Linear Ray and Wave Optics in Phase Space

Bridging Ray and Wave Optics via the Wigner Phase-Space Picture

Linear Ray and Wave Optics in Phase Space

Bridging Ray and Wave Optics via the Wigner Phase-Space Picture

Amalia Torre
ENEA–UTS Tecnologie Fisiche Avanzate
Frascati (Rome), Italy

ELSEVIER

Amsterdam • Boston • Heidelberg • London • New York • Oxford • Paris
San Diego • San Francisco • Singapore • Sydney • Tokyo

ELSEVIER B.V.
Radarweg 29
P.O. Box 211, 1000 AE Amsterdam
The Netherlands

ELSEVIER Inc.
525 B Street, Suite 1900
San Diego, CA 92101-4495
USA

ELSEVIER Ltd
The Boulevard, Langford Lane
Kidlington, Oxford OX5 1GB
UK

ELSEVIER Ltd
84 Theobalds Road
London WC1X 8RR
UK

First edition 2005

Library of Congress Cataloging in Publication Data
A catalog record is available from the Library of Congress.

British Library Cataloguing in Publication Data
A catalogue record is available from the British Library.

ISBN-13: 978 0 444 51799 9
ISBN-10: 0 444 51799 5

♾ The paper used in this publication meets the requirements of ANSI/NISO Z39.48-1992 (Permanence of Paper).
Printed in The Netherlands.

To my mother
whose caress still lingers on my cheek.
To my father
whose smile still shines into my eyes.
To my country
whose colours feed my avid love of life.

Preface

> ...
> Amo i colori, tempi di un anelito
> inquieto, irresolvibile, vitale
> spiegazione umilissima e sovrana
> dei cosmici "perché" del mio respiro.
> ...
>
> A. Merini, *Colori*
> (from A. Merini, *Fiore di poesia*, Einaudi, Torino, 1998)

Ray, wave and quantum concepts are central to diverse and seemingly incompatible models of light. Each model particularizes a specific "manifestation" of light, and then corresponds to adequate physical assumptions and formal approximations, whose domain of applicability is well established. Accordingly each model comprises its own set of geometrical and dynamical postulates with the pertinent mathematical means.

Geometrical optics models the light field as an aggregate of incoherent light rays, naïvely perceived as the trace of the motion of the "luminous corpuscles", which, emitted by the source, move through space in obedience to the usual laws of mechanics. It treats light rays as lines in 3-space dimensions and is accordingly concerned with the dynamical laws establishing how the rays bend when propagating in inhomogeneous media described by the refractive index function. Geometrical optics is not suited to explain interference, diffraction and quantum coherence effects, but, in contrast, it provides a particularly convenient means for the design of optical systems, which is based on the purely geometrical rules of ray tracing.

Geometrical optics has developed its own mathematical framework, which can remarkably be brought into correspondence with that of the Hamiltonian mechanics of point-particles, with "time" corresponding to the arc-length along the ray path and the mechanical "potential" to the refractive index of the optical medium.

Wave optics accounts for the wave characteristics of light. Originating directly from the classical electromagnetic theory, it shares with this theory the same system of theoretical principles and methods, which can notably be put in correspondence with those proper to relativistic quantum mechanics. Then, the geometry of light rays is replaced by the geometry of "luminous" waves, whose propagation is geometrically pictured as the transfer of the interference shaped vibrations from one portion of the medium to the contiguous one.

Wave optics treats the light waves as complex functions of position in 3-space dimensions and is accordingly concerned with the dynamical laws establishing how the wave function changes as the optical wave propagates through inhomogeneous media.

Quantum optics recovers the grainy view of geometrical optics, picturing the light ray as a stream of particle-like entities, the photons. Whereas geometrical optics deals with the influence at a macroscopic level of the medium on the trajectory of the photon streams, quantum optics is typically concerned with the wave-like question relevant to the coherence properties of the photon beams and to the relevance of those properties on the interaction of light with matter, which can correspondingly be treated quantum mechanically. Coherent and squeezed states of light are the building concepts of quantum optics.

Wigner optics bridges between ray and wave optics. It offers the optical phase space as the ambience and the Wigner function based technique as the mathematical machinery to accommodate between the two opposite extremes of light representation: the localized ray of geometrical optics and the unlocalized wave function of wave optics.

Notably quantum optics finds a convenient formulation in the proper phase space with the consequent geometrical view of coherent and squeezed states as circles and ellipses. The Wigner function methods can suitably be applied to quantum optics as well to enable effective analytical means for calculating expectation values and transition probabilities for the aforementioned states.

The purpose of the book is to introduce the reader to the optical phase-space and to the approaches to optics based on the Wigner distribution function, that have been developed over the past 25 years or so in several scientific titles. These yield the formal context, where concepts and methods of both ray and wave optics coalesce into a unifying formalism. In this respect, emphasis is given to the Lie algebra representation of optical systems and accordingly to the Lie algebra view of light propagation through optical systems.

The book is made as self-contained as possible. Chapter 1 presents the Hamiltonian equations of motion, which are basic to the development of both the transfer matrix formalism, appropriate to paraxial ray optics (Chapters 2 and 3), and the transfer operator formalism, suited to paraxial wave optics (Chapters 4 and 5). The relation of both formalisms to the Lie algebra methods is gently displayed.

Chapter 6 introduces the Wigner distribution function, elucidating its origin taken in quantum mechanics and illustrating its properties. A host of diverse optical signals are considered and the relevant Wigner distribution functions are analytically evaluated and graphically shown to help the intuitive perception of the simultaneous account of the signal in the space and spatial frequency domains, conveyed by the Wigner distribution function. Chapter 7 frames the Wigner distribution function within the broad realm of the phase-space signal representations, and illustrates the procedure, and the relevant optical architectures, for displaying the Wigner distribution function of a given signal. In Chapter 8 the laws for the transfer of the Wigner distribution function through linear optical systems are derived. Attention is drawn to the relation between the Wigner distribution function and the fractional Fourier transform, which is a revealing and effective tool for the space-frequency representation of signals (optical or not). Chapter 9 is concerned with the moments of the Wigner distribution function and their propagation laws.

The Wigner representation is presented on the fascinating border-line between quantum mechanics and signal theory.

Chapters are made as self-consistent as possible. Indeed, the Introduction to each chapter is conceived as a summary of the basic results of previous chapters, which are central to those that are going to be presented. A basic role is assigned to the diagrams, which illustrate the syllabus of each chapter, and the figures, which confer physical reality to conceptual architectures. A wide bibliography is given in relation to topics both carefully investigated and briefly mentioned.

Throughout the book the calculations are kept at an accessible level; most mathematical steps are justified. Difficulties might be encountered in connection with the algebra of operators, which do not obey the familiar rules of the algebra of scalars. Careful and illustrative comments on the peculiar behavior of operators are provided in § 1.4.1 in order to help the readers who are not acquainted with the operator algebra.

It is my hope to give the flavour of the fascinating feature of optics that enables a visible account of abstract mathematical entities, like, for instance, symplectic matrices and metaplectic operators, represented through integral transforms. Symplectic matrices and integral transforms, which essentially provide the formal structures for the considerations developed in Chapters 1 to 5, are intimately related, being indeed different representations of the same $Sp(2,\mathbb{R}) \sim Mp(2,\mathbb{R})$ group element. Firstly recognized within a purely quantum mechanical context, this relation has been applied in optics in connection with the fractional Fourier transform. The link between ray matrices and transfer operators from the alternative viewpoint of linear canonical transformations and relevant representations, is elucidated in § 5.6. This is an example of those *parallel paths*, that, explicitly illustrated or implicitly suggested in the text or in the problems, are intended to improve the feeling for the specific topic under consideration and to gain some insight and intuition for unforeseen correspondences and analogies between totally different physical problems.

I am pleased to express my deep gratitude to Professor W.A.B. Evans, whose stimulating discussions, critical comments and technical suggestions have been precious to the completion of the book. I am greatly indebted to Dr. A. De Angelis for his enlightening suggestions, and to Professor A. Reale and Professor A. Scafati for their helpful comments. It is dutiful of me to thank Dr. G. Dattoli, who introduced me to the Lie algebra theory during the stage of our collaboration on the quantum picture of the Free Electron Laser dynamics. I am grateful to Dr. S. Bollanti, Dr. F. Flora and Dr. L. Mezi for their useful comments, and to Mrs. G. Gili, Mr. S. Lupini, Mrs. G. Martoriati, Mrs. M.T. Paolini, Mrs. L. Santonato, Dr. S. Palmerio, Dr. B. Robouch, Dr. N. Sacchetti and Dr. V. Violante for their constant and invaluable sympathy. I thank our librarians, Mrs. C. De Palo and Mrs. M. Liberati, who at certain periods have patiently accepted the role of "my" librarians. It is a pleasure to thank the Optical Society of America for kindly giving me the permission to reproduce material from Applied Optics and Optics Letters, and Einaudi for permitting me to reproduce the lines from Merini's poem, which opened this Preface, my literal translation of which now closes it (below). I express my appreciation to Professor A. Lohmann for his prompt and kind response to my request of reproducing material from papers by himself and his coworkers, which appeared in Optics Communication. I am pleased to express my gratitude to the Fondation Magritte for allowing the reproduction of the evocatively emotional Magritte painting *La corde sensible* for the cover. I am also indebted to the Elsevier production team for expertly implementing my ideas in relation to the cover.

A joyful "Thank you, sorellina" is directed to Dr. F. Mucci, for enthusiastically listening to the description of my "conceptual castles". A bow is for the friends who share my passion for the theatre, for forgiving my absences from the prelims, being forgetfully enraptured in "mie adorate formuline".

It is with intense emotion that my thanks goes also to zia Aida and Maria, who could not see the completion of the book.

...
I love colours, times of a yearning
restless, irresolvable, vital,
very humble and supreme explanation
of the cosmic "why" of my breath.
...

Colours

Contents

1

Hamiltonian Picture of Light Optics. First-Order Ray Optics

1.1 Introduction

The phase space representation of light optics naturally arises from the Hamiltonian formulation of geometrical optics. Geometrical optics gives a simple model for light behaviour, in which the wave character of light is ignored. It is valid whenever light waves propagate through or around objects which are very large compared to the wavelength of the light and when we do not examine too closely what is happening in the proximity of shadows or foci. Accordingly, it does not account for diffraction, interference or polarization effects. Geometrical optics employs the concept of *light ray* [1], which we may give the naïve view as an infinitesimally thin beam of light. Several formal definitions of light ray have been elaborated within both the corpuscular and wave theory to accommodate geometrical abstraction and physical observability. All definitions work well in certain situations, but in others are confronted with intrinsically physical difficulties. Thus, for instance, the corpuscular view of rays as trajectories of "luminous" corpuscles confronts with the problem that the energy density may become infinite. Likewise the wave-like view of rays as orthogonal trajectories to the phase fronts of the light wave confronts with the difficulty of individualizing a defined wave front in the two-wave overlap distribution. Indeed, the ray must be thought of as a convenient and successful model which supports our perception, and hence facilitates the formal description, of a wide class of light phenomena. Geometrical optics establishes the geometrical rules governing the propagation of light rays through optical systems.

The analogy of geometrical optics of light rays to Hamiltonian mechanics of material particles is well established and effectively exploited. The *Hamiltonian formalism* was originally developed by Hamilton for optics in his 1828 paper *Theory of Systems of Rays* and in subsequent papers and brief notes, published during the years from 1830 to 1837 [2.1]. In his papers, Hamilton formulates the problem of studying the geometry of light rays as they pass through optical systems in terms of well-defined relations between the local coordinates of the rays entering and emerging from the system, specified with respect to the optical axis and properly chosen planes across the axis. He shows that, if the ray coordinates are suitably defined, the input-output relations configure as *symplectic* transformations, generated by a function of the ray variables, the *characteristic function*, whose functional form is determined solely by the optical properties of the system. Later, Hamilton realized that the same method could be applied unchanged to mechanical systems, replacing the optical axis by the time axis, the light rays by the particle trajectories and the ray-coordinates by the mechanical phase-space variables [2.2].

The phase space representation is a familiar method within the Hamiltonian formulation of classical mechanics, which describes the dynamics of a mechanical system with m degrees of freedom in terms of m generalized independent coordinates $(q_1, q_2, ..., q_m)$ and the same number of canonically conjugate variables $(p_1, p_2, ..., p_m)$ [3]. The mechanical phase space is the Cartesian space of these $2m$ coordinates. For example, the state of a free particle at a certain time is represented in the proper $6D$ phase space by a *representative point*, specified by the Cartesian coordinates $\mathbf{q} = (q_x, q_y, q_z)$ and the relevant momenta $\mathbf{p} = (p_x, p_y, p_z)$. The motion of the particle in real space corresponds to a trajectory in phase space. Then, the state of an ensemble of identical and noninteracting particles at a given time corresponds to a set of points in the $6D$ phase space. The domain occupied by this set of points moves through phase space as the particles move in real space. However, as the total number of particles remain constant, so will the total number of phase space points. Evidently a real density can be associated with the representative points in phase space, and correspondingly a distribution function of density $\rho(\mathbf{q}, \mathbf{p}, t)$ can be defined so that $\rho(\mathbf{q}, \mathbf{p}, t)dV$ specifies the number of representative points in the element of volume dV in the vicinity of the point $(\mathbf{q}, \mathbf{p})$. Liouville's theorem states the invariance of the density of representative points along the trajectory of any point, and accordingly of the volume of the phase space domain, even though its shape may change considerably during the motion.

Likewise the geometric-optical phase space is the $4D$ Cartesian space of the ray position and momentum coordinates (q_x, q_y, p_x, p_y). However, the phase spaces of classical mechanics and geometrical optics are globally different. The particle momentum of classical mechanics is not restricted in value, whilst the ray momentum of geometrical optics is confined within a circle determined by the local refractive index through the inherent form of the optical Hamiltonian. In the linear approximation the ray momentum is assumed to range well below its natural limit, which then is ignored. Thus the geometric-optical phase space of linear optics comes to be similar to the mechanical phase space.

Section 1.2 reviews the Hamiltonian formulation of geometrical optics and

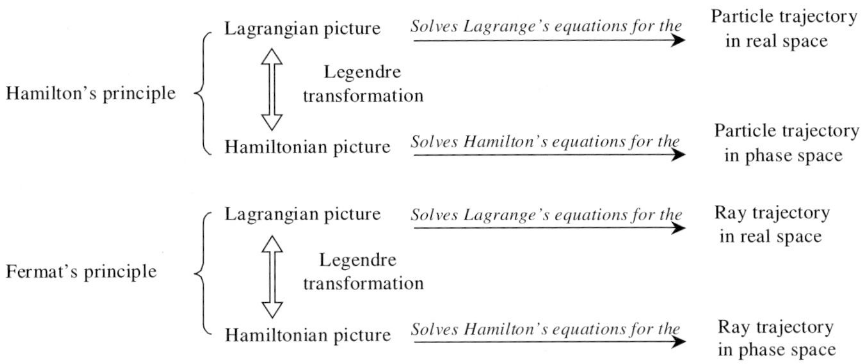

FIGURE 1.1. The Fermat extremal principle based formulation of geometrical optics mirrors that of classical mechanics, based on the Hamilton minimal principle.

introduces the related concept of geometric-optical phase space. Section 1.3 emphasizes the symplectic nature of ray propagation, and details the suited mathematical settings (Poisson brackets and Lie operators) to approach the integration of Hamilton's equations for the light ray. In Sect. 1.4 the ray-transfer operator is introduced and the relevant Lie-transformation based formalism is described. Illustrative examples of phase-space transformations are given in Sect. 1.5. Sections 1.6 and 1.7 illustrate the linear approximation to light-ray propagation, which naturally yields the ray-transfer matrix formalism. Finally, Sect. 1.8 clarifies the link between the ray-matrix approach and the cardinal point (and planes) method.

1.2 Hamiltonian picture of light-ray propagation

We will give a brief account of the Hamiltonian formulation of geometrical optics in order to fix the notations we adopt and to trace the conceptual path towards the phase space representation and the inherent geometry.

Hamiltonian optics develops from Fermat's principle of extremal optical path, which is the optical analog of Hamilton's principle of least action (Fig. 1.1). From Hamilton's principle one can derive both the Lagrangian and Hamiltonian mechanics, related through the Legendre transformation [3]. Likewise from Fermat's principle one can develop the Lagrangian as well as the Hamiltonian formulation of optics [4]. The former yields the equations for the ray variables in real space, while the latter the equations for the ray variables in phase space. We will cursorily illustrate the basic steps leading to both

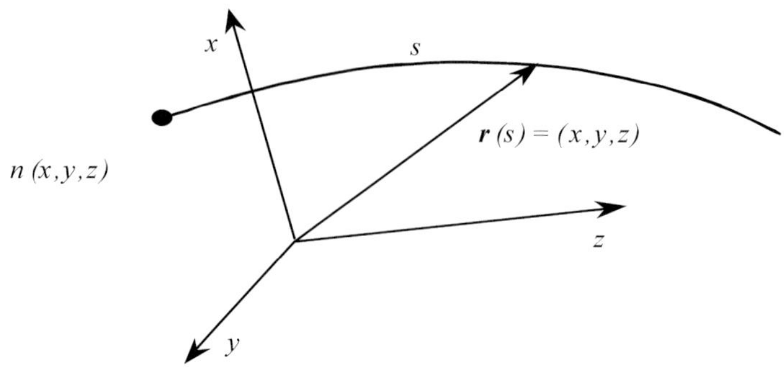

FIGURE 1.2. Geometrical optics describes the medium by the refractive index function $n(x, y, z)$ and the light rays by the 3-vector of functions $\mathbf{r}(s) \equiv (x(s), y(s), z(s))$ of the arc length s measured along the ray path.

pictures, addressing the reader to [4] for a more detailed treatment.

As a natural scenario for introducing Fermat's principle [4, 5], we consider an inhomogeneous medium, occupying a certain region in the $3D$ space, where we suppose a Cartesian system of coordinates (x, y, z) be assigned. The optical properties of the medium are typically described by the refractive index $n(x, y, z)$, given as a scalar function of space[a]. A light ray is propagating in the medium along some trajectory. Regarded as a line in the $3D$ space, the ray can accordingly be described by the position vector $\mathbf{r}(s) \equiv (x(s), y(s), z(s))$ for points on the ray, with the coordinates being functions of the arc length s measured along the ray path with respect to a chosen point (Fig. 1.2).

Fermat's principle combines the geometrical and physical aspects of the ray propagation through the concept of *optical path.*

We recall that given two points P_1 and P_2 and a curve $\mathcal{C}$ connecting them, the *geometrical* path length $\mathfrak{L}(\mathcal{C})$ from P_1 to P_2 along $\mathcal{C}$ is defined as the length of $\mathcal{C}$ and hence is formally given by the line integral

$$\mathfrak{L}(\mathcal{C}) = \int_{P_1}^{P_2} ds, \tag{1.2.1}$$

performed along $\mathcal{C}$ from P_1 to P_2; s denotes the arc length measured along the path and $ds = \sqrt{dx^2 + dy^2 + dz^2}$ is the infinitesimal arc length.

Correspondingly, the *optical* path length $\mathcal{L}(\mathcal{C})$ along the ray trajectory $\mathcal{C}$

---

[a] We will consider only linear spatially nondispersive isotropic media, whose refractive index function is accordingly dependent on position and independent of direction. Hence we will distinguish only between homogeneous and inhomogeneous media, according to whether the scalar index function is uniform or changes from point to point within the medium.

from P_1 to P_2 is defined as the line integral along $\mathcal{C}$ of the refractive index:

$$\mathcal{L}(\mathcal{C}) = \int_{P_1}^{P_2} n(x, y, z) ds. \tag{1.2.2}$$

If the medium is homogeneous, so that $n(x, y, z) = n_0$, the optical path length is the geometrical path length multiplied by the refractive index: $\mathcal{L} = n_0 \mathfrak{L}$.

Slightly correcting the original formulation, as given by Fermat in his Œvreus (1891) [5.1]: "*Je reconnois premièrement...la vérité de ce principe, que la nature agit toujours par les voies les plus courtes,*" Fermat's principle states that, among all the possible paths $\mathcal{C}$ connecting the points P_1 and P_2, the light ray would follow the path $\overline{\mathcal{C}}$ whose optical path length $\mathcal{L}(\overline{\mathcal{C}})$ is an extremum. Therefore it may be a minimum, which is the most frequent case, a maximum, or it may be stationary with respect to the optical path lengths of other paths closely adjacent to $\overline{\mathcal{C}}$. In mathematical terms, the actual ray path $\overline{\mathcal{C}}$ is identified as the extremal of the variational problem

$$\delta\mathcal{L}(\overline{\mathcal{C}}) = \delta \int_{P_1}^{P_2} n(x, y, z) ds = 0, \tag{1.2.3}$$

where the δ variation is intended for small deviations with respect to $\overline{\mathcal{C}}$ of the integration path between the two fixed endpoints P_1 and P_2.

The formal correspondence of Fermat's principle (1.2.3) with Hamilton's principle becomes apparent, once changing the integration variable from s to one of the Cartesian coordinates, say z. Then, evidencing the differential element dz in the line element ds, we may cast Fermat's principle (1.2.3) exactly in the same form as Hamilton's principle, i.e.,

$$\delta \int_{P_1}^{P_2} \{n(x, y, z)\sqrt{1+x'^2+y'^2}\}dz=0, \tag{1.2.4}$$

where the integrand function is naturally identified as the *optical* Lagrangian

$$L(x, y, x', y', z) \equiv n(x, y, z)\sqrt{1+x'^2+y'^2}, \tag{1.2.5}$$

with

$$x' = \frac{dx}{dz}, \quad y' = \frac{dy}{dz}. \tag{1.2.6}$$

The Lagrangian system of equations plainly follow as

$$\begin{aligned} \frac{d}{dz}\left(\frac{\partial L}{\partial x'}\right) - \frac{\partial L}{\partial x} &= 0, \\ \frac{d}{dz}\left(\frac{\partial L}{\partial y'}\right) - \frac{\partial L}{\partial y} &= 0, \end{aligned} \tag{1.2.7}$$

and, on account of (1.2.5) and (1.2.6), can be rearranged into a single vectorial differential equation of second order, known as the *ray equation,*

$$\frac{d}{ds}(n\frac{d\mathbf{r}}{ds}) = \nabla n, \tag{1.2.8}$$

relating the ray trajectory to the optical properties of the medium [6]. Indeed, $\frac{d\mathbf{r}}{ds}$ is the tangent to the ray path, and $\nabla n \equiv (\frac{\partial n}{\partial x}, \frac{\partial n}{\partial y}, \frac{\partial n}{\partial z})$. Interestingly, Eq. (1.2.8) shows a noticeable resemblance to the equation of classical relativistic mechanics $\mathbf{F} = \frac{d\mathbf{p}}{dt}$, that rules the dynamics of a point-particle, the force $\mathbf{F}$ determining the rate of variation in time of the particle momentum $\mathbf{p}$. With time replaced by the arc length s and the mechanical potential by the refractive index n, the gradient ∇n may be interpreted as the "force", acting along the gradient of the potential $n(x,y,z)$, and $n\frac{d\mathbf{r}}{ds}$ as the "momentum of the particle", whose rate of variation being determined by the force ∇n [7].

The transition from the Lagrangian to the Hamiltonian picture is then pursued under Legendre transformation, which takes x' and y' to the *ray momenta* p_x and p_y through

$$p_x = \frac{\partial L}{\partial x'}, \quad p_y = \frac{\partial L}{\partial y'}, \tag{1.2.9}$$

and accordingly the Lagrangian function to the optical Hamiltonian as

$$H(x, y, p_x, p_y, z) = p_x x' + p_y y' - L(x, y, x', y', z). \tag{1.2.10}$$

By (1.2.5) the generalized momenta p_x and p_y explicitly write as

$$\begin{aligned} p_x &= n(x,y,z)\frac{x'}{\sqrt{1+x'^2+y'^2}} = n\frac{dx}{ds} \\ p_y &= n(x,y,z)\frac{y'}{\sqrt{1+x'^2+y'^2}} = n\frac{dy}{ds}, \end{aligned} \tag{1.2.11}$$

in accordance with the mechanical analogy suggested above. As we see, p_x and p_y equal the local value of the refractive index times the direction cosines $(\frac{dx}{ds}, \frac{dy}{ds})$ of the light ray relative to x and y axis, respectively. Hence, they are also termed *optical direction cosines* of the ray.

We express the function (1.2.10) in terms of its proper variables to obtain the Hamiltonian of the ray in the form

$$H(x, y, p_x, p_y, z) = -\sqrt{n^2(x,y,z) - p_x^2 - p_y^2}, \tag{1.2.12}$$

which is just the negative of the optical direction cosine of the ray along the z direction.

Finally, in analogy with mechanics we may write Hamilton's equations for the light ray as

$$\begin{aligned} \frac{dq_x}{dz} &= \frac{\partial H}{\partial p_x}, & \frac{dq_y}{dz} &= \frac{\partial H}{\partial p_y} \\ \frac{dp_x}{dz} &= -\frac{\partial H}{\partial q_x}, & \frac{dp_y}{dz} &= -\frac{\partial H}{\partial q_y}, \end{aligned} \tag{1.2.13}$$

thus completing the Hamiltonian formulation of geometrical optics. We changed $x \to q_x$ and $y \to q_y$ for purpose of future use.

Evidently, within the mechanical analogy the light propagation can be assimilated to a dynamical problem with two degrees of freedom, the length coordinate z being given the distinguished role as the independent (*evolution*) variable. We emphasize that the z direction can be arbitrarily fixed, and hence conveniently chosen to coincide with some privileged direction of the system. Ordinary optical systems are commonly designed to have a plane of symmetry, where an *ideal* ray can be identified, which represents the path of a ray through the system corresponding to adequately assigned input conditions. This *ideal ray* is understood as the *optical axis*. We restrict our attention to *centered* optical systems, where all of the optical components are aligned with their optical axes lying on the same straight line, which then may be designated as the optical axis for the overall system; the z axis may be chosen as this common axis. The optical axis of a system of coaxial lenses, for instance, is just the common axis of the lenses and can conveniently be taken as z axis. Throughout the book we refer to the optical axis as the reference z axis.

Equations (1.2.13) provide the formal answer to the basic problem of geometrical optics of determining the final variables of the ray after passing through an optical system, once assigned the ray variables before propagation and specified the optical properties of the system. In fact, given the values of the position and direction coordinates of the ray at some z_i, i.e., the initial values for the equations of motion (1.2.13), it is possible in principle to solve (1.2.13) for the ray position and direction coordinates at any other z.

Following again the suggestion of Hamiltonian mechanics, we can construct the *geometrical-optical phase space* as the Cartesian space of the four ray-variables (q_x, q_y, p_x, p_y). Notably, according to (1.2.11), the refractive index n represents the normalization factor to turn the geometrical variables dq_x/ds and dq_y/ds into the phase-space variables p_x and p_y. The corresponding step in mechanics is to replace velocity with momentum, the mass being the proper normalization factor.

In the geometrical optical phase space the light ray is individualized by the trajectory of the point whose z-evolving coordinates (q_x, q_y, p_x, p_y) from the specified initial values are determined by the equations of motion. A collection of rays distributed over a range of possible positions and directions fills a certain region in phase space, with which we can associate a density distribution function $\rho(q_x, q_y, p_x, p_y, z)$. The change in the ray bundle as it propagates through optical media reflects into the motion of the representative points as they move through phase space in accord with the equations of motion (1.2.13). While the exact phase-space motion of each representative point (i.e., each ray in the bundle) is uniquely determined by the initial conditions, it is rather impracticable to calculate an exact solution for the whole bundle of rays. It is therefore convenient to provide a statistical description of the behaviour of the assembly of rays, regarded as an ensemble of identical systems differing over a range of initial conditions, and accordingly to follow the evolution of the density distribution function ρ.

Geometric optics is primarily concerned with the behaviour of a single ray when it passes through optical media. We will firstly tackle the general problem of the evolution under the Hamiltonian (1.2.12), introducing the basic tools of the Hamilton-Lie formalism. Then we will focus our attention on linear ray-optics and hence on the related ray-matrix formalism. We will illustrate the ability of the matrix method as a tracing tool for both a single paraxial ray and an ensemble of paraxial rays, thus paralleling the single-particle and particle-ensemble pictures of classical statistical mechanics. In later Chapters, we will illustrate the substantially different tools of linear wave-optics, dealing with light distributions instead of bundles of rays. There we will introduce the *wave-optical phase space*, whose coordinates are customarily set as the spatial coordinate and the spatial frequency. The geometric- and wave-optical phase spaces are basically different, since a "point" in phase space, identified by a definite set of all phase-space coordinates, has no physical meaning in wave-optics, due to the uncertainty relation of Fourier analysis. The wave optical phase space is in a sense "granular", there being a minimum possible phase space volume associated with a source or an optical system in consequence of the wave properties of light. In this Chapter we deal with the geometrical-optical phase space.

Finally we evidence a basic feature of the optical phase space, which globally differentiates from the mechanical phase space. In fact, the functional form (1.2.12) of the optical Hamiltonian limits the optical momenta according to $p_x^2 + p_y^2 \leq n^2$. Thus, the ray-momentum coordinates p_x and p_y are confined within a circle, whereas the ray-position coordinates q_x and q_y are both allowed to range over all the real line. As a consequence, the geometrical-optical phase space configures as the product of a circle for the $\mathbb{R}^2$-plane: $C_2 \times \mathbb{R}^2$, in contrast with the mechanical phase space being the well known $\mathbb{R}^4 = \mathbb{R}^2 \times \mathbb{R}^2$. As we will see, first-order optics ignores the above limit on the ray momenta; it assumes that both the ray coordinates and momenta remain small through the propagation, so that p_x and p_y are well inside their natural range, whose limits can therefore be ignored.

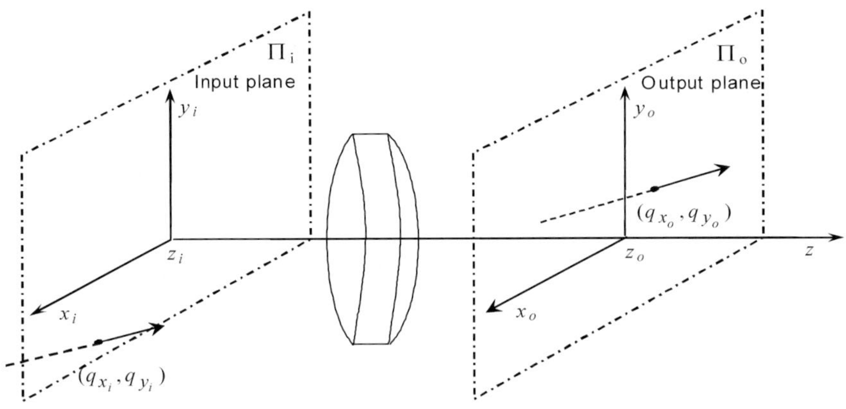

FIGURE 1.3. An optical system and the relevant input and output planes, to which the coordinates of the incoming and outgoing rays are referred.

1.3 Hamiltonian picture of light-ray propagation: formal settings

We consider a light ray propagating along an optical system, which we suppose to be completely characterized by the refractive index function, given at each point before and after the system and in the system itself. The optical z-axis is commonly oriented as the direction of light propagation, assumed to occur from left to right. The light rays therefore enter the system at the left and emerge from it at the right. It is natural to describe the evolution of the ray along the axis by determining at any axial position the intersection point of the ray with a plane orthogonal to the axis, and the optical direction cosines of the ray at that point. Thus, we fix two planes, Π_i and Π_o, across the optical axis, located at positions z_i and z_o, respectively at the left and the right of the system (Fig. 1.3). We will refer to Π_i as the *input* plane and to Π_o as the *output* plane. Also, we will equip each plane with a coordinate system. In the following, unless otherwise specified, rectangular coordinate systems will be taken in both the reference planes. The orientation and the position of the axes can be chosen arbitrarily; it is a bare matter of convenience to choose the origins lying on the optical z-axis and the coordinate axes parallel, and equally scaled, from one to the other plane.

The incoming ray is unambiguously identified by the coordinates (q_{x_i}, q_{y_i}) of the intersection point with the input plane Π_i, and by the optical direction cosines (p_{x_i}, p_{y_i}) at that point. Explicitly, we can write

$$p_{x_i} = n_i \sin \alpha_{x_i} \ , \quad p_{y_i} = n_i \sin \alpha_{y_i} , \tag{1.3.1}$$

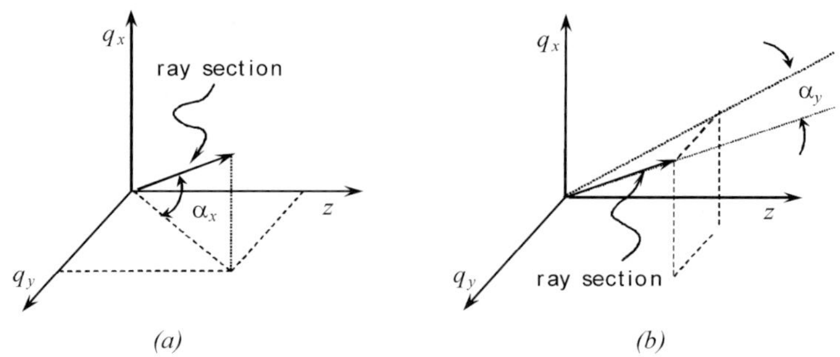

FIGURE 1.4. The angles *(a)* α_x and *(b)* α_y, entering the definition of the optical momenta p_x and p_y, are the complements to $\frac{\pi}{2}$ of the angles of the ray to the x and y axis, respectively.

where $n_i \equiv n(q_{x_i}, q_{y_i}, z_i)$ is the local value of the refractive index at the intersection point and $(\alpha_{x_i}, \alpha_{y_i})$ are the angles, the incoming ray defines with the assigned planes y_i-z and x_i-z (Fig. 1.4). Similarly, the outgoing ray is specified by the coordinates (q_{x_o}, q_{y_o}) of the intersection with the output plane Π_o and by the optical direction cosines (p_{x_o}, p_{y_o}) at this plane:

$$p_{x_o} = n_o \sin \alpha_{x_o} \ , \quad p_{y_o} = n_o \sin \alpha_{y_o} , \tag{1.3.2}$$

where $n_o \equiv n(q_{x_o}, q_{y_o}, z_o)$ and $(\alpha_{x_o}, \alpha_{y_o})$ are the angles of the ray relative to the planes y_o-z and x_o-z.

Evidently, the position coordinates q_x and q_y have the dimensions of length, whilst the direction coordinates p_x and p_y are dimensionless.

As already noted, the problem of geometrical optics is to determine the input-output relations of the optical system,

$$\begin{aligned} q_{x_o} &= q_{x_o}(q_{x_i}, q_{y_i}, p_{x_i}, p_{y_i}; z_o), \qquad & p_{x_o} &= p_{x_o}(q_{x_i}, q_{y_i}, p_{x_i}, p_{y_i}; z_o), \\ q_{y_o} &= q_{y_o}(q_{x_i}, q_{y_i}, p_{x_i}, p_{y_i}; z_o), \qquad & p_{y_o} &= p_{y_o}(q_{x_i}, q_{y_i}, p_{x_i}, p_{y_i}; z_o), \end{aligned} \tag{1.3.3}$$

which relate in a desirable explicit form the position and direction coordinates of the incoming ray at z_i to those of the corresponding outgoing ray at z_o.

The input-output relations (1.3.3) describe a transformation in the optical phase-space of the representative point $P_i \equiv (q_{x_i}, q_{y_i}, p_{x_i}, p_{y_i})$ of the input ray to the representative point $P_o \equiv (q_{x_o}, q_{y_o}, p_{x_o}, p_{y_o})$ of the output ray. Evidently, the transformation occurs through a trajectory from P_i to P_o, ruled by the equations of motion (1.2.13) with the inherent Hamiltonian (1.2.12). The transformation $P_i \longrightarrow P_o$ is *symplectic*.

1.3.1 *Ray propagation as a symplectic transformation*

We clarify the formal and physical significance of the ray propagation as a *symplectic* transformation [3, 9].

To this end we firstly note that, like the mechanical phase space, the optical phase space naturally configures as a $4D$ linear vector space, the 4-vectors $\mathbf{u}$ being formed by the z-depending ray-coordinates (q_x, q_y, p_x, p_y). Similarly it can conveniently be equipped with the euclidean metric defined as usual by the scalar product of vectors:

$$(\mathbf{u}, \mathbf{v}) = \sum_{j=1}^{4} u_j v_j = \mathbf{u}^\top \cdot \mathbf{v}, \tag{1.3.4}$$

where $\mathbf{u}^\top$ is the transpose vector and, for notational convenience, the vector components are indexed by the integers from 1 to 4. Conforming with the common practice, by "vectors" we mean *column* vectors; then, *row* vectors are identified by transposition, i.e., $\mathbf{u} = (q_x, q_y, p_x, p_y)^\top \equiv (u_1, u_2, u_3, u_4)^\top$.

Exploiting the vector notation, we can write Hamilton's equations (1.2.13) in one form

$$\frac{du_j}{dz} = \sum_{l=1}^{4} J_{jl} \frac{\partial H}{\partial u_l}, \qquad j = 1, .., 4, \tag{1.3.5}$$

with the aid as well of the suitably defined coefficients J_{jl}, which are allowed to be only 1, -1 or 0 according to their own indices:

$$J_{jl} = \delta_{l,j+2}, \quad j = 1, 2; \qquad J_{jl} = -\delta_{l,j-2}, \quad j = 3, 4. \tag{1.3.6}$$

Evidently, J_{jl}'s can conveniently be arranged into the 4×4 antisymmetric matrix $\mathbf{J}$ as

$$\mathbf{J} = \begin{pmatrix} \mathbf{0} & \mathbf{I} \\ -\mathbf{I} & \mathbf{0} \end{pmatrix}, \tag{1.3.7}$$

where $\mathbf{I}$ denotes the 2×2 unit matrix as $\mathbf{0}$ the 2×2 null matrix$^{(b)}$. The matrix $\mathbf{J}$ is known as the *symplectic unit* matrix. It has the nice properties

$$\mathbf{J}^{-1} = \mathbf{J}^\top = -\mathbf{J}, \quad \mathbf{J}^2 = -\mathbf{I}, \quad \det \mathbf{J} = 1, \tag{1.3.8}$$

$\mathbf{J}^\top$ denoting the transpose matrix.

We suppose now to take the vector $\mathbf{u}$ to the vector $\mathbf{v} \equiv (v_1, v_2, v_3, v_4)$ under the transformation

$$v_j = v_j(u_1, ..., u_4), \quad j = 1, .., 4, \tag{1.3.9}$$

[b] It is a common convention to represent $2m \times 2m$ matrices as partitioned into four $m \times m$ submatrices [8].

identified by the associated Jacobian matrix $\mathbf{S}$, whose entries $S_{ij} = \partial v_i/\partial u_j$ depend in general on u_j, $j = 1, .., 4$.

It is easy to prove that $\mathbf{v}$ obeys equations

$$\frac{dv_j}{dz} = S_{jl} J_{lk} S_{hk} \frac{\partial H}{\partial v_h}, \qquad j = 1, .., 4, \tag{1.3.10}$$

the summation signs over repeated indices being omitted.

We require that the transformation (1.3.9) leaves the form of Hamilton's equations invariant, i.e., that $\mathbf{v}$ obeys the same equations (1.3.5) as $\mathbf{u}$, namely

$$\frac{dv_j}{dz} = J_{jl} \frac{\partial H}{\partial v_l}, \qquad j = 1, .., 4. \tag{1.3.11}$$

With this request applied to (1.3.10), we end up with the matrix relation:

$$\mathbf{SJS}^{\top} = \mathbf{J}. \tag{1.3.12}$$

It represents the condition for $\mathbf{S}$ to be a symplectic matrix at any point in phase space [9]. As noted, in fact, $\mathbf{S}$ may depend in general on the vector $\mathbf{u}$. In contrast, the product $\mathbf{SJS}^{\mathsf{T}}$ results into the matrix $\mathbf{J}$, independent of $\mathbf{u}$. Evidently the unit matrix $\mathbf{I}$ and the matrix $\mathbf{J}$ are symplectic as well.

We recall that in general nonsingular square matrices form a group under the row-column matrix multiplication, the existence of the identity and the unique inverse as well as the associativity being a plain consequence of the matrix multiplication properties. In particular, the matrices obeying (1.3.12) do form a group due to the closure property of the symplectic condition, which is preserved by matrix multiplication; in fact, if $\mathbf{S}_1$ and $\mathbf{S}_2$ obey (1.3.12), so will the product $\mathbf{S}_1\mathbf{S}_2$. We invite the reader to verify that, if $\mathbf{S}$ is symplectic, the inverse $\mathbf{S}^{-1}$ exists and is symplectic as well as the transpose $\mathbf{S}^{\top}$. The group of the 4×4 symplectic real matrices is denoted as $Sp(4, \mathbb{R})$.

Since $\det \mathbf{J} = 1$ and $\det \mathbf{S} = \det \mathbf{S}^{\mathsf{T}}$, by (1.3.12) it follows that

$$\det \mathbf{S} = \pm 1. \tag{1.3.13}$$

It can be shown that actually for a symplectic matrix it is always $\det \mathbf{S} = +1$ [9.2]. Remarkably, in the 2×2 case the necessary and sufficient condition for a matrix to be symplectic is that it has determinant $+1$. Consequent to the symplectic condition (1.3.12), we also find that the 16 entries of the matrix $\mathbf{S}$ are not all independent; only 10 can be arbitrarily chosen, the others being fixed by (1.3.12). This reflects the general property of the $2m \times 2m$ symplectic matrices, for which only $m(2m + 1)$ of the $4m^2$ elements can arbitrarily be assigned (Problem 1).

Interestingly the Jacobian matrix $\mathbf{S}$ relates the infinitesimal vectors $d\mathbf{v}$ and $d\mathbf{u}$ (see also Problem 2). Indeed, being

$$dv_i = \sum_{j=1}^{4} \frac{\partial v_i}{\partial u_j} du_j = \sum_{j=1}^{4} S_{ij} du_j, \tag{1.3.14}$$

we can concisely write

$$d\mathbf{v} = \mathbf{S}\, d\mathbf{u}. \tag{1.3.15}$$

It is therefore a direct consequence of the symplectic condition (1.3.12) that the quantity $(d\mathbf{u}_1, \mathbf{J}d\mathbf{u}_2)$, for any choice of $d\mathbf{u}_1$ and $d\mathbf{u}_2$, remains unchanged under the symplectic transformation (1.3.15), which takes $\mathbf{u}$ to $\mathbf{v}$ and accordingly $d\mathbf{u}_1$ to $d\mathbf{v}_1$ and $d\mathbf{u}_2$ to $d\mathbf{v}_2$. Explicitly, we can easily prove that

$$(d\mathbf{v}_1, \mathbf{J}d\mathbf{v}_2) = (d\mathbf{u}_1, \mathbf{J}d\mathbf{u}_2). \tag{1.3.16}$$

In particular, if the new and old vectors, $\mathbf{v}$ and $\mathbf{u}$, are linked by a linear transformation as $v_i = \sum_{j=1}^{4} a_{ij} u_j$, $i = 1, .., 4$, the entries of $\mathbf{S}$ are just the $\mathbf{u}$-independent coefficients of the transformation, i.e., $S_{ij} = a_{ij}$. Hence, relations (1.3.15) and (1.3.16) hold true for the finite vectors as well. Explicitly, if

$$\mathbf{v} = \mathbf{S}\, \mathbf{u}, \tag{1.3.17}$$

with $\mathbf{S}$ symplectic and $\mathbf{u}$-independent, then

$$(\mathbf{v}_1, \mathbf{J}\mathbf{v}_2) = (\mathbf{u}_1, \mathbf{J}\mathbf{u}_2). \tag{1.3.18}$$

It is evident that if the transformation which takes the vector $\mathbf{u}$ to $\mathbf{v}$ is z-dependent:

$$v_j = v_j(u_1, ..., u_4; z), \quad j = 1, .., 4, \tag{1.3.19}$$

the simple derivation we have sketched for the symplectic condition (1.3.12) no longer holds. But, indeed, this is just the case of our main interest. In fact, as we said earlier, the propagation of light rays through optical systems is formally described by the solution to Hamilton's equations (1.3.5), which, for assigned input conditions, yield the ray variables, and so the relevant phase space vector $\mathbf{u}(z)$, at each z along the optical axis through the axial interval we are interested in, say from z_i to z_o. Therefore, what we must verify is that the ray propagation from z_i to z_o, we may picture as a transformation of the initial ray-vector $\mathbf{u}(z_i)$ to the final ray-vector $\mathbf{u}(z_o)$:

$$\mathbf{u}(z_i) \longrightarrow \mathbf{u}(z_o) = \mathbf{u}(z_i + \Delta z), \qquad \Delta z = z_o - z_i, \tag{1.3.20}$$

ruled by Hamilton's equations (1.3.5), be symplectic.

To prove this, it is enough to demonstrate that any infinitesimal transformation along the path (1.3.20) is symplectic, i.e., the relevant Jacobian matrix obeys relation (1.3.12). The continuous evolution of the ray vector $\mathbf{u}(z)$ along the finite space interval Δz can be viewed as a succession of infinitesimal symplectic transformations, which, in virtue of the product preservation property mentioned above, result into a single finite symplectic transformation.

Then, let us consider an infinitesimal transformation along the evolution path (1.3.20) taking the initial ray vector $\mathbf{u}(z_i)$ at z_i to the ray vector $\mathbf{u}(z_i+dz)$ at the infinitesimally close axial position $z_i + dz$. For convenience, we set

$$\begin{aligned}\mathbf{w} &= \mathbf{u}(z_i) \\ \mathbf{v} &= \mathbf{u}(z_i + dz) = \mathbf{w} + \left.\tfrac{d\mathbf{u}}{dz}\right|_{z_i} dz\end{aligned} \tag{1.3.21}$$

where, of course, only first-order terms in the elemental displacement dz have been retained.

In accord with Hamilton's equations (1.3.5), one has

$$\left.\frac{du_j}{dz}\right|_{z_i} = J_{jl} \left.\frac{\partial H}{\partial u_k}\right|_{z_i}, \tag{1.3.22}$$

which, once inserted into the second of (1.3.20), yields

$$v_j = w_j + J_{jk} \left.\frac{\partial H}{\partial u_k}\right|_{z_i} dz. \tag{1.3.23}$$

The Jacobian matrix comes to be

$$S_{jl} = \frac{\partial v_j}{\partial w_l} = \frac{\partial w_j}{\partial w_l} + J_{jk} \left.\frac{\partial^2 H}{\partial u_l \partial u_k}\right|_{z_i} dz = \delta_{jl} + J_{jk} H''_{lk}(\mathbf{w}, z_i) dz, \tag{1.3.24}$$

where the second order derivatives of the Hamiltonian function, $H''_{lk} = \left.\frac{\partial^2 H}{\partial u_l \partial u_k}\right|_{z_i}$, which form the so-called Hessian matrix $\mathbf{H}'' = (H''_{lk})_{lk}$, are evaluated at $\mathbf{u}(z_i)$ and z_i. Due to the assumed interchangeability of the derivative order with respect to the vector components in the Hessian entries: $H''_{lk} = H''_{kl}$, relation (1.3.24) can finally be cast in the matrix form

$$\mathbf{S} = \mathbf{I} + \mathbf{J}\mathbf{H}''(\mathbf{w}, z_i) dz. \tag{1.3.25}$$

In order to assess the symplecticity of $\mathbf{S}$ we must take the product $\mathbf{S}\mathbf{J}\mathbf{S}^\top$, which, consistently to first order, is easily proved to be:

$$\begin{aligned}\mathbf{S}\mathbf{J}\mathbf{S}^\top &= (\mathbf{I} + \mathbf{J}\mathbf{H}''(\mathbf{w}, z_i) dz)\mathbf{J}(\mathbf{I} - \mathbf{H}''(\mathbf{w}, z_i)\mathbf{J} dz) \\ &= \mathbf{J} + \mathbf{J}\mathbf{H}''(\mathbf{w}, z_i)\mathbf{J} dz - \mathbf{J}\mathbf{H}''(\mathbf{w}, z_i)\mathbf{J} dz = \mathbf{J}.\end{aligned} \tag{1.3.26}$$

The symplecticity of the infinitesimal transformation mapping the initial ray vector $\mathbf{u}(z_i)$ to the infinitesimally displaced vector $\mathbf{u}(z_i + dz)$ along the finite evolution path (1.3.20) governed by Hamilton's equation (1.3.5) is then demonstrated. Accordingly, we may think of the ray propagation as a continuous sequence of symplectic transformations, under which the ray vectors in phase space are taken one to the other, the axial coordinate z being the stepping parameter. We may say that optical systems operate symplectic transformations on the ray variables q and p [10].

It is worth emphasizing that the foregoing discussion, we have developed with specific reference to the light ray propagation, frames within the general issue of Hamiltonian mechanics [3], stating that any *canonical* transformation (i.e., such to map phase-space vectors obeying equations of motion of Hamiltonian form into phase-space vectors obeying as well equations of motion of Hamiltonian form) is symplectic, whether it depends or not on the evolution variable inherent in the dynamical problem under examination.

Readers interested in a rigorous treatment of symplectic groups are directed to [9]. Additional comments are given in Chapters 2 and 3. The Appendix provides an elementary exposition of the basic notions pertaining to Lie algebras and Lie groups.

1.3.2 *Poisson brackets and Lie operators*

Central to our approach to the geometrical optics problem of determining the transformation (1.3.3) of the light-ray coordinates, generated by the Hamiltonian (1.2.10) through Eqs. (1.2.13), are the strictly related concepts of *Poisson bracket* and *Lie operator*. Poisson brackets and Lie operators represent the formal tools to introduce the Hamiltonian equations of motion into the appropriate Lie algebraic ambience, where their intrinsic symmetry can more clearly be evidenced and effectively exploited. The definitions and main properties of both the Poisson brackets and Lie operators are recalled below. The discussion is developed at a general level. Thus, (q, p) denotes a set of canonically conjugate variables for some Hamiltonian system and z the independent variable, independently of their specific meaning in ray optics.

Poisson brackets

Let f and g be any two continuously differentiable functions of the canonical variables (q, p). The Poisson bracket of f and g, denoted by $\{f, g\}$, is still a function of (q, p), formed according to [3, 10.2, 11]

$$\{f, g\} = \sum_\alpha \left(\frac{\partial f}{\partial q_\alpha}\frac{\partial g}{\partial p_\alpha} - \frac{\partial f}{\partial p_\alpha}\frac{\partial g}{\partial q_\alpha} \right), \tag{1.3.27}$$

where the sum runs over all the pairs of conjugate variables (q_α, p_α) for the problem at hand. Since both the functions f and g may depend on the independent variable z, the Poisson bracket $\{f, g\}$ may depend on z as well.

Clearly the Poisson bracket operation creates a binary relation: $(f, g) \to \{f, g\}$ of every pair of dynamical variables to one dynamical variable, identified through the rule (1.3.27). In particular, the relations

$$\{q_\alpha, q_\beta\} = \{p_\alpha, p_\beta\} = 0; \quad \{q_\alpha, p_\beta\} = \delta_{\alpha\beta}, \tag{1.3.28}$$

characterize the canonically conjugate variables (q_α, p_α), as the following

$$\{q_\alpha, f\} = \frac{\partial f}{\partial p_\alpha}, \qquad \{p_\alpha, f\} = -\frac{\partial f}{\partial q_\alpha}, \tag{1.3.29}$$

establish the rules to form the Poisson brackets of any pair of conjugate variables (q_α, p_α) with a function f of (q, p). In particular, if f is taken as the Hamiltonian function H, the Poisson brackets (1.3.29) reproduce the right hand sides of Hamilton's equations, which accordingly write also as

$$\frac{dq_\alpha}{dz} = \{q_\alpha, H\}, \qquad \frac{dp_\alpha}{dz} = \{p_\alpha, H\}, \tag{1.3.30}$$

for every pair of conjugate variables (q_α, p_α).

Evidently the Poisson bracket of f and the vector-valued function $\mathbf{g} = (g_1(q,p), \ldots, g_m(q,p))^\top$ is the vector function $\{f, \mathbf{g}\} = (\{f, g_1\}, \ldots, \{f, g_m\})^\top$.

The Poisson bracket operation has the four remarkable properties:

$$\begin{array}{ll} \textit{antisymmetry:} & \{f,g\} = -\{g,f\} \\ \textit{linearity}: & \{f, ag+bh\} = a\{f,g\} + b\{f,h\} \\ \textit{derivation property:} & \{f, gh\} = \{f,g\}h + g\{f,h\} \\ \textit{Jacobi identity:} & \{f,\{g,h\}\} = \{\{f,g\},h\} + \{g,\{f,h\}\} \end{array} \tag{1.3.31}$$

holding for any functions f, g, h and complex numbers a, b. These properties directly follow from definition (1.3.27), exploiting the linearity and the Leibnitz rule for differentiation. The antisymmetry implies the identity: $\{f, f\} = 0$. Interestingly, the Poisson bracket operation behaves with respect to the ordinary and Poisson product of functions in the similar fashion as the derivative operation, i.e., through a Leibnitz-like rule.

Let us consider now the set $\{f(q,p)\}$ of all the continuously differentiable functions of the phase space variables (q, p). It is naturally structured as a linear vector space, since any linear combination of functions in the set yields a function in the set. Further, as remarked before, the Poisson bracket takes any pair of functions in the set to a function in the set. Thereby, observing the properties (1.3.31), it becomes clear that the set $\{f(q,p)\}$ configures as a Lie algebra with the Poisson bracket acting as the Lie product. In particular, as the reader may easily verify, the quadratic homogeneous polynomials of the type aq^2+bp^2+cqp are closed under Poisson brackets and so they span a Lie algebra. The quadratic homogeneous polynomials identify a particularly interesting subset within the overall space of dynamical variables $\{f(q,p)\}$, since they are basic to the representation of linear optical systems in both geometrical and wave optics. In fact, as we will see, in the linear approximation the optical Hamiltonian (1.2.12) turns into a quadratic polynomial. We also note that the subset of the inhomogeneous linear polynomials of the type $aq + bp + c$ does form a Lie algebra as well; in particular, the Poisson bracket operation images this subset onto that of constant functions. In Chapter 4 we will see that by the "quantization" procedure the Lie algebra of the inhomogeneous linear polynomials maps to the Heisenberg-Weyl algebra, which will appear to be the basic building unit of wave optics, in accordance with the claimed analogy of wave optics to quantum mechanics. Also, the quadratic Hamiltonian functions of ray optics will map to the quadratic Hamiltonian operators of wave optics and correspondingly the Lie group of the symplectic matrices of ray optics will map to the metaplectic group of the Fresnel-integral operators of wave optics.

Finally, we establish a further characterization of the symplectic transformations in terms of the Poisson brackets. We refer to the $4D$ phase space of geometrical optics and gather the ray variables into the 4-vectors $\mathbf{u}$, as in § 1.3.1. It is easily seen that the symplectic unit matrix $\mathbf{J}$ directly relates to the fundamental Poisson brackets (1.3.28), which synthesize in one form

$$\{u_i, u_j\} = J_{ij}, \tag{1.3.32}$$

as well as to the general Poisson bracket (1.3.27), which writes concisely as

$$\{f, g\} = \frac{\partial f}{\partial u_i} J_{ij} \frac{\partial g}{\partial u_j}. \tag{1.3.33}$$

Composing the derivatives $(\frac{\partial f}{\partial u_i})_{i=1,..,4}$ and $(\frac{\partial g}{\partial u_i})_{i=1,..,4}$ into the 4-vectors $\partial_u f$ and $\partial_u g$, the right-hand side of (1.3.33) can be interpreted as the scalar product of the vectors $\partial_u f$ and $\mathbf{J}\partial_u g$. In symbols we write

$$\{f, g\} = (\partial_u f,\, \mathbf{J}\partial_u g). \tag{1.3.34}$$

Let the phase space vector $\mathbf{u}$ transform to $\mathbf{v}$ under a symplectic transformation, whose associated Jacobian matrix $\mathbf{S}$ obeys thereby the relation (1.3.12). By the chain rule for derivatives, one has that $\partial_v = (\mathbf{S}^\top)^{-1}\partial_u$. Then, regarding f and g as functions of the new variables $\mathbf{v}$, we find

$$(\partial_v f,\, \mathbf{J}\partial_v g) = (\partial_u f, \mathbf{S}^{-1}\mathbf{J}(\mathbf{S}^\top)^{-\mathbf{1}}\partial_u g) = (\partial_u f,\, \mathbf{J}\partial_u g), \tag{1.3.35}$$

which shows that the Poisson bracket $\{f, g\}$ has not changed under the symplectic transformation $\mathbf{u} \to \mathbf{v}(\mathbf{u})$. Conversely, if the identity (1.3.35) holds, then the matrix $\mathbf{S}$ of the transformation $\mathbf{u} \to \mathbf{v}(\mathbf{u})$ is symplectic.

Consequently the necessary and sufficient condition for a transformation to be symplectic is that it preserves the Poisson brackets (1.3.27) of dynamical variables and in particular the fundamental Poisson brackets (1.3.28). The invariance of the Poisson brackets under symplectic transformations conforms with the invariance of the skew product as stated in (1.3.16).

Lie operators

Let $f(q,p,z)$ be some continuously differentiable function of the conjugate variables q, p. We associate with f the *Lie operator* $\widehat{\mathsf{L}}_f$ by the rule [11, 4]

$$\widehat{\mathsf{L}}_f(q,p,z) \equiv \{\cdot, f\} = \sum_\alpha \left(\frac{\partial f}{\partial q_\alpha} \frac{\partial}{\partial p_\alpha} - \frac{\partial f}{\partial p_\alpha} \frac{\partial}{\partial q_\alpha} \right), \tag{1.3.36}$$

where as in (1.3.27) the sum runs over all the pairs of conjugate variables (q_α, p_α) pertinent to the problem at hand[(c)]. Then the Lie operator $\widehat{\mathsf{L}}_f$ generated by f acts on the function g as the right Poisson bracket by f, i.e.,

$$\widehat{\mathsf{L}}_f\, g \equiv \{g, f\}\,. \tag{1.3.37}$$

For notational conciseness, in the following we will omit to show the dependence of $\widehat{\mathsf{L}}_f$ on its own arguments q, p and z.

The powers of $\widehat{\mathsf{L}}_f$ are obtained by repeatedly Poisson bracketing, so that the j-th power of $\widehat{\mathsf{L}}_f$ means j-fold Poisson brackets with f according to

$$\begin{aligned} &\widehat{\mathsf{L}}_f^0 = f,\ \widehat{\mathsf{L}}_f^1 = \{\cdot, f\},\ \widehat{\mathsf{L}}_f^2 = \{\{\cdot, f\}, f\}, \ldots \\ &\widehat{\mathsf{L}}_f^j = \{\{\{..\{\cdot, \underbrace{f\}..\}, f\}, f\}}_{j \text{ times}}, \ldots \end{aligned} \tag{1.3.38}$$

Of course, the zero power yields the identity operator.

Clearly, $\widehat{\mathsf{L}}_f$ can act also on a vector-valued function $\mathbf{g} = (g_1, .., g_m)^\top$ thus giving the vector-valued function $\widehat{\mathsf{L}}_f\mathbf{g} = \{\mathbf{g}, f\} = (\{g_1, f\}, .., \{g_m, f\})^\top$.

The properties of the Lie operators parallel those of the Poisson brackets. In fact, as a consequence of (1.3.31) we find:

$$\begin{array}{ll} \textit{antisymmetry:} & \widehat{\mathsf{L}}_f g = -\widehat{\mathsf{L}}_g f \\ \textit{linearity:} & \widehat{\mathsf{L}}_f(ag + bh) = a\widehat{\mathsf{L}}_f g + b\widehat{\mathsf{L}}_f h \\ \textit{derivation property:} & \widehat{\mathsf{L}}_f(gh) = (\widehat{\mathsf{L}}_f g)\, h + g\, (\widehat{\mathsf{L}}_f h) \\ \textit{bracket property:} & \widehat{\mathsf{L}}_f \{g, h\} = \{\widehat{\mathsf{L}}_f g, h\} + \{g, \widehat{\mathsf{L}}_f h\}. \end{array} \tag{1.3.39}$$

---

[c] In the literature, the Lie operator, variously denoted as : f : [12.1], $\widehat{f}$ [12.5], or $\{f, \circ\}$ [11], is commonly defined by the left Poisson bracket $\{f, \cdot\}$, whilst the operator (1.3.36) defined by the right Poisson bracket is reported as *vector field* [3.2, 10.2]. Evidently the operators $\{f, \cdot\}$ and $\{\cdot, f\}$ differ by a minus sign. We prefer to introduce the Lie operator through the "opposite" definition (1.3.36), in order to avoid the cumbersome presence of the minus sign in the equations, we will encounter in the following.

Like the Poisson bracket, the Lie operator acts on the ordinary and Poisson product of functions in the similar manner as the derivative operator, and hence through a Leibnitz-like rule.

We consider now the set of dynamical variables $\{f(q,p)\}$, which, as previously remarked, configures as a Lie algebra under the Poisson bracket operation. Evidently, by (1.3.36) we can individualize a correspondence of $\{f(q,p)\}$ to the set of the associated Lie operators $\{\widehat{\mathsf{L}}_f\}$:

$$\{f(q,p)\} \;\rightarrow\; \{\widehat{\mathsf{L}}_f\}. \tag{1.3.40}$$

This set can be structured as a linear space; in fact, exploiting the linearity of the Poisson bracket operation, we define linear combinations of Lie operators by the rule

$$a\widehat{\mathsf{L}}_f + b\widehat{\mathsf{L}}_g = \widehat{\mathsf{L}}_{af+bg} = \{\cdot, af + bg\}\,, \tag{1.3.41}$$

to obtain yet a Lie operator.

Notably, the operators $\{\widehat{\mathsf{L}}_f\}$ span a Lie algebra with respect to the commutation between operators. We recall that the commutator of two operators $\widehat{\mathsf{F}}$ and $\widehat{\mathsf{G}}$ is defined by the antisymmetric product

$$[\,\widehat{\mathsf{F}}, \widehat{\mathsf{G}}\,] \;=\; \widehat{\mathsf{F}}\widehat{\mathsf{G}} - \widehat{\mathsf{G}}\widehat{\mathsf{F}}. \tag{1.3.42}$$

The direct calculation shows that the commutator of two Lie operators, $\widehat{\mathsf{L}}_f$ and $\widehat{\mathsf{L}}_g$, is equal to the Lie operator associated with the Poisson bracket of the generating functions g and f; namely

$$[\,\widehat{\mathsf{L}}_f, \widehat{\mathsf{L}}_g\,] = -\,\{\cdot, \{f,g\}\} = -\,\widehat{\mathsf{L}}_{\{f,g\}} = \widehat{\mathsf{L}}_{\{g,f\}}. \tag{1.3.43}$$

The commutator of two Lie operators is therefore a Lie operator as well. This is just the basic requisite for the linear space $\{\widehat{\mathsf{L}}_f\}$ be a Lie algebra under the commutator product (1.3.42). In addition, relation (1.3.43) states that the correspondence (1.3.40) is an homomorphism [9.6], as the algebra-generating operation between elements of $\{f(q,p)\}$, i.e., the Poisson bracket $\{f,g\}$, is imaged to the algebra-generating operation between the corresponding elements of $\{\widehat{\mathsf{L}}_f\}$, i.e., the commutator bracket $[\,\widehat{\mathsf{L}}_f, \widehat{\mathsf{L}}_g\,]$.

As an example, with $f = \frac{q^2}{2}$, $g = \frac{p^2}{2}$ and hence $\{f,g\} = qp$, we find: $f \rightarrow \widehat{\mathsf{L}}_f = -q\frac{\partial}{\partial p}$, $g \rightarrow \widehat{\mathsf{L}}_g = p\frac{\partial}{\partial q}$ and $\{f,g\} \rightarrow \widehat{\mathsf{L}}_{\{f,g\}} = q\frac{\partial}{\partial q} - p\frac{\partial}{\partial p} = -[\,\widehat{\mathsf{L}}_f, \widehat{\mathsf{L}}_g\,]$.

1.4 Hamilton's equations for the light-ray

Hamilton's equations can be cast in terms of Poisson brackets. In fact, on account of (1.3.30), we may rewrite (1.2.13) in the suitable vector form

$$\frac{d}{dz}\mathbf{u}(z) = \{\mathbf{u}(z), H(q,p,z)\}\,, \qquad \mathbf{u}(z_i) = \mathbf{u}_i \tag{1.4.1}$$

the ray-vector $\mathbf{u}(z)$ containing the z-dependent ray variables, whose known values at the entrance plane at z_i are grouped into the vector $\mathbf{u}_i$.

Then, introducing the Lie operator associated with the optical Hamiltonian according to (1.3.37); namely,

$$\widehat{\mathsf{L}}_H(z) = \{\cdot, H\} = \sum_{\alpha=x,y} \left(\frac{\partial H}{\partial p_\alpha} \frac{\partial}{\partial q_\alpha} - \frac{\partial H}{\partial q_\alpha} \frac{\partial}{\partial p_\alpha} \right), \tag{1.4.2}$$

Eqs. (1.4.1) are recast into the concise and elegant form

$$\frac{d}{dz}\mathbf{u}(z) = \widehat{\mathsf{L}}_H(z)\,\mathbf{u}(z), \qquad \mathbf{u}(z_i) = \mathbf{u}_i \tag{1.4.3}$$

which combines both the vector and Lie-operator notation. For notational conciseness, only the possible dependence of $\widehat{\mathsf{L}}_H$ on z is explicitly shown.

As already remarked, the solution of Eq. (1.4.3) is the answer to the fundamental problem of geometrical optics: to determine the input-output relations for a given optical device, which allow us to find out the final vectors $\mathbf{u}_o$ for assigned input vectors $\mathbf{u}_i$, or, inversely, to design an optical system in order that the ray transfer relations $\mathbf{u}_i \to \mathbf{u}_o$ have some desired properties.

Equations of the type (1.4.3) appear in several and differing fields of physics. A large variety of both analytic and numerical methods of integration has been devised, which provide explicit expressions for exact or approximate analytic solutions and define general procedures for numerical algorithms with definite degree of accuracy [12, 13]. Indeed, the integration of (1.4.3) would deserve a proper treatment of its own. Here we will give a little insight into the problem, supporting the proper formal approach by intuitive considerations; emphasis will be put on the operator nature of (1.4.3). The reader is assumed to be familiar with the prime concept of "operator" and with the basic rules of the "arithmetic of operators". The considerations below, and those that will be developed in the next Chapters, will clearly define the link between physical optical systems and abstract operators, which are intended as the mathematical objects conceived to represent those systems in relation with the effect they produce on the light disturbance. We suggest the reader mentally retains throughout a correspondence between the realizable optical system on the one hand and the relevant mathematical representation on the other.

1.4.1 *The ray transfer operator*

The concise form (1.4.3) of the equations of motion resembles that of a homogeneous first-order differential equation, which writes in general as $dy(z)/dz = a(z)y(z)$. The solution is well known to be the exponential function $y(z) = y(z_i)e^{\int_{z_i}^{z} a(z)dz}$, which simplifies into $y(z) = y(z_i)\; e^{a\cdot(z-z_i)}$ in the case of a z-independent coefficient: $a(z) = a$. Clearly the similarity is purely formal, as the 4-component ray vector $\mathbf{u}(z)$ enters (1.4.3) in place of the one function $y(z)$ and correspondingly the z-dependent Lie operator $\widehat{\mathsf{L}}_H(z)$ in place of the ordinary z-function $a(z)$. As is well known, operators behave in a substantially different way than c-functions, i.e., functions that send scalars to scalars. On approaching the integration of Eq. (1.4.3) we are unavoidably confronted with the operator nature of $\widehat{\mathsf{L}}_H$, which may also depend on the axial parameter z. When the Lie operator $\widehat{\mathsf{L}}_H$ is z-independent, the similarity with the ordinary differential equation may be brought a step further, since, as we will see, the solution to (1.4.3) takes an exponential-like form.

In fact, when the Hamiltonian function H does not depend explicitly on the axial variable z, as in the case of propagation in a homogeneous medium where $n(x, y, z) = n_0$, the associated Lie operator $\widehat{\mathsf{L}}_H$ is z-independent as well. The sought solution can then be given the transparent exponential form

$$\mathbf{u}(z) = e^{(z-z_i)\cdot\widehat{\mathsf{L}}_H}\,\mathbf{u}_i. \tag{1.4.4}$$

The propagation of the light ray is seen as the effect of the action of the operator $e^{(z-z_i)\cdot\widehat{\mathsf{L}}_H}$, which in phase space transfers the point $\mathbf{u}_i$ to the point $\mathbf{u}(z)$, and correspondingly in real space the ray at z_i to the ray at z.

Since the exponential function is defined for operators exactly like for scalars, the operator $e^{(z-z_i)\cdot\widehat{\mathsf{L}}_H}$ is understood as the power series

$$e^{(z-z_i)\cdot\widehat{\mathsf{L}}_H} \equiv \sum_{j=0}^{\infty} \tfrac{(z-z_i)^j}{j!}\widehat{\mathsf{L}}_H^{\,j}. \tag{1.4.5}$$

Exponential series of Lie operators, like (1.4.5), are in general known as *Lie transformations* [11]. The theory of Lie transformations develops quite naturally from the theory of Lie operators. In fact, the Lie transformations $\{e^{t\widehat{\mathsf{L}}_f}; t \in \mathbb{R}\}$ constitute the Lie group associated with the Lie algebra $\{\widehat{\mathsf{L}}_f\}$ of the Lie operators; roughly speaking, we may say that this group is obtained by exponentiating the corresponding Lie algebra through a real parameter. It is a common language in the theory of groups to refer to $\{\widehat{\mathsf{L}}_f\}$ as the algebra of the generators of the group of transformations $\{e^{t\widehat{\mathsf{L}}_f}; t \in \mathbb{R}\}$. Additional comments on Lie transformations are given in § 2.2.2.

It is easy to prove that the vector (1.4.4) is a solution of Hamilton's equations (1.4.3) corresponding to the initial values $\mathbf{u}_i$. Differentiating both sides of (1.4.4) with respect to z and summing back to $\mathbf{u}(z)$, we end up with the desired proof

$$\frac{d}{dz}\mathbf{u}(z) = \sum_{j=1}^{\infty} \tfrac{(z-z_i)^{j-1}}{(j-1)!}\widehat{\mathsf{L}}_H^{\,j}\,\mathbf{u}_i = \widehat{\mathsf{L}}_H \cdot e^{(z-z_i)\cdot\widehat{\mathsf{L}}_H}\,\mathbf{u}_i = \widehat{\mathsf{L}}_H\mathbf{u}(z). \tag{1.4.6}$$

However, the above vector provides only the formal answer to the problem of ray evolution, which actually is solved once the action of the operator $e^{(z-z_i)\cdot \widehat{\mathsf{L}}_H}$ on the initial vector $\mathbf{u}_i$ is explicitly defined. The functional form of $e^{(z-z_i)\cdot \widehat{\mathsf{L}}_H}$ is obviously determined by the Hamiltonian function $H(q,p)$, which in turn reflects the geometry and optical features of the system. In other words, expression (1.4.4) is basically a series expansion, whose terms result from repeatedly Poisson bracketing H with $\mathbf{u}_i$. It is the Hamiltonian function that determines the form of the various terms in the series and hence the possibility of summing them to a closed explicit form.

Additional comments deserves the case when the Hamiltonian depends explicitly on z, and so also $\widehat{\mathsf{L}}_H$ does. In that case the relation between the ray variables at the planes z_i and z, as expressed in (1.4.4), is not properly correct. Although an operator-form representation for the evolution of the ray variables can still be guessed, it cannot be given in the form of the single Lie transformation $e^{(z-z_i)\cdot \widehat{\mathsf{L}}_H}$, i.e., as the exponential of the Lie operator $\widehat{\mathsf{L}}_H$ over the whole propagation length from z_i to z.

We may in general relate the ray variables at z_i and z in the operator form

$$\mathbf{u}(z) = \widehat{\mathfrak{M}}\,(z, z_i)\,\mathbf{u}_i, \tag{1.4.7}$$

reflecting the view of the initial vector $\mathbf{u}_i$ being transformed into $\mathbf{u}(z)$ by the operator $\widehat{\mathfrak{M}}(z, z_i)$, which accordingly is called the *ray transfer operator.*

As proved before, when the Hamiltonian is z-independent, the transfer operator is exactly represented by the Lie transformation (1.4.5):

$$\widehat{\mathfrak{M}}\,(z, z_i) = e^{(z-z_i)\cdot \widehat{\mathsf{L}}_H}. \tag{1.4.8}$$

If the Hamiltonian depends on z, this representation no longer holds. This is due to the fact that the arithmetic for operators is not the same as for scalars. Operators are objects that may not obey the commutative law under multiplication, and hence it is not generally true that $\widehat{\mathsf{A}}\widehat{\mathsf{B}} = \widehat{\mathsf{B}}\widehat{\mathsf{A}}$ for operators, whilst it is always true that $ab = ba$ for scalars. Accordingly, exponential operators may not obey the semigroup property; thus it is not generally true that $e^{\widehat{\mathsf{A}}}\cdot e^{\widehat{\mathsf{B}}} = e^{\widehat{\mathsf{A}}+\widehat{\mathsf{B}}}$ for operators, whilst it is always true that $e^a \cdot e^b = e^{a+b}$ for scalars. To gain a feeling for this peculiar aspect of the arithmetic of operators, we arrange Eq. (1.4.3) as

$$d\mathbf{u}(z) = \widehat{\mathsf{L}}_H(z)\,dz \cdot \mathbf{u}(z). \tag{1.4.9}$$

Substituting for $d\mathbf{u}(z) = \mathbf{u}(z+dz) - \mathbf{u}(z)$, we obtain

$$\mathbf{u}(z+dz) = [1 + \widehat{\mathsf{L}}_H(z)\,dz]\,\mathbf{u}(z) = e^{\widehat{\mathsf{L}}_H(z)\,dz}\,\mathbf{u}(z), \tag{1.4.10}$$

where the exponential representation on the right-hand side is in accord with the power series (1.4.5) for infinitesimal dz's. The above describes the ray propagation by the infinitesimal length from z to $z + dz$. As noted in § 1.3.1, the propagation by a finite length can be thought of as an infinite chain of infinitesimal propagations like (1.4.10), sequentially ordered according to the axial variable. Thus, for instance, we have

$$\mathbf{u}[(z+dz)+dz] = e^{\widehat{\mathsf{L}}_H(z+dz)\,dz}\,\mathbf{u}(z+dz) = e^{\widehat{\mathsf{L}}_H(z+dz)\,dz}\,e^{\widehat{\mathsf{L}}_H(z)\,dz}\,\mathbf{u}(z). \quad (1.4.11)$$

The two exponential operators above multiply to one exponential operator $e^{[\widehat{\mathsf{L}}_H(z+dz)+\widehat{\mathsf{L}}_H(z)]\,dz}$ only under the condition that the exponents $\widehat{\mathsf{L}}_H(z)$ and $\widehat{\mathsf{L}}_H(z+dz)$ commute with each other, i.e., $\widehat{\mathsf{L}}_H(z)\widehat{\mathsf{L}}_H(z+dz) = \widehat{\mathsf{L}}_H(z+dz)\widehat{\mathsf{L}}_H(z)$. In that case, sequentially concatenating propagations by all the infinitesimal segments dz's along the desired distance from z_i to z, we end up with the exponential representation

$$\widehat{\mathfrak{M}}\,(z, z_i) = e^{\int_{z_i}^{z}\widehat{\mathsf{L}}_H\,(z)\,dz}, \quad (1.4.12)$$

in a certain analogy with the solution $y(z) = y(z_i)y(z_i)e^{\int_{z_i}^{z}a(z)dz}$ of the scalar homogeneous differential equation. This is not surprising as the commutativity makes operators closer to scalars. Also, we regain the simple form (1.4.8) since, for a z-independent generator $\widehat{\mathsf{L}}_H$, $\int_{z_i}^{z}\widehat{\mathsf{L}}_H\,dz = (z - z_i)\,\widehat{\mathsf{L}}_H$. Evidently the ray evolution is found once the operator (1.4.12) is left to act on the input data to yield the explicit ray-coordinates at z.

In general, as the Lie operators $\widehat{\mathsf{L}}_H(z)$'s at different z do not commute with each other, the semigroup property of exponential scalars is not recovered within the operator context. We are then confronted with the problem of seeking for a proper representation, exact or approximate, for the transfer operator $\widehat{\mathfrak{M}}(z, z_i)$, i.e., with the problem to give the infinite product of z-ordered infinitesimal transformations $e^{\widehat{\mathsf{L}}_H(z)\,dz}$ an explicit, possibly closed, functional form. Readers interested in the detailed aspects concerning the integration of equations of the type (1.4.3) are addressed to [12, 13].

In the forthcoming sections we will consider the ray propagation through *linear* (or *first-order*) optical systems, which can effectively be described by ray-parameter independent 4×4 matrices. More specifically, we will consider systems, whose symmetry properties allow to reduce the dimensionality of the problem from the $4D$ to the $2D$ optical phase space, so that we will need only a 2×2 matrix to describe the ray propagation. We will see that linear optical systems arise from Hamiltonians that are, or are approximated as, quadratic polynomials in the phase space variables; departures from the linearity are identified as *aberrations* [14]. Thus, second-degree polynomials are seen to play a distinguished role within geometrical optics; indeed, linear optical systems stand as the concrete realizations of Hamiltonian dynamics, generated by polynomials quadratic in the phase space variables.

1.5 Lie transformations in the optical phase space

The Hamilton-Lie formalism, described in the previous sections, allows us to identify a path to integrate Hamilton's equations for the light ray to the corresponding symplectic transformation in the geometrical-optical phase space:

$$H \xrightarrow{\text{Poisson bracket operation}} \widehat{\mathsf{L}}_H \xrightarrow{\text{Exponentiation}} e^{(z_o - z_i)\widehat{\mathsf{L}}_H} . \tag{1.5.1}$$

The resulting phase space transformation:

$$\mathbf{u}(z_i) \xrightarrow{\text{by } e^{(z_o - z_i)\widehat{\mathsf{L}}_H}} \mathbf{u}(z_o) = e^{(z - z_i)\cdot\widehat{\mathsf{L}}_H}\, \mathbf{u}(z_i) , \tag{1.5.2}$$

reflects the evolution of the ray along the optical z-axis.

It is evident that, as exemplified in (1.5.1) for optical Hamiltonians, we can generate phase-space transformations, having or not a direct link with the z-evolution of the light ray. We may consider any continuously differentiable phase-space variable $f(q,p)$ and then form the corresponding Lie transformation, whose "validity" as a phase-space transformation is established according to whether it correctly maps the geometric-optical phase-space into itself. We recall in fact that the optical phase space has the peculiar $C_2 \times \mathbb{R}^2$ structure in contrast with the $\mathbb{R}^4$ structure of the mechanical phase space.

We will give some examples of Lie transformations in the optical phase space. In particular, in §§ 1.5.1 and 1.5.2 we will consider as phase-space functions $f(q,p)$ linear and quadratic polynomials in q and p. In that case, we can easily obtain the explicit expressions of the transformed phase space vector, simply using the exponential series expansion (1.4.5). Then in § 1.5.3 we will briefly describe the factorization-based method widely used to deal with the Lie transformations generated by higher-order polynomials.

To simplify notations we will consider the single pair of conjugate variables (q,p), and hence we will account for the proper $[-n,n] \times \mathbb{R}$ structure of the relevant $2D$ phase-space. We will denote by ζ the real parameter of the exponentiation from the Lie operator $\widehat{\mathsf{L}}_f$ to the Lie transformation $e^{\zeta\widehat{\mathsf{L}}_f}$. We will not detail the physical meaning of ζ, which depends, of course, on that of the variable $f(q,p)$ in order that the exponent $\zeta\widehat{\mathsf{L}}_f$ be dimensionless. Thus, if $f(q,p)$ can be interpreted as an optical Hamiltonian, ζ may acquire the meaning as the evolution variable, and accordingly the relevant Lie transformation that as the ray-transfer operator.

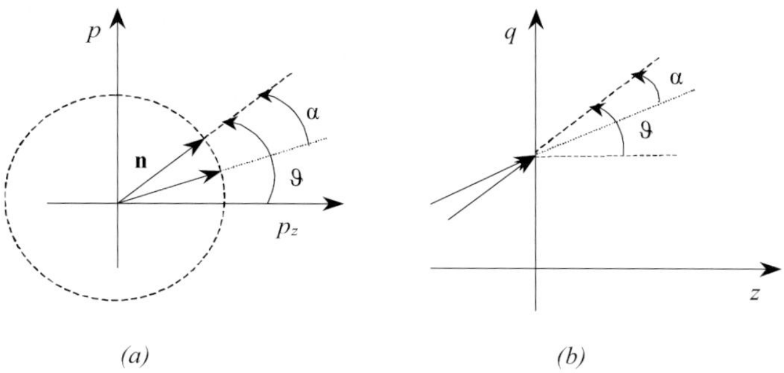

FIGURE 1.5. As the tip of the ray-momentum vector $\mathbf{n} = (p, p_z)$ is confined on the circle of radius n, *(a)* the "momentum translation" is a rotation of the vector, with the tip slipping along the circle, which amounts to *(b)* a variation of the direction of the ray with respect to the q-z reference axes.

1.5.1 *Linear polynomials in q and p*

The momentum variable p generates a shift of the position coordinate of the phase-space vector, without affecting the relative momentum coordinate. In fact, being

$$\widehat{\mathsf{L}}_p = \frac{\partial}{\partial q}, \tag{1.5.3}$$

the pair (q, p) is turned into

$$\begin{pmatrix} q(\zeta) \\ p(\zeta) \end{pmatrix} = e^{\zeta \cdot \partial/\partial q} \begin{pmatrix} q \\ p \end{pmatrix} = \sum_{j=1}^{\infty} \frac{\zeta^j}{j!} \frac{\partial^j}{\partial q^j} \begin{pmatrix} q \\ p \end{pmatrix} = \begin{pmatrix} q + \zeta \\ p \end{pmatrix}, \tag{1.5.4}$$

which may correspond to a shift of the ray position along the reference axis.

Likewise, the position variable q, the associated Lie operator being

$$\widehat{\mathsf{L}}_q = -\frac{\partial}{\partial p}, \tag{1.5.5}$$

would generate a shift in the momentum coordinate of the phase-space vector as

$$\begin{pmatrix} q(\zeta) \\ p(\zeta) \end{pmatrix} = e^{-\zeta \cdot \partial/\partial p} \begin{pmatrix} q \\ p \end{pmatrix} = \sum_{j=1}^{\infty} \frac{(-\zeta)^j}{j!} \frac{\partial^j}{\partial p^j} \begin{pmatrix} q \\ p \end{pmatrix} = \begin{pmatrix} q \\ p - \zeta \end{pmatrix}. \tag{1.5.6}$$

Since by (1.5.6) the momentum coordinate could be carried out of its proper optical range: $p^2 \leq n^2$, the Lie transformation $e^{-\zeta \cdot \partial/\partial p}$ does *not* yield in general a proper transformation in the optical phase-space $[-n, n] \times \mathbb{R}$, which could not be properly mapped into itself.

Actually, the "ray-momentum translation" should signify a variation of the ray direction, which evidently can be realized by a rotation of the ray about its position on the reference axis. As shown in Fig. 1.5, we denote by ϑ the angle the ray forms with the optical axis. Hence, the transverse and longitudinal components of the ray momentum, p and p_z, write as

$$\begin{aligned} p &= n \sin \vartheta, \\ p_z &= n \cos \vartheta. \end{aligned} \tag{1.5.7}$$

Rotating clockwise the ray by the angle α about its position on the reference axis q, the components of the ray momentum become

$$\begin{aligned} \overline{p} &= n \sin(\vartheta - \alpha) = p \cos \alpha - p_z \sin \alpha, \\ \overline{p}_z &= n \cos(\vartheta - \alpha) = p \sin \alpha + p_z \cos \alpha. \end{aligned} \tag{1.5.8}$$

As we will discuss in detail in Sect. 1.6, linear geometrical optics assumes that the direction of the ray is almost parallel to the optical z-axis, so that $p \ll p_z \simeq n$. Then, with both the angles ϑ and α taken enough small to justify the first-order approximation of relations (1.5.7) and (1.5.8), we obtain

$$\overline{p} \simeq p - n\alpha, \qquad \overline{p}_z \simeq n, \tag{1.5.9}$$

being as well $p \simeq n\vartheta$. Evidently, within the linear approximation of first-order optics, a variation of the ray direction may be pictured in the optical phase-space as a shift of the momentum coordinate, which safely remains well inside its proper optical range. Hence, in the linear optics regime the Lie transformation $e^{-\zeta \cdot \partial / \partial p}$, generated by the position variable q, yields a proper transformation in the geometrical-optical phase-space, actually involving a small region around the origin well inside the inherent strip $[-n, n] \times \mathbb{R}$.

1.5.2 *Quadratic polynomials in q and p*

We consider now the Lie transformations, generated by quadratic polynomials, and describe their effect on the phase-space vectors. We disregard the question linked to the proper optical range for the ray momentum, assuming to work in the linear approximation according to which, as said earlier, the rays propagate in the close proximity of the optical axis ($p_x^2 + p_y^2 \ll n^2$ and $q_x^2 + q_y^2 \ll 1$). In other words, we suppose that only small regions in the neighbourhood of the phase-space origin are involved in the transformations we consider. We are therefore enabled to deal with the optical phase space as if it were the $\mathbb{R}^4$-hyperplane or the $\mathbb{R}^2$-plane, when considering, as we will do below, only one transverse coordinate. We will see that the dynamics generated by quadratic polynomials is depicted in phase plane by portions of (nondegenerate or degenerate) conics. The discussion anticipates the more accurate analysis we will develop in Chapters 2 and 3.

We consider firstly

$$f(q,p) = \tfrac{1}{2}p^2, \tag{1.5.10}$$

which within the analogy with mechanics corresponds to the Hamiltonian of a free particle. The associated Lie operator is

$$\widehat{\mathsf{L}}_{p^2/2} = p\frac{\partial}{\partial q}, \tag{1.5.11}$$

which yields the phase plane transformation

$$\begin{aligned}\binom{q(\zeta)}{p(\zeta)} &= e^{\zeta\cdot p\partial/\partial q}\binom{q}{p}_i = \binom{q}{p}_i + \zeta\, p\tfrac{\partial}{\partial q}\binom{q}{p}_i + \\ &+\tfrac{1}{2}\zeta^2\,(p\tfrac{\partial}{\partial q})(p\tfrac{\partial}{\partial q})\binom{q}{p}_i + \ldots = \binom{q_i}{p_i} + \zeta\,\binom{p_i}{0} = \binom{q_i+\zeta\, p_i}{p_i},\end{aligned} \tag{1.5.12}$$

describing a "free" evolution with the momentum coordinate being unchanged: $p(\zeta) = p_i$. It is easily verified that $\{q(\zeta), p(\zeta)\} = \{q_i, p_i\} = 1$.

Correspondingly, with

$$f(q,p) = \tfrac{1}{2}q^2, \tag{1.5.13}$$

which may be put in relation with the potential part of the Hamiltonian of a classical particle experiencing an elastic force, we obtain

$$\widehat{\mathsf{L}}_{q^2/2} = -q\frac{\partial}{\partial p}, \tag{1.5.14}$$

and hence

$$\begin{aligned}\binom{q(\zeta)}{p(\zeta)} &= e^{-\zeta\, q\partial/\partial p}\binom{q}{p}_i = \binom{q}{p}_i - \zeta\, q\tfrac{\partial}{\partial p}\binom{q}{p}_i + \\ &+\tfrac{1}{2}\zeta^2\,(q\tfrac{\partial}{\partial p})(q\tfrac{\partial}{\partial p})\binom{q}{p}_i + \ldots = \binom{q_i}{p_i} + \zeta\,\binom{0}{q_i} = \binom{q_i}{p_i-\zeta\, q_i}.\end{aligned} \tag{1.5.15}$$

In contrast with (1.5.12), the above transformation leaves unchanged the position coordinate $q(\zeta) = q_i$.

The mixed polynomial

$$f(q,p) = qp, \tag{1.5.16}$$

generates the Lie operator

$$\widehat{\mathsf{L}}_{qp} = q\frac{\partial}{\partial q} - p\frac{\partial}{\partial p}, \tag{1.5.17}$$

whose powers $\widehat{\mathsf{L}}^{\,j}_{qp}$ affect the phase-plane vector according to

$$\widehat{\mathsf{L}}_{qp}\begin{pmatrix}q\\p\end{pmatrix} = \begin{pmatrix}q\\-p\end{pmatrix},\quad \widehat{\mathsf{L}}^{2}_{qp}\begin{pmatrix}q\\p\end{pmatrix} = \begin{pmatrix}q\\p\end{pmatrix}, \ldots, \widehat{\mathsf{L}}^{j}_{qp}\begin{pmatrix}q\\p\end{pmatrix} = \begin{pmatrix}q\\(-)^j p\end{pmatrix}, \ldots \tag{1.5.18}$$

Then, summing up all the terms of the exponential series we end up with the transformation

$$\begin{pmatrix} q(\zeta) \\ p(\zeta) \end{pmatrix} = e^{\zeta \cdot \widehat{\mathsf{L}}_{qp}} \begin{pmatrix} q \\ p \end{pmatrix}_i = \begin{pmatrix} e^{\zeta} q_i \\ e^{-\zeta} p_i \end{pmatrix}. \tag{1.5.19}$$

The conjugate variables are scaled by inverse factors so that $q(\zeta)p(\zeta) = q_i p_i$.

The phase-plane trajectories relevant to the "dynamics" described by (1.5.12), (1.5.15) and (1.5.19) are clearly portions of the straight lines $p = p_i$ and $q = q_i$, both originating from the initial point (q_i, p_i), and of the branch of the equilateral hyperbola through the point (q_i, p_i), having the q and p axes as asymptotes.

As a further example, we consider the quadratic polynomial

$$f_{\text{a.o.}}(q, p) = \tfrac{1}{2}p^2 + \tfrac{1}{2}q^2, \tag{1.5.20}$$

which can be regarded as the basic form of the attractive oscillator Hamiltonian. The associated Lie operator is

$$\widehat{\mathsf{L}}_{\text{a.o.}} = p\frac{\partial}{\partial q} - q\frac{\partial}{\partial p}, \tag{1.5.21}$$

whose effect on the phase plane vector is easily determined to be

$$\widehat{\mathsf{L}}_{\text{a.o.}} \begin{pmatrix} q \\ p \end{pmatrix} = [p\frac{\partial}{\partial q} - q\frac{\partial}{\partial p}] \begin{pmatrix} q \\ p \end{pmatrix} = \begin{pmatrix} p \\ -q \end{pmatrix}, \tag{1.5.22}$$

that is a clockwise rotation by $\frac{\pi}{2}$. We can thereby guess to recover the initial vector at the forth power, being $\frac{4\pi}{2} = 2\pi$. Readers can check this by direct calculation of the various powers, which sum finally to give

$$\begin{pmatrix} q(\zeta) \\ p(\zeta) \end{pmatrix} = e^{\zeta \cdot \widehat{\mathsf{L}}_{\text{a.o.}}} \begin{pmatrix} q \\ p \end{pmatrix}_i = \begin{pmatrix} q_i \cos\zeta + p_i \sin\zeta \\ -q_i \sin\zeta + p_i \cos\zeta \end{pmatrix}. \tag{1.5.23}$$

Evidently, the relative trajectory in phase plane lies on the circle $q^2 + p^2 = q_i^2 + p_i^2$, in analogy with the well known phase-space picture of the motion of the mechanical harmonic oscillator.

It is interesting to compare the "dynamics" (1.5.23) relative to the phase-plane function (1.5.20) with that generated by the polynomial

$$f_{\text{r.o.}}(q, p) = \tfrac{1}{2}p^2 - \tfrac{1}{2}q^2, \tag{1.5.24}$$

which can be regarded as the basic form of the Hamiltonian of a repulsive oscillator. The associated Lie operator is

$$\widehat{\mathsf{L}}_{\text{r.o.}} = p\frac{\partial}{\partial q} + q\frac{\partial}{\partial p}, \tag{1.5.25}$$

and the corresponding Lie transformation affects the phase plane vector as

$$\begin{pmatrix} q(\zeta) \\ p(\zeta) \end{pmatrix} = e^{\zeta \cdot \widehat{\mathsf{L}}_{\mathrm{r.o.}}} \begin{pmatrix} q \\ p \end{pmatrix}_i = \begin{pmatrix} q_i \cosh\zeta + p_i \sinh\zeta \\ q_i \sinh\zeta + p_i \cosh\zeta \end{pmatrix}. \tag{1.5.26}$$

The reader may easily see that the phase-plane trajectory relative to the "motion" (1.5.26) lies on the branch of the equilateral hyperbola $q^2 - p^2 = q_i^2 - p_i^2$ through the point (q_i, p_i), with the lines $p = \pm q$ being the asymptotes.

In Chapters 2 and 3 we will illustrate in details the phase plane view of the dynamics generated by all the above considered quadratic dynamical functions, with explicit reference to the optical systems they represent.

Finally, we can consider the polynomial

$$f(q,p) = \tfrac{1}{2}aq^2 + cqp + \tfrac{1}{2}bp^2, \tag{1.5.27}$$

with constant coefficients a, b and c, whose Lie operator is

$$\widehat{\mathsf{L}}_f = a\widehat{\mathsf{L}}_{q^2/2} + c\widehat{\mathsf{L}}_{qp} + b\widehat{\mathsf{L}}_{p^2/2} = (cq + bp)\frac{\partial}{\partial q} - (aq + cp)\frac{\partial}{\partial p}. \tag{1.5.28}$$

The reader can verify by direct calculation that the single-exponential form $e^{\zeta[(cq+bp)\partial/\partial q-(aq+cp)\partial/\partial p]}$ is less comfortably managed when applied to the starting data to determine the ray-coordinates $(q(\zeta), p(\zeta))$ in an explicit and closed form. An appropriate disentanglement technique may usefully be exploited to turn the single exponential operator into a more convenient product of exponential operators, each being suitably determined and placed in order that the inherent calculations become more feasible. Among the possible multiple-exponential forms of the Lie transformation $e^{\zeta[(cq+bp)\partial/\partial q-(aq+cp)\partial/\partial p]}$, we report for illustrative purposes the following ordered product

$$e^{\zeta\{a\widehat{\mathsf{L}}_{q^2/2}+c\widehat{\mathsf{L}}_{qp}+b\widehat{\mathsf{L}}_{p^2/2}\}} = e^{\mathfrak{a}(\zeta)\cdot\widehat{\mathsf{L}}_{q^2/2}}\, e^{\mathfrak{c}(\zeta)\cdot\widehat{\mathsf{L}}_{qp}}\, e^{\mathfrak{b}(\zeta)\cdot\widehat{\mathsf{L}}_{p^2/2}}. \tag{1.5.29}$$

The explicit expressions of the ζ-depending functions $\mathfrak{a}(\zeta)$, $\mathfrak{b}(\zeta)$ and $\mathfrak{c}(\zeta)$ in terms of the coefficients a, b and c can be found in the literature [13]. We invite the reader to work out the phase plane vector $\mathbf{u}(\zeta)$ generated by (1.5.29) from the input data $\mathbf{u}_i$, on account of relations (1.5.12), (1.5.15) and (1.5.19). Also, we suggest to compare the explicit expressions of the coefficients $\mathfrak{a}(\zeta)$, $\mathfrak{b}(\zeta)$ and $\mathfrak{c}(\zeta)$, and then to determine the range of ζ from $\zeta_i = 0$, through which the multiple-exponential representation (1.5.29) may properly be used.

As a conclusion, we evidence a significant feature common to all the ray-variable transformations, generated by second-degree polynomials. As it clearly emerges from the above results, the ray variables at any value of the stepping parameter ζ are determined by the input data through a linear relation with coefficients depending on ζ. The linearity of the mathematical relation between the ray variables during propagation is the distinctive mark of first-order ray-optics, as we will see in Sect. 1.7.

1.5.3 *Beyond the quadratic polynomials*

As noted in the previous paragraph, the homogeneous quadratic polynomials provide the basic mathematical models for first-order optical systems. Notably they represent the only Hamiltonian functions, for which the corresponding transfer operators can be given an exact analytic representation, typically in the form of a finite product of exponential operators. This is true for both z-independent and z-dependent Hamiltonians. Precisely, it is proved that the ray-transfer operator, we may denote by $\widehat{\mathfrak{M}}_2\,(z, z_i)$, associated with quadratic Hamiltonians is exactly equal, in a neighbourhood of the identity, to the product of the Lie transformations, generated by the three basic monomials q^2, p^2 and qp. Explicitly we may write

$$\widehat{\mathfrak{M}}_2\,(z, z_i) = e^{\alpha\,(z,z_i)\cdot \widehat{\mathsf{L}}_{q^2/2}}\, e^{\gamma\,(z,z_i)\cdot \widehat{\mathsf{L}}_{qp}}\, e^{\beta\,(z,z_i)\cdot \widehat{\mathsf{L}}_{p^2/2}}, \tag{1.5.30}$$

where the z-depending coefficients $\alpha\,(z, z_i)$, $\beta\,(z, z_i)$ and $\gamma\,(z, z_i)$ obey a set of nonlinear first-order differential equations, determined by the specific functional form of the Hamiltonian as a quadratic function of q and p.

Deviations from linearity, signified by powers of q or p higher than two in the optical Hamiltonian, can be accommodated by an approximate representation, which, preserving the factored product structure of the transfer operator, adds factors to the basic quadratic-Hamiltonian generated operator $\widehat{\mathfrak{M}}_2$ as [12.1]

$$\widehat{\mathfrak{M}}\,(z, z_i) = \cdots \widehat{\mathfrak{M}}_m\,(z, z_i) \cdots \widehat{\mathfrak{M}}_3\,(z, z_i)\widehat{\mathfrak{M}}_2\,(z, z_i), \quad m > 2, \tag{1.5.31}$$

thus yielding a modular, easily manageable, expression for the transfer operator, in which each accuracy-improving factor links to the previously and independently established operator modulus. Each added factor, $\widehat{\mathfrak{M}}_m\,(z, z_i)$, $m = 3, 4,..,$ in the conceptually infinite product (1.5.31) is in the form of a Lie transformation generated by a polynomial function of m-th degree. Clearly the number and the specific expressions of the added factors are determined by the required degree of accuracy and the functional form of the higher-order terms in the Hamiltonian. The analytical and numerical procedures to elaborate the representation (1.5.31) for $\widehat{\mathfrak{M}}\,(z, z_i)$ at the desired aberration order m are well documented in the literature [12, 14].

1.6 Linear ray optics and quadratic Hamiltonian functions

First-order (or *linear*) ray optics is marked by a linear relationship between the ray coordinates at the assigned end-planes. It is concerned with the behaviour of the *paraxial* rays, whose distances from the axis and inclination angles relative to the axis are supposed to be small. First-order ray optics assumes that such off-axis distances and inclination angles are both small enough to justify the lower order approximation for all the quantities pertaining to the ray propagation with respect to them[d]. The terms paraxial ray optics and first-order ray optics are frequently taken as equivalent.

Thus we can replace the arc element ds along the ray by the straight element dz along the optical axis: $ds \simeq dz$. The ray momenta (1.2.11) can therefore be written as

$$p_x = n\frac{dq_x}{ds} \simeq nq_x', \qquad p_y = n\frac{dq_y}{ds} \simeq nq_y', \tag{1.6.1}$$

i.e., in terms of the slopes q_x' and q_y' of the ray to the z axis, which are further approximated by the angles ϑ_x and ϑ_y or equivalently α_x and α_y (Fig. 1.4).

As a consequence of (1.6.1), we may establish a further characterization of the paraxial propagation in terms of the ray momentum according to

$$p_x^2 + p_y^2 = n^2(q_x'^2 + q_y'^2) \ll n^2, \tag{1.6.2}$$

thus suggesting the quadratic approximation of the optical Hamiltonian as

$$H(q,p,z) = \tfrac{1}{2n(q,z)}(p_x^2 + p_y^2) - n(q,z). \tag{1.6.3}$$

We see that the Hamiltonian of a paraxial ray has the same form as the Hamiltonian of a non relativistic particle with the well recognizable kinetic and potential parts [7]. Accordingly, the phase space of paraxial optics can be thought of as similar to the phase space of classical mechanics; in other words, paraxial optics "lives well within" the proper restriction $|p| \ll n$ (see Sect. 1.2), which then comes to represent no more a restriction but the "status".

Evidently first-order ray optics is concerned with optical systems whose refractive index varies gently in space. In fact, if the refractive index varies drastically in space, we are likely to find rays with large slopes and off-axis distances. We may therefore expand the function $n(q_x, q_y, z)$ with respect to the profile coordinates and retain only the constant, linear and quadratic terms to obtain

$$\begin{aligned} n(q_x, q_y, z) = {} & n_0(z) - n_{1x}(z)q_x - n_{1y}(z)q_y \\ & -\tfrac{1}{2}n_{2x}(z)q_x^2 - \tfrac{1}{2}n_{2y}(z)q_y^2 - n_{2xy}(z)q_xq_y, \end{aligned} \tag{1.6.4}$$

where $n_0(z) = n(0,0,z)$ denotes the refractive index along the optical axis. Needless to say, this approximation is limited to regions across the axis comprising values of the transverse coordinates such that $n(q_x, q_y, z) \geq 1$.

The optical Hamiltonian becomes then quadratic in the ray-variables

$$\begin{aligned} H(q,p,z) = {} & \tfrac{1}{2n_0(z)}(p_x^2 + p_y^2) + n_{1x}(z)q_x + n_{1y}(z)q_y \\ & +\tfrac{1}{2}n_{2x}(z)q_x^2 + \tfrac{1}{2}n_{2y}(z)q_y^2 + n_{2xy}(z)q_xq_y - n_0(z), \end{aligned} \tag{1.6.5}$$

---

[d] The circular functions of angles, for instance, admit the first-order approximation, i.e., $\tan\vartheta \simeq \sin\vartheta \simeq \vartheta$ and $\cos\vartheta \simeq 1$, for angles roughly estimated as $|\vartheta| \leq 15^\circ = 0.262$ rad. In fact, $\sin 15^\circ = 0.259$, $\tan 15^\circ = 0.268$ and $\cos 15^\circ = 0.966$.

with the same formal appearance as the Hamiltonian of the classical motion under a *weak* potential [3.1, 7]. Of course, the zero-order term, $-n_0(z)$, is unessential as its Lie operator is zero. On the other hand, the first-order terms are related to possible tilts and misalignements with respect to the optical axis; they vanish in centered systems as those we consider here.

As anticipated in § 1.5.2, the ray propagation under a quadratic Hamiltonian is described by a linear relation between the ray variables along the optical axis; in other words, quadratic Hamiltonians generate linear symplectic transformations in phase space. Terms of order higher than two in the optical Hamiltonian (identified as *aberrations*) generate nonlinear transformations in phase space (Problem 4). Aberrations can be elegantly classified and effectively computed by use of appropriate group-theoretical tools [12, 13]. The modular scheme (1.5.31) is an exemplary result of the group-theoretical treatment of aberrating systems. As noted, in it deviations of the system response from linearity are represented at every aberration order by symplectic maps suitably concatenated one to the other to calculate aberrations up to the desired order. The conceptual and computational advantages of group-theoretical treatments are then evident.

We consider now optical devices, whose geometry enables us to further simplify expression (1.6.5) of the optical Hamiltonian. Thus, in *orthogonal* systems the linear q_x, q_y and mixed q_xq_y terms in (1.6.5) can be dropped out; also, in *axially symmetric* systems the transverse dependence can be expressed through the radial distance $q_x^2 + q_y^2$ from the optical axis. In both cases, the ray propagation can be treated as a dynamical problem with one-degree of freedom, demanding then for a $2D$ phase space. Figure 1.6 shows the possible realization of an orthogonal and an axisymmetric system; an example of a nonorthogonal system is also depicted.

The presentation below is taken on a cursory and elementary level; it serves to justify the simplification from a $2D$ to a $1D$ model. The interested reader is directed to [4.4] for an exhaustive investigation of the role of "symmetries" in the Hamiltonian formulation, whose power indeed "*lies in the ability to yield over-riding results governing the behaviour of types of systems defined solely with respect to the symmetries they possess.*" In this connection, we emphasize that light propagation occurs inherently in the real $3D$ space. The expression "$2D$ (or $1D$) model" accords with the mechanical-like view, where the evolution variable is the "time" coordinate, but contrasts with the real optical view (for stationary media), where the evolution variable is itself a spatial coordinate. Clearly, when speaking of "$2D$" or "$1D$" picture, we properly refer to the number of profile coordinates suited to describe the propagation possibly in accord with the symmetry properties of the system.

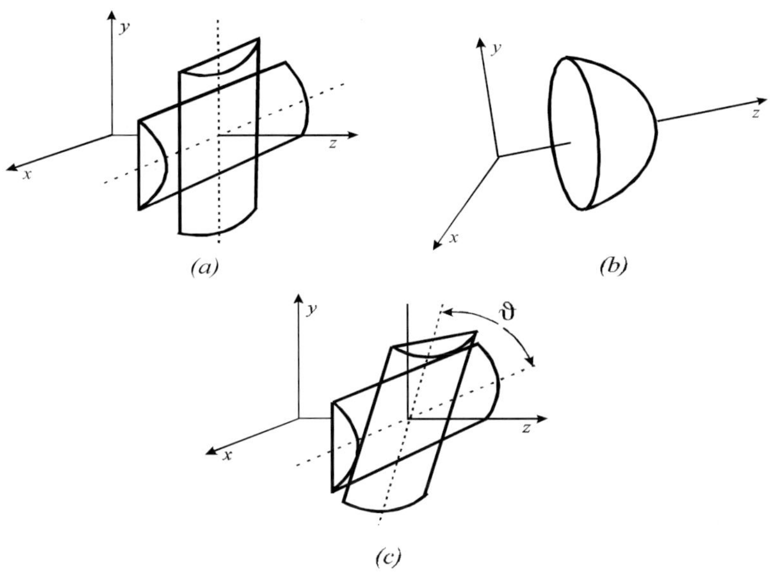

FIGURE 1.6. Examples of *(a)* an orthogonal optical system, *(b)* an axially symmetric optical system and *(c)* a nonorthogonal optical system.

1.6.1 *Orthogonal systems*

The configuration of Fig. 1.6.a) is a basic example of an orthogonal system, being in general characterized by two mutually orthogonal planes of symmetry intersecting at the optical axis. The system sketched in the figure consists, in particular, of two identical cylindrical lenses oriented with their axes orthogonal to one another[(e)]. Clearly the response of such a system to the incoming ray can be investigated by considering separately what happens in the two mutually perpendicular planes of symmetry, conveniently arranged to be the planes x-z and y-z. Systems which are symmetric about two orthogonal planes of symmetry intersecting at the optical axis are named *simple astigmatic*, whereas systems having no planes of symmetry are named *generally astigmatic* [4.2]. Anticipating the methods we will introduce later, we mention that simple astigmatic linear systems are described by two 2×2 matrices,

[e] A cylindrical lens has one or two cylindrical surfaces, their axes being strictly parallel. If the cylinder axis of the lens is aligned with the y axis, as the first in Fig. 1.6.a), the lens has the designed focal length for rays in the x-z plane, and no focusing power for rays in the y-z plane. In fact, a side view of the lens gives the appearance of a plano convex lens, while a top view shows a slab of transparent material. Cylindrical lenses are used to design anamorphic systems, which have different power or magnification between the two ordinarily orthogonal planes of symmetry; one can produce, for instance, a square image of a rectangular object.

accounting for the ray propagation in the two symmetry planes. As an example of a nonorthogonal system, we may think of a sequence of two cylindrical lenses oriented at an angle of, say, 45° to one another (Fig. 1.6.c)).

Dropping out the linear and mixed terms, the optical Hamiltonian (1.6.5) is seen to be the sum of two functions, separately depending on the ray coordinates (q_x, p_x) and (q_y, p_y) and in the same formal fashion, namely

$$\begin{aligned} H(q,p,z) &= \tfrac{1}{2n_0(z)}(p_x^2 + p_y^2) + \tfrac{1}{2}n_{2x}(z)q_x^2 + \tfrac{1}{2}n_{2y}(z)q_y^2 \\ &\equiv H_x(q_x,p_x,z) + H_y(q_y,p_y,z), \end{aligned} \tag{1.6.6}$$

with the $1D$ Hamiltonians H_x and H_y,

$$H_\xi(q_\xi, p_\xi, z) = \tfrac{1}{2n_0(z)}p_\xi^2 + \tfrac{1}{2}n_{2\xi}(z)q_x^2, \qquad \xi = x, y, \tag{1.6.7}$$

ruling the ray evolution in the two planes of symmetry x-z and y-z. Equations of motion (1.4.3) separate then into two independent sets of equations for the ray variables (q_x, p_x) and (q_y, p_y):

$$\frac{d}{dz}\mathbf{u}_\xi(z) = \widehat{\mathsf{L}}_{H_\xi}\,\mathbf{u}_\xi(z), \qquad \mathbf{u}_\xi(z_i) = \mathbf{u}_{\xi i}, \quad \xi = x, y, \tag{1.6.8}$$

where the 2-component vector $\mathbf{u}_\xi(z)$ comprises the ray coordinates (q_x, p_x) or (q_y, p_y) and $\widehat{\mathsf{L}}_{H_\xi}$ is the Lie operator associated with H_x or H_y:

$$\widehat{\mathsf{L}}_{H_\xi} = \tfrac{1}{n_0(z)}p_\xi\frac{\partial}{\partial q_\xi} - n_{2\xi}(z)q_\xi\frac{\partial}{\partial p_\xi} \qquad \xi = x, y. \tag{1.6.9}$$

In conclusion we may say that in the case of linear separable systems the general $2D$ problem of first-order optics, ruled by the Hamiltonian (1.6.5) with the one set of Eqs. (1.4.3) for the overall ray-motion, turns into two $1D$ problems, ruled by the Hamiltonians H_x and H_y with the two sets of Eqs. (1.6.8) for the "partial" ray-motions in the two symmetry planes of the system.

1.6.2 *Axially symmetric systems*

The hemi-spherical lens in Fig. 1.6.b) is an example of an optical system, which is symmetric under rotations around the optical axis. Systems which are circularly symmetric about the optical axis are named *stigmatic* [4.2].

In general, the refractive index in an axially symmetric device, with the optical axis conveniently taken as z axis, is a function of the axial coordinate z and the distance $\rho = \sqrt{q_x^2+q_y^2}$ from the axis: $n(q_x, q_y, z) = n(\rho, z)$ [4.2]. It is convenient to transform the rectangular coordinates q_x and q_y to the polar coordinates ρ and ϕ in the transverse plane, with

$$q_x = \rho\cos\phi, \quad q_y = \rho\sin\phi. \tag{1.6.10}$$

The optical Hamiltonian (1.2.12) writes therefore

$$H(\rho, \phi, p_\rho, p_\phi, z) = -\sqrt{n^2(\rho, z) - p_\rho^2 - \frac{p_\phi^2}{\rho^2}}. \tag{1.6.11}$$

The optical momenta p_ρ and p_ϕ are obtained from the proper Lagrangian function $L(\rho, \phi, \rho', \phi', z) = n(\rho, z)\sqrt{1+\rho'^2+\rho^2\phi'^2}$ by (1.2.9), and hence

$$p_\rho(z) = n(\rho, z)\frac{\rho'}{\sqrt{1+\rho'^2+\rho^2\phi'^2}}, \qquad p_\phi(z) = n(\rho, z)\frac{\rho^2\phi'}{\sqrt{1+\rho'^2+\rho^2\phi'^2}}. \tag{1.6.12}$$

Reflecting the axial symmetry of the system, the Hamiltonian (1.6.11) does not explicitly depend on the angle ϕ, and hence is invariant under rotations of the coordinate system in the transverse plane. In the language of classical mechanics ϕ is said a *cyclic* coordinate. The conjugate momentum p_ϕ is then a constant of the motion: $p_\phi(z) = p_\phi(z_i)$, determined by the initial values. The corresponding coordinate $\phi(z)$ is simply obtained integrating the second of (1.6.12), once the radial motion $\rho(z)$ and $p_\rho(z)$ has been solved from the relevant Hamilton's equations.

It is useful to introduce some definitions. A plane through the optical axis is referred to as a *meridional plane* or a *meridian.* A ray lying on a meridian is said a *meridional ray*, whilst a ray that does not lie on a meridian is a *skew ray.* The *skewness* is defined by the twisted product

$$\mathfrak{s} = q_x p_y - q_y p_x, \tag{1.6.13}$$

whose formal analogy with the angular momentum of a classical particle in a $2D$ motion is evident.

The skewness just equals the ray momentum p_ϕ: $p_\phi = q_x p_y - q_y p_x$, and so the invariance of p_ϕ in axisymmetric systems implies the invariance of $\mathfrak{s}$ as well. Analogously in the motion of a particle under a central force, the angular moment is invariant. Also, as p_ϕ relates to the sine of the angle of the ray with the meridian at the ray position, for a meridional ray we have that $p_\phi(z_i) = 0$, and hence $p_\phi(z) = 0$ at any other axial position. It is a property of axisymmetric systems that a meridian ray remains a meridian ray throughout the entire system, as a skew ray remains a skew ray.

Incidentally, we note that meridional and skew rays "feel" different refractive index functions. In fact, we may recast (1.6.11) in terms of a sort of an effective refractive index $n_{eff}(\rho, z)$ as

$$H(\rho, p_\rho, p_\phi, z) = -\sqrt{n_{eff}^2(\rho, z) - p_\rho^2}. \tag{1.6.14}$$

For a meridional ray n_{eff} is just the refractive index $n(\rho, z)$ of the medium:

$$n_{eff}(\rho, z) = n(\rho, z), \qquad \textit{meridional ray} \tag{1.6.15}$$

whilst for a skew ray

$$n_{eff}(\rho, z) = \sqrt{n^2(\rho, z) - \frac{p_\phi^2(z_i)}{\rho^2}}, \qquad \textit{skew ray} \tag{1.6.16}$$

due to the non vanishing constant value $p_\phi(z_i)$ of the skewness.

It is evident that the propagation of a meridional ray can be described by two variables only. In fact, because the ray remains coplanar with the optical axis, we may trace only two of the four ray variables, following the ray path in its own meridional plane, identified by the intersection of the ray with the input plane. The transfer properties of axisymmetric optical systems can efficaciously be determined by tracing properly chosen meridional rays. The tracing of skew rays is performed in the design of systems with large apertures, that in a sense contrasts with the hypothesis inherent in the first-order approximation [17.2].

Accordingly, as fas as meridional rays are concerned, we are again confronted with the $1D$ problem of tracing the ray path in a plane through the optical axis. Moreover, at the lower order in the ray coordinates ρ and p_ρ, the Hamiltonian (1.6.14) becomes quadratic, i.e.,

$$H(\rho, p_\rho, z) = \frac{1}{2n_0(z)} p_\rho^2 + \frac{1}{2} n_2(z) \rho^2, \tag{1.6.17}$$

in the meridional plane individualized by the input ray.

The optics of meridional rays in linear axisymmetric systems is known as *Gaussian* optics, in honor of K.F. Gauss [15], who approached the analysis of a refracting sphere by using a power series expansion in the off-axis distances and angles. The term Gaussian optics is frequently used in the literature as equivalent to paraxial optics or first-order optics.

1.7 Planar model of first-order optical systems

In accord with the considerations of the previous section, the propagation of paraxial (meridional) rays through simple astigmatic and stigmatic optical devices can be treated as a $1D$ dynamical problem, with the "plane of the motion" being respectively any one of the two planes of symmetry of the system or any one of the infinite meridional planes. In both cases, the relevant Hamiltonian is quadratic,

$$H(q, p, z) = \frac{1}{2n_0(z)} p^2 + \frac{1}{2} n_2(z) q^2, \tag{1.7.1}$$

with the associated Lie operator $\widehat{\mathsf{L}}_H$ being thereby

$$\widehat{\mathsf{L}}_H = \frac{1}{n_0(z)} p \frac{\partial}{\partial q} - n_2(z) q \frac{\partial}{\partial p}, \tag{1.7.2}$$

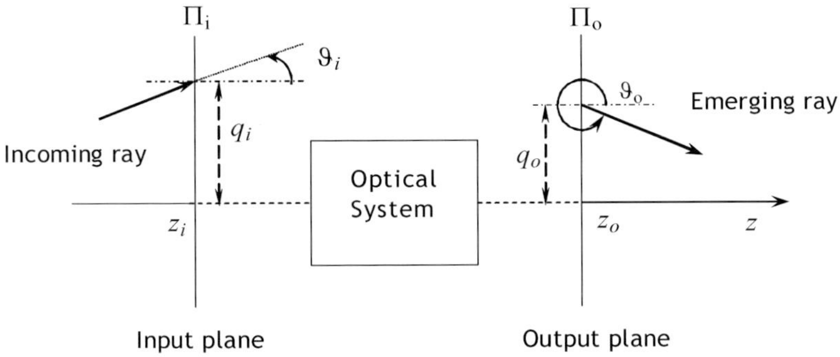

FIGURE 1.7. The ray coordinates in the linear approximation for a 2D optical system specify the height of the ray and the angle it forms relative to the optical axis.

and the ray coordinates q and p having the geometrical meaning as illustrated in Fig. 1.7. At any axial position z, $q(z)$ specifies the ray displacement relative to the axis and $p(z) = n_0(z)\vartheta(z)$ the *reduced* ray slope, given as the inclination angle ϑ of the ray to the axis, multiplied by the on-axis refractive index $n_0(z)$. Following the usual practice, q is taken positive upward from the axis and negative downward as ϑ is reckoned positive if a counterclockwise rotation takes us from axis to ray.

Interestingly, Hamiltonian (1.7.1) resembles that of the classical oscillator with varying mass and frequency. The equations for the ray variables have, in fact, the oscillator-like form

$$\begin{aligned} \tfrac{d}{dz} q &= \tfrac{1}{n_0(z)} p, \\ \tfrac{d}{dz} p &= -n_2(z) q, \quad q(z_i) = q_i\,, \ p(z_i) = p_i\,, \end{aligned} \tag{1.7.3}$$

where q_i and p_i specify the starting data.

A closer look at Eqs. (1.7.3) reveals the possibility of rewriting them into the nice matrix form

$$\frac{d}{dz}\mathbf{u}(z) = \mathbf{H}(z)\,\mathbf{u}(z), \qquad \mathbf{u}(z_i) = \mathbf{u}_i, \tag{1.7.4}$$

involving the phase-plane vector $\mathbf{u}(z) = (q(z), p(z))^{\top}$ and the 2×2 matrix

$$\mathbf{H}(z) = \begin{pmatrix} 0 & 1/n_0(z) \\ -n_2(z) & 0 \end{pmatrix}, \tag{1.7.5}$$

which can be regarded as the matrix representative of the Lie operator $\widehat{\mathsf{L}}_H$.

We discuss this aspect within the proper formal context.

1.7.1 *Quadratic polynomials and matrix representatives*

As a general property, any Poisson bracket based operator associated with a quadratic polynomial defined on a $2m$-dimensional phase space can be given a $2m \times 2m$ matrix representation.

We consider a $2D$ phase space appropriate to the present $1D$ optical model. Let $A(q,p)$ be a quadratic homogeneous polynomial in the conjugate variables q and p, i.e.,

$$A(q,p) = \frac{1}{2}aq^2 + \frac{1}{2}bp^2 + cqp, \tag{1.7.6}$$

with arbitrary real coefficients a, b and c. Here the possible dependence of A and hence of the coefficients a, b and c on the axial variable is unessential. As noted in § 1.3.2, the set $\{A(q,p)\}$ spans a Lie algebra under Poisson bracket.

The Poisson brackets of the conjugate variables q and p with A are linear functions of q and p. In fact, we find

$$\left\{ \begin{pmatrix} q \\ p \end{pmatrix}, A(q,p) \right\} = \widehat{\mathsf{L}}_A \begin{pmatrix} q \\ p \end{pmatrix} \equiv \begin{pmatrix} c & b \\ -a & -c \end{pmatrix} \begin{pmatrix} q \\ p \end{pmatrix} \tag{1.7.7}$$

where explicitly is $\widehat{\mathsf{L}}_A = (cq + bp)\frac{\partial}{\partial q} - (aq + cp)\frac{\partial}{\partial p}$. Evidently, relation (1.7.7) individualizes a correspondence

$$\{A(q,p)\} \longleftrightarrow \{\mathbf{A}\} \tag{1.7.8}$$

of the quadratic homogeneous polynomials $A(q,p)$, and hence of the associated Lie operators $\widehat{\mathsf{L}}_A$, to the 2×2 matrices

$$\mathbf{A} = \begin{pmatrix} c & b \\ -a & -c \end{pmatrix}, \tag{1.7.9}$$

whose entries are given by the coefficients of the quadratic monomials in A.

The matrix $\mathbf{A}$ can be regarded as the matrix representative of the quadratic form (1.7.6), and accordingly of the corresponding Lie operator $\widehat{\mathsf{L}}_A$. $\mathbf{A}$ has null trace being $Tr(\mathbf{A}) = c - c = 0$. Note that the matrix representative (1.7.5) of the quadratic Hamiltonian (1.7.1) is traceless.

The linear space of nonsingular square matrices configures as a Lie algebra with respect to the Lie bracket

$$[\mathbf{A}, \mathbf{B}] = \mathbf{AB} - \mathbf{BA}. \tag{1.7.10}$$

In particular, since $Tr([\mathbf{A}, \mathbf{B}]) = 0$, even the subspace of traceless square matrices is a Lie algebra with respect to the commutation product (1.7.10).

The Lie algebra $\{A(q,p)\}$ of the homogeneous quadratic polynomials is isomorphic to the Lie algebra of the traceless matrices $\{\mathbf{A}\}$, formed by the rule

(1.7.7). The isomorphism between two Lie algebras demands for a one-to-one mapping, which preserves the Lie products associated with the given algebraic structures [9.6]. It is not difficult to prove that if by (1.7.7) the matrices **A** and **B** correspond respectively to the quadratic homogeneous polynomials A and B, then the commutator $\mathbf{C} = [\mathbf{A}, \mathbf{B}]$ corresponds to the Poisson-bracket generated polynomial $\{A, B\}$. Hence, the correspondence (1.7.8), generated through (1.7.7), preserves the Lie products of the two algebras, i.e., the Poisson bracket for the quadratic homogeneous polynomials $\{A(q,p)\}$ and the Lie bracket for the 2×2 traceless matrices $\{\mathbf{A}\}$. In addition, the polynomial $A(q,p)$ can be recovered from the matrix (1.7.9), forming the scalar product

$$A(q,p) = \tfrac{1}{2}(\mathbf{Au}, \mathbf{Ju}) = \tfrac{1}{2}(q,p)\mathbf{A}^\top \mathbf{J} \left(\begin{smallmatrix} q \\ p \end{smallmatrix}\right), \tag{1.7.11}$$

where **J** is the 2×2 symplectic unit matrix and $\mathbf{A}^\top$ is the transpose of **A**. The mapping (1.7.8) is then an isomorphism of $\{A(q,p)\}$ onto $\{\mathbf{A}\}$.

1.7.2 *Input-output relation for first-order optical systems*

Consider the matrix form (1.7.4) of the linearized Hamilton equations (1.7.3).

Suppose to divide the space interval $z_o - z_i$ from the input to the output plane into a number N of segments Δz_j, $j = 1, .., N$ appropriately chosen so that in each the Hamiltonian matrix **H** be almost constant, i.e.,

$$\mathbf{H}(z) = \mathbf{H}(z_j) \equiv \mathbf{H}_j, \qquad z_j \le z < z_{j+1}, \quad j = 1, .., N, \tag{1.7.12}$$

with $z_1 \equiv z_i$ and $z_{N+1} \equiv z_o$. Clearly in each subinterval the solution to (1.7.4) is obtained by the exponential mapping of the respective matrix $\mathbf{H}_j$ as

$$\mathbf{u}(z) = e^{\mathbf{H}_j \cdot \Delta z_j}\, \mathbf{u}(z_j), \quad \Delta z_j = z_{j+1} - z_j, \quad z_j \le z < z_{j+1}, \quad j = 1, .., N. \tag{1.7.13}$$

The transfer of the ray from z_i to z_o, and hence of the phase space vector from $\mathbf{u}(z_i)$ to $\mathbf{u}(z_o)$, is then completed by concatenating the N exponential mappings (1.7.13) in the proper order according to

$$\mathbf{u}(z_o) = e^{\mathbf{H}_N \cdot \Delta z_N} \cdots e^{\mathbf{H}_2 \cdot \Delta z_2}\, e^{\mathbf{H}_1 \cdot \Delta z_1}\, \mathbf{u}(z_i), \tag{1.7.14}$$

and correctly taking the limit for $N \to \infty$, that is for infinitesimal elements $\Delta z_j \to dz$, for which condition (1.7.12) can safely be kept.

The exponential function is defined for matrices exactly as for scalars. Thus, each exponential in (1.7.14) means the power series

$$e^{\mathbf{H}_j \cdot \Delta z_j} \equiv \sum_{h=0}^{\infty} \frac{(\Delta z_j)^h}{h!} \mathbf{H}_j^h, \tag{1.7.15}$$

and so is seen to be a 2×2 matrix, as each power in the series is a 2×2 matrix.

Therefore we can ultimately express the input-output relation from z_i to z_o in the explicit matrix form

$$\begin{pmatrix} q_o \\ p_o \end{pmatrix} = \begin{pmatrix} A & B \\ C & D \end{pmatrix} \begin{pmatrix} q_i \\ p_i \end{pmatrix}, \tag{1.7.16}$$

thus establishing the linear relationship between the ray variables at the assigned planes, we claimed as the distinctive feature of first-order optics.

1.7.3 *The ray transfer matrix*

The 2×2 matrix in (1.7.16), which exemplifies the input-to-output relation for the ray propagation through a "linear" device between two assigned reference screens, is known as the *ray transfer* (or *ABCD*) *matrix* [16]. It is evident that the diagonal elements A and D are dimensionless, while the off-diagonal B and C have dimensions of length and inverse length, respectively. $\mathbf{M}$ is just the Jacobian matrix of the transformation $\binom{q_i}{p_i} \to \binom{q_o}{p_o}$ under the Hamiltonian (1.7.1); it is then symplectic and so unimodular: $\det \mathbf{M} = \mathbf{1}$ (see Problem 6).

The ray transfer matrix is the basic tool for both the diagnostic and the design of first-order optical systems. First-order optics is in fact mainly concerned with the determination of the optical matrices, which completely describe the paraxial ray transformations from the input to the output plane of the system.

Turn back again to Eq. (1.7.4). Following the suggestion of the transfer equation (1.7.16), we express the general solution to (1.7.4) in the form

$$\mathbf{u}(z) = \mathbf{M}(z, z_i)\mathbf{u}(z_i), \quad \mathbf{M}(z, z_i) = \begin{pmatrix} A(z, z_i) & B(z, z_i) \\ C(z, z_i) & D(z, z_i) \end{pmatrix} \tag{1.7.17}$$

letting the ray-matrix, i.e., its entries, vary continuously with the axial variable z, the assigned initial position z_i entering as a fixed parameter. Then, replacing for $\mathbf{u}(z)$ in (1.7.4), we obtain the equation obeyed by $\mathbf{M}(z, z_i)$ as

$$\frac{d}{dz}\mathbf{M}(z, z_i) = \mathbf{H}(z)\mathbf{M}(z, z_i), \quad \mathbf{M}(z_i, z_i) = \mathbf{I}. \tag{1.7.18}$$

The transfer matrix in (1.7.16) is a definite determination of $\mathbf{M}(z, z_i)$ for the specific value z_o of the evolution variable z; as the latter allows us to follow the ray evolution continuously along the propagation distance, the former catches two "instants" of this propagation at the assigned reference planes. If one is not interested in the details of the ray evolution along the optical system, once the relevant transfer matrix has been determined, the real optical system that fills the space between the end-planes becomes unessential; all we need to know about the transformation the system may act on the light rays is

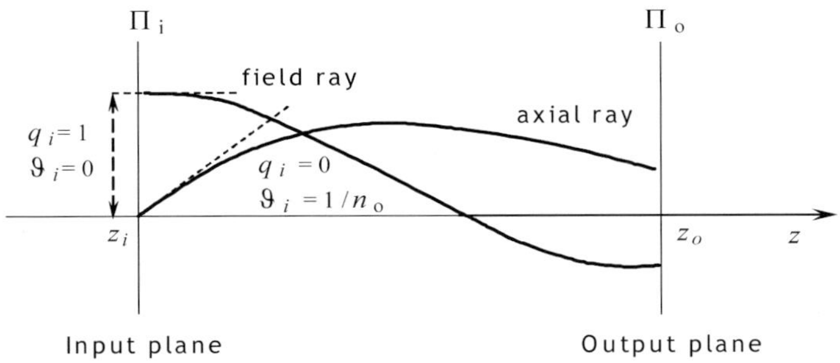

FIGURE 1.8. Cos-like and sine-like solutions of the linearized Hamilton's equations for the light ray.

stored in that matrix. The knowledge of the transfer matrix may be gained by integration of (1.7.18) or by other sort of considerations, as we will see later.

With the form (1.7.17) for the ray matrix $\mathbf{M}(z, z_i)$, Eq. (1.7.18) yields the set of first order differential equations for the entries A, B, C and D as

$$\begin{cases} \dfrac{dA}{dz} = \dfrac{C}{n_0}, & A(z_i) = 1, \\ \dfrac{dC}{dz} = -n_2 A, & C(z_i) = 0. \end{cases} \qquad \begin{cases} \dfrac{dB}{dz} = \dfrac{D}{n_0}, & B(z_i) = 0 \\ \dfrac{dD}{dz} = -n_2 B, & D(z_i) = 1. \end{cases} \tag{1.7.19}$$

Both pairs (A, C) and (B, D) obey Eqs. (1.7.3). They represent two particular rays evolving from the input data $(1, 0)$ and $(0, 1)$, respectively. In fact, the ray with coordinates (A, C) is a paraxial ray leaving the input plane parallel to the axis at the unitary height: $q_i = 1$; it is called *field ray.* On the other hand, the vector (B, D) describes a paraxial ray having null height and slope angle $\vartheta_i = 1/n_0$ at the entrance plane (Fig. 1.8); it is called *axial ray* [4.2]. Equation (1.7.17) suggests therefore a view of the ray propagation as resulting from the superposition of the axial and field ray evolutions with weights fixed by the initial values (q_i, p_i); clearly, superposition is a by-product of linearity.

Eventually we are confronted back with Hamilton's equations (1.7.3). But, we search now for two particular solutions, which, originating from specific sample rays, allow us to define the ray propagation matrix $\mathbf{M}(z, z_i)$. It is a significant consequence of linearity that the knowledge of the behaviour of the axial and field rays yields a global characterization of the ray propagation through the optical system (see Problem 7).

In conclusion we may say that the $1D$ ray-optics of first-order systems is wholly contained into Eqs. (1.7.16) and (1.7.18), the former picturing any first-order optical system as a "black box" fully characterized by its own ray-transfer matrix $\mathbf{M}$, and the latter tracing the procedure to determine $\mathbf{M}$, once specified the geometry and the optical properties of the system through $\mathbf{H}$, the matrix image of the relevant Hamiltonian function.

Alternative approaches to the integration of Eq. (1.7.4) or (1.7.18) make use of the Lie algebra based methods [13]. Matrices behave like operators rather than like scalars; they may not obey the commutative law of multiplication, and hence the rules of calculus for scalars do not apply in general to matrices as they do not apply in general to operators. On integrating Eq. (1.7.4) or (1.7.18), we are confronted with the operator-like nature of the matrix $\mathbf{H}$. Then, the considerations developed in § 1.4.1 regarding the integration of (1.4.3), apply here as well. We shall not detail the specific aspects of the problem. We only mention that, as in the case of the transfer operator, when the Hamiltonian matrix $\mathbf{H}$ does not explicitly depend on the axial variable z, we are enabled to represent $\mathbf{M}(z, z_i)$ in the closed form as the single-exponential of $\mathbf{H}$ through the whole propagation length from z_i to z:

$$\mathbf{M}(z, z_i) = e^{\mathbf{H}\cdot(z-z_i)}. \tag{1.7.20}$$

Readers may check that (1.7.20) solves (1.7.18) by direct calculation. Also, we may note that all the infinitesimal transformations of the type $e^{\mathbf{H}\,dz}$ in the infinite sequence obtained from (1.7.14) for $N \to \infty$ come to have the same exponents, in the case of a z-independent Hamiltonian, and hence combine into the single exponential (1.7.20), just as in the case of the transfer operator.

Notably the procedure, leading to Eqs. (1.7.19), provides a practicable technique to solve (1.7.18), and (1.7.4), turning around the problems of operator ordering and disentanglement arising from the operator nature of $\mathbf{H}$ and $\mathbf{M}$.

Finally, it may be interesting to discuss the link between the transfer matrix $\mathbf{M}(z, z_i)$, defined by the matrix relation (1.7.17), and the transfer operator $\widehat{\mathfrak{M}}(z, z_i)$, defined by the operator relation (1.4.7). Both $\mathbf{M}(z, z_i)$ and $\widehat{\mathfrak{M}}(z, z_i)$ relate the same objects, namely, the ray variables at different positions along the propagation axis, but the transfer operator describes the ray evolution under the general optical Hamiltonian (1.2.10), whilst the transfer matrix is exclusively pertinent to the quadratic Hamiltonian (1.7.1). However, within the linear approximation, the transfer matrix is just the matrix representative of the transfer operator. This is an obvious consequence of the correspondence $\widehat{\mathsf{L}}_H \longleftrightarrow \mathbf{H}$ and the power series representation of exponential functions of both operators and matrices.

1.7.4 *The Lagrange invariant*

Let $(q_1(z), p_1(z))$ and $(q_2(z), p_2(z))$ denote any pair of paraxial rays evolving according to Eqs. (1.7.3). It immediately follows that

$$\frac{d}{dz}(q_1 p_2 - q_2 p_1) = 0, \tag{1.7.21}$$

and accordingly

$$q_1 p_2 - q_2 p_1 = q_{1_i} p_{2_i} - q_{2_i} p_{1_i} = const. \tag{1.7.22}$$

Equations (1.7.21) and (1.7.22) establish the invariance of the cross product

$$\Delta \equiv q_1 p_2 - q_2 p_1 \tag{1.7.23}$$

for any determination of the ray variables (q, p) along the ray trajectory. Δ is called *Lagrange invariant* [4.2]. Within the phase-plane picture, the modulus of the cross product $q_1 p_2 - q_2 p_1$ is the area of the parallelogram in the (q, p) plane, delimited by the vectors with components (q_1, p_1) and (q_2, p_2). Equation (1.7.22) states then the invariance of the area of such a parallelogram, thus expressing Liouville's theorem within the context of $1D$ linear optics.

The invariance of the skew-product (1.7.23), discussed here within the specific context of $1D$ first-order optics, characterizes in general any Hamiltonian system under the linear approximation. In fact, the above result relates directly to that established in § 1.3.1 about the invariance of the product

$$\varpi(\mathbf{u}_1, \mathbf{u}_2) \equiv (\mathbf{u}_1, \mathbf{J}\mathbf{u}_2), \tag{1.7.24}$$

under the linear symplectic transformation (1.3.17). Within the formalism of classical mechanics, $\varpi(\mathbf{u}_1, \mathbf{u}_2)$ is called a *2-form* [3.2, 3.3]. It establishes a correspondence of pairs of phase space vectors to real numbers; in particular, when applied to infinitesimal vectors, it produces the so called *fundamental 2-form* $\varpi(d\mathbf{u}_1, d\mathbf{u}_2) = (d\mathbf{u}_1, \mathbf{J}d\mathbf{u}_2)$, which, as shown in § 1.3.1, remains unchanged under any symplectic transformation.

Evidently the form (1.7.24) is antisymmetric: $\varpi(\mathbf{u}_1, \mathbf{u}_2) = -\varpi(\mathbf{u}_2, \mathbf{u}_1)$. It reduces to the Lagrange invariant (1.7.23) in the case of a $2D$ phase space. Notably, the Poisson bracket of dynamical variables can be understood as a 2-form (Eq. (1.3.34)). Some properties of $\varpi(\mathbf{u}_1, \mathbf{u}_2)$ are listed in Problem 9.

As a conclusion we may say that linear ray-optics deals with a $4D$ (or $2D$ for stigmatic and simple astigmatic optical systems) linear vector space, where the antisymmetric and nondegenerate form (1.7.24) is defined. First-order optical systems generate symplectic transformations of the geometrical-optical phase space, that preserve the *2-form* (1.7.24).

1.7.5 *Square law medium*

We briefly detail the second-order coefficient $n_2(z)$ in the series expansion (1.6.4) of the refractive index $n(q, z)$. The quadratic profile (1.6.4) is a typical model for the *square-law* (also referred to as l*ens-like* or *graded-index*) medium.

We consider the surfaces $n(q_x, q_y, z) = const.$, which intersect the optical axis at right angles. In particular, the refracting surface that intersects the axis at the point z_o is represented by the equation

$$n(q_x, q_y, z) = n(0, 0, z_o) \equiv n_0(z_o). \tag{1.7.25}$$

Expanding the left hand side above at the lower order of q_x, q_y and z, in the assumption of rotational symmetry, we have

$$n(q_x, q_y, z) \simeq n_0(z_o) + n_0'(z_o)(z - z_o) - \tfrac{1}{2}n_2(z_o)(q_x^2 + q_y^2), \tag{1.7.26}$$

the prime denoting as usual differentiation with respect to z. Comparing with (1.7.25), we obtain the surface equation

$$z - z_o = \tfrac{n_2(z_o)}{2n_0'(z_o)}(q_x^2 + q_y^2) \equiv \tfrac{1}{2R_o}(q_x^2 + q_y^2), \tag{1.7.27}$$

which represents a paraboloid having at the vertex z_o the curvature radius

$$R_o = \frac{n_0'(z_o)}{n_2(z_o)}. \tag{1.7.28}$$

The refracting *power* $\mathcal{P}$ of a surface is defined as the ratio of the on-axis variation of the refractive index Δn through the surface to the curvature radius R of the surface, i.e., $\mathcal{P} = \Delta n/R$. Therefore by (1.7.28) we can relate the coefficient $n_2(z_o)$ to the surface power $\mathcal{P}(z_o)$ at z_o as

$$n_2(z_o)dz = \mathcal{P}(z_o), \tag{1.7.29}$$

which conversely suggests to express $\mathcal{P}(z_o)$ as the differential of the area under the curve $n_2(z)$, namely

$$\mathcal{P}(z_o) = d\int_{z_i}^{z_o} n_2(z')dz'. \tag{1.7.30}$$

1.8 ABCD matrix and focal, principal and nodal planes

In the theory of image-forming devices, the optical systems are commonly characterized by the *cardinal points* and their associated *planes* [17].
In this section we outline the relation between the ray-matrix entries A, B, C, D of the optical system, referenced to assigned input and output planes, and the location and role of the relative cardinal points or planes. In this respect, it is convenient to recall the usually adopted nomenclature and sign conventions.

1.8.1 *Object space and image space*

Given an optical system, we take as usual the optical axis as z-axis, oriented from left to right according to the assumed direction of ray-propagation. Supposing the system occupies a certain region of the xyz-space, extending on the axis from $z_{\rm O}$ to $z_{\rm I}$, we call the space $z < z_{\rm O}$ the *object* space and the space $z > z_{\rm I}$ the *image* space.

It is a common situation that the optical system separates two homogeneous media. We assign the refractive indices n_{0_1} to the object space and n_{0_2} to the image space. The light rays in the object and image spaces are sections of straight lines and the optical system determines a transformation of the straight lines of the object space into the straight lines of the image space. The portions of these straight lines that as light rays have a physical reality are called *real* parts of the ray, whilst their extrapolations, which have no physical reality, are called the *virtual* parts of the ray. Referring to the object and image spaces, it is evident that the half space $z > z_{\rm O}$ is the *virtual object* space, as the half space $z < z_{\rm I}$ is the *virtual image* space. A bundle of rays incoming toward the optical device is said to belong to a real or virtual object point, if it originates at or converge to a point located in the real or virtual object space. Similarly, real and virtual image points are understood as the points of intersection of the real and virtual parts of a bundle of rays emerging from the system; outgoing rays then converge to real image points, whilst diverge from virtual image points. In order to specify the character (real or virtual) of the object and image points appropriate reference frames are arranged in the object and image spaces. Denoting indeed by $\Pi_{\rm O}$ and $\Pi_{\rm I}$ the planes normal to the axis at $z_{\rm O}$ and $z_{\rm I}$, in the object reference frame the axial distances are assumed positive when they are measured to the left of $\Pi_{\rm O}$ and negative when they are measured to the right. Likewise in the image reference frame, the axial distances are positive when measured to the right of $\Pi_{\rm I}$ and negative when to the left. In both cases, the transverse distances are positive upward from the axis and negative downward. In the following, the terms object and image spaces will be used to comprise the relative real as well as virtual half-spaces.

Finally, we recall that a point P in the object space and a point P′ in the image space are called *conjugate* if all the rays through P are brought to P′. Due to the principle of the reversibility of light rays, it is evident that all the rays through P′ are brought to P; in other words, P and P′ as object and image points are interchangeable. The planes through P and P′ perpendicular to the optical axis are called *conjugate planes*. It is worth stressing that the expression "all the rays through P" must be understood as all the rays of a finite bundle, whose transverse sizes are determined by the optical system and the approximations inherent in the adopted formalism.

Then, let us assign the input and output planes Π_i and Π_o, whose location along the optical axis can arbitrarily be chosen; in fact, the ray-variables q and p refer to the real as well as the virtual parts of the ray. Clearly, a formal simplification is gained if one assumes the object and image planes $\Pi_{\rm O}$ and $\Pi_{\rm I}$ as input and output planes, respectively.

1.8.2 *Ray-coordinate and momentum magnification*

The *ray-coordinate* (or *linear*) *magnification* of the optical system $\mathcal{M}_q$, relative to the assigned reference planes, is the ratio between the heights of the emerging and incoming rays, namely

$$\mathcal{M}_q = \frac{q_o}{q_i}. \tag{1.8.1}$$

Similarly, the *ray-momentum magnification* is defined in terms of the optical momenta of the incoming and outgoing rays according to

$$\mathcal{M}_p = \frac{p_o}{p_i}. \tag{1.8.2}$$

In terms of the entries of the ray transfer matrix of the system, $\mathcal{M}_q$ and $\mathcal{M}_p$ are expressed as

$$\mathcal{M}_q = A + B\frac{p_i}{q_i}, \quad \mathcal{M}_p = D + C\frac{q_i}{p_i}. \tag{1.8.3}$$

For a field ray, the ray-coordinate magnification takes the simple form

$$\mathcal{M}_q = A\,, \quad (\text{field ray}: \; p_i = 0) \tag{1.8.4}$$

and correspondingly for an axial ray the ray-momentum magnification is

$$\mathcal{M}_p = D\,, \quad (\text{axial ray}: \; q_i = 0) \tag{1.8.5}$$

thus yielding for the entries A and D the meaning as the ray coordinate and momentum magnifications for the field and axial rays.

Interestingly we can relate the angular aperture of a fan of rays emanating from an object point to that of the same fan of rays converging to the image point after traversing a perfect imaging system. Suppose that P_i and P_o are two conjugate points in the object and image space of the system. All the rays from P_i intersect at P_o irrespective of their angles to the optical axis. We may consider, for instance, two rays emanating from P_i with angles ϑ_{1_i} and ϑ_{2_i} to the axis, which are imaged into P_o with angles ϑ_{1_o} and ϑ_{2_o}. According to the phase-plane area preserving rule, which for paraxial propagation takes the form of the Lagrange invariance (1.7.22), we can write

$$n_{0_1} q_i \Delta\vartheta_i = n_{0_2} q_o \Delta\vartheta_o, \tag{1.8.6}$$

where q_i and q_o are the heights to the axis of the object and image points, whilst $\Delta\vartheta_i = \vartheta_{2_i} - \vartheta_{1_i}$ and $\Delta\vartheta_o = \vartheta_{2_o} - \vartheta_{1_o}$ represent the angular separations of the fan of rays before and after propagation through the imaging system. Equation (1.8.6) establishes a clear relation between the linear and the angular magnification of a bundle of rays, whose spatial extension can be magnified only at the expense of the angular divergence, that is by having an image beam with a smaller angular divergence than that of the object beam.

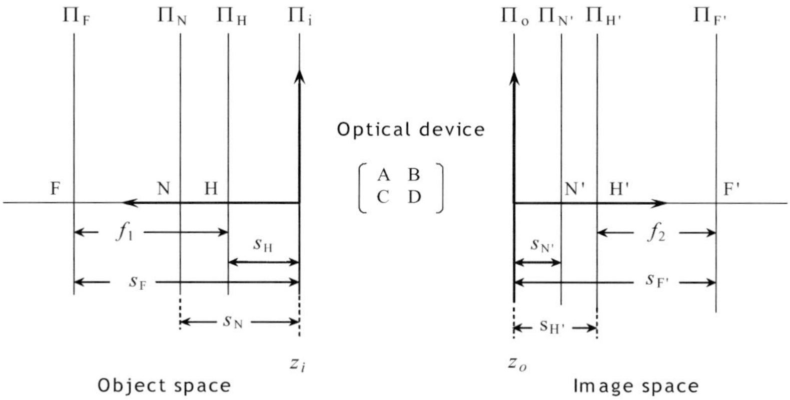

FIGURE 1.9. Cardinal points and associated planes for an optical system represented by the ray matrix $\left(\begin{smallmatrix} A & B \\ C & D \end{smallmatrix}\right)$. The figure is drawn for the case where $C < 0$, $\frac{A-1}{C} > 0$ and $\frac{D-1}{C} > 0$.

1.8.3 *Focal, principal and nodal planes*

Consider now light rays incoming parallel to the optical axis: $p_i = 0$. The output ray variables q_o and p_o depend on q_i only, thus yielding for the ratio q_o/p_o the q_i-independent value

$$\frac{q_o}{p_o} = \frac{A}{C}. \tag{1.8.7}$$

Since $q_o/p_o = -s_2/n_{0_2}$, where s_2 specifies the axial coordinate in the image space of the intersection of the outgoing ray with the optical axis, the above relation states that a bundle of parallel rays, incident on the system, is transformed into a bundle of rays, whose real (virtual) part is directed toward (diverges from) the *back focal point* F′, located on the axis at

$$s_{\mathrm{F}'} = -n_{0_2}\frac{A}{C}, \tag{1.8.8}$$

from Π_o. The axial coordinate $s_{\mathrm{F}'}$ of F′ is named *back focal length* and the plane $\Pi_{\mathrm{F}'}$ through F′ orthogonal to the axis is the *back focal plane* (Fig. 1.9).

The opposite situation corresponds to a ray emerging from the system with zero slope, $p_o = Cq_i + Dp_i = 0$, and arbitrary q_o thus implying

$$\frac{q_i}{p_i} = -\frac{D}{C}. \tag{1.8.9}$$

Being $q_i/p_i = -s_1/n_{0_1}$, where s_1 specifies the axial coordinate in the object space of the intersection of the incident ray with the optical axis, the real (virtual) part of the incident ray diverges from (is directed toward) the *front focal point* F, located on the optical axis at

$$s_{\mathrm{F}} = -n_{0_1}\frac{D}{C}, \tag{1.8.10}$$

from Π_i, which is reported as *front focal length.* The plane Π_{F} through F normal to the optical axis is the *front focal plane* (Fig. 1.9). Evidently the primary and secondary focal points are conjugate to points at an infinitely great distance from the system respectively in the image and object space.

Consider again an incident ray, diverging from the front focal point F. The ray coordinates (q_i, p_i) are related as $q_i/p_i = -D/C$, and the ray emerges from the output plane Π_o with $q_o = -p_i/C$ and zero slope. Extending the incident and refracted rays to their point of intersection, we define the plane Π_{H} through this point normal to the axis as the *primary principal plane*. Similarly, the *secondary principal plane* $\Pi_{\mathrm{F}'}$ is identified by the intersection of the real or virtual part of a parallel input ray and the corresponding output ray, converging to the back focal point F′ ($q_o = Aq_i$, $p_o = Cq_i$). The intersection points H and H′ of Π_{H} and $\Pi_{\mathrm{H}'}$ with the optical axis are the *primary* and *secondary principal points*, located at

$$s_{\mathrm{H}} = n_{0_1}\frac{1-D}{C}, \qquad s_{\mathrm{H}'} = n_{0_2}\frac{1-A}{C}, \tag{1.8.11}$$

respectively in the object and image space.

Furthermore, the position of the front focal point F with respect to Π_{H} is specified by the so called *primary focal length* f_1, given as

$$f_1 = n_{0_1}\frac{q_o}{p_i} = -\frac{n_{0_1}}{C}, \tag{1.8.12}$$

and correspondingly the position of F′ with respect to $\Pi_{\mathrm{H}'}$ is given by

$$f_2 = -n_{0_2}\frac{q_i}{p_o} = -\frac{n_{0_2}}{C}, \tag{1.8.13}$$

reported as the *secondary focal length.*

Notably, the primary and secondary focal lengths are in the same ratio as the refractive indices of the media on the two sides of the system

$$\frac{f_2}{f_1} = \frac{n_{0_2}}{n_{0_1}}. \tag{1.8.14}$$

In particular, if the medium is the same on both sides: $n_{0_1} = n_{0_2}$, then $f_1 = f_2$. Conventionally, the ratio of the primary and secondary focal length to the respective refractive indices, denoted as

$$f = \frac{f_1}{n_{0_1}} = \frac{f_2}{n_{0_2}}, \tag{1.8.15}$$

is simply referred to as the (*reduced*) focal length of the system. The inverse of f defines the (focusing) *power* of the system:

$$\mathcal{P} = \frac{1}{f}, \tag{1.8.16}$$

which is typically measured in diopters (D) if f is given in meters.

As the focal points F and F′ are both real ($f > 0$) or virtual ($f < 0$), the sign of f allows a characterization of the system as a converging or diverging device. When $f > 0$, rays diverging from F are imaged into parallel rays and correspondingly parallel rays are imaged into rays converging to F′. Conversely, when $f < 0$, the system turns rays converging to F into parallel rays and parallel rays into rays diverging from F′. Also, on account of (1.8.12) and (1.8.13), the lower-left entry C of the ray matrix simply relates to f as

$$C = -\frac{1}{f} = \mathcal{P}. \tag{1.8.17}$$

Remarkably, if the locations of the focal and principal points (or planes) relative to the assigned input and output planes are known, the behaviour of the system with respect to these planes is completely determined, being the relevant ray matrix known by (1.8.8), (1.8.10) and (1.8.11). Thus, the system can effectively be modeled by its two focal and principal planes. In the next section, we discuss a characterization of the optical systems, for which the focal or principal planes are assumed as reference planes.

Let us establish now the condition under which an incident ray with slope $\vartheta_i = p_i/\ n_{0_1}$ at Π_i emerges from Π_o with the same slope $\vartheta_o = p_o/\ n_{0_2} = \vartheta_i$. According to the expression (1.8.3) for the ray-momentum magnification $\mathcal{M}_p$, it is evident that the input ray coordinates must satisfy the relation

$$\mathcal{M}_p = D + C\frac{q_i}{p_i} = \frac{n_{0_2}}{n_{0_1}}. \tag{1.8.18}$$

Thereby, any incident ray, whose real (virtual) part originates at (is directed toward) the point N, located in the object space at

$$s_{\rm N} = \frac{n_{0_1}}{C}(\frac{n_{0_2}}{n_{0_1}} - D) = -f_1\,(\frac{f_2}{f_1} - D), \tag{1.8.19}$$

is transformed into a ray with the same slope: $\vartheta_o = \vartheta_i$, whose real (virtual) part is directed toward (originates at) the point N′, located at

$$s_{\rm N'} = \frac{n_{0_2}}{C}(\frac{n_{0_1}}{n_{0_2}} - A) = -f_2(\frac{f_1}{f_2} - A), \tag{1.8.20}$$

in the image space.

The points N and N′ are called *nodal points* and accordingly the *nodal planes* Π_N and $\Pi_{N'}$ are the planes through N and N′ normal to the axis. It is clear that N and N′ are *conjugate points*. If the medium on both sides of the system is the same, the nodal points coincide with the principal points: N = H and N′ = H′. In general, when $n_{0_1} \neq n_{0_2}$, the two set of points are separate, although the distance between the nodal points is exactly equal to the distance between the principal points: NN′ = HH′. In addition, the further relations: FN = F′H′ and F′N′ = FH between the focal and nodal points hold.

Other cardinal points are defined which are of some interest, as the *negative cardinal* and *nodal points*, respectively characterized by a unit negative lateral and angular magnification (see Problem 11).

Finally, to further evidence the analogy between the characterization of the imaging properties of an optical system by the cardinal points (and planes) and that by the ray matrix, we deduce from (1.7.16) the relation

$$\frac{n_{0_1}}{\left[s_1 - \frac{1-D}{C} n_{0_1}\right]} + \frac{n_{0_2}}{\left[s_2 - \frac{1-A}{C} n_{0_2}\right]} = -C, \tag{1.8.21}$$

where again s_1 and s_2 give the axial coordinates of the points P_1 and P_2 at which the incident and refracted rays intersect the optical axis respectively in the object and image space. On account of (1.8.11) and (1.8.15), the above relation can be cast in the form of the well known *thin lens formula*

$$\frac{n_{0_1}}{\overline{s}_1} + \frac{n_{0_2}}{\overline{s}_2} = \frac{1}{f}, \tag{1.8.22}$$

with $\overline{s}_1$ and $\overline{s}_2$ denoting the axial coordinates of the points P_1 and P_2 referenced to the principal planes Π_H and $\Pi_{H'}$.

By a further manipulation, (1.8.21) can be given the *Newtonian form*

$$(s_1 + n_{0_1} \tfrac{D}{C})(s_2 + n_{0_2} \tfrac{A}{C}) = \frac{n_{0_1} n_{0_2}}{C^2}, \tag{1.8.23}$$

or more concisely

$$\widetilde{s}_1 \widetilde{s}_2 = f_1 f_2, \tag{1.8.24}$$

where the axial coordinates $\widetilde{s}_1$ and $\widetilde{s}_2$ of the points P_1 and P_2 are referenced to the focal planes Π_F and $\Pi_{F'}$, respectively.

1.8.4 *Special Optical Systems*

The ray matrix takes specific forms when the focal or the principal planes are taken as input or output planes [16, 17]. In fact, when the principal planes Π_H, $\Pi_{H'}$ are taken as input and output planes the front and primary focal points as well as the back and secondary focal points coincide: $s_F = f_1$ and $s_{F'} = f_2$. The ray matrix from the primary to the secondary principal plane is therefore

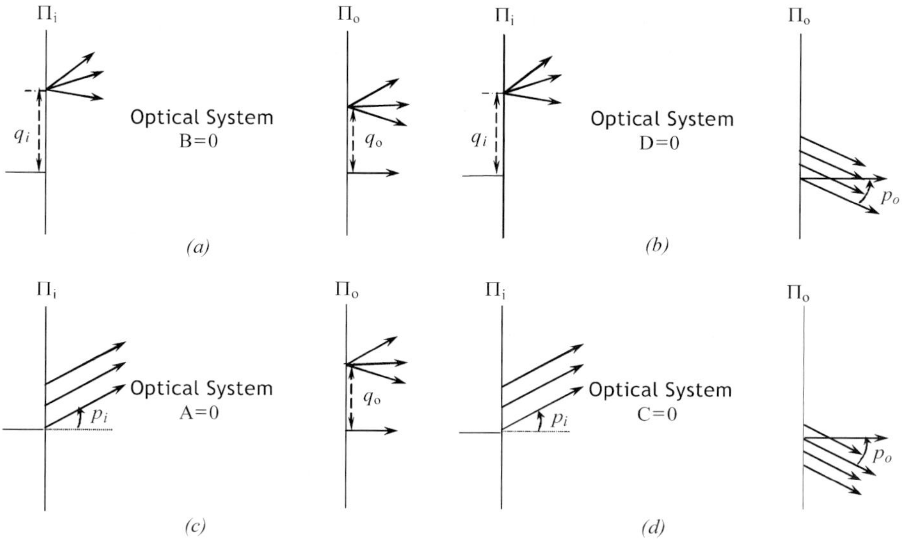

FIGURE 1.10. Optical systems having one vanishing entry in their ray transfer matrix. *(a) imaging* systems ($B = 0$), *(b) collimating* systems ($D = 0$), *(c) focusing* systems ($A = 0$) and *(d) telescopic* systems ($C = 0$).

$$\mathbf{M}_{_{\mathrm{H}\to\mathrm{H}'}} = \begin{pmatrix} 1 & 0 \\ C & 1 \end{pmatrix}, \tag{1.8.25}$$

where by (1.8.17) $C = -1/f$. The form of $\mathbf{M}_{_{\mathrm{H}\to\mathrm{H}'}}$ evidences the property of the principal planes to be conjugate, points of $\Pi_{_\mathrm{H}}$ being imaged into points of $\Pi_{_{\mathrm{H}'}}$. In particular, (1.8.25) produces a unit positive image magnification for any input ray. In fact, the principal planes are also defined as conjugate planes for which $\mathcal{M}_q = 1$, and accordingly called *unit planes.*

Matrix (1.8.25) is a particular form of the most general optical matrix

$$\mathbf{M}_{_{\mathrm{imaging}}} = \begin{pmatrix} A & 0 \\ C & 1/A \end{pmatrix}, \tag{1.8.26}$$

which, having $B = 0$, relates conjugate planes. In fact, as $q_o = Aq_i$, the position of the image ray depends only on the position of the object ray; thus, all the rays through the point at q_i on Π_i are brought to the point at q_o on Π_o (Fig. 1.10.a)). Evidently, $\mathcal{M}_q = A$. Optical systems described by matrices having $B = 0$ are called *imaging* systems (or, *magnifiers*).

Special systems are also obtained when the focal planes are taken as reference planes. In particular, the ray matrix for the propagation from the front focal plane to a given output plane comes to have $D = 0$, namely

$$\mathbf{M}_{\text{collimating}} = \begin{pmatrix} A & -1/C \\ C & 0 \end{pmatrix}, \tag{1.8.27}$$

The sign of $C = -1/f$ establishes if the object and image distributions are real ($f > 0$) or virtual ($f < 0$). Optical systems, described by the ray matrix (1.8.27), image a point into a bundle of parallel rays (Fig. 1.10.b)) since $p_o = Cq_i$ for all p_i. They are referred to as *collimating* systems.

Conversely, the propagation from a given input plane to the back focal plane is characterized by $A = 0$, and hence by the ray matrix

$$\mathbf{M}_{\text{focusing}} = \begin{pmatrix} 0 & -1/C \\ C & D \end{pmatrix}. \tag{1.8.28}$$

Systems (1.8.28) image a bundle of parallel rays into a point (Fig. 1.10.c)) as $q_o = -p_i/C$ for all q_i. They are called *focusing* systems. As we will see later, the condition $A = 0$ is the requisite for the optical system to perform the *imperfect* scaled Fourier transform of the input signal. Note that the positions of the output and the input plane for the systems (1.8.27) and (1.8.28) can be determined from the relative positions (1.8.11) of the principal points.

Also, if both the focal planes are taken as reference planes, the ray-transfer matrix from the front to the back focal plane writes as

$$\mathbf{M}_{\text{F}\to\text{F}'} = \begin{pmatrix} 0 & -1/C \\ C & 0 \end{pmatrix}, \tag{1.8.29}$$

thus combining the features of both (1.8.27) and (1.8.28). Systems like (1.8.29) perform the *perfect* scaled Fourier transform of the input signal; accordingly they are reported as perfect *Fourier transformers*.

Another interesting situation arises in the case where $C = 0$. In that case, $p_o = Dp_i$, and hence the direction of the image ray depends only on the direction of the object ray; thus, a bundle of parallel rays is turned into an other bundle of parallel rays, as in a telescope focused to infinity (Fig. 1.10.d)). The ray matrix takes the form

$$\mathbf{M}_{\text{afocal}} = \begin{pmatrix} A & B \\ 0 & 1/A \end{pmatrix}, \tag{1.8.30}$$

the corresponding optical systems being called *afocal* (or *telescopic*).

Finally, the *pure magnifier* combines the features of the systems (1.8.26) and (1.8.30), being both imaging and telescopic. The relative ray matrix is

$$\mathbf{M}_{\text{pure magnifier}} = \begin{pmatrix} A & 0 \\ 0 & 1/A \end{pmatrix}. \tag{1.8.31}$$

1.9 Summary

Within the Hamiltonian formulation of geometrical optics, light rays are described as points in the $4D$ geometrical-optical phase space, while evolving along the optical axis z of the system. At every $z = const$ plane, the phase space variables (q, p) specify the position and the optical cosines of the ray with respect to the assigned coordinate system in the plane. As a consequence of Fermat's principle, the optical Hamiltonian function turns out to be $H(q, p, z) = -\sqrt{n^2(q, z) - p^2}$, with $n(q, z)$ being the refractive index function. The phase-space dynamics, which parallels the ray propagation in physical space, is determined by the Hamiltonian set of ordinary differential equations.

The ray propagation may be seen as a continuous sequence of symplectic transformations, which map points in phase space into points in phase space as the ray proceeds along the z direction. In this respect, Poisson brackets and Lie operators have a decisive role to form the ray transfer operator, which carries the input ray coordinates to the corresponding values at the desired z.

First-order ray optics, which arises from the quadratic approximation of the optical Hamiltonian in both the position and direction ray variables, describes linear transformations in phase space. The ray transfer matrix $\mathbf{M}$, i.e., the 4×4 symplectic matrix containing the ray-variables independent coefficients of the linear relation of the output to the input ray data, is the mathematical tool to describe the transformation of the ray while going through first-order optical systems. Notably, in the case of stigmatic and simple astigmatic optical systems, the $2D$ Hamiltonian model of ray propagation, which demands for a $4D$ optical phase space, can be reduced to a $1D$ picture, demanding for a $2D$ optical phase space, and hence for a 2×2 optical matrix.

In the next Chapters we will consider the basic optical systems within the framework of the $1D$ model of linear optics. We will derive the relevant ray-matrices and analyse their effect on single light rays in physical space as well as in phase space. The effect on a collection of rays will also be visualized in the phase plane in the case of bundles of rays emitted by idealized light sources, which accordingly are amenable to simple phase-space representations both before and after the propagation through the system.

Problems

1. Consider a $2m \times 2m$ matrix $\mathbf{M} = \begin{pmatrix} \mathbf{A} & \mathbf{B} \\ \mathbf{C} & \mathbf{D} \end{pmatrix}$, where $\mathbf{A}$, $\mathbf{B}$, $\mathbf{C}$ and $\mathbf{D}$ are $m \times m$ submatrices. *(a)* Find the relations obeyed by the submatrices $\mathbf{A}$, $\mathbf{B}$, $\mathbf{C}$ and $\mathbf{D}$ when $\mathbf{M}$ is symplectic. *(b)* Prove that as a consequence of such relations only $m(2m+1)$ entries of $\mathbf{M}$ can be arbitrarily given. *(c)* Show that, in the case of a 2×2 matrix, the symplectic relation turns into the condition that $\det \mathbf{M} = 1$.

2. *(a)* Prove that the Jacobi matrix $\mathbf{S}$ of the Hamiltonian evolution from $\mathbf{u}_i$ to $\mathbf{u}(z)$ obeys the equation of evolution

$$\frac{d}{dz}\mathbf{S} = \mathbf{J}\mathbf{H}''\mathbf{S}$$

involving the Hessian matrix, introduced in § 1.3.1, $\mathbf{H}'' = (H''_{jl})_{jl}$, $H''_{jl} = \partial^2 H/\partial u_j \partial u_l$. (Hint: Slightly changing the initial vector to $\mathbf{u}_i + d\mathbf{u}_i$, which evidently evolves into $\mathbf{u} + d\mathbf{u}$ with $d\mathbf{u} = \mathbf{S}d\mathbf{u}_i$, find the equation obeyed by $d\mathbf{u}$). *(b)* Verify that when the Hamiltonian is quadratic in the phase space variables (Eq. (1.7.1)) the matrix $\mathbf{JH}'' = \mathbf{H}$ (Eq. (1.7.5)), thus proving that in that case $\mathbf{S}$ obeys the same equation as $\mathbf{u}$. *(c)* Prove that in general the representative matrix $\mathbf{A}$, assignable to any quadratic phase space observable $A(q,p)$ by (1.7.7), is also obtainable as $\mathbf{A} = \mathbf{JA}''$, where $\mathbf{A}''$ is the Hessian matrix of the quadratic form $A(q,p)$.

3. Using the power series expansion (1.4.5) for the Lie transformation, prove that the phase-plane motion under polynomial (1.5.27) is described by the expressions $q(z) = [\cos(\alpha z) + c\sin(\alpha z)/\alpha]q_i + [b\sin(\alpha z)/\alpha]p_i$ and $p(z) = [\cos(\alpha z) - c\sin(\alpha z)/\alpha]p_i - [a\sin(\alpha z)/\alpha]q_i$, where $\alpha = \sqrt{ab - c^2}$. Compare the cases *(a)* $\alpha > 0$ and *(b)* $\alpha < 0$.

4. A free-propagation system is described by the Hamiltonian function $H(q,p,z) = -\sqrt{n_0^2 - p^2}$, n_0 being the constant value of the refractive index. Expressing H as a series of powers of the ray momentum p, determine the corrections to the paraxial propagation, essentially described by (1.5.12), which arise from powers higher than two, like for instance p^4, p^6, and so on (generally referred to as spherical aberrations).

5. A skew ray enters an axially symmetric optical system, intersecting the x axis at q_{x_i} with optical direction cosines (p_{x_i}, p_{y_i}). Determine the relation between the subsequent intersections $q_x(z)$ of the ray with the x axis during the propagation through the system and the corresponding optical direction cosines at those intersections.

6. Prove the unimodularity of the ray matrix $\mathbf{M}$ from the requirement that $q_o(q_i, p_i)$ and $p_o(q_i, p_i)$ are canonically conjugate variables, i.e., $\{q_o, p_o\} = \{q_i, p_i\} = 1$.

7. *(a)* Find the field-ray coordinates $(A(z), C(z))$ from the assigned axial ray-coordinates $(B(z), D(z))$ and vice versa. *(b)* Show that it is possible to find any paraxial ray $(q(z), p(z))$ if the field-ray or the axial-ray coordinates are known.

8. Let $\mathbf{M}$ and $\mathbf{H}$ be two $2m \times 2m$ matrices, linked through the group operation $\mathbf{M} = \exp\{\alpha\mathbf{H}\}$, with α arbitrary real parameter. *(a)* Prove that if $\mathbf{M}$ is symplectic, then $\mathbf{H}$ satisfies the relation: $\mathbf{HJ} + \mathbf{JH}^\top = 0$. *(b)* Show that $\mathrm{Tr}(\mathbf{H}) = 0$. *(c)* Show that $\mathbf{HJ}$ is a symmetric matrix. *(d)* Prove that, partitioning $\mathbf{H}$ as $\mathbf{M}$ in Problem 1, i.e., $\mathbf{H} = \left(\begin{smallmatrix}\mathbf{L} & \mathbf{F}\\ \mathbf{G} & \mathbf{N}\end{smallmatrix}\right)$, the submatrices $\mathbf{F}$ and $\mathbf{G}$ are symmetric and $\mathbf{L} = -\mathbf{N}^\top$. *(e)* Verify that only $m(2m+1)$ entries of $\mathbf{H}$ are independent.

9. Prove that: *(a)* $\varpi(a\mathbf{u}_1, b\mathbf{u}_2) = ab\,\varpi(\mathbf{u}_1, \mathbf{u}_2)$ for any complex numbers a and b; *(b)* $\varpi(\mathbf{u}_2, \mathbf{u}_1) = -\varpi(\mathbf{u}_1, \mathbf{u}_2)$; *(c)* $\varpi(\mathbf{u}, \mathbf{u}) = 0$; *(d)* $\varpi(\mathbf{u}_1 + \mathbf{u}_2, \mathbf{v}) = \varpi(\mathbf{u}_1, \mathbf{v}) + \varpi(\mathbf{u}_2, \mathbf{v})$; *(e)* $\varpi(\mathbf{v}, \mathbf{u}_1 + \mathbf{u}_2) = \varpi(\mathbf{v}, \mathbf{u}_1) + \varpi(\mathbf{v}, \mathbf{u}_2)$; *(f)* $\varpi(\mathbf{u}_2, \mathbf{M}\mathbf{u}_1) = \varpi(\mathbf{M}^{-1}\mathbf{u}_1, \mathbf{u}_2)$ for any symplectic matrix $\mathbf{M}$.

10. Show that the integration of the two linearized equations (1.7.3) is equivalent to the integration of one linear differential equation of the Riccati type: *(a)* $s_1' - n_2 s_1^2 - 1/n_0 = 0$ for the function $s_1(z) \equiv q(z)/p(z)$, or *(b)* $s_2' + s_2^2/n_0 + n_2 = 0$ for $s_2(z) \equiv p(z)/q(z)$. *(c)* Find the expression of $q(z)$ and $p(z)$ in terms of both $s_1(z)$ and $s_2(z)$. *(d)* Give a geometrical interpretation of $s_1(z)$ and $s_2(z)$ in the (q,p) plane. *(e)* Express the z-evolution of s_1 and s_2 in terms of the field and axial ray coordinates.

11. Find the positions of *(a)* the negative cardinal points, defined as conjugate points for which $\mathcal{M}_q = -1$ and *(b)* the negative nodal points, for which $\mathcal{M}_p = -1$.

References

[1] Definitions, comments and investigations pertaining to the concept of light ray from the mechanical and electromagnetic views can be found in [4.2], [4.5], and in

[1.1] M. Kline and I.W. Kay, *Electromagnetic Theory and Geometrical Optics* (Interscience Publishers, New York, 1965).

[1.2] G.V. Rozenberg, "The light ray (contribution to the theory of the light field)", Sov. Phys. Usp. **20**, 55-79 (1977).

[1.3] D.S. Goodman, "General Principles of Geometric Optics", in *Handbook of Optics*, Vol. I, M. Bass, E.W. Van Stryland, D.R. Williams and W.L. Wolfe (eds.) (Mc-Graw Hill, New York, 2nd ed., 1995), ch. 1; and references therein.

[2] The original papers by W. R. Hamilton concerning optics and mechanics are collected in

[2.1] *The Mathematical Papers of Sir William Rowan Hamilton*, Vol. I: Geometrical Optics, A.W. Conway and J.L. Synge (eds.) (Cambridge Univ. Press, London, 1931). Also in http://www.maths.tcd.ie/pub/HistMath/People/Hamilton/Optics.html

[2.2] *The Mathematical Papers of Sir William Rowan Hamilton*, Vol. II: Dynamics, A.W. Conway and J.L. Synge (eds.)(Cambridge Univ. Press, London, 1940).

[3] The Hamiltonian formulation of classical mechanics with the related formalism of phase space, and Liouville's theorem, is presented in

[3.1] H. Goldstein, *Classical Mechanics* (Addison-Wesley, Reading, 2nd ed., 1980).

[3.2] V.I. Arnold, *Mathematical Methods of Classical Mechanics* (Springer-Verlag, New York, 1978).

[3.3] R. Abraham and J.E. Marsden, *Foundations of Mechanics* (Addison-Wesley, Redwood City, 2nd ed., 1978).

[3.4] J. Liouville, "Sur la théorie de la variation des constantes arbitraires", Journ. de Math. Pures et Appl. **3**, 342-349 (1838).

[4] As to the Hamiltonian formulation of geometrical optics we quote below a few basic textbooks and papers, in some of which investigations in the optical phase space are also illustrated at various extents. In particular, we address the reader to the book by Wolf [4.11], where through "the tactics of Lie algebras and groups" a complete description of "the geometric model of light" is presented, comprising the paraxial, metaxial and global regimes.

[4.1] J.L. Synge, *Geometrical Optics* (Cambridge Univ. Press, Cambridge, 1937).

[4.2] R.K. Luneburg, *Mathematical Theory of Optics* (California Univ. Press, Berkeley, 1964).

[4.3] M. Born and E. Wolf, *Principles of Optics* (Pergamon Press, Oxford, 1965).

[4.4] H.A. Buchdahl, *An Introduction to Hamiltonian Optics* (Cambridge Univ. Press, Cambridge, 1970).

[4.5] O.N. Stavroudis, *The Optics of Rays, Wavefronts and Caustics* (Academic Press, New York, 1972).

[4.6] D. Marcuse, *Light Transmission Optics* (Van Nostrand-Reinhold, Princeton, 1972).

[4.7] A.K. Ghatak and K. Thyagarajan, *Contemporary Optics* (Plenum Press, New York, 1978).

[4.8] S. Solimeno, B. Crosignani and P. Di Porto, *Guiding, Diffraction and Confinement of Optical Radiation* (Academic Press, Orlando, 1985).

[4.9] H.M. Ozatkas, Z. Zalevsky and M.A. Kutay, *The Fractional Fourier Transform with Applications in Optics and Signal Processing* (Wiley, New York, 2001).

[4.10] V. Lakshminarayanan, A.K. Ghatak and K. Thyagarajan, *Lagrangian Optics* (Kluwer, Boston, 2002).

[4.11] K.B. Wolf, *Geometric Optics on Phase Space* (Springer, Berlin, 2004).

[4.12] R.J. Pegis, "The modern development of Hamiltonian Optics", *Progress in Optics*, E. Wolf (ed.) (North-Holland, Amsterdam, 1963), Vol. I, ch. 1, pp. 1-29.

[4.13] J.A. Arnaud, "Hamiltonian theory of beam propagation", *Progress in Optics*, E. Wolf (ed.) (North-Holland, Amsterdam, 1973), Vol. XI, ch. 6, pp. 247-304.

[4.14] T. Sekiguchi and K.B. Wolf, "The Hamiltonian formulation of optics", Am. J. Phys. **55**, 830-835 (1987).

[4.15] M.A. Alonso and G.W. Forbes, "Generalization of Hamilton's formalism for geometrical optics", J. Opt. Soc. Am. A **12**, 2744-2752 (1995).

[5] A fascinating survey of variational principles and their relation to dynamics and quantum theory is given in [5.6]. The primitive formulation of the Fermat principle, and the pertinent elaborations and investigations extending the principle to anisotropic media can be found in [1.1], chs. 2, 3; in [4.4], chs. 1, 11; and in

[5.1] P. de Fermat, *Œvreus*, (Gauthier-Villars, Paris, 1891), p. 354.

[5.2] H.H. Hopkins, "An extension of Fermat's theorem", Opt. Acta, **17**, 223-225 (1970).

[5.3] W.A. Newcomb, "Generalized Fermat principle", Am. J. Phys. **51**, 338-340 (1983).

[5.4] A.L. Rivera, S.M. Chumakov and K.B. Wolf, "Hamiltonian formulation of geometrical anisotropic optics", J. Opt. Soc. Am. A**12**, 1380-1389 (1995).

[5.5] W. Shen, J. Zhang, S. Wang and S. Zhu, "Fermat's principle, the general eikonal equation, and space geometry in a static anisotropic medium", J. Opt. Soc. Am. A**14**, 2850-2854 (1997).

[5.6] W. Yourgrau and S. Mandelstam, *Variational Principles in Dynamics and Quantum Theory* (Dover Publ., New York, 1979).

[6] A wide and detailed investigation of the ray equation in inhomogneous media for various refractive index profiles is presented in [4.10], and in

[6.1] E.W. Marchand, *Gradient Index Optics* (Academic Press, New York, 1978).

[6.2] S. Cornbleet, *Microwave and Geometrical Optics* (Academic Press, London, 1994).

[7] The optical-mechanical analogy is investigated in [4.2], [4.3], [4.13], and in

[7.1] J.A. Arnaud, "Analogy between optical rays and nonrelativistic particle trajectories: a comment", Am. J. Phys. **44**, 1067-1069 (1976).

[7.2] W.K. Kahn and S. Yang, "Hamiltonian analysis of beams in an optical slab guide", J. Opt. Soc. Am. **73**, 684-690 (1983).

[7.3] J. Evans and M. Rosenquist, "'F=ma' optics", Am. J. Phys. **54**, 876-883 (1986).

[7.4] R.J. Black and A. Ankiewicz, "Fiber-optic analogies with mechanics", Am. J. Phys. **53**, 554-563 (1985).

[7.5] N. Chako and B. B. Blank, "Electron Optics", in [4.2], Suppl. Note # 1, pp. 373-410.

[8] A.S. Deif, *Advanced Matrix Theory for Scientists and Engineers* (Abacus Press, Turnbridge Wells, 1982).

[9] For the general theory of Lie algebras and Lie groups we may suggest among the great variety of textbooks the following:

[9.1] H. Weyl, *The Classical Groups* (Princeton Univ. Press, Princeton, 1946).

[9.2] M. Hamermesh, *Group Theory and its Applications to Physical Problems* (Addison-Wesley, Reading, 1962).

[9.3] S. Helgason, *Differential Geometry and Symmetric Spaces* (Academic Press, New York, 1962).

[9.4] M. Hausner and J.T. Schwartz, *Lie Groups; Lie Algebras* (Gordon and Breach, New York, 1968).

[9.5] W. Miller, Jr., *Symmetry Groups and Their Applications* (Academic Press, New York, 1972).

[9.6] R. Gilmore, *Lie Groups, Lie Algebras and Some of Their Applications* (John Wiley & Sons, New York, 1974).

[10] Examples of applications of the concepts of symplecticity to physics, and in particular to optics, are in

[10.1] Y.S. Kim and M.E. Noz, "Illustrative examples of the symplectic group", Am. J. Phys. **51**, 368-375 (1983).

[10.2] V. Guillemin and S. Sternberg, *Symplectic Techniques in Physics* (Cambridge Univ. Press, New York, 1990).

[11] S. Steinberg, "Lie series, Lie transformations, and their applications", in *Lie Methods in Optics*, J. Sánchez Mondragón and K.B. Wolf (eds.) (Springer-Verlag, Berlin, 1986), ch. 3, pp. 45-104.

[12] Conceptual and applicative analyses of Lie theoretic methods within light and magnetic optics of both linear and nonlinear systems are elaborated in

[12.1] A.J. Dragt and J.M. Finn, "Lie series and invariant functions for analytic symplectic maps", J. Math. Phys. **17**, 2215-2227 (1976).

[12.2] A.J. Dragt, "Lectures on nonlinear orbit dynamics", in *Physics of High Energy Particle Accelerators*, AIP Conf. Proc. ***87***, R. A. Carrigan, F. R. Huson and M. Month (eds.) (AIP, New York, 1982), pp. 147-313; *Lie Methods for Nonlinear Dynamics with Applications to Accelerator Physics*, Univ. Maryland Phys. Dept. Rep. (1998).

[12.3] A.J. Dragt, "Lie algebraic theory of geometrical optics and optical aberrations", J. Opt. Soc. Am. **72**, 372-378 (1982); *Lie Algebraic Methods for Ray and Wave Optics*, Univ. Maryland Phys. Dept. Rep. (1998).

[12.4] A.J. Dragt, E. Forrest and K.B. Wolf, "Foundations of a Lie algebraic theory of geometrical optics", in *Lie Methods in Optics*, J. Sánchez Mondragon and K.B. Wolf (eds.) (Springer-Verlag, Berlin, 1986), ch. 4, pp. 105-157.

[12.5] K.B. Wolf, "Symmetry in Lie optics", Ann. Phys. **172**, 1-25 (1986).

[12.6] M. Navarro Saad and K.B. Wolf, "The group-theoretical treament of aberrating systems. I. Aligned lens systems in third aberration order", J. Math. Phys. **27**, 1449-1457 (1986).

[12.7] K.B. Wolf, "The group-theoretical treament of aberrating systems. II. Axis-symmetric inhomogeneous systems and fiber optics in third aberration order", J. Math. Phys. **27**, 1458-1465 (1986); III. The classification of asymmetric aberrations", J. Math. Phys. **28**, 2498-2507 (1987).

[12.8] K.B. Wolf, "Symmetry-adapted classification of aberrations", J. Opt. Soc. Am. A **5**, 1226-1232 (1988).

[12.9] A.J. Dragt, "Computation of maps for particle and light optics by scaling, splitting, and squaring", Phys. Rev. Lett. **75**, 1946-1948 (1995).

[13] Integration methods of operator differential-equations are well illustrated in

[13.1] J. Wei and E. Norman, "Lie algebraic solution of linear differential equations", J. Math Phys. **4**, 575-581 (1963).

[13.2] R.M. Wilcox, "Exponential operators and parameter differentiation in quantum physics", J. Math. Phys. **8**, 962-982 (1967).

[13.3] G. Dattoli, J.C. Gallardo and A. Torre, "An algebraic view to the operatorial ordering and its applications to optics", Riv. Nuovo Cim. **11**, 11, 1-79 (1988).

[13.4] G. Dattoli and A. Torre, "A general view to Lie algebraic methods in applied mathematics, optics and transport systems for charged beam accelerators", in *Dynamical Symmetries and Chaotic Behaviour in Physical Systems*, G. Maino, L. Fronzoni and M. Pettini (eds.) (World Sc., Singapore, 1991), pp. 21-89.

[13.5] G. Dattoli, P.L. Ottaviani, A. Torre and L. Vázquez, "Evolution operator equations: integration with algebraic and finite-difference methods. Applications to physical problems in classical and quantum mechanics and quantum field theory", Riv. Nuovo Cim. **20**, 2, 1-133 (1997).

[14] Investigations and classifications of optical (light and magnetic) aberrations are elaborated in [4.1]-[4.5], [4.7]-[4.10], [17.1] and [17.2] as well as in the titles below. The group-theoretic treatment of aberrations, pioneered by Dragt's papers [12.1]-[12.3], is illustrated also in [4.10], ch. 8, pp. 161-183, as in [12.4]-[12.8]. An exaustive treatment is presented in [4.11], Part IV, pp. 291-353.

[14.1] H.H. Hopkins, *Wave Theory of Aberrations* (Clarendon Press, Oxford, 1950).

[14.2] H.A. Buchdahl, *Optical Aberration Coefficients* (Dover, New York, 1968).

[14.3] W.T. Welford, *Aberrations of Optical Systems* (A. Higler, Bristol, 1986).

[14.4] V.N. Mahajan, *Aberration Made Simple* (SPIE, Bellingham, 1991).

[14.5] K.B. Wolf, "A group-theoretic treatment of Gaussian optics and 3rd-order aberrations", Lecture Note Phys. **201**, 133-136 (1984).

[14.6] M. Navarro Saad and K.B. Wolf, "Applications of a factorization theorem for 9th-order aberration optics", J. Symb. Comput. **1**, 235-239 (1985).

[14.7] K.B. Wolf and G. Krotzsch, "Group-classified polynomials of phase-space in higher-order aberration expansions", J. Symb. Comput. **12**, 673-693 (1991).

[15] K.F. Gauss, "Dioptrische Untersuchugen", Trans. Roy. Soc. Sci. at Göttingen, vol. I (1843).

[16] The ray matrix approach to linear optics is extensively presented in many textbooks and papers. See, for instance, [12] in ch. 2.

[17] The characterization of optical systems by cardinal planes is illustrated in

[17.1] F.A. Jenkins and H.E. White, *Fundamentals of Optics* (McGraw-Hill, New York, 4th ed., 1976).

[17.2] W.J. Smith, *Modern Optical Engineering* (McGraw-Hill, New York, 1966).

[17.3] R. Kingslake, *Optical System Design* (Academic Press, New York, 1983).

2

1D First-Order Optical Systems: The Ray-Transfer Matrix

2.1 Introduction

We apply the strategy devised in Chapter 1 to construct the optical matrices, describing the propagation of light rays through linear stigmatic and simple astigmatic optical systems. The basic tools of the strategy are the optical Hamiltonian $H(q_x, q_y, p_x, p_y, z)$ and the associated Hamilton's equations for the evolution of the light ray as propagating through the system along the optical z axis. The system is described by the refractive index function $n(q_x, q_y, z)$, as the ray is described by the canonically conjugate variables (q_x, q_y) and (p_x, p_y), specifying the cartesian coordinates and the direction cosines of the ray at its crossing point with the reference screen transversal to the optical axis. More precisely, the optical momenta (p_x, p_y) are the direction cosines of the ray scaled by the local value of the refractive index. The Cartesian space spanned by the ray position and momentum coordinates is the geometrical optical phase space. The ray is represented by points in the $4D$ optical phase space and accordingly the ray path in real space corresponds to a trajectory of the ray representative point in phase space [1].

The optical Hamiltonian has the peculiar form of the square-root function $H(q, p, z) = -\sqrt{n^2(q, z) - p^2}$, having some resemblance to that of a relativistic particle. The associated Lie operator $\widehat{\mathsf{L}}_H(q, p, z)$, obtained from H by Poisson bracket: $\widehat{\mathsf{L}}_H \equiv \{\cdot, H\}$, rules the evolution of the ray variables $q(z)$ and $p(z)$, as functions of the propagation parameter z, according to Hamilton's equations

$$\frac{d}{dz}\begin{pmatrix} q(z) \\ p(z) \end{pmatrix} = \widehat{\mathsf{L}}_H(q, p, z)\begin{pmatrix} q(z) \\ p(z) \end{pmatrix}, \quad q(z_i) = q_i, \;\; p(z_i) = p_i, \tag{2.1.1}$$

z_i denoting the axial position where the ray coordinates are known; q and p here are handy abbreviations of (q_x, q_y) and (p_x, p_y), respectively.

In particular, if propagation in the close proximity of the axis is concerned, the optical Hamiltonian simplifies into the form $\frac{p^2}{2n} - n(q,z)$, resembling the Hamiltonian of a non relativistic particle moving under the potential $-n(q,z)$ [2]. The optical momenta p_x and p_y come to be directly linked to the inclination angles of the ray in x and y directions. In addition, as the refractive index is a smooth function of space, it may be described by a quadratic function of the ray coordinates (q_x, q_y) over the pertinent small region across the axis. The Hamiltonian function becomes then quadratic in both q and p coordinates, thus yielding the linearized form of Hamilton equations (2.1.1) as

$$\frac{d}{dz}\begin{pmatrix} q(z) \\ p(z) \end{pmatrix} = \mathbf{H}(z)\begin{pmatrix} q(z) \\ p(z) \end{pmatrix}, \qquad q(z_i) = q_i, \;\; p(z_i) = p_i, \tag{2.1.2}$$

where the ray-parameter independent matrix $\mathbf{H}(z)$ is obtained from the Hamiltonian $H(q,p,z)$ by Poisson bracketing with the conjugate variables q and p.

Accordingly the propagation manifests as a sequence of linear transformations of the ray variables from the initial values (q_i, p_i), as it is suitably expressed by the concise matrix form of the input-output relationship

$$\begin{pmatrix} q_o \\ p_o \end{pmatrix} = \mathbf{M}(z_o, z_i)\begin{pmatrix} q_i \\ p_i \end{pmatrix}. \tag{2.1.3}$$

The ray variables on the assigned input and output planes, delimiting the system at positions z_i and z_o along the optical axis, are related by the 4×4 ray-transfer matrix $\mathbf{M}(z_o, z_i)$, which is the mathematical recipient for the optical features of the system and hence for its response to the incoming illumination.

The optical matrix is the exhaustive descriptor of ray propagation in geometrical first-order optics, since its specification provides a complete account of the ingoing-to-outgoing ray transformation operated by the optical system. As discussed in § 1.7.3, the representative Hamiltonian matrix $\mathbf{H}(z)$ and the ray transfer matrix $\mathbf{M}(z, z_i)$ are closely linked, the functional forms of the entries of $\mathbf{M}$ being determined by the functional forms of those of $\mathbf{H}$. In the particular case that the Hamiltonian function does not depend explicitly on the axial parameter z, the ray-transfer matrix $\mathbf{M}(z_o, z_i)$ is simply obtained by exponentiating the constant matrix $\mathbf{H}$ through the axial distance $z_o - z_i$ between the reference entrance and exit planes as

$$\mathbf{M}(z_o, z_i) = e^{\mathbf{H}\,(z_o - z_i)}. \tag{2.1.4}$$

Additional simplifications are attained when the propagation occurs through stigmatic or simple astigmatic systems. In the first case, the system is circularly symmetric about the optical axis; then, if the input ray lies on one of the infinite symmetry planes through the axis, we may track only two of the four

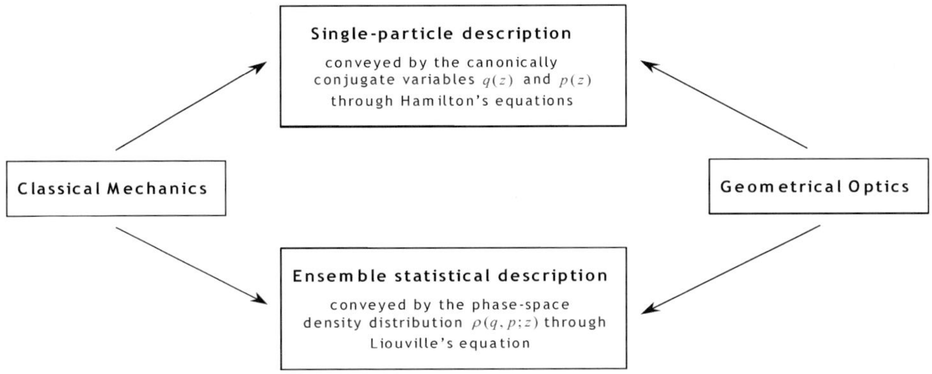

FIGURE 2.1. Exemplificative scheme of the single-particle and ensemble descriptions of a classical dynamical system, e.g. a point particle, which are mirrored into the geometrical-optics approach to light propagation in terms of the single-ray variables $q(z)$, $p(z)$ and the ray-bundle density $\rho(q, p; z)$ in phase space.

ray parameters, following the ray path in that plane, where the propagating ray keeps on lying. The problem becomes one-dimensional, as the motion of the ray representative point takes place in a $2D$ subspace of the $4D$ optical phase space, and accordingly $\mathbf{H}$ and $\mathbf{M}$ come to be 2×2 matrices. Likewise, the propagation through simple astigmatic systems, which are symmetric about two orthogonal planes intersecting the optical axis, can be accounted for by 2×2 matrices, being the motion in the optical phase space decomposable into two independent motions in the two $2D$ subspaces. We will limit our discussion to first-order stigmatic and simple astigmatic optical systems, which, being describable within the $1D$ model, will shortly be referred to as $1D$ *systems*.

The conceptual path, developed in Chapter 1 and briefly synthesized here, pertains to the single-ray picture of light propagation. Evidently within the ray-picture of geometrical optics a light distribution may be seen as an ensemble of uncorrelated rays. As noted in § 1.2, account of the propagation of the light distribution by integration of Hamilton's equations for each ray in the ensemble is, of course, impracticable. An alternative strategy must be envisaged. We may presumably guess that, as the single-ray description of light propagation parallels the Hamiltonian formulation of the single particle dynamics, thus a ray-ensemble description might be elaborated paralleling the statistical formulation of the dynamics of an ensemble of particles [3, 4]. In other words, referring to Fig. 2.1, we may say that, as we have delineated the correspondence between classical mechanics and geometrical optics through the upper path, so we delineate here such a correspondence through the lower path. Indeed, Liouville's equation for the phase-space density distribution $\rho(q, p; z)$ is derived in Sect. 2.2, where are also examined some basic properties of the Lie transformations. The Lie operator $\widehat{\mathsf{L}}_H$ and accordingly the ray matrix $\mathbf{M}$ confirm as the basic tools to describe the light beam propagation within the ray-model of geometrical optics. Some considerations aimed at relating the geometrical and physical interpretations of ρ close the section. Then, inspired by the functional form of the first-order optics Hamiltonian, being quadratic in the ray variables, we start to construct the matrix optics, generated by the quadratic monomials $\frac{1}{2}p^2$, $\frac{1}{2}q^2$ and $\frac{1}{2}qp$, the corresponding optical systems being then identified as a section of homogeneous medium, a refracting surface and an ideal magnifier, respectively (Sect. 2.4). Finally, Sect. 2.5 illustrates the phase-space picture of the transformations effected by the above quoted systems on both a single ray and some exemplary bundles of rays.

2.2 Ray-ensemble description of light propagation

The considerations developed in Chapter 1 have depicted the conceptual and practical procedure for the single-ray tracing, in full analogy with the Hamiltonian dynamics of a single particle [1]. A single ray is described in real space by the four canonically conjugate z-dependent coordinates $(q(z), p(z))$ and visually represented at any z by a mathematical point in the geometrical-optical phase space. The ray path in physical space corresponds to a trajectory of the relative representative point in phase space. The position and direction coordinates of the ray provide an "individual" characterization; their evolution, and accordingly the phase-space trajectory of the ray, is completely and uniquely defined by Hamilton's equations in correspondence with the specified input data. Such a description is no longer viable when dealing with an ensemble of rays. We need identify appropriate parameters to provide a "global" characterization of the ray-ensemble, and then elaborate the relevant evolution equations to account for the propagation of the ray-bundle.

Evidently, within the ray picture of geometrical optics, a light beam may be described as a collection of rays. It covers a certain area on the reference screen and contains rays throughout a certain range of inclination angles. Hence, it may be visually represented in the $4D$ optical phase space by a certain finite volume $\mathcal{V}$; each point in the volume $\mathcal{V}$ is representative of a ray in the bundle. If the bundle propagates through an optical system, the phase-space volume, originally occupied by the rays, changes its shape and moves through phase space. Thus, instead of following the motion of every single ray in the bundle, we may observe the motion of the pertinent phase-space volume.

Paralleling the methods of statistical mechanics [4], we introduce the phase space function $\rho(q, p; z)$, specifying the density of representative points through phase space. Precisely, $\rho(q, p; z)$ is defined in such a way that, at any axial

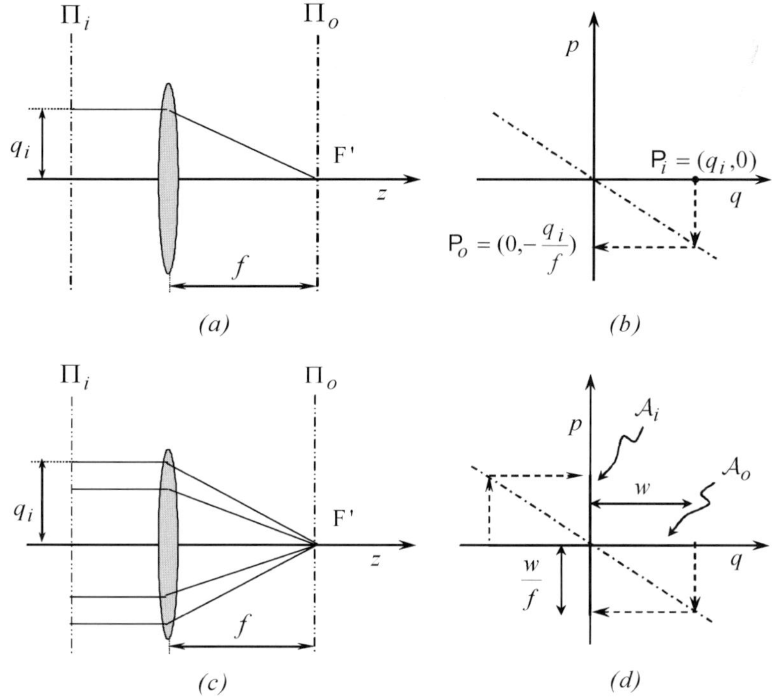

FIGURE 2.2. Naïve view of the phase-plane transformation of the representative point of a single ray and the representative region of a ray-bundle, effected by an optical system. *(a)* A ray propagates to the back focal point of a positive lens and *(b)* the corresponding representative point in the phase plane "moves" from P_i to P_o. Likewise, *(c)* a beam of parallel rays is focused to the lens back focus and *(d)* the corresponding representative segment P_1P_2, lying on the q-axis, changes to the segment P_3P_4, lying on the p-axis.

position z, the number of representative points $d\mathcal{N}(q,p;z)$ in the volume element $dqdp$ around the point (q,p) in phase space is given by the product $d\mathcal{N}(q,p;z) = \rho(q,p;z)dqdp$. Clearly, $\rho(q,p;z)$ takes on values throughout the beam volume $\mathcal{V}$ in phase space, and vanishes outside $\mathcal{V}$.

As an example, we may refer to the simple situations illustrated in Fig. 2.2, inspired by [3.8]. A paraxial ray, incoming parallel to the optical axis of a focusing lens, is deviated to pass through the lens back focus F′ (Fig. 2.2.a)). The input-ray representative point P_i, that lies on the q-axis of the phase plane, "moves" to the output-ray representative point P_o, that in contrast lies on the p-axis (Fig. 2.2.b)). Correspondingly, a beam of parallel rays hinging on the lens is focussed to a spot in the back focus. The incoming beam is represented in the phase plane by a segment, lying on the q-axis, of extension equal to the spatial width $2w$ of the beam (Fig. 2.2.c)). Likewise the fan of the focussed rays passing through F′ identifies a segment, lying on the p-axis, of extension $2\frac{w}{f}$ (Fig. 2.2.d)). It is evident that the region in the optical phase plane representing the beam moved and changed its shape; actually, in the simple case of the $1D$ model pertaining to Fig. 2.2, only the extension of the zero-area phase-plane region has been affected by the lens focusing effect. The density of representative points in phase plane changed as well.

It is therefore natural to inquire whether it is possible to envisage a law for the evolution of the phase-space density $\rho(q, p; z)$, on account of that the evolution of every single ray of the ensemble is governed by Hamilton's equations. We will find that an "evolution" equation for $\rho(q, p; z)$ can be written down, and notably in a form strictly similar to Hamilton's equations for the ray coordinates. Moreover, we will see that the phase space density $\rho(q, p; z)$ is evolved by the ray transfer operator or, when the linear approximation is concerned, by the ray transfer matrix and thus we will regain the single-ray picture as basic to the ray-ensemble picture as well. The discussion presented in the next paragraphs basically parallels the derivation of Liouville's theorem in statistical mechanics [3, 4]. Phase-space based methods are ordinary tools in magnetic optics. Indeed, as is well known, the charged-beam optics manifests a close analogy with the optics of light rays. Such an analogy will frequently be emphasized through the book, even though not specifically analysed; in this connection, we suggest some titles in [3].

Before going on, we note that Figs. 2.b) and 2.d) anticipate in a sense some aspects of the forthcoming discussion. In fact, hypothetical trajectories have been traced in the figure for the representative points to carry the initial point or region to the final point or region. Thus, the passage through the lens is pictured as a q-constant path connecting the zero-momentum points representing the incoming rays to the nonzero-momentum points representing the exiting rays. In a like manner, the propagation from the lens to the back focal point is graphically depicted as a p-constant path to carry the representative points of the redirected rays to the zero off-axis displacement points on the p-axis. Such graphic views of the quoted transformations (transfer by a lens and propagation through a homogeneous medium) will be justified by the later considerations (Sects. 2.4 and 2.5). We will see that the propagation through a homogeneous medium can be depicted as a continuum horizontal segment (a q-shear), signifying indeed a continuous transformation of the ray coordinate. In contrast, the passage through a lens produces a discontinuity of the ray direction, due to the refraction at the shaping surfaces of the lens at which the refractive index function abruptly changes from one side to the other. A continuous path picture of the refraction is a convenient, even though not strictly correct, visualization of the process, which thus is viewed as a p-shear.

2.2.1 *Liouville's equation for the ray-ensemble*

We consider the variation of the density function $\rho(q,p;z)$ with the propagation variable z. It is mathematically given by the total derivative of $\rho(q,p;z)$ with respect to z, namely

$$\frac{d}{dz}\rho(q,p;z) = \frac{\partial}{\partial z}\rho(q,p;z) + \frac{\partial \rho}{\partial q}\frac{dq}{dz} + \frac{\partial \rho}{\partial p}\frac{dp}{dz}, \tag{2.2.1}$$

which, of course, displays contributions from both the explicit and implicit dependence on z, manifested respectively through the partial derivative $\partial\rho/\partial z$ and the derivative-chain terms, involving the ray coordinates q and p [a]. For brevity, we use again the abbreviation (q,p) for (q_x, q_y, p_x, p_y).

Replacing for dq/dz and dp/dz the respective expressions from Hamilton's equations, the chain-like terms simplify into the Poisson bracket of ρ and H:

$$\frac{\partial \rho}{\partial q}\frac{dq}{dz} + \frac{\partial \rho}{\partial p}\frac{dp}{dz} = \frac{\partial \rho}{\partial q}\frac{\partial H}{\partial p} - \frac{\partial \rho}{\partial p}\frac{\partial H}{\partial q} = \{\rho, H\}, \tag{2.2.2}$$

by which the equation for $\rho(q,p;z)$ follows in the concise form

$$\frac{d}{dz}\rho = \frac{\partial}{\partial z}\rho(q,p;z) + \{\rho, H\}. \tag{2.2.3}$$

In general the z-evolution of any dynamical variable pertinent to a system, whose dynamics is governed by Hamilton-like equations of evolution, occurs according to Eq. (2.2.3).

On account of the physical meaning of $\rho(q,p;z)$, we can further specify its variation with z. In fact, we may observe that the total number of representative points in phase space remains constant, even though the shape of the region containing them may presumably change as a consequence of the individual motions of the enclosed points. In particular, let us consider a fixed volume $\mathcal{V}$ in phase space and denote by $\mathcal{S}$ the closed surface delimiting it. The number of ray-representative points inside $\mathcal{V}$ at a certain z is obtained integrating the density $\rho(q,p;z)$ over the volume itself, namely

$$\mathcal{N}_{\mathcal{V}}(z) = \int_{\mathcal{V}} \rho(q,p;z)d\mathcal{V}. \tag{2.2.4}$$

Evidently, $\mathcal{N}_{\mathcal{V}}(z)$ depends on $\mathcal{V}$ and z. Being $\mathcal{V}$ fixed, $\mathcal{N}_{\mathcal{V}}(z)$ changes only with z, due to the motion of the representative points reflecting the propagation of the corresponding rays in the beam. As no rays are created or destroyed,

[a] The derivative $\frac{d}{dz}$ yields the z-derivative of $\rho(q,p;z)$ in a frame moving with the point (q,p) in phase space. In contrast, $\frac{\partial}{\partial z}$ yields the z-derivative of $\rho(q,p;z)$ in a fixed frame.

the net rate at which $\mathcal{N}_\mathcal{V}(z)$ varies, $\frac{\partial \mathcal{N}_\mathcal{V}(z)}{\partial z}$, must equal the net rate of points crossing the boundary surface $\mathcal{S}$. So we write

$$\frac{\partial}{\partial z}\mathcal{N}_\mathcal{V}(z) = -\int_\mathcal{S} \rho\, \dot{\mathbf{u}} \cdot \mathbf{n}\, d\mathcal{S}. \tag{2.2.5}$$

the surface integral representing the rate at which points cross the surface $\mathcal{S}$ outward. In fact, in conformity with the standard conventions, $\mathbf{n}$ is the outward-drawn unit normal to the surface element dS, and $\dot{\mathbf{u}}= (dq/dz, dp/dz)^\top$ is the velocity vector in phase space, specifying the rate by which the ray at the phase space point (q, p) moves to the point $(q + dq, p + dp)$. The minus sign accounts for the fact that, if points leave the volume $\mathcal{V}$, $\mathcal{N}_\mathcal{V}$ decreases.

According to Gauss' divergence theorem [5], the surface integral may be turned into the volume integral as

$$\int_\mathcal{S} \rho\, \dot{\mathbf{u}} \cdot \mathbf{n}\, dS = \int_\mathcal{V} \nabla \cdot (\rho\, \dot{\mathbf{u}}\,)\, d\mathcal{V}, \tag{2.2.6}$$

where $\nabla\cdot(\,)$ denotes the divergence operation, and hence $\nabla\cdot(\,) = \partial/\partial q + \partial/\partial p$. Interestingly we find

$$\nabla \cdot (\rho\, \dot{\mathbf{u}}\,) = \frac{\partial}{\partial q}(\rho \frac{dq}{dz}) + \frac{\partial}{\partial p}(\rho \frac{dp}{dz}) = \{\rho, H\}. \tag{2.2.7}$$

By use of expression (2.2.4), giving $\mathcal{N}_\mathcal{V}$ as the volume integral of ρ, and (2.2.6) in (2.2.5), we end up with

$$\int_\mathcal{V} \{\frac{\partial}{\partial z}\rho + \nabla \cdot (\rho\, \dot{\mathbf{u}}\,)\}\, d\mathcal{V} = 0, \tag{2.2.8}$$

which, being the volume $\mathcal{V}$ arbitrarily chosen, yields the continuity equation for the phase-space density $\rho(q, p; z)$ as

$$\frac{\partial}{\partial z}\rho + \nabla \cdot (\rho\, \dot{\mathbf{u}}\,) = 0. \tag{2.2.9}$$

Moreover, using the above relation into Eq. (2.2.3) for the derivative $\frac{d\rho}{dz}$, on account of (2.2.7), we find that the continuity equation is equivalent to

$$\frac{d}{dz}\rho = 0. \tag{2.2.10}$$

Equation (2.2.10) is the desired equation for the z-evolution of the phase-space density $\rho(q, p; z)$. We rewrite it in the convenient form

$$\frac{\partial}{\partial z}\rho = -\{\rho, H\} = -\,\widehat{\mathsf{L}}_H\,\rho, \tag{2.2.11}$$

with the initial data being conveyed through $\rho(q,p;z_i) \equiv \rho_i(q,p)$. We refer to (2.2.11) as *Liouville's equation* for the ray-ensemble, as it directly follows by *Liouville theorem* (2.2.10). It is evident its similarity to Hamilton's equations (2.1.1) for the light ray.

Evidently, as the phase-space vector $\mathbf{u}(z)=(q(z),p(z))^\top$ may be regarded as the descriptor of the single ray, the phase space density $\rho(q,p,z)$ may be ragarded as the descriptor of the ray-bundle. It is of value that the z-evolution of both $\mathbf{u}$ and ρ is essentially governed by the Lie operator $\widehat{\mathsf{L}}_H$, even though with opposite signs due to the invariance of $\rho(q,p,z)$ along the ray path, on which we will comment later. As a conclusion, we may say that Hamiltonian optics deals with evolution equations of the type

$$\frac{\partial}{\partial z}\mathfrak{v} = \widehat{\mathcal{H}}\mathfrak{v}, \tag{2.2.12}$$

where $\mathfrak{v}$ may denote the phase-space vector $(q(z),p(z))^\top$ or the phase-space density $\rho(q,p;z)$ and accordingly the operator $\widehat{\mathcal{H}}$ comes to be $\widehat{\mathcal{H}} = \pm\widehat{\mathsf{L}}_H$.

The formal similarity of Eq. (2.2.11) to (2.1.1) makes evident that the considerations we developed in § 1.4.1 about the integration of (2.1.1) apply to the integration of (2.2.11) as well. The crucial point of those considerations was concerned with the operator nature of $\widehat{\mathsf{L}}_H$ and its possible dependence on the axial parameter z. We found that the simplest case pertains to a z-independent Hamiltonian, in which case the solution to (2.2.11) takes the form

$$\rho(q,p;z) = e^{-(z-z_i)\widehat{\mathsf{L}}_H}\rho(q,p;z_i), \tag{2.2.13}$$

where $\rho(q,p;z_i)$ denotes the phase-space density of the beam at the input reference screen. Expression (2.2.13) reflects the same view as Eq. (1.4.4); thus, the phase-space density $\rho(q,p;z)$ at z originates from the assigned function $\rho(q,p;z_i)$ at the input z_i-plane by way of the Lie transformation $e^{-(z-z_i)\widehat{\mathsf{L}}_H}$. In Section 1.4 we considered Lie transformations acting directly on the ray variables (q,p), and, as specific examples in § 1.5.2 we described the Lie transformations generated by quadratic functions of q and p. Here, the Lie transformation acts on the phase-space density $\rho(q,p;z)$, which is a function of q and p. To see what happens we need to use some properties of the Lie transformations, on which we will briefly comment in the next paragraph. Firstly we will elaborate a geometrical and physical interpretation of Eq. (2.2.10).

Geometry and Physics related with Liouville's theorem

Equation (2.2.10) shows that the density of phase space points does not change by propagation of the beam through the optical medium. As d/dz represents the "mobile" derivative, we can give (2.2.10) a nice pictorial view in phase space. Suppose to individualize a "point" $\mathrm{P} = (q,p)$ in phase-space, i.e., a ray,

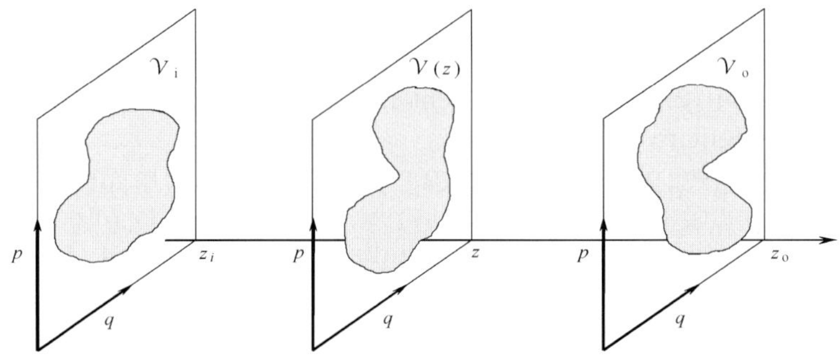

FIGURE 2.3. The phase-space region (shaded areas) occupied by the assembly of the ray-bundle representative points changes as the beam proceeds along the z axis, but the relevant volume remains constant: $\mathcal{V}_i = \mathcal{V}(z) = \mathcal{V}_o$.

by some mark. Let $\rho(\mathrm{P}; z_i)$ be the phase space density at P for a certain z_i. The point P describes a trajectory in phase space with increasing $z > z_i$. For each $z > z_i$ we can evaluate the density $\rho(\mathrm{P}; z)$ at P, moving with it along its own path. We find that it does not change, i.e., $\rho(\mathrm{P}; z) = \rho(\mathrm{P}; z_i)\ \forall\, z > z_i$. The motion of the ray-ensemble representative points in phase space is then like that of an incompressible fluid in real space. In other words, $\rho(q, p; z)$ behaves like a "property" of the point in phase space, and accordingly it is transported by the point along its own trajectory when it is representative of a light ray.

The invariance of the phase-space density $\rho(q, p; z)$ implies the invariance of the phase-space volume $\mathcal{V}$ of the region enclosing the ray-bundle representative points. In fact, suppose to select at a certain $z = z_i$ a certain number of representative points around a chosen point, through which the density takes on the constant value ρ_i. Let $\mathbb{V}_i$ be the volume enclosing the chosen points. The total number of points in $\mathbb{V}_i$ evidently is

$$\mathbb{N}_i = \rho_i \mathbb{V}_i. \tag{2.2.14}$$

Following the points as they move through phase-space, we see that the boundaries of the enclosing volume change with $z > z_i$, but by definition the total number of points contained inside it does not vary: $\mathbb{N}(z) = \mathbb{N}_i$. Also, the density of points keeps on the value $\rho(z) = \rho_i$ by virtue of (2.2.10). Thus, for any $z > z_i$ we can write

$$\mathbb{N}(z) = \rho_i \mathbb{V}(z) = \mathbb{N}_i, \tag{2.2.15}$$

which by comparison with (2.2.14) immediately leads to

$$\mathbb{V}(z) = \mathbb{V}_i. \tag{2.2.16}$$

The volume enclosing a fixed number of representative points in phase space remains constant in z, although, due to the motion of the enclosed points, its shape may sensibly change. So, the total phase-space volume $\mathcal{V}$ of the beam remains constant in z, regardless of the motion of each ray and the corresponding motion of the representative points through phase space: $\mathcal{V}_o = \mathcal{V}(z) = \mathcal{V}_i$ (Fig. 2.3). This result is rather intuitive if we think of a flowing "fluid". If the total quantity of fluid and the density of the fluid do not change through the motion, the volume occupied by the fluid remains constant as well, even though it may deform significantly during the fluid's flow.

The invariance of the phase space volume applies obviously also to the volume elements $d\mathcal{V}$'s: $d\mathcal{V}_o = d\mathcal{V}_i$ along any ray path. In § 2.4 we will elaborate a geometrical interpretation of the volume element $d\mathcal{V}$ and its invariance along the ray path in connection with the conceptual definition of the intuitive perception of the elementary bundle of rays.

Finally, we note that the ray transformation from the phase-space point $\mathrm{P}_i(q_i, p_i)$ to the point $\mathrm{P}_o(q_o, p_o)$ is accompanied by a transformation of the phase-space volume elements from $d\mathcal{V}_i$ to $d\mathcal{V}_o$ [(b)]. We known that if $\mathbf{S}$ denotes the Jacobian matrix associated with the transformation $\mathrm{P}_i \to \mathrm{P}_o$ (see § 1.3.1), the relevant volumes $d\mathcal{V}_i$ and $d\mathcal{V}_o$ are linked by the determinant of $\mathbf{S}$ as $d\mathcal{V}_o = \det \mathbf{S}\, d\mathcal{V}_i$. The invariance of the volume elements under ray propagation implies therefore that $\det \mathbf{S} = 1$, thus solving between the two possibilities offered by Eq. (1.3.13) for the determinant of the symplectic matrix associated with the ray transfer. Viceversa, the unimodularity of the symplectic Jacobian matrix associated with the ray transformation implies the invariance of the phase space volume. This offers a clear geometrical interpretation of the symplectic nature of ray propagation, signifying that elementary volumes along the ray path in phase-space remain constant. Evidently, areas play the role as volumes within a $1D$ picture, we are mainly concerned with (see § 1.7.4).

2.2.2 *Basics of Lie transformations*

As noted in § 1.3.2, given a dynamical variable $f(q,p)$ we can generate the Lie operator $\widehat{\mathsf{L}}_f$ by Poisson bracket operation:

$$\widehat{\mathsf{L}}_f = \{\cdot, f\}, \tag{2.2.17}$$

and the Lie transformation $\widehat{\mathcal{T}}_f$ by computing the power series:

$$\widehat{\mathcal{T}}_f(\zeta) \equiv \sum_{j=0}^{\infty} \frac{\zeta^j}{j!} \widehat{\mathsf{L}}_f^j = e^{\zeta \widehat{\mathsf{L}}_f}, \quad \zeta \in \mathbb{R}, \tag{2.2.18}$$

[b] In the present discussion the concept of volume in phase space has been left on the same intuitive footing as that in physical space. Actually, phase space is a mathematical space in which indeed the infinitesimal volume is defined in terms of the linear antisymmetric nondegenerate *2n-form*, and finite volume are accordingly obtained from it by integration over the domain of interest. For an accurate analysis the reader is directed to [3.3].

where ζ is a real parameter and the j-th power of $\widehat{\mathsf{L}}_f$ is obviously

$$\widehat{\mathsf{L}}_f^j = \{\{\{..\{\cdot, \underbrace{f\}..\}, f\}, f\}}_{j \text{ times}} . \tag{2.2.19}$$

As evidenced by the exponential notation on the right-hand side of (2.2.18), the power series is nothing but the exponential function of the operator $\zeta \widehat{\mathsf{L}}_f$. Evidently, $\widehat{\mathcal{T}}_f(0)$ is the identity operator: $\widehat{\mathcal{T}}_f(0) = \widehat{\mathsf{I}}$ [6].

Lie transformations have many interesting and useful properties, which directly follow from the properties of the Lie operators and the exponentiation. We shall not give a detailed account of all the properties of the Lie transformations, for which the reader is directed to [6], but shall comment in particular on those that are useful to gain insight into the behaviour of the phase-space density $\rho(q, p; z)$ through Eq. (2.2.13). We emphasize that the mathematical setting here is not properly complete; questions relevant, for instance, to the analyticity or the radius of convergence of the power series expansions or of operator identities, we will be concerned with, will not be approached.

The product preservation property

We prove the *product preservation property*, which through the relation

$$e^{\zeta \widehat{\mathsf{L}}_f} hg = (e^{\zeta \widehat{\mathsf{L}}_f} h)\,(e^{\zeta \widehat{\mathsf{L}}_f} g), \tag{2.2.20}$$

states that the Lie transformation acting on the product of functions produces the same result as the product of the Lie transformations acting separately on the single functions in the product. Then, consider the product of two dynamical variables $h(q, p)$ and $g(q, p)$ for which the Poisson brackets with $f(q, p)$ are meaningful:

$$v(q, p) = h(q, p) g(q, p). \tag{2.2.21}$$

We wish to determine the effect of the Lie transformation $\widehat{\mathcal{T}}_f(\zeta)$ on $v(q, p)$. Obviously, we are faced with the evaluation of repeated Poisson brackets of f and v. In fact, in accord with the power series representation (2.2.18), one has

$$e^{\zeta \widehat{\mathsf{L}}_f} v(q, p) = \sum_{j=0}^{\infty} \frac{\zeta^j}{j!} \widehat{\mathsf{L}}_f^j v(q, p) = v + \zeta\{v, f\} + \frac{\zeta^2}{2}\{\{v, f\}, f\} + ..., \tag{2.2.22}$$

We recall from § 1.3.2 that the Lie operator acts on the product of functions in the similar way as differentiation. Higher powers of the Lie operator satisfy an identity analogous to the Liebniz rule for derivatives, and hence

$$\widehat{\mathsf{L}}_f^j hg = \sum_{l=0}^{j} \binom{j}{l} \widehat{\mathsf{L}}_f^l h \widehat{\mathsf{L}}_f^{j-l} g. \tag{2.2.23}$$

We invite the reader to prove this relation by iteration. With this expression inserted in (2.2.22), we obtain

$$e^{\zeta \widehat{\mathsf{L}}_f} v = \sum_{j=0}^{\infty} \sum_{l=0}^{j} \frac{\zeta^{j-l}}{(j-l)!} \frac{\zeta^l}{l!} \widehat{\mathsf{L}}_f^l h \widehat{\mathsf{L}}_f^{j-l} g = [\sum_{l=0}^{\infty} \frac{\zeta^l}{l!} \widehat{\mathsf{L}}_f^l h] [\sum_{j=l}^{\infty} \frac{\zeta^{j-l}}{(j-l)!} \widehat{\mathsf{L}}_f^{j-l} g] = [\sum_{l=0}^{\infty} \frac{\zeta^l}{l!} \widehat{\mathsf{L}}_f^l h] [\sum_{j=0}^{\infty} \frac{\zeta^j}{j!} \widehat{\mathsf{L}}_f^j g] = (e^{\zeta \widehat{\mathsf{L}}_f} h)\,(e^{\zeta \widehat{\mathsf{L}}_f} g), \tag{2.2.24}$$

thus proving the desired property (2.2.20). In the second step of (2.2.24), we have made use of the permutation

$$\sum_{j=0}^{\infty} \sum_{l=0}^{j} = \sum_{l=0}^{\infty} \sum_{j=l}^{\infty} . \tag{2.2.25}$$

Accordingly, the Lie transformation of powers of q and p is

$$e^{\zeta \widehat{\mathsf{L}}_f} q^m p^n = (e^{\zeta \widehat{\mathsf{L}}_f} q)^m \,(e^{\zeta \widehat{\mathsf{L}}_f} p)^n, \tag{2.2.26}$$

as can easily be proved by applying repeatedly the rule (2.2.24).

We consider now a dynamical variable $g(q, p)$, and suppose that it can be expanded in a power series with respect to the conjugate variables q and p around a certain point in phase-space, which for brevity we may asssume to be the origin: $(q, p) = (0, 0)$. We write therefore

$$g(q, p) = \sum_{m=0}^{\infty} \sum_{n=0}^{\infty} g_{m,n} q^m p^n . \tag{2.2.27}$$

We apply $\widehat{\mathcal{T}}_f(\zeta)$ to (2.2.27) and use the linearity of the Lie transformation to pass the exponential operator to each term of the series. Explicitly, we have

$$\widehat{\mathcal{T}}_f(\zeta)\, g(q, p) = \sum_{m=0}^{\infty} \sum_{n=0}^{\infty} g_{m,n} \widehat{\mathcal{T}}_f(\zeta)\, [q^m p^n]. \tag{2.2.28}$$

Then, using the product preserving property (2.2.20) to pass the exponential operator into each variable, we finally obtain

$$e^{\zeta \widehat{\mathsf{L}}_f}\, g(q, p) = \sum_{m=0}^{\infty} \sum_{n=0}^{\infty} g_{m,n} (e^{\zeta \widehat{\mathsf{L}}_f} q)^m \,(e^{\zeta \widehat{\mathsf{L}}_f} p)^n. \tag{2.2.29}$$

It is evident that we can recompose the function g according to the series expansion (2.2.27) at the phase-space point $(e^{\zeta \widehat{\mathsf{L}}_f} q, e^{\zeta \widehat{\mathsf{L}}_f} p)$, namely

$$e^{\zeta \widehat{\mathsf{L}}_f} g(q, p) = g(e^{\zeta \widehat{\mathsf{L}}_f} q, e^{\zeta \widehat{\mathsf{L}}_f} p). \tag{2.2.30}$$

The action of a Lie transformation on a phase-space function is then to perform a Lie transformation on its arguments. In other words, the above result establishes a parallel between the transformation experienced by the phase-space vector $\mathbf{u} = (q,p)^\top$ and that experienced by the phase-space function $g(q,p) = g(\mathbf{u}^\top)$, i.e.,

$$\begin{array}{ccc} \text{phase-space vector} & & \text{phase-space function} \\ \mathbf{u} \rightarrow e^{\zeta \widehat{\mathsf{L}}_f} \mathbf{u} = \mathbf{v} & \Longrightarrow & g(\mathbf{u}^\top) \rightarrow e^{\zeta \widehat{\mathsf{L}}_f} g(\mathbf{u}^\top) = g(\mathbf{v}^\top) \end{array} \tag{2.2.31}$$

The composition property (2.2.30) of the Lie transformations is essential for the later considerations regarding the evolution of the phase-space density.

The Poisson-bracket preservation property

Another remarkable property of the Lie transformations is the *Poisson bracket preserving property*, which in a sense reproduces (2.2.20) with the ordinary product hg being replaced by the Poissonian product $\{h,g\}$, i.e.,

$$e^{\zeta \widehat{\mathsf{L}}_f} \{h,g\} = \{e^{\zeta \widehat{\mathsf{L}}_f} h,\, e^{\zeta \widehat{\mathsf{L}}_f} g\}. \tag{2.2.32}$$

The Lie transformation acting on the Poisson bracket of functions gives the same result as the Poisson bracket of the Lie transformations acting separately on the single functions involved in. Needless to say, the Poisson bracket preservation property has the same proof as (2.2.20), being based on the Leibniz rule for the higher powers of Lie operators acting on a Poisson bracket:

$$\widehat{\mathsf{L}}_f^j \{h,g\} = \sum_{l=0}^{j} \binom{j}{l} \{\widehat{\mathsf{L}}_f^l\, h,\, \widehat{\mathsf{L}}_f^{j-l}\, g\}, \tag{2.2.33}$$

which, paralleling (2.2.23), follows from the derivation-like behaviour of $\widehat{\mathsf{L}}_f$ with respect to the Poisson bracket multiplication.

Evidently the property (2.2.32) establishes that every Lie transformation is symplectic. In fact, the Lie transformation $\widehat{\mathcal{T}}_f(\zeta)$ generates a mapping which takes the conjugate variables q and p to the new variables $\overline{q}$ and $\overline{p}$ as

$$\begin{aligned} q \rightarrow \overline{q} &= e^{\zeta \widehat{\mathsf{L}}_f} q, \\ p \rightarrow \overline{p} &= e^{\zeta \widehat{\mathsf{L}}_f} p. \end{aligned} \tag{2.2.34}$$

The Poisson bracket of h and g as functions of $\overline{q}$ and $\overline{p}$ is then

$$\{h(\overline{q},\overline{p}), g(\overline{q},\overline{p})\} = \{e^{\zeta \widehat{\mathsf{L}}_f} h(q,p),\, e^{\zeta \widehat{\mathsf{L}}_f} g(q,p)\} = \{h,g\}(\overline{q},\overline{p}), \tag{2.2.35}$$

thus evidencing the preservation of the Poisson bracket under the mapping (2.2.34); use has been done of the composition property (2.2.30) to pass the operator $e^{\zeta \widehat{\mathsf{L}}_f}$ from the arguments to the respective functions h and g, and viceversa, after applying (2.2.32), from the function $\{h, g\}$ to the respective arguments. In particular,

$$\{\overline{q}, \overline{p}\} = \{q, p\} = 1. \tag{2.2.36}$$

In § 1.3.2 we proved that the preservation of the Poisson bracket of any two dynamical variables is a necessary and sufficient condition for a transformation be symplectic.

Lie transformations as a Lie group

We now take up the question of the group structure of the set of Lie transformations $\{e^{\zeta \widehat{\mathsf{L}}_f};\ \zeta \in \mathbb{R}\}$ generated by the dynamical variable $f(q, p)$. Accordingly we will linger on some properties of Lie algebras and Lie groups to help the reader to gain some familiarity with these concepts (see also Appendix A). Although the discussion below will be concerned with the Lie algebra of the operators (2.2.17) and the corresponding Lie group of the transformations (2.2.18), it has a general validity, being relevant to the link between the concept of Lie algebra and that of the associated Lie group [7].

As remarked in § 1.3.2, the set of Lie operators $\{\widehat{\mathsf{L}}_f\}$, generated by the dynamical variables $\{f(q, p)\}$ through (2.2.17), do form a Lie algebra with respect to the Lie bracket $[,]$ of operators

$$\left[\widehat{\mathsf{H}}, \widehat{\mathsf{G}}\right] = \widehat{\mathsf{H}}\widehat{\mathsf{G}} - \widehat{\mathsf{G}}\widehat{\mathsf{H}}, \tag{2.2.37}$$

which, as proved by (1.3.43), defines a binary operation taking two Lie operators to a Lie operator as well:

$$[,] : (\widehat{\mathsf{L}}_h, \widehat{\mathsf{L}}_g) \in \{\widehat{\mathsf{L}}_f\} \times \{\widehat{\mathsf{L}}_f\} \to \left[\widehat{\mathsf{L}}_h, \widehat{\mathsf{L}}_g\right] = \widehat{\mathsf{L}}_{\{g,h\}} \in \{\widehat{\mathsf{L}}_f\}. \tag{2.2.38}$$

Evidently, if $\{g, h\} = 0$, the corresponding Lie operators commute with each other: $\left[\widehat{\mathsf{L}}_h, \widehat{\mathsf{L}}_g\right] = 0$.

The notion of group as that of algebra implies the existence of a binary operation sending pairs of elements of the set into elements of the set. When dealing with operators the binary operation to structure a set as a Lie algebra or a Lie group is just the commutation product (2.2.37) between elements in the algebra and the sequential application, i.e. the multiplication, of operators in the group.

It is a notable property of Lie algebras that by exponentiating the elements of the algebra through a real parameter ζ we may form the associated group. An element $\widehat{\mathcal{T}}$ of a Lie group is said to be generated by the element $\widehat{\mathsf{L}}$ of a Lie algebra in the sense that for an infinitesimal value $d\zeta$ of the group parameter we have $\widehat{\mathcal{T}}(d\zeta) = \widehat{\mathsf{I}} + d\zeta\widehat{\mathsf{L}}$. Thus, the Lie algebra of a group is a description of that group near the identity element. Sequential multiplications of infinitesimal-step elements $\widehat{\mathcal{T}}(d\zeta)$ lead by induction to finite-step elements $\widehat{\mathcal{T}}(\zeta)$, for which the exponential representation in terms of the generator $\widehat{\mathsf{L}}$ is consequently obtained. In fact, it is proved by induction that

$$\widehat{\mathcal{T}}(\zeta + d\zeta) = \widehat{\mathcal{T}}(\zeta)\widehat{\mathcal{T}}(d\zeta) = \widehat{\mathcal{T}}(\zeta)(\widehat{\mathsf{I}} + d\zeta\widehat{\mathsf{L}}), \tag{2.2.39}$$

which means

$$\widehat{\mathcal{T}}(\zeta) + d\widehat{\mathcal{T}} = \widehat{\mathcal{T}}(\zeta) + d\zeta\widehat{\mathsf{L}}\widehat{\mathcal{T}}(\zeta), \tag{2.2.40}$$

thus yielding the first-order differential equation for $\widehat{\mathcal{T}}(\zeta)$:

$$\frac{d}{d\zeta}\widehat{\mathcal{T}}(\zeta) = \widehat{\mathsf{L}}\widehat{\mathcal{T}}(\zeta). \tag{2.2.41}$$

The exponential representation of the group element $\widehat{\mathcal{T}}(\zeta)$ in terms of the algebra element $\widehat{\mathsf{L}}$ immediately follows:

$$\widehat{\mathcal{T}}(\zeta) = e^{\zeta\widehat{\mathsf{L}}}, \tag{2.2.42}$$

with being $\widehat{\mathcal{T}}(0) = \widehat{\mathsf{I}}$ for every $\widehat{\mathsf{L}}$ in the algebra.

Then, every operator in the algebra generates a subgroup of exponential-type operators $\{e^{\zeta\widehat{\mathsf{L}}};\, \zeta \in \mathbb{R}\}$, which is continuously parametrized by the real parameter ζ. One typically speaks of a *one-parameter* subgroup. Each value of the parameter ζ individualizes one operator in the subgroup. Operators corresponding to different values of the parameter commute with each other; namely

$$\widehat{\mathcal{T}}(\zeta_1)\widehat{\mathcal{T}}(\zeta_2) = \widehat{\mathcal{T}}(\zeta_2)\widehat{\mathcal{T}}(\zeta_1), \tag{2.2.43}$$

which directly follows from the basic relation (2.2.39) clearly implying in finite terms that

$$\widehat{\mathcal{T}}(\zeta_1)\widehat{\mathcal{T}}(\zeta_2) = \widehat{\mathcal{T}}(\zeta_1 + \zeta_2) = \widehat{\mathcal{T}}(\zeta_2 + \zeta_1) = \widehat{\mathcal{T}}(\zeta_2)\widehat{\mathcal{T}}(\zeta_1). \tag{2.2.44}$$

The subgroup $\{e^{\zeta\widehat{\mathsf{L}}};\, \zeta \in \mathbb{R}\}$ is therefore abelian. Also, (2.2.43) allows us to identify the inverse to $\widehat{\mathcal{T}}(\zeta)$ as the operator individualized by the opposite value $-\zeta$ of the parameter, namely

$$\widehat{\mathcal{T}}^{-1}(\zeta) = \widehat{\mathcal{T}}(-\zeta) = e^{-\zeta\widehat{\mathsf{L}}}. \tag{2.2.45}$$

In the light of these general considerations, the Lie transformations for a given Lie operator $\widehat{\mathsf{L}}_f$ are seen to form the one-parameter subgroup $\{\widehat{\mathcal{T}}_f(\zeta) = e^{\zeta\widehat{\mathsf{L}}_f};\, \zeta \in \mathbb{R}\}$, for which

$$\widehat{\mathcal{T}}_f(\zeta_1)\widehat{\mathcal{T}}_f(\zeta_2) = \widehat{\mathcal{T}}_f(\zeta_1 + \zeta_2) = \widehat{\mathcal{T}}_f(\zeta_2)\widehat{\mathcal{T}}_f(\zeta_1) \tag{2.2.46}$$

and

$$\widehat{\mathcal{T}}_f^{-1}(\zeta) = \widehat{\mathcal{T}}_f(-\zeta) = e^{-\zeta \widehat{\mathsf{L}}_f}. \tag{2.2.47}$$

Each operator in the subgroup produces a symplectic mapping of the conjugate variables q and p to the new variables $\overline{q}$ and $\overline{p}$ as given in (2.2.34), i.e.,

$$\{q, p\} \to \{\overline{q} = e^{\zeta \widehat{\mathsf{L}}_f} q,\, \overline{p} = e^{\zeta \widehat{\mathsf{L}}_f} p\}. \tag{2.2.48}$$

Clearly, successive Lie transformations produce a symplectic transformation as well. Thus, the composite operator $e^{\zeta_1 \widehat{\mathsf{L}}_f} e^{\zeta_2 \widehat{\mathsf{L}}_g}$ generates the two-step symplectic mapping

$$\{q, p\} \to \{\widetilde{q} = e^{\zeta_2 \widehat{\mathsf{L}}_g} q,\, \widetilde{p} = e^{\zeta_2 \widehat{\mathsf{L}}_g} p\} \to \{\overline{q} = e^{\zeta_1 \widehat{\mathsf{L}}_f} \widetilde{q},\, \overline{p} = e^{\zeta_1 \widehat{\mathsf{L}}_f} \widetilde{p}\}. \tag{2.2.49}$$

However it cannot in general be associated with a single Lie transformation, unless, according to (2.2.46), $g = f$, or more in general, as will be proved below, $\{g, h\} = 0$.

Notably, near the identity, i.e., for sufficiently small values of the parameters, Lie transformations multiply into Lie transformations. Namely, we have

$$\widehat{\mathcal{T}}_f(\zeta_1)\widehat{\mathcal{T}}_g(\zeta_2) = \widehat{\mathcal{T}}_h(\zeta), \tag{2.2.50}$$

for ζ_1 and ζ_2 sufficiently small. The explicit expression of the closure relation (2.2.50) for Lie transformations, i.e., for exponential operators, within a sufficiently small neighborhood of the identity is known as the Baker-Campbell-Hausdorff (BCH) formula. It dictates the conditions and the rules to unit two exponential operators into a single exponential operator [8]. The exponents of the factors on one side combine into a single exponent of the product on the other, which is ultimately determined by the underlying algebraic structure, directly involving the Lie products of the exponents of the factors.

Here we will prove (2.2.50) in a rather naïve way, exploiting the power series expansion (2.2.18). Firstly we consider the interesting case that $\{f, g\} = 0$ and so $\left[\widehat{\mathsf{L}}_f, \widehat{\mathsf{L}}_g\right] = 0$. In that case, in fact, the closure relation (2.2.50) of Lie transformations holds for finite values of the parameters ζ_1 and ζ_2, with the multiplication law turning into the additivity of the exponents as

$$\widehat{\mathcal{T}}_f(\zeta_1)\widehat{\mathcal{T}}_g(\zeta_2) = \widehat{\mathcal{T}}_{\zeta_1 f + \zeta_2 g}(1) = \widehat{\mathcal{T}}_g(\zeta_2)\widehat{\mathcal{T}}_f(\zeta_1), \tag{2.2.51}$$

which further becomes merely additive with respect to the parameters ζ_1 and ζ_2 if also $g = f$, i.e.,

$$\widehat{\mathcal{T}}_f(\zeta_1)\widehat{\mathcal{T}}_f(\zeta_2) = \widehat{\mathcal{T}}_f(\zeta_1 + \zeta_2) = \widehat{\mathcal{T}}_f(\zeta_2)\widehat{\mathcal{T}}_f(\zeta_1), \tag{2.2.52}$$

thus recovering the $\widehat{\mathsf{L}}_f$-generated subgroup relation (2.2.46).

It will be evident that the hypothesized commutativity of the operators $\widehat{\mathsf{L}}_f$ and $\widehat{\mathsf{L}}_g$ is crucial to extend the additivity of the exponents from exponential scalars to exponential operators, thus yielding the scalar-like relation (2.2.51).

We consider the exponential operator on the right of (2.2.51), which explicitly writes as

$$\widehat{\mathcal{T}}_{\zeta_1 f+\zeta_2 g}(1) = e^{\zeta_1 \widehat{\mathsf{L}}_f+\zeta_2 \widehat{\mathsf{L}}_g} \equiv \sum_{j=0}^{\infty} \frac{1}{j!}(\zeta_1 \widehat{\mathsf{L}}_f + \zeta_2 \widehat{\mathsf{L}}_g)^j, \tag{2.2.53}$$

and hence demands evaluating powers of the operator $\zeta_1 \widehat{\mathsf{L}}_f + \zeta_2 \widehat{\mathsf{L}}_g$. The square is, for instance,

$$\begin{gathered}(\zeta_1 \widehat{\mathsf{L}}_f + \zeta_2 \widehat{\mathsf{L}}_g)^2 = (\zeta_1 \widehat{\mathsf{L}}_f + \zeta_2 \widehat{\mathsf{L}}_g)\,(\zeta_1 \widehat{\mathsf{L}}_f + \zeta_2 \widehat{\mathsf{L}}_g) = \\ \zeta_1^2 \widehat{\mathsf{L}}_f^2 + \zeta_1\zeta_2 \widehat{\mathsf{L}}_f \widehat{\mathsf{L}}_g + \zeta_2\zeta_1 \widehat{\mathsf{L}}_g \widehat{\mathsf{L}}_f + \zeta_2^2 \widehat{\mathsf{L}}_g^2 = \zeta_1^2 \widehat{\mathsf{L}}_f^2 + 2\zeta_1\zeta_2 \widehat{\mathsf{L}}_f \widehat{\mathsf{L}}_g + \zeta_2^2 \widehat{\mathsf{L}}_g^2.\end{gathered} \tag{2.2.54}$$

It should be noted in the unfolded expression above the two products $\widehat{\mathsf{L}}_f \widehat{\mathsf{L}}_g$ and $\widehat{\mathsf{L}}_g \widehat{\mathsf{L}}_f$, which are in general different. Hence they can be summed to $2\widehat{\mathsf{L}}_f \widehat{\mathsf{L}}_g$ only under the assumption of the commutativity of $\widehat{\mathsf{L}}_f$ and $\widehat{\mathsf{L}}_g$. In that case, the square of the operator $\zeta_1 \widehat{\mathsf{L}}_f + \zeta_2 \widehat{\mathsf{L}}_g$ comes to have the same expression as that of the square of a bynomial of scalars. Accordingly, the j-th power $(\zeta_1 \widehat{\mathsf{L}}_f + \zeta_2 \widehat{\mathsf{L}}_g)^j$ can be calculated by the well known formula for the j-th power of a bynomial:

$$(\zeta_1 \widehat{\mathsf{L}}_f + \zeta_2 \widehat{\mathsf{L}}_g)^j = \sum_{l=0}^{j} \tbinom{j}{l} \zeta_1^l \widehat{\mathsf{L}}_f^l \zeta_2^{j-l} \widehat{\mathsf{L}}_g^{j-l}. \tag{2.2.55}$$

Upon substituting (2.2.55) into the power series (2.2.53), we obtain

$$\begin{aligned} e^{\zeta_1 \widehat{\mathsf{L}}_f+\zeta_2 \widehat{\mathsf{L}}_g} &\equiv \sum_{j=0}^{\infty}\sum_{l=0}^{j} \tfrac{1}{(j-l)!l!} \zeta_1^l \widehat{\mathsf{L}}_f^l \zeta_2^{j-l} \widehat{\mathsf{L}}_g^{j-l} \\ &= \sum_{l=0}^{\infty} \tfrac{1}{l!} \zeta_1^l \widehat{\mathsf{L}}_f^l \sum_{j=l}^{\infty} \tfrac{1}{(j-l)!} \zeta_2^{j-l} \widehat{\mathsf{L}}_g^{j-l} = e^{\zeta_1 \widehat{\mathsf{L}}_f} e^{\zeta_2 \widehat{\mathsf{L}}_g}, \end{aligned} \tag{2.2.56}$$

thus proving relation (2.2.51).

We emphasize again that (2.2.51) grounds on the expression (2.2.55) for powers of a bynomial of operator functions, which reproduces that for powers of a bynomial of scalar functions only for commuting operators (Problem 3).

In general, to prove (2.2.50) we exploit the power series expansion (2.2.18). We suppose now that both the real numbers ζ_1 and ζ_2 are sufficiently small to justify the truncation of the series expansions of both $\widehat{\mathcal{T}}_f(\zeta_1)$ and $\widehat{\mathcal{T}}_g(\zeta_2)$ at the first order of ζ_1 and ζ_2. We write then

$$\begin{aligned}\widehat{\mathcal{T}}_f(\zeta_1 \sim 0) &= \widehat{\mathsf{I}} + \zeta_1 \widehat{\mathsf{L}}_f, \\ \widehat{\mathcal{T}}_g(\zeta_2 \sim 0) &= \widehat{\mathsf{I}} + \zeta_2 \widehat{\mathsf{L}}_g,\end{aligned} \tag{2.2.57}$$

so that the product of $\widehat{\mathcal{T}}_f(\zeta_1)$ and $\widehat{\mathcal{T}}_g(\zeta_2)$ is

$$\widehat{\mathcal{T}}_f(\zeta_1)\widehat{\mathcal{T}}_g(\zeta_2) = \widehat{\mathsf{I}} + \zeta_1 \widehat{\mathsf{L}}_f + \zeta_2 \widehat{\mathsf{L}}_g, \qquad \zeta_1, \zeta_2 \sim 0 \tag{2.2.58}$$

where, of course, only the first order terms in the parameters ζ_1 and ζ_2 have been retained. Resorting again to the power series expansion, we can recover the exponential representation as

$$\widehat{\mathcal{T}}_f(\zeta_1)\widehat{\mathcal{T}}_g(\zeta_2) = e^{\zeta_1 \widehat{\mathsf{L}}_f + \zeta_2 \widehat{\mathsf{L}}_g} = e^{\widehat{\mathsf{L}}_{\zeta_1 f + \zeta_2 g}} = \widehat{\mathcal{T}}_{\zeta_1 f + \zeta_2 g}(1), \tag{2.2.59}$$

for ζ_1 and ζ_2 sufficiently small, and hence near the identity $\widehat{\mathsf{I}}$.

As previously remarked, the validity of the multiplication law (2.2.51) for Lie transformations near the identity grounds on the BCH theorem, which provides the rule to combine the exponents of the factors on the left as

$$e^{\zeta_1 \widehat{\mathsf{L}}_f} e^{\zeta_2 \widehat{\mathsf{L}}_g} = e^{\widehat{\mathsf{L}}_h(\zeta_1, \zeta_2, \widehat{\mathsf{L}}_f, \widehat{\mathsf{L}}_g)}, \tag{2.2.60}$$

giving a complete analytic expression of the exponent on the right as an infinite series of iterated Lie brackets of $\zeta_1 \widehat{\mathsf{L}}_f$ and $\zeta_2 \widehat{\mathsf{L}}_g$. Explicitly,

$$\begin{aligned}\widehat{\mathsf{L}}_h(\zeta_1, \zeta_2, \widehat{\mathsf{L}}_f, \widehat{\mathsf{L}}_g) &= \zeta_1 \widehat{\mathsf{L}}_f + \zeta_2 \widehat{\mathsf{L}}_g + \tfrac{\zeta_1 \zeta_2}{2} [\widehat{\mathsf{L}}_f, \widehat{\mathsf{L}}_g] \\ &+ \tfrac{1}{12}\{\zeta_1^2 \zeta_2 [\widehat{\mathsf{L}}_f, [\widehat{\mathsf{L}}_f, \widehat{\mathsf{L}}_g]] + \zeta_1 \zeta_2^2 [\widehat{\mathsf{L}}_g, [\widehat{\mathsf{L}}_g, \widehat{\mathsf{L}}_f]]\} + \mathcal{O}(\zeta_1^3 \zeta_2, \zeta_1^2 \zeta_2^2, \zeta_1 \zeta_2^3),\end{aligned} \tag{2.2.61}$$

and the series may in general converge only for sufficiently small ζ_1 and ζ_2 [8].

The BCH formula confirms in precise terms that, as noted a number of times, the additivity of exponents does not apply in general to the product of exponential operators, being applicable only to commuting operators. In fact, we see from (2.2.61) that, if $\widehat{\mathsf{L}}_f$ and $\widehat{\mathsf{L}}_g$ commute with each other, the infinite summation on the right-hand side simply reduces to the sum $\zeta_1 \widehat{\mathsf{L}}_f + \zeta_2 \widehat{\mathsf{L}}_g$, thus exactly yielding $e^{\zeta_1 \widehat{\mathsf{L}}_f} e^{\zeta_2 \widehat{\mathsf{L}}_g} = e^{\zeta_1 \widehat{\mathsf{L}}_f + \zeta_2 \widehat{\mathsf{L}}_g}$ just as for scalars. Relation (2.2.51), we proved in the case that $\left[\widehat{\mathsf{L}}_f, \widehat{\mathsf{L}}_g\right] = 0$ using the power series expansions of the operators on both sides, is evidently in accord with the BCH formula. It is needless to emphasize the relevance of the BCH formula within the group theoretical framework, as it establishes a direct relationship between the structure of the Lie algebra of a group (exemplified by the Lie brackets in (2.2.61)) and the local structure of that group (exemplified by the multiplication of operators).

Finally we wish to comment on the physical meaning of the parameter ζ entering the algebra-to-group operation of exponentiation, of which (2.2.18) is an example relatively to the algebra of Lie operators (2.2.17). As remarked also in § 1.5, the meaning of ζ depends on the context we are working in as well as on the meaning of the phase space variable that generates the Lie transformation.

When geometrical optics is concerned and the Lie transformation is associated with the Hamiltonian flow, the exponentiation parameter ζ takes the role as the axial distance $z_o - z_i$ by which the light ray propagates and accordingly the Lie transformation is the optical "propagator" of the ray from z_i to z_o. We note that, as the optical Hamiltonian is unitless, the corresponding Lie operator has the dimension of length^{-1}.

2.2.3 *Ray-bundle propagation: the ray transfer operator and matrix*

Applying the result (2.2.30) to Eq. (2.2.13) we find that $\rho(q,p;z)$ evolves from the initial distribution $\rho(q,p;z_i)$ according to

$$\rho(q,p;z) = \rho(e^{-(z-z_i)\widehat{\mathsf{L}}_H} q, e^{-(z-z_i)\widehat{\mathsf{L}}_H} p; z_i). \tag{2.2.62}$$

We recall that in § 1.4.1 when investigating the z-evolution of the position and direction coordinates of the ray, we introduced the ray-transfer operator $\widehat{\mathfrak{M}}(z,z_i)$ and proved that in the case of a z-independent Hamiltonian it takes the exponentiated form

$$\widehat{\mathfrak{M}}(z,z_i) = e^{(z-z_i)\widehat{\mathsf{L}}_H}. \tag{2.2.63}$$

According to (2.2.47), the operator entering the arguments of ρ in (2.2.62) appears to be the inverse of the ray-transfer operator (2.2.63), namely

$$e^{-(z-z_i)\widehat{\mathsf{L}}_H} = \widehat{\mathfrak{M}}^{-1}(z,z_i). \tag{2.2.64}$$

In fact, if the ray variables $\mathbf{u}(z) = (q(z), p(z))^\top$ evolve from the input data $\mathbf{u}(z_i)$ by $\widehat{\mathfrak{M}}(z,z_i)$, conversely, the input data $\mathbf{u}(z_i)$ can be recovered from $\mathbf{u}(z)$ by the inverse operator $\widehat{\mathfrak{M}}^{-1}(z,z_i)$:

$$\mathbf{u}(z_i) = \widehat{\mathfrak{M}}^{-1}(z,z_i)\mathbf{u}(z). \tag{2.2.65}$$

Relation (2.2.62) suggests therefore the following picture. Let (q,p) be the coordinates of a ray at z, then $(e^{-(z-z_i)\widehat{\mathsf{L}}_H} q, e^{-(z-z_i)\widehat{\mathsf{L}}_H} p)$ may be interpreted as the starting data for the chosen ray (q,p) consistently with the Hamiltonian dynamics, identified by the relevant Lie operator $\widehat{\mathsf{L}}_H$. Evidently, this is a manifestation of the previously established invariance of the phase-space density along the phase-space trajectory.

Remarkably, according to (2.2.62), the Lie transformation $e^{(z-z_i)\widehat{\mathsf{L}}_H}$, which is basic to the single-ray tracing, becomes as well basic to the ray-ensemble picture of light propagation ruling the evolution of the phase-space density.

In § 1.5.3 we mentioned that in general the ray transfer operator $\widehat{\mathfrak{M}}(z,z_i)$ can be represented in the factored form as the product of Lie transformations. Then, applying property (2.2.30) to each term in the product or exploiting the group closure property (2.2.50) of Lie transformations, we may restate relation (2.2.62) in the general form

$$\rho(q,p;z) = \rho(\widehat{\mathfrak{M}}^{-1}(z,z_i)q, \widehat{\mathfrak{M}}^{-1}(z,z_i)p; z_i) = \rho_i(\widehat{\mathfrak{M}}^{-1}(z,z_i)q, \widehat{\mathfrak{M}}^{-1}(z,z_i)p), \tag{2.2.66}$$

involving the ray-transfer operator, of which the single-exponential form (2.2.63) is a specific determination, suited to a z-independent Hamiltonian.

In particular, since in the linear approximation a 4×4 or 2×2 matrix representation is allowed for the ray transfer operator, property (2.2.62) can be formulated in terms of the ray-transfer matrix $\mathbf{M}(z, z_i)$ as well. Thus, using the handy vector notation, we can write

$$\rho(\mathbf{u}^\top; z) = \rho(\mathbf{v}^\top; z_i) = \rho_i(\mathbf{v}^\top), \qquad \mathbf{v} \equiv \mathbf{M}^{-1}(z, z_i)\mathbf{u}, \tag{2.2.67}$$

which in particular shows that

$$\rho(0,0;z) = \rho_i(0,0), \tag{2.2.68}$$

being the phase plane origin mapped into itself by $\mathbf{M}$.

Equation (2.2.66) and the corresponding linear model form (2.2.67) signify the relevance of the ray-transfer operator $\widehat{\mathfrak{M}}(z, z_i)$ and the ray-transfer matrix $\mathbf{M}$ for both the sinlge-ray and ray-ensemble pictures of light propagation pertinently to the geometrical-optics view. Equations (2.2.66) and (2.2.67) clearly show, in fact, that any distribution function is propagated along the ray-trajectories in phase space. In particular, the single-ray picture is easily recovered from (2.2.66) by noting that if the phase-space distribution is initially represented by a single ray, and so mathematically modelled by a δ function,

$$\rho(q,p;z_i) = \delta(q - q_i)\delta(p - p_i), \tag{2.2.69}$$

it remains a δ-function at any $z > z_i$, evolving into

$$\rho(q,p;z) = \delta(q - q(z))\delta(p - p(z)). \tag{2.2.70}$$

Here, $q(z)$ and $p(z)$ are the ray variables at z propagated from the input data (q_i, p_i) through the transfer operator $\widehat{\mathfrak{M}}(z, z_i)$ or, in the linear model, through the transfer matrix $\mathbf{M}(z, z_i)$ (see also Problem 4). Accordingly, throughout the remaining of the chapter we will work within the single-ray picture of linear geometrical optics. In due account of the established significance of the ray-transfer matrix, in the next sections we will proceed deriving the ray matrices for assigned z-independent model Hamiltonians. Their correspondence to recognizable optical systems will then be investigated.

Before closing this section we wish to illustrate the relation of the phase space density $\rho(q,p;z)$ to those measurable quantities that the traditional radiometry uses to characterize the light distribution and its propagation in connection with the radiative properties of the source.

2.2.4 *Phase space density and radiance*

As announced, in this paragraph we elucidate the connecting link between the basic quantity of the phase space representation of geometrical optics, i.e., the phase space density $\rho(q, p; z)$, and the basic quantity of radiometry, i.e., the radiance[(c)]. The discussion is strictly aimed at giving a feeling for the physical concreteness of the phase space density. Accordingly we merely derive the formal relation of ρ to the radiance function, addressing the reader to [9] for a fuller account of both the radiometric model and the relatively recently investigated link with the statistical wave theory.

We emphasize that, as geometrical optics, radiometry and the related theory of radiative energy transport are based on the concept of the light ray, understood as the geometrical trajectory along which the optical energy, radiated by the source, propagates [10]. Thus, according to the radiometric model, the radiant energy is localized both in space and in direction. We will see that the radiance is defined as a function of both position and direction. Position and direction are the building variables of the geometrical-optical phase space, which then seems to offer the natural frame for the radiometric theory as well.

Radiometry has believed to properly describe the behavior of radiation fields generated by spatially incoherent sources, such as the conventional thermal sources. Relatively recent investigations concerning the statistical and wave-optics content of the phenomenological laws of radiometry have revealed an intimate, and in a sense surprising, connection between radiometry and the theory of partial coherence. The laws of classical radiometry have been generalized to pertain to field distributions generated by planar sources of any state of coherence. In this respect, following the original suggestions by Walther [9.6, 9.7], a wide class of generalized radiance functions has been introduced through a certain number of different non-equivalent definitions, linearly involving various second-order correlation functions of the optical distribution across the source. Interestingly, the various expressions elaborated for the radiance exhibit a remarkable similarity with some of the distribution functions arising in the phase space representation of wave optics (and quantum mechanics as well) [9.9, 9.13], as for instance the Wigner distribution function we will examine in Chapter 6. As the phase-space density of the geometric-optical phase-space picture of ray bundles relates to the radiance function pertinent to radiometry with incoherent sources, thus the distribution functions of the

---

[c] "*Radiometry* is the science and the technology of the measurement of electromagnetic radiant energy. Measurements of radiant energy are conducted throughout the electromagnetic spectrum, using methods which are suited to the spectral region of interest." [9.2]. Although we survey the basic concepts of the radiometric model on an elementary level without entering the details of the system configurations or inherent techniques for radiometric measurements, it is worthwhile to emphasize that we refer to the optical region of the spectrum ($\lambda = 100\mathring{A} \div 0.1cm$).

wave-optical phase-space description of light fields relate to the generalized radiance functions pertaining to radiometry with partially coherent sources. Moreover, both the generalized radiance functions of radiometry and the space-frequency distribution functions of wave optics do not retain in general all the properties expected for them in traditional radiometry and in the ordinary probability theory, respectively. This fact indicates that neither the formers neither the latters can in general formulated in a manner that is strictly similar to the traditional radiometric and geometric-optical description of light fields [9.9]. However, it was shown that all the generalized radiance functions regain the properties postulated for the radiance in conventional radiometry, in the case of planar secondary quasi-homogeneous sources [(d)] and in the limit of short wavelengths. This result unequivocally establishes the place and the role of classical radiometry within the statistical wave theory of optics [9.11, 9.12].

Radiance: basic definition

The *radiance* (or *brightness)* is the area and solid angle density of the radiant power. It depends in general on the point in space at which it is evaluated and on the direction through that point. In order to visualize the definition of radiance we consider a planar radiating surface S in a homogeneous medium of refractive index n_0. S may be the surface of a real source or of an optical element like a mirror, a lens or a limiting aperture; also, it may be a fictitious surface situated everywhere in the radiation field. We may locate S in the plane $z = 0$, and suppose it radiates into the half-space $z > 0$. Then, for any point P_S on S, specified by the position vector $\mathbf{q} = (q_x, q_y)$ in the source plane, we denote by $d\mathsf{S}$ the element of area around P_S, and by $d\Omega$ the element of solid angle around a direction specified by the unit vector $\mathbf{v}$ originating from P_S (Fig. 2.4.a)). It is assumed in radiometry that the power $d\mathrm{P}$ $[W]$ radiated by the source element $d\mathsf{S}$ within the solid angle element $d\Omega$ is obtainable according to the elementary law

$$d\mathrm{P} = \mathcal{B}(\mathbf{q}, \mathbf{v}) \cos\theta d\mathsf{S} d\Omega. \tag{2.2.71}$$

Here θ is the angle that the $\mathbf{v}$-direction makes with the positive z axis; evidently, $0 \leq \theta < \pi/2$. Thus, the product $d\mathsf{A} = d\mathsf{S}\cos\theta$ yields the projected element of area perpendicular to $\mathbf{v}$. The function $\mathcal{B}(\mathbf{q}, \mathbf{v})$ is known as the ra-

---

[d] The optical intensity across a planar *quasihomogeneous* source changes so slowly with position in the source plane that is remains essentially constant through distances comparable with the correlation length of the radiation field across the source plane. Also, it is supposed that the linear dimensions of the source are large compared with both the mean wavelength of the radiated light and the correlation length. Evidently quasihomogeneous sources may be spatially coherent and incoherent in the "local" sense, but are always spatially rather incoherent in the "global" sense [9.8, 9.4].

Finally we recall that a *secondary* source radiates after direct or indirect (i.e., through an optical system) illumination by a *primary* source.

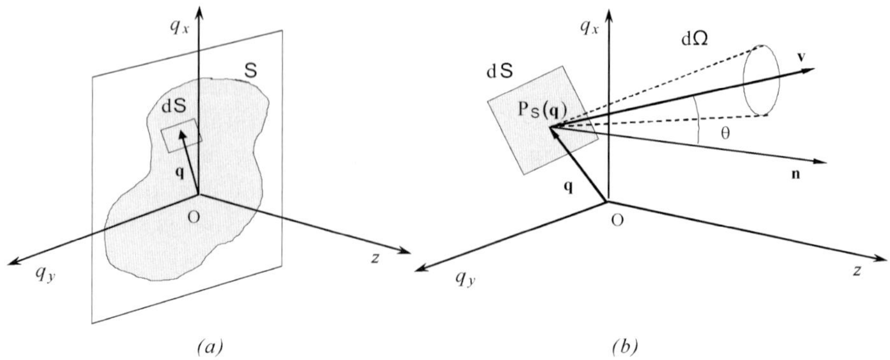

FIGURE 2.4. Basic elements illustrating the definition of radiance. *(a)* The planar source S in the plane $z = 0$ radiating into the half-plane $z > 0$ and *(b)* the geometrical quantities inherent in Eq. (2.2.71), i.e., the area element dS around the source point $P_S(\mathbf{q})$ and the element $d\Omega$ of solid angle around the ray-direction $\mathbf{v}$.

diance or brigthness of the source. It represents the rate at which energy is radiated into the unit solid angle around the specified direction $\mathbf{v}$ by the source element of unit projected area (perpendicular to $\mathbf{v}$) around the specified point $P_S(\mathbf{q})$; hence it is measured in $W \cdot m^{-2} \cdot sr^{-1}$ [(e)].

According to its physical significance, the radiance is necessarily nonnegative,

$$\mathcal{B}(\mathbf{q}, \mathbf{v}) \geq 0 \tag{2.2.72}$$

for all possible values of its arguments, and so for all position vectors $\mathbf{q} = (q_x, q_y)$ in the source plane and directions $\mathbf{v}$ pointing into the half-space $z > 0$. Also, it vanishes at points $\mathbf{q}$ in the source plane lying outside the source area S, i.e.,

$$\mathcal{B}(\mathbf{q}, \mathbf{v}) = 0 \qquad \text{if } \mathbf{q} \notin \mathsf{S}, \tag{2.2.73}$$

for all the possible directions $\mathbf{v}$, and hence for "rays" that do not cross the source surface S.

Evidently, the total power radiated by the source is

$$\mathrm{P} = \int_{2\pi} d\Omega \cos\theta \int_{\mathsf{S}} \mathcal{B}(\mathbf{q}, \mathbf{v}) d\mathsf{S}, \tag{2.2.74}$$

---

[e] On account of the spectral content of the emitted radiation, it is customary to consider the contributions to the radiated power for each frequency component separately. The symbol $d\mathrm{P}_\omega$ $[W \cdot nm^{-1}]$ is typically used with the meaning as the limiting value of the ratio $\Delta\mathrm{P}/\Delta\omega$ for frequency interval $\Delta\omega$ being made infinitesimally small; $\Delta\mathrm{P}$ is clearly intended as the amount of radiant power contained within $\Delta\omega$ around the specific frequency ω. Needless to say, $d\mathrm{P}_\omega$ obeys the law (2.2.71) in terms of the spectral radiance $\mathcal{B}_\omega(\mathbf{q}, \mathbf{p})$ $[W \cdot m^{-2} \cdot sr^{-1} \cdot nm^{-1}]$. We will ignore here these frequency-related aspects of the radiometric model; accordingly we will not consider the possible frequency-dependence of the refractive index.

the integrations extending over the source area S and over the 2π-solid angle formed by all the directions $\mathbf{v}$ pointing into the half-space $z > 0$.

The radiance as a phase-space function

Being a function of both position in space and direction from that position, the brightness comes naturally to be a kind of phase space function. Evidently the pair $(\mathbf{q}, \mathbf{v})$ identifies a ray emerging from the source point $\mathrm{P_S}(\mathbf{q})$ into the $\mathbf{v}$ direction (Fig. 2.4.b)), the ray momentum $\mathbf{p}$ being obtained just scaling the unit vector $\mathbf{v}$ by the refractive index value n_0 of the medium. Thus the radiance can as well be understood as a function of the proper ray-variables $(\mathbf{q}, \mathbf{p})$: $\mathcal{B}(\mathbf{q}, \mathbf{v}) \longleftrightarrow \mathcal{B}(\mathbf{q}, \mathbf{p})$.

Resorting to the phase-space picture of light distributions in terms of the density of rays in phase space, it is evident that the radiating surface S corresponds to a phase-space density function, we may denote as $\rho_\mathrm{S}(\mathbf{q}, \mathbf{p})$. Likewise the sample ray through the source point $\mathrm{P_S}(\mathbf{q})$ extending in the $\mathbf{v}$ direction corresponds to the point $(\mathbf{q}, n_0\mathbf{v})$ in phase space. Then, the rays emitted by the surface element $d\mathrm{S}$ at $\mathrm{P_S}$ within the solid angle $d\Omega$ around $\mathbf{v}$ individualize the element $d\mathcal{V}$ of volume in phase space, which encloses the sample ray. According to the geometric-optical picture of uncorrelated rays, giving independent energy contributions, we can say that each ray radiating from the surface carries a certain amount of power independently of the other emitted rays. Evidently, $d\mathrm{P}$ is the power carried by the phase-space volume $d\mathcal{V}$, and so it is also obtainable as

$$d\mathrm{P} = \delta\mathrm{P}\rho_\mathrm{S}(\mathbf{q}, \mathbf{p})d\mathcal{V}, \tag{2.2.75}$$

Here $\delta\mathrm{P}$ roughly denotes the average power carried by each ray radiated by the surface S. The product $\delta\mathrm{P}\rho_\mathrm{S}(\mathbf{q}, \mathbf{p})$ acquires the meaning as the optical power density in phase space of the light field across the radiating surface. It is worth noting that the linear addition of power represented by the integral (2.2.74) for the total power is consistent with the view of incoherent energy contributions emerging from all elements of the source into all directions, thus implying that there are no interference effects present.

We consider in some detail the phase-space volume $d\mathcal{V} = dq_x dq_y dp_x dp_y$, which we may rewrite as

$$d\mathcal{V} = d\mathrm{S} dp_x dp_y, \tag{2.2.76}$$

in terms of the area element $d\mathrm{S} = dq_x dq_y$ enclosing the chosen point $\mathrm{P_S}(q_x, q_y)$ in the source plane. Recalling from Eq. (1.3.1) the expressions of the optical momenta p_x, p_y:

$$p_x = n_0 \sin\alpha_x, \qquad dp_y = n_0 \sin\alpha_y = n_0 \cos\alpha_x \sin\vartheta_y, \tag{2.2.77}$$

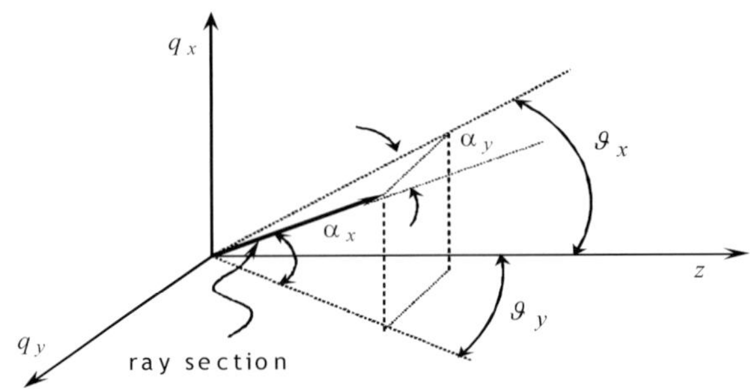

FIGURE 2.5. Geometry of the angles relevant to Eqs. (2.2.77).

where the geometry of the involved angles is clarified in Fig. 2.5, the differentials dp_x and dp_y write as

$$dp_x = n_0 \cos\alpha_x d\alpha_x, \qquad p_y = n_0 \cos\alpha_x \cos\vartheta_y d\vartheta_y. \tag{2.2.78}$$

Accordingly, by successive steps we can elaborate for $d\mathcal{V}$ the expression

$$d\mathcal{V} = n_0^2 d\mathsf{S} \cos^2\alpha_x \cos\vartheta_y d\alpha_x d\vartheta_y = n_0^2 d\mathsf{S} \cos\theta \cos\alpha_x d\alpha_x d\vartheta_y = n_0^2 \cos\theta d\mathsf{S} d\Omega, \tag{2.2.79}$$

after realizing that $\cos\alpha_x \cos\vartheta_y = \cos\theta$ and the elemental solid angle around the $\mathbf{v}$-direction is $d\Omega = \cos\alpha_x d\alpha_x d\vartheta_y$.

Equation (2.2.79) establishes an interesting link between the phase-space volume $d\mathcal{V}$ and some geometrical quantities in physical space, like areas and solid angles.

Inserting (2.2.79) into (2.2.75) and comparing with (2.2.71), we are enabled to relate the radiance $\mathcal{B}(\mathbf{q}, \mathbf{p})$ to the optical power density in phase space; precisely, we can write

$$\mathcal{B}(\mathbf{q}, \mathbf{p}) = n_0^2 \delta \mathrm{P} \rho_{\mathsf{S}}(\mathbf{q}, \mathbf{p}). \tag{2.2.80}$$

The dependence on z of both $\mathcal{B}_\omega$ and ρ_{S} above can tacitly be assumed because, as noted, the radiating surface may be arbitrarily located in space.

Equation (2.2.80) provides the key to turn from the real space representation to the phase space picture. A measurable quantity associated with the optical field in real space, i.e., the distribution of radiant energy at a given plane z, relates to the corresponding density of rays in phase space, thus reflecting as well the intuitive view that the ray density in phase space reproduces the spatial and angular energy distribution of the optical beam. In this connection, the concept of emittance domain in phase space arises quite naturally as the region through the representative phase-space volume of the ray bundle, which encloses most of the beam energy, say 86% or 99% according to the adopted criterion. By Liouville's theorem, this region, or more specifically its boundary, can loosely be taken as a representative region, and accordingly as a representative contour, to provide through its shape evolution a pictorial view in phase space of the propagation of the ray bundle in real space.

Remarkably, by (2.2.80) the invariance of the phase space density ρ_{S} along the ray trajectory implies the invariane of the quantity $\frac{\mathcal{B}}{n_0^2}$ along the ray path in a homogeneous lossless and gainless medium (see also Problem 7); this is a well established result of traditional radiometry [9.1]-[9.3].

As already noted, relation (2.2.80) applies to the classical context of geometrical optics and spatially incoherent sources. Under these conditions, in fact, we can loosely define an energy density distribution in phase space, and interpret it in terms of the radiance of the classical theory of radiometry. In the modern theory various expressions for the radiance have been elaborated, which account for coherent and partially coherent sources, being linearly expressed in terms of various second-order correlation functions of the optical distribution across the source. However, they cannot be interpreted the same as the radiance of the classical theory, i.e., as the radiated power distribution in position and direction. Likewise, wave-optical analogs of the phase space density have been defined in order to combine in a single view, comprising both position and direction, the two wave-optical description of light signals, separately pertaining to the spatial and spatial frequency (i.e., angular) domains. As an example of such phase-space (i.e., space-frequency) distributions, in Chapter 6 we will consider the Wigner distribution function. We will see that it cannot strictly be interpreted as an energy density distribution in phase space, even though a direct relation with a generalized radiance function can be established, thus resembling the classical relation (2.2.80).

For completeness' sake, we report as well the definitions of other two measurable quantities, which are obtained separately integrating the brightness $\mathcal{B}$ over the source area S or over the 2π-solid angle formed by all the possible ray-directions $\mathbf{p}$. Thus, we obtain the *radiant intensity* $\mathcal{J}(\mathbf{p})$ $[W \cdot sr^{-1}]$ in a given direction by the integral

$$\mathcal{J}(\mathbf{p}) = \cos\theta \int_{\mathsf{S}} \mathcal{B}(\mathbf{q}, \mathbf{p}) d\mathsf{S}, \tag{2.2.81}$$

which specifies the angular distribution of the power radiated by the source. Correspondingly the *irradiance* $\mathcal{E}(\mathbf{q})$ $[W \cdot m^{-2}]$, specifying the radiated power by the unit area of the source at any given point $\mathbf{q}$, results from the integral

$$\mathcal{E}(\mathbf{q}) = \int_{2\pi} \mathcal{B}(\mathbf{q}, \mathbf{p}) \cos\theta d\Omega. \tag{2.2.82}$$

As the radiance, the radiant intensity and the irradiance are typically referred to a specific interval of frequency. Also, as the radiance, they were originally defined for rather incoherent thermal sources and later generalized to sources of any state of coherence. The generalization preserves the formal expressions (2.2.81) and (2.2.82) but in terms of the generalized radiance function. Thus, as the radiance, the generalized irradiance does not preserve the physical meaning as the corresponding quantity of traditional radiometry. In contrast, the generalized radiant intensity retains its physical meaning as representing the angular distribution of the radiated power [9.9, 9.10, 9.4].

Elementary beam of rays

Before closing this paragraph we briefly comment on the radiometric concept of *elementary beam* of light. Exploiting some geometrical arguments and related invariances, it is possible to give an "operative" definition of the abstract concept of ray, conciliating the mathematical idea of the ray as a thin pencil of radiation with vanishing solid angle and cross-sectional area, and hence not capable of carrying a finite amount of energy, and the physical need for an "entity" capable, in contrast, of carrying a finite and measurable, even though infinitesimally small, amount of energy.

Resorting to the basic geometry of Fig. 2.4.b), we reconsider the (possibly fictitious) elemental surface $d\mathsf{S}_1$ centered around point P_1 and the light ray emerging from P_1 into the direction $\mathbf{v}$ (Fig. 2.6). Let $d\mathsf{S}_2$ be the (possibly fictitious) elemental surface, that, centered around point P_2 at distance d from P_1 along $\mathbf{v}$, collects all the rays originating from points of $d\mathsf{S}_1$. The elemental solid angle $d\Omega_2$ subtended by $d\mathsf{S}_1$ at P_2 is $d\Omega_2 = d\mathsf{A}_1/d^2$, where $d\mathsf{A}_1 = d\mathsf{S}_1 \cos\theta_1$ is the projected area element. Correspondingly, the elemental solid angle $d\Omega_1$ subtended by $d\mathsf{S}_2$ at P_1 is $d\Omega_1 = d\mathsf{A}_2/d^2$, with $d\mathsf{A}_2 = d\mathsf{S}_2 \cos\theta_2$. It is easy to see that

$$d\mathsf{A}_1 d\Omega_1 = d\mathsf{A}_2 d\Omega_2, \tag{2.2.83}$$

thus implying the invariance along the ray path of the product of the projected area element $d\mathsf{A}_1$ and the solid angle $d\Omega_1$ subtended at this element by another area element along the ray, which collects all the rays coming from $d\mathsf{A}_1$.

Relation (2.2.83) suggests to define an *elementary beam of rays* as that including all the rays passing through any two elements of area along a segment of some central ray path, whose length be very large compared with the linear dimensions of either area element (Fig. 2.6) [9.5, 9.2, 9.3].

The invariant projected-area solid-angle product in (2.2.83) is said the *geometrical extent* of the beam; in a sense it may be regarded as the product of what we may consider the spatial and angular extents of the beam. Interestingly, by (2.2.79), the element of volume $d\mathcal{V}$ in phase space enclosing the ray of interest directly relates to the geometrical extent of the corresponding beam by the constant refractive index of the medium through which propagation occurs. It is therefore evident that Eq. (2.2.83) is another form to express

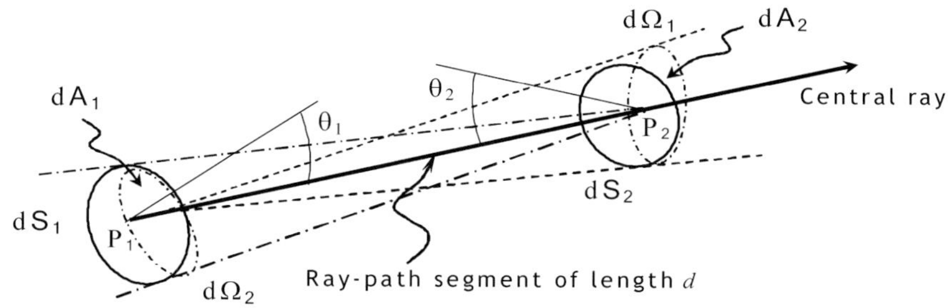

FIGURE 2.6. Illustrating the definition of the geometrical extent of an elementary beam of rays and its invariance under propagation through a homogeneous isotropic medium.

Liouville's invariance of phase-space volumes along the ray path. Also, by (2.2.71), $d\text{P}$ can be interpreted as the power carried by the elementary beam radiated by the source around the ray emerging from the specified point $\text{P}_{\text{S}}(\mathbf{q})$ into the specified direction $\mathbf{v}$. Hence, by virtue of the conservation of power we recover the already established invariance of the quantity $\mathcal{B}/n_0^2$ along the ray path, i.e, within the corresponding elementary beam.

2.2.5 *An example*

As an illustrative example of the application of Liouville equation (2.2.11), we determine the z-evolution under paraxial propagation through a free medium of an initially gaussian phase-space density distribution. We take

$$\rho_{\mathcal{G}}(q,p;z_i) = \frac{1}{2\pi\sigma_q\sigma_p} e^{-\frac{q^2}{2\sigma_q^2} - \frac{p^2}{2\sigma_p^2}}, \tag{2.2.84}$$

where the positive quantities σ_q and σ_p specify the "extension" of the ray density in phase plane, respectively, along the q and p-axis at the input plane z_i. We note that $\rho_{\mathcal{G}}$ is normalized to unity.

Gaussian distributions like (2.2.84) are of noticeable relevance since, as we will see, the Wigner distribution function of a Gaussian signal for which $\sigma_q\sigma_p = \frac{1}{2k_0}$, $k_0 = \frac{2\pi}{\lambda_0}$ being the vacuum wave number, basically takes the same form as (2.2.84), being Gaussian both in the space and spatial frequency domains. Notably, this is the only case where the Wigner distribution function is nonnegative everywhere through the phase space, and so it is interpretable as an energy density distribution.

When propagating through a homogeneous medium of refractive index n_0, the linear approximation gives for the optical Hamiltonian the free-particle like form $H = \frac{1}{2n_0}p^2$. Accordingly, Liouville's equation (2.2.11) gives

$$\frac{\partial}{\partial z}\rho(q,p;z) = -\frac{p}{n_0}\frac{\partial}{\partial q}\rho(q,p;z), \tag{2.2.85}$$

whose general solution is

$$\rho(q,p;z) = \rho(q - \tfrac{p}{n_0}z, p; z_i). \tag{2.2.86}$$

In particular, the input density distribution (2.2.84) propagates into

$$\rho_{\mathcal{G}}(q,p;z) = \frac{1}{2\pi\sigma_q\sigma_p} e^{-\frac{(q-pz/n_0)^2}{2\sigma_q^2} - \frac{p^2}{2\sigma_p^2}}. \tag{2.2.87}$$

Evidently, it remains Gaussian under free propagation, whose effect in the phase plane is pictured as a shear of the initial distribution along the q-axis, as will specifically be evidenced in § 2.5. In general, we will see that Gaussian distributions remain Gaussian under propagation through any first-order optical system, i.e., systems that are described by Lie transformations generated by polynomials quadratic in the optical coordinates q and p.

Finally, it is worth elaborating an example of the geometrical beam-extent invariance (2.2.83). Let us consider a stigmatic Gaussian beam, being circularly symmetric around the propagation z-axis. As is well known, the beam intensity transversally decreases through a gaussian profile to $e^{-2} \simeq 0.135$ of its on-axis value at the beam radius, usually denoted by $w(z)$. Also, the beam diverges as it propagates along the z-axis with an increase of the beam radius, which takes on the minimum value w_0 at the waist, and then far from the waist - i.e., for relevant Fresnel numbers $\pi w_0^2/\lambda z \ll 1$ - follows the simple law with z:

$$w(z) \simeq \frac{\lambda z}{\pi w_0}, \tag{2.2.88}$$

λ being the wavelength in the medium [13].

It is natural to take as an exemplary area A_1 the cross section of the beam at the waist: $\mathsf{A}_1 = \pi w_0^2$, and accordingly as solid angle Ω_1 that subtended at the waist by the cross section of the beam at some z far from the waist: $\Omega_1 = \pi w^2(z)/z^2 = \lambda^2/\pi w_0^2$. The product $\mathsf{A}_1\Omega_1$ is then the z-independent value

$$\mathsf{A}_1\Omega_1 = \lambda^2. \tag{2.2.89}$$

2.3 Quadratic monomials and symplectic matrices

As briefly recalled in the introductory notes, the optical Hamiltonian, suited to the $1D$ picture of geometrical linear optics, takes the quadratic form (§ 1.7)

$$H(q,p) = \tfrac{1}{2n_0}p^2 + \tfrac{n_2}{2}q^2, \tag{2.3.1}$$

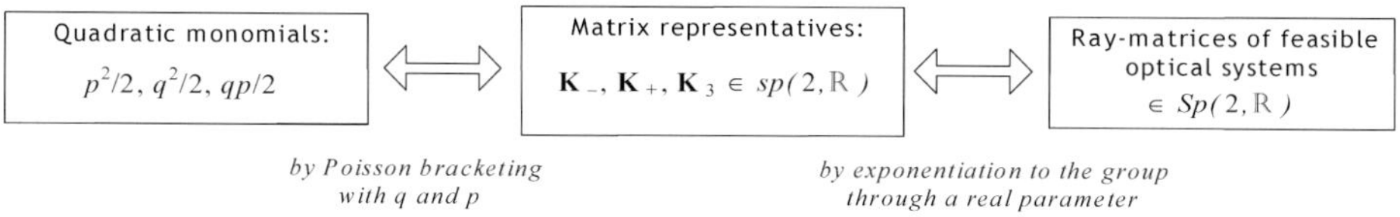

FIGURE 2.7. Conceptual and formal path to single out the basic optical systems pertaining to the linear approximation.

where n_0 denotes the on-axis value of the refractive index and n_2 is linked to the medium refracting power (see § 1.7.5).

We have also recalled from Sect. 1.7 the steps to proceed from the optical Hamiltonian H to the corresponding ray-transfer matrix $\mathbf{M}$:

$$H \quad \rightarrow \quad \mathbf{H} \quad \rightarrow \quad \mathbf{M}, \tag{2.3.2}$$

with $\mathbf{M}$ being locally or globally expressed as the exponential mapping of the matrix representative $\mathbf{H}$ of H by the pertinent (i.e., infinitesimal or finite) propagation distance Δz:

$$\mathbf{M}(z_i + \Delta z, z_i) = e^{\Delta z\, \mathbf{H}}. \tag{2.3.3}$$

The 2×2 traceless matrix $\mathbf{H}$ corresponds to the optical Hamiltonian (2.3.1) under the Poissonian bracket based rule (1.7.7), and so

$$\mathbf{H} = \begin{pmatrix} 0 & 1/n_0 \\ -n_2 & 0 \end{pmatrix}. \tag{2.3.4}$$

Evidently, (2.3.2) retraces the path (1.5.1) within the purely matrix formalism appropriate to first-order ray-optics.

Here we proceed investigating the optical valence of the matrices generated, through the path (2.3.2), by homogeneous quadratic polynomials in the optical coordinates q and p (Fig. 2.7). Since any quadratic homogeneous function of q and p can be expressed as a linear combination of the quadratic monomials p^2, q^2 and qp, we will firstly derive the matrix representatives of these monomials, and then we will exponentiate them through proper real parameters to form the corresponding transformation matrices. Finally, we will analyse the effects that such matrices produce on the ray variables in order to verify whether they may be interpreted as the ray-transfer matrices of constructible optical systems. We will be able to identify the p^2- and q^2-generated optical systems as a section of homogeneous medium and a refracting surface (or, equivalently, a thin lens). In the next chapter we will prove that every symplectic real matrix as well as every first-order optical system can be synthesized by an appropriately designed sequence of homogeneous medium sections and thin lenses. This will confer concreteness to the qp-generated optical system, which, identified as a pure magnifier, will then be used as a basic component for more compact (as well as conceptually and practically interesting) syntheses of symplectic matrices and accordingly first-order optical systems.

The Lie operators associated with the monomials $p^2/2$, $q^2/2$ and $qp/2$, we specifically denote by $\widehat{\mathrm{K}}_-$, $\widehat{\mathrm{K}}_+$ and $\widehat{\mathrm{K}}_3$, write as (see also § 1.5.2)

$$\begin{aligned}
\widehat{\mathrm{K}}_- &\equiv \{\cdot, \tfrac{1}{2}p^2\} = p\frac{\partial}{\partial q}, \\
\widehat{\mathrm{K}}_+ &\equiv \{\cdot, \tfrac{1}{2}q^2\} = -q\frac{\partial}{\partial p}, \\
\widehat{\mathrm{K}}_3 &\equiv \{\cdot, \tfrac{1}{2}qp\} = \tfrac{1}{2}\left(q\frac{\partial}{\partial q} - p\frac{\partial}{\partial p}\right).
\end{aligned} \tag{2.3.5}$$

Correspondingly, by Poisson bracketing with q and p according to (1.7.7) the respective matrix representatives are obtained as

$$\begin{aligned}
\tfrac{1}{2}p^2 &\longrightarrow \mathbf{K}_- = \begin{pmatrix} 0 & 1 \\ 0 & 0 \end{pmatrix}, \\
\tfrac{1}{2}q^2 &\longrightarrow \mathbf{K}_+ = \begin{pmatrix} 0 & 0 \\ -1 & 0 \end{pmatrix}, \\
\tfrac{1}{2}qp &\longrightarrow \mathbf{K}_3 = \tfrac{1}{2}\begin{pmatrix} 1 & 0 \\ 0 & -1 \end{pmatrix}.
\end{aligned} \tag{2.3.6}$$

It is evident that the matrix representative $\mathbf{H}$ of the Hamiltonian (2.3.1) can uniquely be expressed as a linear combination of the above matrices:

$$\mathbf{H} = \frac{1}{n_0}\mathbf{K}_- + n_2\mathbf{K}_+. \tag{2.3.7}$$

Before proceeding further, we briefly comment on matrices (2.3.6). Again we keep the discussion on a relatively elementary level; we shall not detail, in fact, upon some genuinely mathematical peculiarities of the group $Sp(2,\mathbb{R})$ (and the relevant algebra $sp(2,\mathbb{R})$), for which the reader is directed to [11]. Our main interest is in evidencing the relation of the symplectic algebras and groups to the context of geometrical linear optics we are dealing with. The algebra may be seen as providing the "seed" for the evolution, being linked to the equation ruling the ray propagation. Correspondingly, the group may be seen as providing the "agent" of the evolution, being linked to the ingoing-to-outgoing ray relationships. Thus, as the algebra comprises the Hamiltonian-related matrices, the group comprises the propagation describing matrices. We noted before that the structures of the algebra and the associated group may be related by the exponential mapping. Notably, the group $Sp(2,\mathbb{R})$ is not of exponential type; namely, the elements comprised by all the one-parameter subgroups in $Sp(2,\mathbb{R})$, generated by the elements of the relevant algebra $sp(2,\mathbb{R})$, do not exhaust the entire group but leave a good portion out [11.2, 11.5].

2.3.1 *The symplectic Lie algebra* $sp(2,\mathbb{R})$ *and Lie group* $Sp(2,\mathbb{R})$

Firstly we note that $\mathbf{K}_-$, $\mathbf{K}_+$ and $\mathbf{K}_3$ obey the commutator brackets

$$[\mathbf{K}_+, \mathbf{K}_-] = 2\mathbf{K}_3, \quad [\mathbf{K}_3, \mathbf{K}_\pm] = \mp\mathbf{K}_\pm. \tag{2.3.8}$$

The set $\{\mathbf{K}_-, \mathbf{K}_+, \mathbf{K}_3\}$ is then closed with respect to the Lie product, and hence, the linear space, it spans under the ordinary addition of matrices and multiplication of a matrix by (real) scalars, configures as a Lie algebra. Precisely, the commutation relations (2.3.8) are constitutive of the symplectic algebra $sp(2,\mathbb{R})$ [7, 11]. Every matrix $\mathbf{A}$ in the algebra can uniquely be expressed in terms of the basis set $\{\mathbf{K}_-, \mathbf{K}_+, \mathbf{K}_3\}$ through arbitrary real coefficients as

$$\mathbf{A} = a_-\mathbf{K}_- + a_+\mathbf{K}_+ + a_3\mathbf{K}_3 = \begin{pmatrix} a_3/2 & a_- \\ -a_+ & -a_3/2 \end{pmatrix}. \tag{2.3.9}$$

Then, using the general expression above, it is an easy matter to prove that every element in the algebra obeys the relations

$$\mathrm{Tr}(\mathbf{A}) = 0, \qquad \mathbf{AJ} + \mathbf{JA}^\top = \mathbf{0}, \tag{2.3.10}$$

(see also Problem I.8), where $\mathbf{J}$ denotes the antisymmetric matrix

$$\mathbf{J} = \begin{pmatrix} 0 & 1 \\ -1 & 0 \end{pmatrix}, \tag{2.3.11}$$

introduced in § 1.3.1 and further exploited in § 1.7.1. Evidently, the matrix representative $\mathbf{H}$ of the optical Hamiltonian (2.3.1) is an element of $sp(2,\mathbb{R})$.

It was earlier remarked that in general an algebra *generates* the corresponding group by means of the *algebra-to-group operation*, i.e., by the exponentiation of the algebra elements through arbitrary parameters. On account of the operator nature of matrices, we may reiterate the considerations developed in § 2.2.2 about the algebra of Lie operators and the corresponding group of Lie transformations. Hence we may say that every matrix $\mathbf{A}$ in the symplectic algebra $sp(2,\mathbb{R})$ uniquely determines the one-parameter subgroup of real matrices $\{\widehat{\mathcal{T}}_\mathbf{A}(\zeta) = e^{\zeta\mathbf{A}};\ \zeta \in \mathbb{R}\}$ in the symplectic group $Sp(2,\mathbb{R})$, the binary operation being the additive composition with respect to the parameter, i.e.,

$$\widehat{\mathcal{T}}_\mathbf{A}(\zeta_1)\widehat{\mathcal{T}}_\mathbf{A}(\zeta_2) = \widehat{\mathcal{T}}_\mathbf{A}(\zeta_1 + \zeta_2) = \widehat{\mathcal{T}}_\mathbf{A}(\zeta_2)\widehat{\mathcal{T}}_\mathbf{A}(\zeta_1). \tag{2.3.12}$$

Every subgroup comprises both the identity, as $\widehat{\mathcal{T}}_\mathbf{A}(0) = \mathbf{I}$, and the unique inverse $\widehat{\mathcal{T}}_\mathbf{A}^{-1}(\zeta) = \widehat{\mathcal{T}}_\mathbf{A}(-\zeta)$. The symplectic group $Sp(2,\mathbb{R})$ is therefore formed (even though not exclusively) by real matrices $\mathbf{S}$'s obtained by computing the power series expansions

$$\mathbf{S} = \sum_{j=0}^{\infty} \frac{\zeta^j}{j!}\mathbf{A}^j = e^{\zeta\mathbf{A}}, \qquad \mathbf{A} \in sp(2,\mathbb{R}),\ \zeta \in \mathbb{R}, \tag{2.3.13}$$

with $\mathbf{A}$ element in the algebra and ζ real number.

Multiplication of group elements within a sufficiently small neighborhood of the identity matrix $\mathbf{I}$ is ruled by the BCH formula (see § 2.2.2), and hence

$$e^{\zeta_1 \mathbf{A}} e^{\zeta_2 \mathbf{B}} = e^{\mathbf{C}(\zeta_1,\zeta_2,\mathbf{A},\mathbf{B})}. \tag{2.3.14}$$

The matrix $\mathbf{C}$ writes in terms of the matrices $\zeta_1 \mathbf{A}$ and $\zeta_2 \mathbf{B}$ in the similar way as $\widehat{\mathsf{L}}_h$ in (2.2.61) writes in terms of $\zeta_1 \widehat{\mathsf{L}}_f$ and $\zeta_2 \widehat{\mathsf{L}}_g$, and so it is an infinite sum of terms involving only the matrices $\zeta_1 \mathbf{A}$ and $\zeta_2 \mathbf{B}$ and their multiple commutators:

$$\begin{aligned} \mathbf{C}(\zeta_1,\zeta_2,\mathbf{A},\mathbf{B}) &= \zeta_1 \mathbf{A} + \zeta_2 \mathbf{B} + \tfrac{\zeta_1\zeta_2}{2}[\mathbf{A},\mathbf{B}] \\ +\tfrac{1}{12}\{\zeta_1^2\zeta_2[\mathbf{A},[\mathbf{A},\mathbf{B})]] &+ \zeta_1\zeta_2^2[\mathbf{B},[\mathbf{B},\mathbf{A}]]\} + \mathcal{O}(\zeta_1^3\zeta_2, \zeta_1^2\zeta_2^2, \zeta_1\zeta_2^3). \end{aligned} \tag{2.3.15}$$

In general the series may converge only for sufficiently small ζ_1 and ζ_2.

It is a notable aspect of $Sp(2,\mathbb{R})$ that it is not of exponential type. Namely, $Sp(2,\mathbb{R})$ matrices exist that do not arise directly by a unique exponential mapping from the algebra $sp(2,\mathbb{R})$, but in contrast by the composition of appropriate exponential-type elements of the group. The reader interested in this remarkable, and recently investigated, feature of the $Sp(2,\mathbb{R})$ may consult the relevant refs. [11.2, 11.5]. Here we will not further detail this topic.

As a direct consequence of the exponential mapping (2.3.13) from the algebra matrices, obeying (2.3.10), to the group matrices, the matrices in every algebra-generated subgroup can be proved to obey the relation

$$\mathbf{S}\mathbf{J}\mathbf{S}^{\top} = \mathbf{J}, \tag{2.3.16}$$

which of course extends to the non-exponential type elements as well, being preserved under matrix product. Relation (2.3.16) is the identifying property of the symplectic matrices (see also § 1.3.1). The reader is suggested to prove identity (2.3.16), by writing it in the handy form $\mathbf{J}\mathbf{S}^{\top} = \mathbf{S}^{-1}\mathbf{J}$, and then using the series expansions for $\mathbf{S}^{\top}$ and $\mathbf{S}^{-1}$ and the second of relations (2.3.10).

As a further identifying feature of the symplectic matrices, we can say that, as already established in § 1.3.1, they define linear transformations of a $2D$ vector space $\left\{\mathbf{u} = (u_1, u_2)^{\top}\right\}$ under which the antisymmetric bilinear form

$$(\mathbf{u}, \mathbf{J}\mathbf{v}) = u_1 v_2 - u_2 v_1, \tag{2.3.17}$$

(which in the optical context is just the Lagrange invariant Δ of § 1.7.4), is invariant. In fact, if $\mathbf{u} \to \mathbf{u}' = \mathbf{S}\mathbf{u}$ and $\mathbf{v} \to \mathbf{v}' = \mathbf{S}\mathbf{v}$, by (2.3.16) we find that

$$(\mathbf{u}', \mathbf{J}\mathbf{v}') = (\mathbf{u}', \mathbf{S}^{\top}\mathbf{J}\mathbf{S}\mathbf{v}') = (\mathbf{u}, \mathbf{J}\mathbf{v}). \tag{2.3.18}$$

On the other hand, the skew product (2.3.17) can be obtained as the determinant of the 2×2 matrix formed by the components of the involved vectors $\mathbf{u}$ and $\mathbf{v}$; namely

$$(\mathbf{u}, \mathbf{J}\mathbf{v}) = u_1 v_2 - u_2 v_1 = \det \begin{pmatrix} u_1 & v_1 \\ u_2 & v_2 \end{pmatrix}. \tag{2.3.19}$$

Under the linear transformation $\mathbf{u} \to \mathbf{u}' = \mathbf{S}\mathbf{u}$ and $\mathbf{v} \to \mathbf{v}' = \mathbf{S}\mathbf{v}$, the determinant in (2.3.19) is multiplied by $\det \mathbf{S}$, and hence

$$(\mathbf{u}', \mathbf{J}\mathbf{v}') = \det \mathbf{S} \cdot (\mathbf{u}, \mathbf{J}\mathbf{v}) = (\mathbf{u}, \mathbf{J}\mathbf{v}), \tag{2.3.20}$$

thus establishing that the symplectic matrices are unimodular[(f)]

$$\det \mathbf{S} = 1. \tag{2.3.21}$$

Consequently, only three of the four entries of $\mathbf{S}$ are linearly independent. The symplectic group $Sp(2, \mathbb{R})$ is, in fact, 3-dimensional. One generally speaks of $Sp(2, \mathbb{R})$ as a three-parameter Lie group. It is evident that the generating algebra $sp(2, \mathbb{R})$ is 3-dimensional as well, as unequivocally displayed by the general expression (2.3.9) for elements $\mathbf{A}$ in terms of the three algebra generating matrices $\mathbf{K}_-$, $\mathbf{K}_+$ and $\mathbf{K}_3$ (see also Problem 5 in ch.1).

Thus, the path naïvely traced in (2.3.2) can formally be interpreted as developing from the Lie algebra of the quadratic homogeneous polynomials in both q and p coordinates, and then proceeding through the Lie algebra of the corresponding traceless matrices, formed by Poisson bracket operation with q and p, up to the symplectic Lie group of the optical ray matrices, formed by exponentiation of that matrix algebra, as sketched in Fig. 2.7.

2.4 Quadratic monomials and first-order optical systems

We form now the exponential mappings of the matrices (2.3.6) and then inspect the transformations they act on the ray variables, in order to identify the optical systems associable with the monomials $\frac{1}{2}p^2$, $\frac{1}{2}q^2$ and $\frac{1}{2}qp$. We will see that p^2 describes a section of homogeneous medium as q^2 an interface between two media. The mixed monomial qp will be seen to produce an overall scaling of the ray variables, whose optical realization will be clarified in the next chapter. To still preserve the link between the matrix exponentiating procedure, which may appear somewhat "abstract", and the Hamilton equations solving

[f] The proof of the unimodularity of the $Sp(2, \mathbb{R})$ matrices, given here, is based on that presented in [7.1] (already quoted in § 1.3.1), where the general case of n-dimensional symplectic groups and accordingly n-dimensional vector spaces is considered. Evidently, it can as well be interpreted in the light of Liouville's invariance of areas, which transform through $\det \mathbf{S}$ under the transformation of the relevant vector space by $\mathbf{S}$.

procedure, we will resort to both procedures when examining the monomials p^2 and q^2, that directly relate to the model Hamiltonian (2.3.1).

It is worth noting that the matrices $\mathbf{K}_-$, $\mathbf{K}_+$ and $\mathbf{K}_3$ have been formed by rule (1.7.7) from the generic monomials $\frac{1}{2}p^2$, $\frac{1}{2}q^2$ and $\frac{1}{2}qp$, that have not been given an explicit interpretation as optical Hamiltonian functions, for which indeed appropriate normalizing factors, possibly for the position variable q, should be needed. The entries of $\mathbf{K}_-$, $\mathbf{K}_+$ and $\mathbf{K}_3$ have therefore definite units in accord with those of the ray coordinates q [length] and p [unitless]. Accordingly, the parameter ζ of the exponential mapping from the algebra matrices $\{\mathbf{K}_-, \mathbf{K}_+, \mathbf{K}_3\}$ to the group matrices cannot a priori be given the role as a "length" as in (2.3.3); its meaning conforms with the matrix at hand in order that the resulting exponentiated form be interpretable as an optical matrix, for which, as we know, the diagonal entries are dimensionless, whilst the off-diagonal have units of length and length^{-1}.

Finally we emphasize that, on evaluating the exponential functions of the matrices $\mathbf{K}_-$, $\mathbf{K}_+$ and $\mathbf{K}_3$, we compute, and then sum up, the powers in the series (2.3.13). This can easily be carried out due to the simple form of the involved matrices, for which in fact we find

$$\mathbf{K}_-^2 = \mathbf{K}_+^2 = \mathbf{0}, \quad \mathbf{K}_3^{2j-1} = \frac{1}{2^{2(j-1)}}\mathbf{K}_3, \quad \mathbf{K}_3^{2j} = \frac{1}{2^{2j}}\mathbf{I}, \quad j = 1, 2, 3... \tag{2.4.1}$$

Then, we are enabled to disregard the questions concerned with the convergence and the evaluation of matrix functions defined by infinite series like (2.3.13), for which we may address the reader to [14].

2.4.1 *Propagation along a section of homogeneous medium*

As earlier noted, the paraxial propagation in a homogeneous medium of refractive index n_0 is modelled by the quadratic monomial

$$H(q, p) = \tfrac{1}{2n_0}p^2. \tag{2.4.2}$$

In fact, the equations for the ray variables write as

$$\begin{cases} \frac{d}{dz}q = \frac{1}{n_0}p, \\ \frac{d}{dz}p = 0, \qquad q(z_i) = q_i, \; p(z_i) = p_i\,, \end{cases} \tag{2.4.3}$$

showing that, on propagating from the input plane Π_i to the output Π_o, which bound the medium respectively at z_i and z_o, the ray experiences the transformation

$$\begin{pmatrix} q_o \\ p_o \end{pmatrix} = \begin{pmatrix} q_i + \Delta z\, p_i/n_0 \\ p_i \end{pmatrix} = \begin{pmatrix} 1 & \Delta z/n_0 \\ 0 & 1 \end{pmatrix} \begin{pmatrix} q_i \\ p_i \end{pmatrix}, \qquad \Delta z = z_o - z_i. \tag{2.4.4}$$

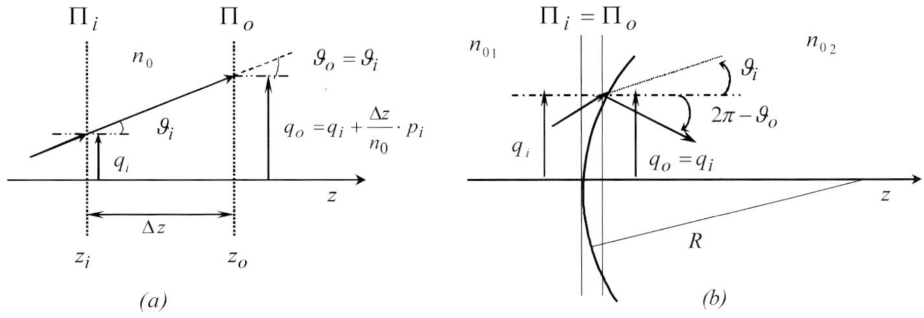

FIGURE 2.8. The optical systems generated by the monomials $\frac{1}{2}p^2$ and $\frac{1}{2}q^2$ are respectively identified as *(a)* a section of homogeneous medium and *(b)* a refracting surface.

The position coordinate changes by $\delta q = \frac{\Delta z}{n_0} p_i$, whilst the optical momentum remains constant. The ray travels in the medium by the axial distance Δz along a straight line in the incoming direction of motion (Fig. 2.8.a)).

Likewise the exponentiation of the matrix representative $\mathbf{H} = \mathbf{K}_-/n_0$ of (2.4.2) yields the matrix on the right of eq. (2.4.4). Consequently, $\mathbf{K}_-$ may be regarded as the generator of the free propagation, and the exponentiated form

$$e^{d\mathbf{K}_-} = \begin{pmatrix} 1 & d \\ 0 & 1 \end{pmatrix}, \quad d \equiv \frac{\Delta z}{n_0} \tag{2.4.5}$$

may be interpreted as the ray matrix of a homogeneous medium of refractive index n_0 extending over the axial length Δz. The ratio $d = \frac{\Delta z}{n_0}$ of Δz to n_0 is usually reported as the *reduced length* [12].

Note that the positions of the reference planes Π_i and Π_o enter the ray matrix (2.4.5) through the relative separation distance $z_o - z_i$. The system is in fact invariant under translations along the z axis, provided that the light ray is equally translated. Clearly this invariance reflects the medium homogeneity.

The above results are of course in accord with those of § 1.5.2, concerning the Lie transformation generated by the operator $\widehat{\mathrm{L}}_{p^2/2} = \widehat{\mathrm{K}}_-$ and its effect on the phase-space variables. Equations (2.4.4) and (1.5.12) are in fact identical, both describing a translation in the unperturbed initial direction of motion.

2.4.2 *Refraction at the interface between two media*

Consider the quadratic monomial

$$H(q,p) = \tfrac{1}{2} n_2 q^2, \tag{2.4.6}$$

which can be roughly regarded as obtained from the model Hamiltonian (2.3.1) in the case that the free-propagation term p^2/n_0 remains negligible through the evolution length of interest.

The relevant Hamilton's equations

$$\begin{cases} \frac{d}{dz}q = 0\,, \\ \frac{d}{dz}p = -n_2 q, \qquad q(z_i) = q_i,\, p(z_i) = p_i. \end{cases} \tag{2.4.7}$$

are easily integrated to the transformation

$$\begin{pmatrix} q_o \\ p_o \end{pmatrix} = \begin{pmatrix} q_i \\ p_i - n_2 \Delta z q_i \end{pmatrix} = \begin{pmatrix} 1 & 0 \\ -n_2 \Delta z & 1 \end{pmatrix} \begin{pmatrix} q_i \\ p_i \end{pmatrix}, \qquad \Delta z = z_o - z_i \tag{2.4.8}$$

which leaves unchanged the position coordinate $q_o = q_i$, whilst changing the optical momentum by $\Delta p = -n_2 \Delta z q_i$. Due to the unavoidable diffraction of light through a finite distance, the above transformation is actuable only if the input and output planes are infinitely close to each other: $\Delta z \rightarrow dz$. Thus, the transformation (2.4.8) represents the *instantaneous* variation of the direction of propagation of the ray, as that occurring in the refraction at the boundary surface between two regions of different refractive indices, with the two reference planes being placed immediately to the left and immediately to the right of the surface (Fig. 2.8.b)). In fact, according to the interpretation of n_2 given in § 1.7.5, the non zero off-diagonal entry in (2.4.8) is minus the refracting power of the surface $\mathcal{P} = \frac{\Delta n}{R}$, if Δn represents the difference between the refractive indices at the output and input planes and R specifies the local curvature radius of the boundary surface. We recall that the usual conventions require to take R positive or negative according to whether the center of curvature is to the right or left of the surface, with the light being as usual assumed to propagate from left to right (see § 1.8 for the sign conventions in the optics of image-forming instruments).

By exponentiating the matrix representative $\mathbf{H} = n_2 \mathbf{K}_+$ of (2.4.6), we obtain the refraction matrix in (2.4.8). Therefore, $\mathbf{K}_+$ can be interpreted as the generator of the refraction at the interface between two different media, and accordingly the matrix for the corresponding ray transformation is

$$e^{\mathcal{P}\mathbf{K}_+} = \begin{pmatrix} 1 & 0 \\ -\mathcal{P} & 1 \end{pmatrix}, \tag{2.4.9}$$

with $\mathcal{P}$ [(length)$^{-1}$] being the surface refracting power. The reference planes are taken at the left and right of the surface.

On account of the results of § 1.8.3 concerning the lower-left entry C of the ray-matrix of an arbitrary optical system (see Eq. 1.8.17), the power $\mathcal{P}$ can be expressed in terms of the reduced focal length f of the refracting surface as

$$\mathcal{P} = \frac{1}{f}, \qquad f = \frac{f_1}{n_{0_1}} = \frac{f_2}{n_{0_2}}, \tag{2.4.10}$$

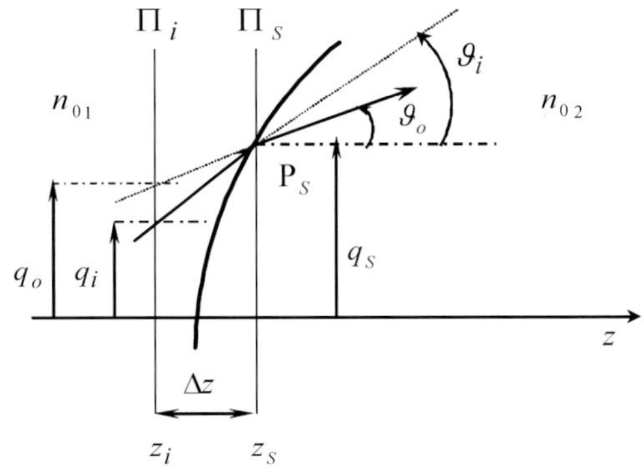

FIGURE 2.9. The refraction at the interface between two media transforms the input ray variables $(q_i, n_{0_1} \sin\vartheta_i)$, referred to the chosen z_i-plane, to the ray variables $(q_S, n_{0_2} \sin\vartheta_o)$, referred to the profile screen at the intersection point $\mathrm{P_S}$ on the surface, which are then projected back to the input plane into the final ray variables $(q_o, n_{0_2} \sin\vartheta_o)$. With $\Delta z \to 0$ and $|q| \ll 1$ we recover the instantaneous view of Fig. 2.8.b).

where f_1 and f_2 denote the primary and secondary focal lengths of the refracting surface, as n_{0_1} and n_{0_2} the on-axis values of the refractive indices of the media on the two sides of the surface.

In particular, the refraction matrix for a planar interface ($R \to \infty$) turns into the identity matrix, with the input-output relations being then

$$\begin{aligned} q_o &= q_i, \\ p_o &= p_i. \end{aligned} \tag{2.4.11}$$

Interestingly the second relation above expresses the conservation law of the optical momentum before and after the refraction at the planar interface. Also, it may be regarded as the paraxial form of the Snell law of refraction, explicitly writing as

$$n_{0_2} \vartheta_o = n_{0_1} \vartheta_i. \tag{2.4.12}$$

Picture (2.4.8) conforms with the result expressed by Eq. (1.5.15), describing the canonical transformation generated by the Lie operator $\widehat{\mathrm{L}}_{q^2/2} = \widehat{\mathrm{K}}_+$.

Before closing this paragraph, we briefly evidence the conceptual difficulties arising when dealing with refracting surface-transformations within the Hamiltonian formalism. Indeed, since the Hamiltonian formalism would not easily accomodate light refraction, the symplecticity of the transformation is not directly assured. By its own nature, the Hamiltonian view to optical propagation would imply that each ray have a unique evolution once specified the refractive index function of the medium through which propagation occurs. In fact, once assigned the initial values of the ray coordinates, i.e, position and angle, the first order differential equations (2.1.1), or their linearized form (2.1.2), uniquely determine the "future" values of the ray coordinates, i.e., position and angle at every subsequent determination of the propagation variable. In contrast, at the refracting surface separating two different media the incident ray ordinarily produces both a refracted and a reflected ray. Moreover, the differential equation formulation of Hamiltonian optics requires a medium with a differentiable refractive index. Evidently, the interface introduces a discontinuity of the index function, i.e., "the potential", entering Hamilton's equations (2.1.1) (or equivalently, the ray equation (1.2.8)). Discontinuous-index systems such as refracting surfaces have been treated through a complementary description, resorting to suitable geometrical contructions and physical considerations about the dynamics of the process rather than to the differential equation formalism; it has been proved that refracting surface transformations factorize into two root symplectic transformations, each depending on the properties of one of the two media separated by the interface (Fig. 2.9, inspired by [15.2]). The reader interested in this refraction-related aspect of the Hamilton-like theory of light propagation may consult [15]. Within the linear approximation we are concerned with, we will use expression (2.4.9) for the refraction ray-matrix, implicitly assuming that the relevant refracting surface has been appropriately treated to reduce the light reflection to the minimum.

2.4.3 *Positive scale transform of the ray-coordinates*

As a completition of the program sketched in Fig. 2.7, we now consider the optical transformation generated by the third of matrices (2.3.6), i.e., $\mathbf{K}_3$. A direct calculation of the exponential function $e^{s\mathbf{K}_3}$, with s being a dimensionless real parameter, yields the scaling relation between the position and momentum coordinates of the incoming and outgoing rays, that is

$$\begin{pmatrix} q_o \\ p_o \end{pmatrix} = \begin{pmatrix} m & 0 \\ 0 & m^{-1} \end{pmatrix} \begin{pmatrix} q_i \\ p_i \end{pmatrix}, \qquad m \equiv e^{s/2}, \tag{2.4.13}$$

in accord as well with the results of § 1.5.2 (Eq. (1.5.19)).

The system represented by the matrix

$$\mathbf{S}(m) \equiv \begin{pmatrix} m & 0 \\ 0 & m^{-1} \end{pmatrix} \tag{2.4.14}$$

is then recognized as an ideal magnifier (see Eq. (1.8.31)), having a positive magnification factor $m > 0$. The system (2.4.14) works as a magnifier for $m > 1$, producing a magnification of the ray-coordinate and conversely a demagnification by the same amount of the ray reduced slope, and correspondingly as a demagnifier for $m < 1$.

Accordingly,

$$q_o p_o = q_i p_i , \tag{2.4.15}$$

and hence the product of the spatial and angular ray-coordinates remains invariant under propagation along the system (2.4.14). The product qp is known as the Smith-Helmholtz invariant, and (2.4.15) is the paraxial form of the Abbe's sine condition holding in general for perfect imaging systems [9.1, 3.4].

The optical realizations of the scale transforming system (2.4.13) will be illustrated in the next chapter.

2.5 Quadratic monomials in phase space

We recall that, according to the linear approximation, small regions across the optical axis in the reference screens as well as small inclination angles with respect to the axis are implicitly assumed to be involved under the propagation of the ray through the system. Therefore, as already noted, the geometrical-optical phase space, which would properly have the $[-n_0, n_0] \times \mathbb{R}$ structure, can safely be viewed as the $\mathbb{R}^2$ phase-plane.

As stated earlier, the variables we are dealing with have different units. Therefore, in order that the expressions we shall obtain be dimensionally meaningful, the phase plane variables q and p must suitably be scaled (or thought of as being suitably scaled), for instance, by some characteristic length for the problem under examination, so that they come to have the same dimensionality. This is implied in the discussion and all the graphs below, even when, for simplicity's sake, the ordinary symbols q and p are still displayed.

Transformations (2.4.4) and (2.4.8) have an immediate pictorial view in the optical phase-plane, both describing a translation of the ray representative point along directions parallel to the coordinate axes. In fact, by (2.4.4) the representative point $P_i = (q_i, p_i)$ of the input ray is translated to the representative point $P_o = (q_i + p_i \Delta z / n_0, p_i)$ of the output ray along the line $p = p_i$, as shown in Fig. 2.10.a). Likewise, the transformation (2.4.8) can conveniently be visualized as a translation of the point $P_i = (q_i, p_i)$ to the point $P_o = (q_i, p_i - \mathcal{P} q_i)$ along a segment parallel to the p axis (Fig. 2.11.a)).

It is interesting to see the effect of the transformations (2.4.4) and (2.4.8) on a collection of rays. We will consider some idealized bundle of rays, whose phase-space representative distributions can easily be depicted. Needless to say, the $1D$ picture, we are working in, confers further astraction to the descriptions to follow; a propagating spherical wave, for instance, is visualized by a longitudinal section of a cone of rays as a laser spot by a segment line.

Firstly, let us consider a fan of rays uniformly emanating from a point source placed at q_i in the plane z_i, which is physically obtainable, for instance, by focusing a He-Ne laser beam. It is represented by a segment in the phase-plane parallel to the p-axis, whose extent Δp_i is determined by the angular width of the beam. As a result of the free propagation from z_i to z_o, the representative segment is horizontally sheared by the angle β, with $\tan\beta = \Delta z/n_0$. Accordingly, the spatial extent of the beam is changed to $\Delta q_o = \Delta p_i \Delta z/n_0$, whilst the angular width remains invariant (Fig. 2.10.b)).

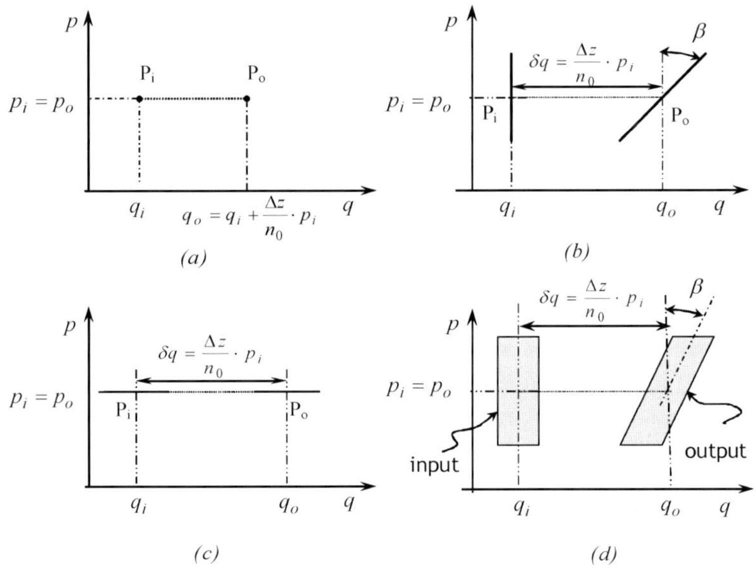

FIGURE 2.10. Light propagation through a homogeneous medium n_0 over the distance Δz. Schematic of the phase plane transformations of *(a)* the representative point of a single ray, and of the representative distributions of *(b)* a bundle of rays emanating from the same point, *(c)* a bundle of parallel rays distributed over a certain spatial extent, and *(d)* a bundle of rays uniformly distributed over different positions and directions.

We may observe that a point source generates a spherical wave, which is described in the linear approximation by a quadratic-phase waveform. The sheared segment of Fig. 2.10.b), representing the fan of rays after propagating by d from the source, can therefore be interpreted as the representative phase-space distribution of a spherical wave with curvature radius just equal to d. Accordingly, we may say that a spherical wave with curvature radius R is represented in the phase plane by the line $p = \frac{q}{R}$. With increasing d the representative segment is sheared more and more within the horizontal strip, determined by the initial angular aperture of the wave, getting parallel to the q axis when $d \to \infty$. At infinite distance from the source, in fact, a spherical wave is fairly like a plane one.

A beam of parallel rays uniformly distributed over a certain spot in the

z_i-plane, obtainable for instance by collimating a He-Ne laser beam into a given direction, may be represented by a segment parallel to the q-axis, whose length Δq_i reproduces the spatial extent of the beam. Effected by (2.4.4), the representative segment is shifted parallel to the q-axis by $\delta q = \frac{\Delta z}{n_0} p_i$, the spatial and angular extents of the beam remaining unchanged (Fig. 2.10.c)).

Finally, we consider a bundle of rays uniformly distributed over different positions and angles about a certain phase-plane point. A line source of given spatial extent emitting rays uniformly within an angular width $\Delta\vartheta_i$ around a fixed direction may be a model source for such a bundle of rays, practically realizable by a thermal source, followed by an appropriate set of long and narrow slits for collimating purposes. The representative distribution in the phase plane covers a rectangular region, whose width and height are determined respectively by the spatial extent and angular divergence of the beam. As sketched in Fig. 2.10.d), the free propagation produces an overall translation and, most importantly, a horizontal shearing of the ray-distribution by the angle $\beta = \tan^{-1}(\frac{\Delta z}{n_0})$. The angular width of the beam remains constant as the rays propagate over the homogeneous section from z_i to z_o: $\Delta p_i = \Delta p_o$, but the spatial extent is increased from Δq_i to $\Delta q_o = \Delta q_i + \Delta p_i \frac{\Delta z}{n_0}$.

It is evident that the transformation (2.4.4) preserves the area of the phase-plane representative distributions: $\mathcal{A}_i = \mathcal{A}_o$, in accord with the Liouville's theorem. In fact, we see that the zero-area distributions in Figs. 2.10.b) and 2.10.c) are turned into zero-area distributions by (2.4.4); likewise, the regions of Fig. 2.10.d), representing the ray bundle before and after the propagation over Δz, have the same area: $\mathcal{A}_i = \mathcal{A}_o = \Delta q_i \Delta p_i$.

Correspondingly, the effect of the transformation (2.4.8) on the above considered bundles of rays is shown in Fig. 2.11.b)-d).

The vertical segment in Fig. 2.11.b), representing a bundle of rays emanating from one point within an angular width $\Delta\vartheta_i$, is translated along the p-axis by $\delta p = -\mathcal{P}\, q_i$, the spatial and angular extents of the bundle being unchanged.

Likewise, the horizontal segment in Fig. 2.11.c), representative of a beam of parallel rays, is vertically shifted by $\delta p = -\mathcal{P}\, q_i$ and sheared by β, being $\tan\beta = -\mathcal{P}$. The angular extent of the beam increases to $\Delta p_o = |\mathcal{P}|\,\Delta\, q_i$. In accordance with the previously developed considerations concerning the phase space representation of a spherical wave, it is evident that Fig. 2.11.c) shows the transformation of a beam of parallel rays into a spherical wave with curvature radius $R = -1/\mathcal{P}$ due to the refraction at a non planar surface. Conversely, due to the invertibility of the ray path, we may say that a spherical wave may be transformed into a bundle of parallel rays by matching the curvature radius of the wave to the refracting power, and so the focal length (Eq. (2.4.10)), of the surface such that $R = \frac{1}{\mathcal{P}}$.

Finally, the rectangular region representing a beam of rays uniformly dis-

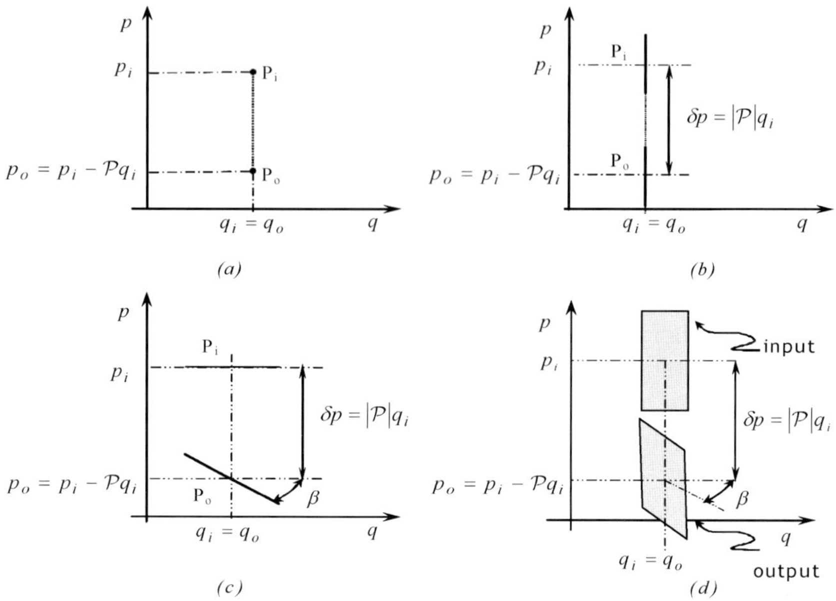

FIGURE 2.11. Refraction at an interface between two media with refracting power $\mathcal{P}$ (in the figures $\mathcal{P} > 0$). Schematic of the phase plane transformations of *(a)* the representative point of a single ray, and of the representative distributions of a bundle of rays with *(b)* angular extent only, *(c)* spatial extent only and *(d)* angular and spatial extents.

tributed over different positions and directions is translated and p-sheared by the angle $\beta = -\tan^{-1}\mathcal{P}$ (Fig. 2.11.d)). The spatial extent of the beam is not changed by the refraction through the interface, $\Delta q_o = \Delta q_i$, whilst the angular width is increased from Δp_i to $\Delta p_o = \Delta p_i + |\mathcal{P}|\,\Delta q_i$.

Notice also that, as in the case of the propagation through a homogeneous medium, the area $\mathcal{A}$ of the representative distributions in phase space is preserved by the refraction at the interface between two media (recall that we are assuming the reflection be negligible at the interface).

Figure 2.12.a) illustrates the scale transformation experienced by a single ray. The coordinates of the incoming and outgoing rays satisfy the relation (2.4.15), which for fixed initial values can be viewed as the equation of an equilateral hyperbola $qp = q_i p_i$, having the coordinate axes as asymptotes. According to the sign of the product $q_i p_i$, the hyperbola branches are in the first and third quadrant ($q_i p_i > 0$) or in the second and forth quadrant ($q_i p_i < 0$). In the case illustrated in the figure, $q_i p_i > 0$. The representative point $P_i = (q_i, p_i)$ of the incoming ray slips along the relevant hyperbola branch to the representative point $P_o = (mq_i, \frac{p_i}{m})$ of the outgoing ray.

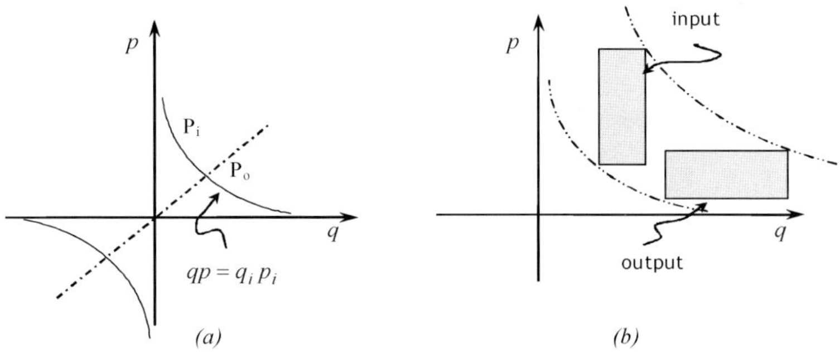

FIGURE 2.12. The scaling matrix in the phase plane. The transformations experienced by *(a)* a single ray and *(b)* a spatially and angularly extended bundle of rays.

In Fig. 2.12.b) we show the transformation acted by (2.4.13) on the phase plane rectangular region, representing a bundle of rays uniformly distributed over position and direction through the respective ranges Δq_i and Δp_i.The spatial and angular extents of the beam are changed to $\Delta q_o = m\Delta q_i$ and $\Delta p_o = \Delta p_i/m$. So, it is possible to reduce the spatial extent of the bundle but only producing a spread in the directions of the rays in the bundle. The behaviour of the system (2.4.13) gives signs, in geometrical optics, of the "uncertainty" principle. Mutuing the wording from the quantum mechanical context, we may say that the pure magnifier (2.4.13) produces a squeeze[(g)] in the phase plane contracting the ray-bundle representative area in one direction and enlarging it in the orthogonal direction.

Finally, we consider the effects of the ray-transformations (2.4.4), (2.4.8) and (2.4.13) on the phase-plane Gaussian distributions of the type (2.2.84). Specifically, we show the deformations experienced by the relative contours of constant density, which are clearly exemplified by the ellipses

$$\frac{q^2}{2\sigma_q^2} + \frac{p^2}{2\sigma_p^2} = \varepsilon \tag{2.5.16}$$

for assigned values of the positive number ε. The value of ε determines the fraction of the total energy enclosed within the relevant elliptical contour; thus, for instance, the contour corresponding to $\varepsilon = 2$ comprises about 86.5% of the optical energy of the beam. Thus, in Figs. 2.13.b) and 2.13.c) we may respectively see the q- and p-shears of the phase-plane contours (being circular

[g] The term *squeeze* is largely used in quantum mechanics to indicate that some of the quantum fluctuations are *squeezed* out of one observable and into the second noncommuting observable.

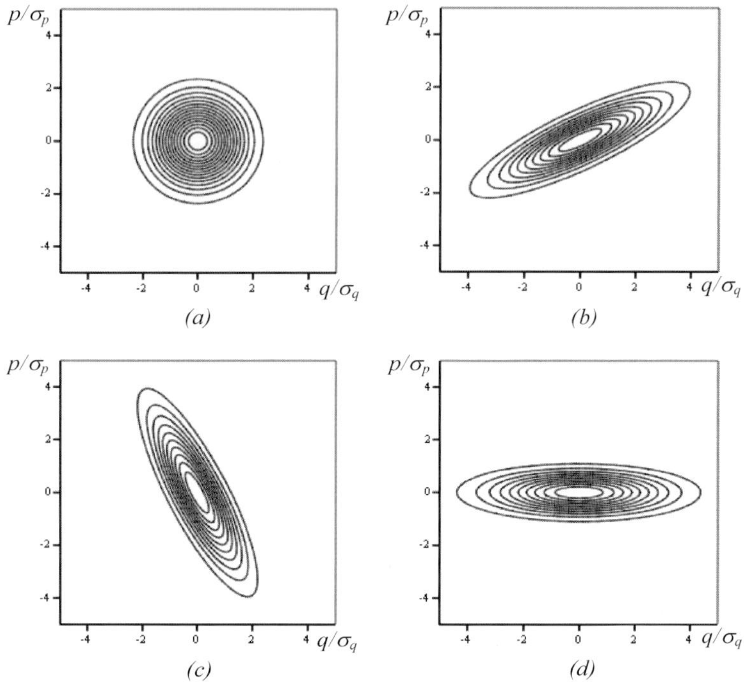

FIGURE 2.13. Transformations of the phase-plane contours of the Gaussian ray-density $\frac{1}{2\pi\sigma_q\sigma_p}e^{-\frac{q^2}{2\sigma_q^2}-\frac{p^2}{2\sigma_p^2}}$. *(a)* The circular contours of the initial distribution are *(b)* q-sheared by the propagation through a homogeneous medium, *(c)* p-sheared by the refraction at a two-media interface, and *(c)* squeezed in the p direction by a scaling transform.

at the input plane) consequent to the propagation through a homogeneous medium and the refraction at an interface. Similarly the squeeze along the p-axis due to the propagation through a magnifier with magnification $m > 1$ is shown in Fig. 2.13.d), the initial circles being transformed into the ellipses with the major axes on the q-axis. We emphasize that circles and ellipses are the pictorial representations adopted in quantum optics for coherent and squeezed states.

It is evident from the above figures that segments are turned into segments, romboidal contours are turned into romboidal contours as finally elliptical contours are turned into elliptical contours [3.10, 3.7, 3.9] (see Problem10). Indeed, the above figures make evident the general property that the functional form of the phase-plane ray distribution is preserved by any linear ray-transformation generated by quadratic functions in q and p. This straightforwardly follows from relation (2.2.67) implying that the arguments of the input ray distribution are replaced by their linear combination according to the specific optical transformation at hand.

As a conclusion, we note that within the framework of charged particle beam dynamics, the area of a closed domain over the phase-plane pertaining to the beam, chosen possibly in conformity with definite criteria, is usually referred to as the beam (*transverse*) *emittance* [3.10]-[3.13], apart from multiplicative constant factors differing among the various slightly different definitions existing in the literature. The emittance remains unchanged if the beam propagates through a non dissipative medium and hence it is an intrinsic parameter of the beam. It corresponds in optics to the second-moment-based M^2 *beam quality* parameter. In full analogy with particle beam optics, a phase-plane domain can as well be associated with optical systems on account of their geometry and optical performance; it represents the largest region in phase plane within which all the incoming rays can propagate through the system reaching then the output plane, and is referred as the system *acceptance*. The investigation of the propagation properties of the beam emittance in relation to the system acceptance is evidently basic to the optimal design of the system in view of the desired optical performance.

2.6 Summary

We have derived Liouville's equation for the phase-space density $\rho(q, p; z)$ appropriate to a "collective" description of light ray propagation. Then, exploiting the properties of Lie transformations we have showed that the phase-space ray density is propagated through the individual phase space ray-trajectories.

We proceeded considering the matrix optics generated by the homogeneous quadratic monomials $\frac{1}{2}p^2$, $\frac{1}{2}q^2$ and $\frac{1}{2}qp$, the associated Lie matrices being the generators of the symplectic group $Sp(2, \mathbb{R})$. The optical systems, generated by $\frac{1}{2}p^2$ and $\frac{1}{2}q^2$, have been recognized as performing the elementary optical operations as propagation through a homogeneous medium and refraction at the interface between two media, whereas the optical system generated by $\frac{1}{2}qp$ acts as a positive scaler, which may enlarge the spatial extent of the beam at the expense of the angular spread.

Finally the phase-plane picture have allowed us to recognize the transformations acted by the aforementioned optical systems on light beam as a horizontal shear, a vertical shear and an overall scale (or, "squeezing").

In the next chapter we will prove that every linear optical system can be understood as a sequence of a certain number of translations and refractions. We will establish two interesting decompositions of optical matrices, realizable under specific conditions. One of these decompositions involves the optical systems generated by the algebra basis matrices associated with the monomials

$\frac{1}{2}p^2$, $\frac{1}{2}q^2$ and $\frac{1}{2}qp$. Accordingly, optical systems may be described in phase plane by a sequence of a horizontal shear, a vertical shear and a squeezing. The other decomposition involves the optical systems generated by the algebra basis matrices associated with the polynomials $\frac{1}{2}q^2 + \frac{1}{2}p^2$, $\frac{1}{2}q^2$ and $\frac{1}{2}qp$, the former being proved to generate rotations in phase-plane.

Problems

1. Prove that through the symplectic mapping (2.2.34) generated by the Lie transformation $e^{\zeta \widehat{L}_f}$ associated with the dynamical variable $f(q,p)$, f is an invariant function, namely $f(\overline{q},\overline{p}) = f(q,p)$ over all range of the evolution paramer ζ in (2.2.34). Verify that the results of § 1.5.2 are in accord with this invariance.

2. Consider a perfect imaging system. Applying the invariance of the phase-plane areas, prove the sine condition

$$n_{0_1}\Delta q_i \sin\vartheta_i = n_{0_2}\Delta q_o \sin\vartheta_o,$$

where Δq_i and Δq_o denote the transverse extensions of the object and the image, whilst ϑ_i and ϑ_o represent the limiting angles (to the optical axis) of the fan of rays respectively accepted and transmitted by the system.

3. Verify that

$$(\widehat{A}+\widehat{B})^2=\widehat{A}^2+2\widehat{A}\widehat{B}+\widehat{B}^2-[\widehat{A},\widehat{B}],$$
$$(\widehat{A}+\widehat{B})^3=\widehat{A}^3+3\widehat{A}^2\widehat{B}+3\widehat{A}\widehat{B}^2+\widehat{B}^3-3\widehat{A}[\widehat{A},\widehat{B}]-3\widehat{B}[\widehat{A},\widehat{B}]+[\widehat{A},[\widehat{A},\widehat{B}]]-2[\widehat{B}[\widehat{B},\widehat{A}],$$

which clearly evidence the role of the commutator $[\widehat{A},\widehat{B}]$ to establish the difference of the algebra of operators to that of scalars.

4. Write down the composition rule for the product $e^{\widehat{A}}\, e^{\widehat{B}}$ of two exponential operators, whose exponents commute to a scalar: $[\widehat{A},\widehat{B}] = c$. Verify that $e^{\widehat{A}}\, e^{\widehat{B}}$ is not commutative.

5. Prove that inserting the form (2.2.70) for the ray density distribution in the Liouville equation (2.2.11) yields Hamilton's equations for the light ray.

6. Exploiting the invariance of the phase-space volume and the relation between the volume element $d\mathcal{V}$ and the geometrical extent $dAd\Omega$ of the elementary bundle of rays, deduce the transformation law of the geometrical extent of the elementary beam under refraction at the interface between two media.

7. Exploiting the invariance of the phase-space volume, deduce the transformation law of the radiance under refraction at the interface between two media, on account of the relation between the volume element $d\mathcal{V}$ and the geometrical factor $d\mathsf{S}\cos\theta d\Omega$ as well as of the conservation of the radiant energy.

8. Prove that in the case of propagation through a homogeneous medium, which is described by the Hamiltonian function $H = -\sqrt{n_0^2 - p^2}$, the exact ray evolution obtains

$$q(z) = q_i + \frac{p_i}{\sqrt{n_0^2 - p_i^2}}(z - z_i),$$
$$p(z) = p_i,$$

where the term $\frac{p_i}{\sqrt{n_0^2 - p_i^2}}$ is the tangent of the angle of the input ray to the axis.

9. Analyse the effect of the spherical aberrations, i.e., deviations from the paraxial propagation, on the Gaussian density distribution (2.2.84), using the results of Problem 4 in ch. 1. Verify the accord with the general result of Problem 8.

10. Verify that *(a)* straight lines are mapped into straight lines under the linear ray-optics transformation (2.1.3); in particular, verify that parallel straight lines are turned into parallel straight lines, the relevant angular coefficients η being related by the hyperbolic Möbius transformation

$$\eta_o = \frac{D\eta_i + C}{A\eta_i + B};$$

and that *(b)* ellipses are mapped into ellipses with the same area.

References

[1] For the Hamiltonian formulation of geometrical optics see [4] in ch. 1.

[2] See [7] in ch.1 for a discussion of the analogy between geometrical optics and classical, and relativistic, mechanics of a point-particle. The light-magnetic optics analogy is as well evidenced in [3.10], [3.11], [3.13].

[3] In relation with the Liouville's theorem we quote the original paper by Liouville, and the classical textbooks on mechanics by Arnold [3.3] and Goldstein [3.2], or on statistical mechanics by Tolman [4.1], ch. 3, pp. 48-52, Pathria [4.2], ch. 2, pp. 34-36, Balescu [4.3], ch. 2, pp. 37-44, and Toda [4.4], ch. 1, 17-22. A derivation of Liouville's theorem with an enlightening discussion of its implications in geometrical optics is presented by Marcuse [3.4]. Examples of the applicative implications of Liouville's theorem are in the papers [3.5]-[3.8]. For a discussion of the phase space method, and in particular of Liouville's theorem, within the context of the charged-beam optics, we suggest the books by Lichtenberg [3.10], Wollnik [3.11], and Hawkes and Kasper [3.13]. The latter exhaustively covers the whole of electron optics from both the geometrical and wave optics viewpoints. The paper by Lejeune and Aubert [3.11] provides one of the first systematic accounts of the multivariate issues concerned with the concepts of emittance and brightness in magnetic optics. See also the recent title by Dragoman and Dragoman [3.14].

[3.1] J. Liouville, "Sur la théorie de la variation des constantes arbitraires", Journ. de Math. Pure et Appl. **3**, 342-349 (1838).

[3.2] H. Goldstein, *Classical Mechanics* (Addison Wesley, Reading, 2nd ed., 1980), ch. 9, pp. 426-428.

[3.3] V.I. Arnold, *Mathematical Methods of Classical Mechanics* (Springer-Verlag, New York, 4th ed., 1984), ch. 3, pp. 68-74.

[3.4] D. Marcuse, *Light Transmission Optics* (Van Nostrand-Reinhold, Princeton, New Jersey, 1972), ch. 3, pp. 112-124.

[3.5] D. Marcuse, "Compression of a bundle of light rays", Appl. Opt. **10**, 494-497 (1971).

[3.6] M.C. Hudson, "Calculation of the maximum optical coupling efficiency into multimode optical waveguides", Appl. Opt. **13**, 1029-1033 (1974).

[3.7] G. Nemes, I.E. Teodorescu and M. Nemes, "Phase space treatment of optical beams", Preprint LOP-43-1984, CIP, Bucharest (1984).

[3.8] W.A.E. Goethals, "Laser beam analysis by geometrical optics", in *The Physics and Technology of Laser Resonators*, D.R. Hall and P.E. Jackson (eds.)(Adam Hilger, Bristol, 1989), pp. 143-153.

[3.9] G. Nemes, "A phase space approach to beam-pulse representations and measurements", Preprint LOP-76-1990, IAP, Bucharest (1990).

[3.10] A.J. Lichtenberg, *Phase-Space Dynamics of Particles* (J. Wiley & Sons, New York, 1969).

[3.11] C. Lejeune and J. Aubert, "Emittance and Brightness: Definitions and Measurements" in *Applied Charged Particle Optics*, Adv. Electron. & Electr. Physics, Suppl. **13A**, A. Septier (ed.) (Academic Press, New York, 1980), pp. 159-259.

[3.11] H. Wollnik, *Optics of Charged Particles* (Academic Press, Orlando, Florida, 1987).

[3.13] P.W. Hawkes and E. Kasper, *Principles of Electron Optics*, 3 vols. (Academic Press, London, Vol. 1-2, 1989, Vol. 3, 1994).

[3.14] D. Dragoman and M. Dragoman, *Quantum-Classical Analogies* (Springer, Berlin, 2004), ch. 6, pp. 143-186.

[4] As basic textbooks on statistical mechanics we quote

[4.1] R.C. Tolman, *The Principles of Statistical Mechanics* (Oxford Univ. Press, New York, 1938).

[4.2] R.K. Pathria, *Statistical Mechanics* (Pergamon Press, Oxford, 1972).

[4.3] R. Balescu, *Equilibrium and Nonequilibrium Statistical Mechanics* (John Wiley & Sons, New York, 1975). Here, Lie operators and Lie transformations are introduced and exploited; also, both the classical and the quantum phase-space pictures are presented, the latter being conveyed through the Wigner distribution function.

[4.4] M. Toda, R. Kubo and N. Saitô, *Statistical Physics I. Equilibrium Statistical Mechanics* (Springer-Verlag, Berlin 1983). Liouville's theorem of classical statistical mechanics is discussed, whose correspondence to quantum statistical mechanics is as well elucidated through the Wigner distribution function.

[5] G.C. Oates, "Vector Analysis", in *Handbook of Applied Mathematics - Selected results and methods,* C. E. Pearson (ed.) (Van Nostrand-Reinhold Company, New York, 1983), ch. 3, pp. 129-178.

[6] S. Steinberg, "Lie serie, Lie transformations, and their applications," in *Lie Methods in Optics*, J. Sánchez Mondragón and K.B. Wolf (eds.) (Springer-Verlag, Berlin, 1986), ch. 3, pp. 45-104.

[7] For the general theory of Lie algebras and groups we address the reader to [9] in ch. 1. Specifically, we quote below the textbook by Hamermesh being of concern to the proof of the unimodularity of the symplectic matrices afforded in § 2.3.1.

[7.1] M. Hamermesh, *Group Theory and its Applications to Physical Problems* (Addison-Wesley, Reading, MA, 1962), pp. 403-406.

[8] We list below the original papers by Baker [8.2], Campbell [8.1], and Hausdorff [8.3], along with a few related titles from the wide relevant literature. Comments on the BCH formula are given in [6], where a list of references is provided. The derivation of the BCH formula can be found also in the textbooks [8.7] and [8.8].

[8.1] J.E. Campbell, "On a law of combination of operators", Proc. London Math. Soc. **29**, 14-32 (1897).

[8.2] H.F. Baker, "On the exponential theorem for a simply transitive continuous group, and the calculation of the finite equations from constants of structure," Proc. London Math. Soc. **3**, 91-129 (1901).

[8.3] F. Hausdorff, "Die symbolische exponentialformel in der gruppentheorie", Ber. Verh. Saechs. Akad. Wiss. Leipzig Math.-Phys. Klasse **58**, 19-48 (1906).

[8.4] W. Magnus, "On the exponential solution of differential equations for a linear operator", Comm. Pure Appl. Math. **7**, 649-673 (1954).

[8.5] R.D. Richtmeyer and S. Greenspan, "Espansions of the Campbell-Baker-Hausdorff formula by computer", Comm. Pure Appl. Math. **18**, 107-108 (1965).

[8.6] R. M. Wilcox, "Exponential operators and parameter differentiation in quantum physics", J. Math. Phys. **8**, 962-982 (1967).

[8.7] S. Helgason, *Differential Geometry and Symmetric Spaces* (Academic Press, New York, 1962), pp. 36-40.

[8.8] M. Hausner and J.T. Schwartz, *Lie Groups; Lie Algebras* (Gordon and Breach, New York, 1968), pp. 65-71.

[9] The basic concepts of radiometry and radiative transfer theory, and their connection to partial coherence theory are illustrated in a variety of titles; a few are quoted below. See [3.11] and [3.13], ch. 47, pp. 971-988, ch. 78-79, pp. 1719-1767, for a discussion of the concept of brightness, and coherence and brightness from the viewpoint of electron optics; the close analogy to light optics is enlightened and effectively supported by numerous references.

[9.1] M. Born and E. Wolf, *Principles of Optics* (Pergamon, Oxford, 1965), pp.181-183.

[9.2] F. Grum and R.J. Becherer, *Optical Radiation Measurements. I. Radiometry* (Academic Press, New York, 1979), chs. 1-3, pp. 1-97.

[9.3] W.R. McCluney, *Introduction to Radiometry and Photometry* (Artech House, Boston, 1994).

[9.4] L. Mandel and E. Wolf, *Optical Coherence and Quantum Optics* (Cambridge University Press, Cambridge, 1995), ch. 5, pp. 287-307 and references therein.

[9.5] F. Nicodemus, "Radiance", Am. J. Phys. **31**, 368-377 (1963).

[9.6] A. Walther, "Radiometry and Coherence", J. Opt. Soc. Am. **58**, 1256-1259 (1968).

[9.7] A. Walther, "Radiometry and Coherence", J. Opt. Soc. Am. **63**, 1622-1623 (1973).

[9.8] E. Wolf, "Coherence and Radiometry", J. Opt. Soc. Am. **68**, 6-17 (1978).

[9.9] A. T. Friberg, "On the existence of a radiance function for finite planar sources of arbitrary states of coherence", J. Opt. Soc. Am. **69**, 192-197 (1979).

[9.10] A. T. Friberg, "Phase-space methods for partially coherent wavefields", in *Optics in four dimensions*, M.A. Machado and L.M. Narducci (eds.) (Amer. Inst. Phys., New York, 1981), pp. 313-330.

[9.11] L.A. Apresyan and Yu.A. Kravtsov, "Photometry and coherence: wave aspects of the theory of radiation transport", Sov. Phys. Usp. **27**, 301-313 (1984).

[9.12] J.T. Foley and E. Wolf, "Radiometry as a short-wavelength limit of statistical wave theory with globally incoherent sources", Opt. Comm. **55**, 236-241 (1985).

[9.13] G.S. Agarwal, J.T. Foley and E. Wolf, "The radiance and phase-space representations of the cross-spectral density operator", Opt. Comm. **62**, 67-72 (1987).

[9.14] R.G. Littlejohn and R. Winston, "Corrections to classical radiometry", J. Opt. Soc. Am. A **10**, 2024-2037 (1993).

[10] A reading list about the concept of light ray is provided in [1] of ch. 1.

[11] Specific investigations about the symplectic group, in relation as well to geometrical optics, are presented in

[11.1] V. Guillemin and S. Sternberg, *Symplectic Techniques in Physics* (Univ. Press, New York, 1990).

[11.2] R. Simon and N. Mukunda, "The two-dimensional symplectic and metaplectic groups and their universal cover", in *Symmetries in Science VI*, B. Gruber (ed.) (Plenum Press, New York, 1993), pp. 659-689.

[11.3] Arvind, B. Dutta, N. Mukunda and R. Simon, "The Real symplectic groups in quantum mechanics and optics", Pramana **45**, 471-513 (1995); also in arXiv:quant-ph/9509002 v3.

[11.4] J.F. Cariñena and J. Nasarre, "On the symplectic structures arising in geometric optics", Fortschr. Phys. **44**, 181-198 (1996).

[11.5] R. Simon and K.B. Wolf, "Structure of the set of paraxial optical systems", J. Opt. Soc. Am. **17**, 342-355 (2000).

[11.6] K.B. Wolf, *Geometric Optics on Phase Space* (Springer, Berlin, 2004).

[12] As to the ray matrix approach to linear optics we suggest [13.3], [13.5] and

[12.1] W. Brouwer, *Matrix Mehods in Optical Instrument Design* (Benjamin, New York, 1964).

[12.2] A. Nussbaum and R.A. Phillips, *Contemporary Optics for Scientists and Engineers* (Prentice-Hall, Englewood Cliffs, 1976).

[12.3] M.V. Klein and T. Furtak, *Optics* (John Wiley & Sons, New York, 2nd ed., 1986).

[12.4] K. Halbach, "Matrix representation of Gaussian optics", Am. J. Phys. **32**, 90-108 (1964).

[12.5] H. Kogelnik, "Imaging of optical modes - Resonators with internal lenses", Bell System Tech. J. **44**, 455-494 (1965).

[13] Detailed descriptions of the properties of Gaussian beams are presented in [3.4], ch.6, pp.230-262, and in

[13.1] A. Yariv, *Quantum Elctronics* (John Wiley & Sons, New York, 2nd ed, 1975), ch. 6, pp. 99-129.

[13.2] J.A. Arnaud, *Beam and Fiber Optics* (Academic Press, London, 1976), ch. 2, pp. 54-79.

[13.3] A.E. Siegman, *Lasers* (Univ. Book, Mill Valley, 1986), chs. 16-17, pp. 626-697.

[13.4] P.W. Milonni and J.H. Eberly, *Lasers* (John Wiley & Sons, New York, 1988), pp. 484-508.

[13.5] B.E.A. Saleh and M.C. Teich, *Fundamentals of Photonics* (John Wiley & Sons, New York, 1991), ch. 3, pp. 80-107.

[13.6] N. Hodgson and H. Weber, *Optical Resonators - Fundamentals, Advanced Concepts and Applications* (Springer-Verlag, London, 1997), ch. 2, pp. 76-104.

[14] A. S. Deif, *Advanced Matrix Theory for Scientists and Engineers* (Abacus Press, Turnbridge Wells, 1982), ch. 5; in particular, for the methods to compute the exponential function of matrices see pp. 176-182.

[15] For an investigation of the refracting surface transformations we address the reader to [11.6], ch. 4, pp. 47-56 and to

[15.1] H. Raszillier and W. Schempp, "Fourier optics from the perspective of the Heisenberg group", in *Lie Methods in Optics*, J. Sánchez Mondragón and K.B. Wolf (eds.) (Springer-Verlag, Berlin, 1986), ch. 2, pp. 29-43.

[15.2] M. Navarro-Saad and K.B. Wolf, "Factorization of the phase-space transformation produced by an arbitrary refracting surface", J. Opt. Soc. Am. A **3**, 340-346 (1986).

[15.3] T. Sekiguchi and K.B. Wolf, "The Hamiltonian formulation of optics", Am. J. Phys. **55**, 830-835 (1987).

[15.4] K.B. Wolf, "Symmetry in Lie optics", Ann. Phys. **172**, 1-25 (1986).

[15.5] K.B. Wolf and G. Krötzsch, "Geometry and dynamics in refracting systems", Eur. J. Phys. **16**, 14-20 (1995).

[15.6] O.N. Stavroudis, *The Optics of Rays, Wavefronts and Caustics* (Academic Press, New York, 1972).

3

The Group of 1D First-Order Optical Systems

3.1 Introduction

In the previous chapter we have traced a line of approach for investigating the relation between the ray-optics description of paraxial propagation through $1D$ optical systems and the symplectic Lie group of 2×2 real matrices, associated with the quadratic polynomials in the conjugate variables q and p under Poisson bracket operation and matrix exponentiation.

When considering rays with small inclination angles to the axis ($p^2 \ll p_z^2 \simeq n^2$) and accordingly objects and images extending across the axis well within all the pupil-apertures of the system ($|q| \ll$ all system pupil-apertures), the optical Hamiltonian becomes quadratic in both q and p, i.e.,

$$H(q, p, z) = \frac{1}{2n_0} p^2 + \frac{n_2}{2} q^2, \tag{3.1.1}$$

showing separate "kinetic" and "potential"-like contributions.
Hamiltonian (3.1.1) is a model function in several fields of physics (Fig. 3.1), both in the classical-like context, where q and p are interpreted as canonically conjugate variables for which $\{q, p\} = 1$, and in the quantum-like context, where q and p are interpreted as noncommuting operators for which $[q, p] = \aleph i$, $\aleph$ signifying the "quantization constant" inherent in the specific framework.

Accordingly, we have examined the transformations of the position and momentum ray-variables, generated by the quadratic monomials $\frac{1}{2}p^2$, $\frac{1}{2}q^2$ and $\frac{1}{2}qp$ in the real space as well as in the optical phase-space, in order to identify the possible optical systems that may produce such transformations. We have drawn in fact a definite correspondence of each monomial to a traceless matrix in the symplectic Lie algebra $sp(2, \mathbb{R})$ and then to a unimodular transformation matrix in the corresponding symplectic Lie group $Sp(2, \mathbb{R})$. The $\frac{1}{2}p^2$-,$\frac{1}{2}q^2$-, and $\frac{1}{2}qp$-generated optical systems are thereby recognized as a free-medium section, a refracting interface and a positive scaler, as sketched in the scheme:

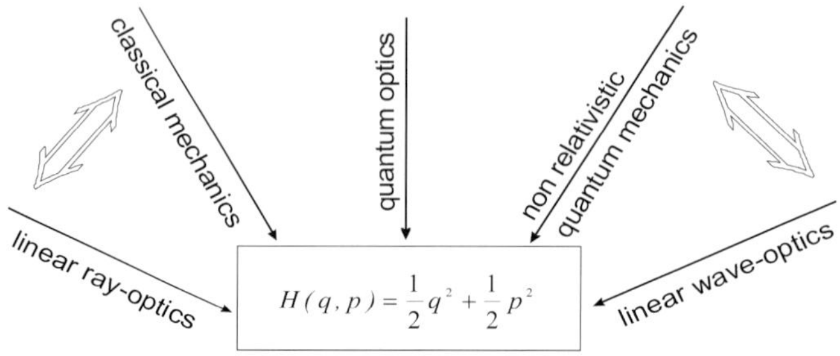

FIGURE 3.1. The harmonic oscillator-like Hamiltonian $H(q,p) = \frac{1}{2}p^2 + \frac{1}{2}q^2$ models several physical systems in many branches of physics.

generating function			generating matrix		transformation matrix
free-medium section	$\frac{1}{2}p^2$	$\rightarrow$	$\mathbf{K}_- = \begin{pmatrix} 0 & 1 \\ 0 & 0 \end{pmatrix}$	$\rightarrow$	$e^{d\mathbf{K}_-} = \begin{pmatrix} 1 & d \\ 0 & 1 \end{pmatrix}$
thin lens	$\frac{1}{2}q^2$	$\rightarrow$	$\mathbf{K}_+ = \begin{pmatrix} 0 & 0 \\ -1 & 0 \end{pmatrix}$	$\rightarrow$	$e^{\frac{1}{f}\mathbf{K}_+} = \begin{pmatrix} 1 & 0 \\ -1/f & 1 \end{pmatrix}$
positive magnifier	$\frac{1}{2}qp$	$\rightarrow$	$\mathbf{K}_3 = \frac{1}{2}\begin{pmatrix} 1 & 0 \\ 0 & -1 \end{pmatrix}$	$\rightarrow$	$e^{s\mathbf{K}_3} = \begin{pmatrix} e^{s/2} & 0 \\ 0 & e^{-s/2} \end{pmatrix}$.

(3.1.2)

Also, we have individualized the respective area-preserving transformations in the q-p plane as a q-shear, a p-shear and a reciprocal scaling in q and p.

Here we further explore the correspondence between quadratic polynomials and first-order optical systems. Section 3.2 describes the product rule of ray matrices for obtaining the overall ray-matrix of a composite system. Basic considerations on free medium section and thin lens matrices are then elaborated in Sect. 3.3. In particular, as regards the free propagation matrices, the difference in their "behaviour" when understood as abstract matrices or optical matrices is remarked. Section 3.4 shows that every symplectic matrix can be synthesized as an ordered product of a finite number of $\frac{1}{2}p^2$- and $\frac{1}{2}q^2$-generated matrices. From the optical viewpoint, this means that every first-order optical system is equivalent to a properly designed arrangement of a finite number of thin lenses separated by uniform index media. Consequently every phase-plane transformation mapping the ingoing ray coordinates to the outgoing ray coordinates can be pictured as a sequence of horizontal and vertical shears. The $\frac{1}{2}p^2$- and $\frac{1}{2}q^2$-generated matrices look thereby as the building blocks for the whole domains of the real symplectic matrices and first-order optical systems. As a conclusion, a one-to-one correspondence between the set of $1D$ linear optical systems and the symplectic group $Sp(2,\mathbb{R})$ of 2×2 real matrices is established. Section 3.6 illustrates the link of the attractive and repulsise oscillator-like Hamiltonian dynamics with fractional Fourier transformers and beam expanders (or reductors), which individualize in the (q, p) plane scaled rotations and squeezes. Two interesting realizations of optical matrices are described. One realization, which applies locally near the group identity, involves the $\frac{1}{2}q^2$-, $\frac{1}{2}qp$-, and $\frac{1}{2}p^2$-generated optical systems, and accordingly is visualized in the optical phase plane as a q-shear followed by a scaling and then by a p-shear (§ 3.5). The other realization, which in contrast applies to the entire group, involves the $\frac{1}{2}q^2$-, $\frac{1}{2}qp$-, and $\frac{1}{2}(p^2 + q^2)$-generated optical systems; it is pictured in the (q, p) plane as a rotation followed by a scaling and a vertical shear (§ 3.7). Conclusive considerations about the link between quadratic polynomials and paraxial ray matrices are given in Sect. 3.8. Finally, Sect. 3.9 describes the method to integrate Hamilton's equation for the optical ray matrix, detailing separately focusing and defocusing quadratic index media.

3.2 Ray matrix of composite optical systems

Optical systems, consisting of several components connected together in cascade, can be handled symply by multiplying the ray matrices of the individual optical elements, arranged in the reverse order [1]. In fact, each optical element has its own input and output planes and acts on the ray incident on its input plane in a certain way to produce the ray at its output plane. Thus the ray propagation through a sequence of N optical elements with ray-matrices $\mathbf{M}_1$, $\mathbf{M}_2$, ..., $\mathbf{M}_N$ is accounted for by the following chain of matrix relations

$$\begin{aligned} \mathbf{u}_1 &= \mathbf{M}_1 \mathbf{u}_0, \\ \mathbf{u}_2 &= \mathbf{M}_2 \mathbf{u}_1, \\ &\dots \\ \mathbf{u}_N &= \mathbf{M}_N \mathbf{u}_{N-1}, \end{aligned} \tag{3.2.1}$$

where as usual $\mathbf{u}_j$ denotes the ray-coordinate vector $\mathbf{u}_j = (q_j, p_j)^\top$ at the fixed reference plane Π_j, $j = 0, .., N$ (Fig. 3.2). The intermediate planes Π_j, $j = 1, .., N-1$ play the double role as the output and input planes respectively for the j-th and $(j+1)$-th optical element, whilst Π_0 and Π_N are the input and output planes for the first and last element as well as for the overall system.

Combining the input-output relations (3.2.1), we obtain the transfer law for the ray coordinates from Π_0 to Π_N in the form

$$\mathbf{u}_N = \mathbf{M}_N \mathbf{M}_{N-1} \cdots \mathbf{M}_2 \mathbf{M}_1 \mathbf{r}_0 \equiv \mathbf{M} \mathbf{u}_0. \tag{3.2.2}$$

Accordingly, the ray matrix $\mathbf{M}$ of the overall system comes to be the product

$$\mathbf{M} = \mathbf{M}_N \mathbf{M}_{N-1} \cdots \mathbf{M}_2 \mathbf{M}_1 \tag{3.2.3}$$

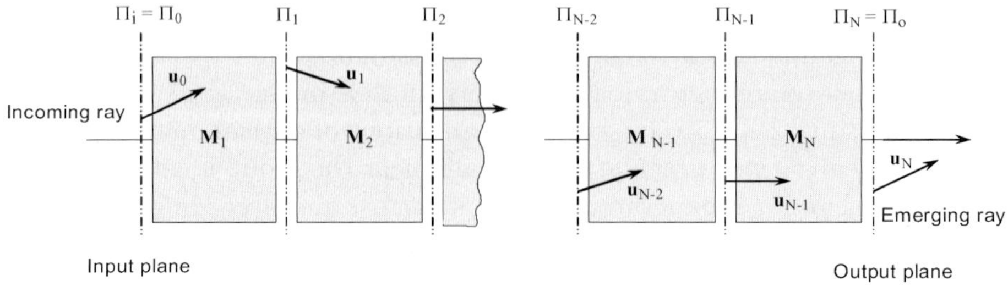

FIGURE 3.2. Propagation of a light ray through a sequence of optical elements. The optical matrix of the overall system is the product of the individual ray matrices from Π_0 to Π_N.

with the matrices $\mathbf{M}_1$, .., $\mathbf{M}_N$ being arranged in the inverse order from that in which the corresponding elements are placed on the axis with respect to the direction of the incoming ray.

3.2.1 *Thick and thin lenses*

As an example of a compound optical system, we consider a *thick* lens, which basically consists of two spherical surfaces separated by an appreciable distance. The ray propagation through a *thick* lens may be seen as comprising the following sequence of ray-transformations: 1) refraction at the front surface with refracting power $\mathcal{P}_1$, 2) free propagation through the lens of axial thickness d and 3) refraction at the back surface with power $\mathcal{P}_2$ (Fig. 3.3.a)). The thick lens matrix is then obtained from the ordered product of refraction-free propagation-refraction matrices, which finally gives

$$\mathbf{M}_{\text{thick lens}} = \begin{pmatrix} 1 - \mathfrak{d}\mathcal{P}_1 & \mathfrak{d} \\ -\mathcal{P} & 1 - \mathfrak{d}\mathcal{P}_2 \end{pmatrix}, \tag{3.2.4}$$

where the focal power $\mathcal{P}$ of the lens is $\mathcal{P} = \mathcal{P}_1 + \mathcal{P}_2 - \mathfrak{d}\mathcal{P}_1\mathcal{P}_2$, and $\mathfrak{d}$ denotes the axial thickness of the lens divided by the relative refractive index.

As a further example, we consider a *thin* lens, formally defined as consisting of two refracting surfaces with negligible separation in between, so that a ray incident at a given point on one surface emerges approximately at the same height from the other surface. The reference planes Π_i and Π_o are taken to coincide with the plane of the lens, although Π_i relates to incident rays whilst Π_o relates to emerging rays (Fig. 3.3.b)). The ray matrix of the thin lens is then obtained as the product of two refraction matrices relevant to the two lens interfaces. The resulting matrix looks like a refraction matrix, i.e.,

$$\mathbf{M}_{\text{thin lens}} = \begin{pmatrix} 1 & 0 \\ -\mathcal{P} & 1 \end{pmatrix}. \tag{3.2.5}$$

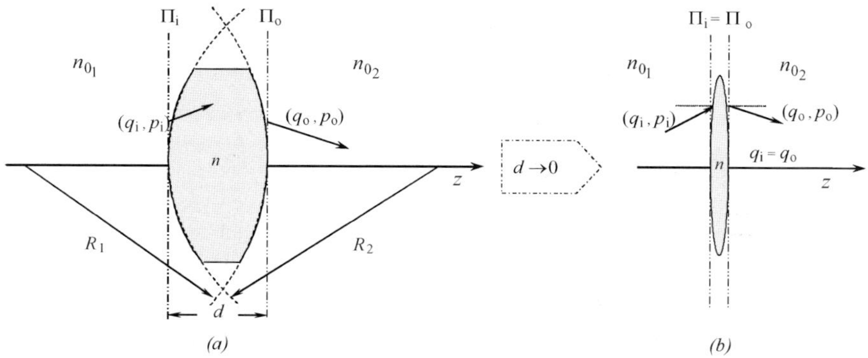

FIGURE 3.3. *(a)* Propagating through a thick lens, the light ray experiences the refraction at the front surface, the propagation through the lens and another refraction at the back surface. *(b)* Being the refracting surfaces of a thin lens separated by a negligible distance, the light ray experiences only the refractions at the lens front and back surfaces.

the focal power $\mathcal{P}$ of the lens being the sum of the refracting powers of the two surfaces: $\mathcal{P} = \mathcal{P}_1 + \mathcal{P}_2$. The lens power $\mathcal{P}$ is commonly expressed in terms of the reduced focal length f as $\mathcal{P} = 1/f$, with $f = f_1/n_{0_1} = f_2/n_{0_2}$, n_{0_1} and n_{0_2} denoting the refractive indices of the media in the object and image spaces.

The formal analogy between the refraction and thin lens matrices establishes the optical equivalence between a refracting interface with a given refracting power $\mathcal{P}$ and a thin lens with the same focusing power.

Disregarding the real design, in the following we will refer to a thin lens as the optical element described by the ray matrix (3.2.5). Also, according to the usual practice, we will refer to *converging*, or *positive*, lenses, when $f > 0$, and to *diverging*, or *negative*, lenses, when $f < 0$. As is well known, positive lenses cause converging spherical waves to converge more rapidly, and diverging spherical waves to diverge less rapidly or to converge; in contrast, negative lenses cause converging spherical waves to converge less rapidly or to diverge, and diverging waves to diverge more rapidly.

3.3 The subgroup of free propagation and thin lens matrices

Free-medium sections and thin lenses are the basic components of every first-order optical system. As it is proved, in fact, every optical transformation can be realized by a finite sequence of free propagations and lens operations, or equivalently every 2×2 real symplectic matrix can optically be synthesized properly arranging free propagation sections and thin lenses.

We denote by $\mathbf{T}(d)$ and $\mathbf{L}(f)$ the optical matrices relevant respectively to portions of a uniform refractive index medium,

$$\mathbf{T}(d) \equiv e^{d\mathbf{K}_-} = \begin{pmatrix} 1 & d \\ 0 & 1 \end{pmatrix}, \tag{3.3.1}$$

and thin lenses,

$$\mathbf{L}(f) \equiv e^{\frac{1}{f}\mathbf{K}_+} = \begin{pmatrix} 1 & 0 \\ -1/f & 1 \end{pmatrix}. \tag{3.3.2}$$

We recall that d and f denote the section length and the lens focal length, both scaled by the relevant refractive index. Also, thin lens matrices are equivalently denoted by $\mathbf{L}(f)$ or $\mathbf{L}(\mathcal{P})$, $\mathcal{P}$ being the focal power: $\mathcal{P} = 1/f$. The limit cases of $d \to 0$ (section of negligible length) and $f \to \pm\infty$ (lens of negligible focal power) identify the unity matrix $\mathbf{I}$: $\mathbf{T}(0) = \mathbf{I}$ and $\mathbf{L}(\pm\infty) = \mathbf{I}$.

Interestingly, $\mathbf{T}$ and $\mathbf{L}$ are unimodular and, respectively, upper- and lower-triangular matrices. They are the representatives of the two Abelian one-parameter subgroups of the symplectic group $Sp(2, \mathbb{R})$, generated respectively by the algebra elements $\mathbf{K}_-$ and $\mathbf{K}_+$ as illustrated in Chapter 2 (§§ 2.3, 2.4.1 and 2.4.2) [2]. The subgroup additivity property (2.3.12) with respect to the parameter of the algebra-to-group exponentiation writes as

$$\mathbf{T}(d_1)\mathbf{T}(d_2) = \mathbf{T}(d_2)\mathbf{T}(d_1) = \mathbf{T}(d), \tag{3.3.3}$$

with

$$d = d_1 + d_2, \tag{3.3.4}$$

and similarly

$$\mathbf{L}(f_1)\mathbf{L}(f_2) = \mathbf{L}(f_2)\mathbf{L}(f_1) = \mathbf{L}(f), \tag{3.3.5}$$

with

$$\frac{1}{f} = \frac{1}{f_1} + \frac{1}{f_2}, \quad \text{or} \quad \mathcal{P} = \mathcal{P}_1 + \mathcal{P}_2. \tag{3.3.6}$$

The above relations express in formal terms the well known property that the sequence of two free space sections d_1 and d_2 is equivalent to one section of length $d = d_1 + d_2$ and, similarly, two cascaded lenses f_1 and f_2 (with no separation in between) are equivalent to the lens $f = \frac{f_1 f_2}{f_1+f_2}$.

As an immediate consequence of (3.3.3) and (3.3.5), we also have

$$\mathbf{T}(d)\mathbf{T}(-d) = \mathbf{T}(-d)\mathbf{T}(d) = \mathbf{I}, \tag{3.3.7}$$

and

$$\mathbf{L}(f)\mathbf{L}(-f) = \mathbf{L}(-f)\mathbf{L}(f) = \mathbf{I}, \tag{3.3.8}$$

which state the existence and unicity of the inverse to every $\mathbf{T}$ and $\mathbf{L}$-type matrix according to

$$[\mathbf{T}(d)]^{-1} = \mathbf{T}(-d) = e^{-d\mathbf{K}_-} \tag{3.3.9}$$

and

$$[\mathbf{L}(f\,)]^{-1} = \mathbf{L}(-f\,) = e^{-\frac{1}{f}\mathbf{K}_+}. \tag{3.3.10}$$

Formally, since both $\mathbf{T}(-d)$ and $\mathbf{L}(-f\,)$ reproduce the typology of the matrices (3.3.1) and (3.3.2), relations (3.3.9) and (3.3.10) confirm that $\mathbf{T}$ and $\mathbf{L}$-type matrices do form two proper one-parameter subgroups of $Sp(2,\mathbb{R})$, when regarded as real upper and lower triangular matrices with unit diagonal entries. Optically, relation (3.3.10) establishes that the inverse of a lens system should be a lens with opposite value of the focal length. This is a physically realizable system; relation (3.3.8) can therefore be implemented by the simple sequence of two lenses with opposite focal lengths. As the lens f alters the direction of propagation of the ingoing ray, the lens $-f$ acts oppositely on the intermediate ray, which then regains the original direction of propagation. Hence matrices (3.3.2) do form a subgroup even when interpreted as the optical matrices of thin lenses.

On the contrary, by (3.3.9) the inverse of the free propagation by the distance d should be the free propagation by the distance $-d$, which in principle may appear not physically realizable since distances occur naturally as positive quantities. Therefore matrices (3.3.1), when interpreted as optical matrices for propagation through portions of a homogeneous medium, do not strictly form a subgroup. In the next paragraph we will see that the propagation corresponding to "negative" values of d can optically be realized by a suitable sequence of free propagation sections and thin lenses. The inverse of the free propagation matrix, i.e. the matrix $\mathbf{T}(-d)$, acquires then a physical concreteness, although unlike $\mathbf{T}(d)$ it does not correspond to a single and unique optical element. Accordingly, relation (3.3.7) becomes optically synthesizable, the effect of the free propagation by d being completely vanished by the propagation through the system $\mathbf{T}(-d)$, as the ray eventually recovers the initial position and direction.

Apart from their physical realizability, "negative" free propagations, when framed within the common conventions regarding the optical systems, can be given a practical interpretation which relates to the character, real or virtual, of the rays we are dealing with. Let us write d in terms of the axial coordinate as $d = z_o - z_i$ (the refractive index being omitted for simplicity), so that

$$\mathbf{T}^{-1}(z_o - z_i) = \mathbf{T}(-(z_o - z_i)) = \mathbf{T}(z_i - z_o). \tag{3.3.11}$$

Suppose that the ray coordinates (q_o, p_o) are known at the plane z_o. Evidently the ray coordinates (q_i, p_i) at the preceding plane z_i may be calculated by the inverse matrix $\mathbf{T}^{-1}(z_o - z_i) = \mathbf{T}(z_i - z_o)$. In symbols, we have

$$\begin{pmatrix} q_i \\ p_i \end{pmatrix} = \mathbf{T}(z_i - z_o) \begin{pmatrix} q_o \\ p_o \end{pmatrix}, \quad z_i - z_o < 0. \tag{3.3.12}$$

If the portion of space from z_i to z_o is not free, but occupied by some optical system, the ray at z_i, we might obtain through $\mathbf{T}(z_i - z_o)$, is not the actual ray at z_i, but the virtual ray (see § I.8.1), namely the ray that would produce the given ray at z_o, if the intervening space between z_i and z_o were free. Exemplary to this view are some optical systems, we will describe below as, for instance, the Fourier transforming configuration realized by diverging thin lenses.

The above discussion gives a physical concreteness to free propagation and thin lens matrices, $\mathbf{T}(d)$ and $\mathbf{L}(f)$, with negative values of d and f. The present treatment acquires indeed a general validity, comprising within the same formalism the cases of diverging lenses, virtual objects and virtual images.

3.3.1 *Free propagation through "negative" distances*

Free propagations corresponding to negative values of the distance can be realized by using a finite number of thin lenses separated by finite distances. This unequivocally establishes the physical realizability of the inverse of free propagation by a constructible optical system, and hence composes the contrast between the "mathematical" and "optical" behaviour of the set of matrices (3.3.1) in favour of the group structure of the set of linear optical systems.

Arranging three lenses appropriately chosen and spaced one from the others one may synthesize a "negative" free propagation. In fact, the formal identity

$$\begin{pmatrix} A & B \\ C & D \end{pmatrix} = \mathbf{L}(f_3)\mathbf{T}(d_2)\mathbf{L}(f_2)\mathbf{T}(d_1)\mathbf{L}(f_1), \tag{3.3.13}$$

can be solved for the parameters of the optical arrangement on the right, i.e. the focal lenghts f_1, f_2, f_3 and the distances d_1, d_2, in order to produce a "negative" free propagation matrix, for which indeed we must have

$$A = D = 1, \quad C = 0, \quad B < 0. \tag{3.3.14}$$

Substituting the various $\mathbf{T}$ and $\mathbf{L}$ matrices entering (3.3.13), we obtain for the entries of the matrix on the left the expressions

$$\begin{aligned} A &= 1 - \tfrac{d_2}{f_2} - \tfrac{B}{f_1}, \quad & B &= d_2 + d_1\left(1 - \tfrac{d_2}{f_2}\right), \\ C &= -\tfrac{1}{f_2} - \tfrac{1}{f_3} + \tfrac{d_2}{f_2 f_3} - \tfrac{D}{f_1}, \quad & D &= 1 + d_1\left(-\tfrac{1}{f_2} - \tfrac{1}{f_3} + \tfrac{d_2}{f_2 f_3}\right) - \tfrac{d_2}{f_3}, \end{aligned} \tag{3.3.15}$$

which by (3.3.14) provide the relations

$$\frac{1}{f_1} = -\frac{1}{f_2} - \frac{1}{f_3} + \frac{d_2}{f_2 f_3}, \quad \frac{d_1}{f_1} = \frac{d_2}{f_3}, \quad \frac{1}{f_2} > \frac{1}{d_1} + \frac{1}{d_2}. \tag{3.3.16}$$

It is immediately evident that f_2 must be positive. Also, it can easily be checked that f_1 and f_3 as well must be positive. Therefore, identity (3.3.13) satisfying requirements (3.3.14) is concevaible with all the three lenses positive:

$$f_1, f_2, f_3 > 0. \tag{3.3.17}$$

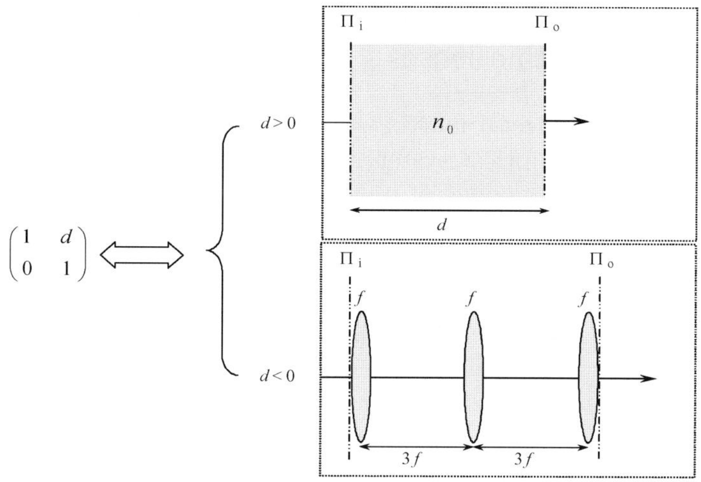

FIGURE 3.4. A possible three-lens configuration implementing the free propagation by a "negative" distance.

Clearly the choice of the parameters, i.e., focal lenghts and distances, satisfying the above relations is not unique. However, in order to prove the existence of some optical configuration being in accord with (3.3.16), we choose

$$f_2 = f_3 \equiv f. \tag{3.3.18}$$

Consequently, we have

$$\frac{1}{f_1} = -\frac{2}{f} + \frac{d_2}{f^2} > 0, \tag{3.3.19}$$

which can be satisfied, for instance, by setting

$$d_2 = 3f. \tag{3.3.20}$$

It follows that

$$f_1 = f, \; d_1 = 3f = d_2, \tag{3.3.21}$$

thus yielding a symmetric configuration with three identical lenses equally spaced from each other (Fig. 3.4). The B-entry of the ray matrix comes to be

$$B = -3f < 0. \tag{3.3.22}$$

The value of f will be fixed according to some specific request on the value of B. Thus, from $B = -d, \; d > 0$, we obtain $f = \frac{d}{3}$. The reader may work out other configurations satisfying (3.3.16).

3.4 Optical matrices factorized in terms of free propagation sections and thin lenses

Every 2×2 real symplectic matrix can be understood as a ray matrix representing a first-order optical system. Hence, first-order optical systems can be regarded as optical analogs of physical processes described by 2×2 real symplectic matrices. It can be proved, in fact, that every 2×2 real symplectic matrix is expressible as the product of a finite number of $\mathbf{T}$ and $\mathbf{L}$-like matrices, thus also proving that every optical system can be realized as an appropriate (finite) sequence of free propagation sections and thin lenses [3].

Here we examine the realizability of every $Sp(2,\mathbb{R})$ system by an optical arrangement of thin lenses and free-medium sections. We shall not dwell on the question regarding the minimum number of lenses needed to realize a particular system. This question is analyzed in detail in [3.5]; there, as a result of an accurate investigation of the various regions of $Sp(2,\mathbb{R})$, that can be reached by one, two and three lens configurations, it is shown that every element in the group can be realized by an optical configuration involving no more than three lenses.

We firstly consider optical configurations, demanding for three parameters to be defined, and prove that they yield suitable optical syntheses for elements in specific regions of the $Sp(2,\mathbb{R})$ group. Then, more complex arrangements with increasing numbers of lenses and free propagation sections are configured in order to show that unimodular purely diagonal matrices are optically realizable. This is basic to the conclusive proof as elaborated in § 3.4.4, concerning the possibility of representing every real symplectic matrix within the connected identity component of the group in the form as the factored product of $\mathbf{T}$, $\mathbf{L}$ and purely diagonal matrices. In the forthcoming discussion the matrix $\mathbf{T}(d)$ is understood as describing truly free propagation sections, and hence the parameter d is limited to nonnegative values. This is in order to evidence the relevance of the inherent *semi*group (instead of group) structure of the set of the free-propagation matrices to prevent the simple three parameter configurations, comprising one or two lenses, from exhausting all the three-parameter $Sp(2,\mathbb{R})$ systems. The restriction on the matrix $\mathbf{T}(d)$ is relaxed in § 3.4.4, where "negative" distances are allowed.

3.4.1 *TLT and LTL configurations*

The simplest realization of a 2×2 symplectic matrix by means of thin lenses and free propagation sections, one may envisage, should involve an arrangement allowing for three parameters, which then might be two distances and one focal length or conversely two focal lengths and one distance. The resulting configurations will be denoted as TLT and LTL, respectively.

Then, let $\mathbf{M} = \left(\begin{smallmatrix} A & B \\ C & D \end{smallmatrix}\right)$ be a symplectic matrix. Supposing $\mathbf{M}$ has non null entries, the following sequence of optical elements can be hypothesized

$$\left(\begin{smallmatrix} A & B \\ C & D \end{smallmatrix}\right) = \mathbf{T}(d_2)\mathbf{L}(f)\mathbf{T}(d_1) \tag{3.4.1}$$

having $\mathbf{M}$ as the relevant ray matrix. The lengths d_1, d_2 and f are determined according to the relations

$$1 - \tfrac{d_2}{f} = A, \quad 1 - \tfrac{d_1}{f} = D, \quad -\tfrac{1}{f} = C, \quad d_1 + d_2 - \tfrac{d_1 d_2}{f} = B. \tag{3.4.2}$$

As f may range from $-\infty$ to $+\infty$, whilst d_1 and d_2 are thought as nonnegative quantities, not all the possible values of the matrix entries A, B, C, D can be reached. In fact, only the two families of symplectic matrices:

$$\left(\begin{smallmatrix} A>1 & B>0 \\ C>0 & D>1 \end{smallmatrix}\right) \quad \text{and} \quad \left(\begin{smallmatrix} A<1 & B \\ C<0 & D<1 \end{smallmatrix}\right), \tag{3.4.3}$$

are synthesizable by the one lens configuration (3.4.1), according to whether the central lens is chosen as negative or positive.

The complementary two lens configuration

$$\left(\begin{smallmatrix} A & B \\ C & D \end{smallmatrix}\right) = \mathbf{L}(f_2)\mathbf{T}(d)\mathbf{L}(f_1) \tag{3.4.4}$$

is also possible, with the choice

$$1 - \frac{d}{f_1} = A, \quad 1 - \frac{d}{f_2} = D, \quad d = B, \quad -\frac{1}{f_1} - \frac{1}{f_2} + \frac{d}{f_1 f_2} = C, \tag{3.4.5}$$

which, on account of the restriction $d \geq 0$, allows to implement matrices having

$$\left(\begin{smallmatrix} A & B>0 \\ C & D \end{smallmatrix}\right). \tag{3.4.6}$$

As a simple example for the optical realization (3.4.4), we may consider a thick lens. On account of the equivalence between a refracting surface and a thin lens, having the same, respectively, refracting and focal power $\mathcal{P}$, it is evident that a thick lens can be realized by a pair of thin lenses spaced by the lens thickness; the ray matrix (3.2.4) can then be factorized as in (3.4.4).

3.4.2 *Special systems as TLT and LTL configurations*

It is interesting to investigate the potential of the previously described single and double lens configurations to implement matrices with one or both vanishing off-diagonal elements, of course, within the inherent restrictions. So doing, we gain a useful visualization of the special systems, which, abstractly introduced in § 1.8.4 as corresponding to ray matrices with one vanishing entry, are now seen to be realizable by thin lenses and free-propagation sections.

Telescopic and imaging systems

The LTL configuration (3.4.4) is appropriate to synthesize symplectic matrices having $C = 0$. In fact, the identity

$$\begin{pmatrix} A & B>0 \\ 0 & A^{-1} \end{pmatrix} = \mathbf{L}(f_2)\mathbf{T}(d)\mathbf{L}(f_1) \qquad (3.4.7)$$

is actuable with $d = f_1 + f_2 = B$ and $A = -f_2/f_1$.
We recall from § 1.8.4 that optical systems described by ray-matrices with $C = 0$ are named *afocal* or *telescopic* systems, as the direction of the emergent ray depends only on the direction of the incident ray. Thus any ray coming into the system parallel to the optical axis emerges parallel to the axis as well.

If $B = 0$, the single lens configuration (3.4.1) can be designed with $1/f = -C$, $d_2/d_1 = -A$ and $1/f = 1/d_1 + 1/d_2$ (*focal* condition) in order to give

$$\begin{pmatrix} A<0 & 0 \\ C<0 & A^{-1} \end{pmatrix} = \mathbf{T}(d_2)\mathbf{L}(f)\mathbf{T}(d_1), \qquad (3.4.8)$$

limited, of course, to negative values of the entries A and C. According to the focal relation, the intermediate lens is positive and can be understood as composed by two positive lenses with focal lenghts d_1 and d_2.
Optical systems with $B = 0$ are called *imaging* systems (or *magnifiers*), as the position of the outcoming ray depends only on the position of the incoming ray. Thus all the rays from the point q_i in the input plane are imaged into the point $q_o = Aq_i$ in the output plane (§ 1.8.4).

Symplectic upper and lower triangular matrices, as those in (3.4.7) and (3.4.8), span two subgroups of $Sp(2, \mathbb{R})$, with the diagonal matrices belonging to both. **T** and **L**-like matrices represent proper subsets of these subgroups.

As a conclusion, we may say that the TLT and LTL realizations (3.4.1) and (3.4.4) are appropriate to synthesize unimodular upper and lower triangular matrices, although the respective non null elements A, B, D and A, C, D can not arbitrarily be valued, as indicated by the restrictions in (3.4.7) and (3.4.8). Also, one and two lens configurations can not describe diagonal matrices, since both sides of (3.4.1) and (3.4.4) turn into the unit matrix when $B = 0$ and $C = 0$. In that case more complex configurations can be envisaged involving additional free-propagation sections and thin lenses besides those entering configurations (3.4.1) and (3.4.4) [3.5].

Collimating and focusing systems

Symplectic matrices having one zero along the diagonal, when regarded as optical matrices, are representative of collimating and focusing systems. In general, matrices with $D = 0$ describe the propagation from the primary focal plane of the system to the plane located by the axial position of the relative secondary principal plane, i.e. $\frac{s_{H'}}{n_{0_2}} = \frac{1-A}{C}$. Correspondingly, matrices

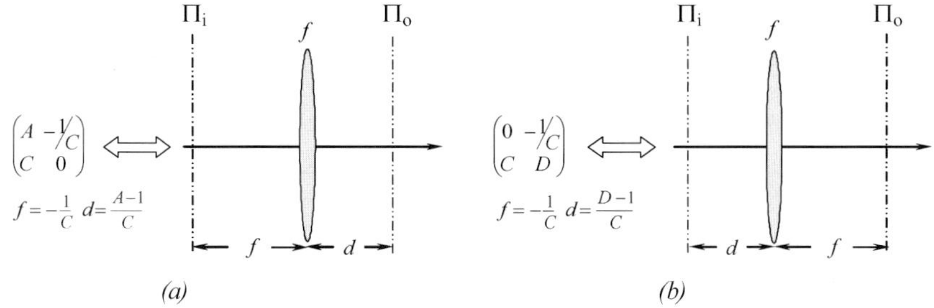

FIGURE 3.5. Optical realization of a symplectic matrix having *(a)* $D = 0$ (*collimating* system) and *(b)* $A = 0$ (*imperfect Fourier* transforming system).

with $A = 0$ account for the propagation from the plane located by the axial position of the primary principal plane as $\frac{s_H}{n_{0_1}} = \frac{1-D}{C}$ to the secondary focal plane of the system (§ 1.8.4).

Evidently, a $D = 0$ matrix can be realized by the one lens setup

$$\begin{pmatrix} A & -C^{-1} \\ C & 0 \end{pmatrix} = \mathbf{T}(d)\mathbf{L}(f)\mathbf{T}(f) \tag{3.4.9}$$

with taking

$$f = -\frac{1}{C}, \quad d = \frac{A-1}{C}, \tag{3.4.10}$$

which then demand for $C < 0$ and $A < 1$ in order that both $\mathbf{T}(d)$ and $\mathbf{T}(f)$ represent positive free propagation sections.

Likewise, to produce an $A = 0$ matrix, the reverse configuration can be arranged according to

$$\begin{pmatrix} 0 & -C^{-1} \\ C & D \end{pmatrix} = \mathbf{T}(f)\mathbf{L}(f)\mathbf{T}(d) \tag{3.4.11}$$

with

$$f = -\frac{1}{C}, \quad d = \frac{D-1}{C}, \tag{3.4.12}$$

thus requiring that $C < 0$ and $D < 1$.

The single lens decompositions (3.4.9) and (3.4.11) establish, within the specified restrictions, the equivalence of collimating and focusing systems to thin lenses, whose input and output planes be properly placed. In (3.4.9) the input is taken at the primary focal plane of the lens as the output at distance $d = \frac{A-1}{C}$ from the secondary principal plane, which for a thin lens is the lens plane (Fig. 3.5.a)). Reversely, in (3.4.11) the input is taken at the distance $d = \frac{D-1}{C}$ from the lens plane and the output at the secondary focal plane (Fig. 3.5.b)). Focusing systems perform the *imperfect* Fourier transform of the

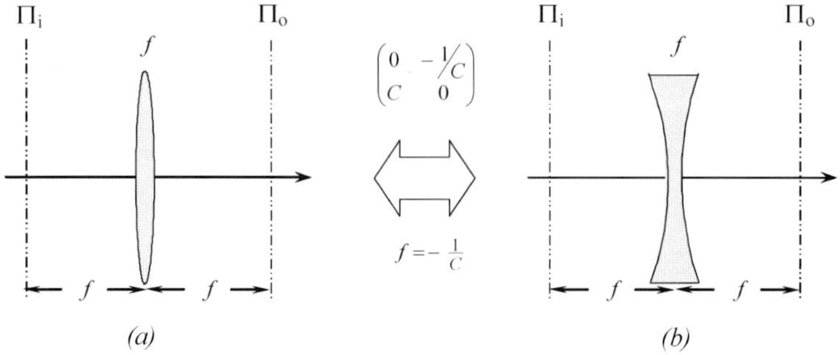

FIGURE 3.6. Single lens realization of an antidiagonal symplectic matrix (*perfect Fourier* transforming system). The "$2f$" system with *(a)* a positive lens and *(b)* a negative lens.

input signal, the amplitude distribution of the output signal being interpreted as the Fourier transform of the input multiplied by a quadratic phase term.

The further request $A = 0$ in (3.4.9), or $D = 0$ in (3.4.11), yields $d = f$, thus providing for the *perfect* Fourier transforming system the single lens realization

$$\mathbf{F}(f) = \begin{pmatrix} 0 & -1/C \\ C & 0 \end{pmatrix} = \mathbf{T}(f)\mathbf{L}(f)\mathbf{T}(f), \tag{3.4.13}$$

with $f = -\frac{1}{C}$. The antidiagonal matrix above describes in general the optical propagation from the primary focal plane of the system to the secondary one, on which the perfect Fourier transform of the ingoing signal is formed (§ 1.8.4). According to (3.4.13), the simplest realization of such a system consists of a positive thin lens, with the light rays being propagated from the front to the back focal plane (see the scheme in Fig. 3.6.a) for the exemplary case of a bi-convex lens). In principle, the TLT design (3.4.13) should strictly admit only positive lenses, corresponding to negative values of C in the given matrix. However, negative values of f are allowed as well. The resulting optical configuration is depicted in Fig. 3.6.b) for the illustrative case of a bi-concave lens. Taking into account the meaning of the sign of the focal length f, which, in accord with the common conventions (§§ 1.8.1 and 3.3), establishes the character, real or virtual, of the object and image rays, the reader may easily verify that the formal representation for the optical setup in the figure turns to be just the TLT product (3.4.13), with f retaining its own sign in the $\mathbf{T}$-matrices. Thus, unimodular antidiagonal matrices can be realized by use of one lens of focal length $f = -\frac{1}{C}$, placed midway between two reference planes separated by $2\,|f|$. This configuration is reported as the $2f$ *system*.

Finally we note that the double lens configuration (3.4.4) can also be valued to correspond to a ray matrix with $D = 0$, $A = 0$, or $A = D = 0$. In particular,

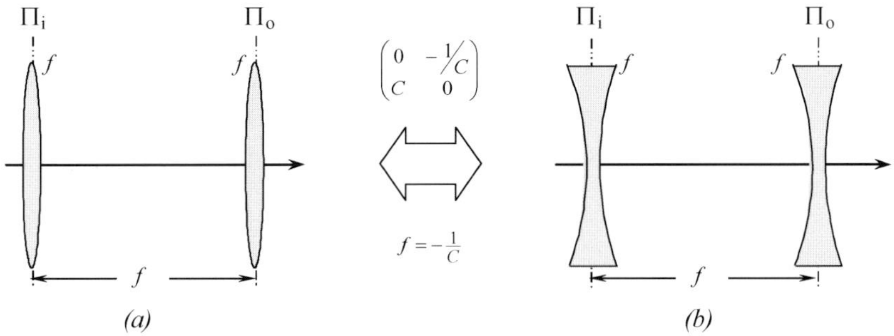

FIGURE 3.7. Double lens realization of an antidiagonal symplectic matrix (*perfect Fourier transforming system*). The *Fourier tube* with *(a)* positive lenses and *(b)* negative lenses.

in Fig. 3.7 we show the double lens setups (also known as *Fourier tubes*), suited to perform the perfect Fourier transform of the object distribution.

3.4.3 *Pure magnifiers*

The basic one and two lens configurations (3.4.1) and (3.4.4) are not suited to realize unimodular purely diagonal matrices, which within the optical context are interpreted as the ray transfer matrices of *pure magnifiers* (§ 1.8.4). Either additional thin lenses or additional free propagation sections or both must be added to the optical arrangements (3.4.1) and (3.4.4), thus introducing further parameters, which can be adjusted in order that both the off-diagonal elements in the resulting ray matrix may vanish.

We consider the optical identity (3.4.8) for the imaging matrix, which for convenience is rewritten as

$$\begin{pmatrix} A{<}0 & 0 \\ C{<}0 & A^{-1} \end{pmatrix} = \mathbf{T}(d_2)\mathbf{L}(f_1)\mathbf{T}(d_1), \tag{3.4.14}$$

with

$$\frac{1}{f_1} = \frac{1}{d_1} + \frac{1}{d_2} = -C, \qquad A = -\frac{d_2}{d_1}. \tag{3.4.15}$$

Concatenating the TLT sequence (3.4.14) with a thin lens on the left or right, the lower-left entry of the resulting optical matrix may vanish. Accordingly, we may realize the pure magnifier matrix by the optical arrangement

$$\mathbf{D}(\mathcal{M}) \equiv \begin{pmatrix} \mathcal{M} & 0 \\ 0 & \mathcal{M}^{-1} \end{pmatrix} = \mathbf{L}(f_2)\mathbf{T}(d_2)\mathbf{L}(f_1)\mathbf{T}(d_1), \tag{3.4.16}$$

the magnification $\mathcal{M}$ being determined by the parameters f and d as

$$\mathcal{M} = -\frac{d_2}{d_1} = -\frac{f_2}{f_1}, \quad \frac{1}{f_1} = \frac{1}{d_1} + \frac{1}{d_2} \tag{3.4.17}$$

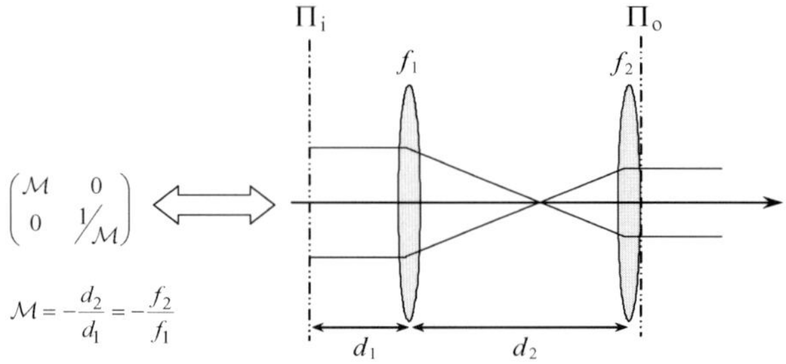

FIGURE 3.8. Two-lens realization of a pure magnifier.

which also yield $d_2 = f_1 + f_2$ (Fig. 3.8).
The magnification $\mathcal{M}$ is clearly negative, thus signifying an inversion of the ray position with respect to the optical axis. We can obtain a positive magnification by cascading two magnifiers like (3.4.16). In fact, being a striking property of diagonal matrices that they multiply into diagonal matrices as

$$\begin{pmatrix} \mathcal{M}_1 & 0 \\ 0 & \mathcal{M}_1^{-1} \end{pmatrix} \begin{pmatrix} \mathcal{M}_2 & 0 \\ 0 & \mathcal{M}_2^{-1} \end{pmatrix} = \begin{pmatrix} \mathcal{M} & 0 \\ 0 & \mathcal{M}^{-1} \end{pmatrix} \tag{3.4.18}$$

with $\mathcal{M} = \mathcal{M}_1 \mathcal{M}_2$, we can have $\mathcal{M} > 0$ with both $\mathcal{M}_1 < 0$ and $\mathcal{M}_2 < 0$.

Equations (3.4.16) and (3.4.18) are basic to our purposes, because they represent the proof that purely diagonal symplectic matrices can be synthesized by concatenating thin lenses and free propagation sections.

The reader can easily verify that the same optical configuration as in (3.4.16) may be obtained adding a free propagation section to the right of the double lens realization (3.4.7) of an afocal matrix.

Pure magnifiers and Fourier transforming systems

Remarkably, a purely magnifying configuration can also be reached by enlarging the two lens configuration (3.4.16) to comprise an additional free propagation section, thus yielding

$$\mathbf{D}(\mathcal{M}) = \mathbf{T}(d_2)\mathbf{L}(f_2)\mathbf{T}(d)\mathbf{L}(f_1)\mathbf{T}(d_1), \tag{3.4.19}$$

where the involved lengths are related through

$$\mathcal{M} = -\tfrac{f_2}{f_1} = 1 - \tfrac{d}{f_1}, \qquad d + d_1\mathcal{M} + \tfrac{d_2}{\mathcal{M}} = 0. \tag{3.4.20}$$

Preserving the restriction of $0 \leq d_1, d, d_2 < \infty$ in order that the $\mathbf{T}$-matrices strictly correspond to free propagation sections, the magnification $\mathcal{M}$, satisfying (3.4.20), comes to be negative and hence the focal lenghts f_1 and f_2 are required to be both positive or both negative.

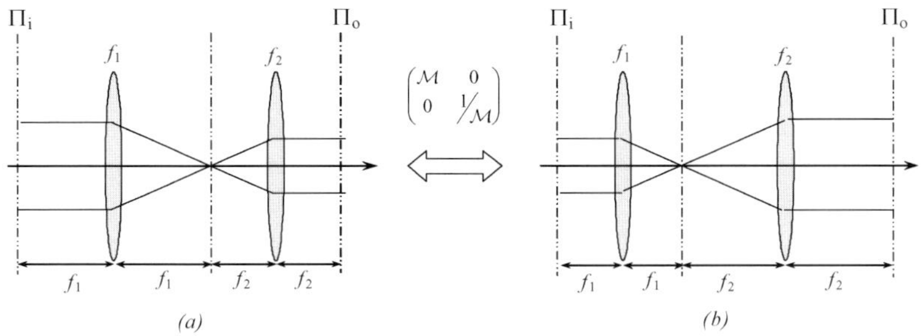

FIGURE 3.9. Optical realization of a *pure magnifier* with $\mathcal{M} < 0$ by cascading two Fourier transforming systems. *(a)* $|\mathcal{M}| < 1$ *(b)* $|\mathcal{M}| > 1$.

We observe that the second of (3.4.20) may be satisfied by setting $d_1 = f_1$, $d_2 = f_2$ and accordingly $d = f_1 + f_2$. The resulting two lens configuration turns to be the sequence of two $2f$ systems, described in § 3.4.2 and pictured in Fig. 3.6.a). Therefore we write

$$\mathbf{D}(\mathcal{M}) = \mathbf{F}(f_2)\mathbf{F}(f_1), \quad \mathcal{M} \equiv -\tfrac{f_2}{f_1}. \tag{3.4.21}$$

On account of the considerations of § 3.4.2 concerning the interpretation of the sign of the focal lengths in the implementation of the Fourier transform by thin lenses, the focal lengths f_1 and f_2 above can be let to be arbitrarily positive or negative, thus allowing for both negative and positive magnifications.

According to (3.4.21), a pure magnifier can be realized by using two lenses, spaced so as to have a common focal point, and embraced by two free-medium sections, whose lengths are adjusted according to the focal length of the corresponding adjacent lens. Thus, in the whole system both the effective length and the optical power vanish. As an example, Fig. 3.9 shows two possible realizations of an $\mathcal{M} < 0$ configuration by concatenating two Fourier transformers, both having positive lenses; the respective performances are pictured for an incoming bundle of parallel rays. The propagation through both systems manifests an inversion of the ray-coordinate with respect to the optical axis along with a magnifying or demagnifying effect according to whether $\frac{f_2}{f_1} > 1$ or $\frac{f_2}{f_1} < 1$. In particular, we emphasize that two identical Fourier transformers in sequence produce a pure inversion of the ray coordinates, in accord with the general property of the Fourier transform. Then, a ray emerging from a sequence of four identical Fourier transforming systems has the same coordinates (position and inclination) as when entering to (see also § 3.6.1).

In contrast, $\mathcal{M} > 0$ configurations demand for lenses of opposite signs. Figure 3.10 shows two possible realizations. The path of an ingoing bundle of

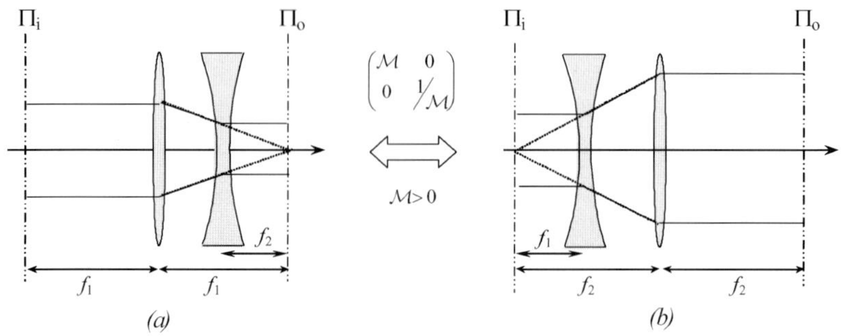

FIGURE 3.10. Optical realization of a *pure magnifier* with $\mathcal{M} > 0$ by cascading two Fourier transforming systems. *(a)* $\mathcal{M} < 1$ *(b)* $\mathcal{M} > 1$.

parallel rays is traced, showing the contrasting effects of the two systems, the first producing a reduction ($\mathcal{M} < 1$) of the transverse size of the bundle, the second, on the contrary, an expansion ($\mathcal{M} > 1$).

The optical configurations pictured in Figs. 3.9 and 3.10 may then be understood as the optical analogs of physical processes, described by 2×2 symplectic purely diagonal matrices with negative and positive entries, respectively. Also, as a positive magnifier can be arranged by cascading two negative magnifying systems, four positive Fourier transforming configurations can be concatenated to synthesize the optical analog of any physical process described by a 2×2 symplectic diagonal positive matrix.

Finally, we note that the configuration (3.4.16) can also be enlarged to comprise an additional thin lens thus leading to the three-lens setup

$$\mathbf{D}(\mathcal{M}) = \mathbf{L}(f_2)\mathbf{T}(d_2)\mathbf{L}(f)\mathbf{T}(d_1)\mathbf{L}(f_1). \tag{3.4.22}$$

We invite the reader to find out the explicit expressions linking the involved parameters d_1, d_2, f_1, f_2 and f. Also, we suggest to prove the realizability of the configuration with $f_1 = d_1$, $f_2 = d_2$, $\frac{1}{f} = \frac{1}{f_1} + \frac{1}{f_2}$ and accordingly $\mathcal{M} = -\frac{f_2}{f_1}$, which is interpreted as the sequence of two Fourier tubes.

The subgroup of pure magnifiers

The ideal magnifiers form a proper subgroup of the set of $1D$ linear optical systems. This straightforwardly follows from the properties of the unimodular diagonal matrices, we have denoted by $\mathbf{D}(\mathcal{M})$. In fact, as already shown, the product of two matrices in the set $\{\mathbf{D}(\mathcal{M}), \mathcal{M} \text{ real}\}$ belongs to the set as well, and hence just rewriting (3.4.18) we have

$$\mathbf{D}(\mathcal{M}_1)\mathbf{D}(\mathcal{M}_2) = \mathbf{D}(\mathcal{M}_2)\mathbf{D}(\mathcal{M}_1) = \mathbf{D}(\mathcal{M}), \quad \mathcal{M} = \mathcal{M}_1\mathcal{M}_2. \tag{3.4.23}$$

The inverse of a pure magnifier is the magnifier with reciprocal magnification:

$$[\mathbf{D}(\mathcal{M})]^{-1} = \mathbf{D}(\mathcal{M}^{-1}), \tag{3.4.24}$$

and the unit matrix **I** can be interpreted as the optical matrix of the magnifier with $\mathcal{M} = 1$:

$$\mathbf{D}(1) = \mathbf{I}. \tag{3.4.25}$$

We note that, according to (3.3.7) and (3.3.8), the unit matrix can be optically synthesized by cascading two, positive and "negative", free propagation sections or two thin lenses having opposite focal lenghts. Due to (3.4.25), the identity matrix can also be implemented by cascading two positive or two negative pure magnifiers with $\mathcal{M}_2 = \frac{1}{\mathcal{M}_1}$, and hence it can be realized as two, positive and negative, Fourier tubes nested one into the other or as four positive Fourier transforming systems concatenated one to the other with the relative focal distances being related by $f_1 f_3 = f_2 f_4$.

Pure magnifiers without inversion

Positive ideal magnifiers constitute a proper subgroup of the group of unimodular purely diagonal matrices $\{\mathbf{D}(\mathcal{M}), \mathcal{M} \text{ real}\}$, because cascading two positive magnifiers produces a positive magnifier. In § 2.4.3 we have elaborated for the pure magnifiers without inversion the exponential representation, rewritten in (3.1.2), involving the traceless matrix $\mathbf{K}_3$, representative of the mixed quadratic monomial $\frac{1}{2}pq$. As we will see later, under specific restrictions the scale matrix $\mathbf{S}(m)$ plays a crucial role as a building "block" for the representation of optical matrices in a factored product form involving as well the **T** and **L** matrices. Therefore, we further comment here on the system $\mathbf{S}(m)$, suggesting in particular some possible optical realizations of it.

We may rephrase relatively to the scaling matrix

$$\mathbf{S}(m) \equiv e^{s\mathbf{K}_3} = \begin{pmatrix} m & 0 \\ 0 & m^{-1} \end{pmatrix}, \qquad m \equiv e^{s/2} \tag{3.4.26}$$

the formal considerations developed in § 3.3 in respect to the free propagation section and thin lens matrices, $\mathbf{T}(d)$ and $\mathbf{L}(f)$, thus completing the view of the three one parameter subgroups of $Sp(2, \mathbb{R})$, generated by the algebra basis $\{\mathbf{K}_-, \mathbf{K}_+, \mathbf{K}_3\}$ by exponentiation through a real parameter.

Evidently, **S**-like matrices constitute a proper subgroup of $Sp(2, \mathbb{R})$; more precisely, the set $\{\mathbf{S}(m), m > 0\}$ is included into the subgroup $\{\mathbf{D}(\mathcal{M}), \mathcal{M} \text{ real}\}$ of unimodular diagonal matrices $\mathbf{D}(\mathcal{M})$. Then, as particular cases of (3.4.23) and (3.4.24), we can state the closure property

$$\mathbf{S}(m_1)\mathbf{S}(m_2) = \mathbf{S}(m_2)\mathbf{S}(m_1) = \mathbf{S}(m), \qquad m = m_1 m_2 > 0 \tag{3.4.27}$$

for matrices (3.4.26), and hence the existence and unicity of the inverse being

$$[\mathbf{S}(m)]^{-1} = \mathbf{S}(m^{-1}). \tag{3.4.28}$$

In addition, the previously developed considerations concerning the optical realization of pure-magnification devices with $\mathcal{M} > 0$ apply to the system (3.4.26) as well. Thus, the symplectic diagonal matrices, generated by the algebra matrix $\mathbf{K}_3$, can optically be realized by concatenating two systems like (3.4.16) or two Fourier transformers (3.4.13), the relevant focal lengths being related to the magnification m as

$$f_2 = -mf_1 \, . \tag{3.4.29}$$

An example is given in Fig. 3.10. Also, we can cascade two negative magnifiers to synthesize the matrix (3.4.26), the resulting configuration may be regarded as optical analogs of the matrix $e^{s\mathbf{K}_3}$, the involved focal lengths f_1, f_2, f_3, f_4 being suitably chosen to satisfy the relation

$$f_2 f_4 = e^{s/2} f_1 f_3. \tag{3.4.30}$$

3.4.4 *Symplectic matrices as optical matrices*

Appropriate conjuctions of $\mathbf{T}, \mathbf{L}$ and $\mathbf{D}$-like matrices can be arranged to be equivalent to any given symplectic matrix, excluding the antidiagonal matrices, for which, as discussed in § 3.4.2, the $2f$ systems and the Fourier tubes provide the suitable optical realizations. In this connection, it is worth emphasizing once again the difference, relatively to the group property, of the matrices $\mathbf{T}$'s according to whether they are regarded as purely mathematical entities or interpreted as formal tools for modelling optical systems. Mathematically, the 2×2 unimodular real upper triangular matrices (3.3.1) do form a subgroup of the symplectic group $Sp(2, \mathbb{R})$. In contrast, optically they represent ray-propagation through portions of a homogeneous medium, and hence do not form a subgroup because the inverse of a free-propagation section cannot be synthesized as a free-propagation section as well; it should demand in fact for sections of "negative" length and thereby is not physically realizable. In the following we regard $\mathbf{T}(d)$ just as the upper triangular matrix (3.3.1), optically synthesizable by portions of a homogeneous medium or by more complex arrangements of lenses separated by uniform medium sections according to whether the parameter d takes on a positive or negative value.

It is easily proved that every symplectic matrix with non vanishing upper-left entry can be realized as the ray matrix of an optical system, composed by the sequence of a free propagation, a pure magnifier and a thin lens, namely

$$\begin{pmatrix} A\neq 0 & B \\ C & D \end{pmatrix} = \mathbf{L}(f)\mathbf{D}(\mathcal{M})\mathbf{T}(d) \tag{3.4.31}$$

with the parameters d, $\mathcal{M}$ and f being related to the entries of the matrix as

$$d = \tfrac{B}{A}\,, \quad \mathcal{M} = A\,, \quad f = -\tfrac{A}{C}. \tag{3.4.32}$$

Needless to say, the optical arrangement on the right of (3.4.31) is basically composed by thin lenses and free propagation sections.

Interestingly, since **D** represents an optical system with vanishing effective length and focal power, the presence of the matrices **T** and **L** in the synthesis (3.4.31) directly relates to the entries B and C, respectively. Thus, if $B = 0$ or $C = 0$, correspondingly **T** or **L** disappears from (3.4.31).

If $A = 0$, and hence $D \neq 0$, the reader may easily verify that the reverse ordering TDL may be effectively applied:

$$\begin{pmatrix} 0 & B \\ C & D\neq 0 \end{pmatrix} = \mathbf{T}(\tfrac{B}{D})\mathbf{D}(\tfrac{1}{D})\mathbf{L}(-\tfrac{D}{C}), \qquad B = -\tfrac{1}{C}. \tag{3.4.33}$$

However, following the suggestion put forward in § 3.4.2 we can verify the feasibility of the optical equivalence

$$\begin{pmatrix} 0 & -C^{-1} \\ C & D \end{pmatrix} = \mathbf{F}(-\tfrac{1}{C})\mathbf{T}(\tfrac{D}{C}), \tag{3.4.34}$$

which unlike (3.4.33) comprises the case $D = 0$ as well. On account of the optical realization of $\mathbf{F}(f)$ by a $2f$ system, it is evident that the product in (3.4.34) describes the propagation from an input plane, placed at distance d from the primary focal plane of the lens, to the secondary focal plane. The presence of $\mathbf{T}(d)$ reveals in fact that system (3.4.34) performs the imperfect Fourier transform of the input signal, and so marks the difference with respect to a perfect Fourier transforming system, corresponding to $d = 0$, i.e. $D = 0$.

Equation (3.4.31) and all others, obtainable changing the order of the optical components in the product configuration on the right, represent the conclusive step of the investigation presented in this section. We may accordingly state that every 2×2 symplectic matrix can arise as an optical matrix; in other words, every $Sp(2,\mathbb{R})$ system can be realized as an optical system composed by a suitably arranged and characterized set of thin lenses separated by (positive) free propagation sections. This establishes the homomorphism between the symplectic group $Sp(2,\mathbb{R})$ and the group of the $1D$ linear optical systems.

3.5 Wei-Norman representation of optical elements: LST synthesis

In the previous section it has been shown that every real symplectic matrix with non vanishing diagonal is expressible in a product form involving the matrices **T**, **L** and **D**. In general, the matrix **T** is to be considered from a purely mathematical viewpoint as a real unimodular upper-triangular matrix, admitting the exponential representation in terms of the algebra matrix $\mathbf{K}_-$, and hence describing from the optical viewpoint a single free-propagation section or a suitable arrangement of free-medium sections and thin lenses.

Here we will see that, resorting to the well known Wei-Norman theorem of Lie group theory [4.1], symplectic matrices having a positive A or D entry admit an ordered product form representation in terms of $\mathbf{T}$, $\mathbf{L}$ and $\mathbf{S}$-like matrices; namely in terms of a upper triangular matrix with unit diagonal entries, a unimodular positive-definite diagonal matrix, and a lower triangular matrix with unit diagonal entries. In fact, the optical arrangement

$$\begin{pmatrix} A & B \\ C & D \end{pmatrix} = \mathbf{L}(\mathcal{P})\mathbf{S}(m)\mathbf{T}(d) \tag{3.5.1}$$

demands for the parameters $\mathcal{P}$, m and d be related to the matrix elements by

$$\mathcal{P} = -\tfrac{C}{A}, \quad m = A, \quad d = \tfrac{B}{A}, \tag{3.5.2}$$

just as (3.4.32) with $\mathcal{M}$ replaced by m. As $m > 0$, identity (3.5.1) is admissible only for matrices with positive diagonal entry A, while D is in general allowed to range from $-\infty$ to ∞. The reader may verify that the factorizations where the matrix $\mathbf{L}$ is on the left of the matrix $\mathbf{T}$ (i.e., the LST, SLT and LTS configurations) are suitable to synthesize symplectic matrices having $A > 0$ and $-\infty < D < \infty$. In contrast, the factorizations where the matrix $\mathbf{L}$ is on the right of $\mathbf{T}$ (which lead to the TSL, STL and TLS configurations) can account for $D > 0$ and A arbitrarily ranging through the real line.

The equivalence expressed in (3.5.1) as well as in all the other factored forms, obtainable changing the order of the component matrices in the product on the right, states the possibility of representing every $Sp(2,\mathbb{R})$ element with positive upper-left or lower-right entry as the ordered product of elements drawn from the one-parameter subgroups generated by the corresponding algebra matrices $\mathbf{K}_-$, $\mathbf{K}_+$, and $\mathbf{K}_3$. Thus, on account of the exponential representations of the matrices involved in the product, we can write

$$\begin{pmatrix} A & B \\ C & D \end{pmatrix} = e^{\mathcal{P}\mathbf{K}_+} e^{s\mathbf{K}_3} e^{d\mathbf{K}_-}, \qquad A > 0 \tag{3.5.3}$$

with $\mathcal{P}$, s and d defined by (3.5.2); namely, $\mathcal{P} = -\frac{C}{A}$, $s = 2\ln A$, $d = \frac{B}{A}$.

Accordingly, every 2×2 symplectic matrix with positive upper-left or lower-right entry can be interpreted as an optical matrix arising from the appropriate conjuctions of (positive or negative) free-propagations, thin lenses and reciprocal scaling systems. Within the context of $1D$ linear optics the decomposition (3.5.3) acquires an utmost relevance when approaching propagation problems. In that case the initial values of the ray matrix entries are set in order to reproduce the identity matrix. The continuity of the process assures that in a

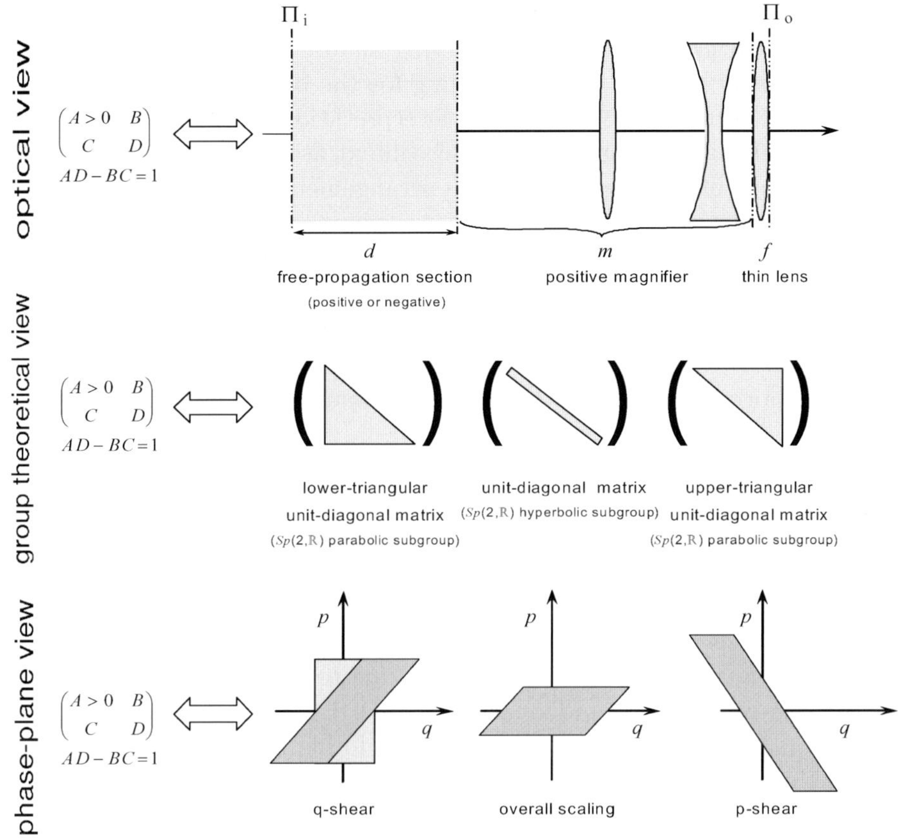

FIGURE 3.11. Wei-Norman synthesis of an optical matrix.

neighbourhood of the initial position, i.e. in a neighbourhood of the identity, the entries A and D remain greater than zero. The representation (3.5.3) is therefore "locally" allowed in that neighbourhood, where we can write

$$\mathbf{M}(z, z_i) = e^{\mathcal{P}(z)\mathbf{K}_+} e^{s(z)\mathbf{K}_3} e^{d(z)\mathbf{K}_-}, \qquad \mathbf{M}(z_i, z_i) = \mathbf{I}, \tag{3.5.4}$$

thus decomposing the overall problem of determining the transfer matrix $\mathbf{M}(z, z_i)$ into the "partial" problems of determining the parameters $\mathcal{P}(z)$, $s(z)$ and $d(z)$. Notably the factors in (3.5.4) directly relate to the "dynamics" associated respectively with the quadratic monomials $\frac{1}{2}q^2$, $\frac{1}{2}qp$ and $\frac{1}{2}p^2$, which can easily be investigated and possibly expressed through definite functional forms of the relevant characteristic parameters $\mathcal{P}(z)$, $s(z)$ and $d(z)$ (see §§ 1.5.2 and 2.4). As an example, when considering the propagation in a parabolic index profile medium, we will see that the differential equations for the matrix elements A, B, C, D turn into differential equations for the parameters $\mathcal{P}$, s, d of the decomposition (3.5.4), which, offering an alternative parametrization of the problem, may provide a deeper feeling for the behaviour of the system.

Figure 3.11 summarizes the various interpretations of the LST decomposition (3.5.1) within the group theoretical context as well as within the optical context in terms of both optical system arrangements and phase-plane transformation sequences.

Moreover, as in the linear approximation the ray transfer matrix $\mathbf{M}$ is the matrix representative of the ray transfer operator $\widehat{\mathfrak{M}}$, relation (3.5.4) also states that the ray transfer operator of any one first-order $1D$ optical system is factorizable in a neighbourhood of the entrance plane into the product of the Lie transformations generated by the monomials $\frac{p^2}{2}$, $\frac{qp}{2}$ and $\frac{q^2}{2}$ as

$$\widehat{\mathfrak{M}}(z, z_i) = e^{\mathcal{P}(z)\widehat{\mathrm{K}}_+} e^{s(z)\widehat{\mathrm{K}}_3} e^{d(z)\widehat{\mathrm{K}}_-}, \qquad \widehat{\mathfrak{M}}(z_i, z_i) = \widehat{\mathrm{I}}, \tag{3.5.5}$$

where $\widehat{\mathrm{K}}_-$, $\widehat{\mathrm{K}}_+$ and $\widehat{\mathrm{K}}_3$ are the Lie operators associated with the quoted monomials (see §§ 1.5.2 and 2.3) [4].

Restrictly to the linear approximation the matrix and operator identities, (3.5.4) and (3.5.5), provide the answer to the problem of z-ordering, which unavoidably arises when approaching the integration of Hamilton's equations in the case of a z-depending Hamiltonian, due to the in general noncommutative nature of matrices and operators. In § 1.5.3 we mentioned that the factorization of $\widehat{\mathfrak{M}}$ can be continued to include higher order terms, accounting for the aberrations. Each of the additional factors takes the form of a Lie transformation, although it cannot be given a 2×2 matrix representation.

3.6 Rotations and squeezes in the phase plane

In § 1.5.2, when discussing the phase-space dynamics generated by the Lie transformations associated with dynamical functions quadratic in the position and momentum ray variables, we have considered the polynomials $\frac{1}{2}(p^2 + q^2)$ and $\frac{1}{2}(p^2 - q^2)$. We have showed that they give rise to definitely different motions in the phase plane, the corresponding trajectories lying respectively along a circle and a branch of hyperbola. Here we reconsider in detail the above bynomials in order to identify the optical systems they may describe.

3.6.1 *Attractive oscillator, phase-plane rotation and fractional Fourier transform*

We denote as $H_{\text{a.o.}}$ the quadratic homogeneous polynomial

$$H_{\text{a.o.}} = \tfrac{1}{2}p^2 + \tfrac{1}{2f_S^2}q^2, \tag{3.6.1}$$

which typically models the attractive oscillator-like dynamics. The space coordinate q has been explicitly scaled by an appropriate length factor f_S, whose optical interpretation is clarified below.

The corresponding Lie operator $\widehat{\mathsf{L}}_{\text{a.o.}}$ is

$$\widehat{\mathsf{L}}_{\text{a.o.}} = p\frac{\partial}{\partial q} - \frac{q}{f_S^2}\frac{\partial}{\partial p}, \tag{3.6.2}$$

which we rewrite in terms of the operators $\widehat{\mathrm{K}}_-$ and $\widehat{\mathrm{K}}_+$, defined by (2.3.5), in the convenient form

$$\widehat{\mathsf{L}}_{\text{a.o.}} = \tfrac{1}{f_S}(f_S\widehat{\mathrm{K}}_- + \tfrac{1}{f_S}\widehat{\mathrm{K}}_+). \tag{3.6.3}$$

Accordingly the matrix representative $\mathbf{H}_{\text{a.o.}}$ linearly relates to the $sp(2,\mathbb{R})$ basis matrices $\mathbf{K}_-$ and $\mathbf{K}_+$ as

$$\mathbf{H}_{\text{a.o.}} = \begin{pmatrix} 0 & 1 \\ -1/f_S^2 & 0 \end{pmatrix} = \frac{2}{f_S}\mathbf{K}_1(f_S), \tag{3.6.4}$$

where we have denoted by $\mathbf{K}_1(f_S)$ the algebra element

$$\mathbf{K}_1(f_S) = \tfrac{1}{2}(f_S\mathbf{K}_- + \tfrac{1}{f_S}\mathbf{K}_+) = \tfrac{1}{2}\begin{pmatrix} 0 & f_S \\ -1/f_S & 0 \end{pmatrix}. \tag{3.6.5}$$

The convenience of such a seemingly odd definition will become apparent later.

In order to identify the optical system described by the harmonic oscillator-like Hamiltonian (3.6.1), we exponentiate the traceless matrix (3.6.4) through the axial distance $\Delta z = z - z_i$. As a result, we obtain the "transfer matrix" for the ray propagation from z_i to z in the form[a]

$$\mathbf{F}^{\alpha}(f_S) = e^{\Delta z\,\mathbf{H}_{\text{a.o.}}} = e^{2\phi\mathbf{K}_1(f_S)} = \begin{pmatrix} \cos\phi & f_S\sin\phi \\ -\frac{1}{f_S}\sin\phi & \cos\phi \end{pmatrix}, \tag{3.6.6}$$

having interpreted the dimensionless argument $\frac{\Delta z}{f_S}$ of the circular functions as the angle ϕ, commonly measured in unity of $\frac{\pi}{2}$ through the parameter α:

$$\phi(z) = \frac{z - z_i}{f_S} \equiv \alpha(z)\frac{\pi}{2}. \tag{3.6.7}$$

Note that, as f_S has the dimension of length, the off-diagonal entries in (3.6.6) have the dimensions of length and $(\text{length})^{-1}$ respectively, as it should be for an optical matrix in the ray-coordinate and momentum representation.

---

[a] As in the case of the matrices $\mathbf{K}_-$, $\mathbf{K}_+$ and $\mathbf{K}_3$ (see § 2.4), we may compute the exponential function of the matrix $\mathbf{K}_1(f_S)$ (as well of $\mathbf{K}_2(f_S)$ below) by simply applying the exponential power series, since powers of $\mathbf{K}_1(f_S)$ (and $\mathbf{K}_2(f_S)$) are easily computed. The interested reader may consult [14] in ch. 2 for a general account of the methods of computing exponential functions of matrices.

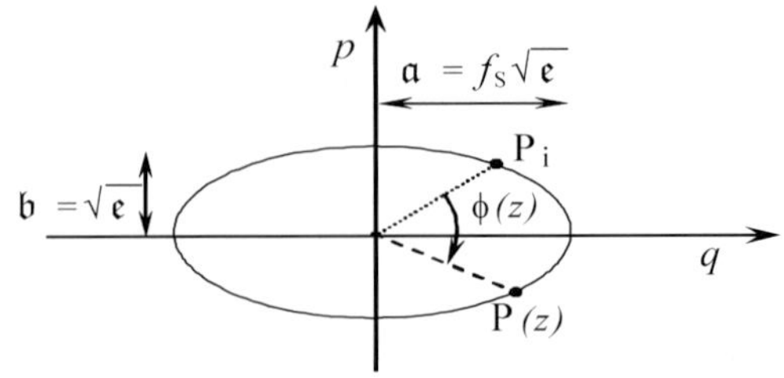

FIGURE 3.12. Phase-plane dynamic generated by the attractive oscillator Hamiltonian (3.6.1). The ray representative point slips along the ellipse determined by the initial data and the family parameter f_S.

The transformation acted by the matrix $\mathbf{F}^\alpha(f_S)$ on the phase-plane variables (q, p) from the initial data (q_i, p_i) is described by

$$\begin{aligned} q(z) &= \quad q_i \cos\phi + p_i f_S \sin\phi \,, \\ p(z) &= -\tfrac{1}{f_S} q_i \sin\phi + p_i \cos\phi \,, \end{aligned} \tag{3.6.8}$$

which is of course in accord with result (1.5.23).

The cyclic nature of the phase plane motion (3.6.8) is apparent. As the ray propagates through the "optical" system (3.6.6), the representative point P in the q-p plane moves along the ellipse

$$\frac{q(z)^2}{f_S^2} + p(z)^2 = \mathfrak{e}, \qquad \mathfrak{e} \equiv \frac{q_i^2}{f_S^2} + p_i^2 . \tag{3.6.9}$$

The constant $\mathfrak{e}$ is determined by the parameter f_S and the initial values; $\phi(z)$ specifies for each value of the axial parameter z the angle of the rotation of the corresponding representative point $P(q(z), p(z))$ with respect to the initial point $P_i(q_i, p_i)$ (Fig. 3.12). The rotation is clockwise or counterclockwise according to the sign of f_S.

Evidently the matrix $\mathbf{F}^\alpha(f_S)$ relates to a rotation by $\phi = \alpha\frac{\pi}{2}$ along an in general elliptical contour in the phase-plane. In fact, as will be clarified later, it can basically be interpreted as a phase-plane rotator. In particular, with $f_S = 1$, $\mathbf{F}^\alpha(1)$ takes the appearance of a pure rotation matrix by the angle ϕ:

$$\mathbf{F}^\alpha = \begin{pmatrix} \cos\phi & \sin\phi \\ -\sin\phi & \cos\phi \end{pmatrix} = e^{2\phi \mathbf{K}_1}, \tag{3.6.10}$$

where $\mathbf{K}_1 \equiv \mathbf{K}_1(1) = \frac{\mathbf{K}_- + \mathbf{K}_+}{2}$, the unit argument being omitted. Accordingly the representative point moves along a circle centered around the origin with radius $r_C = \sqrt{q_i^2 + p_i^2}$.

Interestingly, with $\phi = \frac{\pi}{2}$ (i.e. $\alpha = 1$) we recover the perfect Fourier transform matrix (3.4.13),

$$\mathbf{F}(f_S) = \begin{pmatrix} 0 & f_S \\ -1/f_S & 0 \end{pmatrix} = e^{\pi \mathbf{K}_1(f_S)}, \tag{3.6.11}$$

(the unit order being omitted as well), which then is seen to relate to the phase plane rotation by $\pi/2$.

The matrix $\mathbf{F}^{\alpha}(f_S)$ is reported in the literature as the *fractional Fourier transform* matrix of order α and family parameter f_S.

The term "fractional" evokes the charming vision of the "extension to continuum" of a process that occurs through finite-size steps. It is in fact in conformity with this "idea of fragmentation" that the fractional Fourier transform has been introduced within different contexts of both mathematics and physics. Condon, for instance, investigated about a continuous group of functional transformations isomorphic with the group of rotations of a plane about a fixed point by multiples of an arbitrary angle, thus generalizing the property of the Fourier transform group, which in fact corresponds to rotations by multiples of the right angle [5.1]. Likewise, Namias elaborated the functional form of Fourier-like operators, admitting, as the Fourier transform, the Hermite-Gauss eigenfunctions, but with eigenvalues evenly separated by a *fraction* of the imaginary unit [5.2].

The definition of the Fourier transform of fractional order within the optical context has been inspired by the *performance* of an optical system perceived as obtainable by fragmenting the Fourier transform configuration into a larger and larger number of shorter and shorter free-space sections interleaved with weaker and weaker lenses [5.3], or equivalently as capable of extending to arbitrary angles the property of the Fourier transforming system to rotate by a right angle the Wigner chart in the inherent Wigner phase plane [5.4].

The rotation by a right angle in the optical phase plane is the underlying canonical transformation acted by the Fourier operation on the conjugate variables q and p. Letting the rotation angle vary continuously leads to the fractional Fourier operation. Thus, the finite-step transformations marked by multiples of the right angle, corresponding to Fourier transform systems possibly concatenated one to the other, become specific events within the smooth evolution of the process governed by the continuously varying angle $\phi(z)$ over the 2π range, and so by the continuously varying axial parameter z. Such a phase plane rotation by a continuously varying angle may be achieved by sectioning the basic Fourier transforming configurations into an increasing number of similar configurations appropriately designed, as we will see later. The relation, and the relevant visualization in the optical phase plane, between the Fourier transform and the Wigner distribution function is clarified in § 8.4.1.

After its introduction into the field of optics, the concept of fractional Fourier transform and the relevant formalism gave rise to a huge variety of applications, investigations and new formulations in an increasingly enriched optics scenario. In the forthcoming paragraphs we will simply clarify the concept of fractional Fourier transform at a very basic level, addressing the reader to the related literature for a wider and deeper treatment [5].

Fractional Fourier transform matrices as a symplectic subgroup

Remarkably, for a fixed value of the family parameter f_S, the matrices $\mathbf{F}^\alpha$ span a proper Abelian subgroup of the symplectic group $Sp(2,\mathbb{R})$ with respect to the order α, which may range in principle through all the real line. In fact, with $\alpha = 0$ we obtain the identity matrix; namely,

$$\mathbf{F}^0 = \mathbf{I}. \tag{3.6.12}$$

Also, by the addition formulas of the circular functions, we have

$$\begin{aligned}\mathbf{F}^{\alpha_1}(f_S)\,\mathbf{F}^{\alpha_2}(f_S) &= \begin{pmatrix} \cos(\phi_1+\phi_2) & f_S\sin(\phi_1+\phi_2) \\ -\frac{1}{f_S}\sin(\phi_1+\phi_2) & \cos(\phi_1+\phi_2) \end{pmatrix} \\ &= \mathbf{F}^{\alpha_1+\alpha_2}(f_S) = \mathbf{F}^{\alpha_2}(f_S)\mathbf{F}^{\alpha_1}(f_S)\end{aligned} \tag{3.6.13}$$

which reflects the well known property that the composition of two rotations by the angles ϕ_1 and ϕ_2 is equivalent to the rotation by the angle $\phi = \phi_1 + \phi_2$.

The additivity property (3.6.13) with respect to the order α leads to identify the inverse to $\mathbf{F}^\alpha(f_S)$ as the matrix corresponding to the opposite value of α,

$$[\mathbf{F}^\alpha(f_S)]^{-1} = \mathbf{F}^{-\alpha}(f_S) = \begin{pmatrix} \cos\phi & -f_S\sin\phi \\ \frac{1}{f_S}\sin\phi & \cos\phi \end{pmatrix}. \tag{3.6.14}$$

which then relates to the rotation by the angle ϕ in the reverse direction.

Also, the matrix $\mathbf{F}^\alpha(f_S)$ can be interpreted as the α-th power of $\mathbf{F}^1(f_S)$, i.e., of the "ordinary" Fourier transform:

$$\mathbf{F}^\alpha(f_S) = [\mathbf{F}(f_S)]^\alpha. \tag{3.6.15}$$

In Problem 5 we suggest a direct proof of the above relation in the case of $\alpha = \frac{1}{m}$, $m = \pm1, \pm2, ...$ We invite the reader to interpret the exponential representation (3.6.6) of $\mathbf{F}^\alpha(f_S)$ from the viewpoint of the power relation (3.6.15).

Interestingly, with $\phi = 2j\pi$ for any integer value $j = 0, \pm1, \pm2, ..$, we recover the identity matrix:

$$\mathbf{F}^{4j}(f_S) = \mathbf{F}^0(f_S) = \mathbf{I}. \tag{3.6.16}$$

The cyclic behaviour of the fractional Fourier transforming systems with respect to the order α is then easily established in the general form

$$\mathbf{F}^{\alpha+4j}(f_S) = \mathbf{F}^\alpha(f_S), \tag{3.6.17}$$

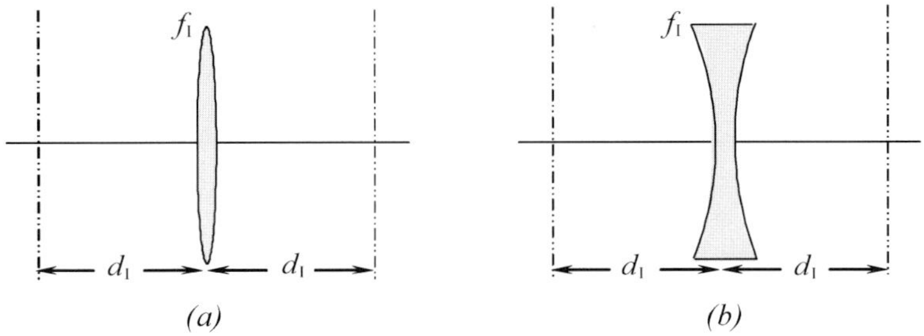

FIGURE 3.13. Lohmann type I optical realization of the fractional Fourier transform matrix $\mathbf{F}^\alpha$.*(a)* Positive lens. *(b)* Negative lens.

which is in accord of course with the well known periodicity of the ordinary Fourier transforming system. Hence the angle ϕ can conveniently be limited to anyone interval of amplitude 2π, as $\phi \in (-\pi, \pi]$ for instance, and accordingly the order α to the interval $\alpha \in (-2, 2]$.

Fragmenting the optical Fourier transforming system. The type I and type II optical setups

We investigate now the possibility of optically synthesizing the fractional Fourier transform matrix $\mathbf{F}^\alpha(f_S)$. Taking as model the single lens geometry of the Fourier transform, described in § 3.4.2, we may think of the $\mathbf{F}^\alpha$ system as formed by a thin lens of focal length f_I placed between two free-medium sections of the same length d_I. It is easy to verify that the symmetric single lens geometry

$$\mathbf{F}^\alpha(f_S) = \mathbf{T}(d_I)\,\mathbf{L}(f_I)\mathbf{T}(d_I), \tag{3.6.18}$$

can be designed with the involved parameters d_I and f_I being determined by the angle ϕ and the family parameter f_S according to the relations

$$d_I = f_S \tan(\phi/2), \qquad f_I = \frac{f_S}{\sin\phi}, \qquad -\pi < \phi < \pi. \tag{3.6.19}$$

Positive values of ϕ in the inherent range, namely $0 < \phi < \pi$, guarantee that both d_I and f_I are positive, thus yielding a one-lens configuration like that pictured in Fig. 3.13.a) with a bi-convex lens. In contrast, negative values of ϕ, $-\pi < \phi < 0$, yield negative values of d_I and f_I as well; the optical configuration demands therefore for a negative lens as that sketched in Fig. 3.13.b), virtual objects and virtual images being involved. The single lens realization (3.6.18) of the $\mathbf{F}^\alpha$ system is reported as *Lohmann's type I geometry* [5.4].

Likewise, a double lens geometry can be designed to optically implement the $\mathbf{F}^\alpha$ system, consisting of two identical lenses of focal length f_{II} spaced by

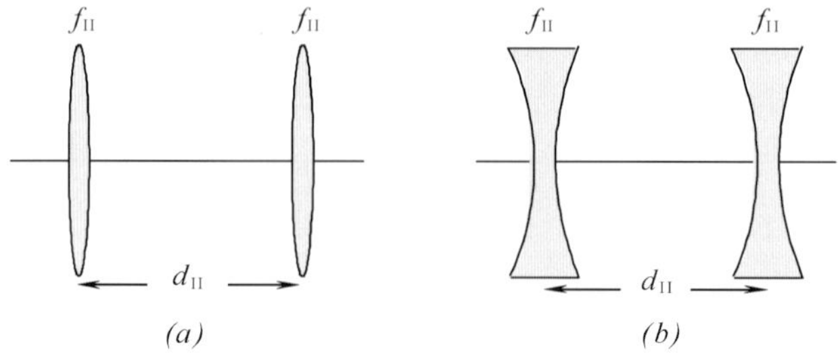

FIGURE 3.14. Lohmann type II optical realization of the fractional Fourier transform matrix $\mathbf{F}^\alpha$. *(a)* Positive lenses. *(b)* Negative lenses.

a free-propagation section of length d_{II}. Explicitly, we obtain

$$\mathbf{F}^\alpha(f_S) = \mathbf{L}(f_{II})\mathbf{T}(d_{II})\mathbf{L}(f_{II}), \tag{3.6.20}$$

with the defining relations for the relevant parameters d_{II} and f_{II}

$$d_{II} = f_S \sin\phi, \quad f_{II} = \frac{f_S}{\tan(\phi/2)}, \quad -\pi < \phi < \pi. \tag{3.6.21}$$

The optical configurations are shown in Fig. 3.14. Positive lenses, and so real object and image rays, are involved for positive values of the rotation angle ϕ, $0 < \phi < \pi$, whilst negative lenses, and hence virtual object and image rays, are involved for negative values of ϕ, $-\pi < \phi < 0$. The double lens realization (3.6.20) of the $\mathbf{F}^\alpha$ system is reported as *Lohmann's type II geometry* [5.4].

The optical configurations (3.6.18) and (3.6.20) are described as well by specific ray-matrices involving the respective parameters d_I, f_I and d_{II}, f_{II}, and of course the order α through the angle ϕ, instead of the general ray-matrix (3.6.6) involving the family parameter f_S. The reader is asked to write down the ray matrices for the fractional type I and type II units.

Evidently the value $\phi = \pi$ cannot be accounted for by the optical representations (3.6.18) and (3.6.20), as the relevant relations (3.6.19) and (3.6.21) become undefined in that case. However, with $\phi = \pi$ the rotation matrix has the form of the scaling matrix with magnification factor $\mathcal{M} = -1$, namely

$$\mathbf{F}^2(f_S) = \mathbf{D}(-1) = -\mathbf{I}, \tag{3.6.22}$$

thus signifying reflection with respect to the optical axis, whatever be the value of f_S. According to the results of § 3.4.3, the sequence of two identical (i.e., with lenses having the same focal distances) Fourier transforming systems produces a π-rotation of the ray vector in the q-p plane. This is clearly in accord with the additivity property (3.6.13), by which we can write

$$\mathbf{F}(f_S)\mathbf{F}(f_S) = \mathbf{F}^2(f_S). \tag{3.6.23}$$

With $\phi = \frac{\pi}{2}$ (i.e., $\alpha = 1$) we regain from (3.6.18) and (3.6.20) the one and two lens configurations of the perfect Fourier transforming system. With decreasing values of α, $|\alpha| < 1$, we can "continuously" fragment the single or double lens Fourier transforming configuration into an increasing number, $|\alpha|^{-1}$, of similar single or double lens configurations with the relevant free-propagation sections and thin lenses, respectively, becoming increasingly shorter and weaker.

With $\alpha = \frac{1}{2}$, for instance, the one-lens geometry (3.6.18) obtains with

$$d_I = (\sqrt{2} - 1) f_S\,, \qquad f_I = \sqrt{2} f_S. \tag{3.6.24}$$

The relevant ray matrix is

$$\mathbf{F}^{\frac{1}{2}}(f_S) = \begin{pmatrix} 1/\sqrt{2} & f_S/\sqrt{2} \\ -1/\sqrt{2}f_S & 1/\sqrt{2} \end{pmatrix}, \tag{3.6.25}$$

whose square is readily verified to yield $\mathbf{F}(f_S)$; namely

$$\mathbf{F}^{\frac{1}{2}}(f_S) \cdot \mathbf{F}^{\frac{1}{2}}(f_S) = \mathbf{F}(f_S). \tag{3.6.26}$$

In other words, $\mathbf{F}^{\frac{1}{2}}(f_S)$ represents the optical system that operates the *square root* of the Fourier transform.

Thus, with the order α ranging within the interval $(-2, 2]$, we obtain a "family" of optical systems with the geometries depicted in Figs. 3.13 and 3.14. The family is uniquely individualized by the real parameter f_S, also reported as the *standard focal length.* It represents the scale of the ordinary Fourier transform, which corresponding to $\alpha = 1$ results by appropriately cascading fractional transforms of lower orders belonging to the same family f_S. In other words, the family parameter f_S identifies the basic Fourier transform unit, which can be fragmented and reconstructed by the opposite operations of sectioning into fractional units down to the required α-th root, and of cascading such fractional units up to the required α-th power. It is clear that both the fragmentation and the reconstruction of the basic Fourier transforming unit can be achieved with fractional configurations of different orders which, of course, sum up to unity. In Fig. 3.15, for instance, we show three possible realizations of an imaging system producing a perfect inverted image of the object ($\alpha = 2$), by cascading fractional order units, the relevant orders summing up to 2 in all cases. Evidently the fragmentation of a system is not unique.

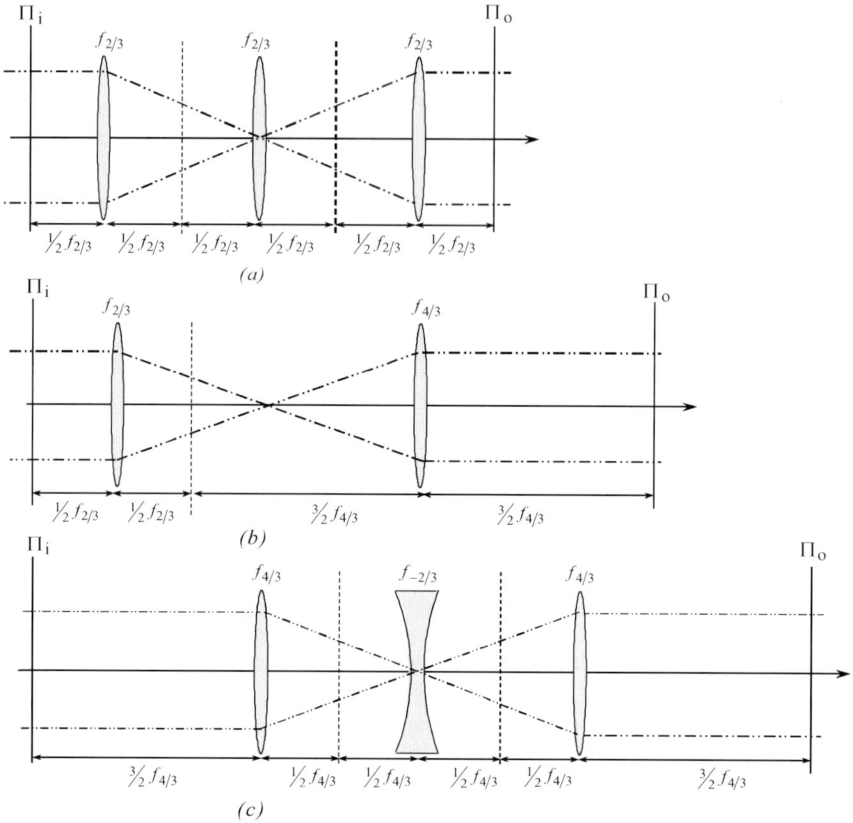

FIGURE 3.15. Perfect imaging system ($\alpha = 2$) obtained by cascading fractional Fourier units of different orders. *(a)* three units with $\alpha_1 = \alpha_2 = \alpha_3 = \frac{2}{3}$; *(b)* two units with $\alpha_1 = \frac{2}{3}$ and $\alpha_2 = \frac{4}{3}$ ($f_{2/3} = f_{4/3}$); *(c)* three units with $\alpha_1 = \frac{4}{3}$, $\alpha_2 = -\frac{2}{3}$ and $\alpha_3 = \frac{4}{3}$ ($f_{4/3} = -f_{-2/3}$).

The concept of *family*, where the fractional transforms with the same standard focal length f_S group together, naturally arises from the rule, by which the systems (3.6.6) are cascaded. As displayed by the semigroup property (3.6.13), only cascading fractional units having the same absolute value of f_S, the overall system is again a fractional Fourier transform, and also belongs to the same family. Notably, as already evidenced in (3.6.10), with $f_S = 1$ we obtain the family of the pure rotations in the q-p plane:

$$\mathbf{F}^{\alpha} = \mathbf{R}(\phi), \qquad \phi = \alpha\tfrac{\pi}{2}. \tag{3.6.27}$$

Finally, we wish to comment on the optical "performance" of a thin lens as it emerges from the considerations here developed. A lens is able to perform two basic operations in optical processing: imaging and Fourier transform. The concept of fractional Fourier transform allows to frame both operations within the same context, where the imaging can be seen as a fractional Fourier transformation and correspondingly the ordinary Fourier transform as a particular realization of the fractional transformation. We know, in fact, that the elementary optical system consisting of a thin lens of focal length f placed midway between two reference planes separated by $2d$ can operate as an imaging or a Fourier transforming system according to the relative values of f and d. For $d = 2f$ the systems acts as an imaging system, whilst with $d = f$ it operates as a Fourier transformer. It is evident that in both the type I and type II geometries one can range from one operation to the other through the intermediate steps of fractional Fourier transformation by continuously adjusting the order parameter α, and accordingly the involved lengths d_{I} and f_{I}, or d_{II} and f_{II}.

It may appear however that the claimed "fractionalization to the continuum" does not have a practical consistency due to the impossibility of realizing an infinitely short free-propagation section as an infinitely weak thin lens. We will see in § 3.9.1 that a quadratic graded-index focusing medium behaves in respect to the (paraxial) ray propagation as a fractional Fourier transforming system, thus offering the possibility of physically defining the process of a continuous fractionalization of the Fourier transforming operation.

Rotations and squeezes in the fractional Fourier transform operation

It is evident that the phase plane operation of the $\mathbf{F}^\alpha$ system results from rotations and scaling transformations, appropriately designed and concatenated one to the other. In fact, as said earlier, with $f_{\text{S}} = 1$ the fractional Fourier transform matrix takes the form as that of a pure rotation matrix by the associated angle $\phi = \alpha\frac{\pi}{2}$. Then we are naturally led to decompose the fractional Fourier transform matrix $\mathbf{F}^\alpha(f_{\text{S}})$ into scale and rotation matrices according to the similarity transformation

$$\mathbf{F}^\alpha(f_{\text{S}}) = \mathbf{S}(f_{\text{S}}^{1/2})\mathbf{F}^\alpha\mathbf{S}(f_{\text{S}}^{-1/2}). \tag{3.6.28}$$

The pure rotation by $\phi = \alpha\frac{\pi}{2}$ is sandwiched between two scale transformations with reciprocal magnifications, $f_{\text{S}}^{-1/2}$ and $f_{\text{S}}^{1/2}$. Note that $\mathbf{S}(f_{\text{S}}^{1/2}) = [\mathbf{S}(f_{\text{S}}^{-1/2})]^{-1}$. Of course, the off-diagonal entries in the pure rotation matrix $\mathbf{F}^\alpha$ above are understood having their own dimensions of length and $(\text{length})^{-1}$, although not evidenced by a visible parameter. Also, the magnification factors $f_{\text{S}}^{-1/2}$ and $f_{\text{S}}^{1/2}$ of the scaling matrices must be regarded as numerical values, as those obtainable, for instance, by cascading two Fourier transforming systems as described in § 3.4.3. In particular, with $\alpha = 1$ we obtain on one side the ordinary Fourier transform whilst on the other the pertinent decomposition in terms of scaling and rotations by the right angle, i.e.,

$$\begin{pmatrix} 0 & f_S \\ -1/f_S & 0 \end{pmatrix} = \begin{pmatrix} \sqrt{f_S} & 0 \\ 0 & 1/\sqrt{f_S} \end{pmatrix} \begin{pmatrix} 0 & 1 \\ -1 & 0 \end{pmatrix} \begin{pmatrix} 1/\sqrt{f_S} & 0 \\ 0 & \sqrt{f_S} \end{pmatrix}. \tag{3.6.29}$$

The above discussion confirms the relation of the Fourier transform to the phase-plane scaled rotations by the right angle and correspondingly of the fractional Fourier transform to scaled rotations by arbitrary, and continuously adjustable, fractions of the right angle. In accord with decomposition (3.6.28) we may regard the fractional Fourier transformers as *phase-plane rotators.*

Resorting to the optical configuration, described in the previous paragraph, we may therefore say that the fragmentation of the Fourier transforming configuration into similar configurations with increasingly shorter free-propagation sections interleaved by increasingly weaker thin lenses corresponds in the optical phase plane to a fragmentation of the rotation angle into increasingly smaller fractions of the right angle.

Finally, paralleling the single point phase-plane view of Fig. 3.12, it is interesting to visualize the effect of the fractional Fourier transform operation on a ray distribution in phase-plane. We consider the Gaussian density distribution

$$\rho_{\mathcal{G}}(q,p;z_i) = \frac{1}{2\pi\sigma_q\sigma_p} e^{-\frac{q^2}{2\sigma_q^2} - \frac{p^2}{2\sigma_p^2}}, \tag{3.6.30}$$

already used for some illustrative examples in Chapter 2; as noted there, the positive quantities σ_q and σ_p specify the rms extensions of the density distribution in the phase plane along the q and p-axis, respectively.

On account of relation (2.2.67), it is easy to elaborate the analytical expression for the effect produced on the distribution (3.6.30) by the optical Fourier transformation (3.6.8). However, it is helpful to give a pictorial view. Figure 3.16 shows the transformation acted by $\mathbf{F}^{2/3}(\frac{1}{2})$ on the representative phase-plane area of the distribution (3.6.30), identified by any one of the elliptical contours (2.5.16), as, for instance, that corresponding to the value $\varepsilon = 2$. The proper sequence of scaling, rotation and scaling, to which according to (3.6.28) the desired fractional Fourier transformation, i.e., $\mathbf{F}^{2/3}(\frac{1}{2})$, can be thought as equivalent, is exemplified. In the graphs the scaled variables $\frac{q}{\sigma_q}$ and $\frac{p}{\sigma_p}$ have been used, thus obtaining as initial phase plane areas circles of radius $\mathfrak{r} = \sqrt{2\varepsilon}$.

3.6.2 *Repulsive oscillator, phase-plane squeeze and Lorentz transformations*

We consider now the quadratic homogeneous polynomial

$$H_{\text{r.o.}} = \tfrac{1}{2}p^2 - \tfrac{1}{2f_S^2}q^2, \tag{3.6.31}$$

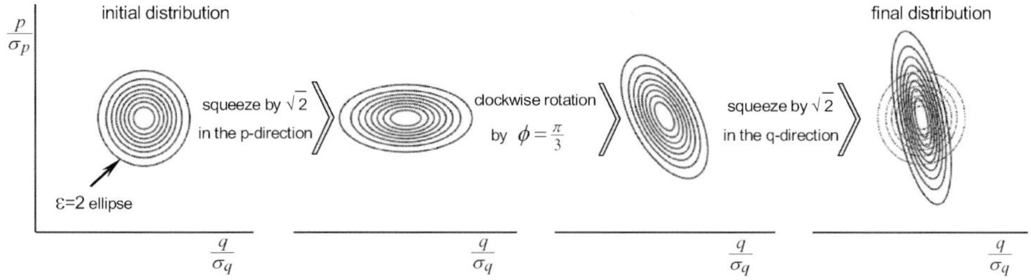

FIGURE 3.16. Phase-plane view of the decomposition (3.6.28) of the fractional Fourier transform matrix $\mathbf{F}^{2/3}(\frac{1}{2})$ into the proper sequence of scaling, rotation and scaling, through the relevant effects produced on the Gaussian ray density (3.6.30).

which typically models the repulsive oscillator dynamics. The Lie operator $\widehat{\mathsf{L}}_{\text{r.o.}}$ is

$$\widehat{\mathsf{L}}_{\text{r.o.}} = p\frac{\partial}{\partial q} + \frac{q}{f_{\text{S}}^2}\frac{\partial}{\partial p}, \tag{3.6.32}$$

which, in the similar way as $\widehat{\mathsf{L}}_{\text{a.o.}}$, relates to the operators $\widehat{\mathrm{K}}_-$ and $\widehat{\mathrm{K}}_+$ as

$$\widehat{\mathsf{L}}_{\text{r.o.}} = \frac{1}{f_{\text{S}}}(f_{\text{S}}\widehat{\mathrm{K}}_- - \frac{1}{f_{\text{S}}}\widehat{\mathrm{K}}_+). \tag{3.6.33}$$

Likewise, the matrix representative $\mathbf{H}_{\text{r.o.}}$writes in terms of the matrices $\mathbf{K}_-$ and $\mathbf{K}_+$ according to

$$\mathbf{H}_{\text{r.o.}} = \begin{pmatrix} 0 & 1 \\ 1/f_{\text{S}}^2 & 0 \end{pmatrix} = \frac{2}{f_{\text{S}}}\mathbf{K}_2(f_{\text{S}}), \tag{3.6.34}$$

where $\mathbf{K}_2(f_{\text{S}})$ is the algebra element

$$\mathbf{K}_2(f_{\text{S}}) = \tfrac{1}{2}(f_{\text{S}}\mathbf{K}_- - \tfrac{1}{f_{\text{S}}}\mathbf{K}_+) = \tfrac{1}{2}\begin{pmatrix} 0 & f_{\text{S}} \\ 1/f_{\text{S}} & 0 \end{pmatrix}. \tag{3.6.35}$$

Exponentiating the traceless matrix (3.6.34) through the real parameter $\Delta z = z - z_i$, we obtain the "transfer matrix" for the ray propagation from z_i to z as

$$\mathbf{B}(\tau, f_{\text{S}}) = e^{\Delta z\,\mathbf{H}_{\text{r.o.}}} = e^{2\tau\,\mathbf{K}_2(f_{\text{S}})} = \begin{pmatrix} \cosh\tau & f_{\text{S}}\sinh\tau \\ \frac{1}{f_{\text{S}}}\sinh\tau & \cosh\tau \end{pmatrix}, \tag{3.6.36}$$

having denoted the dimensionless argument $\frac{\Delta z}{f_{\text{S}}}$ of the hyperbolic sine and cosine functions by

$$\tau(z) = \frac{z - z_i}{f_{\text{S}}}. \tag{3.6.37}$$

The transformation acted by the matrix $\mathbf{B}(\tau, f_S)$ on the phase-plane variables (q, p) from the initial data (q_i, p_i) is then

$$\begin{aligned} q(z) &= \quad q_i \cosh\tau + p_i f_S \sinh\tau\,, \\ p(z) &= -\frac{1}{f_S} q_i \sinh\tau + p_i \cosh\tau, \end{aligned} \tag{3.6.38}$$

in accord with Eq. (1.5.26). Evidently, the phase-plane motion is no more cyclic as in the case of the system (3.6.6). In fact, as the ray propagates through the "optical" system (3.6.36), the representative point in the (q, p) plane moves along one of the branches of the hyperbola

$$\frac{q(z)^2}{f_S^2} - p(z)^2 = \mathfrak{h}, \qquad \mathfrak{h} \equiv \frac{q_i^2}{f_S^2} - p_i^2, \tag{3.6.39}$$

whose definite shape is determined by the initial data, which along with the parameter f_S fix the value of the constant $\mathfrak{h}$.

For a given value of f_S, the matrices $\mathbf{B}(\tau, f_S)$ form an Abelian subgroup of the symplectic group $Sp(2, \mathbb{R})$ with respect to the parameter τ. In fact, the addition formulae for the hyperbolic functions yield the semigroup property

$$\mathbf{B}(\tau_1, f_S)\mathbf{B}(\tau_2, f_S) = \mathbf{B}(\tau_1 + \tau_2, f_S) = \mathbf{B}(\tau_2, f_S)\mathbf{B}(\tau_1, f_S), \tag{3.6.40}$$

according to which we also have

$$\mathbf{B}(\tau, f_S)\mathbf{B}(-\tau, f_S) = \mathbf{B}(-\tau, f_S)\mathbf{B}(\tau, f_S) = \mathbf{I}, \tag{3.6.41}$$

the identity matrix being indeed $\mathbf{I} = \mathbf{B}(0, f_S)$, whatever be f_S.

The inverse to $\mathbf{B}(\tau, f_S)$ is thereby the matrix

$$[\mathbf{B}(\tau, f_S)]^{-1} = \mathbf{B}(-\tau, f_S) = \begin{pmatrix} \cosh\tau & -f_S \sinh\tau \\ -\frac{1}{f_S}\sinh\tau & \cosh\tau \end{pmatrix}. \tag{3.6.42}$$

The nature of the phase plane dynamics described by (3.6.38) is basically determined by the presence of the hyperbolic functions so as the circular functions entering (3.6.8) determine the cyclic behaviour of the fractional Fourier transform generated dynamics. Then, we set $f_S = 1$ in (3.6.36) in order to evidence the intrinsic character of the **B**-generated phase-plane dynamics.

B-*generated phase-plane dynamics: rotations and squeezes*

With $f_S = 1$, the matrix (3.6.36) takes the symmetric form

$$\mathbf{B}(\tau) = \begin{pmatrix} \cosh\tau & \sinh\tau \\ \sinh\tau & \cosh\tau \end{pmatrix} = e^{2\tau \mathbf{K}_2}, \tag{3.6.43}$$

paralleling in terms of the matrix $\mathbf{K}_2 = \frac{\mathbf{K}_- - \mathbf{K}_+}{2}$, and so of the hyperbolic functions, expression (3.6.10) for the fractional Fourier transform matrix $\mathbf{F}^\alpha$ relative to a unit value of the family parameter. Notably, $\mathbf{F}^\alpha$ is an orthogonal matrix: $[\mathbf{F}^\alpha]^{-1} = [\mathbf{F}^\alpha]^\top$, whereas $\mathbf{B}(\tau)$ is symmetric and positive definite[(b)].

Equation (3.6.39) identifies in the q-p plane the equilateral hyperbola

$$q(z)^2 - p(z)^2 = \mathfrak{h}, \qquad \mathfrak{h} \equiv q_i^2 - p_i^2, \tag{3.6.44}$$

whose branches intersect the q or p-axis according to whether the initial values $(q_i\,, p_i)$ are such that $\mathfrak{h} > 0$ or $\mathfrak{h} < 0$. Either cases, the asymptotes are the straight lines $p = \pm q$.

The dynamics of the ray vector in the phase plane under the action of $\mathbf{B}(\tau)$ is similar to that under $\mathbf{S}(m)$, which, as noted in § 2.5, identifies the hyperbola $qp = q_i p_i$. It is evident, in fact, that the hyperbolas $qp = q_i p_i$ and $q^2 - p^2 = q_i^2 - p_i^2$ can be turned one into the other by a proper rotation by $\frac{\pi}{4}$. Then, as phase-plane rotations are actuated by fractional Fourier transforming systems, we may guess for $\mathbf{B}(\tau)$ a three-step realization in terms of pure rotations and squeezes as

$$\mathbf{B}(\tau) = \mathbf{F}^{-1/2}\,\mathbf{S}(e^\tau)\mathbf{F}^{1/2}. \tag{3.6.45}$$

Notably, (3.6.45) allows us to identify an optical realization of the symplectic matrix $\mathbf{B}(\tau)$, on account of the optical realizations of the fractional Fourier transforming and pure magnifying systems discussed in § 3.6.1 and § 3.4.3. We invite the reader to prove that a symmetric single-lens realization of matrix (3.6.45) is actuable with $d = \tanh(\frac{\tau}{2})$ and $f = -\sinh^{-1}(\tau)$ when $\tau > 0$, whilst a more complex configuration demanding for more than two lenses is needed to implement negative values of τ.

In order to clearly show that the matrix $\mathbf{B}(\tau)$ operates in the phase plane as a scale transformation in a similar fashion as $\mathbf{S}(m)$, but with respect to a reference direction making an angle $\theta = \frac{\pi}{4}$ with the q-axis, we consider again the Gaussian distribution (3.6.30) and the relevant representative $\varepsilon = 2$ phase-plane ellipse. For illustrative purposes, we set $\sigma_q = \sigma_p = \sigma$, and accordingly consider the circle $\mathcal{C}$ of radius $\mathfrak{r} = 2\sigma$ centered around the origin,

$$q^2 + p^2 = \mathfrak{r}^2, \tag{3.6.46}$$

---

[b] A matrix $\mathbf{A}$ is *positive definite* if the scalar product $(\mathbf{u}, \mathbf{Au})$ is positive for any vector $\mathbf{u}$ in the linear space acted by $\mathbf{A}$. It is evident that in the case of matrix (3.6.43) we have

$$(p^2 + q^2)\cosh\tau + 2pq\sinh\tau > 0$$

on account of the inequalities $|\sinh\tau| < \cosh\tau$ and $2pq \le p^2 + q^2$ holding through all the domain of the real parameter τ as well as of the ray variables q and p.

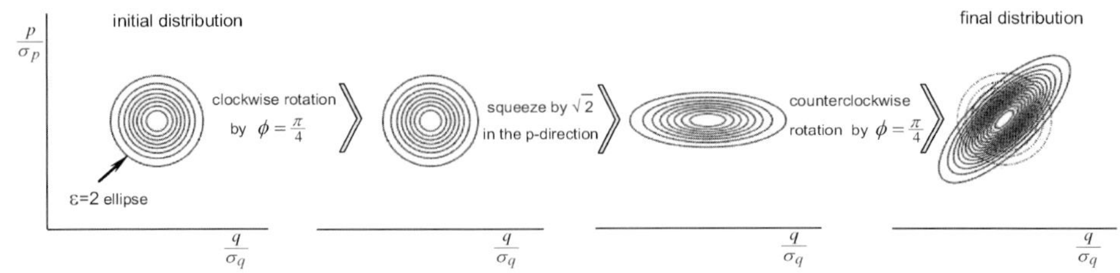

FIGURE 3.17. Phase-plane view of the decomposition (3.6.45) of the squeeze matrix $\mathbf{B}(\frac{1}{2})$ into the proper sequence of rotation, scaling, and rotation, through the relevant effects produced on the Gaussian ray density (3.6.30).

Acting on the beam by $\mathbf{S}(m)$ (we may take $m = e^\tau$), the representative circle $\mathcal{C}$ is transformed into the ellipse

$$e^{-2\tau}q^2 + e^{2\tau}p^2 = \mathfrak{r}^2, \tag{3.6.47}$$

symmetrically positioned in the q-p plane with semiaxis $\mathfrak{a} = e^\tau \mathfrak{r}$ and $\mathfrak{b} = e^{-\tau}\mathfrak{r}$. Then, supposing $\tau > 0$, we may see that the beam-transformation acted by the optical matrix $\mathbf{S}(m)$ produces an elongation of the q-size of the circle by e^τ and correspondingly a contraction of the p-size by the same factor. In other words, the beam representative circle $\mathcal{C}$ is squeezed along the p axis (see third graph in Fig. 3.17). The area of ellipse (3.6.47) is obviously the same as that of the original circle: $\mathcal{A} = 4\pi\sigma^2$.

Likewise, transferring the beam from z_i to z through $\mathbf{B}(\tau)$, the phase-plane circle $\mathcal{C}$ is transformed into the ellipse

$$\cosh(2\tau)\, q^2 + \cosh(2\tau)\, p^2 - 2\sinh(2\tau)\, pq = \mathfrak{r}^2. \tag{3.6.48}$$

The major axis is in the $\frac{\pi}{4}$ or $-\frac{\pi}{4}$ direction according to whether τ is positive or negative. If $\tau > 0$, for instance, the major axis comes to lie along the direction making a $\frac{\pi}{4}$ angle with the q-axis (see the last graph in Fig. 3.17), specifically being $\mathsf{a} = e^\tau\mathfrak{r}$ and $\mathsf{b} = e^{-\tau}\mathfrak{r}$. The initial phase-plane area is then stretched up to the factor $m = e^\tau$ in the direction at $\frac{\pi}{4}$ to the q-axis and orthogonally squeezed down by the same factor. The system $\mathbf{B}(\tau)$ is indeed referred to as *hyperbolic expander* or *reductor* according to whether $\tau > 0$ or $\tau < 0$. Figure 3.17 shows the rotation-scaling-rotation sequence, whose overall effect is the expected squeeze of the phase plane area in the direction at $\frac{\pi}{4}$ to the q-axis. As in the graphs of Fig. 3.16, the scaled variables $\frac{q}{\sigma_q}$ and $\frac{p}{\sigma_p}$ have been used.

It is possible to produce squeeze in any direction θ, combining scaling and rotation matrices through a similarity transformation analogous to (3.6.45) with $\mathbf{F}^{1/2}$ replaced by $\mathbf{F}^\alpha$. In fact, the reader can easily verify that the matrix

$$\mathbf{B}(\tau,\theta) = \mathbf{F}^{-\alpha}\,\mathbf{S}(e^{\tau})\mathbf{F}^{\alpha} = \begin{pmatrix} \cosh\tau + \cos(2\theta)\sinh\tau & \sin(2\theta)\sinh\tau \\ \sin(2\theta)\sinh\tau & \cosh\tau - \cos(2\theta)\sinh\tau \end{pmatrix}, \tag{3.6.49}$$

produces a scale transformation with respect to the direction making the angle $\theta = \alpha\frac{\pi}{2}$ with the q-axis. Like (3.6.43), the matrix $\mathbf{B}(\tau,\theta)$ is symmetric and positive definite. Also, it is characterized by two parameters specifying the "entity" and the "direction" of the squeeze. In accord with the previously adopted symbology, we identify the specific determinations of matrix (3.6.49) corresponding to $\theta = 0$ and $\theta = \frac{\pi}{4}$ as $\mathbf{S}(m)$ and $\mathbf{B}(\tau)$ respectively; namely, $\mathbf{B}(\tau,0) \equiv \mathbf{S}(m)$, $m = e^{\tau}$ and $\mathbf{B}(\tau,\frac{\pi}{4}) \equiv \mathbf{B}(\tau)$.

Interestingly, although the matrices $\mathbf{S}$ and $\mathbf{F}^{\alpha}$ separately form proper subgroups of the group $Sp(2,\mathbb{R})$, the matrices $\mathbf{B}(\tau,\theta)$ do not span a subgroup. In fact, successive squeeze transformations along different directions do not compose into one scale transformation, but into a squeeze preceded or followed by a rotation (see Problem 10).

It is worth evidencing the formal similarity between the squeeze matrix $\mathbf{B}(\tau)$ and the Lorentz transformation matrix in the Cartesian system of z and ct, positive and negative values of τ respectively corresponding to boosts and antiboosts. Likewise, the pure magnifier matrix $\mathbf{S}(m)$ has the form of the Lorentz transformation matrix in the Dirac's light-cone coordinate system, which is rotated in the Euclidean plane (z, ct) by $\frac{\pi}{4}$. Accordingly, we may note that as successive non collinear squeezes do not combine into one squeeze, Lorentz boosts in different directions do not multiply into one Lorentz boost, but into a Lorentz transformation preceded or followed by the Wigner rotation. The Lie algebra $sp(2,\mathbb{R})$ is in fact isomorphic to the underlying algebra $so(2,1)$ of the group $SO(2,1)$ of the Lorentz transformations in three (two space and one time) dimensions. This isomorphism, and in particular the analogy in the behaviour of optical scale transformations and Lorentz boosts, has suggested optical experiments to test the abstract properties of the Lorentz group [6].

Finally, we briefly comment on the factored representation of the squeeze matrix (3.6.36), which is naturally suggested by (3.6.45) in the form

$$\mathbf{B}(\tau, f_{\mathrm{S}}) = \mathbf{F}^{-1/2}(f_{\mathrm{S}})\,\mathbf{S}(e^{\tau})\mathbf{F}^{1/2}(f_{\mathrm{S}}) = \mathbf{S}(f_{\mathrm{S}}^{1/2})\mathbf{B}(\tau)\mathbf{S}(f_{\mathrm{S}}^{-1/2}), \tag{3.6.50}$$

paralleling expression (3.6.28) for the fractional Fourier transform matrix $\mathbf{F}^{\alpha}(f_{\mathrm{S}})$. The parameter f_{S} retains its role as the focal length of the Fourier transform unit associated with the fractional Fourier transforming systems with $\alpha = \pm\frac{1}{2}$, entering (3.6.50). The transformation acted by $\mathbf{B}(\tau, f_{\mathrm{S}})$ produces therefore the further scalings by $f_{\mathrm{S}}^{-1/2}$ and $f_{\mathrm{S}}^{1/2}$ with respect to that acted by $\mathbf{B}(\tau)$.

3.6.3 Generator matrices of pure rotations and squeezes

We frame the matrices $\mathbf{K}_1$ and $\mathbf{K}_2$ within the proper algebraic context. Defined according to relations (3.6.5) and (3.6.35) for a unit value of the length parameter f_S, the matrices $\mathbf{K}_1$ and $\mathbf{K}_2$ have been recognized as the generators of pure rotations and squeezes in the phase plane, and associated with optical systems displaying respectively a definite focusing and defocusing behaviour.

Let us rewrite the pertinent expressions, i.e.

$$\mathbf{K}_1 = \tfrac{1}{2}(\mathbf{K}_- + \mathbf{K}_+), \qquad \mathbf{K}_2 = \tfrac{1}{2}(\mathbf{K}_- - \mathbf{K}_+), \tag{3.6.51}$$

with the explicit matrix forms of $\mathbf{K}_-$ and $\mathbf{K}_+$ being reproduced (3.1.2).

As a synthesis of the considerations developed in §§ 3.6.1 and 3.6.2, we may complete the table of correspondences (3.1.2) adding the following:

generating function		generating matrix		transformation matrix
$\frac{1}{2}p^2 + \frac{1}{2}q^2$	$\rightarrow$	$\mathbf{K}_1 = \frac{1}{2}\begin{pmatrix} 0 & 1 \\ -1 & 0 \end{pmatrix}$	$\rightarrow$	$\mathbf{F}^\alpha = e^{2\phi \mathbf{K}_1} = \begin{pmatrix} \cos\phi & \sin\phi \\ -\sin\phi & \cos\phi \end{pmatrix}$
$\frac{1}{2}p^2 - \frac{1}{2}q^2$	$\rightarrow$	$\mathbf{K}_2 = \frac{1}{2}\begin{pmatrix} 0 & 1 \\ 1 & 0 \end{pmatrix}$	$\rightarrow$	$\mathbf{B}(\tau) = e^{2\tau \mathbf{K}_2} = \begin{pmatrix} \cosh\tau & \sinh\tau \\ \sinh\tau & \cosh\tau \end{pmatrix}$,

(3.6.52)

which allow us to characterize the $\frac{1}{2}(p^2+q^2)$-generated systems by orthogonal unimodular matrices and correspondingly the $\frac{1}{2}(p^2-q^2)$-generated systems by symmetric definite positive unimodular matrices.

As noted in § 2.3.1, the matrices $\{\mathbf{K}_-, \mathbf{K}_+, \mathbf{K}_3\}$ do form a closed set with respect to the Lie products, being in fact

$$[\mathbf{K}_+, \mathbf{K}_-] = 2\mathbf{K}_3, \quad [\mathbf{K}_\pm, \mathbf{K}_3] = \pm\mathbf{K}_\pm. \tag{3.6.53}$$

In a like manner, the set of matrices $\{\mathbf{K}_1, \mathbf{K}_2, \mathbf{K}_3\}$ also close under commutation according to

$$[\mathbf{K}_1, \mathbf{K}_2] = \mathbf{K}_3, \quad [\mathbf{K}_1, \mathbf{K}_3] = -\mathbf{K}_2, \quad [\mathbf{K}_2, \mathbf{K}_3] = -\mathbf{K}_1, \tag{3.6.54}$$

Therefore, like $\{\mathbf{K}_-, \mathbf{K}_+, \mathbf{K}_3\}$, the set $\{\mathbf{K}_1, \mathbf{K}_2, \mathbf{K}_3\}$ is a basis for $sp(2, \mathbb{R})$ as well; it is, in fact, easy to verify that any matrix in the algebra can uniquely be written as a linear real combination of $\mathbf{K}_1$, $\mathbf{K}_2$ and $\mathbf{K}_3$.

Finally we mention that the one-parameter subgroups generated by $\mathbf{K}_2$ and $\mathbf{K}_3$, formed respectively by the matrices $\mathbf{B}(\tau)$ and $\mathbf{S}(m)$, are reported in the group theory as *hyperbolic subgroups*, whilst that generated by $\mathbf{K}_1$, formed by the pure rotation matrices $\mathbf{F}^\alpha$, is called *elliptic subgroup*. Also, the subgroups generated by $\mathbf{K}_-$ and $\mathbf{K}_+$, respectively formed by the matrices $\mathbf{T}(d)$ and $\mathbf{L}(f)$, are referred to as *parabolic subgroups*. The reader may gain a feeling for the difference between matrices in these subgroups, by noting that, unlike for the elliptic subgroup (see Eq. (3.6.15)), powers of the matrices in the hyperbolic and parabolic subgroups grow indefinitely and never reproduce the identity; the hyperbolic and parabolic subgroups are in fact noncompact [2].

3.6.4 *Rotations and squeezes as building blocks of any symplectic transformation*

As a conclusion of the present investigation of the rotation and squeeze generating matrices, we prove that any optical (symplectic) matrix $\mathbf{M} \in Sp(2, \mathbb{R})$ is uniquely expressible as the product of a **B**- and an $\mathbf{F}^\alpha$-matrix. In symbols,

$$\mathbf{M} = \mathbf{B}(\tau, \theta)\mathbf{F}^\alpha, \tag{3.6.55}$$

with the entries A, B, C, D on one side being related to the two squeeze-parameters τ, θ, and the rotation angle $\phi = \alpha\pi/2$ on the other side as

$$\begin{aligned} &\cosh\tau = \tfrac{1}{2}\sqrt{A^2 + B^2 + C^2 + D^2 + 2} \\ &\tan(2\theta) = \tfrac{2(BD+AC)}{A^2+B^2-C^2-D^2}, \qquad \tan\phi = \tfrac{B-C}{A+D}, \end{aligned} \tag{3.6.56}$$

The inverse-order factorization, namely

$$\mathbf{M} = \mathbf{F}^{\alpha'}\mathbf{B}(\tau', \theta'), \tag{3.6.57}$$

is as well actuable with

$$\tau' = \tau, \quad \phi' = \phi, \quad \theta' = \theta + \phi. \tag{3.6.58}$$

Equations (3.6.55) and (3.6.57) yield the *polar* decomposition of symplectic matrices, in general involving a real symmetric positive-definite unimodular matrix and a real skew-symmetric unimodular matrix, corresponding to $\mathbf{B}(\tau, \theta)$ and $\mathbf{F}^\alpha$ in our case [3.6].

3.7 Iwasawa representation of optical elements: $\mathbf{LSF}^\alpha$ synthesis

In § 3.5 we have described the representation of optical matrices as ordered products of elements drawn from the $\mathbf{K}_+$-, $\mathbf{K}_3$- and $\mathbf{K}_-$-generated subgroups. Inspired by the original suggestion by Wei and Norman [4.1], the LST decomposition is widely exploited in quantum mechanics in connection with the solution of the Schrödinger equation with an underlying $SU(2)$ or $SU(1,1)$, i.e. angular momentum and harmonic oscillator-like, symmetry. However, as remarked, the LST optical arrangement is not global as it is not applicable to every optical matrix, although it is definitely suitable to propagation problems.

In contrast, the representation of optical matrices as ordered products of three elements drawn from the $\mathbf{K}_1$-, $\mathbf{K}_3$- and $\mathbf{K}_+$-generated subgroups is global in the sense that it applies to every symplectic matrix and is unique in the sense that the parameters of the representation are uniquely determined in terms of the entries of the given matrix. In fact, the LSF$^\alpha$ arrangement involving thin lenses, pure magnifiers and phase-plane rotators, we will describe here, resort to the Iwasawa decomposition theorem, which is a basic result of the Lie group theory valid for every noncompact semisimple Lie group.

For a specific discussion of the Iwasawa decomposition theorem within the general context of the Lie group theory, we address the reader to [7]. We will show here the feasibility of the Iwasawa representation in the case of interest to us, i.e., for the optical matrices belonging to the symplectic group $Sp(2, \mathbb{R})$.

Every ray-transfer matrix can be decomposed into the factored product of a lens, a positive-definite pure magnifier and a phase-plane rotator as

$$\begin{pmatrix} A & B \\ C & D \end{pmatrix} = \mathbf{L}(\mathcal{P})\mathbf{S}(m)\mathbf{F}^\alpha. \tag{3.7.1}$$

The relations between the various parameters obtain in the form

$$\begin{gathered} A = m\cos\phi, \qquad B = m\sin\phi, \\ C = -m\mathcal{P}\cos\phi - \tfrac{1}{m}\sin\phi, \quad D = -m\mathcal{P}\sin\phi + \tfrac{1}{m}\cos\phi, \end{gathered} \tag{3.7.2}$$

which are solved for $\mathcal{P}$, m and ϕ to give

$$m^2 = A^2 + B^2, \quad \mathcal{P} = -\tfrac{AC+DB}{A^2+B^2}, \quad \tan\phi = \tfrac{B}{A}. \tag{3.7.3}$$

It may be convenient to employ a non unit value of the standard focal length $f_{\rm S}$ in the fractional Fourier transform matrix entering (3.7.1). In that case, $f_{\rm S}$ may be left as a free parameter, whose magnitude can freely be chosen according to our convenience relatively to the specific problem under examination [8].

Taking due account of the exponential representations of the matrices $\mathbf{L}$, $\mathbf{S}$ and $\mathbf{F}^\alpha$, we can express the product in (3.7.1) into the functional form

$$\begin{pmatrix} A & B \\ C & D \end{pmatrix} = e^{\mathcal{P}\mathbf{K}_+}e^{s\mathbf{K}_3}e^{2\phi\mathbf{K}_1}, \qquad m = e^{s/2}. \tag{3.7.4}$$

Equation (3.7.1) expresses in the optical context the general property of every noncompact semisimple Lie group (in particular, of $Sp(2, \mathbb{R})$), whose elements can uniquely be decomposed into the product of three factors, taken each one from a maximal nilpotent group $\mathcal{N}$, a maximal Abelian subgroup $\mathcal{A}$ and the maximal compact subgroup $\mathcal{K}$. For $Sp(2, \mathbb{R})$ the subgroup $\mathcal{K}$ is formed by the orthogonal matrices $\mathbf{F}^\alpha$, whilst $\mathcal{N}$ and $\mathcal{A}$ can be chosen to comprise respectively the lower-triangular matrices $\mathbf{L}$ with unit entries on the diagonal and the mutually commuting positive diagonal matrices $\mathbf{S}$. Hence, the "$\mathcal{NAK}$" decomposition takes the form (3.7.1). We note that the three factors belong respectively to the parabolic, hyperbolic and elliptic subgroup of $Sp(2, \mathbb{R})$.

Figure 3.18 summarizes the group-theoretical and optical content of the LSF$^\alpha$ decomposition (3.7.1), mirroring Fig. 3.11 relative to the LST decomposition. Notably, as evidenced in the figure, the ray transformation acted by any first order optical system can be interpreted as a definite sequence of a phase-plane rotation, an overall scaling and a p-shear.

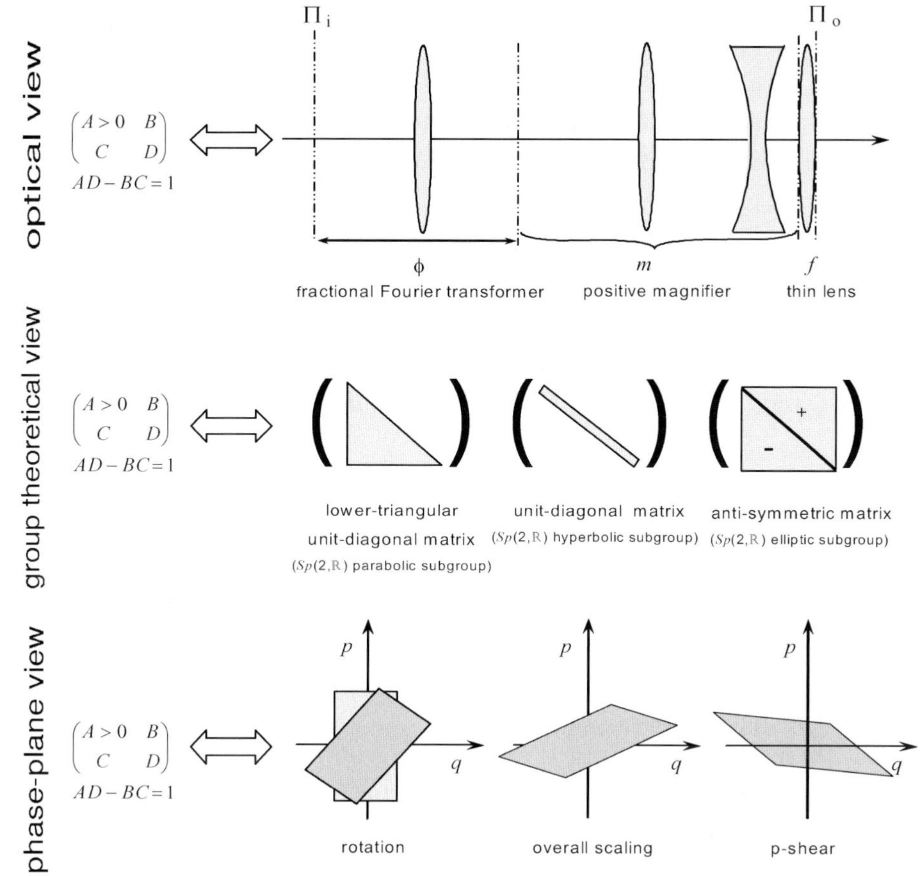

FIGURE 3.18. Iwasawa synthesis of an optical matrix.

3.8 Canonical and noncanonical representations of symplectic matrices

We complete here the investigation started in the previous chapter concerning the relation between quadratic polynomials and optical systems. We consider indeed the general quadratic function in q and p (see also §§ 1.5.2 and 2.3)

$$Q(q,p) = \tfrac{1}{2}bp^2 + cqp + \tfrac{1}{2}aq^2. \tag{3.8.1}$$

The representative traceless matrix is

$$\mathbf{Q} = \begin{pmatrix} c & b \\ -a & -c \end{pmatrix}, \tag{3.8.2}$$

which is expressible in terms of the algebra basis $\{\mathbf{K}_-, \mathbf{K}_+, \mathbf{K}_3\}$ as

$$\mathbf{Q} = a\mathbf{K}_+ + 2c\mathbf{K}_3 + b\mathbf{K}_- . \tag{3.8.3}$$

It is easy to compute the exponential function of $\mathbf{Q}$, thus comprising all the correspondences established in (3.1.2) and (3.6.52) into one form:

$$\begin{array}{ccccc} \text{generating function} & & \text{generating matrix} & & \text{transformation matrix} \\ \frac{1}{2}bp^2 + cqp + \frac{1}{2}aq^2 & \rightarrow & \mathbf{Q} = \begin{pmatrix} c & b \\ -a & -c \end{pmatrix} & \rightarrow & e^{\zeta\,\mathbf{Q}} = \begin{pmatrix} A(\zeta) & B(\zeta) \\ C(\zeta) & D(\zeta) \end{pmatrix}. \\ a, b, c, \zeta \in \mathbb{R} & & & & \end{array} \tag{3.8.4}$$

The entries of the group element $e^{\zeta\,\mathbf{Q}}$, depending on both the exponentiation parameter ζ and the coefficients (a, b, c) of the quadratic form, explicitly write as

$$\begin{array}{ll} A(\zeta) = \cosh(\Delta\zeta) + \frac{c}{\Delta}\sinh(\Delta\zeta), & B(\zeta) = \frac{b}{\Delta}\sinh(\Delta\zeta), \\ C(\zeta) = -\frac{a}{\Delta}\sinh(\Delta\zeta), & D(\zeta) = \cosh(\Delta\zeta) - \frac{c}{\Delta}\sinh(\Delta\zeta), \end{array} \tag{3.8.5}$$

where Δ is the square root of $\det\mathbf{Q}$: $\Delta = \sqrt{c^2 - ab}$.

In the group theory the real numbers (a, b, c) are known as the "coordinate" of the *canonical* representation of group elements of exponential type, i.e., representable in the form $e^{\zeta\,\mathbf{Q}}$ with the generating matrix $\mathbf{Q}$ being expressed in terms of the algebra basis $\{\mathbf{K}_-, \mathbf{K}_+, \mathbf{K}_3\}$, as shown in (3.8.3). As mentioned in § 2.3.1, there are portions of the symplectic group $Sp(2, \mathbb{R})$, whose elements can not be represented as exponential mappings of elements in its algebra $sp(2, \mathbb{R})$.

On the other hand, in § 3.5 we have described the LST representation of symplectic matrices, which locally applies to both exponential and non exponential type matrices. It involves three factors taken each from the one-parameter subgroups generated by the algebra basis elements $\{\mathbf{K}_-, \mathbf{K}_+, \mathbf{K}_3\}$. In the group theory, the LST-like realization is known as a *noncanonical* representation of symplectic matrices. The relevant parameters $(\mathcal{P}, m, d)$ can therefore be interpreted as the coordinates of the (local) non canonical LST representation of the exponential-type group element.

Evidently, the traceless matrix $\mathbf{Q}$ is equally expressible in terms of the algebra elements $\{\mathbf{K}_+, \mathbf{K}_3, \mathbf{K}_1\}$, which pertain to the Iwasawa representation discussed in § 3.7. In fact, one has

$$\mathbf{Q} = \mathfrak{a}\mathbf{K}_+ + 2\mathfrak{c}\mathbf{K}_3 + 2\mathfrak{b}\mathbf{K}_1, \tag{3.8.6}$$

with, of course,

$$\mathfrak{a} = a - b, \quad \mathfrak{b} = b, \quad \mathfrak{c} = c. \tag{3.8.7}$$

Then paralleling the relation between the coefficients (a, b, c) and the parameters $(\mathcal{P}, m, d)$ of the Wei-Norman representation, the coefficients $(\mathfrak{a}, \mathfrak{b}, \mathfrak{c})$ and the parameters $(\mathcal{P}, m, \phi)$ of the global LSF$^{\alpha}$ representation (3.7.1) are similarly interpretable as canonical and noncanonical (global) coordinates of the exponential-type group elements. The reader may deduce the relations between the canonical and noncanonical coordinates (a, b, c) and $(\mathcal{P}, m, d)$, as well as $(\mathfrak{a}, \mathfrak{b}, \mathfrak{c})$ and $(\mathcal{P}, m, \phi)$, using (3.8.5), (3.5.2) and (3.7.2) [4.3, 4.6].

3.8.1 *A bit on the non-exponential type elements of $Sp(2, \mathbb{R})$*

Let us summarize the basic results we have estabished up to now.

The symplectic group $Sp(2, \mathbb{R})$ is a three-parameter Lie group, which consists of all 2×2 real unimodular matrices $\mathbf{S}$. The unimodularity of elements in $Sp(2, \mathbb{R})$ is equivalent to the symplectic condition

$$\mathbf{S}\mathbf{J}\mathbf{S}^{\top} = \mathbf{J}, \tag{3.8.8}$$

where $\mathbf{J} = \left(\begin{smallmatrix} 0 & 1 \\ -1 & 0 \end{smallmatrix}\right)$ is the unit symplectic matrix.

The Lie algebra $sp(2, \mathbb{R})$ consists of all 2×2 real traceless matrices. It is the algebra of $Sp(2, \mathbb{R})$ in the sense that the exponential of any matrix in $sp(2, \mathbb{R})$ is a matrix of $Sp(2, \mathbb{R})$; namely

$$\mathbf{E} = e^{\zeta \mathbf{Q}} \in Sp(2, \mathbb{R}) \quad \forall\, \mathbf{Q} \in sp(2, \mathbb{R}) \ \text{ and } \ \zeta \in \mathbb{R}. \tag{3.8.9}$$

However, not all elements in $Sp(2, \mathbb{R})$ can be interpreted as exponential of elements in the relevant algebra $sp(2, \mathbb{R})$.

We may convince ourselves of this by a simple reasoning. Let us consider the trace of any exponential-type matrix in $Sp(2, \mathbb{R})$. Being

$$\mathbf{E} = e^{\zeta \mathbf{Q}} = \sum_{j=0}^{\infty} \frac{\zeta^j}{j!} \mathbf{Q}^j, \tag{3.8.10}$$

the linearity of the trace allows us to write

$$Tr\mathbf{E} = \sum_{j=0}^{\infty} \tfrac{\zeta^j}{j!} Tr\, \mathbf{Q}^j = \sum_{j=0}^{\infty} [\tfrac{\zeta^{2j}}{(2j)!} Tr\, \mathbf{Q}^{2j} + \tfrac{\zeta^{2j+1}}{(2j+1)!} Tr\, \mathbf{Q}^{2j+1}], \tag{3.8.11}$$

where the even and odd power terms have been separated. Noting that

$$\mathbf{Q}^{2j} = (-\det \mathbf{Q})^j \mathbf{I}, \qquad \mathbf{Q}^{2j+1} = (-\det \mathbf{Q})^j \mathbf{Q}, \qquad j = 0, 1, 2, ... \tag{3.8.12}$$

on due account of that $Tr\mathbf{I} = 2$ and $Tr\mathbf{Q} = 0$, we end up with

$$Tr\mathbf{E} = 2 \sum_{j=0}^{\infty} (-)^j \tfrac{\zeta^{2j}}{(2j)!} (\det \mathbf{Q})^j = 2 \cos(\zeta \sqrt{\det \mathbf{Q}}). \tag{3.8.13}$$

Then comprising the case that $\det \mathbf{Q} > 0$, yielding the circular cosine ($-1 \leq \cos x \leq 1$), as well as the case that $\det \mathbf{Q} < 0$, yielding the hyperbolic cosine ($\cosh x \geq 1$), we can conclude that all the single-exponential type matrices $\mathbf{E}$ in $Sp(2, \mathbb{R})$ satisfy the relation

$$Tr\, \mathbf{E} \geq -2. \tag{3.8.14}$$

Evidently, symplectic matrice with traces < -2 exist, and so symplectic matrices that do not arise from exponential mappings of elements in the algebra exist. As basic examples, we may report the matrices (see also Problem 11)

$$\begin{pmatrix} -1 & B\neq 0 \\ 0 & -1 \end{pmatrix} \quad \text{and} \quad \begin{pmatrix} -1 & 0 \\ C\neq 0 & -1 \end{pmatrix}, \tag{3.8.15}$$

which in fact can be obtained as the product of two exponential type matrices:

$$\begin{pmatrix} -1 & B \\ 0 & -1 \end{pmatrix} = e^{2\pi \mathbf{K}_1} e^{-B\mathbf{K}_-} \quad \text{and} \quad \begin{pmatrix} -1 & 0 \\ C & -1 \end{pmatrix} = e^{2\pi \mathbf{K}_1} e^{C\mathbf{K}_+}. \tag{3.8.16}$$

3.9 Integrating the equation for the ray transfer matrix

We reconsider here the problem of solving Hamilton's equations for the optical matrix $\mathbf{M}(z, z_i)$,

$$\mathbf{M}(z, z_i) = \begin{pmatrix} A(z, z_i) & B(z, z_i) \\ C(z, z_i) & D(z, z_i) \end{pmatrix}, \tag{3.9.1}$$

for the propagation under the quadratic Hamiltonian (3.1.1), where the medium parameters $n_0(z)$ and $n_2(z)$ are in general z-dependent functions.

It has been proved in § 1.7.3 that the entries A, B, C, D of $\mathbf{M}(z, z_i)$ are obtained as solutions to the set of first-order differential equations

$$\begin{aligned} A' &= \frac{C}{n_0}, \quad A(z_i, z_i) = 1, \qquad B' = \frac{D}{n_0}, \quad B(z_i, z_i) = 0, \\ C' &= -n_2 A, \quad C(z_i, z_i) = 0; \qquad D' = -n_2 B, \quad D(z_i, z_i) = 1; \end{aligned} \tag{3.9.2}$$

the prime denoting derivative with respect to the axial coordinate z.

Interestingly, the optical equivalences established in the previous sections, in particular the LST parametrization, suggest alternative, and possibly further enlightening and effective, integration approaches to the set (3.9.2).

Since $A(z_i, z_i) = 1$ and $D(z_i, z_i) = 1$, the continuity of the elements of $\mathbf{M}(z, z_i)$ as functions of z, assures that $A(z, z_i) > 0$ and similarly $D(z, z_i) > 0$ in a neighbourhood of the initial position z_i. We can therefore resort to the factorization (3.5.1) in terms of the matrices $\mathbf{L}$, $\mathbf{S}$ and $\mathbf{T}$, thus expressing $\mathbf{M}(z, z_i)$ in terms of the z-dependent parameters $\mathcal{P}$, s and d as

$$\mathbf{M}(z, z_i) = e^{\mathcal{P}(z)\mathbf{K}_+} e^{s(z)\mathbf{K}_3} e^{d(z)\mathbf{K}_-}. \tag{3.9.3}$$

As prescribed by relations (3.5.2), the magnification factor $m(z) = e^{s(z)/2}$, the focal power $\mathcal{P}(z)$ and the reduced distance $d(z)$ are linked to the matrix entries A, B, C, D through

$$A = e^{s/2}, \quad C = -\mathcal{P}\, e^{s/2}\ , \quad B = d\, e^{s/2}. \tag{3.9.4}$$

Therefore, exploiting Eqs. (3.9.2) it is easy to obtain the set of differential equations for $\mathcal{P}$, s and d as

$$\begin{aligned} &\mathcal{P}' - \tfrac{1}{n_0}\mathcal{P}^2 - n_2 = 0, && \mathcal{P}(z_i) = 0, \\ &s' + \tfrac{2}{n_0}\mathcal{P} = 0, && s(z_i) = 0, \\ &d' - \tfrac{1}{n_0}e^{-s} = 0, && d(z_i) = 0. \end{aligned} \tag{3.9.5}$$

Equation (3.9.4) represents the link between the parametrizations (3.9.1) and (3.9.3) of the optical matrix $\mathbf{M}$. Indeed, once the explicit expressions for the overall matrix entries are known from (3.9.2), the decomposition (3.9.3) is uniquely determined through the algebraic relations (3.9.4). Conversely, solving (3.9.5) for $\mathcal{P}$, s, d allows us to evaluate the matrix entries by (3.9.4).

Moreover, limiting ourselves to consider in (3.9.2) the equations for (A, C), as the entries (B, D) obey formally similar set of equations, we see that they yield two independent homogeneous linear equations of the second order for A and C; namely,

$$\begin{aligned} &A'' + \tfrac{n_0'}{n_0}A' + \tfrac{n_2}{n_0}A = 0, && A(z_i) = 1,\ A'(z_i) = 0, \\ &C'' - \tfrac{n_2'}{n_2}C' + \tfrac{n_2}{n_0}C = 0, && C(z_i) = 0,\ C'(z_i) = -n_2(z_i), \end{aligned} \tag{3.9.6}$$

The entries B and D obey the same equations as A and C, respectively, with the proper initial values being $B(z_i) = 0$, $B'(z_i) = \frac{1}{n_0(z_i)}$ and $D(z_i) = 1$, $D'(z_i) = 0$. Solving, for instance, the first of (3.9.6) for A and B, allows us to obtain C and D simply by integration of the relevant equations in (3.9.2).

Likewise, inspecting Eqs. (3.9.5), it is evident that solving for $\mathcal{P}$ allows us to integrate for s and then for d. In this connection, the equation for $\mathcal{P}$ deserves some comments. It is structured as the *generalized Riccati equation* [9]

$$\frac{dw}{dz} = a_0(z) + a_1(z)w + a_2(z)w^2, \quad w(z_i) = w_i. \tag{3.9.7}$$

whic, in fact, with $a_0(z) = n_2(z)$, $a_1(z) = 0$ and $a_2(z) = \frac{1}{n_0(z_i)}$ turns into the equation for $\mathcal{P}$. Setting

$$\mathcal{P} = -n_0(z)\frac{v'}{v}, \tag{3.9.8}$$

the equation for $\mathcal{P}$ is transformed into the homogeneous linear equation of the second order for the function $v(z)$

$$v'' + \frac{n_0'}{n_0}v' + \frac{n_2}{n_0}v = 0, \tag{3.9.9}$$

the same as the equation for A in (3.9.6).

The above expresses a general property of the Riccati equation (3.9.7), which is turned into a homogeneous linear equation of the second order by the substitution $w = -\frac{v'}{a_2(z)v}$. Conversely, any linear second order equation can be given the Riccati form by a suitable transformation of the dependent variable.

As a conclusion, we may say that the tracking of a paraxial ray propagating through an inhomogeneous medium depends on the solution of a homogeneous linear equation of the second order ((3.9.6) or (3.9.9)). From it all the parameters, which completely characterize the relevant optical matrix, i.e., the entries A, B, C, D or the functions $\mathcal{P}$, s, d of the possibile factorization in terms of the matrices $\mathbf{L}$, $\mathbf{S}$, $\mathbf{T}$, can straigthforwardly be obtained. We emphasize yet again that the parameters $\mathcal{P}$, s, d pertain to a local representation of the optical matrix $\mathbf{M}(z, z_i)$, whilst the entries A, B, C, D to a global representation.

In the next paragraph we discuss in detail the simple case where the medium parameters $n_0(z)$ and $n_2(z)$ are z-independent. In that case, the solutions to Eqs. (3.9.6) can be given in explicit form.

3.9.1 *Ray-tracing in a quadratic graded-index medium*

The ray propagation along a quadratic graded-index medium, described by the parabolic profile of the refractive index [10]

$$n(q, z) = n_0 - \tfrac{1}{2}n_2 q^2 \tag{3.9.10}$$

where the on-axis value n_0 and the transverse derivative n_2 are both constant with the axial coordinate z, is explicitly solvable[c]. In fact, with z-independent medium parameters, Eqs. (3.9.6) take the form

$$\frac{d^2}{dz^2}y + \frac{n_2}{n_0}y = 0, \tag{3.9.11}$$

with the initial values being assigned according to the specific entry, A or C, we are solving for.

The character of the solution of the above equation, and hence the behaviour of the light ray in the medium, depends on the sign of n_2. Accordingly, we may distinguish between focusing and defocusing media.

---

[c] As already noted, the index function (3.9.10) is valid over a limited range of the transverse coordinate q, appropriate to the linear approximation of our concern, as it becomes unphysical if n becomes less than the refractive index in the vacuum, generally taken as unity.

Focusing quadratic medium

A focusing index profile, for which $n_2 > 0$, exhibits a maximum on the optical axis and decreases radially from it. The general solution of Eq. (3.9.11) writes in terms of the harmonic functions as

$$y(z) = c_1 \cos\zeta + c_2 \sin\zeta, \tag{3.9.12}$$

where c_1 and c_2 denote two arbitrary constants, determined by the initial values, and ζ is the dimensionless axial coordinate

$$\zeta(z) = \sqrt{\frac{n_2}{n_0}}(z - z_i). \tag{3.9.13}$$

Determining c_1 and c_2 according to the initial values appropriate to the entries A, B, C or D, we obtain

$$\begin{aligned} A &= \cos\zeta, & B &= \frac{1}{\sqrt{n_0 n_2}}\sin\zeta\,, \\ C &= -\sqrt{n_0 n_2}\sin\zeta\,, & D &= \cos\zeta. \end{aligned} \tag{3.9.14}$$

It is apparent that the field ray corresponds to the cosine solution of Eq. (3.9.11) as the axial ray to the sine solution.

The optical matrix $\mathbf{M}(z, z_i)$ writes therefore

$$\mathbf{M}_{\text{foc}}(z, z_i) = \begin{pmatrix} \cos\zeta & \frac{1}{\sqrt{n_0 n_2}}\sin\zeta \\ -\sqrt{n_0 n_2}\sin\zeta & \cos\zeta \end{pmatrix}, \tag{3.9.15}$$

and the ray variables change with the axial coordinate according to

$$\begin{aligned} q(z) &= \quad q_i \cos\zeta + \frac{p_i}{\sqrt{n_0 n_2}}\sin\zeta, \\ p(z) &= -\, q_i \sqrt{n_0 n_2}\sin\zeta + p_i \cos\zeta. \end{aligned} \tag{3.9.16}$$

The ray oscillates back and forth across the axis with the spatial period

$$\mathcal{Z} = 2\pi\sqrt{\frac{n_0}{n_2}}, \tag{3.9.17}$$

known as the *pitch* (Fig. 3.19). The maximum transverse excursion of the ray is calculated to be $q_{\text{max}} = \sqrt{q_i^2 + p_i^2/n_0 n_2}$ and the maximum angle the ray forms with the axis along its path is $\vartheta_{\text{max}} = p_{\text{max}}/n_0 = \sqrt{n_2/n_0}\,q_{\text{max}}$. If the width of the material is greater than $2q_{\text{max}}$, the ray remains confined within the medium, which then acts as a light guide.

The matrix (3.9.18) has the same form as the fractional Fourier transform matrix (3.6.6). In fact, we can write

$$\mathbf{M}_{\text{foc}}(z, z_i) = \mathbf{F}^{\alpha(z)}(f_S), \tag{3.9.18}$$

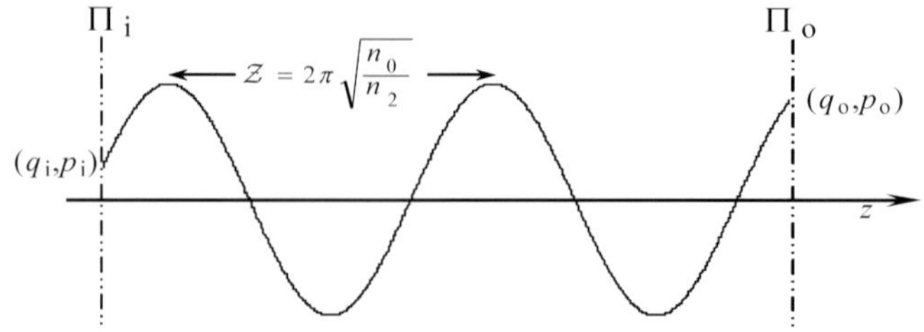

FIGURE 3.19. Ray trajectory in a focusing medium.

where the standard focal length f_S is identified as

$$f_S = \frac{1}{\sqrt{n_0 n_2}} \tag{3.9.19}$$

and the order α ranges continuously with the axial coordinate z as expected according to relation (3.6.7), which gives then

$$\zeta(z) = \alpha(z)\tfrac{\pi}{2} = \tfrac{z-z_i}{n_0 f_S}.$$

As already noted, in fact, a quadratic graded-index focusing medium provides the optical system to implement the fractional Fourier transforming operation with a continuously ranging order. The oscillations of the ray path across the optical axis correspond to the continuous rotation of the representative point in the phase plane, which describes the ellipse (3.6.9) determined by the medium parameters n_0 and n_2, and the initial data.

As the order α is directly linked to the propagation variable z, the system may act as a Fourier transforming system at the proper distance L_F from the input screen. In fact, if L_F is such that $\alpha(L_F + z_i) = 1$, and hence

$$L_F = \tfrac{\pi}{2}\sqrt{\frac{n_0}{n_2}} = \frac{\mathcal{Z}}{4}, \tag{3.9.20}$$

the propagation through the medium over the distance L_F from the input plane results in the ordinary Fourier transformation with focal lenth f_S. This establishes the possibility of optically realizing a Fourier transformation as well by a piece of a quadratic medium of proper length L_F.

At distances αL_F from z_i, the system behaves as a Fourier transformer of fractional order α. Then we can imagine the graded-index medium as formed by a continuum of planes from z_i along the optical axis; on each plane at z we can observe the fractional Fourier transform of order $\alpha = \frac{z-z_i}{L_F}$ of the object at z_i. At planes regularly placed from the input plane at distances L_F, $2L_F$,

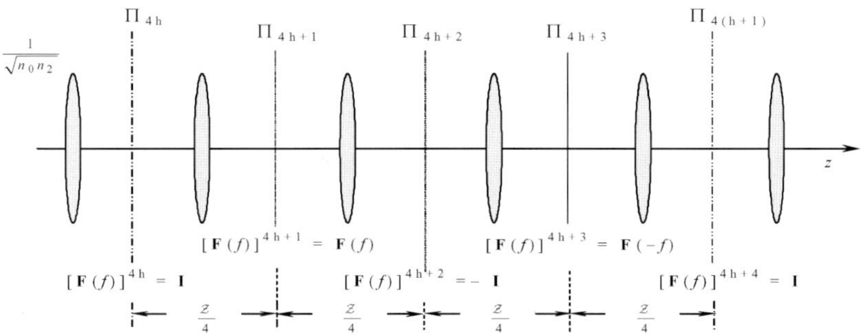

FIGURE 3.20. A quadratic focusing medium is like a sequence of Fourier transformers.

$3L_F$, and so on, we may detect the successive powers of the Fourier transform of the object; specifically, at distance jL_F we observe the j-th power of the Fourier transform of the object (Fig. 3.20). For $j = 4h$, we recover the identity matrix and hence the input signal, according to the periodicity of the Fourier transform operation[d]. The overall medium can be understood as a sequence of identical Fourier transforming configurations, L_F fixing the z-size of the Fourier transform units, by which the medium may be thought as composed.

Defocusing quadratic medium

For a defocusing index profile: $n_2 < 0$, which exhibits a minimum on the axis and increases far from it, Eq. (3.9.11) is solved by hyperbolic sine and cosine functions. Hence the optical matrix writes as

$$\mathbf{M}_{\text{def}}(z, z_i) = \begin{pmatrix} \cosh\xi & \frac{1}{\sqrt{-n_0 n_2}}\sinh\xi \\ \sqrt{-n_0 n_2}\sinh\xi & \cosh\xi \end{pmatrix}, \quad \xi(z) \equiv \sqrt{-\frac{n_2}{n_0}}(z - z_i) \tag{3.9.21}$$

which has evidently the same form as the squeeze matrix (3.6.36), being indeed

$$\mathbf{M}_{\text{def}}(z, z_i) = \mathbf{B}(\xi, f_S), \quad f_S = \frac{1}{\sqrt{-n_0 n_2}}. \tag{3.9.22}$$

As the height of the ray with respect to the axis increases while propagating from the input plane, the representative point in the phase plane moves along the branch of the hyperbola (3.6.39). The propagation along a defocusing quadratic medium physically realizes the three-step phase-plane transformation (3.6.50) involving squeezes and scaled rotations.

---

[d] More precisely, as it will be shown in § 5.4, the Fourier transformed signal is affected by a phase factor, which obviously cannot be accounted for within the ray-optics description of the performance of the optical system.

Being $\cosh\xi > 0$, the Wei-Norman decomposition can safely be applied through the whole ξ (i.e., z) domain. This is left as an exercise to the reader.

3.10 Summary

It has been proved that every first-order optical system can be understood as a sequence of thin lenses separated by free-propagation sections, visualizable a sequence of vertical and horizontal shears in the phase plane view.

The optical systems, generated by the attractive and repulsive oscillator-like Hamiltonians $\frac{1}{2}p^2 + \frac{1}{2}q^2$ and $\frac{1}{2}p^2 - \frac{1}{2}q^2$, have been examined. The former is recognized as performing scaled rotations in the phase plane, whereas the latter acts as a squeezing transformation.

Two interesting realizations of optical matrices have been described in correspondence with the two algebra basis $\{\mathbf{K}_-, \mathbf{K}_+, \mathbf{K}_3\}$ and $\{\mathbf{K}_-, \mathbf{K}_1, \mathbf{K}_3\}$. One realization, which applies locally near the entrance screen, synthesises the optical matrix as a sequence of a (positive or negative) free-propagation section, a positive-definite pure magnifier and a thin lens, and accordingly is visualized in the optical phase plane as a horizontal shear followed by a scaling and a vertical shear. The other realization allows us to describe the system as a sequence of a fractional Fourier transforming system, a positive-definite pure magnifier and a thin lens, and hence is pictured in the phase plane as a rotation followed by a scaling and then by a vertical shear.

In the next Chapter, the light propagation will be reconsidered within the framework of wave optics. We will investigate the ray-wave connection, working out the operator representation of the linear $1D$ optical systems through a sort of quantization procedure, similar to that applied to classical mechanics to yield quantum mechanics. Ray vectors will be replaced by wave functions and the unimodular matrices of the symplectic group $Sp(2,\mathbb{R})$ will be replaced by the unitary operators of the metaplectic group $Mp(2,\mathbb{R})$.

Problems

1. Consider a thick lens, consisting of two refracting surfaces with curvature radius R_1 and R_2 separated by a distance d on the optical axis. Denote by n_{0_1} and n_{0_2} the refractive indices in the object and image space, and by n the refractive index of the lens material. Take the reference planes immediately at the left and at the right of the lens (Fig. 3.3.a)). *(a)* Find the positions of the principal, focal and nodal points of a the lens. *(b)* Express the thick lens matrix (3.2.4) in terms of the coordinates of the principal points s_{H} and $s_{\mathrm{H}'}$. *(c)* Write down the explicit expression of the reduced focal length of a thin lens in terms of the relevant parameters, i.e., n_{0_1}, n_{0_2}, n, R_1 and R_2.

2. Enlarge configuration (3.3.13) by *(a)* an additional free-propagation section placed on the right or the left, and *(b)* two additional free-propagation sections placed on both sides. Write down in both cases the relations for the relevant parameters and individualize some configuration in accord with the obtained relations.

3. Specialize the parameters of the LTL configurations, which may synthesize symplectic matrices having *(a)* $D = 0$, *(b)* $A = 0$, and *(c)* $A = D = 0$. Specify in every case the restrictions on the non null matrix elements.

4. Specify the parameters $\mathcal{P}$, m and d for the "dual" decomposition:

$$\left(\begin{smallmatrix} A & B \\ C & D \end{smallmatrix}\right) = \mathbf{T}(d)\mathbf{S}(m)\mathbf{L}(\mathcal{P}), \qquad D > 0,$$

which reproduces what in the group theory is called the antinormal ordering. In turn, the LST realization (3.5.1) reproduces the so called normal ordering.

5. Prove relation (3.6.15) for $\alpha = \frac{1}{m}$, $m = 1, 2, ..$(Hint: Rewrite (3.6.15) as $[\mathbf{F}^{\alpha}(f_S)]^{1/\alpha} = \mathbf{F}(f_S)$ and use Sylvester's theorem according to which

$$\left(\begin{smallmatrix} A & B \\ C & D \end{smallmatrix}\right)^m = \frac{1}{\sin(m\Theta)} \begin{pmatrix} A\sin(m\Theta) - \sin(m-1)\Theta & B\sin(m\Theta) \\ C\sin(m\Theta) & D\sin(m\Theta) - \sin(m-1)\Theta \end{pmatrix}, \quad \cos\Theta = \frac{A+D}{2}.)$$

6. Decompose the fractional Fourier transforming matrix $\mathbf{F}^{\alpha}(f_S)$ into the form $\mathrm{S}^{-1}\mathrm{F}^{\alpha}\mathrm{S}$ for negative values of f_S.

7. Write down the relation between the matrix entries (A, B, C, D) and the the parameters $(\mathcal{P}, m, \phi)$ of the $\mathcal{KAN}$ decomposition:

$$\left(\begin{smallmatrix} A & B \\ C & D \end{smallmatrix}\right) = \mathbf{F}^{\alpha}\mathbf{S}(m)\mathbf{L}(\mathcal{P}).$$

8. Consider two symplectic matrices $\mathbf{M}_1$ and $\mathbf{M}_2$, whose relative $\mathcal{NAK}$ decompositions are given with the appropriate parameters being $\mathcal{P}_1$, m_1, ϕ_1 and $\mathcal{P}_2$, m_2, ϕ_2. Prove that the parameters $\mathcal{P}$, m, ϕ of the LSF$^{\alpha}$ synthesis of the product matrix

$$\mathbf{M}_2\mathbf{M}_1 = \mathbf{L}(\mathcal{P})\mathbf{S}(m)\mathbf{F}^{\alpha}$$

are

$$m = \overline{m}m_2, \quad \phi = \phi_1 + \overline{\phi}, \quad \mathcal{P} = \mathcal{P}_2 + \frac{\overline{\mathcal{P}}}{m_2^2},$$

where

$$\begin{aligned} \overline{m}^2 &= m_1^2\cos^2\phi_2 + (m_1^{-2} + m_1^2\mathcal{P}_1^2)\sin^2\phi_2 - m_1^2\mathcal{P}_1\sin(2\phi_2), \\ \overline{\mathcal{P}} &= \frac{1}{2\overline{m}^2}[2m_1^2\mathcal{P}_1\cos(2\phi_2) + (m_1^2 - m_1^{-2} - m_1^2\mathcal{P}_1^2)\sin(2\phi_2)], \\ \tan\overline{\phi} &= \frac{\tan\phi_2}{m_1^2(1-\mathcal{P}_1\tan\phi_2)}. \end{aligned}$$

(Hint: Write the Iwasawa decomposition of the internal sequence $\mathbf{F}^{\alpha_2}\mathbf{L}(\mathcal{P}_1)\mathbf{S}(m_1) = \mathbf{L}(\overline{\mathcal{P}})\mathbf{S}(\overline{m})\mathbf{F}^{\overline{\alpha}}$ in the product $\mathbf{M}_2\mathbf{M}_1$; then combine the resulting sequential rotations, lenses and magnifiers).

9. Identify the parameters $(\mathcal{P}, m, \phi)$ of the $\mathcal{NAK}$ decomposition of the squeeze matrix (3.6.49):

$$\mathbf{B}(\tau, \theta) = \mathbf{L}(\mathcal{P})\mathbf{S}(m)\mathbf{F}^{\alpha},$$

in terms of the squeeze parameters τ, θ.

10. Prove that two squeezes into different directions do not multiply into a single squeeze. (Hint: Consider as an example the product $\mathbf{B}(\tau_2, 0)\mathbf{B}(\tau_1, \theta_1)$; identify the parameters of the relevant $\mathcal{NAK}$ decomposition; then, verify that they do not reproduce the results of Problem 9, thus proving that $\mathbf{B}(\tau_2, 0)\mathbf{B}(\tau_1, \theta_1) = \mathbf{B}(\tau, \theta)\mathbf{F}^{\beta}$. It is also possible to express $(\tau, \theta, \psi = \beta\frac{\pi}{2})$ in terms of $(\tau_2, \tau_1, \theta_1)$.)

11. Verify that the matrices

$$\begin{pmatrix} -\cos\phi & \sin\phi \\ -\sin\phi & -\cos\phi \end{pmatrix} \quad \text{and} \quad \begin{pmatrix} -\cosh\tau & \sinh\tau \\ \sinh\tau & -\cosh\tau \end{pmatrix},$$

are not of exponential type. (Hint: Write down the above matrices in the neighbourhood of the identity, i.e., respectively as $\mathbf{I}+\phi\mathbf{Q}_1+\mathcal{O}(\phi^2)$ and $\mathbf{I}+\tau\mathbf{Q}_2+\mathcal{O}(\tau^2)$. Verify that $\mathbf{Q}_1$ and $\mathbf{Q}_2$ do not belong to $sp(2, \mathbb{R})$.)

12. Infer the equations for the parameters $\mathcal{P}$, m and ϕ relevant to the Iwasawa decomposition of the optical matrix $\mathbf{M}(z, z_i)$ of the parabolic index medium.

References

[1] For the ray matrix approach to linear optics see [12] of ch. 2.

[2] For theoretical and applicative considerations on the symplectic Lie algebra and group, see [9] and [10] in ch. 1. More recent investigations, in relation as well to geometrical optics, can be found in [11] of ch. 2. The non exponential nature of the symplectic group $Sp(2, \mathbb{R})$ is analysed in

[2.1] R. Simon and N. Mukunda, "The two-dimensional symplectic and metaplectic groups and their universal cover", in *Symmetries in Science VI*, B. Gruber (ed.) (Plenum Press, New York, 1993), pp. 659-689.

[3] The relation between "mathematics" and "optics" of the 2×2 real symplectic matrix is investigated in

[3.1] H. H. Arsenault, "Generalization of the principal plane concept in matrix optics", Am. J. Phys. **48**, 397-399 (1980).

[3.2] M. Nazarathy and J. Shamir, "First-order optics - a canonical operator representation: lossless systems", J. Opt. Soc. Am. **72**, 356-364 (1982).

[3.3] H. H. Arsenault and B. Macukow, "Factorization of the transfer matrix for symmetrical optical systems", J. Opt. Soc. Am. **73**, 1350-1359 (1983).

[3.4] B. Macukow and H. H. Arsenault, "Matrix decomposition for nonsymmetrical optical systems", J. Opt. Soc. Am. **73**, 1360-1366 (1983).

[3.5] E.C.G. Sudarshan, N. Mukunda and R. Simon, "Realization of first order optical systems using thin lenses", Optica Acta **32**, 855-872 (1985).

[3.6] R. Simon and K.B. Wolf, "Structure of the set of paraxial optical systems", J. Opt. Soc. Am. A **17**, 342- 355 (2000).

[3.7] K.B. Wolf, *Geometric Optics on Phase Space* (Springer, Berlin, 2004), ch. 10, pp. 179-204.

[4] The operator (or matrix) decomposition in terms of the algebra generators $\widehat{\mathrm{K}}_{+}$, $\widehat{\mathrm{K}}_{-}$ and $\widehat{\mathrm{K}}_{3}$ (or $\mathbf{K}_{+}$, $\mathbf{K}_{-}$, $\mathbf{K}_{3}$) is illustrated in several papers in relation with both the unitary evolution operators associated with quantum systems and the symplectic maps associated with Hamiltonian systems. We quote

[4.1] J. Wei and E. Norman, "Lie algebraic solution of linear differential equations", J. Math Phys. **4**, 575-581 (1963).

[4.2] R. M. Wilcox, "Exponential operators and parameter differentiation in quantum physics", J. Math. Phys. **8**, 962-982 (1967).

[4.3] D.T. Truax, "Baker-Campbell-Hausdorff relations and unitarity of $SU(2)$ and $SU(1,1)$ squeeze operators", Phys. Rev. D **31**, 1988-1991 (1985).

[4.4] G. Dattoli, J. C. Gallardo and A. Torre, "An algebraic view to the operatorial ordering and its applications to optics", Riv. Nuovo Cim. **11**, 11, 1-79 (1988).

[4.5] G. Dattoli and A. Torre, "A general view to Lie algebraic methods in applied mathematics, optics and transport systems for charged beam accelerators", in *Dynamical Symmetries and Chaotic Behaviour in Physical Systems*, G. Maino, L. Fronzoni and M. Pettini (eds.) (World Sc. Publ., Singapore, 1991), pp. 21-89.

[4.6] M. Ban, "Decomposition formulas for $SU(1,1)$ and $SU(2)$ Lie algebras and their applications in quantum optics", J. Opt. Soc. Am. B **10**, 1347-1359 (1993).

[4.7] A. Torre, "Lie algebra treatment of two and three-mode interaction Hamiltonians", Recent Res. Devel. Quantum Electronics, 2, 111-131 (2000).

[4.8] A. J. Dragt and J.M. Finn, "Lie series and invariant functions for analytic symplectic maps", J. Math. Phys. **17**, 2215-2227 (1976).

[4.9] A. J. Dragt, "Lectures on nonlinear orbit dynamics", *Physics of High Energy Particle Accelerators*, AIP Conference Proceedings, R. A. Carrigan, F. R. Huson and M. Month (eds.) 87 (Amer. Inst. Phys., New York, 1982), pp. 147-313.

[4.10] A. J. Dragt, "Lie algebraic theory of geometrical optics and optical aberrations", J. Opt. Soc. Am. **72**, 372-378 (1982).

[4.11] A. J. Dragt, E. Forrest and K. B. Wolf, "Foundations of a Lie algebraic theory of geometrical optics", *Lie Methods in Optics*, J. Sánchez Mondragon and K.B. Wolf (eds.) (Springer-Verlag, Berlin, 1986), ch. 4, pp. 105-157.

[5] The bibliography about the fractional Fourier transform and its various applications in optics is extremely wide. Here we quote the seminal papers by Condon and Namias as well as the basic papers by Mendlovic and Ozaktas, and by Lohmann, through which the concept of fractional Fourier transform has been introduced in optics. A wider view on the subject can be found in the review papers [5.7]-[5.9], and in the comprehensive book [5.10]. Fractional Fourier transformers are also investigated in [3.7] (see, in particular, ch. 15, pp. 341-353).

[5.1] E. U. Condon, "Immersion of the Fourier transform in a continuous group of functional transformations", Proc. Natl. Acad. Sci. USA **23**, 158-164 (1937).

[5.2] V. Namias, "The fractional order Fourier transform and its application to quantum mechanics", J. Inst. Maths. Applics. **25**, 241-265 (1980).

[5.3] D. Mendlovic and H.M. Ozaktas, "Fourier transforms of fractional order and their optical interpretation: I", J. Opt. Soc. Am. A **10**, 1875-1881 (1993); II, *ibid.*, 2522-2531.

[5.4] A.W. Lohmann, "Image rotation, Wigner rotation, and the fractional order Fourier transform", J. Opt. Soc. Am. A **10**, 2181-2186 (1993).

[5.5] D. Mendlovic, H.M. Ozaktas and A.W. Lohmann, "Graded-index fibers, Wigner distribution functions, and the fractional Fourier transform", Appl. Opt. **33**, 6188-6193 (1994).

[5.6] G.S. Agarwal and R. Simon, "A simple realization of fractional Fourier transformation and relation to harmonic oscillator Green's function", Opt. Commun. **110**, 23-26 (1994).

[5.7] A.W. Lohmann, D. Mendlovic and Z. Zalevsky, "Fractional transformations in optics", in *Progress in Optics*, Vol. XXXVIII, E. Wolf, ed. (Elsevier, Amsterdam, 1997), ch. 4, pp. 263-342.

[5.8] H. M. Ozaktas, M.A. Kutay and D. Mendlovic, "Introduction to the fractional Fourier transform and its applications", in *Advances in Imaging and Electron Physics*, Vol. 106, P.W. Hawkes (ed.) (Academic Press, San Diego, 1999), ch. 4, pp. 239-291.

[5.9] A. Torre, "The fractional Fourier transform and some of its applications to optics", in *Progress in Optics*, Vol. 43, E. Wolf (ed.) (Elsevier, Amsterdam, 2002), ch. 7, pp. 521-586.

[5.10] H.M. Ozatkas, Z. Zalevsky and M.A. Kutay, *The Fractional Fourier Transform with Applications in Optics and Signal Processing* (Wiley, New York, 2001).

[6] The link between the symplectic group $Sp(2,\mathbb{R})$ and the little Lorentz group $SO(2,1)$ is investigated in [3.7], ch. 5, pp. 69-90, and in

[6.1] Y.S. Kim and M.E. Noz, "Dirac's light-cone coordinate system", Am. J. Phys. **50**, 721-724 (1982).

[6.2] Y.S. Kim and M.E. Noz, "Illustrative examples of the symplectic group", Am. J. Phys. **51**, 368-375 (1983).

[6.3] D. Han, Y.S. Kim and M.E. Noz, "Linear canonical transformations of coherent and squeezed states in the Wigner phase space", Phys. Rev. A **37**, 807-814 (1988);II Quantitative analysis", Phys. Rev. A **40**, 902-912 (1989).

[6.4] Y.S. Kim and M.E. Noz, *Phase Space Picture of Quantum Mechanics* (World Sc. Publ., Singapore, 1991).

[6.5] P.K. Aravind, "Simulating the Wigner angle with a parametric amplifier", Phys. Rev. A**42**, 4077-4084 (1990).

[6.6] F. Ciocci, G. Dattoli, C. Mari and A. Torre, "Measuring the Wigner rotation with electron beams", Phys. Rev. A **46**, 5149-5153 (1992).

[6.7] D. Han, Y.S. Kim and M.E. Noz, "Wigner rotations and Iwasawa decompositions in polarization optics", Phys. Rev. E **60**, 1036-1041 (1999).

[6.8] J.J. Monzón and L.L. Sánchez-Soto, "Multilayer optics as an analog computer for testing special relativity", Phys. Lett. A**262**, 18-26 (1999).

[6.9] T. Yonte, J.J. Monzón, L.L. Sánchez-Soto, J.F. Cariñena and C. López-Lacasta, "Understanding multilayers from a geometrical viewpoint", J. Opt. Soc. Am. **19**, 603-609 (2002).

[7] For a discussion of the Iwasawa decompostion theorem within the general group theory context we may suggest

[7.1] K. Iwasawa, "On the representations of Lie algebras", Jpn. J. Math. **19**, 513-523 (1948).

[7.2] K. Iwasawa, "On some types of topological groups", Ann. Math. **50**, 507-558 (1948).

[7.3] S. Helgason, *Differential Geometry and Symmetric Spaces* (Academic Press, New York, 1962), pp. 219-225.

[7.4] S. Lang, *"$SL_2(R)$"* (Addison Wesley, New York, 1993), pp. 659-689.

[8] Applications of the Iwasawa decomposition to optical problems are illustrated in [3.6], [6.7] [6.9], [3.7], ch. 9, pp. 173-177, and in

[8.1] R. Simon and N. Mukunda, "Iwasawa decomposition for $SU(1,1)$ and the Güoy effect for squeezed states", Opt. Commun. **95**, 39-45 (1993).

[8.2] R. Simon and N. Mukunda, "Iwasawa decomposition in first-order optics: universal treatment of shape-invariant propagation for coherent and partially coherent beams", J. Opt. Soc. Am. A **15**, 2146-2155 (1998).

[9] E.L. Ince, *Ordinary Differential Equations* (Dover Publ., New York, 1956).

[10] For a detailed ray tracing in a parabolic-index medium we suggest

[10.1] D. Marcuse, *Light Transmission Optics* (Van Nostrand-Reinhold, Princeton, 1972).

[10.2] E.W. Marchand, *Gradient Index Optics* (Academic Press, New York, 1978).

[10.3] S. Cornbleet, *Microwave and Geometrical Optics* (Academic Press, London, 1994).

[10.4] V. Lakshminarayanan, A.K. Ghatak and K. Thyagarajan, *Lagrangian Optics* (Kluwer Academic Publ., Boston, 2002).

4

Wave-Optical Picture of First-Order Optical Systems

4.1 Introduction

We elaborate the wave optical picture of light propagation through first-order optical systems within the context of an operator treatment, which parallels the matrix treatment of linear ray optics, as presented in the previous chapters.

In accord with the scalar wave model of light, by which all the components of the electric or magnetic field associated with the optical wave can be treated independently, we deal with a single scalar function of position and time. Further we suppose that such a scalar function be the complex-valued amplitude of a single monochromatic component of the wave field at a certain temporal frequency ω, and so only its dependence on the spatial coordinates will be object of interest. Accordingly, the scalar Helmholtz equation and the parabolic wave equation, resulting from the former in the paraxial approximation, come to be the basic equations in our discussion.

As earlier emphasized, geometrical optics and classical mechanics display a structural similarity, both developing through the Hamiltonian formalism from analogous principles; the optics of light rays develops from the extremal Fermat principle as the mechanics of material particles from the minimal Hamilton principle [1]. Likewise, the structural similarity between wave optics and quantum mechanics manifests into the formal identity of the Helmholtz wave equation for the scalar light field to the Klein-Gordon equation for the quantum mechanical wave function of a spinless free particle. Thus the parabolic wave equation for slowly varying envelope wave fields looks like the Schrödinger equation for non relativistic quantum particles [2].

Guided by these similarities, we will translate the matrix formalism of ray optics into an operator formalism, plainly retracing the line from classical to quantum mechanics. Since linear ray optics is in a sense like the particle dynamics under quadratic Hamiltonians, we will operate a kind of "waviza-

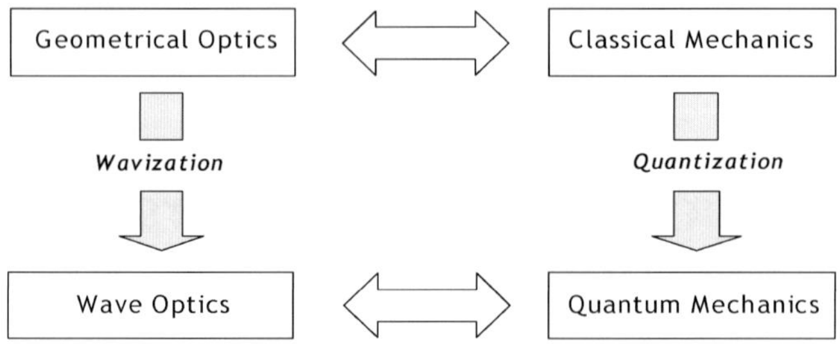

FIGURE 4.1. From geometrical optics to wave optics by a kind of a *wavization* procedure.

tion" [9.7] of the Hamiltonian dynamics under quadratic polynomials (Fig. 4.1). Thus we will elaborate the formalization of paraxial wave optics from that of paraxial ray optics, by properly replacing ray vectors by wave functions and accordingly ray-transfer operators by wave-transfer operators. More precisely, mirroring the ray optics picture of light propagation in terms of symplectic transformations acted by optical systems on points in the $4D$ (or, $2D$) phase space, we will describe the light propagation in terms of unitary transformations acted by optical systems on complex-valued functions belonging to the Hilbert space of square-integrable functions on the real line.

We will maintain the picture of the optical system as a sort of "black box between two reference screens" and that of light propagation as a "transformation" of the light disturbance between the two end-planes. The incoming light disturbance, whose distribution in the input plane is supposed to be known, constitutes the "excitation" to which the system responds producing another light disturbance, whose distribution in the output plane is to be determined. Thus the ray-optics picture of optical systems as "operators" mapping phase space points into phase space points, is paralleled by the wave-optics picture of optical systems as "operators" mapping square-integrable complex-valued functions into square-integrable complex-valued functions.

Linearity is a basic system property. It permits one to decompose any propagation problem into "pieces", which can easily be "analyzed" or plainly "solved"; then one must be able to compose the individual solved "pieces" back together to indicate the solution of the original problem. We are accordingly allowed to represent or to look for representing optical signals as weighted sums (or, integrals) of a set of pre-selected waveforms, which are so chosen that the responses of the system to any one of them be easily obtained and easily composed back to reconstruct the response of the system to the original signal. Delta functions and complex exponentials naturally arise as convenient mathematical waveforms for such a signal and system performance piecewise decomposition. Evoked as eigenstates of the position and momentum operators, delta functions and complex exponentials are the mathematical description of the intuitive view of idealized point-like and unidirectional disturbances. They will be used to elaborate two equivalent signal representations, as delta functions are functional to the spatial representation of the signal and complex exponentials are functional to the spectral representation of the signal. Notably, delta functions and complex exponentials are found to be the interchangeable object-image pair of a physically realizable optical system - two identical free-space sections intertwined by a proper thin lens, i.e. the Fourier transformer. In fact, whilst in general systems Fourier transforms do not represent physical signals, the optical Fourier transformer produces on the output plane the detectable (scaled) Fourier transform of the input waveform.

As a conclusion, we will see that a wave-optical phase space can be conceived as formed by the Fourier conjugate variables. Such variables span the precise domains where the two equivalent signal descriptions, spatial and spectral, may be elaborated, i.e., the domain of the spatial coordinates, over which the signal extends, and the domain of the propagation angles of the plane waves components contained in the signal. However, analysis, computation and visualization of the light propagation, and hence of the optical system behaviour, in this wave-optical phase space need a joint space-angle description of the signal and, of course, a knowledge of the action of the various elements of paraxial optics on such a joint signal representation.

It is assumed that the reader is familiar with the basics of wave optics, which therefore is only briefly reviewed in Sect. 4.2. In Sect. 4.3 the wave-transfer operator and the associated point-spread function are introduced as basic tools for the wave-optical description of optical systems. The formalization of paraxial wave optics by a kind of a quantization procedure of paraxial ray optics is elaborated in Sect. 4.4. The basic waveforms, i.e., unit impulses and spatial harmonics, for signal decompositions are discussed in Sect. 4.5. Finally, in Sect. 4.6 the spatial Fourier representation of the signal is described, and the conceptual introduction of the wave-optical phase space as a counterpart of the symplectic geometric-optical phase space is commented on.

4.2 Essentials of the scalar wave model of light. The paraxial wave equation in a quadratic medium

In accordance with the scalar wave model of light, the optical wave is described by a single scalar function of position and time $\psi(x, y, z; t)$, called the *optical wave function*. It may be related to any one of the rectangular components of the electric or magnetic field associated with the optical wave [3].

In a linear homogeneous and isotropic medium of refractive index n, the optical wave function obeys the *wave equation*

$$(\frac{\partial^2}{\partial x^2} + \frac{\partial^2}{\partial y^2} + \frac{\partial^2}{\partial z^2})\psi - \frac{n^2}{c^2}\frac{\partial^2}{\partial t^2}\psi = 0, \tag{4.2.1}$$

where c is the speed of light in the vacuum. Equation (4.2.1) is applicable also to inhomogeneous media as long as the relative refractive index $n(x, y, z)$ changes but slightly over distances compared with the wavelength of the electromagnetic field, as it is the case when optical fields are involved. In fact, the approximately scalar character of most optical systems stems from the fact, that usually the characteristic physical dimensions of the system components greatly exceed the wavelength of light. Moreover, it is consistent with the paraxial approximation that amounts to saying that the wave field is purely transverse. This is the inherent wave-optical characterization of light propagation through first-order optical systems, which are of primary interest here.

In general, an infinite number of solutions to the wave equation (4.2.1) exists. In fact, all the waveforms of the type

$$\psi(x, y, z; t) = \psi(k_x x + k_y y + k_z z - \omega t), \tag{4.2.2}$$

for any choice of the parameters k_x, k_y, k_z and ω satisfying the relation

$$k^2 \equiv k_x^2 + k_y^2 + k_z^2 = n^2\frac{\omega^2}{c^2}, \tag{4.2.3}$$

are solutions of the wave equation. The unique solution is individualized by the specific boundary conditions for the propagation problem at hand. This means that at a certain time the wave field over a certain region in the space is given, and the propagation in space and time is then calculated by (4.2.1).

It is a common practice to investigate the behaviour of the single monochromatic components of the light wave, separating the rapidly time-varying factor $e^{-i\omega t}$ at the temporal frequency ω from the stationary amplitude as follows

$$\psi(x, y, z; t) = \phi(x, y, z)e^{-i\omega t}. \tag{4.2.4}$$

Then the position dependent *complex amplitude* $\phi(x, y, z)$ assumes the role of the descriptor of the wave; it obeys the *reduced* or *time-free wave equation* (also known as the *scalar Helmholtz equation*)

$$(\frac{\partial^2}{\partial x^2} + \frac{\partial^2}{\partial y^2} + \frac{\partial^2}{\partial z^2})\phi + k^2\phi = 0, \tag{4.2.5}$$

the dispersion relation (4.2.3) linking the wave number k in the medium to the temporal frequency ω of the wave. Since the refractive index is allowed to vary in space, the wave number may accordingly be regarded as a position dependent function: $k(x, y, z) = k_0 n(x, y, z)$, with a space scale large compared with the mean wavelength λ of the wave; $k_0 = \frac{\omega}{c}$ is the vacuum wave number.

As we solely consider monochromatic waves, the terms wave function and complex amplitude are interchangeably used to denote a complex function of position only, obeying the time-free wave equation (4.2.5). Clearly the wave function and the complex amplitude are just mathematical entities; it is the real part of the wave function, i.e., of the complex amplitude times the harmonic exponential $e^{-i\omega t}$, that may represent a physical quantity like, as noted, the amplitude of any component of the electric or magnetic field vectors.

The plane waves and the spherical waves represent two sets of elementary solutions to the Helmholtz equation (4.2.5) in a homogeneous medium. Described respectively by the complex amplitudes

$$\begin{aligned} \phi_{\mathrm{P}}(x,y,z) &= e^{i(k_x x + k_y y + k_z z)} = e^{i\mathbf{k}\cdot\mathbf{r}}, \\ \phi_{\mathrm{S}}(x,y,z) &= \tfrac{\lambda}{r} e^{ikr}, \quad r > \lambda \end{aligned} \tag{4.2.6}$$

the plane and the spherical waves can pictorially be visualized by the relative sets of constant phase surfaces, separated by $\lambda = \frac{2\pi}{k}$ and extending through the whole $3D$ space with no spatial confinement. Thus the plane waves are depicted as an unlimited sequence of parallel planes orthogonal to the position-independent wave vector $\mathbf{k} = (k_x, k_y, k_z)$, which individualizes the direction of propagation of the wave, whilst the spherical waves are viewed as a set of far to infinite reaching concentric spheres. The plane waves are unidirectional, with the intensity of the wave (and so the energy flow), uniformly distributed over the infinite constant-phase planes, being constant along the direction of propagation, i.e. along $\mathbf{k}$. In contrast, the spherical waves have multidirectional features, with the intensity of the wave decreasing with the distance r from the central radiating source according to the inverse square law.

The hypothesis of paraxiality implies that the propagation occurs mostly in one direction, the z direction, for instance [4]. Then we can reasonably set

$$k_z \simeq k, \tag{4.2.7}$$

and accordingly write the complex amplitude for the general solution to (4.2.5) in the nearly plane-wave form

$$\phi(x,y,z) = \psi(x,y,z)e^{ikz}, \quad k = k_0 n_0 \tag{4.2.8}$$

where the modulating amplitude $\psi(x,y,z)$ varies slowly with z in comparison with the leading plane wave e^{ikz}. More in general, when propagation occurs in a weakly inhomogeneous medium the carrier phase factor is replaced as

$$e^{ik_0 n_0 z} \to e^{ik_0 \int_{z_i}^{z} n_0(z')dz'}, \tag{4.2.9}$$

thus accounting for the proper phase shift $ik_0 \int_{z_i}^{z} n_0(z')dz'$ experienced by the wave field along the z axis from the assigned reference plane at z_i; $n_0(z) \equiv$

$n(0, 0, z)$ denotes indeed the on-axis value of the refractive index. The slowly varying envelope $\psi(x, y, z)$ then obeys the *parabolic wave equation*

$$\frac{i}{k_0}\frac{\partial}{\partial z}\psi = -\frac{1}{2k_0^2 n_0(z)}\left(\frac{\partial^2}{\partial x^2} + \frac{\partial^2}{\partial y^2}\right)\psi + \frac{1}{2n_0(z)}\left[n_0^2(z) - n^2 - i\frac{n_0'(z)}{k_0}\right]\psi, \tag{4.2.10}$$

with the potential-like term $\frac{1}{2n_0(z)}[n_0^2(z) - n^2 - i\frac{n_0'(z)}{k_0}]\psi$ disappearing when propagation takes place in a homogeneous medium: $n(x, y, z) = n_0$. The prime denotes derivative with respect to z. On going from (4.2.5) to (4.2.10) through (4.2.8) the diffusive term $\left|\frac{\partial^2\psi}{\partial z^2}\right|$ has been neglected with respect to $\left|k\frac{\partial\psi}{\partial z}\right|$, in conformity with the assumption that significative variations of ψ occur over distances considerably exceeding the wavelength of light.

The hypothesis of paraxiality (4.2.7) amounts to turning the sphere of wave vectors (4.2.3) into a paraboloid. In fact, (4.2.7) is also equivalent to

$$k_x, k_y \ll k_z \approx k, \qquad \text{or} \qquad k_\perp \ll k_z \approx k, \tag{4.2.11}$$

where $k_\perp$ represents the transverse component of the wave vector comprising the x and y components as $k_\perp^2 = k_x^2 + k_y^2$. Thus the dispersion surface, which by (4.2.3) individualizes a sphere in the $3D$ space of the Cartesian components k_x, k_y and k_z of $\mathbf{k}$, or a circle in the $2D$ space of the transverse and longitudinal wave vector components $k_\perp$ and k_z, turns into

$$k_z = k - \frac{k_x^2 + k_y^2}{2k} = k - \frac{k_\perp^2}{2k}, \tag{4.2.12}$$

which can be viewed as a paraboloid in the $3D$ space spanned by (k_x, k_y, k_z) or a parabola in the $2D$ space spanned by $(k_\perp, k_z)$ (Fig. 4.2).

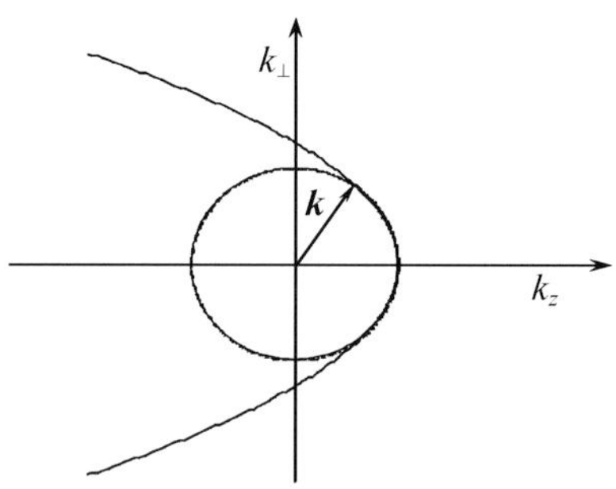

FIGURE 4.2. The paraxial approximation amounts to replacing the circle $k_\perp^2 + k_z^2 = k^2$ by the parabola $k_z = k - k_\perp^2/2k$. (Inspired by [3.3])

Evidently, the same occurs in geometric optics where the sphere of ray momenta $p_z^2 + p_x^2 + p_y^2 = n^2$ (Eq. (1.2.12)) is approximated by the paraboloid $p_z = n - \frac{p_x^2 + p_y^2}{2n}$, when paraxial propagation is assumed.

We can further shape the parabolic equation (4.2.10) for the wave amplitude $\psi(x, y, z)$, assuming that ψ changes smoothly through the transverse plane and is mostly concentrated near the axis so that the second-order approximation of the refractive index with respect to the transverse coordinates is allowed. Replacing $n^2(x, y, z)$ by the second-order expansion in x and y,

$$n^2(x, y, z) \simeq n_0^2(z) - n_0(z)n_{2x}(z)x^2 - n_0(z)n_{2y}(z)y^2, \tag{4.2.13}$$

one obtains the parabolic wave equation appropriate to describe the propagation of a wave field in a quadratic-like medium; namely,

$$\frac{i}{k_0}\frac{\partial}{\partial z}\psi = -\frac{1}{2k_0^2 n_0}\left(\frac{\partial^2}{\partial x^2} + \frac{\partial^2}{\partial y^2}\right)\psi + \frac{1}{2}\left[n_{2x}x^2 + n_{2y}y^2\right]\psi - i\frac{n_0'}{2k_0 n_0}\psi. \tag{4.2.14}$$

with the z-dependence of the refractive-index related functions n_0, n_{2x}, n_{2y} and n_0' being implied.

The term $-\frac{in_0'\psi(x,y,z)}{2k_0 n_0}$ above can be absorbed into a convenient redefinition of the field amplitude from $\psi(x, y, z)$ to $\psi = \frac{\Phi(x,y,z)}{\sqrt{n_0(z)}}$. The resulting equation for Φ formally resembles the Schrödinger equation for a quantum oscillator with two independent degrees of freedom and "time"-dependent "mass" and "frequencies". In fact, by a simple separation of the variables x and y as

$$\Phi(x, y, z) = \varphi_x(x, z)\varphi_y(y, z), \tag{4.2.15}$$

the $2D$ parabolic equation (4.2.14) turns into two $1D$ Schrödinger-like equations of the form

$$i\hbar\frac{\partial}{\partial t}\varphi(q, t) = \frac{1}{2}\left[\frac{1}{m(t)}\hat{\mathsf{p}}^2 + m(t)\Omega^2(t)\hat{\mathsf{q}}^2\right]\varphi(q, t), \tag{4.2.16}$$

with the mechanics-to-optics correspondences:

$$\hbar \to \frac{\lambda_0}{2\pi}, \quad t \to z, \quad m(t) \to n_0(z) \quad \Omega^2(t) \to \frac{n_2(z)}{n_0(z)}, \tag{4.2.17}$$

and the usual quantum mechanical definition of the position and momentum operators: $\hat{\mathsf{q}} = q$ and $\hat{\mathsf{p}} = -i\hbar\frac{\partial}{\partial q}$ (see § 4.4.1), q symbolizing any one of the profile coordinates x or y.

On deriving the parabolic wave equation (4.2.10) from the time-free wave equation (4.2.5) on the basis of the paraxial approximation, the questions concerning the compatibility of the inherent assumptions with the exact Maxwell's equations as well as the description of nonparaxial propagation have been ignored. The paraxial description of light propagation may safely be applied when the variations in the transverse profile of the optical field per wavelength are small and the propagation distance is not too long compared with the depth of focus of the optical field. Naturally, there are situations where the paraxial description is no longer adequate as it is the case of tightly focused light beams, semiconductor laser beams with large divergence angles and non-paraxial eigenmodes of stable resonators. A great variety of approaches has been devised to extend the treatment of scalar wave propagation beyond the paraxial approximation as well as to extend the treatment of scalar waves to vector waves, namely to electromagnetic waves satisfying Maxwell's equations. In [5.1], for instance, an expansion scheme of the formal solution to the exact Maxwell's equations for the electric field vector in powers of a suitably defined parameter is presented, and accordingly the recurrence equations for the correction terms of different orders are derived. The expansion parameter, which compares the space scales of the transverse and longitudinal variations of the wave amplitude, is understood as a measure of the beam divergence angle and so of the departure from the nearly parallel propagation of the beam to the axis. Further investigations have been performed in order to obtain exact rather than perturbative solutions to the Helmholtz wave equation. In [5.13] two families of exact solutions of the Helmholtz equation are presented: one in the oblate spheroidal coordinate system and the other in the prolate spheroidal coordinate system, the fundamental modes for the two families being respectively the propagating Gaussian beams and the spherical waves. The traditional Gaussian beam is then recovered within a nonparaxial description of wave-propagation, re-interpreting the beam wave front as a section of an oblate ellipsoid rather than of a sphere. Also, the potential of the complex point source generated fields [5.10], which are rigorous solutions of Maxwell's equations, to describe physical nonparaxial beam-like waves has deeply been investigated, thus suggesting a certain number of models to avoid the non-physical singularities inherently carried by this kind of solutions [5.12]. For a detailed investigation, the reader may consult [5].

4.3 Ray and wave optics

Geometric and wave optics are intimately connected, with the results of geometrical optics being recoverable as a limiting case of wave optics, but not the other way around. In the short wavelength limit, in fact, (or, more precisely, in the asymptotic approximation for large wave numbers) the solution of the time-independent wave equation yields the eikonal equation of geometric optics [6]. In contrast, wave optics cannot be recovered from geometrical optics by purely algebraic transformations, but by a sort of quantization procedure. Wave optics can be regarded as a *quantized* geometrical optics. A *quantum* theory of light rays has been elaborated from both a differential and integral equation approach, applying the same rules for transition from classical to quantum mechanics, according to which variables are replaced by linear operators [7]. Notably, quantization of the equations of geometric optics leads to the scalar Helmholtz equation, and accordingly the paraxial limit of wave optics corresponds to the non relativistic limit of quantum mechanics. Thus, the relation between wave optics and ray optics is similar to the relation between quantum mechanics and classical mechanics, the light wavelength in the vacuum being like the Planck constant. We recall also that the axial coordinate of optics is like the time coordinate of mechanics.

Geometrical optics grounds on the concept of *ray of light*, whose trajectories are uniquely determined by the ray-equation, or equivalently by the optical Hamiltonian equations, supplemented with the proper initial values for the ray position and direction. Correspondingly, wave optics is based on the concept of *optical wave function,* whose behaviour is uniquely determined by the reduced wave equation, supplemented with the appropriate boundary conditions for the optical field. In both cases, the optical properties of the medium, in which the wave is propagating, are accounted for by the refractive index, assigned (in the case of a nondispersive medium of concern here) as a scalar function of space position only. On the other hand, instead with the properties and the tracing of the ray paths, geometrical optics may alternatively be concerned with the properties and the construction of the surfaces to which the ray trajectories are normal, the so called wave-surfaces or *wave fronts*. The eikonal is representative of such a "wave front" approach of geometrical optics to light propagation; it is a scalar potential that can be interpreted as an optical path length from some reference wavefront surface. As Fermat's principle is the building block for the ray-tracing procedure, the eikonal equation is the basic tool for the wavefront-evolving investigations, since it determines the transformations experienced through propagation by the equilevel surfaces of the eikonal. The eikonal is recovered in wave optics as the phase of the quasi-plane wave expression of the light field as it results from the zero-order term of the asymptotic representation of the solution to the Helmholtz equation in powers of the inverse of the wave number. In wave optics the concept of *ray* is not elementary at all; its place is taken by the simplest solutions to the Helmholtz equation in a homogeneous medium: the spatial harmonic plane waves and the spherical waves, which, originating respectively from a harmonic disturbance distributed over a plane and from a light point, retain in the respective sources the basic attributes of the ray, the direction and the location. The concept of light ray contrasts with the uncertainty relation, as it inherently implies the precise specification of both the location and the direction of the ray, which is precluded by the classical wave properties. In wave optics, light rays are individualized as the trajectories perpendicular to the wave fronts of the propagating optical disturbance. Figure 4.3 offers a naïve illustration of the ray-to-wave relation in the case we are considering of plane and spherical

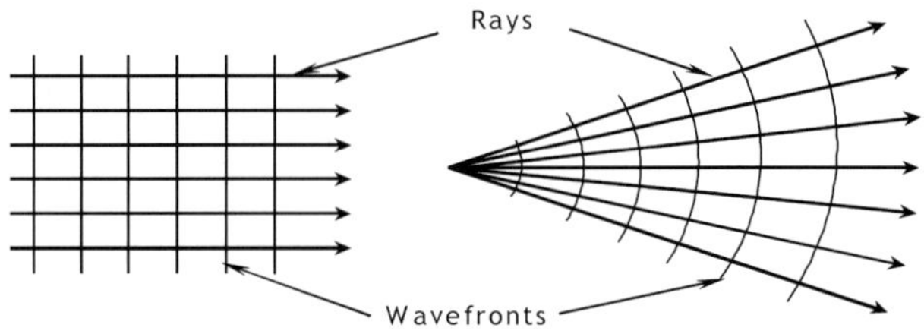

FIGURE 4.3. The ray of geometrical optics are orthogonal to the wave fronts of wave optics. The wave fronts are the surfaces along which the eikonal remains constant.

waves. The constant phase surfaces are parallel planes and concentric spheres respectively, and accordingly the ray trajectories trace straight lines, reciprocally parallel in the direction of propagation or emanating outward in all direction from the central source.

However, as the "ray", both the plane and spherical waves are idealized entities, that can hardly be realized in practice demanding for an infinite extent planar source or in contrast a point-like source, respectively. Nevertheless, they are useful mathematical models for various situations. Plane waves are the basic tools of Fourier analysis, which represents the wave field as a sum of purely harmonic waves, and so, resting on the unidirectional feature of such waves, determines the directional content of the wave field. On the other hand, the spherical waves are instrumental to Huygens' construction of advancing the wave fronts of the optical disturbance. We also observe that far away from the source small segments of a spherical wave may be considered as approximately plane waves. The optical wave fields that minimize the uncertainty product, thus manifesting a ray-like feature, and are both mathematically representable and physically realizable are the real Gaussian waveforms. Therefore they can be assimilated on the proper q-κ scale of the *wave-optical phase space* to a product of delta functions like $\delta(q-q_0)\delta(\kappa-\kappa_0)$, just as the light ray can be represented on the proper q-p scale of the geometric-optical phase space by a single point, and hence by the product $\delta(q-q_0)\delta(p-p_0)$ [a].

As we will see in Chapter 6, the Wigner representation elaborates a picture of the light disturbance in both space and frequency (i.e. direction), as an

[a] The symbol κ denotes the *spatial frequency*, we will introduce later as the Fourier conjugate variable to the spatial coordinate q; it relates to the direction of propagation of the wave. Thus, q_0 and κ_0 localize the "centres" of the distributions of the beam in space and direction of propagation respectively, whose relative spreads multiply to the minimum value $\frac{1}{2}$, allowed by the uncertainty relation of Fourier analysis. It is in the light of the uncertainty product based "granular texture" of the wave-optical phase space that the product $\delta(q-q_0)\delta(\kappa-\kappa_0)$ must be interpreted as signifying the "granule" in that space.

alternative to the separate space and frequency descriptions of wave optics, yielded respectively by the wave function and its Fourier transform. Thus the ray-like model of geometrical optics, and so the feasibility to precisely detail an "optical entity" by both position and direction attributes, is recovered to a certain extent within the Wigner space, composing information concerning both the space and angular distributions of the signal into a single entity, the Wigner distribution function. However, due to its intrinsic nonpositiveness, the Wigner distribution function cannot be interpreted as the wave energy distribution in space and frequency. Notably, Gaussian signals are the only waveforms, for which such an interpretation of the relevant Wigner distribution function is feasible, being it everywhere positive through the phase space.

4.3.1 *Reviewing the operator model of ray optics*

From the viewpoint of ray optics the optical system is seen to operate a mapping of the ray position and momentum coordinates from the values, pertaining to the reference screen assigned at the entrance to the system, to the values, pertaining to the reference screen assigned at the exit from the system (Fig. 4.4.a)). Evidently such a view is not concerned with the internal workings of the system, but only with the relation between the system inputs and outputs. It finds the proper formulation in the operator relation (§ 1.4.1)

$$\mathbf{u}_o = \widehat{\mathfrak{M}}\left(z_o, z_i\right)\mathbf{u}_i, \tag{4.3.1}$$

where the column vector $\mathbf{u} = \left(q_x, q_y, p_x, p_y\right)^{\top}$ comprises the four ray coordinates. The ray-transfer operator $\widehat{\mathfrak{M}}\left(z_o, z_i\right)$ is the mathematical descriptor to account for the "finite-step" transformations of the ray-variables due to propagation through the system from the input plane to the output plane, located along the optical axis at the axial positions conventionally denoted as z_i and z_o. In fact, $\mathbf{u}_i$ and $\mathbf{u}_o$ localize the representative points of the incoming and emerging ray in the $4D$ geometrical optics phase space, where the bilinear antisymmetric metric

$$\varpi\left(\mathbf{u}, \mathbf{v}\right) = u_j J_{jl} v_l \tag{4.3.2}$$

is defined (§ 1.6.4). For notational convenience the ray vector components are indexed by the integers 1,.., 4, summation over repeated indices being implied. As noted in § 1.3.1, the numbers J_{jl} above, that can only be 0, 1, and -1 according to the rules $J_{jl} = \delta_{l,j+2},\ j = 1,2$ and $J_{jl} = -\delta_{l,j-2},\ j = 3,4$, form the symplectic unit matrix $\mathbf{J}$. When dealing with a $2D$ optical phase space as in the case of stigmatic and separable astigmatic optical systems, whose behaviour can effectively be depicted on the phase plane spanned by only one pair of conjugate ray variables, the product (4.3.2) is related to the oriented area of the parallelogram spanned by the two phase-plane vectors $\mathbf{u}$ and $\mathbf{v}$.

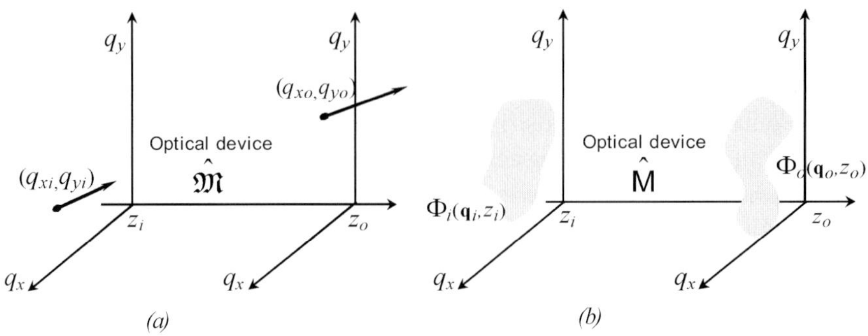

FIGURE 4.4. *(a)* Ray vectors and ray transfer operators pertain to the ray-optics formalism as *(b)* wave functions and wave transfer operators pertain to the wave-optics formalism.

The mapping (4.3.1) is symplectic as it preserves the metric (4.3.2) in the neighborhood of the ray trajectory from $\mathbf{u}_i$ to $\mathbf{u}_o$. Thus, for any pair of infinitesimal vectors $d\mathbf{v}_i$ and $d\mathbf{r}_i$ originating at $\mathbf{u}_i$, which are mapped to $d\mathbf{v}_o$ and $d\mathbf{r}_o$ originating at $\mathbf{u}_o$, the number obtained by (4.3.2) for $d\mathbf{v}_i$ and $d\mathbf{r}_i$ is the same as the number obtained by (4.3.2) for $d\mathbf{v}_o$ and $d\mathbf{r}_o$, i.e. $\varpi\,(d\mathbf{v}_o,\, d\mathbf{r}_o) = \varpi\,(d\mathbf{v}_i,\, d\mathbf{r}_i)$. This implies that elementary phase space volumes along the ray path remain constant. Equivalently, considering the 4×4 Jacobian matrix $S_{jl} = \partial u_{o_j}/\partial u_{i_l}$ of the transformation (4.3.1) so that $d\mathbf{v}_o = \mathbf{S}d\mathbf{v}_i$ and $d\mathbf{r}_o = \mathbf{S}d\mathbf{r}_i$, it follows that $\mathbf{S}^\top(\mathbf{u}_i)\mathbf{J}\mathbf{S}(\mathbf{u}_i) = \mathbf{J}$, which formally expresses the symplecticity of the ray-variables dependent Jacobian matrix.

In the case that the ray propagates nearly close to the optical axis, the first-order approximation of relations (4.3.1) with respect to the ray position and momentum is permitted, thus yielding the linear (or paraxial[b]) form of the transfer relations as

$$\mathbf{u}_o = \mathbf{M}(z_o, z_i)\,\mathbf{u}_i, \tag{4.3.3}$$

with the Jacobian matrix $\mathbf{M}(z_o, z_i)$ comprising now the ray-variables independent coefficients of the transformation $\mathbf{u}_i \to \mathbf{u}_o$ (§ 1.7.2). Also, as the direction cosines of the ray can be approximated by the respective angles, the paraxial optical momenta (p_x, p_y) turn to be the product of the angles of the ray relative respectively to the planes y-z and x-z at the intersection point with the reference screen times the on-axis value of the refractive index at that point.

The matrix relation (4.3.3) is both conceptually and practically efficacious,

[b] As remarked in § 1.6, in its basic acception, the term *paraxial* implies rays whose distances from the axis and angles to the axis are small. However, it is commonly used with a more stringent valence to signify the first-order approximation of the optical Hamiltonian of the system and accordingly of the input-output relations (4.3.1), with respect to both the ray off-axis distances and inclination angles through all the input-to-output distance.

as it reflects the mapping view of the general operator relation (4.3.1) through a descriptor, the ray-transfer matrix $\mathbf{M}(z_o, z_i)$, which is determined exclusively by the geometry and optical features of the system under consideration. Linear ray optics can, in fact, depict the optical system as a "black box" equipped with the input plane, the output plane, and the 4×4 (or 2×2 if a $1D$ picture is allowed) symplectic ray-matrix $\mathbf{M}$. The knowledge about $\mathbf{M}$ enables us to send rays into the system, with assigned transverse coordinates and inclination angles with respect to the entrance plane, and calculate the coordinates and the angles at which they will emerge with respect to the exit plane after passing through the system.

As a basic result of linear ray optics, the evolution law obeyed by the optical matrix $\mathbf{M}$ writes in the form of the linear first-order differential equation

$$\frac{d}{dz}\mathbf{M}(z, z_i) = \mathbf{H}(z)\mathbf{M}(z, z_i), \quad \mathbf{M}(z_i, z_i) = \mathbf{I}, \tag{4.3.4}$$

where the matrix $\mathbf{H}(z)$ obtains from the paraxial form of the Hamiltonian of the system under Poisson bracket with the conjugate variables q, p (§ 1.7.3).

The matrix $\mathbf{M}$ belongs to the Lie group $Sp(4,\mathbb{R})$, or $Sp(2,\mathbb{R})$ in the case of $1D$ systems, of 4×4 (or, 2×2) symplectic matrices, and so $\mathbf{M}^\top \mathbf{J}\mathbf{M} = \mathbf{J}$. Accordingly the generator matrix $\mathbf{H}$ can be placed in the symplectic Lie algebra $sp(4,\mathbb{R})$, or $sp(2,\mathbb{R})$, $\mathbf{M}$ and $\mathbf{H}$ being related by an exponential mapping (§ 2.3) [8]. In fact, by (4.3.4) the ray matrix is locally expressible in the exponential form $\mathbf{M}(z_i + dz, z_i) = e^{\mathbf{H}(z_i)\,dz}$ for any elemental distance dz from z_i. The propagation of the ray through the optical system progresses by multiplying the infinitesimal ray-matrices $e^{\mathbf{H}(z)\,dz}$, thus obtaining for any desired propagation distance $z_o - z_i$ the infinite-term product $\mathbf{M}(z_o, z_i) = e^{\mathbf{H}(z_o - dz)\,dz} \cdots e^{\mathbf{H}(z_i + dz)\,dz} e^{\mathbf{H}(z_i)\,dz}$. The one-exponential form over the entire propagation distance $z_o - z_i$ is allowed when $\mathbf{H}$ does not depend explicitly on the axial parameter z. In which event, the exponents of the infinite-exponential form plainly sum up to yield the global representation of $\mathbf{M}(z_o, z_i)$ as the single exponential

$$\mathbf{M}(z_o, z_i) = e^{\mathbf{H}\,(z_o - z_i)}. \tag{4.3.5}$$

The exponential form (4.3.5) has been basic to the considerations developed in Chapters 2 and 3. It will be recovered here, formulated within the operator representation we will elaborate to describe the light wave propagation.

4.3.2 *Introducing the operator model of wave optics. The wave-optical transfer operator*

From the viewpoint of wave optics, the optical system is seen to operate a mapping of a set of functions given in the input plane into a set of functions in the output plane. The "objects" of such a mapping are not the position and momentum coordinates of the incoming and outgoing rays as in (4.3.1), but the light distributions before and after the passage through the system, and hence the wave functions in the assigned profile planes (Fig. 4.4.b)). Accordingly, the tool of the mapping is the *wave transfer operator* $\widehat{\mathsf{M}}(z_o, z_i)$, which acts on the wave function $\Phi_i(\mathbf{q}_i, z_i)$ of the incoming light to yield the wave function $\Phi_o(\mathbf{q}_o, z_o)$ of the outgoing light, as formally expressed by the operator relation

$$\Phi_o(\mathbf{q}_o, z_o) = \widehat{\mathsf{M}}(z_o, z_i)\Phi_i(\mathbf{q}_i, z_i). \tag{4.3.6}$$

The position vectors $\mathbf{q}_i = (q_{x_i}, q_{y_i})$ and $\mathbf{q}_o = (q_{x_o}, q_{y_o})$ identify points respectively in the input and output plane by pairs of Cartesian coordinates with respect to the assigned reference frames in the respective planes. The origins of these reference frames are conveniently taken on the optical z axis, and the axes disposed parallel from one plane to the other and equally scaled.

The point-spread function

Evidently the "finite step" picture for light propagation pertinent to the operator relation (4.3.6) is equivalent to the "elemental step" view pertinent to the wave equation. Then, as the wave equation is linear, the mapping (4.3.6) is linear as well. The response of the system to a linear combination of various independent inputs is simply the same linear combination of the responses of the system to the individual inputs. In formal terms, the wave-transfer operator $\widehat{\mathsf{M}}(z_o, z_i)$ comes to be linear, and hence the relation

$$\begin{aligned}\widehat{\mathsf{M}}\,[a\Phi_i(\mathbf{q}_i, z_i) + b\Psi_i(\mathbf{q}_i, z_i)] &= \\ a\widehat{\mathsf{M}}\Phi_i(\mathbf{q}_i, z_i) + b\widehat{\mathsf{M}}\Psi_i(\mathbf{q}_i, z_i) &= a\Phi_o(\mathbf{q}_o, z_o) + b\Psi_o(\mathbf{q}_o, z_o),\end{aligned} \tag{4.3.7}$$

holds for all complex constants a and b and wave functions Φ_i and Ψ_i.

The transfer relation (4.3.6) can then be recast as the $2D$ integral transform

$$\Phi_o(\mathbf{q}_o, z_o) = \int \mathfrak{G}(\mathbf{q}_o, \mathbf{q}_i; z_o, z_i)\Phi_i(\mathbf{q}_i, z_i)d\mathbf{q}_i \tag{4.3.8}$$

which is indeed the continuous version of (4.3.7) expressing the wave function at position $\mathbf{q}_o$ in the output plane as the weighted superposition of the contributions at different positions $\mathbf{q}_i$ in the input plane. Naturally, a $2D$ integration is implied by (4.3.8), the expression $\int ...d\mathbf{q}_i$ representing $\iint ...dq_{x_i}dq_{y_i}$. As in the Chapter all integrals without limits of integrations are to be taken from $-\infty$ to $+\infty$, integration in (4.3.8) is intended to extend over the entire plane.

The weighting function $\mathfrak{G}(\mathbf{q}_o, \mathbf{q}_i; z_o, z_i)$ is referred to as the *point-spread (or impulse response) function* of the system. It is the response of the system to a true point object, mathematically represented by the delta function as $\delta(\mathbf{q} - \mathbf{q}_i, z_i) = \delta(q_x - q_{x_i}, z_i)\delta(q_y - q_{y_i}, z_i)$. As mathematical representations of point sources or unit impulsive signals, the functions $\delta(\mathbf{q}, z)$ provide, in fact, a convenient set of *elementary* waveforms, into which any optical wave can be decomposed. Exploiting the sifting property of the δ function, we can write the wave function $\Phi_i(\mathbf{q}_i, z_i)$ in the input plane as the weighted superposition

$$\Phi_i(\mathbf{q}_i, z_i) = \int \Phi_i(\mathbf{q}_i', z_i)\delta(\mathbf{q}_i' - \mathbf{q}_i, z_i)d\mathbf{q}_i', \tag{4.3.9}$$

the weighting factors being given by the numbers $\Phi_i(\mathbf{q}_i', z_i)$.

Inserting the integral representation (4.3.9) into the input-output relation (4.3.6) and invoking the linearity of $\widehat{\mathsf{M}}$, which then can separately operate on the individual unit impulses, we end up with the integral transform (4.3.8). Also, the formal expression for the spread function $\mathfrak{G}(\mathbf{q}_o, \mathbf{q}_i; z_o, z_i)$ obtains as

$$\mathfrak{G}(\mathbf{q}_o, \mathbf{q}_i; z_o, z_i) = \widehat{\mathsf{M}}(z_o, z_i)\delta(\mathbf{q}_i' - \mathbf{q}_i, z_i), \tag{4.3.10}$$

thus confirming that it represents the image distribution generated by the system responding to the point object at $(\mathbf{q}_i, z_i)$. An *ideal* optical system is expected to transfer a point object into a point image between two planes determined by the system; thereby, the response function of an ideal system with respect to suitably assigned profile planes may be a delta function.

An optical system is then completely characterized if its responses to unit impulses placed at all possible points in the input plane are known. The response to any optical wave is obtained by composing such responses with suitable weighting factors as just provided by the incoming wave. Of course, the spread function depends on points in both the object and image planes; hence the form of the response in the image plane generally varies with the position of the unit impulse over the object plane. In this respect, it is straightforward to recognize the practical relevance of the *space-invariant* systems, for which the shape of the image for a given object does not depend on the specific location of the object in the input plane; only the location of the image over the output plane changes as the point object explores the input screen. Space invariant systems are discussed in § 4.6.3. Notably, they have complex exponentials as their eigenfunctions, and hence the Fourier-transform description is valuable for their analysis.

It will be proved in Chapter 5 that, in the frame of situations allowing for the first-order approximation and hence for the matrix representation of the optical systems, the impulse response $\mathfrak{G}(\mathbf{q}_o, \mathbf{q}_i; z_o, z_i)$ is expressible in terms of the entries of the optical matrix $\mathbf{M}$ of linear ray optics. This establishes a relation between the geometric- and wave-optics pictures (4.3.3) and (4.3.8) of light propagation for the case of first-order optical systems. Evidently such a relation has both a conceptual charm, connecting different mathematical structures, and practical advantages, enabling calculations with operators be replaced by calculations with 4×4 matrices, or even more simply with 2×2 matrices if the single transverse coordinate depiction is allowed.

One-dimensional model: the line-spread function

In order to maintain the correspondence with the $1D$ picture of the previous chapters, we limit the dependence of the optical wave function to a single profile coordinate. Extension to two transverse dimensions is straightforward, apart from the unavoidable increase in formal complexity. Naturally the description in terms of a single transverse Cartesian coordinate is consistent with a simple astigmatic optical system, for which, in fact, the $2D$ transfer operator can be split into two $1D$ operators: $\widehat{\mathsf{M}} = \widehat{\mathsf{M}}_x\widehat{\mathsf{M}}_y$, each operating on a single transverse dimension. Thus, supposing the input wave function be separable as well, i.e., $\varphi_i(q_{x_i}, q_{y_i}, z_i) = \varphi_{x_i}(q_{x_i}, z_i)\varphi_{y_i}(q_{y_i}, z_i)$, one can analyze the behaviour of the system one dimension at a time (see § 4.2). Of course, the $2D$ description of the system, resulting by properly composing the obtained $1D$ descriptions, is valid also for non-separable type inputs. Axis-symmetric systems, being evidently separable, are describable by a single transverse Cartesian coordinate as well; however, in that case the circular symmetry is not formally accounted for since the inherent operators have a different representation in terms of the proper radial coordinate in the profile screen.

Then we deal in general with complex-valued functions $\varphi(q, z)$, regarded as functions of the profile coordinate q, the axial coordinate z being like a parameter. Naturally, $\varphi(q, z)$ is supposed to belong to the Hilbert space $\mathcal{L}^2(\mathbb{R})$ of the square integrable functions defined on the real line. For any two functions ψ, φ in $\mathcal{L}^2(\mathbb{R})$ the inner product is defined by the well-known integral

$$(\psi, \varphi)(z) = \int \psi^*(q, z)\varphi(q, z)dq, \tag{4.3.11}$$

the norm of any function being accordingly given as

$$\|\varphi\|^2(z) = (\varphi, \varphi)(z) = \int |\varphi(q, z)|^2 dq. \tag{4.3.12}$$

The asterisk as usual denotes complex conjugation. As is well known, two functions having vanishing inner product, $(\psi, \varphi) = 0$, are reported as orthogonal, and functions with unit norm, $(\varphi, \varphi) = 1$, are said to be normalized to unity.

In conformity with the single transverse coordinate picture, the input-output relation (4.3.8) turns into the $1D$ integral transform

$$\varphi_o(q_o, z_o) = \int \mathsf{g}(q_o, q_i; z_o, z_i)\varphi_i(q_i, z_i)dq_i, \tag{4.3.13}$$

and correspondingly the inherent spread function, formally given as

$$\mathsf{g}(q_o, q_i; z_o, z_i) = \widehat{\mathsf{M}}(z_o, z_i)\delta(q_i' - q_i, z_i), \tag{4.3.14}$$

comes to represent the response of the system to a line source in the input plane at q_i represented by the δ-distribution $\delta(q_i' - q_i, z_i)$. Hence it should properly be called the *line spread* function. As previously mentioned, as far as propagation through first-order optical systems is concerned, an explicit relation between the spread function and the ray matrix of geometrical optics can be established. In fact, in the next chapter we will be able to relate the line spread function $\mathsf{g}(q_o, q_i; z_o, z_i)$ to the ray-matrix entries A, B, D by a quadratic phase function.

For ease of writing, we may simply denote the spread function as $\mathsf{g}(q_o, q_i)$, omitting to display the axial coordinates z_i and z_o referring to the location of the input and the output planes, unless of specific relevance to the topic under discussion.

As earlier noted, the complex amplitude is related to any one component of the electric or magnetic field of the wave. Therefore, disregarding the precise dimensions, the magnitude squared of the complex amplitude can be interpreted as the optical intensity or the optical energy density, the two quantities differing by the light speed in the medium. Then, ignoring the further limits arising from the single profile-coordinate description of the light disturbance, we are dealing with, we may consider the quantity $|\varphi(q, z)|^2$ as the *optical intensity*, or the *optical energy density*. Accordingly, the non negative integral in (4.3.12) relates to the optical power or energy of the wave field.

We will consider lossless and gainless optical systems, which may reshape the amplitude and phase of the input wave function without affecting however the energy content. The conservation of the optical power or energy through propagation is mathematically expressed by the identity of the norms of the incoming and outgoing wave functions, namely

$$\|\varphi_o\|\,(z_o) = \|\varphi_i\|\,(z_i). \tag{4.3.15}$$

As $\varphi_o = \widehat{\mathsf{M}}\varphi_i$, this identity signifies the property of the transfer operator $\widehat{\mathsf{M}}(z, z_i)$ to preserve the norm of the wave function, which is reported as *unitarity*. Clearly, relation (4.3.15) does not apply to optical systems having limiting apertures, like optical resonators, for instance, or media with gain or loss profiles, which are in fact described by non unitary transfer operators.

Needless to say, "physical" signals are represented by wave functions which are smooth and negligible outside a definite region of space, and so the energy-integral (4.3.12) converges to a finite value. We will consider as well "idealized" signals described by wave functions that are not normalizable in the common sense of the energy-like integral (4.3.12), as, for instance, the impulse and the monofrequency signal, respectively represented by the delta and the linear phase function. Such idealized signals can be handled mathematically as limiting processes of normalizable signals or by resorting to the generalized function method. We shall not dwell on the mathematical details of such procedures, and use the delta and linear-phase functions as powerful tools in a kind of interplay between physical intuition and mathematical rigor.

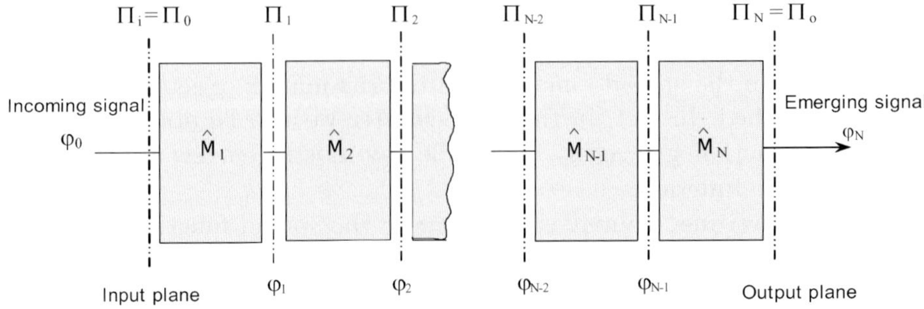

FIGURE 4.5. The wave transfer operator of a composite optical system is the product of the transfer operators of the various optical elements: $\widehat{\mathsf{M}} = \widehat{\mathsf{M}}_N ... \widehat{\mathsf{M}}_2 \widehat{\mathsf{M}}_1$.

Composition rule for the transfer operator and the spread function

In order to elaborate the composition rule for the wave-transfer operator, we resort to the system described in § 3.2, consisting of N optical elements connected together in sequence, from z_0 to z_1, from z_1 to z_2 and so on up to z_N (Fig. 4.5). Denoting by $\widehat{\mathsf{M}}_j(z_j, z_{j-1})$ the transfer operator accounting for the propagation from z_{j-1} to z_j through the j-th optical element, $j = 1, .., N$, we may follow the input signal $\varphi_0(q_0, z_0)$ whilst propagating through the various components. Then, we write

$$\begin{aligned} \varphi_1(q_1, z_1) &= \widehat{\mathsf{M}}_1(z_1, z_0)\varphi_0(q_0, z_0), \\ \varphi_2(q_2, z_2) &= \widehat{\mathsf{M}}_2(z_2, z_1)\varphi_1(q_2, z_1), \\ &\cdots\cdots \\ \varphi_N(q_N, z_N) &= \widehat{\mathsf{M}}_N(z_N, z_{N-1})\varphi_{N-1}(q_{N-1}, z_{N-1}), \end{aligned} \tag{4.3.16}$$

which sound as the wave optical versions of the ray-vector relations (3.2.1). The optical propagator $\widehat{\mathsf{M}}(z_N, z_0)$ for the overall system from z_0 to z_N results from multiplying the transfer operators of the various optical elements, and so

$$\widehat{\mathsf{M}}(z_N, z_0) = \widehat{\mathsf{M}}_N(z_N, z_{N-1}) \;....\widehat{\mathsf{M}}_2(z_2, z_1)\widehat{\mathsf{M}}_1(z_1, z_0), \tag{4.3.17}$$

in the same way as the ray matrix for the composed system is obtained multiplying the ray matrices of the various components (Eq. (3.2.3)). As the matrices, the transfer operators must be arranged in the product in the reverse order to that the wave hits the corresponding elements, and so from right to left, being the propagation conventionally assumed to occur from left to right.

It is easy to turn the above composition rule for $\widehat{\mathsf{M}}$ into the composition rule for the impulse response of the system, which writes as

$$\mathsf{g}(q_N, q_0) = \int \cdots \int dq_{N-1} ... dq_1 \, \mathsf{g}_N(q_N, q_{N-1}) ... \mathsf{g}_1(q_1, q_0). \tag{4.3.18}$$

The spread function of a cascade of N systems is then calculated as an $(N-1)$-fold integral of the product of the spread functions of the various components, the integration being performed over all the intermediate planes from 1 to $N-1$, if one numbers the input and output planes respectively by 0 and N.

We learned in the foregoing chapters that in general optical systems are "noncommuting" objects, and so a change of the order of the single optical elements in a cascade may change the properties of the overall system. This manifests into the in general non commutativity of the transfer operators corresponding to different optical systems.

Unitarity and inverse operator

The unitarity of the transfer operator, which as noted before expresses the condition that the system be lossless, turns into a specific relation obeyed by the corresponding spread function. In fact, let $\varphi_i(q_i, z_i)$ and $\varphi_o(q_o, z_o) = \widehat{\mathsf{M}}(z_o, z_i)\varphi_i(q_i, z_i)$ designate as usual the complex amplitudes of the input and output signals, respectively. The conservation of the norm

$$(\varphi_o, \varphi_o) = (\varphi_i, \varphi_i) \tag{4.3.19}$$

requires that the equality

$$\iiint dq_i dq_i' dq_o \mathsf{g}^*(q_o, q_i')\, \mathsf{g}(q_o, q_i)\varphi_i^*(q_i', z_i)\varphi_i(q_i, z_i) = \int dq_i \varphi_i^*(q_i, z_i)\varphi_i(q_i, z_i) \tag{4.3.20}$$

be satisfied for any input signal $\varphi_i(q_i, z_i)$, thus yielding for g the relation

$$\int dq_o \mathsf{g}^*(q_o, q_i')\, \mathsf{g}(q_o, q_i) = \delta(q_i' - q_i). \tag{4.3.21}$$

In this connection, it is interesting to think, from a purely conceptual viewpoint, of a cascade of two optical systems whose output distribution be exactly equal to the input distribution. Namely, given an optical system described by the transfer operator $\widehat{\mathsf{M}}$, we question about the existence of an associated optical system, the relevant transfer operator being denoted as $\widehat{\mathsf{M}}_{\text{inv}}$, such that cascaded with $\widehat{\mathsf{M}}$ produces an optical system that has no effect on the input signal whatever it be, apart from a trivial translation along the optical axis. This is mathematically expressed by the relation

$$\widehat{\mathsf{M}}(z_1, z_i)\widehat{\mathsf{M}}_{\text{inv}}(z_o, z_1) = \widehat{\mathsf{M}}_{\text{inv}}(z_i, z_2)\widehat{\mathsf{M}}(z_o, z_2) = \widehat{\mathsf{I}}, \tag{4.3.22}$$

z_1 and z_2 denoting some intermediate points between z_i and z_o determined for each sequence, i.e., $\widehat{\mathsf{M}}\widehat{\mathsf{M}}_{\text{inv}}$ and $\widehat{\mathsf{M}}_{\text{inv}}\widehat{\mathsf{M}}$, according to the axial extension of the system and its inverse.

Relation (4.3.22) allows us to identify $\widehat{\mathsf{M}}_{\text{inv}}$ as the inverse of $\widehat{\mathsf{M}}$: $\widehat{\mathsf{M}}_{\text{inv}} = \widehat{\mathsf{M}}^{-1}$, and the unitarity of $\widehat{\mathsf{M}}$ assures the existence of $\widehat{\mathsf{M}}^{-1}$, in the same way as the symplecticity of the ray transfer matrix guarantees the existence of the (unique) inverse matrix $\mathbf{M}^{-1}$. This means that the inverse of any optical system is conceptually conceivable, its physical realizability being however a matter of system synthesis. The discussion, we presented in § 3.3.1, about the realizability of the inverse of the free-medium section makes clear the difference between the mathematical existence and the physical feasibility of a system.

4.4 From the ray-optical matrix to the wave-optical operator

We translate now the matrix formalism, previously illustrated within the context of first-order ray-optics, into an operator formalism, specifically adequate to the wave optical description of light propagation, thus setting a "bridge" between the linear ray-optics formalism, reviewed in § 4.3.1, and the linear wave-optics formalism, introduced in § 4.3.2. To this end, we will exploit the quantization procedure, turning all the variables of ray optics into linear operators [7]. As a natural by-product, the ray transfer matrices will be turned into definite wave transfer operators, accounting for the transformations of the wave complex-amplitude when going through the optical system of concern [9].

We shall not detail the quantization of ray optics, but merely outline the basic steps leading to the evolution equation for the wave-transfer operator in the paraxial régime. This will naturally suggest us the strategy to obtain the operator representation of first-order optical systems, that will be used in the next chapter to deduce the functional form of the spread function of any paraxial optical system in terms of the entries of the relative ray matrix.

As said earlier, paraxial ray optics is generated by quadratic polynomials under Poisson brackets and matrix exponentiation. Accordingly we may expect that paraxial wave optics be generated by *quantized* quadratic polynomials under Lie brackets and operator exponentiation. In fact, as we will see, the relation between paraxial ray optics and paraxial wave optics can be interpreted in the light of the relation between the symplectic group of the ray transfer matrices and the metaplectic group of wave transfer operators.

The presentation below is kept on a nontechnical level, refining later representations firstly used invoking intuition from, and analogy with, classical and quantum mechanics.

4.4.1 *Wave-optical position and momentum operators. Hamiltonian operator for paraxial wave optics*

Extrapolating from geometrical to wave optics, the canonical variables q and p of ray-optics are replaced by the canonical Hermitian operators $\widehat{\mathsf{q}}$ and $\widehat{\mathsf{p}}$ [10]. In the spatial coordinate representation, the position operator $\widehat{\mathsf{q}}$ acting on the optical wave function has the effect of multiplying by the coordinate value q:

$$\widehat{\mathsf{q}} = q, \tag{4.4.1}$$

whereas the momentum operator $\widehat{\mathsf{p}}$ acts by the derivative with respect to q:

$$\widehat{\mathsf{p}} = -i\tfrac{\lambda_0}{2\pi}\frac{\partial}{\partial q} = -\tfrac{i}{k_0}\frac{\partial}{\partial q}. \tag{4.4.2}$$

We emphasize again that the vacuum wavelength $\lambda_0 = \frac{2\pi}{k_0}$ corresponds to Planck's constant $\hbar$ of the quantum theory.

As the Poisson bracket of the classical functions q and p is

$$\{q, p\} = 1, \tag{4.4.3}$$

the Lie bracket of the corresponding operators $\widehat{\mathsf{q}}$ and $\widehat{\mathsf{p}}$ is

$$[\widehat{\mathsf{q}}, \widehat{\mathsf{p}}] = \widehat{\mathsf{q}}\widehat{\mathsf{p}} - \widehat{\mathsf{p}}\widehat{\mathsf{q}} = i\widehat{\mathcal{I}}, \qquad \widehat{\mathcal{I}} \equiv \frac{\widehat{\mathsf{I}}}{k_0}, \tag{4.4.4}$$

$\widehat{\mathsf{I}}$ denoting the identity operator.

The operators $\left\{\widehat{\mathsf{q}}, \widehat{\mathsf{p}}, \widehat{\mathcal{I}}\right\}$, with the commutation brackets

$$[\widehat{\mathsf{q}}, \widehat{\mathsf{p}}] = i\widehat{\mathcal{I}}, \quad [\widehat{\mathsf{q}}, \widehat{\mathcal{I}}\,] = 0, \quad [\widehat{\mathsf{p}}, \widehat{\mathcal{I}}\,] = 0, \tag{4.4.5}$$

form the Heisenberg-Weyl algebra [11], which parallels the algebra spanned by the linear monomials $\{q, p, 1\}$ under Poisson bracket operation, on which we have briefly commented in § 1.3.2 and § 1.5.1.

The quantization procedure proceeds then on turning the scalar equation of geometrical optics (Eq. (1.2.12)):

$$H(q, p, z) = -p_z, \tag{4.4.6}$$

into the proper operator form

$$\widehat{\mathcal{H}}(\widehat{\mathsf{q}}, \widehat{\mathsf{p}}, z) = -\widehat{\mathsf{p}}_z, \tag{4.4.7}$$

substituting for the Hamiltonian function $H(q, p, z)$ and the z-component of the optical moment p_z the respective operator realizations, $\widehat{\mathcal{H}}$ and $\widehat{\mathsf{p}}_z$.

Since, of course,

$$\widehat{\mathsf{p}}_z = -\tfrac{i}{k_0}\frac{\partial}{\partial z}, \tag{4.4.8}$$

squaring both sides of (4.4.7) yields the operator form of the exact (i.e., circular) dispersion relation of the optical momenta (see § 4.2), which then, once applied to the optical wave function, results into the $2D$ Helmholtz equation.

Similarly, taking on the left side of (4.4.7) the paraxial form of the Hamiltonian function yields the operator form of the approximate (i.e., parabolic) dispersion relation of the optical momenta (see § 4.2), which then, once applied to the optical wave function, results into the $2D$ parabolic wave equation.

Needless to say, the Hamiltonian operator $\widehat{\mathcal{H}}(\widehat{\mathsf{q}}, \widehat{\mathsf{p}}, z)$ is obtained from the Hamiltonian function $H(q, p, z)$ of geometrical optics by means of the suitable correspondence rule. We shall not dwell here on the intrinsic difficulty of going from a classical function of the canonical position and momentum variables q and p to an operator function of the quantum observables $\widehat{\mathsf{q}}$ and $\widehat{\mathsf{p}}$, consequent to the non commutative relation (4.4.4), which allows indeed for different non equivalent correspondences. The interested reader is directed to the references quoted in [13]. Here we simply say that each classical phase-space function $f(q,p)$ is transcribed into a Hermitian operator $\widehat{f}(\widehat{\mathsf{q}}, \widehat{\mathsf{p}})$, once the specific rule of association has been chosen. In fact, there is no unique quantization scheme, that could be deduced from the basic principles, for assigning a Hermitian operator function of the noncommuting $\widehat{\mathsf{q}}$ and $\widehat{\mathsf{p}}$ to a given c-number function of q and p. As an example, we may consider the simple mixed monomial qp. In principle it could equally well be turned into the nonequivalent operator forms $\widehat{\mathsf{q}}\widehat{\mathsf{p}}$, $\widehat{\mathsf{p}}\widehat{\mathsf{q}}$ or $(\widehat{\mathsf{q}}\widehat{\mathsf{p}}+\widehat{\mathsf{p}}\widehat{\mathsf{q}})/2$. Since the first two forms do not lead to Hermitian operators as desired, the monomial qp is uniquely carried to the symmetric operator form $(\widehat{\mathsf{q}}\widehat{\mathsf{p}}+\widehat{\mathsf{p}}\widehat{\mathsf{q}})/2$. In contrast the monomial q^2p^2 could be quantized among others as the non equivalent Hermitian operators $(\widehat{\mathsf{q}}^2\widehat{\mathsf{p}}^2 + \widehat{\mathsf{p}}^2\widehat{\mathsf{q}}^2)/2$ or $\widehat{\mathsf{q}}\widehat{\mathsf{p}}^2\widehat{\mathsf{q}}$ or $\widehat{\mathsf{p}}\widehat{\mathsf{q}}^2\widehat{\mathsf{p}}$; in that case, indeed, the requirement for a Hermitian operator is not a sufficient criterion for uniquely determining the operator function to which the classical monomial q^2p^2 might be carried to. Uniqueness of the correspondence $f(q,p) \rightleftharpoons \widehat{f}(\widehat{\mathsf{q}}, \widehat{\mathsf{p}})$ is assured only within a particular rule of association. Here we apply the Weyl rule because of its intimate relation to the Wigner representation, that is the main topic of the forthcoming Chapters (see in particular § 8.2.3, and the references there addressed). Very succinctly we recall that the Weyl rule takes "plane waves" in the space of the canonical scalar pairs (q,p) to "plane waves" in the space of the canonical operator pairs $(\widehat{\mathsf{q}}, \widehat{\mathsf{p}})$, namely, $e^{i\xi q+i\eta p} \rightleftharpoons e^{i\xi\widehat{\mathsf{q}}+i\eta\widehat{\mathsf{p}}}$. As a consequence, in the Weyl scheme, to $q^m p^n$ corresponds the operator formed by summing up all the permutations of m $\widehat{\mathsf{q}}$'s and n $\widehat{\mathsf{p}}$'s, treated as individual objects, and then dividing by $(m+n)!$.

Remarkably, in the paraxial régime in quadratic index media, with which we are mainly concerned, the geometric/wave correspondence is uniquely determined. In fact, as far as propagation through first-order optical systems is concerned, the basic geometrical phase-space functions, being quadratic in q and p, have a unique transcription as operator functions of the wave optical observables $\widehat{\mathsf{q}}$ and $\widehat{\mathsf{p}}$, since all the quantization rules give the same result on quadratic polynomial functions in q and p. It is not so for higher-order polynomials, except for the special classes of functions which are linear in one of the two canonical variables, and so are like $qf(p) + g(p)$ or $pf(q) + g(q)$.

We remind the reader that the geometrical optical Hamiltonian $H(q, p, z)$ pertinent to the propagation through first-order optical systems displays separate quadratic contributions from the ray variables as (§§ 1.6-1.7)

$$H(q, p, z) = \tfrac{1}{2n_0(z)}p^2 + \tfrac{n_2(z)}{2}q^2, \tag{4.4.9}$$

the meaning of the medium parameters n_0 and n_2 being elucidated in § 1.7.5. The quantized form of the "classical" Hamiltonian (4.4.9) is then uniquely obtained according to the correspondences

$$p^2 \rightleftharpoons \widehat{\mathsf{p}}^2, \qquad q^2 \rightleftharpoons \widehat{\mathsf{q}}^2, \qquad qp \rightleftharpoons \tfrac{1}{2}(\widehat{\mathsf{q}}\widehat{\mathsf{p}}+\widehat{\mathsf{p}}\widehat{\mathsf{q}}), \tag{4.4.10}$$

which hence yield the Hamiltonian operator appropriate to first-order wave optics in the Hermitian form

$$\widehat{\mathcal{H}}_{\text{paraxial}}(\widehat{\mathsf{q}}, \widehat{\mathsf{p}}, z) \equiv \widehat{\mathsf{H}}(\widehat{\mathsf{q}}, \widehat{\mathsf{p}}, z) = \tfrac{1}{2n_0(z)}\widehat{\mathsf{p}}^2 + \tfrac{n_2(z)}{2}\widehat{\mathsf{q}}^2. \tag{4.4.11}$$

4.4.2 *Evolution equation for the wave transfer operator in the paraxial regime*

The evolution equation we will deal with takes therefore the form

$$\frac{i}{k_0}\frac{\partial}{\partial z}\varphi(q, z) = \widehat{\mathsf{H}}(\widehat{\mathsf{q}}, \widehat{\mathsf{p}}, z)\varphi(q, z), \qquad \varphi(q, z_i) = \varphi_i(q), \tag{4.4.12}$$

where $\varphi_i(q)$ specifies the light wave distribution in the input reference screen. On account of the explicit functional form (4.4.11) of $\widehat{\mathsf{H}}(\widehat{\mathsf{q}}, \widehat{\mathsf{p}}, z)$, the reader may easily verify the correspondence of the above equation with the $2D$ parabolic wave equation (4.2.16) in quadratic media.

Accordingly, by (4.3.6), Eq. (4.4.12) yields the equation for the wave transfer operator $\widehat{\mathsf{M}}(z, z_i)$ in the paraxial regime, i.e.,

$$\frac{i}{k_0}\frac{\partial}{\partial z}\widehat{\mathsf{M}}(z, z_i) = \widehat{\mathsf{H}}(\widehat{\mathsf{q}}, \widehat{\mathsf{p}}, z)\widehat{\mathsf{M}}(z, z_i), \qquad \widehat{\mathsf{M}}(z_i, z_i) = \widehat{\mathsf{I}}, \tag{4.4.13}$$

which represents the *wave-optical* version of the ray-matrix equation (4.3.4).

As already noted, matrices and operators are strictly similar entities. Within the context of the correspondence between (4.3.4) and (4.4.13), matrices are intended to act on the geometrical phase space of the ray vectors, as operators are intended to act on the Hilbert space $\mathcal{L}^2(\mathbb{R})$ of the square integrable optical wave functions. Therefore, the analysis elaborated in §§ 1.4.1 and 1.7.2-3 in relation to the solving techniques of the ray-matrix equation (4.3.4) is adequate to the wave-operator equation (4.4.13) as well. In particular, if the Hamiltonian operator $\widehat{\mathsf{H}}(\widehat{\mathsf{q}}, \widehat{\mathsf{p}}, z)$ does not explicitly depend on the axial coordinate z: $\widehat{\mathsf{H}}(\widehat{\mathsf{q}}, \widehat{\mathsf{p}}, z) = \widehat{\mathsf{H}}(\widehat{\mathsf{q}}, \widehat{\mathsf{p}})$, exponentiation of $\widehat{\mathsf{H}}$ over the finite propagation distance $z - z_i$ can loosely be performed, thus giving the solution to (4.4.13) in the form

$$\widehat{\mathsf{M}}(z, z_i) = e^{-ik_0 \widehat{\mathsf{H}}(z - z_i)}. \tag{4.4.14}$$

In fact, retracing the discussion presented in §§ 1.4.1 and 1.7.2-3 (recalled in § 4.3.1) on the evolution equation (4.3.4) and the exponential mapping (4.3.5) as relative solving form, we may say that by (4.4.13) the transfer operator $\widehat{\mathsf{M}}(z, z_i)$ from z_i to z arises from an infinite string of unitary infinitesimal operators of the type $e^{-ik\widehat{\mathsf{H}}(z)dz}$ for successive elemental increments dz of the evolution parameter from z_i up to z. Hence if $\widehat{\mathsf{H}}$ does not depend on z, the ordered product of all those infinitesimal unitary operators combines to the unitary single-exponential form (4.4.14). Evidently, expression (4.4.14) completes the correspondences with ray optics, being it the *wave-optical* version of (4.3.5). One needs only substitute $-ik_0\widehat{\mathsf{H}}$ for $\mathbf{H}$ in the expression for the ray-transfer matrix, whether we consider propagation over infinitesimal or finite distance.

The one-to-one correspondences (4.4.10) and the exponential law (4.4.14) represent the very basic dictionary to translate the ray optics description of first-order optical systems into the language and formalism of wave optics.

As a final comment, we note that a ray-optics based description of light propagation through optical systems can account only for the amplitude information but lacks the phase information. In fact, the homogeneous term $n_0(z)$ arising in the quadratic approximation of the optical Hamiltonian is merely dropped out in ray optics, in accord with the fact, that as a local theory geometrical optics can shift the origin of the energy. Indeed, this term is uninfluential to the ray path evolution, since its Lie operator is zero. If retained in the "classical" Hamiltonian function and then passed into the "quantum" Hamiltonian operator, it would yield the additional term $\frac{1}{2}n_0(z)\psi(q, z)$ in the $1D$ evolution equation (4.4.12). Of course, this term could be absorbed into the wave amplitude through the phase factor $e^{i\frac{k_0}{2}\int_{z_i}^{z} n_0(z)dz}$, which, once properly composed according to the decomposition scheme (4.2.15) with the identical phase factor associated with the other profile coordinate, would yield the paraxial representation (4.2.8) of the wave field. The axial phase shift is not accounted for by the ray optics description of light propagation, and hence is not recoverable from the ray matrix formalism. In situations where phase information are required in addition to amplitude information, as, for instance, when studying holography or optical resonators, the on-axis phase shift $e^{ik_0\int_{z_i}^{z} n_0(z)dz}$ must be separately calculated and added to the complex amplitude $\varphi(q_x, q_y, z)$ to obtain a complete wave optical representation of the paraxially propagating disturbance.

4.4.3 The wave transfer operator from the perspective of quantum mechanics

The correspondence of the transfer operator $\widehat{\mathsf{M}}(z, z_i)$ to the time evolution operator $\widehat{\mathsf{U}}(t, t_i)$ of quantum mechanics is naturally suggested by the transfer relation (4.3.6) as well as by the evolution equation (4.4.13).

As the propagation operator $\widehat{\mathsf{M}}(z, z_i)$ transfers the optical wave function in the plane z_i to the optical wave function in the plane z on account of the optical properties of the medium, likewise the *time evolution operator* $\widehat{\mathsf{U}}(t, t_i)$ transfers the system wave function at time t_i to the system wave function at time t on account of the dynamical features of the problem:

$$\psi(x, t) = \widehat{\mathsf{U}}(t, t_i)\psi(x, t_i). \tag{4.4.15}$$

where x shortly signifies the appropriate set of system coordinates [10].

It is well known that $\widehat{\mathsf{U}}(t, t_i)$ obeys the Schrödinger equation

$$i\hbar\frac{\partial}{\partial t}\widehat{\mathsf{U}}(t, t_i) = \widehat{\mathsf{H}}(x, t)\widehat{\mathsf{U}}(t, t_i), \quad \widehat{\mathsf{U}}(t_i, t_i) = \widehat{\mathsf{I}} \tag{4.4.16}$$

where $\widehat{\mathsf{H}}(x, t)$ is the Hamiltonian operator of the system.

As said earlier, Eq. (4.4.13) for $\widehat{\mathsf{M}}(z, z_i)$ is in a sense the "Schrödinger equation" of wave optics. The similarity between $\widehat{\mathsf{M}}(z, z_i)$ and $\widehat{\mathsf{U}}(t, t_i)$ involves other mathematical aspects; indeed, it reflects the similarity of the physical problems the two operators describe: propagation of light waves and dynamical evolution of quantum states.

In this connection, we recall that for any linear operator $\widehat{\mathsf{O}}$ acting on the Hilbert space $\mathcal{L}^2(\mathbb{R})$ of the square integrable functions $\varphi(q)$ defined on the real line, the *hermitian conjugate* (or, *adjoint*) operator $\widehat{\mathsf{O}}^\dagger$ obeys the relation

$$(\psi, \widehat{\mathsf{O}}^\dagger\varphi) = (\widehat{\mathsf{O}}\psi, \varphi), \tag{4.4.17}$$

involving the inner product (4.3.11) between functions.

The relation between the operator $\widehat{\mathsf{O}}$ and its hermitian conjugate $\widehat{\mathsf{O}}^\dagger$ turns into specific properties of utmost relevance to the physical interpretation of $\widehat{\mathsf{O}}$ itself, and hence to its role within the physical context of concern.

It is well known that *Hermitian (or, self-adjoint)* operators, for which the formal identity

$$\widehat{\mathsf{O}} = \widehat{\mathsf{O}}^\dagger, \tag{4.4.18}$$

holds, may correspond to physical observables, as their eigenvalues being real can be interpreted as possible values of dynamical variables. Strictly speaking, self-adjoint operators are characterized by the further condition that the domain of $\widehat{\mathsf{O}}^\dagger$ is the same as that of $\widehat{\mathsf{O}}$; throughout the book we will consider the terms Hermitian and self-adjoint as equivalent.

On the other hand, unitary operators, characterized by the relation

$$\widehat{\mathsf{O}}^{-1} = \widehat{\mathsf{O}}^{\dagger}, \tag{4.4.19}$$

have the distinguished property of preserving the inner product between functions. Hence the norm of the transformed function $\widehat{\mathsf{O}}\varphi$ is the same as the norm of φ. In symbols we may write

$$(\widehat{\mathsf{O}}\psi, \widehat{\mathsf{O}}\varphi) = (\psi, \varphi), \qquad (\widehat{\mathsf{O}}\varphi, \widehat{\mathsf{O}}\varphi) = (\varphi, \varphi). \tag{4.4.20}$$

As remarked earlier, the wave transfer operator $\widehat{\mathsf{M}}(z, z_i)$ for gainless and lossless systems is unitary. The time evolution operator $\widehat{\mathsf{U}}(t, t_i)$ for non dispersive quantum states is unitary as well. The unitarity of $\widehat{\mathsf{M}}$ and $\widehat{\mathsf{U}}$ is the mathematical expression of the conservation of the optical power flux and the probability to find the quantum system in any one of its possible dynamical states. It directly follows from the relevant evolution equations (4.4.13) and (4.4.16), according to which both $\widehat{\mathsf{M}}$ and $\widehat{\mathsf{U}}$ result from the exponential of a Hermitian operator by a purely imaginary parameter. In fact, if in general $\widehat{\mathsf{O}} = e^{i\tau\widehat{\mathsf{H}}}$, with τ real and $\widehat{\mathsf{H}}$ Hermitian, it follows that $\widehat{\mathsf{O}}^{\dagger} = e^{-i\tau\widehat{\mathsf{H}}^{\dagger}} = e^{-i\tau\widehat{\mathsf{H}}} = \widehat{\mathsf{O}}^{-1}$.

Evidently, unitarity plays within the wave-optical framework the same role as symplecticity plays within the geometrical-optical context. In particular, concerning propagation through first-order optical systems, the traceless matrix $\mathbf{H}$ corresponds to the Hermitian operator $\widehat{\mathsf{H}}$, respectively characterized by the pertinent relations

$$\mathbf{JH} + \mathbf{H}^{\top}\mathbf{J} = \mathbf{0} \quad \rightleftharpoons \quad \widehat{\mathsf{H}} = \widehat{\mathsf{H}}^{\dagger}. \tag{4.4.21}$$

Also, as exponentiation of $\mathbf{H}$ by a real parameter generates the ray-propagating matrix, likewise exponentiation of $\widehat{\mathsf{H}}$ by a purely imaginary parameter generates the wave-propagating operator. As an example, for an optically homogeneous medium along the z axis, we have the correspondence

$$e^{\mathbf{H}(z-z_i)} \quad \rightleftharpoons \quad e^{-ik_0\widehat{\mathsf{H}}(z-z_i)} \tag{4.4.22}$$

between the ray-optics and wave-optics transfer operators from z_i to z.

4.4.4 *Further exploring the correspondence between ray and wave optics pictures. I*

According to the matrix relation (4.3.5), the evolution of the ray variables is ruled by the Lie operator $\widehat{\mathsf{L}}_H = \{\cdot, H\}$, with which the matrix representative $\mathbf{H}$ is associated via Poisson bracket operation. In contrast, as displayed by (4.4.13), the evolution of the wave function is ruled by the Hamiltonian operator $\widehat{\mathsf{H}}(\widehat{\mathsf{q}}, \widehat{\mathsf{p}}, z)$. This seeming incongruence between the ray and wave-optical description of light propagation originates from the fact that ray optics is concerned with the evolution of the ray variables (q, p) whereas wave optics is concerned with the evolution of the wave function $\varphi(q, z)$, i.e., with the Schrödinger-like representation of the light disturbance.

The correspondence between the ray and wave optics pictures looks straightforward once investigated from the viewpoint of the Heisenberg operators $\widehat{\mathsf{q}}(z)$ and $\widehat{\mathsf{p}}(z)$, whose evolution is ruled by

$$\begin{aligned} \tfrac{i}{k_0}\tfrac{d}{dz}\widehat{\mathsf{q}}(z) &= [\widehat{\mathsf{q}}, \widehat{\mathsf{H}}(\widehat{\mathsf{q}}, \widehat{\mathsf{p}}, z)], \\ \tfrac{i}{k_0}\tfrac{d}{dz}\widehat{\mathsf{p}}(z) &= [\widehat{\mathsf{p}}, \widehat{\mathsf{H}}(\widehat{\mathsf{q}}, \widehat{\mathsf{p}}, z)], \end{aligned} \tag{4.4.23}$$

which stand out as the *wave-optical* form of Hamilton's equations of ray optics. Hence, in the light of the optical Schrödinger equation (4.4.12), the (paraxial) "propagation" of $\widehat{\mathsf{q}}(z)$ and $\widehat{\mathsf{p}}(z)$ through an optical system is formally interpreted in the sense of the similarity transformations (by the inverse operator $\widehat{\mathsf{M}}^{-1}(z, z_i)$),

$$\begin{aligned} \widehat{\mathsf{q}}(z) &= \widehat{\mathsf{M}}^{-1}(z, z_i)\widehat{\mathsf{q}}(z_i)\widehat{\mathsf{M}}(z, z_i), \\ \widehat{\mathsf{p}}(z) &= \widehat{\mathsf{M}}^{-1}(z, z_i)\widehat{\mathsf{p}}(z_i)\widehat{\mathsf{M}}(z, z_i), \end{aligned} \tag{4.4.24}$$

of the Schrödinger operators (4.4.1) and (4.4.2), accordingly regarded as the "input" operators $\widehat{\mathsf{q}}(z_i)$ and $\widehat{\mathsf{p}}(z_i)$, i.e. $\widehat{\mathsf{q}}(z_i) = \widehat{\mathsf{q}}$ and $\widehat{\mathsf{p}}(z_i) = \widehat{\mathsf{p}}$ [10].

We have learned in § 1.3.2 that operators can be associated with functions by Poisson bracket. In an exactly similar manner, operators can as well be associated with operators by Lie product. Indeed, the operator

$$ad\widehat{\mathsf{O}} = [\cdot, \widehat{\mathsf{O}}]. \tag{4.4.25}$$

generated by an operator $\widehat{\mathsf{O}}$, which acts as the right commutator bracket with $\widehat{\mathsf{O}}$, is called the *adjoint* operator of $\widehat{\mathsf{O}}$ [14].

Introducing the operator-valued vector $\widehat{\mathbf{w}}(z) = (\widehat{\mathsf{q}}(z), \widehat{\mathsf{p}}(z))^\top$, Heisenberg's equations are then recast in a form similar to (1.4.3) for the ray vector $\mathbf{u}$; namely,

$$\frac{i}{k_0}\frac{d}{dz}\,\widehat{\mathbf{w}}(z) = ad\widehat{\mathsf{H}}(z)\,\widehat{\mathbf{w}}(z), \quad \widehat{\mathbf{w}}(z_i) = \widehat{\mathbf{w}}_i, \tag{4.4.26}$$

where, of course, $\widehat{\mathbf{w}}_i = (\widehat{\mathsf{q}}(z_i), \widehat{\mathsf{p}}(z_i))^\top$. Therefore, the evolution relations (4.4.24) for $\widehat{\mathsf{q}}$ and $\widehat{\mathsf{p}}$ can compactly be expressed through the solving form of (4.4.26). In particular, if $\widehat{\mathsf{H}}$ is z-independent, one has

$$\widehat{\mathbf{w}}(z) = e^{-ik_0(z-z_i)\,ad\widehat{\mathsf{H}}}\,\widehat{\mathbf{w}}_i. \tag{4.4.27}$$

The reader may prove that (4.4.27) is equivalent to the similarity transformations (4.4.24) with the transfer operator $\widehat{\mathsf{H}}$ being given the pertinent functional form (4.4.14) (Problem 2). Naturally the exponential operator $e^{-ik_0(z-z_i)\,ad\widehat{\mathsf{H}}}$ corresponds to the power series

$$e^{-ik_0(z-z_i)\,ad\widehat{\mathsf{H}}} = \sum_{j=0}^{\infty} \frac{(-ik_0)^j}{j!}(z - z_i)^j (ad\widehat{\mathsf{H}})^j, \tag{4.4.28}$$

the power $(ad\widehat{\mathsf{H}})^j = [[...[[\cdot, \widehat{\mathsf{H}}], \widehat{\mathsf{H}}], ...], \widehat{\mathsf{H}}]$ demanding for the j-fold commutators with $\widehat{\mathsf{H}}$. Therefore the evolution relation (4.4.27) explicitly means

$$\widehat{\mathbf{w}}(z) = \widehat{\mathbf{w}}_i - ik_0(z - z_i)\,[\widehat{\mathbf{w}}_i, \widehat{\mathsf{H}}] - \tfrac{k_0^2}{2}(z - z_i)^2[[\widehat{\mathbf{w}}_i, \widehat{\mathsf{H}}], \widehat{\mathsf{H}}] + ... \tag{4.4.29}$$

The analogy with the evolution relation (1.4.4) where the Lie transformation (1.4.5) acts on the ray variables of geometrical optics is evident.

The direct analogy with the formalism of ray optics remains in some sense hidden, since wave optics deals with the wave function $\varphi(q, z)$ rather than with the position and momentum observables $\widehat{\mathsf{q}}$ and $\widehat{\mathsf{p}}$. The wave function does not have a direct counterpart within ray optics, the local character inherent in the ray picture contrasting with the non local character of the wave function picture. *Locality* is partially recovered in the mean values $\langle\, \widehat{\mathsf{q}}\,\rangle$ and $\langle\, \widehat{\mathsf{p}}\,\rangle$ of the operators $\widehat{\mathsf{q}}$ and $\widehat{\mathsf{p}}$ in the state $\varphi(q, z)$, which, as we will see in Chapter 5, evolve through the laws of ray optics in the considered case of quadratic Hamiltonian, and so relatively to the paraxial propagation. In addition, in § 5.5.1 it is shown that for quadratic Hamiltonian operators like (4.4.11) matrix representatives can be associated with the adjoint operators (4.4.25) just applying a rule similar to (1.7.7) for Lie operators, with functions being replaced by operators and Poisson brackets by Lie brackets.

4.5 Eigenfunctions of $\widehat{\mathsf{q}}$ and $\widehat{\mathsf{p}}$: point-like and spatial harmonic waveforms

We give a closer look at the eigenfunctions of the position and momentum operators $\widehat{\mathsf{q}}$ and $\widehat{\mathsf{p}}$. This is a genuine quantum mechanical problem, that we review within the optical context. So, the position eigenstates $\delta(q)$, as we said earlier, are interpreted as representing unit *impulsive* signals, generated by sources concentrated into a point. Correspondingly, the momentum eigenstates e^{ik_0pq} are interpreted as representing plane waves, the relative momentum eigenvalues specifying the direction of propagation of the wave. Both the delta and linear phase functions can be regarded as "elementary" signals, into which any optical disturbance can be decomposed; a property already exploited in § 4.3.2 to express the wave-optical input-output relationships into the integral form (4.3.13). Thus, we are naturally led to introduce the Fourier integral as an alternative representation of the optical disturbance. Both $\widehat{\mathsf{q}}$ and $\widehat{\mathsf{p}}$ have a continuous spectrum of eigenvalues [10].

4.5.1 *The impulse function: $\delta(q)$*

The eigenfunction $\varphi_{q'}(q)$ of $\widehat{\mathsf{q}}$ belonging to the coordinate eigenvalue q', for which

$$\widehat{\mathsf{q}}\varphi_{q'}(q) = q'\varphi_{q'}(q), \tag{4.5.1}$$

is represented by the delta function: $\varphi_{q'}(q) = \delta(q - q')$.

We will not detail the mathematical aspects of the theory of the distributions, or generalized functions, of which the δ-function is an example [15]. We will simply use the defining properties of the δ-function, whose intuitive understanding rests on the rough description as a "function" having an infinite height and zero width such that its area is equal to unity, as formalized by

$$\delta(q - q') = 0, \qquad q \neq q'; \qquad \int \delta(q)dq = 1, \tag{4.5.2}$$

which evidently cannot describe an "ordinary" function. In conformity with the view of the δ-function being highly concentrated in one point, it is frequently used as mathematical representation of *unphysical* idealized quantities exhibiting "point" or "line" characteristics, as point- or line-like sources, or any other quantities highly localized in some coordinate system.

The δ-function is as well defined by the remarkable property of sifting out the value of any suitably continuous function $\varphi(q)$ in a single desired point, according to the integral relation

$$\varphi(q_0) = \int \varphi(q)\delta(q - q_0)dq, \tag{4.5.3}$$

which suggests the conceptually and practically pleasing possibility of representing $\varphi(q)$ as a superposition of a large number of suitably positioned and weighted point-like functions.

The derivatives of the delta function exhibit the same sifting property. In fact, being

$$\int \varphi(q)\delta^{(m)}(q - q_0)dq = (-)^m\varphi^{(m)}(q_0), \tag{4.5.4}$$

the m-th derivative $\delta^{(m)}(q)$ sifts out the value $\varphi^{(m)}(q_0)$ of the mth derivative of φ in the desired point q_0, provided of course this value exists. The delta function is an even function: $\delta(-q) = \delta(q)$, whilst the derivatives are alternatively odd or even: $\delta^{(m)}(-q) = (-)^m\delta^{(m)}(q)$.

Also worth mention is the relation

$$\int \delta(q - q')\delta(q - q'')dq = \delta(q' - q''), \tag{4.5.5}$$

which along with the superposition integral (4.5.3) allows us to consider the ensemble of the eigenfunctions $\{\varphi_{q'}(q), q' \in \mathbb{R}\}$ as a complete orthonormal set.

Interestingly the δ-function can be visualized, and even formally defined, as the limit of a sequence of appropriate functions of decreasing width, increasing height and unit area. In particular, we recall the definition of the delta function as limit of a sequence of real or complex Gaussian functions according to

$$\delta(q) = \lim_{\sigma \to 0} \frac{1}{\sqrt{\pi\sigma}} e^{-\frac{q^2}{\sigma}}, \tag{4.5.6}$$

the parameter σ being allowed to be real or purely imaginary.

Finally, it is noteworthy the integral representation

$$\delta(q) = \frac{1}{2\pi} \int e^{\pm i\eta q} d\eta, \tag{4.5.7}$$

establishing a relation of Fourier conjugation between the delta function and the harmonic function $e^{\pm i\eta q}$ (see Eqs. (4.6.1) and (4.6.2)). Equation (4.5.7) can be deduced from the representation of $\delta(q)$ as limit of a sequence of sinc functions: $\delta(q) = \lim_{K \to \infty} \frac{\sin(Kq)}{\pi q}$ (Problem 3). As a plain consequence of (4.5.7), one finds

$$\delta^{(m)}(q) = \frac{(\pm i)^m}{2\pi} \int \eta^m e^{\pm i\eta q} d\eta. \tag{4.5.8}$$

4.5.2 *The spatial harmonic waveform:* $e^{i\kappa q}$

The eigenvalue problem for the momentum operator $\widehat{\mathsf{p}} = -\frac{i}{k_0}\frac{\partial}{\partial q}$ is expressed by the equation

$$-\frac{i}{k_0}\frac{\partial}{\partial q}\varphi_p(q) = p\varphi_p(q), \tag{4.5.9}$$

which is solved by the $1D$ harmonic waveforms $e^{ik_0 pq}$. The eigenfunctions $\varphi_\kappa(q)$ of $\widehat{\mathsf{p}}$ are then conveniently written as

$$\varphi_\kappa(q) = \frac{1}{\sqrt{2\pi}} e^{i\kappa q}, \qquad \kappa = k_0 p, \tag{4.5.10}$$

absorbing the eigenvalue p into the product $k_0 p \equiv \kappa$. Clearly the eigenvalue p, and so κ, is real, otherwise $\varphi_\kappa(q)$ will blow up either as $q \to -\infty$ or as $q \to +\infty$. The factor $\frac{1}{\sqrt{2\pi}}$ ensures that the functions $\{\varphi_\kappa(q), \kappa \in \mathbb{R}\}$ fulfil the orthogonality relation

$$\int \varphi_\kappa^*(q)\varphi_{\kappa'}(q) dq = \delta(\kappa - \kappa'). \tag{4.5.11}$$

Comparison of (4.5.10) with (4.2.6) suggests to interpret the momentum eigenfunction $\varphi_\kappa(q)$ for any κ as the profile of a plane wave along a chosen transversal coordinate axis, let us say q_x, in a designated reference plane at some $z = z_i$.

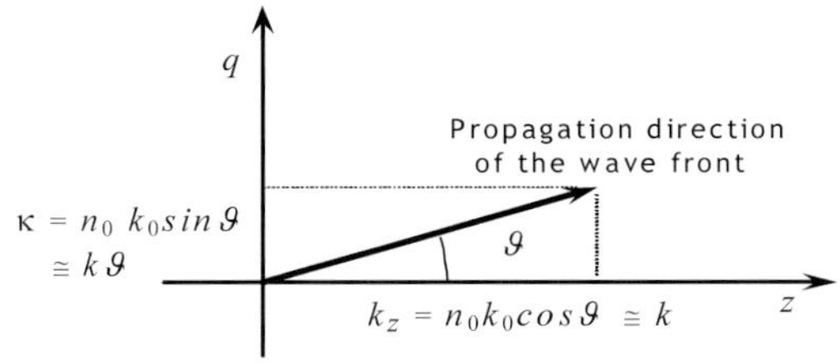

FIGURE 4.6. The spatial frequency as transverse wave number.

The harmonic profile (4.5.10) can be continued into a $3D$ plane wave, which may be supposed to propagate through some medium of uniform index n_0 in a direction uniquely identified by the eigenvalue κ for a given vacuum wave number k_0. The amplitude of the wave can in fact be written as (Eq. (4.2.6))

$$\psi_\kappa(q_x, q_y, z) = \frac{1}{\sqrt{2\pi}} e^{i[\kappa q_x + k_z(z-z_i)]}, \tag{4.5.12}$$

with then (Eq. (4.2.3))

$$\kappa^2 + k_z^2 = n_0^2 k_0^2 = k^2. \tag{4.5.13}$$

The eigenvalue κ identifies the transverse component of the wave vector, so that, if ϑ is the angle of the wave vector relative to the z axis, κ is (Fig. 4.6)

$$\kappa = k \sin\vartheta = k_0 n_0 \sin\vartheta, \qquad \text{or} \qquad p = n_0 \sin\vartheta, \tag{4.5.14}$$

thus yielding a direct relation of the eigenvalues κ to the directions of propagation in any chosen meridian plane through the z-axis in conformity with the $2D$ picture we are considering. Also, as the direction of the wave vector is the direction of the light ray in the geometrical optics picture, we are enabled to establish a direct correspondence between the eigenvalue p of the momentum operator $\widehat{\mathsf{p}}$ and the ray momentum $p = n_0 \sin\vartheta$ of geometrical optics.

Interestingly we see from (4.5.13) that

$$k_z = \pm\sqrt{n_0^2 k_0^2 - \kappa^2}. \tag{4.5.15}$$

In order that k_z be real the range of κ turns out to be bounded by $n_0 k_0$, i.e., $\kappa^2 \le n_0^2 k_0^2$ or $p^2 \le n_0^2$. The spectrum of $\widehat{\mathsf{p}}$ cannot extend beyond the interval $[-n_0, n_0]$, fixed by the refractive index n_0 of the medium, just as the range of the "classical" variable. However, in conformity with first-order ray optics, paraxial wave optics assigns small values to moment observable $\widehat{\mathsf{p}}$. The eigenvalues of $\widehat{\mathsf{p}}$ are supposed to remain well confined within the natural range $[-n_0, n_0]$, so that one can consider the spectrum as it extended over all the real line, thus avoiding the difficulties inherent in a limited spectrum. In fact, according to the Stone-von Neumann theorem, the Schrödinger realizations (4.4.1) and (4.4.2) in $\mathcal{L}^2(\mathbb{R})$ of the canonically self-adjoint operators $\widehat{\mathsf{q}}$ and $\widehat{\mathsf{p}}$, fulfilling the Heisenberg-Weyl commutation relation (4.4.4), holds only if both operators are unbounded. Account of the limited spectrum of the optical momentum, and hence of the global range of the ray directions over the full sphere beyond both the paraxial and metaxial regimes, has led to the formulation of the Euclidean model of optics, the underlying algebra being the Euclidean algebra $iso(3)$ ($iso(2)$) instead of the Heisenberg-Weyl algebra [16].

In the paraxial approximation the expression of κ or p in terms of the direction of the ray simplifies as

$$\kappa = k_0 p \simeq k_0 n_0 \vartheta, \qquad \text{or} \qquad p \simeq n_0 \vartheta, \tag{4.5.16}$$

by the allowed approximation $\sin\vartheta \simeq \vartheta$ for $\vartheta \ll 1$ (Fig. 4.6).

4.6 Spatial Fourier representation of optical wave fields

Paralleling representation (4.5.3) of the wave amplitude in terms of the δ-functions, we can decompose $\varphi(q,z)$ in terms of the frequency eigenwaves $\varphi_\kappa(q)$ for any fixed value of the axial parameter z. We write then the integral

$$\varphi(q,z) = \int \widetilde{\varphi}(\kappa,z)\varphi_\kappa(q)d\kappa = \tfrac{1}{\sqrt{2\pi}} \int \widetilde{\varphi}(\kappa,z)e^{i\kappa q}d\kappa, \tag{4.6.1}$$

which, in the light of the physical significance of $\varphi_\kappa(q)$, has the meaning of a decomposition of the $2D$ optical wave function $\varphi(q,z)$ across any assigned z-plane into $2D$ plane waves travelling along slightly different directions, definitely specified by the values of κ for a given wave number $k = k_0 n_0$.

The weighting factors $\widetilde{\varphi}(\kappa,z)$, regarded as a function of κ, form the *Fourier transform* of $\varphi(q,z)$, which by (4.5.11) shapes in turn as the integral

$$\widetilde{\varphi}(\kappa,z) = \int \varphi^*_\kappa(q)\varphi(q,z)dq = \tfrac{1}{\sqrt{2\pi}} \int \varphi(q,z)e^{-i\kappa q}dq. \tag{4.6.2}$$

The $\widetilde{\ }$-sign over a function symbol will be used throughout to identify the Fourier transform of the function. In the common language of Fourier analysis, integral (4.6.1) is referred to as the *inverse* Fourier transform. Note the symmetry of expressions (4.6.1) and (4.6.2) with respect to the variables q, κ.

Thus the eigenvalue κ acquires the significance as the Fourier conjugate variable to q; indeed, it is reported as *spatial frequency*. In addition, the Fourier transform (4.6.2) is also referred to as the *angular spectrum*, since it describes the directional content of the signal amplitude across any transverse plane in terms of plane wave components, each of which, according to relations (4.5.14) or (4.5.16), can equivalently be labelled by either the spatial frequency κ or the angle ϑ of the direction of propagation of the wave relative to the z axis.

Remarkably both $\varphi(q,z)$ and $\widetilde{\varphi}(\kappa,z)$ provide a complete representation of the light disturbance separately in the spatial coordinate and the spatial frequency domains, respectively. Paralleling the interpretation of $|\varphi(q,z)|^2$ as indicative of the energy density in space, we can relate the squared modulus of $\widetilde{\varphi}(\kappa,z)$ to the energy density per unit frequency, thus understanding $|\widetilde{\varphi}(\kappa,z)|^2$ as indicative of the *optical energy density spectrum.* Evidently, $|\widetilde{\varphi}(\kappa,z)|^2$ tells us which spatial frequencies (i.e. directions) are present through the total extent of the waveform at every z plane. Of course, no indication can be drawn from $\widetilde{\varphi}(\kappa,z)$ about precisely where in space such frequencies occur. The physical ideas, that have suggested to, and the mathematical machinery, that has been developed in order to elaborate a joint space-frequency description of the wave field will be the subject of Chapters 6 to 9.

Finally, we like to note that the angular spectrum $\widetilde{\varphi}(\kappa,z)$ mirrors the momentum wave function of quantum mechanics, the Fourier transform operation being as well the intertwining tool between the position and momentum representations of the quantum system state. It is in fact obvious that we can equally well describe the optical wave field in the momentum space, with the relevant relations connecting the q- and p-space being, of course,

$$\begin{aligned}\varphi(q,z) &= \sqrt{\tfrac{k_0}{2\pi}}\int \widetilde{\varphi}(p,z)e^{ik_0pq}dq,\\ \widetilde{\varphi}(p,z) &= \sqrt{\tfrac{k_0}{2\pi}}\int \varphi(q,z)e^{-ik_0pq}dp.\end{aligned} \tag{4.6.3}$$

To facilitate the comparison with the quantum mechanical relations we recall once again that $\frac{1}{k_0}$ corresponds to $\hbar$.

Integrals (4.5.3) and (4.6.1) realize the decomposition of the optical disturbance into impulsive and monofrequency signals. The choice of one with respect to the other is essentially suggested by the symmetry properties of the system, as reflected by the relative point spread function, as we will see in §§ 5.2.1 and 5.2.2 when describing the propagation through a free-medium section and a thin lens.

4.6.1 *Basics of the Fourier transform*

As the Fourier representation of the optical disturbance finds wide use in the later discussion, we list without proof a few of the basic mathematical properties of the Fourier transform, addressing the interested reader to [17] for detailed proofs as well as wider comments.

For notational convenience, we regard $\widetilde{\varphi}(\kappa,z)$ as resulting from the action of the Fourier transform operator $\widehat{\mathcal{F}}$ on the complex amplitude $\varphi(q,z)$:

$$(\widehat{\mathcal{F}}\varphi)(\kappa,z) = \widetilde{\varphi}(\kappa,z) = \tfrac{1}{\sqrt{2\pi}}\int \varphi(q,z)e^{-i\kappa q}dq, \tag{4.6.4}$$

and inversely, φ results from the action of the inverse operator $\widehat{\mathcal{F}}^{-1}$on $\widetilde{\varphi}$:

$$(\widehat{\mathcal{F}}^{-1}\widetilde{\varphi})(q,z) = \varphi(q,z) = \frac{1}{\sqrt{2\pi}} \int \widetilde{\varphi}(\kappa,z)e^{i\kappa q}d\kappa. \tag{4.6.5}$$

We shall not dwell on the "existence conditions" that confer a mathematical sense to the above integrals, directing the reader to the quoted references. Indeed, as we are interested in the Fourier transform from a physical rather than mathematical viewpoint, we may justifiably consider the physical existence of a waveform as a sufficient condition for defining its Fourier transform.

Enumerating the properties of the Fourier operator, we omit to explicitly display the now irrelevant dependence of φ and $\widetilde{\varphi}$ on the axial coordinate z.

The *linearity property* allows to calculate the Fourier transform of a linear combination of functions as the same linear combination of the Fourier transforms of the individual components, i.e.,

$$\widehat{\mathcal{F}}\{a\varphi_1 + b\varphi_2\} = a\widehat{\mathcal{F}}\varphi_1(q) + b\widehat{\mathcal{F}}\varphi_2(q) = a\widetilde{\varphi}_1(\kappa) + b\widetilde{\varphi}_2(\kappa), \tag{4.6.6}$$

for any complex numbers a, b, and wave functions φ_1, φ_2. In fact, confirming the analysis of § 3.4.2, we will see that the Fourier transform can be performed by simple one or two-lens optical systems. The Fourier operator may then be regarded as a wave-transfer operator, and as such comes to be linear.

The *similarity property* specifies the effect on the Fourier transform of a scaling of the input variable. Scaling the q-variable by the real parameter $\mathfrak{a}$ causes the κ-variable to scale accordingly by the inverse factor $\frac{1}{\mathfrak{a}}$ and produces also a change in the overall amplitude of the Fourier transform as

$$\widehat{\mathcal{F}}\varphi(\mathfrak{a}q) = \frac{1}{|\mathfrak{a}|}\widetilde{\varphi}(\tfrac{\kappa}{\mathfrak{a}}). \tag{4.6.7}$$

As scale transformations are performed optically by ideal magnifying systems, relation (4.6.7) describes the performance of a Fourier transformer-ideal magnifier composite system.

The *shift property* is concerned with the effect on the spectrum of a translation of the origin of the coordinate axis, which in fact produces a linear modulation of the Fourier transform, being

$$\widehat{\mathcal{F}}\varphi(q - q_0) = e^{-i\kappa q_0}\widetilde{\varphi}(\kappa). \tag{4.6.8}$$

The *modulation property* in turn accounts for the shift of the Fourier transform in response to a linear frequency modulation of the waveform; namely,

$$\widehat{\mathcal{F}}[e^{i\kappa_0 q}\varphi(q)] = \widetilde{\varphi}(\kappa - \kappa_0). \tag{4.6.9}$$

The *Parseval relation* establishes the conservation of the signal energy, that can intuitively justified considering the performance of a lossless optical Fourier transformer. Thus, we can write

$$\int |\widetilde{\varphi}(\kappa)|^2\, d\kappa = \int |\varphi(q)|^2\, dq. \tag{4.6.10}$$

The *convolution property* expresses the particularly interesting and useful result that the Fourier transform of the convolution of two functions equals the product of the Fourier transforms of the convolved functions. In formal terms, given the wave function $\varphi(q)$ as the convolution integral

$$\varphi(q) = \int \varphi_1(q - q')\varphi_2(q')\, dq' \equiv \varphi_1 * \varphi_2, \tag{4.6.11}$$

the Fourier transform is proved to be simply the product

$$\widetilde{\varphi}(\kappa) = \sqrt{2\pi}\widetilde{\varphi}_1(\kappa)\,\widetilde{\varphi}_2(\kappa). \tag{4.6.12}$$

The *multiplication property* mirrors the convolution property in the spatial frequency domain, stating that the Fourier transform of the product of two functions equals the convolution of the respective Fourier transforms. Thus, the spectrum of the product

$$\varphi(q) = \varphi_1(q)\,\varphi_2(q), \tag{4.6.13}$$

comes to be the convolution integral

$$\widetilde{\varphi}(\kappa) = \frac{1}{\sqrt{2\pi}} \int \widetilde{\varphi}_1(\kappa - \kappa')\widetilde{\varphi}_2(\kappa')\, d\kappa'. \tag{4.6.14}$$

Remarkably, multiplication of $\varphi(q)$ times q is turned by Fourier transformation into the differentiation of $\widetilde{\varphi}(\kappa)$ with respect to κ, as

$$\widehat{\mathcal{F}}\,[q\,\varphi(q)] = i\frac{d}{d\kappa}\widetilde{\varphi}(\kappa). \tag{4.6.15}$$

Conversely, differentiation of $\varphi(q)$ with respect to q produces under Fourier transform the multiplication of $\widetilde{\varphi}(\kappa)$ times $i\kappa$:

$$\widehat{\mathcal{F}}\,[\frac{d}{dq}\varphi] = i\kappa\widetilde{\varphi}(\kappa). \tag{4.6.16}$$

Relations (4.6.15) and (4.6.16) show that the position and momentum operators $\widehat{\mathsf{q}}$ and $\widehat{\mathsf{p}}$ are Fourier imaged respectively into (i times) differentiation and ($\frac{1}{k_0}$ times) multiplication in the frequency domain:

$$\begin{aligned} \widehat{\mathsf{q}} &\xrightarrow{\widehat{\mathcal{F}}} i\frac{\partial}{\partial\kappa}, \\ \widehat{\mathsf{p}} &\xrightarrow{\widehat{\mathcal{F}}} \frac{1}{k_0}\kappa \end{aligned} \tag{4.6.17}$$

The relations above are used in § 4.6.6 to identify the momentum representation of the operators $\widehat{\mathsf{q}}$ and $\widehat{\mathsf{p}}$, and accordingly the frequency representation of the position and frequency operators $\widehat{\mathsf{q}}$, $\widehat{\kappa}$ (there introduced).

Anticipating the results of later sections, we mention that the convolution (4.6.11) and the product (4.6.13) may be regarded as the transformation laws for light signals propagating respectively through free-space sections and thin lenses. One of the functions in (4.6.11) and (4.6.13) can be interpreted as the input waveform and the other as the impulse response of the free-space section or correspondingly the phase-modulation function of the lens. Accordingly, the product (4.6.12) and the convolution (4.6.14) represent the transfer laws for the above quoted systems in the spatial-frequency domain. As a convolutive-like relation in the q-domain turns into a multiplicative-like relation in the Fourier conjugate domain, i.e., the κ-domain, similarly a multiplicative-like relation in the q-domain turns into a convolutive-like relation in the κ-domain. Convolution and multiplication are conjugate operations with respect to Fourier transformation. Free propagation and lens transfer (and, more in general, space-invariant and spreadless systems) are just the concrete realization of such a mathematical relation.

In this connection, we note that delta and harmonic functions are conjugate with respect to Fourier transformation as proved by the relations

$$\begin{aligned} &\delta(q'-q) = \tfrac{1}{2\pi}\int e^{i\kappa(q'-q)}d\kappa, \qquad \delta(\kappa'-\kappa) = \tfrac{1}{2\pi}\int e^{iq(\kappa'-\kappa)}dq, \\ &e^{i\kappa q} = \int e^{i\kappa q'}\delta(q'-q)dq' = \int e^{i\kappa' q}\delta(\kappa'-\kappa)d\kappa', \end{aligned} \tag{4.6.18}$$

expressing the formal link between these two idealized optical disturbances, whose behaviours in the q and κ domains are complementary to one another. Relations (4.6.18) furnish the mathematical ground for the performance of the optical Fourier transformer, which images a bundle of parallel rays into a point (i.e., spatial harmonics into δ-functions) in the Fourier plane and conversely a point in the input plane into a bundle of parallel rays (i.e., δ-functions into spatial harmonics).

4.6.2 *Input-output relations in the spatial frequency representation: the wave-spread function*

The integral transform (4.3.13) is the mathematical model for the light wave propagation through optical systems, exemplified in terms of a relation between the optical disturbances entering to and emerging from the system. In it both the input and output signals are described in the space domain and hence are represented by the complex amplitudes $\varphi_i(q, z_i)$ and $\varphi_o(q, z_o)$ at the assigned input and output reference planes.

Equivalent input-output relations can be elaborated, in which the input and output signals are described in different domains. One can relate, for instance, the frequency spectra of the input and output signals $\widetilde{\varphi}_i(\kappa, z_i)$ and $\widetilde{\varphi}_o(\kappa, z_o)$, or the field distribution at one plane and the Fourier transform at the other plane, namely $\varphi_i(q, z_i)$ and $\widetilde{\varphi}_o(\kappa, z_o)$, or inversely $\widetilde{\varphi}_i(\kappa, z_i)$ and $\varphi_o(q, z_o)$. In any case the transfer relation takes the form of an integral transform similar to (4.3.13), with the inherent spread functions being specific of the involved domains. Evidently, the existence of four equivalent forms of input-output relations with the respective spread functions is a direct consequence of the existence of four pairs of input-output domains in which the signals can be represented, i.e., space-space, frequency-frequency, space-frequency and frequency-space. This can be brought into relation with the existence of exactly four Hamilton's characteristics, i.e., the point characteristic, the angle characteristic and the two mixed characteristics. In fact, a definite relation can be shown to hold between the four system functions of wave optics and the corresponding Hamilton's characteristics of geometrical optics. As it is beyond the scope of this book to present a more detailed analysis of the subject, the interested reader is referred to [18].

Here we deduce the relation between the Fourier transforms of the object and image distributions $\widetilde{\varphi}_i$ and $\widetilde{\varphi}_o$. Fourier transforming both sides of (4.3.13) with respect to q_o and then expressing the input amplitude $\varphi_i(q, z_i)$ in terms of the relevant frequency spectrum $\widetilde{\varphi}_i(\kappa, z_i)$ by (4.6.1), we end up with the integral relation

$$\widetilde{\varphi}_o(\kappa_o, z_o) = \int d\kappa_i \mathsf{h}(\kappa_o, \kappa_i; z_o, z_i)\widetilde{\varphi}_i(\kappa_i, z_i), \tag{4.6.19}$$

formally identical to (4.3.13). The weighting function h is explicitly given by

$$\mathsf{h}(\kappa_o, \kappa_i; z_o, z_i) = \tfrac{1}{2\pi} \iint dq_i dq_o \, e^{-i(\kappa_o q_o - \kappa_i q_i)} \, \mathsf{g}(q_o, q_i; z_o, z_i), \tag{4.6.20}$$

with the reciprocal expression of g versus h,

$$\mathsf{g}(q_o, q_i; z_o, z_i) = \tfrac{1}{2\pi} \iint d\kappa_i d\kappa_o \, e^{i(\kappa_o q_o - \kappa_i q_i)} \, \mathsf{h}(\kappa_o, \kappa_i; z_o, z_i). \tag{4.6.21}$$

It is evident that h and g are related by a Fourier transform-like relation; however, due to the minus sign in the exponent $(\kappa_o q_o - \kappa_i q_i)$, they do not constitute a Fourier-integral pair[c].

For ease of writing, as in the case of $\mathsf{g}(q_o, q_i)$ we simply write $\mathsf{h}(\kappa_o, \kappa_i)$ instead of $\mathsf{h}(\kappa_o, \kappa_i; z_o, z_i)$.

Interestingly, if the input to the system is a spatial harmonic $\varphi_{\kappa_i}(q_i, z_i) = \frac{e^{i\kappa_i q_i}}{\sqrt{2\pi}}$, the system responds producing the output

$$\varphi_o(q_o, z_o) = \tfrac{1}{\sqrt{2\pi}} \int dq_i \mathsf{g}(q_o, q_i) e^{i\kappa_i q_i}, \tag{4.6.22}$$

[c] The Fourier transform of a $2D$ function $f(q_1, q_2)$ is defined by the $2D$ integral

$$\widetilde{f}(\kappa_1, \kappa_2) = \tfrac{1}{2\pi} \iint dq_2 dq_1 \, e^{-i(\kappa_1 q_1 + \kappa_2 q_2)} f(q_1, q_2).$$

When the inner product $(\kappa_1 q_1 + \kappa_2 q_2)$ in the exponent is replaced by the skew-product $(\kappa_1 q_1 - \kappa_2 q_2)$, the resulting transform is also reported as *symplectic* Fourier transform.

which first by (4.6.21) and then by (4.6.18) is seen to be

$$\varphi_o(q_o, z_o) = \frac{1}{\sqrt{2\pi}} \int d\kappa_o \mathsf{h}(\kappa_o, \kappa_i)\, e^{i\kappa_o\, q_o}. \tag{4.6.23}$$

The function h yields therefore the spectral representation of the system response to a spatial harmonic waveform, which is represented by a δ-function in κ-space: $\widetilde{\varphi}_{\kappa_i}(\kappa_i, z_i) = \delta(\kappa - \kappa_i)$. Hence, h is reported as the *wave-spread* (or *wave response*) *function* of the system. The input-output relation (4.6.19) has the same significance as (4.3.13), since the response of the system in the κ-domain is reconstructed from the superposition of the responses to elementary δ-stimuli $\delta(\kappa - \kappa_i)$, corresponding to frequency eigenwaves in the q-domain.

As a conclusion of this account about the wave spread function, we say that being h and g linked by (4.6.20) and (4.6.21), all relations for g correspond to analogous relations for h, and vice versa. In particular, relation (4.3.21) expressing the losslessness of the optical system is mirrored into

$$\int d\kappa_o \mathsf{h}^*(\kappa_o, \kappa_i') \,\mathsf{h}(\kappa_o, \kappa_i) = \delta(\kappa_i' - \kappa_i). \tag{4.6.24}$$

The reader can easily deduce the composition rule for h, which parallels, of course, relation (4.3.18) for g.

4.6.3 *Space-invariant and spreadless systems*

Space-invariant systems, which are characterized by the remarkable property that the image distribution does not depend on the specific location of the object in the input plane [17]. Thus, variations in the location of the input distribution over the object plane do not affect the structure of the output distribution, but only its location in the image plane, being indeed shifted by the same amount as the object distribution[d].

Shifting the input over the object plane as $\varphi_i(q_i - Q, z_i)$, the output is correspondingly shifted through the image plane into $\varphi_o(q_o - Q, z_o)$ for any shift amount Q. This implies that the spread function does not depend separately on its arguments but only through the difference $q_o - q_i$:

$$\mathsf{g}(q_o, q_i) = \mathfrak{g}(q_o - q_i). \tag{4.6.25}$$

The input-output relation for shift-invariant systems writes then in the form

$$\varphi_o(q_o, z_o) = \int dq_i \mathfrak{g}(q_o - q_i)\varphi_i(q_i, z_i) = \mathfrak{g} * \varphi_i, \tag{4.6.26}$$

[d] In practice, optical systems are seldom space-invariant over all the object plane, but it is usually possible to divide the input plane into small regions through which the system is approximately invariant. Thus, within the linear approximation of our concern, the condition for space-invariance may be checked over a small region of the object plane near the axis.

thus showing that the image distribution is the *convolution* of the object distribution with the point-spread function of the system.

Taking advantage of the convolution property of the Fourier transform, the above relation takes the simple multiplicative form

$$\widetilde{\varphi}_o(\kappa, z_o) = \mathfrak{h}(\kappa)\widetilde{\varphi}_i(\kappa, z_i). \tag{4.6.27}$$

in the κ-space. The function $\mathfrak{h}(\kappa)$ is, apart from a $\frac{1}{\sqrt{2\pi}}$ factor, the Fourier transform of the point-spread function $\mathfrak{g}(q)$ according to

$$\mathfrak{h}(\kappa) = \int dq\, e^{-i\kappa q}\, \mathfrak{g}(q) = \sqrt{2\pi}\widetilde{\mathfrak{g}}(\kappa). \tag{4.6.28}$$

It is reported as the *optical transfer function.*

The "local" character of the input-output relation (4.6.27) is evident: no mixing of the spatial frequencies is produced by the system, each spectral component of the input being transmitted into the output signal though modulated by the transfer function $\mathfrak{h}$. In particular, spatial harmonics are simply transferred to the output plane. Naturally, by an appropriate design of the optical system, i.e., of the relative $\mathfrak{h}$-function, one may remove or preferentially pass certain desired spatial frequency component of the signal; in other terms, one may operate a spatial frequency filtering of the input signal.

Equation (4.6.27) relates the Fourier transform of the output signal $\widetilde{\varphi}_o$ to the Fourier transform of the input $\widetilde{\varphi}_i$ through a multiplication by the system function $\mathfrak{h}$. This suggests an alternative procedure to that indicated by the space-domain convolution (4.6.26) to find the system output, requiring the often simpler sequence with a detour through the frequency domain:

space-invariant systems: $\mathfrak{g}(q_o, q_i) = \mathfrak{g}(q_o - q_i)$

q-space		κ-space
φ_i	Fourier transformation $\longrightarrow$	$\widetilde{\varphi}_i$
convolution by $\mathfrak{g}$ $\downarrow$		multiplication by $\mathfrak{h} \propto \widetilde{\mathfrak{g}}$ $\downarrow$
φ_o	inverse Fourier transformation $\longleftarrow$	$\widetilde{\varphi}_o$

In contrast with the non local character of the transfer relation (4.6.26) for space-invariant systems, *spreadless* systems are characterized by a local relation connecting the output to the input signal as

$$\varphi_o(q, z_o) = \mathfrak{f}(q)\varphi_i(q, z_i). \tag{4.6.29}$$

The input signal is simply modulated by the function $\mathfrak{f}(q)$. Then, in full analogy with (4.6.27), one may suitably reshape the input disturbance as desired, removing or preferentially passing certain desired portions of the distribution; in other words, one may perform a filtering operation in the space domain.

In terms of the spread function, spreadless systems are identified by

$$\mathfrak{g}(q_o, q_i) = \mathfrak{f}(q_0)\delta(q_i - q_o). \tag{4.6.30}$$

It is evident that on account of the multiplication property (4.6.14) of the Fourier transform, locality is lost in the frequency domain, where the input-output relation takes the convolutive structure as (4.6.26), namely,

$$\widetilde{\varphi}_o(\kappa, z_o) = \frac{1}{\sqrt{2\pi}} \int \widetilde{\mathfrak{f}}(\kappa - \kappa')\widetilde{\varphi}_i(\kappa', z_i)\, d\kappa' = \frac{1}{\sqrt{2\pi}} \widetilde{\mathfrak{f}} * \widetilde{\varphi}_i. \tag{4.6.31}$$

Therefore, the picture above sketched relatively to the performance of a space-invariant system can be reproduced relatively to the behaviour of a spreadless system as well with the nature, and hence the mathematical representation, of the responses of the two systems being interchanged between the spatial and frequency domains:

spreadless systems: $\mathfrak{g}(q_o, q_i) = \mathfrak{f}(q_0)\delta(q_i - q_o)$

$$\begin{array}{ccc}
q\text{-space} & & \kappa\text{-space} \\
\varphi_i & \xrightarrow{\text{Fourier transformation}} & \widetilde{\varphi}_i \\
\text{multiplication by } \mathfrak{f} \downarrow & & \downarrow \text{convolution by } \widetilde{\mathfrak{f}}/\sqrt{2\pi} \\
\varphi_o & \xleftarrow{\text{inverse Fourier transformation}} & \widetilde{\varphi}_o
\end{array}$$

Relations like (4.6.29) and (4.6.27), where the input and the output of a linear system are related in the spatial or spatial frequency domain by a simple multiplicative relation through the system function, are representative of what are reported as *modulators* or *spatial frequency filters,* according to whether the multiplicative relation holds in the space or frequency domain. Evidently, any space invariant system operates as a spatial frequency filter. A region of homogeneous space furnishes a realizable example of a space-invariant system, whilst a thin lens is an example of a modulator, i.e., a filter in the space domain. Linear filters are basic to the (optical and not) signal processing, which comprises, for instance, spectral analysis, pattern recognition as well as signal recovery and detection [15.3, 17].

It is easily seen that cascading appropriate modulating and filtering devices one obtains a Fourier transformer. In fact, given a waveform $\varphi_i(q, z_i)$, we suppose to firstly multiply it by the linear *chirp* $e^{-i\beta\frac{q^2}{2}}$ [e], i.e.,

$$\varphi_i(q, z_i) \longrightarrow \varphi_i(q, z_i)e^{-i\beta\frac{q^2}{2}}, \tag{4.6.32}$$

---

[e] Here, by linear chirps we signify the pure imaginary Gaussians $e^{i\beta q^2}$, β real.

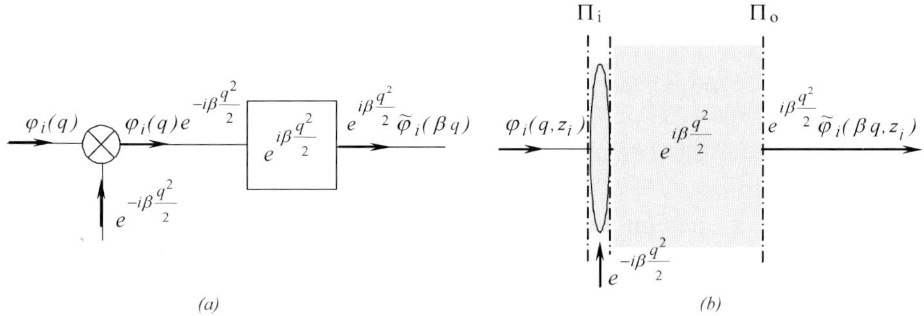

FIGURE 4.7. *(a)* Electrical and *(b)* optical systems implementation of the sequence of chirp multiplication and chirp convolution, producing the Fourier transform of the input signal.

which is then input into a filter with a chirp-like impulse response $\mathfrak{g}(q) = \frac{e^{i\beta \frac{q^2}{2}}}{\sqrt{2\pi}}$, and hence a chirp-like transfer function $\mathfrak{h}(\kappa) = \sqrt{\frac{i}{\alpha}}\, e^{-i\frac{\kappa^2}{2\beta}}$. The output,

$$\varphi_o(q, z_o) = \frac{1}{\sqrt{2\pi}} \int \varphi_i(q', z_i) e^{-i\beta \frac{q'^2}{2}} e^{i\beta \frac{(q-q')^2}{2}} = e^{i\beta \frac{q^2}{2}} \widetilde{\varphi}_i(\beta q, z_i), \qquad (4.6.33)$$

is a chirp-modulated signal whose envelope is the Fourier transform of the input. Since in the linear approximation thin lenses act as modulators by chirps and free-space sections as frequency filters by chirps, the above sequence of processes can be implemented by a lens followed by a free-space section, as sketched in Fig. 4.7.b), which of course supports the basic scheme of a Fourier transformer drawn in Fig. 3.7. In Fig. 4.7.a) the analogous electrical cascade is depicted in order to evidence the optical-electrical devices analogy [17.2].

4.6.4 *The optical uncertainty relation. I*

It is generally observed that spatially "narrow" signals have "wide" frequency spectra, and conversely. The scaling property (4.6.7) is a manifestation of this reciprocal relation between the spreads of any Fourier transform pair $(\varphi, \widetilde{\varphi})$.

The uncertainty product gives a quantitative account of this empiric observation, relating the "width" of the signal to the "width" of its Fourier transform through the inequality

$$\sigma_q \sigma_\kappa \geq \tfrac{1}{2}. \qquad (4.6.34)$$

Here the variance of $|\varphi|^2$ is used as a measure of the width of φ and similarly the variance of $|\widetilde{\varphi}|^2$ is used as measure of the width of $\widetilde{\varphi}$. Accordingly, the symbols σ_q and σ_κ above mean the square roots of

$$\begin{aligned} \sigma_q^2(z) &= \frac{1}{\|\varphi\|^2} \int (q-\overline{q}(z))^2 \, |\varphi(q,z)|^2 \, dq, \\ \sigma_\kappa^2(z) &= \frac{1}{\|\widetilde{\varphi}\|^2} \int (\kappa-\overline{\kappa}(z))^2 \, |\widetilde{\varphi}(\kappa,z)|^2 \, d\kappa. \end{aligned} \tag{4.6.35}$$

On a geometrical account, σ_q and σ_κ are the radii of gyration of the areas under the curves $|\varphi(q,z)|^2$ and $|\widetilde{\varphi}(\kappa,z)|^2$ about an axis passing through the relevant mean values $\overline{q}(z)$ and $\overline{\kappa}(z)$, respectively given as

$$\overline{q}(z) = \frac{1}{\|\varphi\|^2} \int q \, |\varphi(q,z)|^2 \, dq, \qquad \overline{\kappa}(z) = \frac{1}{\|\widetilde{\varphi}\|^2} \int \kappa \, |\widetilde{\varphi}(\kappa,z)|^2 \, d\kappa \, . \tag{4.6.36}$$

For formal convenience, the wave function is supposed to be normalized to unity; also, the dependence of φ and $\widetilde{\varphi}$, as well as of $\overline{q}$, $\overline{\kappa}$, σ_q and σ_κ, on the axial coordinate z will no more be displayed.

A further simplification of expressions (4.6.35) is attained by taking as zero the centroid of $|\varphi|^2$ and $|\widetilde{\varphi}|^2$, i.e., $\overline{q} = 0$ and $\overline{\kappa} = 0$. This condition can easily be realized by shifting φ by $-\overline{q}$ and $\widetilde{\varphi}$ by $-\overline{\kappa}$. The reader may easily verify that σ_q and σ_κ are insensitive to such space and frequency shifts of φ and $\widetilde{\varphi}$.

The width principle (4.6.34) grounds on the variance measure of "width" according to (4.6.35) and on the Fourier transform relation holding between φ and $\widetilde{\varphi}$. In fact, it applies to any Fourier integral pair with definitions (4.6.35) for the inherent spreads around the respective centroids.

Let us consider the frequency variance σ_κ^2. As κ is real, unfolding the square modulus $|\widetilde{\varphi}(\kappa)|^2 = \widetilde{\varphi}^*(\kappa)\widetilde{\varphi}(\kappa)$ we can write

$$\sigma_\kappa^2 = \int |\kappa\widetilde{\varphi}(\kappa)|^2 d\kappa = \|\kappa\widetilde{\varphi}\|^2 \, , \tag{4.6.37}$$

which then can be interpreted as the norm of $\kappa\widetilde{\varphi}$. We have learned in § 4.6.1 that the frequency multiplication of the transform arises from the transform of the derivative of the signal; namely,

$$\kappa\widetilde{\varphi}(\kappa) = \widehat{\mathcal{F}}\,[-i\frac{d}{dq}\,\varphi]. \tag{4.6.38}$$

As Parseval's theorem says that the norm of the Fourier transform equals the norm of the function, one can identify σ_κ^2 as the norm of the derivative of φ. (The imaginary unit i disappears, of course, when multiplying $-id\,\varphi/dq$ by the complex conjugate $id\varphi^*/dq$). Therefore, we can write

$$\sigma_\kappa^2 = \int |\frac{d}{dq}\,\varphi\,|^2 dq = \|\frac{d}{dq}\,\varphi\,\|^2. \tag{4.6.39}$$

Likewise, the spatial variance σ_q^2 is seen to be the norm of $q\varphi$, i.e.,

$$\sigma_q^2 = \int |q\varphi(q)|^2 dq = \|q\varphi\|^2 . \tag{4.6.40}$$

Then, upon multiplying σ_q^2 and σ_κ^2:

$$\sigma_q^2\sigma_\kappa^2 = \int |q\varphi(q)|^2 dq \cdot \int |\frac{d}{dq}\,\varphi\,|^2 dq, \tag{4.6.41}$$

we resort to Schwarz's inequality[f] to compose the product of the two integrals into a single integral, thus finally obtaining

$$\sigma_q^2\sigma_\kappa^2 \geq \left|\int q\varphi^* \frac{d}{dq}\,\varphi\, dq\right|^2 \geq \frac{1}{2}. \tag{4.6.42}$$

The conclusive inequality above follows from the estimation of the intermediate quantity through the steps:

$$\begin{aligned} \left|\int q\varphi^* \tfrac{d}{dq}\,\varphi\, dq\right| &= \tfrac{1}{2}\left\{\left|\int q\varphi^* \tfrac{d}{dq}\,\varphi\, dq\right| + \left|\int q\varphi [\tfrac{d}{dq}\,\varphi]^*\, dq\right|\right\} \\ &\geq \tfrac{1}{2}\left|\int q\varphi^* \tfrac{d}{dq}\,\varphi\, dq + \int q\varphi [\tfrac{d}{dq}\,\varphi]^*\, dq\right| = \tfrac{1}{2}\,\|\varphi\|^2 = \tfrac{1}{2}, \end{aligned} \tag{4.6.43}$$

the last equality resulting from an integration by parts of the two integrals embraced by the modulus symbol under the assumption that the complex amplitude $\varphi(q)$ goes to zero at infinity more rapidly than $\frac{1}{\sqrt{q}}$.

Relation (4.6.34) with definitions (4.6.35) for the signal variances in the space and frequency domains is strongly reminiscent of Heisenberg's uncertainty relation of quantum mechanics. Of course, the analogy is formal only, although the use of the same word "uncertainty" may be misleading. In this respect, we further remark that (4.6.34) expresses a definite relation between the energy density of the signal in space and in spatial frequency, respectively, represented by $|\varphi|^2$ and $|\widetilde{\varphi}\,|^2$. Such a relation takes the ultimate form (4.6.34) as a consequence of the mathematical definitions (4.6.35) being adopted as measures of the space and frequency extents of the signal, and above all as a consequence of the Fourier relation holding between φ and $\widetilde{\varphi}$. The plain

---

[f] We recall that the Schwarz inequality states that

$$\left|\int_a^b f(x)g^*(x)dx\right|^2 \leq \int_a^b |f(x)|^2 \cdot \int_a^b |g(x)|^2\, dx = \|f\|^2 \cdot \|g\|^2 ,$$

for any pair of complex-valued functions $f(x)$ and $g(x)$, which are supposed to be integrable on the interval $[a, b]$. The equality holds if and only if f and g are linearly dependent functions.

content of (4.6.34) is that the energy densities in space and frequency of any signal cannot both be made narrow.

It is customary to take $2\sigma_q$ as representative of the linear extent of the signal: $S \equiv 2\sigma_q$, and correspondingly σ_κ as representative of the signal bandwidth (i.e., the interval of frequency $|\kappa| \leq \sigma_\kappa$, over which the frequency spectrum of the signal differs significantly from zero). Accordingly, inequality (4.6.34), which so turns into

$$SB \geq 1 \tag{4.6.44}$$

is also reported as the *space-bandwidth* product relation.

Alternative, and more convenient, definitions for the waveform extent and bandwidth have been adopted within the signal and information theory, with which definitions the analogy with quantum mechanics vanishes.

Schwarz's inequality turns into an equality when the two functions f and g are proportional to each other. Under this hypothesis, in the case of the functions entering the integrals in (4.6.41), one should have:

$$\frac{d}{dq}\varphi(q) = -cq\varphi(q), \tag{4.6.45}$$

which yields the normalized complex amplitude,

$$\varphi(q) = \left(\tfrac{c_r}{\pi}\right)^{\frac{1}{4}} e^{-(c_r + ic_i)\frac{q^2}{2}}, \tag{4.6.46}$$

c_r and c_i identifying the real and imaginary parts of the constant c: $c = c_r + ic_i$.

Inserting (4.6.46) into the product of integrals in (4.6.41) or equivalently in the single integral in (4.6.42), one obtains

$$\int |q\varphi(q)|^2 dq \cdot \int \left|\frac{d}{dq}\varphi\right|^2 dq = \frac{1}{4}\frac{c_r}{\pi}(c_r^2 + c_i^2)\frac{\pi}{c_r^3} = \frac{1}{4}\left(1 + \frac{c_i^2}{c_r^2}\right), \tag{4.6.47}$$

which turns into the value $\frac{1}{4}$ if and only if $c_i = 0$. Accordingly, the waveforms, which minimize the uncertainty product to $\frac{1}{2}$, shape as real Gaussians, i.e.,

$$\varphi_{\sigma_q\sigma_\kappa = 1/2}(q) = \left(\tfrac{1}{\pi a^2}\right)^{\frac{1}{4}} e^{-\frac{q^2}{2a^2}}, \qquad a \text{ real.} \tag{4.6.48}$$

An example of a minimum uncertainty product waveform, accounting also for possible non zero centroids of φ and $\widetilde{\varphi}$, is provided by the *Gabor function*

$$\varphi_{\mathsf{G}}(q) = \left(\tfrac{1}{\pi a^2}\right)^{\frac{1}{4}} e^{-\frac{1}{2a^2}(q-q_0)^2 + i\kappa_0(q - \frac{q_0}{2})}, \tag{4.6.49}$$

which is of great interest within both optical and not strictly optical contexts. The parameter a is a positive quantity. As the reader may easily verify, $\overline{q} = q_0$, $\overline{\kappa} = \kappa_0$, $\sigma_q^2 = \frac{a^2}{2}$, $\sigma_\kappa^2 = \frac{1}{2a^2}$ and so $\sigma_q\sigma_\kappa = \frac{1}{2}$. Waveform (4.6.49) consists of a sinusoid modulated by a Gaussian form; its intensity is concentrated around the centroid at q_0 through a Gaussian profile, whose width is measured by a. Real-valued functions of space and time, formally similar to (4.6.49), have originally been introduced and studied by Gabor in his investigation of the *essence of the information*, conveyed by channels of communications [20]. Gabor elaborated a "quantum theory" of information, dominated by uncertainty relations analogous to those formulated by Heisenberg in quantum mechanics; accordingly, rephrasing (4.6.34) in the time-frequency context, the simultaneous specificity of a signal in time and temporal frequency is limited by a lower bound of the signal bandwidth and time duration. In Gabor's theory, Gaussian modulated sinusoids like (4.6.49) represent the elementary signals, i.e., the *quanta of information*, which minimize the indeterminacy product. As the original Gabor functions of time, complex-valued space functions of the type (4.6.49) retain the attractive feature of being compact in both space and spatial frequency as they minimize the product of their variances in space and spatial frequency. Since this product specifies roughly the amount of information conveyed by a signal, it is evident its relevance for research on information encoding and transmission in various fields of physical interest. In this respect, we mention that Gabor's signal expansion in terms of Gaussian-weighted complex exponentials like $\varphi_{_G}(q)$ represents a basic example of a joint space-frequency signal representation (see § 7.2.2). In optical terms, waveforms like (4.6.49) may be interpreted as the q-profiles at the waist of Gaussian beams, as those produced, for instance, by stable resonator laser sources.

The frequency spectrum $\widetilde{\varphi}_G(\kappa)$ has the same analytical form as $\varphi_{_G}(q)$, with the Fourier conjugate variables q and κ interchanged; explicitly we find

$$\widetilde{\varphi}_G(\kappa) = (\tfrac{a^2}{\pi})^{\frac{1}{4}} e^{-\frac{a^2}{2}(\kappa-\kappa_0)^2 - iq_0(\kappa - \frac{\kappa_0}{2})}, \tag{4.6.50}$$

which displays a Gaussian profile centered around the mean frequency κ_0.

A comparison between $\varphi_{_G}(q)$ and $\widetilde{\varphi}_G(\kappa)$ manifests the general property of Gaussian forms being invariant under Fourier transform, as the reader may prove by using suitable normalized coordinates: $q \to \frac{q}{a}$, $\kappa \to \kappa a$. Specifically, focussing on the strictly Gaussian factors in (4.6.49) and (4.6.50), we see that

$$\widehat{\mathcal{F}}[e^{-\frac{x^2}{2}}] = e^{-\frac{x^2}{2}}. \tag{4.6.51}$$

In Chapter 9 we will obtain a more precise form of the optical uncertainty relation (4.6.34) showing indeed an additional term besides $\frac{1}{2}$.

4.6.5 *The wave-optical phase space*

From the discussion above it clearly emerges that a wave-optical phase space is conceivable with coordinates naturally suggested by the space and spatial frequency variables q, κ.

The geometric-optical phase space, which has been the ambience of our description of first-order optical systems in the foregoing chapters, is identified by the position and angle ray-variables. The "point" in the geometrical-optics phase-space is the geometric object suitable to the ray based, and hence local, description of light elaborated by geometrical optics. The position and momentum coordinates are in fact intended as simultaneously and precisely measurable attributes of the ray. As shown in Chapters 2 and 3, one can represent a ray in physical space by a point in phase space, and accordingly describe the light propagation by curves in phase space, since points are transformed into points as the ray propagates in real space. The "point" in phase space still remains central to the geometric-optics description of light propagation even when a ray density distribution is involved, since the ray density can be evolved through the trajectories of the single rays in the bundle.

Evidently a wave-optical phase-space representation of light propagation is still not feasible in terms of the hitherto elaborated descriptions of optical wave fields as they involve separately the spatial waveform, signified by $\varphi(q)$, and the frequency spectrum, signified by $\widetilde{\varphi}(\kappa)$. Therefore, a joint representation of optical wave fields together in space and frequency must be elaborated, and accordingly the relevant transfer laws identified, in order to enable a phase-space description of optical signals and their propagation through optical systems.

However, we may guess that even when the desired joint space-frequency representation of optical waveforms will be elaborated, "points" in wave-optical phase space will still remain physically meaningless. Since the simultaneous and precise specificity of any signal in space and frequency is inherently limited by the uncertainty relation, we may expect that any joint signal representation will reflect such a limitation, namely, such an inaccessibility of "points" in the wave-optical phase space. Thus, whilst a phase space distribution like $\delta(q - q_0)\delta(p - p_0)$ is legitimately conceivable in geometric optics, representing a single ray at q_0 in the direction p_0, the analogous phase-space distribution $\delta(q-q_0)\delta(\kappa-\kappa_0)$ is not conceivable in wave optics. In this respect, the minimum uncertainty product waveforms, i.e., the Gaussian wave packets like (4.6.49), may be intended as the signals closest to the ray-concept of geometric optics, having the maximum allowed space and frequency localization.

Finally, we note that the geometry of a space is of course determined by the invariance laws under the inherent transformations of that space into itself. The geometric-optical phase space is indeed symplectic since areas are preserved by the propagating ray transformations in accord with the underlying Hamiltonian dynamics. Likewise the geometry of the wave-optical phase space will be determined by the invariance laws of the individualized space-frequency descriptor when evolved by propagation through optical systems. Specifically it has been proved that in the Wigner representation the wave optical phase space (also called *Mock space*) is an affine space [21].

As a conclusion, it is interesting to remark that, on account of the definite relation between the optical momentum p and the spatial frequency κ, the matrix formalism of ray optics may also be considered to yield relations between the spatial coordinates and the spatial frequencies of the light waveform from the input to the output plane. In fact, taking account of (4.5.14), the input-output relations of ray optics, involving the ray position and slope (q, p) and the $ABCD$ matrix, can be turned into analogous relations, involving the space and frequency variables q and κ and a slightly different transfer matrix, where the off-diagonal elements are scaled by $\frac{1}{k_0}$ and k_0 with respect to those in the ray-optics matrix. In fact, it is easy to verify the correspondence:

$$\underset{\text{canonically conjugate variables } (q,p)}{\begin{pmatrix} q_o \\ p_o \end{pmatrix} = \begin{pmatrix} A & B \\ C & D \end{pmatrix} \begin{pmatrix} q_i \\ p_i \end{pmatrix}} \quad \rightleftharpoons \quad \underset{\text{Fourier conjugate variables } (q,\kappa)}{\begin{pmatrix} q_o \\ \kappa_o \end{pmatrix} = \begin{pmatrix} A & B/k_0 \\ k_0 C & D \end{pmatrix} \begin{pmatrix} q_i \\ p_i \end{pmatrix}} \tag{4.6.52}$$

Of course, the matrix relation on the left is visualizable in phase space identifying the meaningful trajectory of the ray under paraxial propagation. In contrast, the matrix relation on the right has only a mathematical valence linking the conjugate variables q, κ from the input to the output plane. As earlier said, it may acquire a physical content only if it is brought into relation with the transfer laws for some signal representative in the wave-optical phase space. Indeed, we will see in § 5.5 that the centroids $\overline{q}$ and $\overline{\kappa}$ of any waveform $\varphi(q)$ and its spectrum $\widetilde{\varphi}(\kappa)$ evolve under paraxial propagation through the matrix relation on the left of (4.6.52), i.e., just as the ray-coordinates of geometrical optics. Also, such a relation comes to be crucially functional to the paraxial propagation law for the Wigner distribution function, whose arguments, in fact, transform just through that relation (see § 8.3).

4.6.6 *Coordinate and frequency operators in the q and κ representations*

Expressions (4.4.1) and (4.4.2) define the operators $\widehat{\mathsf{q}}$ and $\widehat{\mathsf{p}}$ in the coordinate representation, i.e., through their action on the spatial waveform $\varphi(q)$; namely,

$$\widehat{\mathsf{q}}\varphi(q) = q\varphi(q), \qquad \widehat{\mathsf{p}}\varphi(q) = -\frac{i}{k_0}\frac{\partial}{\partial q}\varphi(q). \tag{4.6.53}$$

The corresponding expressions in the momentum representation identify the operators $\widehat{\mathsf{q}}$ and $\widehat{\mathsf{p}}$ through their action on the momentum wave function $\widetilde{\varphi}(p)$.

We recall that if the operator $\widehat{\mathsf{A}}$ defines a mapping of $\mathcal{L}^2(\mathbb{R})$ into itself, every operator $\widehat{\mathsf{O}}$ is changed to $\widehat{\mathsf{O}}' = \widehat{\mathsf{A}}\widehat{\mathsf{O}}\widehat{\mathsf{A}}^{-1}$, thus bearing the same relation to the transformed wavefunctions $\widehat{\mathsf{A}}\varphi$ as the original operator $\widehat{\mathsf{O}}$ to the original wavefunctions φ, i.e.,

$$\text{if} \quad \varphi \xrightarrow{\widehat{\mathsf{O}}} \psi \quad \text{then} \quad \widehat{\mathsf{A}}\varphi \xrightarrow{\widehat{\mathsf{O}}'=\widehat{\mathsf{A}}\widehat{\mathsf{O}}\widehat{\mathsf{A}}^{-1}} \widehat{\mathsf{A}}\psi. \tag{4.6.54}$$

Then, confirming relations (4.6.17), one can give the operators $\widehat{\mathsf{q}}$, $\widehat{\mathsf{p}}$ in the p-representation as

$$\widehat{\mathsf{q}} = \frac{i}{k_0}\frac{\partial}{\partial p}, \qquad \widehat{\mathsf{p}} = p, \tag{4.6.55}$$

It is convenient in the present context to introduce the spatial frequency operator $\widehat{\kappa}$, just translating into operator form the scalar relation $\kappa = k_0 p$, i.e.,

$$\widehat{\kappa} = k_0 \widehat{\mathsf{p}}. \tag{4.6.56}$$

We will mostly deal in the following with the pair $(\widehat{\mathsf{q}}, \widehat{\kappa})$ instead of $(\widehat{\mathsf{q}}, \widehat{\mathsf{p}})$. Evidently, the operator $\widehat{\kappa}$ can be identified as the differentiation operator

$$\widehat{\kappa} = -i\frac{\partial}{\partial q}, \tag{4.6.57}$$

when acting on $\varphi(q)$, and as the multiplication operator

$$\widehat{\kappa} = \kappa, \tag{4.6.58}$$

when acting on $\widetilde{\varphi}(\kappa)$. Likewise, the coordinate operator $\widehat{\mathsf{q}}$ in the κ-representation writes as

$$\widehat{\mathsf{q}} = i\frac{\partial}{\partial \kappa}, \tag{4.6.59}$$

and hence the commutation bracket between $\widehat{\mathsf{q}}$ and $\widehat{\kappa}$ comes to be

$$[\widehat{\mathsf{q}},\, \widehat{\kappa}] = i\widehat{\mathsf{I}}. \tag{4.6.60}$$

The Hermitian operators $\left\{\widehat{\mathsf{q}},\, \widehat{\kappa}, \widehat{\mathsf{I}}\right\}$, with the relevant Lie products $[\widehat{\mathsf{q}},\, \widehat{\kappa}] = i\widehat{\mathsf{I}}$, $[\widehat{\mathsf{q}},\, \widehat{\mathsf{I}}] = 0$, $[\widehat{\kappa}, \widehat{\mathsf{I}}] = 0$, span the Heisenberg-Weyl algebra in the same way as the operator set $\left\{\widehat{\mathsf{q}},\, \widehat{\mathsf{p}}, \widehat{\mathcal{I}}\right\}$, having indeed the same commutators (4.4.5).

4.7 Summary

We have presented a formalization of wave optics, similar to that of quantum mechanics à la Schrödinger-Dirac, namely in terms of wave functions $\varphi(q)$ belonging to $\mathcal{L}^2(\mathbb{R})$, and unitary transformations $\widehat{\mathsf{M}}$ acting on $\mathcal{L}^2(\mathbb{R})$. As $\varphi(q)$ is the mathematical representation of light distributions, which are subject to an evolution under propagation through optical systems, $\widehat{\mathsf{M}}$ is the mathematical representation of the "agents" of this evolution, i.e., the optical systems.

The representation of optical systems in terms of unitary operators $\widehat{\mathsf{M}}$ has been obtained by means of a quantization-like procedure applied to ray optics, by turning variables into appropriately defined operators. As a first step, the ray variables q and p have been replaced by the position and momentum operators $\widehat{\mathsf{q}}$ and $\widehat{\mathsf{p}} = -\frac{i}{k_0}\frac{\partial}{\partial q}$, whose Lie bracket $[\widehat{\mathsf{q}}, \widehat{\mathsf{p}}] = \frac{i}{k_0}$ mirrors the Poisson bracket of the classical variables $\{q, p\} = 1$.

The complete sets of eigenstates of $\widehat{\mathsf{q}}$ and $\widehat{\mathsf{p}}$ have been seen to be basic to two complementary representations of optical wavefields, carrying information about the relevant energy distributions in space and spatial frequency (i.e., direction). Owing to the Fourier relation linking these two signal representations, one can establish an inequality for the product of the space and frequency variances of the signal, only formally similar to Heisenberg's inequality.

Problems

1. Consider the set of up-to-second degree polynomials in q and p: $\mathcal{P}_{\leq 2} = \{q, p, q^2, p^2, qp\}$ and the set of the corresponding Hermitian operators, having the unique transcription as $\mathsf{P}_{\leq 2} = \{\widehat{\mathsf{q}}, \widehat{\mathsf{p}}, \widehat{\mathsf{q}}^2, \widehat{\mathsf{p}}^2, \frac{\widehat{\mathsf{q}}\widehat{\mathsf{p}}+\widehat{\mathsf{p}}\widehat{\mathsf{q}}}{2}\}$. Verify that the Poisson bracket of any pair of polynomials in $\mathcal{P}_{\leq 2}$ is taken to the Lie bracket of the corresponding operators in $\mathsf{P}_{\leq 2}$ by the factor $\frac{i}{k_0}$. It is not so for higher-order polynomials. Verify, for instance, that the Poisson bracket of the quartic polynomial q^2p^2, which by the Weyl rule is carried into $(q^2p^2)_{\mathrm{W}} = \widehat{\mathsf{q}}^2\widehat{\mathsf{p}}^2 - \frac{2i}{k_0}\widehat{\mathsf{q}}\widehat{\mathsf{p}} + \frac{1}{2k_0^2}$, with any one of the polynomials in $\mathcal{P}_{\leq 2}$ does not equal the commutator bracket of $(q^2p^2)_{\mathrm{W}}$ with the corresponding operator in $\mathsf{P}_{\leq 2}$.

2. Prove the identity: $e^{-\widehat{\mathsf{A}}}\,\widehat{\mathsf{B}}\,e^{\widehat{\mathsf{A}}} = e^{ad\widehat{\mathsf{A}}}\,\widehat{\mathsf{B}}$, for arbitrary operators $\widehat{\mathsf{A}}$, $\widehat{\mathsf{B}}$. (Hint: Introduce parameter-depending operators such as $\widehat{\mathsf{F}}(\tau) = e^{-\tau\widehat{\mathsf{A}}}\,\widehat{\mathsf{B}}\,e^{\tau\widehat{\mathsf{A}}}$, with τ real. Write down the differential equation obeyed by $\widehat{\mathsf{F}}(\tau)$. Then take the solution for $\tau = 1$.)

3. Consider a sequence of frequency spectra being flat over increasing frequency intervals:

$$\widetilde{\varphi}_n(\kappa) = \begin{cases} \frac{1}{\sqrt{2\pi}}, & |\kappa| \leq n\kappa_0 \\ 0, & |\kappa| > n\kappa_0 \end{cases}$$

with n integer and κ_0 arbitrary. Denoting by $\varphi_n(q)$ the corresponding waveforms, prove that $\lim_{n\to\infty} \varphi_n(q) = \lim_{K\to\infty} \frac{\sin(Kq)}{\pi q}$, with $K = n\kappa_0$.

4. Ientify the realness and symmetry properties of the Fourier transform in relation with properties of the signal like *(a)* realness: $\varphi(q) = \varphi^*(q)$, *(b)* imaginariness: $\varphi(q) = \varphi^*(q)$, *(c)* eveness: $\varphi(q) = \varphi(-q)$, *(d)* oddness: $\varphi(q) = -\varphi(-q)$, *(e)* hermiticity: $\varphi(q) = \varphi^*(-q)$, and *(f)* antihermiticity: $\varphi(q) = -\varphi^*(-q)$.

5. Write down the hybrid transfer relations with the object and image waveforms being represented the q and κ space, and vice versa. Deduce the relations between the response functions relevant to the four forms of the input-output relations.

6. Show that *(a)* for a system, in which the output is a replica of the input, the transfer function is $\mathfrak{h}(\kappa) = 1$, whereas *(b)* for a system, in which the output is a replica of the input shifted by Q, the transfer function is $\mathfrak{h}(\kappa) = e^{i\kappa Q}$.

7. Prove that the signal $\varphi(q)$, which is supposed to be centered at $\overline{q}$ in space and at $\overline{\kappa}$ in frequency, and its space and frequency shifted version $\psi(q) = e^{-i\overline{\kappa}(q+\overline{q})}\varphi(q+\overline{q})$ have the same spatial and frequency variances.

8. Expressing the complex amplitude as $\varphi(q) = f(q)e^{-\phi(q)}$, with f and ϕ real and $\int f^2 dq = 1$, prove that

$$\left|\int q\varphi^* \tfrac{d\varphi}{dq}\, dq\right|^2 = \left| -\tfrac{1}{2} + i\int qf^2 \tfrac{d\phi}{dq} dq\right|^2,$$

yielding the general form of the width principle as

$$\sigma_q^2\sigma_\kappa^2 \geq \tfrac{1}{4} + \left|\int qf^2 \tfrac{d\phi}{dq} dq\right|^2.$$

References

[1] The collections of Hamilton's papers specifically concerning light optics and particle dynamics are quoted respectively in [1] and [2] of ch. 1.

[2] For the mechanical-optical analogy see [6] in ch. 1. A detailed investigation of the correspondence between quantum mechanics and classical optics in its varied aspects, as in particular that related to the phase-space view, is presented in

[2.1] D. Dragoman, "Phase space correspondence between classical optics and quantum mechanics", in *Progress in Optics*, E. Wolf (ed.) (Elsevier, Amsterdam, 2002), Vol. 43, ch. 5, 433-496.

[2.2] D. Dragoman and M. Dragoman, *Quantum-Classical Analogies* (Springer, Berlin, 2004).

[3] The wave theory of light is illustrated in many textbooks as, for instance,

[3.1] M. Born and E. Wolf, *Principles of Optics* (Pergamon Press, Oxford, 6th ed., 1980).

[3.2] D. Marcuse, *Light Transmission Optics* (Van Nostrand-Reinhold, Princeton, 1972).

[3.3] J.A. Arnaud, *Beam and Fiber Optics* (Academic Press, New York, 1976).

[3.4] A.K. Ghatak and K. Thyagarajan, *Contemporary Optics* (Plenum Press, New York, 1978).

[3.5] B.E.A. Saleh and M.C. Teich, *Fundamentals of Photonics* (John Wiley & Sons, New York, 1991).

[3.6] S. Solimeno, B. Crosignani and P. Di Porto, *Guiding, Diffraction and Confinement of Optical Radiation* (Academic Press, Orlando, 1985).

[4] The procedure to replace the full wave equation by the parabolic wave equation, on the line of the original suggestion by Fock and Leontovich, is illustrated in the basic textbooks in [3]; in particular, we address the reader to [3.3], ch. 3, pp. 173-177. In Fock's paper [4.3] the similarity between the parabolic wave equation and the time-dependent Schrödinger equation was observed for the first time.

[4.1] M.A. Leontovich, "A method of solution of problems of electromagnetic wave propagation along the earth's surface", Bulletin de l'Académie des Sciences de l'URSS, sér. phys **8**,16-22 (1944) (in Russian).

[4.2] M.A. Leontovich and V.A. Fock, "Solution of the problem of propagation of electromagnetic waves along the earth's surface by the method of parabolic equations", Sov. J. Phys. **10**, 13-20 (1946); see also V.A. Fock, *Electromagnetic Diffraction and Propagation Problems* (Pergamon Press, Oxford, 1965), ch. 11, pp. 213-234.

[4.3] V.A. Fock, "Theory of radiowave propagation in an inhomogeneous (stratified) atmosphere for a raised source", in *Electromagnetic Diffraction and Propagation Problems*, (Pergamon Press, Oxford, 1965), ch. 14, pp. 276-307.

[5] The description of nonparaxial propagation of beam-like wave fields has been investigated in several papers by using rather different approaches, ranging from the pertubation approach to the Maxwell's equation in a beam divergence-related

parameter to the angular spectrum representation of the electric field and to the spheroidal coordinate system representation of the Helmholtz wave equation as well as to the complex point-source model. Below only a few of them are listed.

[5.1] M. Lax, W.H. Louisell and W.B. McKnight, "From Maxwell to paraxial wave optics", Phys. Rev. A **11**, 1365-1370 (1975).

[5.2] L.W. Davis, "Theory of electromagnetic fieds", Phys. Rev. A **19**, 1177-1179 (1979).

[5.3] G.P. Agrawal and D.N. Pattanayak, "Gaussian beam propagation beyond the paraxial approximation", J. Opt. Soc. Am. **69**, 575-578 (1979).

[5.4] W.H. Louisell, "From Maxwell to optical resonators", in *Laser Induced Fusion and X-ray Laser Studies*, S.F. Jacobs, M.O. Scully, M. Sargent III and C.D. Cantrell III (eds.) (Addison-Wesley, Reading, 1976), pp. 369-485.

[5.5] S.R. Seshadri, "Electromagnetic Gaussian beam", J. Opt. Soc. Am. A**15**, 2712-2719 (1998).

[5.6] H.-C. Kim and Y.H. Lee, "Higher-order corrections to the electric field vector of a Gaussian beam", J. Opt. Soc. Am. A **16**, 2232-2238 (1999).

[5.7] Q. Cao, "Corrections to the paraxial approximation solutions in transversely nonuniform refractive-index media", J. Opt. Soc. Am. A **16**, 2494-2499 (1999).

[5.8] A. Ciattoni, B. Crosignani and P. Di Porto, "Vectorial free-space optical propagation: a simple approach for generating all-order non paraxial corrections", Opt. Comm. **177**, 9-13 (2000).

[5.9] G.A. Deschamps, "Gaussian beam as a bundle of complex rays", Electron. Lett. **7**, 684-685 (1971).

[5.10] S.Y. Shin and L.B. Felsen, "Gaussian beam modes by multipoles with complex source points", J. Opt. Soc. Am. **67**, 699-700 (1977).

[5.11] E. Heyman and L.B. Felsen, "Complex-source pulsed-beam fields", J. Opt. Soc. Am. A **6**, 806-817 (1989).

[5.12] C.J.R. Sheppard and S. Saghafi, "Electromagnetic Gaussian beams beyond the paraxial approximation", J. Opt. Soc. Am. A **16**, 1381-1386 (1999).

[5.13] B.T. Landesman and H.H. Barrett, "Gaussian amplitude functions that are exact solutions to the scalar Helmholtz equation", J. Opt. Soc. Am. A **5**, 1610-1619 (1988).

[5.14] Z. Ulanowski and I.K. Ludlow, "Scalar field of nonparaxial Gaussian beams", Opt. Lett. **25**, 1792-1794 (2000).

[5.15] G. Rodríguez-Morales and S. Chávez-Cerda, "Exact nonparaxial beams of the scalar Helmholtz equation", Opt. Lett. **29**, 430-432 (2004).

[6] An exhaustive analysis of the relation of geometrical to wave optics, and in general to Maxwell's electromagnetic theory, is presented in [3.1], [3.6] as well as in the refs. below. Keller's paper [6.5], introducing the "geometrical theory of diffraction", had signified a great increase of the domain of application of ray methods.

[6.1] R.K. Luneburg, *Mathematical Theory of Optics* (California Univ. Press, Berkeley, 1964).

[6.2] M. Kline and I.W. Kay, *Electromagnetic Theory and Geometrical Optics* (Interscience, New York, 1965).

[6.3] G.A. Deschamps, "Ray techniques in electromagnetics", Proc. IEEE **60**, 1022-1035 (1972).

[6.4] S.W. McDonald, "Phase-space representations of wave equations with applications to the eikonal approximation for short-wavelength waves", Phys. Rep. **158**, 337-416 (1988).

[6.5] J.B. Keller, "Geometrical theory of diffraction", J. Opt. Soc. Am. **52**, 116-119 (1962).

[7] The formalization of wave optics à la quantum mechanics is illustrated in [3.2], pp. 103-111, and [3.4], pp. 29-30, on the basis of the original elaboration by Gloge and Marcuse [7.2]. Preliminary suggestions and wider investigations are in the additional titles below. We also suggest [3.3], pp. 44-46, 51-52.

[7.1] N. Chako, "Sur le passage de l'Optique géométrique à l'Optique ondulatoire", Compt. Rend. **251**, 852-853 (1960).

[7.2] D. Gloge and D. Marcuse, "Formal quantum theory of light rays", J. Opt. Soc. Am. **59**, 1629-1631 (1969).

[7.3] G. Eichmann, "Propagation of light beams in continuously focusing lens-like media", Proc. IRE **58**, 837-839 (1970).

[7.4] G. Eichmann, "Quasi-geometric optics of media with inhomogeneous index of refraction", J. Opt. Soc. Am. **61**, 161-168 (1971).

[7.5] M. Feldmann, "Modes in general gaussian optical systems using quantum-mechanics formalism", J. Opt. Soc. Am. **61**, 446-449 (1971).

[7.6] T. Pradhan, "Maxwell's equations from geometrical optics", Phys. Lett. A **122**, 397–398 (1987).

[8] For a general treatment of the symplectic group within the context of the group theory see [9] of ch. 1, whilst for a specific treatment see [11] of ch. 2.

[9] Some titles concerned with investigations related to the operator algebra structure of paraxial wave optics as arising from a quantization-like process of paraxial ray optics, based on the Heisenberg-Weyl algebra, are quoted here. See also [16].

[9.1] A. Vander Lugt, "Operational notation for the analysis and synthesis of optical data-processing systems", Proc. IEEE, **54**, 1055-1063 (1966).

[9.2] H.J. Butterweck, "General theory of linear, coherent, optical data-processing systems", J. Opt. Soc. Am. **67**, 60-70 (1977).

[9.3] M. Nazarathy and J. Shamir, "Fourier optics described by operator algebra", J. Opt. Soc. Am. **70**, 150-159 (1980).

[9.4] H.J. Butterweck, "Principles of optical data-processing", in *Progress in Optics*, XIX, E. Wolf (ed.) (North-Holland, Amsterdam, 1981), ch. 4, pp. 211-280.

[9.5] M. Nazarathy and J. Shamir, "First-order optics - a canonical operator representation: lossless systems", J. Opt. Soc. Am. **72**, 356-364 (1982). "First-order optics: operator representation for systems with loss or gain", *ibid.*, 1398-1408.

[9.6] H. Raszillier and W. Schempp, "Fourier optics from the perspective of the Heisenberg group", in *Lie Methods in Optics*, J. Sánchez Mondragón and K.B. Wolf (eds.) (Springer-Verlag, Berlin, 1986), ch. 2, pp. 29-43.

[9.7] O. Castaños, E. López-Moreno and K.B. Wolf, "Canonical transforms for paraxial wave optics", in *Lie Methods in Optics*, J. Sánchez Mondragón and K.B. Wolf (eds.) (Springer-Verlag, Berlin, 1986), ch. 5, pp. 159-182.

[9.8] K. Wittig, A. Giesen and H.Hügel, "Algebraic approach to characterizing paraxial optical systems", Appl. Opt. **33**, 3837-3848 (1994).

[9.9] R. Simon and N. Mukunda, "Optical phase space, Wigner representation, and invariant quality parameters", J. Opt. Soc. Am. A **17**, 2440-2463 (2000).

[9.10] V. Guillemin and S. Sternberg, *Symplectic Techniques in Physics* (Cambridge Univ. Press, New York, 1990)

[9.11] J. Shamir, *Optical Systems and Processes* (SPIE, Bellingham, 1999).

[10] The reading list on nonrelativistic and/or relativistic quantum mechanics should be extremely long. Among the rich variety of titles, only a few are listed below. In particular, we signalize [10.5] in relation to the theory of linear operators in the Hilbert space of the square integrable functions.

[10.1] A. Messiah, *Quantum Mechanics* (North-Holland, Amsterdam, 1961).

[10.2] A.S. Davydov, *Quantum Mechanics* (Pergamon Press, Oxford, 1976).

[10.3] L.E. Ballantine, *Quantum Mechanics* (Prentice Hall, Englewood Cliffs, 1990).

[10.4] W. Greiner, *Relativistic Quantum Mechanics - Wave Equations* (Springer-Verlag, Berlin, 1990).

[10.5] T.F. Jordan, *Linear Operators for Quantum Mechanics* (John WIley & Sons, New York, 1969).

[11] A presentation of the Heisenberg-Weyl group from the perspective of quantum mechanics, to which it is intimately related, is given in

[11.1] H. Weyl, *The Theory of Groups and Quantum Mechanics* (Dover, New York, 1931).

[11.2] K.B. Wolf, "The Heisenberg-Weyl ring in quantum mechanics", in *Group Theory and Its Applications*, Vol. 3, E.M. Loebl (ed.) (Academic Press, New York, 1975), pp. 189-247.

[13] The methods for obtaining the operators that correspond to particular dynamical variables are illustrated in [11], and in

[13.1] J.R. Shewell, "On the formation of quantum-mechanical operators", Am. J. Phys. **27**, 16-21 (1959).

[13.2] H. Daughaday and B.P. Nigam, "Function in Quantum Mechanics which corresponds to a given function in Classical Mechanics", Phys. Rev. B **139**, 1436-1442 (1964).

[13.3] G.S. Agarwal and E. Wolf, "Calculus for functions of non-commuting operators and general phase-space methods in quantum mechanics. I. Mapping theorems and ordering of funtions of non commuting operators", Phys. Rev. D **2**, 2161-2186 (1970).

[14] S. Steinberg, "Lie serie, Lie transformations, and their applications", in *Lie Methods in Optics*, J. Sánchez Mondragón and K.B. Wolf (eds.) (Springer-Verlag, Berlin, 1986), ch. 3, pp. 45-104.

[15] For an enlightening physics-oriented introduction to the theory of the distributions we suggest [17.3], pp. 50-66; a rigorous mathematical treatment is in

[15.1] R. Bracewell, *The Fourier Transform and Its Applications* (McGraw-Hill, New York, 2nd ed., 1986), ch. 5, pp. 69-97.

[15.2] M. J. Lighthill, *Introduction to Fourier Analysis and Generalised Functions* (Cambridge Univ. Press, London, 1958).

[15.3] A. Papoulis, *The Fourier Integral and its Applications* (McGraw-Hill, New York, 1962), pp. 269-282.

[15.4] K.B. Wolf *Integral Transforms in Science and Engineering* (Plenum Press, New York, 1979).

[15.5] R. Courant and D. Hilbert, *Methods of Mathematical Physics* (Interscience, New York, 1962), Vol. II, pp. 766–798.

[16] The Euclidean model of optics, as arising from a quantization-like process of paraxial ray optics, based on the Euclidean algebra, is presented in the titles below; preliminary suggestions can be found in [9.7].

[16.1] K.B. Wolf, "Elements of Euclidean optics", in *Lie Methods in Optics II*, K.B. Wolf (ed.) (Springer-Verlag, Berlin, 1989), ch. 6, pp. 115-162.

[16.2] K.B. Wolf, *Geometric Optics on Phase Space* (Springer, Berlin, 2004), pp.135-151.

[17] The mathematics of the Fourier transform is described in the devoted textbooks [15.1]-[15.4], whilst an optics-flavoured investigation is offered in

[17.1] J.W. Goodman, *Introduction to Fourier Optics* (McGraw-Hill, New York, 1968).

[17.2] A. Papoulis, *Systems and Transforms with Applications in Optics* (McGraw-Hill, New York, 1968). Here, the electrical-optical devices analogy is expressely evidenced.

[17.3] J.D. Gaskill, *Linear Systems, Fourier Transforms and Optics* (Wiley & Sons, New York, 1978).

[18] The basics of the Hamilton characteristics method are presented in the textbooks [3.1] and [3.6]; a more exhaustive presentation is offered in the books by Synge [18.1], Luneburg [6.1] and Buchdahl [18.2]. The relation of Hamilton's characteristics of geometrical optics to the spread functions of wave optics is investigated in [9.5], and in the paper by Bastiaans [18.3], where the Wigner distribution function is proved to bridge ray and wave optics.

[18.1] J.L. Synge, *Geometrical Optics. An introduction to Hamilton's method* (Cambridge Univ. Press, London, 1937).

[18.2] H.A. Buchdahl, *An Introduction to Hamiltonian Optics* (Cambridge Univ. Press, New York, 1970).

[18.3] M.J. Bastiaans, "The Wigner distribution function and Hamilton's characteristics of a geometric-optical system", Opt. Comm. **30**, 321-326 (1979).

[19] Formal investigations of various inequalities pertaining to the Fourier analysis can be found in [15.1]-[15.3]. See also [17]. For a discussion of the uncertainty principle within the quantum mechanical framework we direct the reader to [10]. Additional refs. are given in ch. 9.

[20] We quote below the seminal Gabor's papers [20.1] on the theory of information and the later paper [20.2] on "optics from the viewpoint of information theory":

[20.1] D. Gabor, "Theory of communication. I: The analysis of information", J. of the IEE **93**, 429-441 (1946); II: The analysis of hearing", *ibid.*, 442-445; III: Frequency compression and expansion", *ibid.*, 445-457.

[20.2] D. Gabor, "Light and Information", in *Progress in Optics*, E. Wolf (ed.) (North-Holland, Amsterdam, 1961), Vol. I , ch. 4, pp. 109-153.

[21] For an exemplificative account of affine space geometry we suggest

[21.1] H.J. Wolters, "Affine and projective geometries. A tutorial.", in http://www.hpl.hp.com/research/mmsl/presentations/3d/GeometryTutorial.pdf

5

1D First-Order Optical Systems: The Huygens-Fresnel Integral

5.1 Introduction

It has been emphasized many times that the quadratic Hamiltonian model for paraxial propagation is central to several fields of physics in its "classical" as well as "quantum" form, as pictorially sketched in Fig. 3.1.

The presentation elaborated in Chapters 2 and 3 was based on the classical-like Hamiltonian model, which had provided the basic algebra for the formalization of paraxial ray-optics.

The presentation started in Chapter 4 is based on the quantum-like Hamiltonian model, which in turns provides the basic algebra suitable to the formalization of paraxial wave-optics. Such a formalization emerges from a quantization procedure, by which the description of the (monochromatic) optical disturbance in terms of rays of light is replaced by the description in terms of optical wave functions, and correspondingly the phase-space variables of geometrical optics are replaced by linear Hermitian operators acting on the Hilbert space $\mathcal{L}^2(\mathbb{R})$ of the square-integrable wave functions $\varphi(q)$:

$$\begin{array}{l} \textsf{light ray} \quad \longrightarrow \quad \textsf{optical wavefunction} \in \mathcal{L}^2(\mathbb{R}) \\ \textsf{phase-space variables} \quad \longrightarrow \quad \textsf{linear Hermitian operators on } \mathcal{L}^2(\mathbb{R}) \end{array} \tag{5.1.1}$$

Accordingly, the position and momentum variables are promoted into the operators $\widehat{\mathsf{q}}$ and $\widehat{\mathsf{p}}$, which act on $\mathcal{L}^2(\mathbb{R})$ according to the Schrödinger realization (strictly admissible only under paraxial approximation):

$$\widehat{\mathsf{q}}\varphi(q) = q\varphi(q), \quad \widehat{\mathsf{p}}\varphi(q) = -\frac{i}{k_0}\frac{\partial}{\partial q}\varphi(q). \tag{5.1.2}$$

Their Lie bracket:

$$[\widehat{\mathsf{q}}, \widehat{\mathsf{p}}] = i\widehat{\mathcal{I}}, \quad \widehat{\mathcal{I}} \equiv \frac{\widehat{\mathsf{I}}}{k_0}, \tag{5.1.3}$$

is in a sense the "quantized" form of the Poisson bracket of the classical variables: $\{q, p\} = 1$.

Also, the ray-transfer matrix $\mathbf{M}(z_o, z_i)$, which characterizes first-order optical systems identifying the relevant input-output relations in terms of the ray-coordinates at the end-planes, is replaced by the wave-transfer operator $\widehat{\mathsf{M}}(z_o, z_i)$, which correspondingly identifies the input-output relations in terms of the "excitation" $\varphi_i(q)$ in the input plane and the system "response" $\varphi_o(q)$ in the output plane, i.e.

$$\varphi_o(q) = \widehat{\mathsf{M}}(z_o, z_i)\varphi_i(q). \tag{5.1.4}$$

As a natural by-product of the quantization process, the equation of motion for $\mathbf{M}$, and the finite-step functional expression of $\mathbf{M}$ as a single exponential find their wave-optical transcriptions respectively in the optical Schrödinger equation for $\widehat{\mathsf{M}}$:

$$\frac{i}{k_0}\frac{\partial}{\partial z}\widehat{\mathsf{M}}(z, z_i) = \widehat{\mathsf{H}}(\widehat{\mathsf{q}}, \widehat{\mathsf{p}}, z)\widehat{\mathsf{M}}(z, z_i), \qquad \widehat{\mathsf{M}}(z_i, z_i) = \widehat{\mathsf{I}}, \tag{5.1.5}$$

and in the single-exponential functional form of $\widehat{\mathsf{M}}$:

$$\widehat{\mathsf{M}}(z_o, z_i) = e^{-ik_0\widehat{\mathsf{H}}(z_o - z_i)}, \tag{5.1.6}$$

arising, of course, as a solution to (5.1.5) in the case of a z-independent Hamiltonian operator: $\widehat{\mathsf{H}}(\widehat{\mathsf{q}}, \widehat{\mathsf{p}}, z) = \widehat{\mathsf{H}}(\widehat{\mathsf{q}}, \widehat{\mathsf{p}})$.

Since the phase space variables of geometrical optics, being quadratic in q and p, have unique wave-optical transcriptions as Hermitian operators, on the basis of the primary correspondences

$$p^2 \rightleftharpoons \widehat{\mathsf{p}}^2, \qquad q^2 \rightleftharpoons \widehat{\mathsf{q}}^2, \qquad qp \rightleftharpoons \tfrac{1}{2}(\widehat{\mathsf{q}}\widehat{\mathsf{p}}+\widehat{\mathsf{p}}\widehat{\mathsf{q}}), \tag{5.1.7}$$

the Hamiltonian operator $\widehat{\mathsf{H}}(\widehat{\mathsf{q}}, \widehat{\mathsf{p}}, z)$ entering (5.1.5) and (5.1.6) is uniquely defined as a plain transcription of the paraxial form of the ray-optics Hamiltonian function (see §§ 4.4.1-2).

Accordingly, relations (5.1.7) and (5.1.6) are central to the present analysis. They will be used, in fact, to elaborate the operator description of first-order optical systems, which eventually will converge into the well known Huygens-Fresnel integral in a form that, establishing the link with ray optics in the linear approximation, allows us to benefit from the effectiveness of the ray matrix formalism [1]. Specifically, paralleling the analysis in Chapter 2, in Sect. 5.2 we consider the quadratic operators $\frac{1}{2}\widehat{\mathsf{p}}^2$, $\frac{1}{2}\widehat{\mathsf{q}}^2$ and $\frac{1}{4}(\widehat{\mathsf{q}}\widehat{\mathsf{p}}+\widehat{\mathsf{p}}\widehat{\mathsf{q}})$, whose exponentiated forms provide the operator representations of a free-medium section, a thin lens, and a positive scaler. Besides, each of these operator representations is

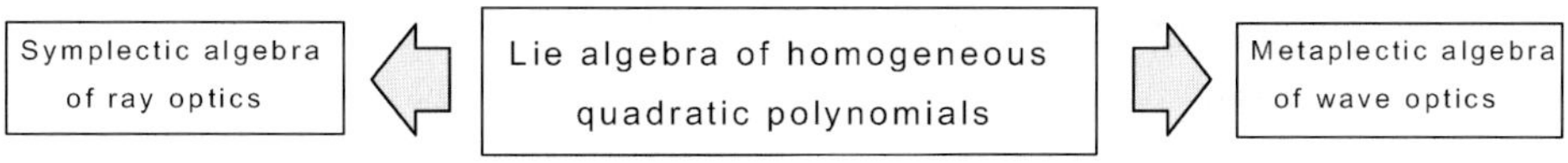

FIGURE 5.1. The quadratic homogeneous polynomials provide the basic algebraic structure of optics, yielding the Lie algebra of the symplectic ray matrices of ray optics as well as the Lie algebra of the metaplectic transfer operators of wave optics.

given an explicit functional form, individualizing for each the definite expression of the relevant line-spread function $\mathsf{g}(q, q')$, which represents the kernel of the integral transform expression of the transfer relations (5.1.4), i.e.

$$\varphi_o(q) = \int \mathsf{g}(q, q')\varphi_i(q')dq', \tag{5.1.8}$$

holding in general as an obvious by-product of the linearity of $\widehat{\mathsf{M}}$.

The operator representations for optical systems, we will deduce in the light of the quantization procedure, can as well be inferred from the parabolic wave equation for quadratic media, once the underlying operator algebra has been recognized, in full analogy with the group theoretical approach to the quantum mechanical Schrödinger equation [1]. The procedure, we will follow, reveals more clearly the link of paraxial wave optics to paraxial ray optics, and hence, within the group theoretical framework, of the metaplectic group $Mp(2, \mathbb{R})$ to the symplectic group $Sp(2, \mathbb{R})$, both having a common root in the algebra of quadratic polynomials (Fig. 5.1).

In Sect. 5.3 the obtained transfer relations for free-propagation and lens transformation are exploited to infer the input-output relationships for an arbitrary $ABCD$ system. This naturally conveys the link between the Huygens-Fresnel integral of the diffraction theory and the ray-matrix picture of geometrical optics. In fact, the line-spread function is found to be a purely imaginary exponential of a quadratic polynomial in the profile coordinates at the end-planes, the relevant coefficients being defined in terms of the entries of the ray-matrix.

The position-momentum identification of light rays in geometrical optics is mirrored in wave-optics by the space and spatial frequency representations of optical fields, respectively conveyed by the wave function $\varphi(q)$ and the Fourier transform $\widetilde{\varphi}(\kappa)$. Remarkably, the Fourier transform of an optical signal is not an abstract entity; in fact, it turns out to be physically detectable by use of plain bulk optics tools, and hence its properties have a definite physical interpretation and specific utilization in practical applications, as, for instance, in the design of devices for signal analysis and synthesis. However, whilst the position-momentum pair in ray-optics yields a "joint" (i.e., local) representation of the optical disturbance, the Fourier transform pair $(\varphi, \widetilde{\varphi})$ yield separate (global) information about the signal. In the forthcoming chapters we will address the issue to elaborate a space-frequency representation of optical signals being as close as possible to the inherently local representation of ray optics. Section 5.4 presents a rather detailed discussion on the optical Fourier transform and the optical fractional Fourier transform.

In Sect. 5.5 the z-evolution of the expectation values of the operators $\widehat{\mathsf{q}}$ and $\widehat{\mathsf{p}}$ is considered. They are seen to evolve in similar fashion as the ray coordinates q and p in paraxial ray-optics.

Finally in Sect. 5.6 it is shown that the integral representation of the metaplectic operators (i.e., the $ABCD$ integral form of the wave-transfer operators $\widehat{\mathsf{M}}$) arises as a unitary representation on the Hilbert space $\mathcal{L}^2(\mathbb{R})$ of linear canonical transformations of $\widehat{\mathsf{q}}$ and $\widehat{\mathsf{p}}$, represented by the ray matrices $\mathbf{M}$. The resulting correspondence of $\widehat{\mathsf{M}}$ to $\mathbf{M}$ is seen to be defined up to a sign. Hence, whilst an isomorphism can be established between the underlying algebraic structures of paraxial ray and wave optics, i.e., the symplectic and metaplectic algebras $sp(2,\Re)$ and $mp(2,\Re)$, an isomorphism up to a sign holds between the corresponding groups $Sp(2,\Re)$ and $Mp(2,\Re)$.

5.2 Quadratic Hamiltonians and metaplectic Lie algebra

As just reviewed, the quadratic monomials of ray optics are uniquely taken to the quadratic Hermitian operators of wave optics, we respectively denote by $\widehat{\mathsf{K}}_-$, $\widehat{\mathsf{K}}_+$, and $\widehat{\mathsf{K}}_3$, i.e.,

$$\begin{array}{ccl} \frac{1}{2}p^2 & \longrightarrow & \widehat{\mathsf{K}}_- = \frac{1}{2}\widehat{\mathsf{p}}^2 \\ \frac{1}{2}q^2 & \longrightarrow & \widehat{\mathsf{K}}_+ = \frac{1}{2}\widehat{\mathsf{q}}^2 \\ \frac{1}{2}qp & \longrightarrow & \widehat{\mathsf{K}}_3 = \frac{1}{4}(\widehat{\mathsf{q}}\widehat{\mathsf{p}}+\widehat{\mathsf{p}}\widehat{\mathsf{q}}), \end{array} \tag{5.2.1}$$

which are indeed the *wave-optical* transcription of (2.3.6). The dimensionality of the above operators directly follows from that of $\widehat{\mathsf{q}}$, having the dimension of length, and $\widehat{\mathsf{p}}$, being dimensionless.

The Lie bracket of any two operators on the right of (5.2.1) corresponds to the Poisson bracket of the corresponding c-number functions on the left (see Problem 4.1), being in fact

$$[\,\widehat{\mathsf{K}}_+,\widehat{\mathsf{K}}_-] = \tfrac{i}{k_0}(2\,\widehat{\mathsf{K}}_3), \quad [\,\widehat{\mathsf{K}}_3,\,\widehat{\mathsf{K}}_+] = \tfrac{i}{k_0}(-\,\widehat{\mathsf{K}}_+), \quad [\,\widehat{\mathsf{K}}_3,\widehat{\mathsf{K}}_-\,] = \tfrac{i}{k_0}(\widehat{\mathsf{K}}_-). \tag{5.2.2}$$

In § 1.3.2 it has been proved that the homogeneous quadratic polynomials span a Lie algebra with respect to the Poisson bracket operation. Then in § 2.3 the isomorphism of this algebra with the Lie algebra $sp(2,\mathbb{R})$ of the real traceless 2×2 matrices $\{\mathbf{K}_-,\mathbf{K}_+,\mathbf{K}_3\}$, which provide the underlying algebraic structure of first-order ray optics, has been evidenced. The correspondences (5.2.1) identify an operator realization of the Lie algebra of the monomials $\frac{1}{2}p^2$, $\frac{1}{2}q^2$ and $\frac{1}{2}qp$. The commutation relations (5.2.2) prove that such an operator realization spans a Lie algebra as well, isomorphic to the former, since the commutator between operators just retraces the Poisson bracket between the corresponding polynomials; the factor $\frac{i}{k_0}$ marks the passage from the ray- to the wave-optical context, as evidenced by the correspondence of the Lie brackets of the q-numbers $\widehat{\mathsf{q}}, \widehat{\mathsf{p}}$ to the Poisson brackets of the c-numbers q, p. Moreover, the exponential rule (5.1.6) mirrors the similar rule (4.3.5) that takes the traceless generating matrices in the symplectic algebra $sp(2,\mathbb{R})$ to the ray-optical matrices in the symplectic group $Sp(2,\mathbb{R})$, whose correspondence to physically realizable optical systems has been elucidated in the previous chapters. Accordingly, by (5.1.6) the Hermitian operators in the metaplectic Lie algebra $mp(2,\mathbb{R})$ spanned by the set $\{\widehat{\mathsf{K}}_-, \widehat{\mathsf{K}}_+, \widehat{\mathsf{K}}_3\}$ are carried to the unitary wave-optical operators in the metaplectic group $Mp(2,\mathbb{R})$, which gives a two-valued representation of the symplectic group $Sp(2,\mathbb{R})$ [2].

We can therefore rephrase within the wave-optical framework the path (3.1.2) from the quadratic monomials $\frac{1}{2}p^2$, $\frac{1}{2}q^2$, $\frac{1}{2}qp$ of first-order ray-optics to the exponential mappings $e^{-ik_0 d\widehat{\mathsf{K}}_-}$, $e^{-ik_0 \mathcal{P}\widehat{\mathsf{K}}_+}$ and $e^{-ik_0 s\widehat{\mathsf{K}}_3}$ of first-order wave optics, which then can justifiably be recognized as the paraxial propagators respectively for a free-medium section, a thin lens and a positive magnifier:

	generating function		generating operator		optical propagator
free-medium section	$\frac{1}{2}p^2$	$\longrightarrow$	$\widehat{\mathsf{K}}_- = \frac{1}{2}\widehat{\mathsf{p}}^2$	$\longrightarrow$	$e^{-ik_0 d\widehat{\mathsf{K}}_-}$
thin lens	$\frac{1}{2}q^2$	$\longrightarrow$	$\widehat{\mathsf{K}}_+ = \frac{1}{2}\widehat{\mathsf{q}}^2$	$\longrightarrow$	$e^{-ik_0 \mathcal{P}\widehat{\mathsf{K}}_+}$
positive magnifier	$\frac{1}{2}qp$	$\longrightarrow$	$\widehat{\mathsf{K}}_3 = \frac{1}{4}(\widehat{\mathsf{q}}\,\widehat{\mathsf{p}}+\widehat{\mathsf{p}}\,\widehat{\mathsf{q}})$	$\longrightarrow$	$e^{-ik_0 s\widehat{\mathsf{K}}_3}$.

(5.2.3)

Needless to say, in the q-domain representation (5.1.2) of $\widehat{\mathsf{q}}$ and $\widehat{\mathsf{p}}$, the generating operators $\{\widehat{\mathsf{K}}_-, \widehat{\mathsf{K}}_+, \widehat{\mathsf{K}}_3\}$ turn out to signify

$$\widehat{\mathsf{K}}_- = -\frac{1}{2k_0^2}\frac{\partial^2}{\partial q^2}, \quad \widehat{\mathsf{K}}_+ = \frac{1}{2}q^2, \quad \widehat{\mathsf{K}}_3 = -\frac{i}{4k_0}(1+2q\frac{\partial}{\partial q}). \tag{5.2.4}$$

Accordingly, the meaning of the real parameters d, $\mathcal{P}$ and s is the same as in the ray matrices of Chapters 2 and 3.

We shall not discuss further the algebra and group related aspects implied by the sketched path (5.2.3), we are going to follow. As we go on, insight will gradually be gained. So, we will inspect the action of the wave-transfer operators $e^{-ik_0 d\widehat{\mathsf{K}}_-}$, $e^{-ik_0 \mathcal{P}\widehat{\mathsf{K}}_+}$ and $e^{-ik_0 s\widehat{\mathsf{K}}_3}$ on a given signal in order to infer the explicit forms of the input-output relation (5.1.8), and hence the functional expressions of the spread functions $\mathsf{g}(q, q')$ of the systems they represent.

5.2.1 Propagation through a free-medium section

Firstly, we consider the operator

$$\widehat{\mathsf{T}}(z_o, z_i) = e^{-i\frac{k_0}{n_0}(z_o - z_i)\widehat{\mathsf{K}}_-} = e^{-i\frac{k_0}{2n_0}(z_o - z_i)\widehat{\mathsf{p}}^2}. \quad (5.2.5)$$

which, as earlier noted, describes the propagation through a homogeneous medium of refractive index n_0 extending from z_i to z_o. The object and image amplitudes are then related by

$$\varphi_o(q) = \widehat{\mathsf{T}}(z_o, z_i)\varphi_i(q) = e^{-i\frac{k_0}{2n_0}(z_o - z_i)\widehat{\mathsf{p}}^2}\varphi_i(q). \quad (5.2.6)$$

Transfer relation in the q-representation: free-medium line-spread function g_d

In order to give an explicit functional form to the above operator relation, it is convenient to express the complex amplitude $\varphi_i(q)$ in terms of the eigenfunctions (4.5.10) of $\widehat{\mathsf{p}}$, i.e. $\varphi_\kappa(q) = \frac{1}{\sqrt{2\pi}}e^{i\kappa q}$, $\kappa = k_0 p$. This is because of a notable property of the eigenfunctions of an operator $\widehat{\mathsf{O}}$ belonging to some eigenvalue μ, which remain eigenfunctions of any operator function $f(\widehat{\mathsf{O}})$ of $\widehat{\mathsf{O}}$ with eigenvalue $f(\mu)$, the functional relation expressed by f being transferred on the eigenvalue μ. Thus, applying the operator $\widehat{\mathsf{T}}(z_o, z_i)$ to the φ_κ's, we obtain

$$\widehat{\mathsf{T}}(z_o, z_i)\varphi_\kappa(q) = e^{-i\frac{(z_o - z_i)}{2n_0 k_0}\kappa^2}\varphi_\kappa(q), \quad (5.2.7)$$

showing that the propagation through a uniform refractive index medium does not modify the propagation direction of a plane wave. Similarly, the ray slope is not altered by the ray matrix $\mathbf{T}(d)$ (Eq. (2.4.3)). However, the propagation imparts to the spatial harmonic $e^{i\kappa q}$ a phase shift proportional to κ^2.

Then, expressing the input amplitude $\varphi_i(q)$ in terms of the Fourier transform $\widetilde{\varphi}_i(\kappa)$, and letting the linear operator $\widehat{\mathsf{T}}(z_o, z_i)$ act separately on each spatial harmonic according to (5.2.7), the transfer relation (5.2.6) recasts into

$$\varphi_o(q) = \frac{1}{\sqrt{2\pi}}\int \widetilde{\varphi}_i(\kappa)e^{-i\frac{(z_o - z_i)}{2n_0 k_0}\kappa^2}e^{i\kappa q}d\kappa \quad (5.2.8)$$

Turning back to $\varphi_i(q)$ by the inverse Fourier transform and carrying out the integration over κ, the explicit form of the image-to-object relation obtains

$$\varphi_o(q) = \sqrt{\frac{k_0}{2\pi i d}}\int e^{i\frac{k_0}{2d}(q-q')^2}\varphi_i(q')dq', \quad d = \frac{(z_o - z_i)}{n_0} \quad (5.2.9)$$

where d denotes the section length scaled by the refractive index, in conformity to the notation used for the ray-matrix $\mathbf{T}(d)$ [a]. By comparison with the

---

[a] We recall that

$$\int e^{-i\zeta\frac{x^2}{2} - i\xi x}dx = e^{-i\frac{\pi}{4}\mathrm{sgn}(\zeta)}\sqrt{\frac{2\pi}{|\zeta|}}e^{i\frac{\xi^2}{2\zeta}}, \ \zeta \text{ real}$$

general form (5.1.8), we may individualize the line-spread function $\mathsf{g}_d(q, q')$ for free-propagation over the reduced distance d as the quadratic phase function

$$\mathsf{g}_d(q, q') = \sqrt{\frac{k_0}{2\pi i d}}\, e^{i\frac{k_0}{2d}(q-q')^2}. \tag{5.2.10}$$

The argument of the square root is taken in the interval $(-\frac{\pi}{2}, \frac{\pi}{2}]$; this means that the phase of $\sqrt{d}$, with d being allowed to be both positive and negative (see § 3.3.1), is either 0 or $\frac{\pi}{2}$.

Like in the matrix relation (2.4.4) for the ray variables (q, p), the positions of the input and output planes enter (5.2.9) through the distance $z_o - z_i$. The system is then invariant for translations along the optical axis, provided the input signal be equally translated. So, the operator $\widehat{\mathsf{T}}(z_o, z_i)$ may equally be denoted as $\widehat{\mathsf{T}}(z_o - z_i)$, or $\widehat{\mathsf{T}}(d)$.

Finally, resorting to the general considerations in § 4.4.2, we note that the transfer relation (5.2.9) does not account for the phase accumulation as the light propagates from the plane at z_i to z_o, the spread function g_d modeling only the transverse profile of the propagating wave. Therefore, the axial phase-shift $e^{in_0 k_0 (z_o - z_i)/2}$ should be added to the propagated signal amplitude $\varphi_o(q)$ in (5.2.9) for a complete account of the propagation effects.

The line-spread function g_d *as the response of the system to a line-like signal*

The line-spread function represents the response of the system to the unit impulse $\delta(q - q')$, as formally expressed by Eq. (4.3.14), which in terms of $\widehat{\mathsf{T}}$ and g_d writes

$$\mathsf{g}_d(q, q') = \widehat{\mathsf{T}}(z_o, z_i)\delta(q - q'), \tag{5.2.11}$$

from which expression (5.2.10) for $\mathsf{g}_d(q, q')$ is easily obtained, on account of the Fourier expansion (4.5.7) of the δ function.

Within the original $2D$ model introduced in § 4.3.2, the δ function yields the mathematical representation of the point source of a spherical wave, and hence Eq. (5.2.11) should describe the propagation over the distance $z_o - z_i$ of a spherical wave originated at point q' in the input plane. In a homogeneous (isotropic) medium a spherical wave radiates uniformly through concentric spherical wave fronts with increasing radii. Within the planar model of optical propagation of our concern here, we must properly consider a cross section of the emanated spherical wave in a plane through the center, i.e. a circular wave. In fact, the picture suited to a planar geometry is that of a line source and an emanated cylindrical wave. Either case, however, the cross section of the wave in any meridional plane through the point source is represented by

---

conforming to the choice of the argument of the square root in the interval $(-\frac{\pi}{2}, \frac{\pi}{2}]$. The symbol $\mathrm{sgn}(\cdot)$ means the sign function, which gives ± 1 according to the sign of the argument.

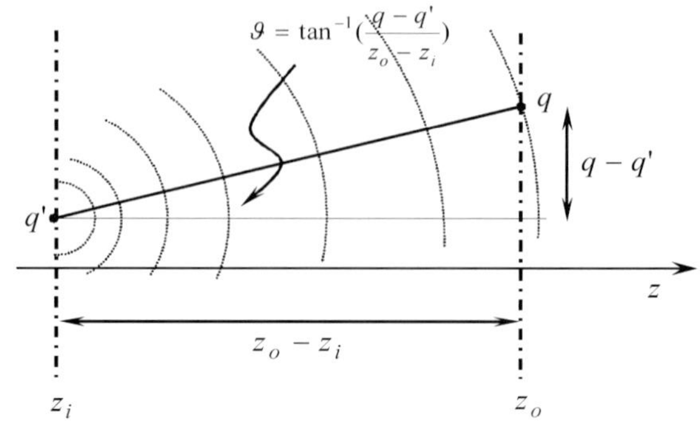

FIGURE 5.2. The line spread function $g_d(q, q')$ can be understood as the complex amplitude at point q in the output plane of a cilindrical wave emanating from point q' in the input plane, with both q and q' ranging through object and image regions across the end-planes which are supposed to be much smaller than the plane-to-plane distance.

$$\varphi(q, z_o) \propto \sqrt{\frac{n_0 k_0}{i}} \frac{e^{i n_0 k_0 \rho}}{\sqrt{\rho}} \tag{5.2.12}$$

with ρ denoting the distance from the source in the q-z plane (Fig. 5.2). If the point source is located at (q', z_i), as supposed in (5.2.11), the signal at (q, z_o) should have the form as (5.2.12), with $\rho = \sqrt{(q-q')^2 + (z_o - z_i)^2}$.

Under the hypothesis that only small regions across the optical axis in both the input and output planes are of interest, their maximum lateral extensions being much smaller of the overall axial distance $z_o - z_i$, we can expand the image-to-object distance ρ with respect to the small angle $\left|\frac{q-q'}{z_o - z_i}\right| \ll 1$, and retain only second order terms as

$$\rho \simeq z_o - z_i + \frac{1}{2} \frac{(q-q')^2}{z_o - z_i}. \tag{5.2.13}$$

We replace ρ by the zero-order term $z_o - z_i$ in the magnitude of (5.2.12) and by the second-order expansion (5.2.13) in the phase, being the phase much sensitive to errors. Then the circular wavefront disturbance (5.2.12) turns into the parabolic wavefront disturbance

$$\varphi(q, z_o) \propto \sqrt{\frac{n_0 k_0}{i(z_o - z_i)}} e^{i \frac{n_0 k_0}{2(z_o - z_i)} (q-q')^2} e^{i n_0 k_0 (z_o - z_i)}, \tag{5.2.14}$$

which, apart from the plane wave factor, has the very functional form of the line spread function g_d. The object-image relation (5.2.9) relevant to free-medium sections acquires therefore the meaning of the composition on the output plane of the parabolic waves emanated from each point on the input plane with strengths fixed by the input signal, the linear sizes of the involved regions over both the input and output planes being much smaller than the axial distance through which the waves are propagated. This is the Fresnel approximation to the Huygens integral, which in its general form involves the composition of the propagated spherical (or circular) waves with appropriate weighting factors [3, 6]. The Fresnel (or paraxial) approximation to the general diffraction theory, which allows us to replace spherical profiles by parabolic profiles, implies roughly that

$$\frac{(w_i+w_o)^4}{4(\lambda_0/n_0)(z_o-z_i)^3} < \varepsilon, \qquad \text{or} \qquad \frac{(\vartheta_i+\vartheta_o)^4(z_o-z_i)}{4(\lambda_0/n_0)} < \varepsilon \tag{5.2.15}$$

if ε is the maximum angle error, in units of π, tolerated in the phase in (5.2.12). Here, w_i and w_o denote the largest radial extensions of the regions in the two end-planes within which the input and output signals may be confined. Correspondingly, ϑ_i and ϑ_o are the angles subtended by w_i and w_o at z_o and z_i. The above reasoning shows that the Fresnel condition (5.2.15) of paraxial wave optics corresponds, and give a quantitative account, to the small angles and off-axis distances condition of paraxial ray optics.

Transfer relation in the κ-representation: free-medium transfer function $\mathfrak{h}_d$

It is evident that the impulse response g_d of a homogeneous-medium section depends on the combination $q-q'$. With the notation of § 4.6.3, we may write

$$\mathsf{g}_d(q,q') = \mathfrak{g}_d(q-q'), \tag{5.2.16}$$

where, according to (5.2.10), $\mathfrak{g}_d$ is

$$\mathfrak{g}_d(q) = \sqrt{\frac{k_0}{2\pi i d}}\, e^{i\frac{k_0}{2d}q^2}. \tag{5.2.17}$$

The system is shift-invariant under transverse translations, as expected due to the uniformity of the refractive index. Translations of the object over the input plane turn into translations of the image over the output plane by the same amount. The same occurs in the simple picture of ray optics, where a shift of the ray height turns into an equal shift in the height of the propagated ray without any alteration of the ray inclination. Accordingly, the transfer relation in the spatial frequency domain writes in the form (see Eq. (4.6.27))

$$\tilde{\varphi}_o(\kappa) = \mathfrak{h}_d(\kappa)\,\tilde{\varphi}_i(\kappa). \tag{5.2.18}$$

The transfer function $\mathfrak{h}_d(\kappa)$, relating to the Fourier transform of $\mathfrak{g}_d$, writes as

$$\mathfrak{h}_d(\kappa) = e^{-i\frac{d}{2k_0}\kappa^2}, \tag{5.2.19}$$

i.e., just as the quadratic phase function entering (5.2.7), which is in fact a manifestation of relation (5.2.18) for the specific case of plane waves.

Notably both the spread and transfer function $\mathfrak{g}_d$, $\mathfrak{h}_d$ have the functional form of a quadratic-phase function, which as noted in § 4.6.3 is typically reported in the signal theory as a (linear) *chirp*. The similarity of the functional forms of $\mathfrak{g}_d$ and $\mathfrak{h}_d$ is due to the basic property of the chirp, which being an imaginary Gaussian form is basically turned into its complex conjugate by Fourier transformation. In fact, one has

$$\widehat{\mathcal{F}}[e^{i\frac{x^2}{2}}] = e^{i\frac{\pi}{4}} e^{-i\frac{x^2}{2}}. \tag{5.2.20}$$

which parallels relation (4.6.51) for real Gaussian forms.

So we see that the transfer relation for free propagation mathematically shapes as a chirp convolution in the space domain and as a chirp multiplication in the frequency domain:

Propagation through a homogeneous medium

spatial domain	frequency domain
$\varphi_i \xrightarrow{\text{convolution by the chirp } e^{i\frac{k_0}{2d}q^2}} \varphi_o$	$\widetilde{\varphi}_i \xrightarrow{\text{multiplication by the chirp } e^{-i\frac{d}{2k_0}\kappa^2}} \widetilde{\varphi}_o$

Convolution and multiplication by a chirp are Fourier conjugate operations.

The chirp convolution is known mathematically as the Fresnel transform. Therefore, the free propagation between two planes separated by a certain distance d provides a practicable tool to physically implement the Fresnel transform of a mathematical function, that is properly turned into an optical signal to be propagated. Typically, the function is mapped to the complex amplitude transmittance of a transparency, which is placed in the object plane and then illuminated by a unit-amplitude plane wave. In accord with the κ-domain transfer relation (5.2.18), the Fresnel transform is equivalent to a spatial frequency filter, with phase-only filtering function $\mathfrak{h}_d$. All the frequency components of the input are propagated without attenuation, but with a phase dispersion quadratic in the frequency.

5.2.2 *Propagation through a thin lens*

As earlier remarked, the operator

$$\widehat{\mathsf{L}}(f) = e^{-i\frac{k_0}{f}\widehat{\mathsf{K}}_+} = e^{-i\frac{k_0}{2f}\widehat{\mathsf{q}}^2} \tag{5.2.21}$$

describes the effect of a thin lens with (reduced) focal length f (and power $\mathcal{P} = \frac{1}{f}$).

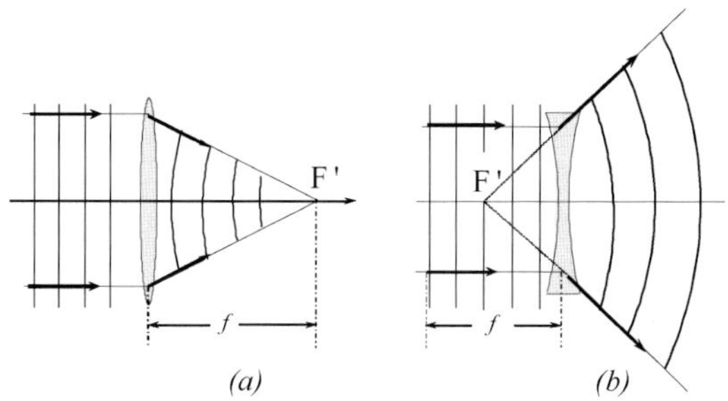

FIGURE 5.3. A thin lens transforms a normally incident plane wave into a spherical wave *(a)* converging to the real secondary focal point F′ (if $f > 0$) or *(b)* diverging from the virtual secondary focal point F′ (if $f < 0$).

Transfer relation in the q-representation: lens line-spread function g_f

As in the space-coordinate representation the operator $\widehat{\mathsf{q}}$ has the effect of multiplying the wave function by the coordinate value q, the thin lens operator $\widehat{\mathsf{L}}(f)$ acts as a phase-only transformer, affecting the incident wave simply by the quadratic-phase factor $e^{-ik_0 q^2/2f}$. The complex amplitudes $\varphi_i(q)$ and $\varphi_o(q)$ of the optical waves incident on and emerging from the lens are then linked by

$$\varphi_o(q) = e^{-i\frac{k_0}{2f}q^2}\varphi_i(q), \tag{5.2.22}$$

the reference planes being as usual located immediately in front of and behind the lens. Unlike a section of homogeneous medium, a thin lens acts as a modulator with the phase-only modulating function $e^{-ik_0 q^2/2f}$.

Since, as previously noted, the chirp factor in (5.2.22) may be interpreted as the paraxial approximation to the circular profile of a spherical wave, the above relation means that a thin lens transforms a normally incident unit amplitude plane wave ($\varphi_i(q) = 1$) into a spherical wave centered about a point on the lens axis. In particular, as illustrated in Fig. 5.3, if the lens is convergent ($f > 0$), the emerging wave is convergent toward the real secondary focal point F′; correspondingly, if the lens is divergent ($f < 0$), the emerging wave is divergent about the virtual secondary focal point F′ (§ 1.8.3). Accordingly a thin lens maps linear profiles into circular profiles (and hence planar wave fronts into spherical wave fronts, within a proper $2D$ picture of light propagation) in the proximity of the optical axis. Under non paraxial conditions the emerging wave fronts exhibit deviations from perfect sphericity, as an effect of the aberrations.

The input-output relation (5.2.22) can easily be cast in the integral form

$$\varphi_o(q) = \int dq' \mathsf{g}_\mathsf{f}(q, q')\varphi_i(q'), \tag{5.2.23}$$

reproducing the general form (5.1.8). The impulse response $\mathsf{g}_{\mathsf{f}}(q,q')$ writes as

$$\mathsf{g}_{\mathsf{f}}(q,q') = \widehat{\mathsf{L}}(f)\delta(q-q') = e^{-i\frac{k_0}{2f}q^2}\delta(q-q'). \tag{5.2.24}$$

The above relation can also be regarded as the eigenvalue equation for the thin lens operator $\widehat{\mathsf{L}}(f)$, with the δ functions being the eigenfunctions of $\widehat{\mathsf{q}}$. The decomposition of the input signal in terms of δ functions provides therefore a convenient representation of the signal, involving the very eigenfunctions of the system. In this respect, we recall that, according to the ray optics picture, a thin lens does not affect the ray position, being $q_o = q_i$.

Clearly the input output relation (5.2.22) does not account for the axial phase shift as the wave propagates through the lens, due for instance to the finite, although negligible, thickness of the lens. Then, if needed, the axial phase delay $e^{ik_0 n_l \mathrm{d}}$ must be added to the output signal amplitude in (5.2.22). Here, n_l denotes the refractive index of the lens material, whilst d is related to the maximum lens thickness t; specifically, d is taken as t, if a cylindrical lens is concerned, and as $\frac{\mathsf{t}}{2}$ in the case of a spherical lens.

Transfer relation in the κ-representation: lens wave-spread function h_{f}

According to the general rules (4.6.19) and (4.6.20), the lens transfer in the spatial frequency domain is described by the superposition integral

$$\widetilde{\varphi}_o(\kappa) = \int d\kappa'\, \mathsf{h}_{\mathsf{f}}(\kappa,\kappa')\widetilde{\varphi}_i(\kappa'), \tag{5.2.25}$$

where the wave-spread function h_{f} is

$$\mathsf{h}_{\mathsf{f}}(\kappa,\kappa') = \sqrt{\frac{f}{2\pi i k_0}}\, e^{i\frac{f}{2k_0}(\kappa-\kappa')^2}, \tag{5.2.26}$$

the phase of the square root being yet again taken in the interval $(-\frac{\pi}{2},\frac{\pi}{2}]$.

Evidently, in the κ-domain the lens behaves as a shift-invariant system, and so the transfer relation in the q-domain is simply multiplicative.

Relation (5.2.25) with the relevant kernel (5.2.26) resembles the q-domain transfer relation (5.2.9) for the propagation through a homogeneous medium, the wave-spread function h_{f}, as a function of κ, shaping like the line-spread function g_{d}, as a function of q. The focusing power $\mathcal{P} = \frac{1}{f}$ is placed in (5.2.26) as the reduced distance d in (5.2.10). Correspondingly, the κ-domain transfer relation (5.2.18) for propagation through a homogeneous medium has the same multiplicative structure of the q-domain transfer relation (5.2.22) for lens focusing. In both cases, the input is simply affected through the chirps $e^{-id\kappa^2/2k_0}$ and $e^{-ik_0q^2/2f}$, the correspondence of $\mathcal{P}$ to d being yet again evident. This is a manifestation of the duality relationship holding between free propagation and lens transfer. Chirp multiplication and chirp convolution are in fact the mathematical operation acted by a thin lens respectively in the space and frequency domain through an inverted scheme with respect to that pertaining to the propagation through a homogeneous medium:

Propagation through a thin lens					
	spatial domain			frequency domain	
φ_i	$\xrightarrow{\text{multiplication by the chirp } e^{-i\frac{k_0}{2f}q^2}}$	φ_o	$\widetilde{\varphi}_i$	$\xrightarrow{\text{convolution by the chirp } e^{i\frac{f}{2k_0}\kappa^2}}$	$\widetilde{\varphi}_o$

5.2.3 *Propagation through a positive pure magnifier*

We describe now the wave-optical representation of a positive pure magnifier, whose wave transfer operator writes as

$$\widehat{\mathsf{S}}(m) = e^{-ik_0 s \widehat{\mathsf{K}}_3} = e^{-i\frac{k_0 s}{4}(\widehat{\mathsf{q}}\,\widehat{\mathsf{p}}+\widehat{\mathsf{p}}\,\widehat{\mathsf{q}})}, \tag{5.2.27}$$

where the real parameters m and s are linked by $m = e^{s/2}$.

Taking the explicit expressions of $\widehat{\mathsf{K}}_3$ in the coordinate representation, the effect of the operator $\widehat{\mathsf{S}}(m)$ on the object wave function $\varphi_i(q)$ follows from

$$\varphi_o(q) = \widehat{\mathsf{S}}(m)\varphi_i(q) = e^{-\frac{s}{4}} e^{-\frac{s}{2}q\frac{\partial}{\partial q}} \varphi_i(q). \tag{5.2.28}$$

It is convenient to decompose the input wave $\varphi_i(q)$ into δ-like components. Then, after analysing the action of $\widehat{\mathsf{S}}(m)$ on each component $\delta(q)$ by inspection of the spread function

$$\mathsf{g}_{\mathsf{m}}(q, q') = e^{-\frac{s}{4}} e^{-\frac{s}{2}q\frac{\partial}{\partial q}} \delta(q' - q), \tag{5.2.29}$$

we recompose the output wave according to the superposition integral

$$\varphi_o(q) = \int dq' \mathsf{g}_{\mathsf{m}}(q, q')\varphi_i(q'). \tag{5.2.30}$$

Changing from q to q with $q = e^{\mathsf{q}}$, so that $q\frac{\partial}{\partial q} = \frac{\partial}{\partial \mathsf{q}}$, the scaling operator $e^{-\frac{s}{2}q\frac{\partial}{\partial q}}$ with respect to q is carried to the shift operator $e^{-\frac{s}{2}\frac{\partial}{\partial \mathsf{q}}}$ with respect to the new variable. Thus, being

$$e^{-\frac{s}{2}q\frac{\partial}{\partial q}} \delta(q' - q) = e^{-\frac{s}{2}\frac{\partial}{\partial \mathsf{q}}} \delta(q' - e^{\mathsf{q}}) = \delta(q' - e^{\mathsf{q}-\frac{s}{2}}) = \delta(q' - \tfrac{q}{m}), \tag{5.2.31}$$

the spread function $\mathsf{g}_{\mathsf{m}}(q, q')$ takes the explicit expression

$$\mathsf{g}_{\mathsf{m}}(q, q') = \sqrt{\tfrac{1}{m}}\,\delta(q' - \tfrac{q}{m}), \tag{5.2.32}$$

which conforms to the ray-optics description of the ideal magnifier transferring point objects at q_i into point images at $q_o = mq_i$ (§ 2.4.3).

Accordingly, the integral transform (5.2.30) yields the input-output relation for the positive ideal magnifier in the simple form

$$\varphi_o(q) = \sqrt{\tfrac{1}{m}}\varphi_i(\tfrac{q}{m}), \tag{5.2.33}$$

showing that the outgoing signal from the system $\widehat{\mathsf{S}}(m)$ is a scaled replica of the incoming signal. The overall magnitude is changed by the factor $\sqrt{\frac{1}{m}}$, thus preserving the energy.

Relation (5.2.33) is as well obtainable from (5.2.28) by the change of variable $q \longrightarrow \mathsf{q}$ and the consequent transformation $q\frac{\partial}{\partial q} \longrightarrow \frac{\partial}{\partial \mathsf{q}}$ as in (5.2.31). The decomposition of the wave function $\varphi_i(q)$ into line-like objects has been aimed at emphasizing the analogy with the ray-optics description of the system.

Finally, we consider the effect of the operator $\widehat{\mathsf{S}}(m)$ in the κ-domain. The frequency spectra of the object and image distributions are related by

$$\widetilde{\varphi}_o(\kappa) = \int d\kappa' \, \mathsf{h}_{\mathsf{m}}(\kappa, \kappa')\widetilde{\varphi}_i(\kappa'), \tag{5.2.34}$$

where according to (4.6.20) the wave response h_{m} is

$$\mathsf{h}_{\mathsf{m}}(\kappa, \kappa') = \sqrt{m}\delta(\kappa' - m\kappa), \tag{5.2.35}$$

which parallels the ray optics description of the system, according to which the ray momentum p_i is transformed into $p_o = \frac{p_i}{m}$ (§ 2.4.3).

The integral transform (5.2.34) turns then into the scaling relation

$$\widetilde{\varphi}_o(\kappa) = \sqrt{m}\widetilde{\varphi}_i(m\kappa). \tag{5.2.36}$$

5.3 Wave-optical transfer relations for an ABCD system

We approach the input-output problem for a 1D first-order optical system, generally described by the ray-transfer matrix $\mathbf{M} = \left(\begin{smallmatrix} A & B \\ C & D \end{smallmatrix}\right)$.

We resort to the equivalences established in Chapter 3 of any optical system to appropriate sequences of a finite number of free-medium sections and thin lenses. In particular, we consider the simple configuration consisting of two free-medium sections and one thin lens, described by Eqs. (3.4.1-2).

Within the present formalism, the equivalence in (3.4.1) can be restated as

$$\widehat{\mathsf{M}}(\mathbf{M}) = \widehat{\mathsf{T}}(d_2)\widehat{\mathsf{L}}(f)\widehat{\mathsf{T}}(d_1), \tag{5.3.1}$$

with the relevant parameters being specified by

$$d_1 = \tfrac{A-1}{C} > 0, \quad d_2 = \tfrac{D-1}{C} > 0, \quad f = -\tfrac{1}{C}, \quad B = d_1 + d_2 - \tfrac{d_1 d_2}{f} > 0. \tag{5.3.2}$$

5.3.1 *Transfer relation in the q-representation: line-spread function of an ABCD system*

Using relations (5.2.9) and (5.2.22) to account for the effects respectively of the operators $\widehat{\mathsf{T}}(d_1)$, $\widehat{\mathsf{T}}(d_2)$, and $\widehat{\mathsf{L}}(f)$, we easily obtain the complex amplitude $\varphi_o(q)$ of the emerging signal from the single lens setup (5.3.1) in the form

$$\varphi_o(q) = \frac{k_0}{2\pi i\sqrt{d_1 d_2}} \int dq' \int dq'' e^{-i\frac{k_0}{2f}q'^2} e^{i\frac{k_0}{2d_2}(q-q')^2} e^{i\frac{k_0}{2d_1}(q'-q'')^2} \varphi_i(q''). \tag{5.3.3}$$

We perform the integration over the dummy variable q' and exploit relations (5.3.2), thus finally obtaining the integral transform

$$\varphi_o(q) = \widehat{\mathsf{M}}(\mathbf{M})\varphi_i(q) = \int dq' \mathsf{g}_{\mathsf{ABCD}}(q, q')\varphi_i(q'), \tag{5.3.4}$$

where the relevant line spread $\mathsf{g}_{\mathsf{ABCD}}(q, q')$ explicitly writes as

$$\mathsf{g}_{\mathsf{ABCD}}(q, q') = \sqrt{\frac{k_0}{2\pi i B}} e^{i\frac{k_0}{2B}(Aq'^2 + Dq^2 - 2qq')} \tag{5.3.5}$$

involving the entries of the ray-matrix of the overall TLT configuration.

Although the TLT configuration (5.3.1) comprises a limited set of ABCD systems, if one regards d_1 and d_2 as positive section lengths, we assume relation (5.3.5) as a general prescription to construct the representing kernel for the integral transform (5.3.4), which defines the wave-optical operator $\widehat{\mathsf{M}}$ associable to the ray-optical matrix $\mathbf{M}$, with the already mentioned convention for the phase of the square root being taken in the interval $(-\frac{\pi}{2}, \frac{\pi}{2}]$. We will address in § 5.6 the question of the uniqueness of the operator $\widehat{\mathsf{M}}$ associable to $\mathbf{M}$ through (5.3.4), and accordingly of $\varphi_o(q)$ resulting from $\varphi_i(q)$ through $\widehat{\mathsf{M}}$.

The spread function $\mathsf{g}_{\mathsf{ABCD}}$ of any first-order optical system is evidently an imaginary Gaussian exponential of the profile coordinates at the end planes, with coefficients determined by the ray-matrix entries. The factor in front of the Gaussian exponential guarantees the losslessness of the sytem. For ease of writing, the subscript ABCD to g will be dropped; our discussion is in fact concerned exclusively with first-order, i.e., "*ABCD*", optical systems.

Remarkably the integral transform in (5.3.4) is the Huygens-Fresnel integral of diffraction theory. In fact, it can also be inferred from the Huygens principle through the Fresnel approximation to the Huygens wavelets, with the relevant eikonal function being expressed in terms of the entries of the overall *ABCD* matrix of the system [4]. This calculation automatically includes the phase shift factor, that on the contrary, as noted many times and further stressed below, is not accounted for by a ray-matrix related approach.

Expression (5.3.4) of the input-output relation for an arbitrary optical system represents the cardinal result of the chapter. It states that, within the paraxial approximation, an arbitrary optical wave can be propagated through an optical device, on account as well of all the diffraction effects, using knowledge only of the overall ray matrix of the system, which synthesizes the detailed structure of the device. Also, it establishes the full correspondence between the ray-optical description of optical systems in terms of ray matrices and the wave-optical description in terms of wave transfer operators, the functional form of the latter being expressed in terms of the entries of the former ones. Thus, an optical system can be analyzed by using the ray matrix formalism or the wave operator formalism, whichever is more convenient. The group of first-order optical systems is the physical realization of the symplectic group of unimodular matrices as well as of the metaplectic group of the corresponding unitary operator representations; the properties of the former are a direct manifestation of the abstract algebraic properties of the latter. Additionally, we emphasize the basic conceptual content of relations (5.3.4) and (5.3.5), which provide the link between the inherently local picture of light propagation in ray optics and the inherently non local picture in wave optics.

To further appreciate the significance of Eqs. (5.3.4) and (5.3.5), we may compare the simple functional form they suggest for the spread function g with the involved expression (4.3.18), yielding the composition rule for composite system. The formal simplicity of rule (4.3.17), demanding for the product of the transfer operators of the various optical components, is only apparent indeed, since the product of operators may yield rather involved functional forms, as proved, for instance, by the double integral form in (5.3.3) accounting for the simple three-component sequence $\widehat{\mathsf{T}}(d_2)\widehat{\mathsf{L}}(f)\widehat{\mathsf{T}}(d_1)$. In contrast, the product of matrices can easily be performed. Thus, the ray-to-wave correspondence established by relations (5.3.4) and (5.3.5) allows to replace the complicate calculation inherent in (4.3.18) by simple multiplications of 2×2 matrices.

From the group theoretical viewpoint, relations (5.3.4)-(5.3.5) signify a correspondence of unimodular matrices in the symplectic Lie group $Sp(2, \mathbb{R})$ to unitary operators in the metaplectic Lie group $Mp(2, \mathbb{R})$:

$$\mathbf{M} \in Sp(2, \mathbb{R}) \longrightarrow \widehat{\mathsf{M}} \in Mp(2, \mathbb{R}), \tag{5.3.6}$$

the latter ones acting on the Hilbert space $\mathcal{L}^2(\mathbb{R})$ of wave functions through the integral transform (5.3.4) with kernel determined by $\mathbf{M}$ according to (5.3.5).

It is in the virtue of such a correspondence that all the considerations developed in Chapter 3 relatively to propagation problems expressed in terms of ray matrices apply to propagation problems expressed in terms of wave-optical propagators. This is to say, we can solve the operator equation (5.1.5) ruling the paraxial propagation through a quadratic medium by solving the corresponding matrix equation (4.3.4), with rules (4.4.10) and (5.3.5) providing the key to turn from the operator to the matrix formalism and vice versa.

However, the correspondence between symplectic matrices and metaplectic operators is not one-to-one, as straightforwardly will emerge in § 5.6, where we will frame the wave-optical relation between ray-matrices and wave-propagators within the general context of unitary representations of linear canonical transformations. We will see that the unitary operator $\widehat{\mathsf{M}}$ associable to $\mathbf{M}$ is specified to within an arbitrary phase. Even though the choice of the phase can be narrowed by requiring that the operators reproduce the multiplication rule of the matrices, a sign ambiguity still remains.

In fact, the mapping (5.3.6) is, globally speaking, 2-to-1, there being two unitary operators, $\widehat{\mathsf{M}}(\mathbf{M})$ and $-\widehat{\mathsf{M}}(\mathbf{M})$, for every $\mathbf{M}$. The operators do not compose exactly as the matrices, the difference being in a sign, i.e.

$$\widehat{\mathsf{M}}_1(\mathbf{M}_1)\widehat{\mathsf{M}}_2(\mathbf{M}_2) = \pm\widehat{\mathsf{M}}(\mathbf{M}_1\mathbf{M}_2), \tag{5.3.7}$$

and accordingly,

$$[\widehat{\mathsf{M}}(\mathbf{M})]^{-1} = \pm\widehat{\mathsf{M}}(\mathbf{M}^{-1}), \quad \mathbf{M}^{-1} = \begin{pmatrix} D & -B \\ -C & A \end{pmatrix}. \tag{5.3.8}$$

Clearly this sign ambiguity implies a sign ambiguity on the propagated wave function as well. So, whilst the ray-variable evolution is uniquely determined through the optical matrix $\mathbf{M}$, the wave function evolution is determined through $\widehat{\mathsf{M}}$ up to a sign. If one is interested in the energy distribution of the propagated wave, such a sign, which is indeed observable [10], can be ignored.

Finally, we stress yet again that the input-output relation (5.3.4), being derived from the ray matrix picture of optical systems, does not include the on-axis phase shift factor $e^{ik_0\mathcal{L}}$, experienced by the optical wave as it travels through the system. This factor must be separately evaluated and added to the outgoing amplitude, if it is of relevance for the problem under study. Naturally the optical path length $\mathcal{L}$ along the axis of the system is the sum of the individual on-axis optical path lengths $\mathcal{L}_j$ for all optical elements in the system: $\mathcal{L} = \sum_j \mathcal{L}_j$, each $\mathcal{L}_j$ being corrected by the factor 1/2 when required.

5.3.2 *Transfer relation relation in the κ-representation: wave-spread function of an ABCD system*

The relation between the spatial frequency spectra of the incoming and outgoing field distributions $\widetilde{\varphi}_i(\kappa)$, $\widetilde{\varphi}_o(\kappa)$ for an $ABCD$ system writes (see § 4.6.2)

$$\widetilde{\varphi}_o(\kappa) = \int d\kappa' \mathsf{h}(\kappa,\kappa')\widetilde{\varphi}_i(\kappa'). \tag{5.3.9}$$

The wave spread function $\mathsf{h}(\kappa,\kappa')$, resulting from the double Fourier transform of the impulse response (5.3.5), has the explicit expression

$$\mathsf{h}(\kappa,\kappa') = \sqrt{\frac{i}{2\pi k_0 C}}\, e^{-i\frac{1}{2k_0 C}\left[D\kappa'^2 + A\kappa^2 - 2\kappa\kappa'\right]}. \tag{5.3.10}$$

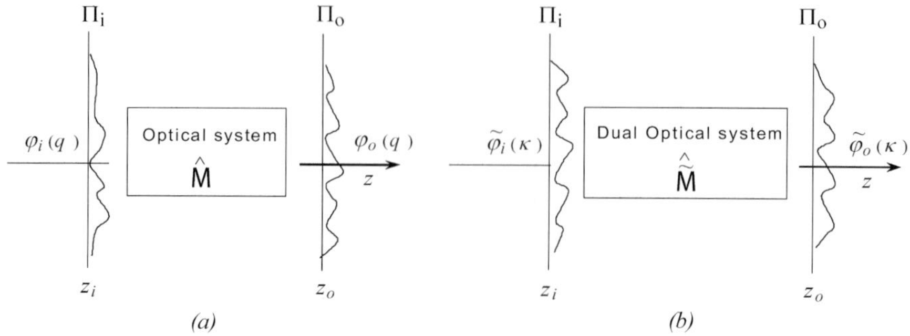

FIGURE 5.4. Relation between *(a)* an optical system and *(b)* its dual. As $\widehat{\mathsf{M}}$ acts in the real space on the light distribution $\varphi_i(q)$, $\widehat{\widetilde{\mathsf{M}}}$ acts in the κ-domain on the Fourier transform $\widetilde{\varphi}_i(\kappa)$.

with the usual convention for the argument of the square root.

Notably, $\mathsf{g}(q,q')$ and $\mathsf{h}(\kappa,\kappa')$ have similar imaginary Gaussian forms with energy-preserving complex factors, that can be synthesized into

$$\mathsf{y}(y,y') = \sqrt{\frac{1}{2\pi i\mathcal{B}}}e^{i\frac{1}{2\mathcal{B}}\left[\mathcal{A}y'^2 + \mathcal{D}y^2 - 2yy'\right]}. \tag{5.3.11}$$

The parameters $\mathcal{A}$, $\mathcal{B}$ and $\mathcal{D}$ entering the above can be understood as the entries of the unimodular matrix $\left(\begin{smallmatrix}\mathcal{A} & \mathcal{B}\\ \mathcal{C} & \mathcal{D}\end{smallmatrix}\right)$, which is alternatively identified with the ray-optics matrix $\mathbf{M} = \left(\begin{smallmatrix}A & B/k_0\\ k_0C & D\end{smallmatrix}\right)$ for the pair (q,κ) when considering the impulse response $\mathsf{g}(q,q')$ or with the matrix $\widetilde{\mathbf{M}} = \left(\begin{smallmatrix}D & -k_0C\\ -B/k_0 & A\end{smallmatrix}\right)$ for the pair $(\kappa,-q)$ when considering the wave response $\mathsf{h}(\kappa,\kappa')$ [b].

As we know from § 3.6.1, the pair (q,κ) can be taken to the pair $(\kappa,-q)$ (apart from scaling factors that can be absorbed by suitable normalizations) by a Fourier transformation, producing a clockwise rotation by $\frac{\pi}{2}$ in the phase plane. It is evident from (5.3.11) that if $\varphi_i(q)$ is the input signal to the optical system, described by the ray matrix $\mathbf{M}$, and $\widetilde{\varphi}_i(\kappa)$ is the input signal to the system, described by the ray matrix $\widetilde{\mathbf{M}}$, then the output signal from $\widetilde{\mathbf{M}}$ is just $\widetilde{\varphi}_o(\kappa)$, that is the Fourier transform of the output signal $\varphi_o(q)$ from $\mathbf{M}$ (Fig. 5.4). In that sense, the system $\widetilde{\mathbf{M}}$ is referred to as the *dual* system of $\mathbf{M}$ [5].

It has been earlier remarked that the single space-angle description of wave fields, conveyed through the ray variables (q,p) of geometrical optics, is replaced by two (Fourier-transform linked) descriptions in wave optics. The

[b] As noted in § 4.6.5, the same symbol $\mathbf{M}$ will be used to signify transformations of both the ray-variables (q,p) and the Fourier conjugate pair (q,κ), in the latter case being implied that the off-diagonal elements of $\mathbf{M}$ be scaled to B/k_0 and k_0C.

q-domain representation, conveyed through the complex amplitude $\varphi(q)$, is naturally associated with the (q, κ) phase space, as the κ-domain representation conveyed through the frequency spectrum $\widetilde{\varphi}(\kappa)$, is naturally associated with the $(-\kappa, q)$ phase space, in which the role of $\varphi(q)$ and $\widetilde{\varphi}(\kappa)$ are symmetrically reversed. Accordingly, the optical matrix $\mathbf{M}$ of ray optics is involved into the wave-optical transfer relation in the q-domain, whilst the dual matrix

$$\widetilde{\mathbf{M}} = \left(\begin{smallmatrix} 0 & 1 \\ -1 & 0 \end{smallmatrix}\right)\mathbf{M}\left(\begin{smallmatrix} 0 & -1 \\ 1 & 0 \end{smallmatrix}\right) \tag{5.3.12}$$

is involved into the wave-optical transfer relations in the κ-domain. The similarity transformation by the pure $\frac{\pi}{2}$-rotation matrix in (5.3.12) reflects the fact that, the phase-plane axes (q, κ) and the dual phase-plane axes $(-\kappa, q)$ are geometrically linked by a $\frac{\pi}{2}$ counterclockwise rotation.

In accord with the prescriptions of § 4.6.6, it is evident that in the frequency domain the operators $\widehat{\mathsf{K}}_-$, $\widehat{\mathsf{K}}_+$, and $\widehat{\mathsf{K}}_3$ have the meaning as in the scheme below, where, for purpose of comparison, the corresponding q-domain realizations (5.2.4) are also reported:

	q-domain	κ-domain	
$\widehat{\mathsf{K}}_- = \frac{1}{2}\widehat{\mathsf{p}}^2$	$-\frac{1}{2k_0^2}\frac{\partial^2}{\partial q^2}$	$\frac{1}{2k_0^2}\kappa^2$	
$\widehat{\mathsf{K}}_+ = \frac{1}{2}\widehat{\mathsf{q}}^2$	$\frac{1}{2}q^2$	$-\frac{1}{2}\frac{\partial^2}{\partial \kappa^2}$	(5.3.13)
$\widehat{\mathsf{K}}_3 = \frac{1}{4}(\widehat{\mathsf{q}\mathsf{p}} + \widehat{\mathsf{p}\mathsf{q}})$	$-\frac{i}{4k_0}(1 + 2q\frac{\partial}{\partial q})$	$\frac{i}{4k_0}(1 + 2\kappa\frac{\partial}{\partial \kappa})$	

The reader will then find intuitive the following correspondences of the basic optical systems to their dual systems:

Optical system	Dual System	
free-space section, d	thin lens, $f=1/d$	
thin lens, f	free-space section, $d=1/f$	(5.3.14)
positive scaler, m	positive scaler, $1/m$	

which of course are another way to express the frequently remarked dual relation between free-space propagation (i.e., chirp convolution) and lens transfer (i.e., chirp modulation), and the transfer relations (5.2.33) and (5.2.36), relevant to the positive scaler.

In relation with the correspondences above, note that in the space domain one deals with a length-unit variable whilst in the frequency domain one deals with a length^{-1}-unit variable. The correspondences in (5.3.14) are intended in the light of the structure of the involved ray-matrices, being lower-triangular, upper-triangular or diagonal, which structure ultimately determines the functional form of the system response, whichever be the input, φ or $\widetilde{\varphi}$.

5.3.3 *Exponential representation of the optical propagator. Operator equivalent of the Wei-Norman matrix synthesis*

Expression (5.3.11), tailored to represent the response of an arbitrary first-order optical system in the spatial or frequency domain according to the specific settings of the optical parameters $\mathcal{A}$, $\mathcal{B}$ and $\mathcal{D}$, enables us to write down the transfer relations for an optical system in the q or κ domain in one form

$$\phi_o(y) = \sqrt{\frac{1}{2\pi i \mathcal{B}}} e^{\frac{i}{2}\frac{\mathcal{A}}{\mathcal{C}}y^2} \int dy' e^{\frac{i}{2}\frac{\mathcal{A}}{\mathcal{B}}(y'-\frac{y}{\mathcal{A}})^2} \phi_i(y'). \tag{5.3.15}$$

It is therefore easy to recognize the possibility of interpreting the transfer relations for an optical system in both the spatial and frequency domain as resulting from the mathematical sequence of

$$\phi_i(y) \xrightarrow{\text{chirp convolution}} \otimes \xrightarrow{\text{scale transformation}} \otimes \xrightarrow{\text{chirp modulation}} \phi_o(y) \tag{5.3.16}$$

each transformation being properly parameterized.

The explicit operator form of the transformations, entering the general transfer scheme (5.3.16), in terms of the group generators $\{\widehat{\mathsf{K}}_-, \widehat{\mathsf{K}}_+, \widehat{\mathsf{K}}_3\}$ depends on whether it concerns the wave function $\varphi_i(q)$ or its frequency spectrum $\widetilde{\varphi}_i(\kappa)$. Looking in particular at the space domain, if the ray-matrix entry $A > 0$, which allows us to link the scale transformation solely to $\widehat{\mathsf{K}}_3$, we obtain for the inherent wave-optical operator the ordered product form [1.5]-[1.7]

$$\widehat{\mathsf{M}}(\mathbf{M}) = e^{ik_0\frac{C}{A}\widehat{\mathsf{K}}_+} e^{-i2k_0\ln(A)\widehat{\mathsf{K}}_3} e^{-ik_0\frac{B}{A}\widehat{\mathsf{K}}_-}, \tag{5.3.17}$$

which evidently translates in terms of metaplectic operators the Wei-Norman decomposition of simplectic matrices in terms of a free propagation, a positive scaling and a lens transfer, subject to the condition that $A > 0$ (§ 3.5):

$$\mathbf{M} = \mathbf{L}(-\frac{C}{A})\mathbf{S}(A)\mathbf{T}(\frac{B}{A}). \tag{5.3.18}$$

It has been noted in § 3.5 that decomposition (5.3.18) is allowed in a neighborhood of the input plane in any propagation problem. This means that any propagation problem can be mathematically thought of as a suitably parameterized sequence of a Fresnel transform, a scale transform and a chirp modulation, and optically as a suitably parameterized cascade of a free-medium section, a positive scaler and a thin lens.

The exponential operator transcription of the transfer scheme (5.3.16) in the frequency domain needs, of course, to represent the basis operators $\widehat{\mathsf{K}}_-$, $\widehat{\mathsf{K}}_+$, $\widehat{\mathsf{K}}_3$ in the Fourier space. Using indeed the appropriate expressions given in (5.3.13), the reader may easily obtain the dual form of (5.3.17) and verify the transfer relation in the κ-domain having the form (5.3.9) with (5.3.10).

5.3.4 *Imaging systems* ($B = 0$)

The case $B = 0$ deserves some comments since expression (5.3.5) of the line spead function seems to be meaningful only for $B \neq 0$. We recall from § 1.8.4 that the vanishing B-element in the optical matrix characterizes *imaging* systems, with the matrix entry A specifying the scaling factor by which the object is expanded or contracted in the image plane. A thin lens is the basic example of an imaging system, with unit magnification.

In terms of wave optics, *imaging* means that the amplitude distribution of the input signal is reproduced on the output plane, apart from possible scaling, folding about the origin and phase shift effects. The image is *perfect* when the corresponding amplitude distribution has not phase factors other than those present in the amplitude distribution of the input signal. A thin lens always produces an *imperfect* image, since, according to (5.2.22), the image amplitude has a quadratic phase term that is not present in the object amplitude.

It is convenient to recast the spread function (5.3.5) in the form

$$\mathsf{g}(q, q') = \sqrt{\frac{k_0}{2\pi i B}}\, e^{i\frac{k_0 A}{2B}(q' - \frac{q}{A})^2} e^{i\frac{k_0 C}{2A} q^2}, \tag{5.3.19}$$

which, on account of the limit relation (4.5.6), immediately yields

$$\mathsf{g}_{\mathsf{B=0}}(q, q') = \sqrt{\tfrac{1}{A}}\, e^{i\frac{k_0 C}{2A} q^2} \delta(q' - \tfrac{q}{A}). \tag{5.3.20}$$

Imaging systems are therefore characterized by input-output relations like

$$\varphi_o(q) = \sqrt{\tfrac{1}{A}}\, e^{i\frac{k_0 C}{2A} q^2} \varphi_i\left(\frac{q}{A}\right) = \sqrt{\tfrac{1}{A}}\, e^{-i\frac{k_0}{2f_2} q^2} \varphi_i(\tfrac{q}{A}), \quad (B = 0), \tag{5.3.21}$$

the back focal length $f_2 = -\frac{A}{C}$ being introduced in the last expression.

Equation (5.3.21) generalizes the imaging relation (5.2.22) of a thin lens to arbitrary imaging systems. The input signal is scaled into the output signal by the matrix entry A; the factor $\sqrt{\frac{1}{A}}$ preserves normalization. As for a thin lens, the image is not perfect due to the presence of the quadratic phase term $e^{-ik_0 q^2/2f_2}$. It represents (in the parabolic approximation) a circular wave converging toward (if $f_2 > 0$) or diverging from (if $f_2 < 0$) the secondary focal point F′ (Fig. 5.5), whose wavefront profile $\mathcal{S}_o$ with curvature radius $\mathcal{R} = f_2$ is then tangent to the output plane Π_o, where the emerging signal is observed. The quadratic phase term in (5.3.19) is compensated if the output signal is observed on $\mathcal{S}_o$, which thereby represents the profile where a perfect image may be detected. If $C = 0$ (i.e., $\mathcal{R} = f_2 \to \infty$), the phase factor disappears and the perfect image is observed on the straight profile of the end-plane Π_o. In that case, (5.3.21) yields the transfer relation for the pure magnifier,

$$\varphi_o(q) = \sqrt{\tfrac{1}{A}}\varphi_i(\tfrac{q}{A}), \tag{5.3.22}$$

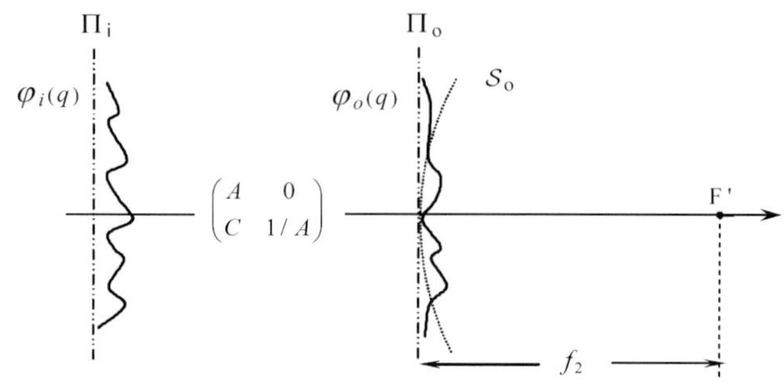

FIGURE 5.5. Imaging system ($B = 0$). The *perfect* image of the input signal may be observed on the circular profile $\mathcal{S}_o$.

which has, of course, the same form as (5.2.33), but may account as well for a folding of the signal about the origin.

5.4 The optical Fourier transform

It has been earlier remarked that the Fourier transform of an optical signal is not an abstract entity. In fact, it is observable, for instance, on the back focal plane of a thin lens. The Fourier transform is the mathematical tool to turn from the space-domain representation to the spatial-frequency domain representation, displaying the directional content of the signal. The observability of the Fourier transform implies the observability of the κ-domain representation of a signal, the focal planes of a lens providing the elementary physical planes where the two Fourier reciprocal descriptions of the signal may be observed. Remarkably, the "gap" between these planes is found to be filled by a continuum of planes, *fractionally* Fourier conjugate to each other, where an intermediate representation is observable, displaying both the space and frequency features of the signal. This is the object of the forthcoming discussion.

5.4.1 *Fourier transform*

The general condition for transferring the input signal to the focal plane of an optical system is that the ray-matrix entry A be zero (§ 3.4.2). With $A = 0$, the transfer relation (5.3.4) becomes

$$\varphi_o(q) = \sqrt{\frac{k_0}{2\pi i f}} e^{i\frac{k_0}{2f} Dq^2} \int dq' e^{-i\frac{k_0}{f} qq'} \varphi_i(q') = \sqrt{\frac{k_0}{if}} e^{i\frac{k_0}{2f} Dq^2} \widetilde{\varphi}_i(\tfrac{k_0}{f} q), \qquad (5.4.1)$$

where use has been done of the relation $B = -\frac{1}{C} = f$.

The amplitude of the output signal, detected at the point q on the secondary focal plane, is directly linked to the Fourier transform of the input signal at the spatial frequency $\kappa = k_0 q/f$. However, the Fourier transform relation between the object and the image is not an exact one, due to the presence of the quadratic phase factor $e^{ik_0 Dq^2/2f}$, which specifies the circular profile, of radius $\mathcal{R} = \frac{f}{D}$, on which the exact Fourier transform of the signal can be observed. If $D = 0$, the quadratic phase term in (5.4.1) disappears; thus, the amplitude distribution at the secondary focal plane is the exact Fourier transform of the input amplitude at the primary focal plane. As mentioned in § 3.4.2, the system is said to respectively perform the *imperfect* or *perfect* Fourier transform of the input illumination. Either case, the intensity distribution on the focal plane reproduces the spatial frequency spectrum of the object signal:

$$|\varphi_o(q)|^2 = \tfrac{k_0}{f} \left|\widetilde{\varphi}_i(\tfrac{k_0}{f}q)\right|^2. \tag{5.4.2}$$

Remarkably, relation (5.4.1) gives a physical content to the rather formal treatment of the spatial Fourier transform presented in § 4.6. On the back focal plane of any optical system, one may "see" the Fourier transform of the input signal, apart from a possible quadratic phase factor (which can be avoided looking on an output circular profile instead of a planar one). However, the frequency domain of the "mathematical" Fourier transform is not directly mapped into the output space domain of the optical system, the scaling constant $\frac{k_0}{f}$ [length^{-2}] being indeed required:

$$\kappa = \frac{k_0}{f} q. \tag{5.4.3}$$

Also, the additional constant complex factor $\sqrt{\frac{k_0}{if}}$ appears. Hence, the relation between the "mathematical" Fourier transform $\widehat{\mathcal{F}}$ (§ 4.6) and the "optical" Fourier transform $\widehat{\mathsf{F}}[\mathbf{F}(f)] \equiv \widehat{\mathsf{F}}_f$ writes in the form

$$\widehat{\mathsf{F}}_f \varphi(q) = \sqrt{\tfrac{k_0}{f}} e^{-i\frac{\pi}{4}} [\widehat{\mathcal{F}}\varphi](\tfrac{k_0}{f}q). \tag{5.4.4}$$

As an elementary example of the tranfer relation (5.4.1), we may think of a thin lens, whose back focal plane is connected by a Fourier transforming relation like (5.4.1) to any plane, placed at a certain distance d against the lens (see Fig. 3.7.b) and the scheme of Fig. 4.7). Indeed, the lens produces on its back focal plane the imperfect Fourier transform of the object, the quadratic phase factor in (5.4.1) specifically being $e^{ik_0(1-\frac{d}{f})q^2/2f}$. It disappears, of course, if the object is placed at the front focal plane ($d = f$) of the lens, which so produces the perfect Fourier transform of the object (see Fig. 3.8).

Either case, the lens images plane waves with different spatial frequencies, i.e. different directions of propagation, into different points of the focal plane, and hence separates the contributions of the various plane waves which compose the input signal. In particular, the plane wave propagating at an angle ϑ with respect to the optical axis is focused into the point $q = n_0 f \vartheta$ (supposing that the media on the lens sides have the same refractive index n_0).

5.4.2 *Fractional Fourier transform*

In § 3.6.1 we have considered the quadratic homogeneous polynomial

$$H_{\text{a.o.}} = \tfrac{1}{2}p^2 + \tfrac{1}{2\,f_{\text{S}}^2}q^2, \tag{5.4.5}$$

where the subscript a.o. signalizes the analogy with the Hamiltonian of an attractive oscillator, having in particular unit mass and frequency $\frac{1}{f_{\text{S}}}$.

The optical matrix, describing the ray propagation ruled by (5.4.5), is

$$\mathbf{F}^{\alpha}(f_{\text{S}}) = \begin{pmatrix} \cos\phi & f_{\text{S}}\sin\phi \\ -\frac{1}{f_{\text{S}}}\sin\phi & \cos\phi \end{pmatrix}, \quad -\pi < \phi \leq \pi. \tag{5.4.6}$$

The angle ϕ is defined in terms of the propagation length $z - z_i$ along the axis and the focal length parameter f_{S}: $\phi = \frac{z - z_i}{f_{\text{S}}}$. In the light of the analogy with the harmonic oscillator, it may be seen to correspond to the phase $\omega(t - t_0)$ of the oscillator. Also, ϕ is usually given as a "fraction" of $\frac{\pi}{2}$ through the parameter α setting $\phi = \alpha\frac{\pi}{2}$. The value $\phi = \frac{\pi}{2}$ signifies the ordinary Fourier transform: $\mathbf{F}^1(f_{\text{S}}) = \mathbf{F}(f_{\text{S}})$, which then can be understood as a specific determination of the continuous evolution, governed by (5.4.6), at definite profile screens identified by $\phi_m = \frac{\pi}{2} + 4m\frac{\pi}{2}$, $m = 0, 1, 2...$ Conversely, by values $0 < \phi < \frac{\pi}{2}$ one may inspect the ray-evolution at intermediate planes within the basic Fourier transform unit, which is so fragmented in *fractional* units [7], and, according to the periodicity of the harmonic functions, at any other plane "separated" from the former by $\Delta\phi = 2m\pi, m = 1, 2...$ Opposite values of ϕ signify the inverse system, being indeed $[\mathbf{F}^{\alpha}(f_{\text{S}})]^{-1} = \mathbf{F}^{-\alpha}(f_{\text{S}})$; also, we recall the remarkable relation: $\mathbf{F}^{\alpha}(f_{\text{S}}) = [\mathbf{F}(f_{\text{S}})]^{\alpha}$.

In § 3.6.1 we have individualized the optical implementation of the system $\mathbf{F}^{\alpha}(f_{\text{S}})$ in terms of both refractive lenses and graded-index media. In fact, mirroring the single- and double-lens geometries, which perform the ordinary Fourier transform, the basic bulk optics realizations of the system (5.4.6) are obtained interleaving free-medium sections and thin lenses according to the schemes:

$$\begin{aligned} &\text{Lohmann's type I:} \quad \mathbf{F}^{\alpha}_{\text{I}}(f_{\text{S}}) = \mathbf{T}(f_{\text{S}}\tan(\tfrac{\phi}{2}))\mathbf{L}(\tfrac{f_{\text{S}}}{\sin\phi})\mathbf{T}(f_{\text{S}}\tan(\tfrac{\phi}{2})) \\ &\text{Lohmann's type II:} \quad \mathbf{F}^{\alpha}_{\text{II}}(f_{\text{S}}) = \mathbf{L}(\tfrac{f_{\text{S}}}{\tan(\phi/2)})\mathbf{T}(f_{\text{S}}\sin\phi)\mathbf{L}(\tfrac{f_{\text{S}}}{\tan(\phi/2)}). \end{aligned} \tag{5.4.7}$$

Also, a quadratic graded-index focusing medium is appropriate to implement the fractional Fourier transforming operation with a continuously ranging order $\alpha(z)$. The bulk optics realizations (5.4.7) are in fact paralleled by the graded-index medium realization synthesized into the scheme:

$$\begin{array}{c}\text{quadratic-index medium of length } L\\ (n(q,z)=n_0-n_2q^2/2,\quad n_2>0)\end{array} \iff \mathbf{F}^{\alpha}(f_{\rm S})\ \left(\alpha(L)=\tfrac{2}{\pi}\sqrt{\tfrac{n_2}{n_0}}L,\ f_{\rm S}=\tfrac{1}{\sqrt{n_0 n_2}}\right). \tag{5.4.8}$$

In particular, with $\alpha = 1$, the resulting optical systems perform the ordinary Fourier transform of the object. Specifically, it amounts to choosing all the lengths involved in (5.4.7) equal to $f_{\rm S}$, and correspondingly the length of the quadratic medium in (5.4.8) as (or, as quartic multiples of) $L_{\rm F} = \frac{\pi}{2}\sqrt{\frac{n_0}{n_2}}$.

Finally, we recall from § 3.6.1 the single exponential representation of the matrix (5.4.6). It writes in terms of the matrix $\mathbf{K}_1(f_{\rm S}) = \frac{1}{2}(f_{\rm S}\mathbf{K}_- + \frac{1}{f_{\rm S}}\mathbf{K}_+) = \frac{1}{2}\begin{pmatrix} 0 & f_{\rm S} \\ -1/f_{\rm S} & 0\end{pmatrix}$, which is nothing but one half the Fourier transform matrix with focal length $f_{\rm S}$. Explicitly, one finds

$$\mathbf{F}^{\alpha}(f_{\rm S}) = e^{2\phi\mathbf{K}_1(f_{\rm S})}. \tag{5.4.9}$$

All the above reviewed results can be read from the wave-optical perspective just replacing the inherent scalar functions of q and p by operator functions of $\widehat{\mathsf{q}}$ and $\widehat{\mathsf{p}}$, and accordingly the inherent matrices by integral transforms. We can therefore say that the quantum harmonic oscillator-like Hamiltonian

$$\widehat{\mathsf{H}}_{\rm a.o.} = \frac{1}{2}\widehat{\mathsf{p}}^2 + \frac{1}{2f_{\rm S}^2}\widehat{\mathsf{q}}^2, \tag{5.4.10}$$

generates the optical propagator

$$\widehat{\mathsf{F}}^{\alpha}_{f_{\rm S}} = e^{-i\frac{\pi}{2}k_0\alpha[f_{\rm S}\widehat{\mathsf{K}}_- + \frac{1}{f_{\rm S}}\widehat{\mathsf{K}}_+]}. \tag{5.4.11}$$

As $\mathbf{F}^{\alpha}(f_{\rm S})$, $\widehat{\mathsf{F}}^{\alpha}_{f_{\rm S}}$ is intended to connect two planes separated by a finite or infinitesimal distance according to whether the order α is given a definite specification within the proper interval $\alpha \in (-2, 2]$ or is allowed to range continuously in conformity with the propagation variable z. Of course,

$$\left[\widehat{\mathsf{F}}^{\alpha}_{f_{\rm S}}\right]^{-1} = \widehat{\mathsf{F}}^{-\alpha}_{f_{\rm S}}, \qquad \widehat{\mathsf{F}}^{\alpha}_{f_{\rm S}} = \left[\widehat{\mathsf{F}}_{f_{\rm S}}\right]^{\alpha}. \tag{5.4.12}$$

The formal input-to-output relation

$$\varphi_o(q) = \widehat{\mathsf{F}}^{\alpha}_{f_{\rm S}}[\varphi_i(q)], \tag{5.4.13}$$

for the fractional Fourier transforming systems (5.4.7) finds its own integral representation through rule (5.3.5), which indeed yields

$$\varphi_o(q) = \sqrt{\frac{k_0}{2\pi i f_{\rm S}\sin\phi}}\int e^{i\frac{k_0}{2f_{\rm S}\sin\phi}[q'^2\cos\phi + q^2\cos\phi - 2qq']}\varphi_i(q')dq', \quad \phi = \alpha\frac{\pi}{2}. \tag{5.4.14}$$

Of course, the above equation may describe as well the paraxial propagation of a light disturbance through a quadratic index medium, if α and f_S are given as in (5.4.8), and so directly linked to the propagation length $z - z_i$ and the medium parameters n_0, n_2. Therefore, it provides the explicit solution to the parabolic wave equation for a quadratic focusing medium, whose ray optics account has been detailed in § 3.9. Equation (5.4.14) is just the wave-optical transcription of the ray transfer relations (3.9.16). It represents an example of the interplay between matrix and operator methods that can effectively been exploited to solve evolutionary differential equations.

The κ-domain representation of the input-output relation (5.4.14) is easily obtained through the rule (5.3.10), by noting that the dual matrix $\widetilde{\mathbf{F}}^{\alpha}(f_S)$ writes just as (5.4.6) with family parameter $\frac{1}{f_S}$; namely, $\widetilde{\mathbf{F}}^{\alpha}(f_S) = \mathbf{F}^{\alpha}(f_S^{-1})$.

The single-exponential form (5.4.11) of the operator $\widehat{\mathsf{F}}^{\alpha}_{f_S}$ is paralleled by the multi-exponential forms, directly obtained from (5.4.7) by a mere replacement of matrices by operators. In fact, the wave-optical transcriptions of (5.4.7) yield the following ordered products of metalectic operators and corresponding sequences of mathematical/optical transformations:

$$\text{type I}\quad\begin{cases} \widehat{\mathsf{F}}^{\alpha}_{f_S} = e^{-ik_0 f_S \tan(\frac{\phi}{2})\widehat{\mathsf{K}}_-}\, e^{-i\frac{k_0}{f_S}\sin\phi\,\widehat{\mathsf{K}}_+}\, e^{-ik_0 f_S \tan(\frac{\phi}{2})\widehat{\mathsf{K}}_-} \\ \varphi_i(q) \xrightarrow{\text{chirp convolution}} \otimes \xrightarrow{\text{chirp modulation}} \otimes \xrightarrow{\text{chirp convolution}} \varphi_o(q) \end{cases}$$

$$\text{type II}\quad\begin{cases} \widehat{\mathsf{F}}^{\alpha}_{f_S} = e^{-i\frac{k_0}{f_S}\tan(\frac{\phi}{2})\widehat{\mathsf{K}}_+}\, e^{-ik_0 f_S \sin\phi\,\widehat{\mathsf{K}}_-}\, e^{-i\frac{k_0}{f_S}\tan(\frac{\phi}{2})\widehat{\mathsf{K}}_+} \\ \varphi_i(q) \xrightarrow{\text{chirp modulation}} \otimes \xrightarrow{\text{chirp convolution}} \otimes \xrightarrow{\text{chirp modulation}} \varphi_o(q) \end{cases} \tag{5.4.15}$$

The reader is invited to write down the bulk optics realization of the dual operator $\widetilde{\widehat{\mathsf{F}}}^{\alpha}_{f_S}$ (corresponding to the dual matrix $\widetilde{\mathbf{F}}^{\alpha}(f_S)$), whose type I and type II implementations turn out to be just the duals respectively of the type II and type I realizations of $\widehat{\mathsf{F}}^{\alpha}_{f_S}$ (and so $\mathbf{F}^{\alpha}(f_S)$).

We conclude this section resorting to the frequently outlined analogy of the ray- and wave-optics *space*-propagation under the quadratic Hamiltonians (5.4.5) and (5.4.10) with the *time*-evolution of a mechanical (respectively, classical and quantum) harmonic oscillator. In the light of this analogy, realizations (5.4.7) and (5.4.8) can be interpreted as optical analogs of the harmonic oscillator dynamics, which is indeed optically reproduced at "sampled times" and at "continuous times". The analogy with the quantum harmonic oscillator dynamics is further evidenced by the eigenvalue equation for the fractional Fourier transform operator $\widehat{\mathsf{F}}^{\alpha}_{f_S}$,

$$\widehat{\mathsf{F}}^{\alpha}_{f_S}\varphi_n(q) = \lambda_n \varphi_n(q). \tag{5.4.16}$$

Physically, it identifies light distributions $\varphi_n(q)$, whose transverse profile is not altered (apart from the complex factor λ_n, $|\lambda_n|^2 = 1$) when propagating through a fractional Fourier transforming system.

Since $\widehat{\mathsf{F}}^{\alpha}_{f_S}$ can equally be interpreted as the *time*-evolution operator of the quantum harmonic oscillator described by the Hamiltonian (5.4.10), it is natural to look for φ_n's in the form of the oscillator energy eigenstates, and hence satisfying the equation

$$\widehat{\mathsf{H}}_{\text{a.o.}}\varphi_n(q) = \mathcal{E}_n\varphi_n(q). \tag{5.4.17}$$

Resorting to the well known oscillator energy eigenstates, we then write

$$\varphi_n(q) = \left(\sqrt{\tfrac{2}{\pi}}\tfrac{1}{2^n n! w}\right)^{1/2} H_n\!\left(\sqrt{2}\tfrac{q}{w}\right)e^{-\frac{q^2}{w^2}} \tag{5.4.18}$$

where H_n denotes the Hermite polynomial of degree n. Also, on account of the already mentioned correspondences:

$$\begin{aligned} &\text{oscillator mass} && \to 1 \\ &\text{oscillator frequency} && \to \tfrac{1}{f_S} \\ &\text{Planck's constant} && \to \lambda_0, \end{aligned} \tag{5.4.19}$$

the parameter w in (5.4.18) and the "energy" eigenvalue $\mathcal{E}_n$ specialize as

$$w = \sqrt{\tfrac{2f_S}{k_0}}, \qquad \mathcal{E}_n = \tfrac{1}{k_0 f_S}\left(n + \tfrac{1}{2}\right). \tag{5.4.20}$$

Then, applying $\widehat{\mathsf{F}}^{\alpha}_{f_S}$ to φ_n we obtain the explicit expression of the propagation factor λ_n; namely,

$$\widehat{\mathsf{F}}^{\alpha}_{f_S}\varphi_n(q) = e^{-i\alpha\frac{\pi}{2}(n+\frac{1}{2})}\varphi_n(q) = (-i)^{\alpha(n+\frac{1}{2})}\varphi_n(q). \tag{5.4.21}$$

The fractional Fourier transform operator $\widehat{\mathsf{F}}^{\alpha}_{f_S}$ can equally be identified through the eigenvalue equation (5.4.21) being required to be solved by the Hermite-Gauss wave functions (5.4.18) with eigenvalues $\lambda_n = (-i)^{\alpha(n+\frac{1}{2})}$. In particular, with $\alpha = 1$ one recovers the well-known property of the Hermite-Gauss functions being eigenfunctions of the ordinary Fourier transform operator:

$$\widehat{\mathsf{F}}_{f_S}\varphi_n(q) = (-i)^{n+\frac{1}{2}}\varphi_n(q). \tag{5.4.22}$$

Relation (5.4.22) for $n = 0$ is in accord with (4.6.51) on account of (5.4.4).

It is worth noting that in the original formulation by Namias [7] an eigenvalue equation similar to (5.4.21) (see Eq. (5.4.42) below) was proposed as defining equation of the fractional Fourier transform operator, which is then conceived to have the same eigenfunctions of the ordinary Fourier transform but with eigenvalues separated by a fraction of the imaginary unit.

Equation (5.4.21) merely reproduces the time evolution of the oscillator energy eigenstates, which in fact occurs through the phase factor $e^{-i\mathcal{E}_n(t-t_0)/\hbar} = (-i)^{\alpha(n+\frac{1}{2})}$. It however states something new, suggesting to interpret the time-evolution operator of the quantum oscillator as a fractional Fourier transform operator. This is a rather general result regarding indeed any quadratic Hamiltonian and hence any paraxial wave propagation, since, as we will see in the next paragraph, any $ABCD$ integral, i.e, any paraxial optical propagator, is related to the fractional Fourier transform.

Additionally, since the q-profile of the eigenmodes of a quadratic focusing medium model just as (5.4.18), Eq. (5.4.21) can as well be interpreted in the sense of the propagation through a focusing medium of the relative eigenmodes, the corresponding propagation constants being so $\mathcal{E}_n = \frac{(n+\frac{1}{2})}{k_0\sqrt{n_0 n_2}}$. This, of course, confirms the earlier mentioned view of the fractional Fourier transform operator as the optical propagator in a quadratic focusing medium.

Relation (5.4.21) could equally be obtained from (5.4.22) on account of the second of (5.4.12); the discussion above emphasizes the role of $\widehat{\mathsf{F}}^{\alpha}_{f_S}$ as an evolution operator and the wave optics-quantum mechanics analogy.

5.4.3 *ABCD integral and fractional Fourier transform. Operator equivalent of the Iwasawa matrix synthesis*

According to the Iwasawa decomposition of symplectic matrices (§ 3.7), every first-order optical system can be synthesized by a sequence of a phase-plane rotator, a positive scaler and a thin lens, as explicitly expressed by

$$\mathbf{M} = \mathbf{L}(\mathcal{P})\mathbf{S}(m)\mathbf{F}^{\alpha}(f_s), \tag{5.4.23}$$

the real parameters m, $\mathcal{P}$ and ϕ being

$$m^2 = A^2 + \frac{B^2}{f_S^2}, \quad \tan\phi = \frac{B}{A\, f_S}, \quad \mathcal{P} = -\frac{1}{m^2}(AC + D\frac{B}{f_S^2}). \tag{5.4.24}$$

Here an in general non unit value of f_S has been considered with respect to the parameterization reported in Eq.(3.7.3).

Thus, simply translating in operator form the above matrix product, we see that every optical propagator can be represented as the ordered product of a lens operator, a scale operator and a fractional Fourier transform operator:

$$\widehat{\mathsf{M}}(\mathbf{M}) = \widehat{\mathsf{L}}(\mathcal{P})\,\widehat{\mathsf{S}}(m)\widehat{\mathsf{F}}^{\alpha}_{f_S}, \tag{5.4.25}$$

with relations (5.4.24) for the involved parameters. The exponential representations of the three components of the product above in terms of the algebra basis $\{\widehat{\mathsf{K}}_-, \widehat{\mathsf{K}}_+, \widehat{\mathsf{K}}_3\}$ are respectively given by (5.2.21), (5.2.27) and (5.4.11).

Clearly while $\widehat{\mathsf{F}}^{\alpha}_{f_{\mathrm{S}}}$ generates the diffraction integral (5.4.14), $\widehat{\mathsf{S}}(m)$ and $\widehat{\mathsf{L}}(\mathcal{P})$ manifest on the incoming signal $\varphi_i(q)$, respectively, through the scaling operation (5.2.33) and the chirp-modulation (5.2.22). Any $ABCD$ system can therefore be represented by the transfer relation

$$\varphi_o(q) = \tfrac{1}{\sqrt{m}} e^{-ik_0 \mathcal{P} \frac{q^2}{2}} [\widehat{\mathsf{F}}^{\alpha}_{f_{\mathrm{S}}} \varphi_i](\tfrac{q}{m}), \tag{5.4.26}$$

which expresses the amplitude distribution $\varphi_o(q)$ over the output plane as the scaled fractional Fourier transform of the amplitude distribution $\varphi_i(q)$ over the input plane with a further quadratic phase term, and hence as the *imperfect* scaled fractional Fourier transform of the input[(c)].

The above offers an equivalent representation of the $ABCD$ integral, which then can alternatively be characterized by the scale factor m, the curvature parameter $\mathcal{P}$ and the angle ϕ, and hence the order α of the fractional Fourier transform. Evidently, $\mathcal{P}$ specifies the curvature radius $\mathcal{R} = -\frac{1}{\mathcal{P}}$ of the circular profile on which the perfect (scaled) fractional Fourier transform can be observed. When $\mathcal{P} = 0$, the quadratic phase factor disappears and the perfect (scaled) fractional Fourier transform is observed on a planar profile ($\mathcal{R} = \infty$). The pure fractional Fourier transform corresponds to $m = 1$ and $\mathcal{P} = 0$.

The $ABCD$ integral represents the explicit functional solution to any paraxial propagation problem ruled by a polynomial Hamiltonian quadratic in q and p. Accordingly, the LST and LSF$^{\alpha}$ decompositions (5.3.17) and (5.4.25) yield a valuable solving tool permitting, in the same way as their matrix counterparts (5.3.18) and (5.4.23) do for the relevant matrix equation (4.3.4), to decompose the overall problem into three subproblems, whose dynamics might be more easily investigated. This may simplify the solving procedure and/or suggest interesting and enlightening analogies. It is therefore evident the physical and mathematical relevance of decomposition rules in relation with evolution problems ruled by second-order differential equations.

In particular, relation (5.4.26) will be used in § 5.4.5 to relate the free-medium propagator to the quadratic medium propagator, i.e. the Fresnel transform to the fractional Fourier transform.

5.4.4 *Fresnel and Fourier transform*

It is interesting to investigate the link between the Fourier transform of a wave field and the relevant far-field pattern. To this end, we transform back to the q-domain the frequency spectrum $\widetilde{\varphi}_{o}(\kappa, z_{o})$ of the free-propagated signal

---

[c] One may talk of *imperfect* fractional Fourier transform in the same sense as one talks of imperfect Fourier transform. An *imperfect* fractional Fourier transform displays a residual quadratic phase factor in the output variable with respect to the basic performance (5.4.14).

amplitude $\varphi_o(q, z_o)$ over the distance $z_o - z_i$. Then, using (5.2.18), we obtain

$$\varphi_o(q, z_o) = \frac{1}{\sqrt{2\pi}} \int d\kappa\, e^{i\kappa q} \widetilde{\varphi}_o(\kappa, z_o) = \frac{1}{\sqrt{2\pi}} \int d\kappa\, e^{i\kappa q}\, e^{-i\frac{(z_o - z_i)}{2n_0 k_0}\kappa^2}\, \widetilde{\varphi}_i(\kappa, z_i). \tag{5.4.27}$$

Writing $z_o = z_i + \ell$ and rehandling the argument of the exponential function, the above integral can be recast in the form

$$\varphi_o(q, z_i + \ell) = \frac{1}{\sqrt{2\pi}} e^{i\frac{n_0 k_0}{2\ell} q^2} \int d\kappa e^{-i\frac{\ell}{2n_0 k_0}(\kappa - \frac{n_0 k_0 q}{\ell})^2} \widetilde{\varphi}_i(\kappa, z_i), \tag{5.4.28}$$

which, on account of the relation (4.5.6), in the limit $\ell \to \infty$ and finite extent of the signal, turns into

$$\varphi_o(q, z_i + \ell) \underset{\ell \to \infty}{\approx} e^{i\frac{n_0 k}{2\ell} q^2} \sqrt{\frac{n_0 k_0}{i\ell}}\, \widetilde{\varphi}_i(\tfrac{n_0 k_0 q}{\ell}, z_i). \tag{5.4.29}$$

The above relation simply says that the far-field pattern, i.e., the freely propagated field distribution at large distances from the input plane ($\ell \to \infty$), is equivalent, apart from a quadratic phase factor, to the spatial Fourier transform of the input signal (the near-field distribution), evaluated at frequencies $\kappa = \frac{n_0 k_0 q}{\ell}$. Accordingly, the observed signal across the back focal plane of a thin lens can be understood as the far field pattern of the input signal, if freely propagated over large distances.

Relation (5.4.29) expresses the well-known result of diffraction theory according to which for large distances and finite extent of the signal, the Fresnel diffraction passes into the Fraunhofer diffraction [6]. We will see in the next subsection that the "gap" between the Fresnel and Fraunhofer transforms is filled by a continuum of fractional Fourier transforms, thus reproducing for a homogeneous medium the view of propagation through fractional transform planes suggested for a quadratic focusing medium.

5.4.5 *Fresnel and fractional Fourier transform*

Interestingly, the expression (5.2.9) of the Fresnel integral connecting two planar surfaces can be adapted to the expression (5.4.14) of the fractional Fourier transform, thus paralleling the relation of the Fraunhofer integral to the standard Fourier transform, previously discussed. In fact, it is easy to see that the Fresnel transform can be interpreted as a magnified fractional Fourier transform with a residual phase-curvature. Explicitly, by (5.4.26) one may decompose the free-medium propagator $\widehat{\mathsf{T}}(d)$ into

$$\widehat{\mathsf{T}}(d) = \widehat{\mathsf{L}}(\mathcal{P})\, \widehat{\mathsf{S}}(m) \widehat{\mathsf{F}}^{\alpha}_{f_S}, \tag{5.4.30}$$

with the parameters m, $\mathcal{P}$, ϕ specifically given as

$$m^2 = 1 + \frac{d^2}{f_S^2}, \quad \mathcal{P} = -\frac{d}{m^2 f_S^2}, \quad \tan\phi = \frac{d}{f_S}. \tag{5.4.31}$$

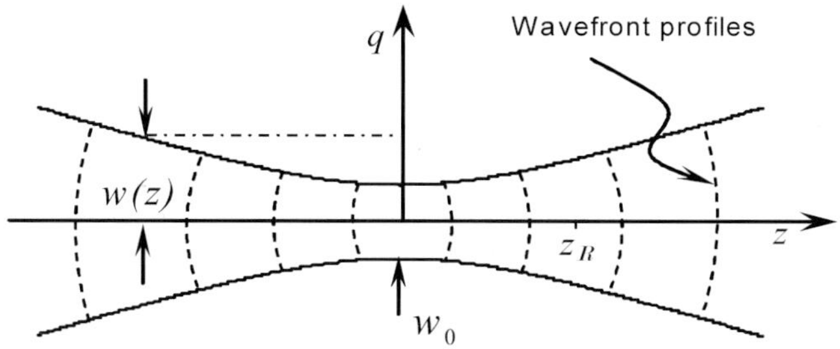

FIGURE 5.6. Schematic of a Gaussian beam, showing the parabolic profiles of the wave fronts and the hyperbolic profiles of the beam width.

Here, the scale parameter can be conveniently set to $f_S = k_0$.

The Fresnel diffraction pattern can therefore be interpreted as the scaled fractional Fourier transform of the signal diffracting from the entrance plane Π_i. The transform is observed on the spherical surface Σ_o with curvature radius

$$\mathcal{R}_o = d(1 + \tfrac{f_S^2}{d^2}), \tag{5.4.32}$$

at distance d from Π_i. The order of the transform increases monotonically with the distance d. As $d \to \infty$, the angle ϕ approaches $\frac{\pi}{2}$ and hence the fractional transform turns into the ordinary Fourier transform. Correspondingly, since $m \to \frac{d}{f_S}$ and $\mathcal{R}_o \to d$, one recovers the far-field diffraction pattern, which according to (5.4.29) is the Fourier transform of the incoming signal. A unifying mathematical description of the diffraction in a free medium is therefore provided by the fractional Fourier transform formalism, which embraces the ordinary Fourier formulation as limiting case.

Notably, the amplitude distributions on the plane Π_i and on the spherical surface Σ_o of curvature radius $\mathcal{R}_o$ are directly connected by a fractional Fourier transformation. The expression (5.4.32) of $\mathcal{R}_o$ closely resembles that for the curvature radius R of the wavefront at d of a Gaussian beam having its waist at the plane Π_i, i.e. $R(d) = d(1+\frac{z_R^2}{d^2})$ ([8], see also § 9.5.1). The scale parameter f_S takes then the meaning as the Rayleigh length z_R of the beam: $z_R = \frac{1}{2}kw_0^2$, w_0 denoting the waist radius ([d]). Besides, the magnification factor m has the form of the normalized beam width $\overline{w}$ at distance d from the waist, which, according to the law $\overline{w}^2(z) \equiv \frac{w^2(z)}{w_0^2} = 1 + \frac{z^2}{z_R^2}$, yields m^2 as in (5.4.31) with $z = d$ and $z_R = f_S$. Finally, the order α of the fractional transform is proportional to the

[d] We recall that when dealing with the propagation laws of Gaussian beam parameters, the z coordinate is commonly measured with respect to the beam waist.

Gouy phase ζ of the beam, which writing

$$\zeta(z) = \tan^{-1}(\frac{z}{z_R}), \tag{5.4.33}$$

turns into the angle ϕ of the transform by properly setting $z = d$ and $z_R = f_S$.

Thus, we can say that the propagation of a Gaussian beam in free space naturally occurs through scaled fractional Fourier transformations, whose order is directly linked to the Gouy phase shift whereas the scale factor accounts for the variation of the beam width with propagation.

5.4.6 *The ABCD integral as a scaled fractional Fourier transform of a scaled input*

Under definite conditions, the ray matrix $\mathbf{M}$ can be decomposed as

$$\mathbf{M} = \mathbf{S}(m_2)\mathbf{F}^{\alpha}(f_s)\mathbf{S}(m_1), \tag{5.4.34}$$

conveying the view of the optical sytem described by $\mathbf{M}$ as performing the fractional Fourier transform of the input signal with two scale factors, one for the input and the other for the output.

Solving (5.4.34) for the relevant parameters, provides the general condition under which decomposition (5.4.34) is feasible. We obtain, in fact,

$$\cos^2\phi = AD, \tag{5.4.35}$$

which yields the inequality

$$0 \leq AD \leq 1, \tag{5.4.36}$$

as a necessary and sufficient condition to interpret the optical transform by $\mathbf{M}$ of the input into the output plane as a fractional Fourier transform of real order.

Correspondingly, the scale factors m_1 and m_2 specialize as

$$m_1^{-4} = \frac{B^2}{f_s^2}\frac{D}{A(1-AD)}, \qquad m_2^4 = \frac{B^2}{f_s^2}\frac{A}{D(1-AD)}. \tag{5.4.37}$$

In the general case of complex order transforms, the condition (5.4.36) can be relaxed; the scale parameters m_1 and m_2 are then allowed to be complex.

Relation (5.4.34) has suggested a generalized form of the fractional Fourier transform, as depending on the order and on two scale factors, which can as well be complex to account for possible field curvatures [7.6].

In a sense, the decomposition (5.4.34) generalizes to the Huygens-Fresnel integral the three factor decomposition (3.6.28), with the scale factors for the input and the output being allowed to be complex and not necessarily reciprocal to each other.

5.4.7 *Basic mathematics of the fractional Fourier transform*

Although introduced within a closely optical context, the fractional Fourier transform has its pure mathematical countepart in the same way as the optical Fourier transform $\widehat{\mathsf{F}}_{f_S}$ is paralleled by the "mathematical" operator $\widehat{\mathcal{F}}$, to which is linked by relation (5.4.4). A similar relation is naturally suggested between the "optical" fractional transform $\widehat{\mathsf{F}}^{\alpha}_{f}$ and the "mathematical" fractional transform $\widehat{\mathcal{F}}^{\alpha}$. Explicilty, one may write:

$$\widehat{\mathsf{F}}^{\alpha}_{f_S}\varphi(q) = \sqrt{\tfrac{k_0}{f}}\, e^{-i\alpha\frac{\pi}{4}}\,[\widehat{\mathcal{F}}^{\alpha}\varphi](\tfrac{k_0}{f}q), \tag{5.4.38}$$

which can be recast in a further "abstract" form changing from the dimensional variable q to the dimensionless $x = \frac{k_0}{f}q$, as we will do below. In fact, conforming to the common mathematical practice, we use here the same symbol to denote the arguments of both the function $\varphi(x)$ and its transform $\widetilde{\varphi}_{\alpha}(x)$. The pertinent physical context gives, of course, specific meaning, role and dimension to the arguments of the tilded and untilded functions.

Accordingly, $\widehat{\mathcal{F}}^{\alpha}$ is defined to signify the following transformation

$$[\widehat{\mathcal{F}}^{\alpha}\varphi](x) \equiv \widetilde{\varphi}_{\alpha}(x) = \sqrt{\tfrac{1-i\cot\phi}{2\pi}} \int e^{\frac{i}{2\sin\phi}[x'^2\cos\phi + x^2\cos\phi - 2xx']}\varphi(x')dx', \tag{5.4.39}$$

which corresponds to the exponential operator representation

$$\widehat{\mathcal{F}}^{\alpha} = e^{i\frac{\phi}{2}}e^{-i\phi\widehat{\mathcal{H}}}, \tag{5.4.40}$$

$\widehat{\mathcal{H}}$ being the "mathematical" generator reproducing the harmonic-oscillator Hamiltonian:

$$\widehat{\mathcal{H}} = -\tfrac{1}{2}\tfrac{\partial^2}{\partial x^2} + \tfrac{1}{2}x^2 = \tfrac{1}{2}\widehat{\mathrm{P}}^2 + \tfrac{1}{2}\widehat{\mathrm{X}}^2 \tag{5.4.41}$$

The argument of the square root in (5.4.39) is as usual taken in the interval $(-\frac{\pi}{2}, \frac{\pi}{2}]$. Of course, $\phi = \alpha\frac{\pi}{2}$, with $\alpha \in (-2, 2]$. The position- and momentum-like operators $\widehat{\mathrm{X}}$, $\widehat{\mathrm{P}}$ in (5.4.41) signify $\widehat{\mathrm{X}} = x$ and $\widehat{\mathrm{P}} = -i\frac{d}{dx}$.

The mathematical definition (5.4.39) of the fractional Fourier transform is properly obtained, within the purely formal context of linear canonical transforms, as unitary representation of the canonical transformation, which produces a rotation, governed by the order parameter α, in the plane of the Fourier conjugate variables (see also § 5.6). It is required to have the basic properties:

i. $\widehat{\mathcal{F}}^{\alpha}$ is continuous with respect to the order parameter α, i.e. $\widehat{\mathcal{F}}^{\beta}$ tends to $\widehat{\mathcal{F}}^{\alpha}$ as β tends to α,

ii. $\widehat{\mathcal{F}}^{\alpha}$ obeys the semigroup property, so that composing two fractional transforms of order α_1 and α_2 yields the fractional transform of order $\alpha_1 + \alpha_2$:

$$\widehat{\mathcal{F}}^{\alpha_1}\widehat{\mathcal{F}}^{\alpha_2} = \widehat{\mathcal{F}}^{\alpha_2}\widehat{\mathcal{F}}^{\alpha_1} = \widehat{\mathcal{F}}^{\alpha_1+\alpha_2},$$

iii. $\widehat{\mathcal{F}}^{\alpha}$ reduces to the identity operator with $\alpha = 0$ and to the ordinary Fourier transform with $\alpha = 1$: $\widehat{\mathcal{F}}^{0} = \widehat{1}$ and $\widehat{\mathcal{F}}^{1} = \widehat{\mathcal{F}}$.

The reader can easily recognize properties *i.-iii.* among those discussed in relation with the optical transform within both the ray-optics (§ 3.6.1) and wave-optics (§ 5.4.2) frameworks. We will discuss in the next section some other properties of the fractional Fourier transform just paralleling those described in § 4.6.1 in regards to the ordinary Fourier transform.

Firstly, it may be interesting to look at the effect of the operator $\widehat{\mathcal{F}}^{\alpha}$ on some basic optical signals. In this connection, we may immediately note that the eigenvalue equation (5.4.21) says that the Hermite-Gaussian functions (5.4.18) are reproduced under the action of $\widehat{\mathcal{F}}^{\alpha}$. In the present context, Eq. (5.4.21) can in fact be rewritten as

$$\widehat{\mathcal{F}}^{\alpha}u_n(x) = (-i)^{\alpha n}u_n(x), \quad u_n(x) = \left(\sqrt{\tfrac{2}{\pi}}\tfrac{1}{2^n n!}\right)^{1/2}H_n(\sqrt{2}x)e^{-x^2}. \qquad (5.4.42)$$

Also, it is instructive to consider the fractional Fourier transform of the line-like, monofrequency and chirp signals, respectively represented by a δ function, and by a linear and a quadratic phase function. The reader is invited to carry on the relevant calculations, yielding as final expressions the following:

Function	Fractional Fourier transform
$\delta(x)$	$\sqrt{\frac{1-i\cot\phi}{2\pi}}e^{\frac{i}{2}x^2\cot\phi}$
$e^{i\gamma x}$	$\sqrt{1+i\tan\phi}\,e^{-\frac{i}{2\cos\phi}(x^2\sin\phi+\gamma^2\sin\phi-2\gamma x)}$
$e^{i\beta x^2}$	$\sqrt{\frac{1+i\tan\phi}{1+\beta\tan\phi}}e^{\frac{i}{2}x^2\left(\frac{\beta-\tan\phi}{1+\beta\tan\phi}\right)}$

(5.4.43)

We see that the chirp function $e^{i\beta\frac{x^2}{2}}$ can be fractionally transformed into a delta function if the order of the transform is chosen to $\alpha = -\frac{2}{\pi}\tan^{-1}(\frac{1}{\beta})$; this property is used to remove chirp-type noises from signals.

We finally mention that fractional Fourier transforms of complex order have been introduced as well. In the light of the interpretation in terms of rotations in the Fourier plane, complex-order transforms are seen to correspond to complex-angle rotations, from which the associated integral transforms emerge in the base form (5.4.39) and with the basic properties *i.* and *ii.*, but with substantial differences with respect to the real-order transforms as far as variable domains, function spaces and operator properties, like boundness and unitarity, are concerned. Complex orders naturally arise when dealing with complex optical systems; gaussian lenses, characterized by complex focal lengths, and gaussian apertures, modelled by imaginary focal lengths, are examples of complex optical systems, described by complex $ABCD$ matrices [4.2].

Some properties of the fractional Fourier transform

We will discuss in particular the similarity, the multiplication and the derivative properties of the fractional Fourier transform, that can interestingly be brought into relation with the corresponding properties of the ordinary Fourier transform, expressed through (4.6.7) and (4.6.17). We will mostly make use of the exponential representation (5.4.40) and the LSF$^\alpha$ decomposition (5.4.25) of $\widehat{\mathcal{F}}^{\alpha}$, whose "mathematical" content will then be emphasized.

Let us start deducing the *similarity property* of the fractional Fourier transform $\widehat{\mathcal{F}}^{\alpha}$, recalling that for the Fourier transform it writes as

$$\widehat{\mathcal{F}}\ \{\varphi(\mathfrak{a}x)\} = \tfrac{1}{|\mathfrak{a}|}\widetilde{\varphi}(\tfrac{x}{\mathfrak{a}}). \tag{5.4.44}$$

with $\mathfrak{a}$ being an arbitrary real parameter.
The function $\varphi(\mathfrak{a}x)$ can be thought of as resulting from a scale transformation of $\varphi(x)$ acted by the operator $\widehat{\mathsf{S}}(\frac{1}{\mathfrak{a}}) = e^{-\frac{\mathfrak{s}}{2}(x\frac{\partial}{\partial x}+\frac{1}{2})}$, with $\mathfrak{a} = e^{-\frac{\mathfrak{s}}{2}}$. So, resorting to the optical analogy (5.4.38) we can write

$$\widehat{\mathcal{F}}^{\alpha}\{\varphi(\mathfrak{a}x)\} = \tfrac{1}{\sqrt{\mathfrak{a}}}e^{i\frac{\phi}{2}}\widehat{\mathsf{F}}^{\alpha}\,\widehat{\mathsf{S}}(\tfrac{1}{\mathfrak{a}})\varphi(x). \tag{5.4.45}$$

On account of the Iwasawa decomposition (5.4.25) with the associated relations (5.4.24), the optical propagator $\widehat{\mathsf{F}}^{\alpha}\,\widehat{\mathsf{S}}(\frac{1}{\mathfrak{a}})$ can alternatively be realized by the ordered product

$$\widehat{\mathsf{F}}^{\alpha}\,\widehat{\mathsf{S}}(\tfrac{1}{\mathfrak{a}}) = \widehat{\mathsf{L}}(\mathcal{P})\,\widehat{\mathsf{S}}(m)\widehat{\mathsf{F}}^{\alpha'},$$

with the parameters being specifically identified through

$$\tfrac{\tan\phi'}{\tan\phi} = \mathfrak{a}^2,\ \ \mathcal{P} = -\cot\phi(1-\tfrac{\cos^2\phi'}{\cos^2\phi}),\ \ m = \mathfrak{a}\tfrac{\sin\phi}{\sin\phi'},\ \ \phi' = \alpha'\tfrac{\pi}{2}. \tag{5.4.46}$$

Then turning back to the "mathematical" context, we end up with

$$\widehat{\mathcal{F}}^{\alpha}\{\varphi(\mathfrak{a}x)\} = \tfrac{1}{|\mathfrak{a}|}\sqrt{\tfrac{1-i\cot\phi}{1-i\cot\phi'}}\,e^{i\frac{x^2}{2}(1-\frac{\cos^2\phi'}{\cos^2\phi})\cot\phi}\ \widetilde{\varphi}_{\alpha'}(\tfrac{x}{\mathfrak{a}}\tfrac{\sin\phi'}{\sin\phi}). \tag{5.4.47}$$

The α-th order fractional Fourier transform of $\varphi(\mathfrak{a}x)$ is not the scaled version of $\widetilde{\varphi}_{\alpha}(x)$, as in (5.4.44); indeed, it is the scaled and chirp-modulated version of $\widetilde{\varphi}_{\alpha'}(x)$ for a different order α', determined by the order α and the scale factor $\mathfrak{a}$ as prescribed by the first of (5.4.46).

The *multiplication property* is concerned with the fractional Fourier transform of the product $x\varphi(x)$ that comes to be a rotational mixture of products by x in the normal and Fourier sense. In fact, we write

$$\widehat{\mathcal{F}}^{\alpha}\{x\varphi(x)\} = [\widehat{\mathcal{F}}^{\alpha}x\widehat{\mathcal{F}}^{-\alpha}]\,\widehat{\mathcal{F}}^{\alpha}\{\varphi(x)\} = \widehat{\mathrm{X}}(\alpha)\widetilde{\varphi}_{\alpha}(x). \tag{5.4.48}$$

The operator $\widehat{\mathrm{X}}(\alpha)$, defined through the similarity transformation

$$\widehat{\mathrm{X}}(\alpha) \equiv \widehat{\mathcal{F}}^{\alpha}x\widehat{\mathcal{F}}^{-\alpha} = \widehat{\mathcal{F}}^{\alpha}\widehat{\mathrm{X}}\widehat{\mathcal{F}}^{-\alpha}, \tag{5.4.49}$$

is evidently the operator into which $\widehat{\mathrm{X}}$ is transformed under $\widehat{\mathcal{F}}^{\alpha}$. Using the exponential representation (5.4.40) of $\widehat{\mathcal{F}}^{\alpha}$, we can give $\widehat{\mathrm{X}}(\alpha)$ the explicit form

$$\widehat{\mathrm{X}}(\alpha) = e^{-i\phi\widehat{\mathcal{H}}}\,\widehat{\mathrm{X}}\,e^{i\phi\widehat{\mathcal{H}}} = \cos\phi\,\widehat{\mathrm{X}} - \sin\phi\,\widehat{\mathrm{P}}, \tag{5.4.50}$$

thus providing for $\widehat{\mathcal{F}}^{\alpha}\{x\varphi(x)\}$ the expression

$$\widehat{\mathcal{F}}^{\alpha}\{x\varphi(x)\} = \left[\cos\phi\,x + i\sin\phi\frac{d}{dx}\right]\widetilde{\varphi}_{\alpha}(x). \tag{5.4.51}$$

Similarly, the *derivative property* establishes the rule to calculate the fractional Fourier transform of the derivative $\frac{d}{dx}\varphi(x)$ in terms of the fractional Fourier transform of φ. As before, we write

$$\widehat{\mathcal{F}}^{\alpha}\left\{\frac{d}{dx}\varphi(x)\right\} = [\widehat{\mathcal{F}}^{\alpha}\frac{d}{dx}\widehat{\mathcal{F}}^{-\alpha}]\,\widehat{\mathcal{F}}^{\alpha}\{\varphi(x)\} = i\widehat{\mathrm{P}}(\alpha)\widetilde{\varphi}_{\alpha}(x), \tag{5.4.52}$$

where the operator $\widehat{\mathrm{P}}(\alpha)$ defined by

$$\widehat{\mathrm{P}}(\alpha) \equiv \widehat{\mathcal{F}}^{\alpha}(-i\frac{d}{dx})\widehat{\mathcal{F}}^{-\alpha} = \widehat{\mathcal{F}}^{\alpha}\,\widehat{\mathrm{P}}\widehat{\mathcal{F}}^{-\alpha}, \tag{5.4.53}$$

is the operator into which $\widehat{\mathrm{P}}$ is transformed under $\widehat{\mathcal{F}}^{\alpha}$. It specializes into

$$\widehat{\mathrm{P}}(\alpha) = e^{-i\phi\widehat{\mathcal{H}}}\,\widehat{\mathrm{P}}\,e^{i\phi\widehat{\mathcal{H}}} = \sin\phi\,\widehat{\mathrm{X}} + \cos\phi\,\widehat{\mathrm{P}}, \tag{5.4.54}$$

thus leading to

$$\widehat{\mathcal{F}}^{\alpha}\left\{\frac{d}{dx}\varphi(x)\right\} = \left[i\sin\phi\,x + \cos\phi\frac{d}{dx}\right]\widetilde{\varphi}_{\alpha}(x). \tag{5.4.55}$$

which looks like a rotational mixture of derivatives with respect to x in the normal and Fourier sense.

As relations (4.6.17) synthesize the rules by which the "position" and "momentum" operators $\widehat{\mathrm{X}}$, $\widehat{\mathrm{P}}$ are transformed between the Fourier conjugate domains, likewise relations (5.4.50) and (5.4.54) synthesize the rules by which $\widehat{\mathrm{X}}$ and $\widehat{\mathrm{P}}$ are transformed between the fractional Fourier conjugate domains. Schematically, we have

x-domain	$\overset{\widehat{\mathcal{F}}}{\longrightarrow}$	Fourier domain	x-domain	$\overset{\widehat{\mathcal{F}}^{\alpha}}{\longrightarrow}$	Fractional Fourier domain
$\widehat{\mathrm{X}}$	$\rightarrow$	$-\widehat{\mathrm{P}}$	$\widehat{\mathrm{X}}$	$\rightarrow$	$\cos\phi\,\widehat{\mathrm{X}} - \sin\phi\,\widehat{\mathrm{P}}$
$\widehat{\mathrm{P}}$	$\rightarrow$	$\widehat{\mathrm{X}}$	$\widehat{\mathrm{P}}$	$\rightarrow$	$\sin\phi\,\widehat{\mathrm{X}} + \cos\phi\,\widehat{\mathrm{P}}$

The explicit calculation of the fractional-Fourier domain representations (5.4.50) and (5.4.54) of the operators $\widehat{\mathrm{X}}$ and $\widehat{\mathrm{P}}$ follow from rule (4.4.25) (see also Problem 4.2) on account of the commutator brackets of $\widehat{\mathcal{H}}$ with $\widehat{\mathrm{X}}$ and $\widehat{\mathrm{P}}$ being $[\widehat{\mathcal{H}}, \widehat{\mathrm{X}}] = -i\,\widehat{\mathrm{P}}$ and $[\widehat{\mathcal{H}}, \widehat{\mathrm{P}}] = i\widehat{\mathrm{X}}$.

We may guess from the expressions in the above table, showing that $\widehat{\mathrm{X}}$ and $\widehat{\mathrm{P}}$ are transformed by the fractional Fourier transform operator into a mixture of $\widehat{\mathrm{X}}$ and $\widehat{\mathrm{P}}$, the potential of the fractional Fourier transform to yield a representation of the signal not purely in space as $\varphi(q)$, or in frequency as $\widetilde{\varphi}(\kappa)$, but in both space and frequency with a continuous, and adjustable, degree of emphasis on space or frequency features. In later chapters, in fact, it will emerge the intimate link of the fractional Fourier transform to the Wigner distribution function (and the ambiguity function), the exemplary joint representation of optical signals.

5.5 Recovering the ray-optical description

As expected, we may recover from the $ABCD$ integral (5.3.4) the transfer law of ray optics, according to which the ray variables from one plane to another plane are linked by a linear relation formally expressed through the ray matrix pertaining to the involved planes.

To this end, we consider the expectation values of the operators $\widehat{\mathsf{q}}$ and $\widehat{\mathsf{p}}$ in the state $\varphi(q,z)$, which are defined by the inner products

$$\begin{aligned} \langle\,\widehat{\mathsf{q}}\,\rangle(z) &= \int dq \varphi^*(q,z)\widehat{\mathsf{q}}\,\varphi(q,z), \\ \langle\,\widehat{\mathsf{p}}\,\rangle(z) &= \int dq \varphi^*(q,z)\widehat{\mathsf{p}}\,\varphi(q,z), \end{aligned} \tag{5.5.1}$$

with the complex amplitude being assumed to be normalized to one: $\|\varphi\| = 1$.

Using the input-output relation for the optical system to express the signal amplitude at different planes along the optical axis, we can relate the expectation values at the plane z, $\langle\,\widehat{\mathsf{q}}\,\rangle(z)$, $\langle\,\widehat{\mathsf{p}}\,\rangle(z)$, to the expectation values at the input plane z_i, $\langle\,\widehat{\mathsf{q}}\,\rangle(z_i)$, $\langle\,\widehat{\mathsf{p}}\,\rangle(z_i)$. It is evident that the expectation values (5.5.1) equal the mean values $\overline{q}$ and $\overline{\kappa}/k_0$ introduced in Eq. (4.6.36), i.e.,

$$\langle\,\widehat{\mathsf{q}}\,\rangle(z) = \overline{q}(z), \qquad \langle\,\widehat{\mathsf{p}}\,\rangle(z) = \tfrac{1}{k_0}\overline{\kappa}(z). \tag{5.5.2}$$

The first identity is obvious. As to the second, using the inverse Fourier transform, one obtains

$$\begin{aligned} \langle\,\widehat{\mathsf{p}}\,\rangle(z) &= \tfrac{1}{2\pi}\iint d\kappa' d\kappa \widetilde{\varphi}^*(\kappa',z)\,\widetilde{\varphi}(\kappa,z)\int dq e^{-i\kappa' q}(-\tfrac{i}{k_0}\tfrac{\partial}{\partial q})e^{i\kappa q} \\ &= \tfrac{1}{k_0}\int \widetilde{\varphi}^*(\kappa,z)\kappa\,\widetilde{\varphi}(\kappa,z)d\kappa = \tfrac{1}{k_0}\overline{\kappa}(z). \end{aligned} \tag{5.5.3}$$

Accordingly, the z-evolution of the expectation values $\langle\,\widehat{\mathsf{q}}\,\rangle(z)$, $\langle\,\widehat{\mathsf{p}}\,\rangle(z)$ yields information on the "motion" (position and direction) of the centroid of the wave field.

In particular, for the propagation through a homogeneous medium over the reduced distance d we have

$$\langle \widehat{\mathsf{q}} \rangle_o = \tfrac{k_0}{2\pi d} \int dq \int dq' \int dq'' q\, e^{-i\frac{k_0}{2d}(q-q')^2} e^{i\frac{k_0}{2d}(q-q'')^2} \varphi_i^*(q')\varphi_i(q''), \quad (5.5.4)$$

which, by use of relations (4.5.8) and (4.5.4) concerning the first derivative of the δ function, yields

$$\langle \widehat{\mathsf{q}} \rangle_o = \langle \widehat{\mathsf{q}} \rangle_i + d\langle \widehat{\mathsf{p}} \rangle_i. \quad (5.5.5)$$

Similarly the expectation value of $\widehat{\mathsf{p}}$ is obtained from the integral

$$\langle \widehat{\mathsf{p}} \rangle_o = \tfrac{k_0}{2\pi d} \int dq \int dq' \int dq'' q\, e^{-i\frac{k_0}{2d}(q-q')^2} \varphi_i^*(q)(-\frac{i}{k_0}\frac{\partial}{\partial q}) e^{i\frac{k_0}{2d}(q-q'')^2} \varphi_i(q''), \quad (5.5.6)$$

which upon calculation gives

$$\langle \widehat{\mathsf{p}} \rangle_o = \langle \widehat{\mathsf{p}} \rangle_i. \quad (5.5.7)$$

Relations (5.5.5) and (5.5.7) have the same form as the matrix relation for the ray variables (q, p) between the end-planes enclosing the medium.

Likewise, from the transfer relation (5.2.22) for a thin lens, it easily follows for the expectation values the same formal relation linking the ray variables in matrix optics, i.e.

$$\begin{aligned} \langle \widehat{\mathsf{q}} \rangle_o &= \langle \widehat{\mathsf{q}} \rangle_i\,, \\ \langle \widehat{\mathsf{p}} \rangle_o &= -\tfrac{1}{f}\langle \widehat{\mathsf{q}} \rangle_i + \langle \widehat{\mathsf{p}} \rangle_i\ . \end{aligned} \quad (5.5.8)$$

In general, using the input-output relation (5.3.4) for an arbitrary $ABCD$ system, after some algebra we end up with

$$\begin{pmatrix} \langle \widehat{\mathsf{q}} \rangle_o \\ \langle \widehat{\mathsf{p}} \rangle_o \end{pmatrix} = \begin{pmatrix} A & B \\ C & D \end{pmatrix} \begin{pmatrix} \langle \widehat{\mathsf{q}} \rangle_i \\ \langle \widehat{\mathsf{p}} \rangle_i \end{pmatrix} \quad (5.5.9)$$

thus proving that the mean values $\langle \widehat{\mathsf{q}} \rangle(z)$ and $\langle \widehat{\mathsf{p}} \rangle(z)$ transform as the ray variables of ray optics.

Equations (5.5.9) may be regarded as the wave-optical version (in the linear approximation) of the Ehrenfest theorem of quantum mechanics, which is well known to provide the law of motion of the mean values of the conjugate observables $\widehat{\mathsf{q}}$ and $\widehat{\mathsf{p}}$ [9]. The Ehrenfest theorem states that the equations of motion obeyed by $\langle \widehat{\mathsf{q}} \rangle$ and $\langle \widehat{\mathsf{p}} \rangle$ are formally identical to Hamilton's equations of the classical theory, except that the quantities which occur on both sides of the classical equations must be replaced by their expectation values. This does not mean in general that the expectation values follow the classical motion. In fact, relations (5.5.9), which establish indeed that $\langle \widehat{\mathsf{q}} \rangle$ and $\langle \widehat{\mathsf{p}} \rangle$ follow the ray motion of geometrical optics, hold only in the first-order approximation, and hence only for Hamiltonians, which are quadratic functions of $\widehat{\mathsf{q}}$ and $\widehat{\mathsf{p}}$.

To see this, consider that the z-evolution of the expectation value of any wave-optical operator $\widehat{\mathsf{O}}$ is ruled by the equation

$$\frac{i}{k_0}\frac{d}{dz}\langle\,\widehat{\mathsf{O}}\,\rangle = \langle\,[\,\widehat{\mathsf{O}},\,\widehat{\mathsf{H}}]\,\rangle + \frac{i}{k_0}\langle\frac{\partial}{\partial z}\,\widehat{\mathsf{O}}\,\rangle, \tag{5.5.10}$$

the second term in the right hand side accounting for the possible explicit dependence of $\widehat{\mathsf{O}}$ on the propagation variable z.

When the canonical operators $\widehat{\mathsf{q}}$ and $\widehat{\mathsf{p}}$ are concerned, Eq. (5.5.10) gives

$$\begin{aligned}\tfrac{i}{k_0}\tfrac{d}{dz}\langle\,\widehat{\mathsf{q}}\,\rangle &= \langle\,[\,\widehat{\mathsf{q}},\,\widehat{\mathsf{H}}\,]\rangle,\\ \tfrac{i}{k_0}\tfrac{d}{dz}\langle\,\widehat{\mathsf{p}}\,\rangle &= \langle\,[\,\widehat{\mathsf{p}},\,\widehat{\mathsf{H}}\,]\rangle.\end{aligned} \tag{5.5.11}$$

As the commutators on the right hand side yield the derivatives of $\widehat{\mathsf{H}}$ with respect to $\widehat{\mathsf{q}}$ and $\widehat{\mathsf{p}}$, we can write

$$\begin{aligned}\tfrac{i}{k_0}\tfrac{d}{dz}\langle\,\widehat{\mathsf{q}}\,\rangle &= \langle\,\tfrac{\partial\,\widehat{\mathsf{H}}}{\partial p}\,\rangle,\\ \tfrac{i}{k_0}\tfrac{d}{dz}\langle\,\widehat{\mathsf{p}}\,\rangle &= -\langle\,\tfrac{\partial\,\widehat{\mathsf{H}}}{\partial q}\,\rangle,\end{aligned} \tag{5.5.12}$$

which represent the definite form of Ehrenfest's theorem.

Evidently, although Eqs. (5.5.12) resemble Hamilton's equations of geometrical optics, one can not say that the expectation values $\langle\,\widehat{\mathsf{q}}\,\rangle$ and $\langle\,\widehat{\mathsf{p}}\,\rangle$ follow the laws of ray optics, unless the following identities

$$\begin{aligned}\langle\,\tfrac{\partial}{\partial p}\,\widehat{\mathsf{H}}(\,\widehat{\mathsf{q}},\,\widehat{\mathsf{p}},z)\rangle &= \tfrac{\partial}{\partial\langle\,\widehat{\mathsf{p}}\,\rangle}\widehat{\mathsf{H}}(\langle\,\widehat{\mathsf{q}}\,\rangle,\langle\,\widehat{\mathsf{p}}\,\rangle,z),\\ \langle\,\tfrac{\partial}{\partial q}\,\widehat{\mathsf{H}}(\,\widehat{\mathsf{q}},\,\widehat{\mathsf{p}},z)\rangle &= \tfrac{\partial}{\partial\langle\,\widehat{\mathsf{q}}\,\rangle}\widehat{\mathsf{H}}(\langle\,\widehat{\mathsf{q}}\,\rangle,\langle\,\widehat{\mathsf{p}}\,\rangle,z),\end{aligned} \tag{5.5.13}$$

can be stated. It is easy to verify that the above identities hold true only if the Hamiltonian operator is a quadratic function of $\widehat{\mathsf{q}}$ and $\widehat{\mathsf{p}}$, and hence only in the paraxial approximation. In that case only, the expectation values of the position and momentum operators follow the "classical" motion.

5.5.1 *Further exploring the correspondence between the ray and wave optics pictures. II*

Continuing the line of the presentation in § 4.4.4, we reconsider again the evolution equations for the Heisenberg operators $\widehat{\mathsf{q}}(z)$ and $\widehat{\mathsf{p}}(z)$, we rewrite in the slightly different form

$$\begin{aligned}\tfrac{d}{dz}\widehat{\mathsf{q}}(z) &= [\widehat{\mathsf{q}},\widehat{\mathcal{H}}],\\ \tfrac{d}{dz}\,\widehat{\mathsf{p}}(z) &= [\widehat{\mathsf{p}},\widehat{\mathcal{H}}],\end{aligned} \tag{5.5.14}$$

with $\widehat{\mathcal{H}} = -ik_0\widehat{\mathsf{H}}$.

Rephrasing the presentation of § 1.7.1, concerning the quadratic polynomials and the associated matrix representatives under Poisson brackets, we consider as a general Hamiltonian operator the bilinear form in $\widehat{\mathsf{q}}$ and $\widehat{\mathsf{p}}$:

$$\widehat{\mathsf{H}}(\widehat{\mathsf{q}}, \widehat{\mathsf{p}}, z) = \tfrac{1}{2}a\widehat{\mathsf{q}}^2 + \tfrac{1}{2}b\,\widehat{\mathsf{p}}^2 + \tfrac{1}{2}c(\widehat{\mathsf{q}}\widehat{\mathsf{p}} + \widehat{\mathsf{p}}\widehat{\mathsf{q}}). \tag{5.5.15}$$

The Lie brackets $[\widehat{\mathsf{q}}, \widehat{\mathcal{H}}]$ and $[\widehat{\mathsf{p}}, \widehat{\mathcal{H}}]$ come to be linear in $\widehat{\mathsf{q}}$ and $\widehat{\mathsf{p}}$, being, in fact,

$$\begin{aligned} [\widehat{\mathsf{q}}, \widehat{\mathcal{H}}] &= \quad c\widehat{\mathsf{q}} + b\,\widehat{\mathsf{p}}, \\ [\widehat{\mathsf{p}}, \widehat{\mathcal{H}}] &= -a\widehat{\mathsf{q}} - \,c\,\widehat{\mathsf{p}}, \end{aligned} \tag{5.5.16}$$

which allows us to identify the matrix representative of $\widehat{\mathcal{H}}$ as

$$\mathbf{H} = \begin{pmatrix} c & b \\ -a & -c \end{pmatrix}. \tag{5.5.17}$$

Clearly the traceless matrix $\mathbf{H}$ belongs to the symplectic algebra $sp(2, \mathbb{R})$.

The Heisenberg equations (5.5.14) can then be recast into a matrix relation similar to Eq. (1.7.4) for the ray vector $\mathbf{u}$; namely,

$$\frac{d}{dz}\widehat{\mathbf{w}}(z) = \mathbf{H}\widehat{\mathbf{w}}(z), \quad \widehat{\mathbf{w}}(z_i) = \widehat{\mathbf{w}}_i \tag{5.5.18}$$

involving the operator-valued vector $\widehat{\mathbf{w}}(z) = (\widehat{\mathsf{q}}(z), \widehat{\mathsf{p}}(z))^{\top}$ (see § 4.4.4).

It is inherent in the above equation a linear transformation for $\widehat{\mathsf{q}}$ and $\widehat{\mathsf{p}}$, which reproduces that for the c-number ray variables q and p. In fact, since the change of the vector $\widehat{\mathbf{w}}(z)$ over an infinitesimal distance dz from the initial plane z_i is effected by the 2×2 symplectic matrix $e^{\mathbf{H}(z_i)dz}$according to

$$\widehat{\mathbf{w}}(z_i + dz) = e^{\mathbf{H}(z_i)dz}\,\widehat{\mathbf{w}}_i, \tag{5.5.19}$$

the evolution of $\widehat{\mathbf{w}}(z)$ over the finite distance $z - z_i$ is effected by an infinite ordered string of 2×2 matrices like that above, which eventually yields for the operators $(\widehat{\mathsf{q}}, \widehat{\mathsf{p}})$ the same finite linear symplectic transformation ruling the evolution of the ray-variables (q, p).

Interestingly, if we conveniently define the operators

$$\widehat{\mathcal{K}}_{-} = -ik_0\widehat{\mathsf{K}}_{-}, \quad \widehat{\mathcal{K}}_{+} = -ik_0\widehat{\mathsf{K}}_{+}, \quad \widehat{\mathcal{K}}_{3} = -ik_0\widehat{\mathsf{K}}_{3} \tag{5.5.20}$$

we can rewrite the commutation relations (5.2.2) as

$$\left[\widehat{\mathcal{K}}_{+}, \widehat{\mathcal{K}}_{-}\right] = 2\,\widehat{\mathcal{K}}_{3}, \quad \left[\widehat{\mathcal{K}}_3, \widehat{\mathcal{K}}_{\pm}\right] \; = \mp\widehat{\mathcal{K}}_{+}, \tag{5.5.21}$$

which, of course, pertain to the metaplectic Lie algebra $mp(2, \mathbb{R})$, and exactly mirror the commutation relations (2.3.8) of the $sp(2, \mathbb{R})$ matrix basis

$\{\mathbf{K}_{-}, \mathbf{K}_{+}, \mathbf{K}_{3}\}$. Also, by rule (5.5.16) the expected correspondences between the $mp(2, \mathbb{R})$ operator basis, and the $sp(2, \mathbb{R})$ matrix basis are recovered, being

$$\widehat{\mathcal{K}}_{-} \rightleftharpoons \mathbf{K}_{-} = \begin{pmatrix} 0 & 1 \\ 0 & 0 \end{pmatrix}, \quad \widehat{\mathcal{K}}_{+} \rightleftharpoons \mathbf{K}_{+} = \begin{pmatrix} 0 & 0 \\ -1 & 0 \end{pmatrix}, \quad \widehat{\mathcal{K}}_{3} \rightleftharpoons \mathbf{K}_{3} = \frac{1}{2}\begin{pmatrix} 1 & 0 \\ 0 & -1 \end{pmatrix}, \tag{5.5.22}$$

The inverse arrows, from matrices to operators, work through the operator form of (1.7.11), which writes

$$\widehat{\mathsf{H}}(\widehat{\mathsf{q}}, \widehat{\mathsf{p}}) = \tfrac{1}{2}(\mathbf{H}\widehat{\mathbf{w}}, \mathbf{J}\widehat{\mathbf{w}}) = \tfrac{1}{2}\widehat{\mathbf{w}}^{\top}\mathbf{H}^{\top}\mathbf{J}\widehat{\mathbf{w}}. \tag{5.5.23}$$

An isomorphism can evidently be established between the symplectic and metaplectic algebras $sp(2, \mathbb{R})$ and $mp(2, \mathbb{R})$. We will see later that the respective groups are isomorphic up to a sign.

5.6 Wave-optical propagators as unitary representations of linear canonical transformations

It is noteworthy that the wave-optical propagators $\widehat{\mathsf{M}}(z, z_i)$ in their integral transform representation (5.3.5) arise as unitary operators acting on the Hilbert space $\mathcal{L}^2(\mathbb{R})$ of wave functions, which correspond to unitary transformations mapping the Heisenberg-Weyl algebra operators $\widehat{\mathsf{q}}$ and $\widehat{\mathsf{p}}$ into real linear combinations of themselves [10].

As said in § 4.4.4, the propagation of a wave field through an optical system described by the propagator $\widehat{\mathsf{M}}$ reflects into the the similarity transformations (by $\widehat{\mathsf{M}}^{-1}$) of the Heisenberg operators $\widehat{\mathsf{q}}(z)$ and $\widehat{\mathsf{p}}(z)$, which then evolve from the Schrödinger operators (5.1.2) according to

$$\begin{aligned} \widehat{\mathsf{q}}(z) &= \widehat{\mathsf{M}}^{-1}(z, z_i)\widehat{\mathsf{q}}(z_i)\widehat{\mathsf{M}}(z, z_i), \\ \widehat{\mathsf{p}}(z) &= \widehat{\mathsf{M}}^{-1}(z, z_i)\widehat{\mathsf{p}}(z_i)\widehat{\mathsf{M}}(z, z_i), \end{aligned} \tag{5.6.24}$$

where, of course, $\widehat{\mathsf{q}}(z_i) = \widehat{\mathsf{q}}$ and $\widehat{\mathsf{p}}(z_i) = \widehat{\mathsf{p}}$.

On the other hand, guided by the considerations developed in the previous section, showing the formal identity of the evolution laws for the q-number pair $(\widehat{\mathsf{q}}(z), \widehat{\mathsf{p}}(z))$ to those of the c-number variables $(q(z), p(z))$, we may relate the propagated operators $\widehat{\mathsf{q}}(z)$ and $\widehat{\mathsf{p}}(z)$ to the input operators $\widehat{\mathsf{q}}(z_i)$ and $\widehat{\mathsf{p}}(z_i)$ through the optical matrix $\mathbf{M}$ of ray-optics. So, we explicitly write

$$\begin{aligned} \widehat{\mathsf{q}}(z) &= A(z, z_i)\widehat{\mathsf{q}}(z_i) + B(z, z_i)\widehat{\mathsf{p}}(z_i), \\ \widehat{\mathsf{p}}(z) &= C(z, z_i)\widehat{\mathsf{q}}(z_i) + D(z, z_i)\widehat{\mathsf{p}}(z_i). \end{aligned} \tag{5.6.25}$$

The unimodularity of $\mathbf{M}$ guarantees that the evolved operators $\widehat{\mathsf{q}}(z)$ and $\widehat{\mathsf{p}}(z)$, as $\widehat{\mathsf{q}}(z_i)$ and $\widehat{\mathsf{p}}(z_i)$, obey the Lie bracket (5.1.3): $[\widehat{\mathsf{q}}(z), \widehat{\mathsf{p}}(z)] = [\widehat{\mathsf{q}}(z_i), \widehat{\mathsf{p}}(z_i)] = \frac{i}{k_0}$.

The question arises about the specific functional form of the transfer operator $\widehat{\mathsf{M}}(z, z_i)$, which through (5.6.24), leads to the linear canonical transformation (5.6.25) for $\widehat{\mathsf{q}}$ and $\widehat{\mathsf{p}}$, namely

$$\begin{aligned}\widehat{\mathsf{M}}^{-1}(z, z_i)\widehat{\mathsf{q}}(z_i)\widehat{\mathsf{M}}(z, z_i) &= A(z, z_i)\widehat{\mathsf{q}}(z_i) + B(z, z_i)\widehat{\mathsf{p}}(z_i), \\ \widehat{\mathsf{M}}^{-1}(z, z_i)\widehat{\mathsf{p}}(z_i)\widehat{\mathsf{M}}(z, z_i) &= C(z, z_i)\widehat{\mathsf{q}}(z_i) + D(z, z_i)\widehat{\mathsf{p}}(z_i)\ .\end{aligned} \tag{5.6.26}$$

In other words, we are trying to define a mapping of symplectic matrices $\mathbf{M}$ into unitary operators $\widehat{\mathsf{M}}$:

$$\mathbf{M} \in Sp(2, \mathbb{R}) \longrightarrow \widehat{\mathsf{M}} \in Mp(2, \mathbb{R}), \tag{5.6.27}$$

based on the requirement that the operators $\widehat{\mathsf{q}}(z)$, $\widehat{\mathsf{p}}(z)$ follow the same matrix law of the ray variables $q(z)$, $p(z)$, when the wave field propagates through the system, and hence the corresponding wave function is acted by $\widehat{\mathsf{M}}$.

Evidently, since both $\widehat{\mathsf{M}}$ and $\widehat{\mathsf{M}}^{-1}$ enter (5.6.26), any operator $\widehat{\mathsf{M}}$, satisfying (5.6.26), is definable up to an arbitrary phase factor. Our analysis below is aimed at exploring the possibility to uniquely define such a phase in accord with the desired preservation of the group structure of the set of $\widehat{\mathsf{M}}$'s. We will see that the phase indeterminacy can be reduced up to a sign.

As suggested by (5.1.8), we assume that $\widehat{\mathsf{M}}(z, z_i)$ concretizes into the integral transform

$$\widehat{\mathsf{M}}(z, z_i)\varphi(q, z_i) = \int dq' \mathbb{G}(q, q', z, z_i)\varphi(q', z_i), \tag{5.6.28}$$

so that the problem of determining $\widehat{\mathsf{M}}$ is turned into that of determining the functional form of the kernel $\mathbb{G}(q, q', z, z_i)$ assuring that relations (5.6.25) hold.

We rewrite relations (5.6.26) in the equivalent form

$$\begin{aligned}\widehat{\mathsf{q}}\,\widehat{\mathsf{M}} &= \widehat{\mathsf{M}}(A\widehat{\mathsf{q}} + B\widehat{\mathsf{p}}), \\ \widehat{\mathsf{p}}\widehat{\mathsf{M}} &= \widehat{\mathsf{M}}(C\widehat{\mathsf{q}} + D\widehat{\mathsf{p}}),\end{aligned} \tag{5.6.29}$$

the dependence of the various functions and operators on the axial coordinate z being implied. Acting on the wave function $\varphi(q)$ by the operators on both sides of (5.6.29), we obtain the identities

$$\begin{aligned}\int dq' \mathbb{G}(q, q')(A\widehat{\mathsf{q}} + B\widehat{\mathsf{p}})\varphi(q') &= \widehat{\mathsf{q}} \int dq' \mathbb{G}(q, q')\varphi(q'), \\ \int dq' \mathbb{G}(q, q')(C\widehat{\mathsf{q}} + D\widehat{\mathsf{p}})\varphi(q') &= \widehat{\mathsf{p}} \int dq' \mathbb{G}(q, q')\varphi(q').\end{aligned} \tag{5.6.30}$$

On account of (5.1.2), we may integrate by parts the left-hand sides of both equations. Then, from the resulting identities, that must hold for any $\varphi(q)$, we obtain the following set of linear partial differential equations for $\mathbb{G}(q, q')$:

$$\begin{aligned}(Aq' + \tfrac{i}{k_0} B \tfrac{\partial}{\partial q'})\mathbb{G}(q, q') &= q\mathbb{G}(q, q'), \\ (Cq' + \tfrac{i}{k_0} D \tfrac{\partial}{\partial q'})\mathbb{G}(q, q') &= -\tfrac{i}{k_0}\tfrac{\partial}{\partial q}\mathbb{G}(q, q').\end{aligned} \tag{5.6.31}$$

We may propose a solution in the form

$$\mathbb{G}(q,q') = \gamma e^{i(\mu q^2+\nu q'^2+\tau qq')}, \tag{5.6.32}$$

with the constant parameters (i.e. independent from q and q') γ, μ, ν, τ being to be appropriately defined. Indeed, inserting (5.6.32) into (5.6.31), we obtain for μ, ν and τ the definite expressions

$$\mu = k_0 \tfrac{D}{2B}, \qquad \nu = k_0 \tfrac{A}{2B}, \qquad \tau = -k_0 \tfrac{1}{B}. \tag{5.6.33}$$

In order to specialize γ, we resort to relation (4.3.21) (adapted to $\mathbb{G}$), which reflects the unitarity of $\widehat{\mathsf{M}}$. With (5.6.32) and (5.6.33), Eq. (4.3.21) yields

$$|\gamma|^2 = \frac{k_0}{2\pi|B|}, \tag{5.6.34}$$

by which γ comes to be defined up to a phase factor: $\gamma = e^{i\zeta}\sqrt{\frac{k_0}{2\pi|B|}}$.

Accordingly, we can establish for $\mathbb{G}(q,q')$ the functional form

$$\mathbb{G}(q,q') = e^{i\zeta}\sqrt{\frac{k_0}{2\pi i|B|}}\, e^{i\frac{k_0}{2B}(Aq'^2+Dq^2-2qq')}, \tag{5.6.35}$$

with an undefined phase ζ.

It could seem convenient, in order to form a one-to-one mapping of symplectic matrices $\mathbf{M}$ into operators $\widehat{\mathsf{M}}$, to envisage a rule to assign a unique phase $\theta(\mathbf{M})$ to every matrix $\mathbf{M}$. With such a choice, however, the resulting set of operators $\widehat{\mathsf{M}}(\mathbf{M})$ do not close under multiplication, and hence it is not a group, as it would be desirable. In fact, let $\mathbf{M}_1 = \left(\begin{smallmatrix} A_1 & B_1 \\ C_1 & D_1 \end{smallmatrix}\right)$ and $\mathbf{M}_2 = \left(\begin{smallmatrix} A_2 & B_2 \\ C_2 & D_2 \end{smallmatrix}\right)$ be two symplectic matrices, which are imaged respectively into $\widehat{\mathsf{M}}_1$ and $\widehat{\mathsf{M}}_2$, with representing kernels $\mathbb{G}_1$ and $\mathbb{G}_2$. Let $\mathbf{M} = \left(\begin{smallmatrix} A & B \\ C & D \end{smallmatrix}\right)$ denote the product matrix $\mathbf{M} = \mathbf{M}_1\mathbf{M}_2$; of course,

$$\mathbf{M} = \begin{pmatrix} A_1A_2 + B_1C_2 & A_1B_2 + B_1D_2 \\ A_2C_1 + C_2D_1 & B_2C_1 + D_1D_2 \end{pmatrix}. \tag{5.6.36}$$

Finally, let $\widehat{\mathsf{M}}_{12}$ and $\widehat{\mathsf{M}}$ respectively denote the operator resulting from the product $\widehat{\mathsf{M}}_{12} = \widehat{\mathsf{M}}_1\widehat{\mathsf{M}}_2$ and that corresponding to the product matrix $\mathbf{M}$ through (5.6.35). From (5.6.28) it follows that the representative kernels $\mathbb{G}_1$ and $\mathbb{G}_2$ of $\widehat{\mathsf{M}}_1$ and $\widehat{\mathsf{M}}_2$ compose as (see also Eq. (4.3.18))

$$\mathbb{G}_{12}(q,q') = \int dq''\mathbb{G}_1(q,q'')\mathbb{G}_2(q'',q'). \tag{5.6.37}$$

In order to verify that the product of operators exactly reproduces the multiplication of matrices according to

$$\widehat{\mathsf{M}}(\mathbf{M}) = \widehat{\mathsf{M}}_1(\mathbf{M}_1)\widehat{\mathsf{M}}_2(\mathbf{M}_2), \tag{5.6.38}$$

for any choice of $\widehat{\mathsf{M}}_1$ and $\widehat{\mathsf{M}}_2$, we will work out the explicit expression of $\mathbb{G}_{12}(q,q')$ in terms of the entries of the product matrix (5.6.36), and then we will compare it with the expression of the representing kernel $\mathbb{G}(q,q')$ of $\widehat{\mathsf{M}}$.

By (5.6.35), the integral in (5.6.37) yields

$$\begin{aligned}\mathbb{G}_{12}(q,q') &= e^{i\zeta_1}e^{i\zeta_2}\frac{k_0}{2\pi\sqrt{|B_1B_2|}}e^{i\frac{k_0}{2}(\frac{A_2}{B_2}q'^2+\frac{D_1}{B_1}q^2)}\int dq'' e^{i\frac{k_0}{2}[\frac{B}{B_1B_2}q''^2-2(\frac{q}{B_1}+\frac{q'}{B_2})]} \\ &= e^{i\zeta_1}e^{i\zeta_2}e^{i\frac{\pi}{4}\mathrm{sgn}(\frac{B}{B_1B_2})}\sqrt{\frac{k_0}{2\pi|B|}}e^{i\frac{k_0}{2B}(Aq'^2+Dq^2-2qq')},\end{aligned} \tag{5.6.39}$$

where sgn(·) is the sign function being ±1 according to whether the argument is positive or negative (see footnote on p. 226).

On the other hand, the representing kernel $\mathbb{G}(q,q')$ of $\widehat{\mathsf{M}}$ is

$$\mathbb{G}(q,q') = e^{i\zeta}\sqrt{\frac{k_0}{2\pi|B|}}e^{i\frac{k_0}{2B}(Aq'^2+Dq^2-2qq')}. \tag{5.6.40}$$

Evidently, for the identity (5.6.38) to hold, one should have

$$\zeta(\mathbf{M}) = \zeta(\mathbf{M}_1)+\zeta(\mathbf{M}_2)+\frac{\pi}{4}\mathrm{sgn}(\frac{B}{B_1B_2}). \tag{5.6.41}$$

A possible definition of the phase is naturally suggested as

$$\zeta(\mathbf{M}) = -\frac{\pi}{4}\mathrm{sgn}(B), \tag{5.6.42}$$

by which we evidently form a one-to-one association between symplectic matrices and metaplectic operators. But with this choice the phase of the product operator $\widehat{\mathsf{M}}$ comes to differ from the sum of the phases of the component operators by the term

$$\zeta(\mathbf{M}) - [\zeta(\mathbf{M}_1)+\zeta(\mathbf{M}_2)] = \frac{\pi}{4}[\mathrm{sgn}(B) - \mathrm{sgn}(B_1) - \mathrm{sgn}(B_2) + \mathrm{sgn}(\frac{B}{B_1B_2})], \tag{5.6.43}$$

which indeed, as the reader may verify by direct check, turns into 0 or $\pm\pi$ according to the signs of the involved B-entries. This means that the set of operators $\widehat{\mathsf{M}}(\mathbf{M})$ we obtain with this choice do not form a group, since on multiplying two operators, $\widehat{\mathsf{M}}_1(\mathbf{M}_1)$ and $\widehat{\mathsf{M}}_2(\mathbf{M}_2)$, it could happen that, depending on $\mathbf{M}_1$ and $\mathbf{M}_2$, the result be $-\widehat{\mathsf{M}}(\mathbf{M})$, which is not in the set. Therefore in order to obtain a group, we must let the correspondence (5.6.27) be not one-to-one, allowing that both $\widehat{\mathsf{M}}(\mathbf{M})$ and $-\widehat{\mathsf{M}}(\mathbf{M})$ be associable to each symplectic matrix $\mathbf{M}$. The resulting set of operators closes under multiplication, and yields a two-valued representation[e] of $Sp(2,\mathbb{R})$, imaging linear

[e] Note that this is not strictly a representation, which is intended to admit only one "image" corresponding to possibly more than one "object".

canonical transformations of the Heisenberg-Weyl algebra operators $\widehat{\mathsf{q}}$, $\widehat{\mathsf{p}}$ into transformations of the Hilbert space $\mathcal{L}^2(\mathbb{R})$ of wave functions $\varphi(q)$.

The integral kernel $\mathbb{G}(q, q')$ for the operator $\widehat{\mathsf{M}}$ corresponding to $\mathbf{M}$ is then,

$$\mathbb{G}(q, q') = \pm\sqrt{\frac{k_0}{2\pi i B}} e^{i\frac{k_0}{2B}(Aq'^2+Dq^2-2qq')}, \tag{5.6.44}$$

thus evidencing that $\widehat{\mathsf{M}}$ is uniquely identified by $\mathbf{M}$ along with the choice of the sign. The phase of the square root is taken within the interval $(-\frac{\pi}{2}, \frac{\pi}{2}]$.

It is a rather common convention to write the kernel $\mathbb{G}$ without the sign, letting implied the possibility that two unitary operators, $\widehat{\mathsf{M}}(\mathbf{M})$ and $-\widehat{\mathsf{M}}(\mathbf{M})$, may correspond to every symplectic matrix $\mathbf{M}$. Evidently, the line spread function $\mathsf{g}_{\mathsf{ABCD}}(q, q')$ in (5.3.5) has been written in conformity with this convention. In the next paragraph, we illustrate an example of this sign ambiguity.

As a conclusion, we may say that, whilst an isomorphism holds between the symplectic and metaplectic algebras $sp(2,\mathbb{R})$ and $mp(2,\mathbb{R})$, the respective groups $Sp(2,\mathbb{R})$ and $Mp(2,\mathbb{R})$ are globally homomorphic (precisely, isomorphic up to a sign).

An example

Let us consider the optical Fourier transform, which is represented in ray optics by the symplectic matrix

$$\mathbf{F}(f) = \begin{pmatrix} 0 & f \\ -1/f & 0 \end{pmatrix}, \tag{5.6.45}$$

and in wave optics by the unitary operator $\widehat{\mathsf{F}}_f$, discussed in § 5.4.1.

We have seen in § 3.6.1 that the Fourier transform matrix is cyclic by 4. A cascade of four identical Fourier transformers compose to the identity; namely,

$$\mathbf{F}(f)\mathbf{F}(f)\mathbf{F}(f)\mathbf{F}(f) = \mathbf{I} \tag{5.6.46}$$

In contrast, the operator $\widehat{\mathsf{F}}_f$ is not cyclic by 4 but by 8, thus differing from the "mathematical" operator $\widehat{\mathcal{F}}$, which is indeed cyclic by 4. Evidently the different periodicity of the "optical" operator $\widehat{\mathsf{F}}_f$ with respect to $\widehat{\mathcal{F}}$ arises from the phase factor $e^{-i\frac{\pi}{4}}$, which basically distinguishes between the optical and the mathematical Fourier transforms.

In fact, propagating an optical signal $\varphi(q)$ through the cascade (5.6.46), one obtains at the intermediate planes, firstly

$$\varphi_1(q) = \widehat{\mathsf{F}}_f\,\varphi(q) = \sqrt{\frac{k_0}{if}}\widetilde{\varphi}(\frac{k_0}{f}q). \tag{5.6.47}$$

and then

$$\varphi_2(q) = \widehat{\mathsf{F}}_f\,\varphi_1(q) = -i\varphi(q). \tag{5.6.48}$$

So, optically Fourier transforming again leads to

$$\varphi_3(q) = \widehat{\mathsf{F}}_f\,\varphi_2(q) = -i\sqrt{\tfrac{k_0}{if}}\,\widetilde{\varphi}(-\tfrac{k_0}{f}q). \tag{5.6.49}$$

Finally after a further transform we end up at the exit plane with

$$\varphi_4(q) = \widehat{\mathsf{F}}_f\,\varphi_3(q) = -\varphi(q). \tag{5.6.50}$$

This means that

$$\widehat{\mathsf{F}}_f^4 = -\widehat{\mathsf{I}}, \tag{5.6.51}$$

and hence, within the formalism of the above paragraph, that

$$\widehat{\mathsf{M}}(\mathbf{F})\widehat{\mathsf{M}}(\mathbf{F})\widehat{\mathsf{M}}(\mathbf{F})\widehat{\mathsf{M}}(\mathbf{F}) = -\widehat{\mathsf{M}}(\mathbf{F}^4), \tag{5.6.52}$$

the minus sign signifying a phase shift by π of the optical wave going through the 4-Fourier transformers cascade [10.10].

Therefore, on one side relation (5.6.46) signifies that a 2π rotation in phase plane suffices to turn the ray vector into itself. On the other side, relation (5.6.51) signifies that a 4π rotation is needed to recover the identity operator $\widehat{\mathsf{I}}$, amounting to an eigth power of the metaplectic operator $\widehat{\mathsf{F}}_f$. The metaplectic group $Mp(2,\mathbb{R})$ realizes a double cover of the symplectic group $Sp(2,\mathbb{R})$ [2].

5.7 Summary

We have elaborated the wave-optical representation of the basic optical systems, identified within the ray-optics context by the quadratic monomials $\frac{1}{2}p^2$, $\frac{1}{2}q^2$ and $\frac{1}{2}qp$, in terms of unitary operators belonging to the metaplectic group $Mp(2,\mathbb{R})$. This is synthesized in the table below, which completes Fig. 5.1 paralleling the ray-optics representation of first-order optical systems in terms of unimodular 2×2 matrices $\mathbf{M}$, with the wave optics representation in terms of unitary operators $\widehat{\mathsf{M}}$, both departing from the quoted monomials:

Ray Optics $Sp(2,\mathbb{R})$	$\Leftarrow$ Quadratic Monomials as generators of $\Rightarrow$	Wave Optics $Mp(2,\mathbb{R})$
$\mathbf{T}(d) = \begin{pmatrix} 1 & d \\ 0 & 1 \end{pmatrix}$	free-medium propagation $\frac{1}{2}p^2$	$\widehat{\mathsf{T}}(d) = \sqrt{\frac{k_0}{2\pi i d}}\int dq' e^{i\frac{k_0}{2d}(q-q')^2}$
$\mathbf{L}(f) = \begin{pmatrix} 1 & 0 \\ -1/f & 1 \end{pmatrix}$	lens transfer $\frac{1}{2}q^2$	$\widehat{\mathsf{L}}(f) = e^{-i\frac{k_0}{2f}q^2}$
$\mathbf{S}(m) = \begin{pmatrix} m & 0 \\ 0 & 1/m \end{pmatrix}$	positive scaling $\frac{1}{2}qp$	$\widehat{\mathsf{S}}(m) = \frac{1}{\sqrt{m}}e^{-\ln(m)\,q\frac{\partial}{\partial q}}$

The operator representation of an arbitrary $ABCD$ system has then been obtained in the form of an integral transform whose kernel is determined by the entries of the ray-matrix of the system, thus yielding the operator

$$\mathbf{M} = \begin{pmatrix} A & B \\ C & D \end{pmatrix} \quad \rightarrow \quad \widehat{\mathsf{M}}(\mathbf{M}) = \sqrt{\frac{k_0}{2\pi i B}} \int dq' e^{i\frac{k_0}{2B}\left[Aq'^2+Dq^2-2qq'\right]},$$

with an implicit possibility that both $\widehat{\mathsf{M}}(\mathbf{M})$ and $-\widehat{\mathsf{M}}(\mathbf{M})$ are associable to each matrix $\mathbf{M}$.

This means that the symplectic and the metaplectic groups, $Sp(2,\mathbb{R})$ and $Mp(2,\mathbb{R})$, are isomporphic up to a sign.

In the forthcoming chapters we will describe the Wigner representation, which composes both the ray and the wave optics approaches to first-order optical systems, retaining the wave optical description in its kinematic structure and displaying the ray optical features in its dynamics.

Problems

1. Prove that the fractional Fourier transform of *(a)* the space-shifted function $\varphi(x - x_0)$ is

$$\widehat{\mathcal{F}}^{\alpha}\left\{\varphi(x-x_0)\right\} = e^{-ix_0(x-\frac{x_0}{2}\cos\phi)\sin\phi}\,\widetilde{\varphi}_{\alpha}(x - x_0\cos\phi),$$

and of *(b)* the frequency-shifted function $e^{-ix_0 x}\varphi(x)$ is

$$\widehat{\mathcal{F}}^{\alpha}\left\{e^{-ix_0 x}\varphi(x)\right\} = e^{ix_0(x-\frac{x_0}{2}\cos\phi)\cos\phi}\,\widetilde{\varphi}_{\alpha}(x - x_0\sin\phi),$$

both manifesting an α-dependent *shift variance.*

2. Write down the evolution law under $\widehat{\mathcal{F}}^{\alpha}$ of the Heisenberg-Weyl operators $e^{ib\widehat{\mathsf{P}}}$ and $e^{ia\widehat{\mathsf{X}}}$, a and b being real parameters, which are known to produce respectively space and frequency shifts of the wave function.

3. Prove that the fractional Fourier transform of the convolution

$$\varphi(x) = \int \varphi_1(x')\varphi_2(x-x')dx',$$

is

$$\widehat{\mathcal{F}}^{\alpha}\left\{\varphi(x)\right\} = \frac{e^{-i\frac{x^2}{2}\tan\phi}}{|\cos\phi|} \int e^{i\frac{x'^2}{2}\tan\phi}[\widehat{\mathcal{F}}^{\alpha}\varphi_2](x')\,\varphi_1(\tfrac{x-x'}{\cos\phi})dx'.$$

It obtains chirp-modulating the fractional Fourier transform of one function, convolving with the scaled version of the other function and finally multiplying by a chirp-function and a scale factor. (Hint: use the result of Problem 2 to specialize the effect of $\widehat{\mathcal{F}}^{\alpha}$ on $\varphi_2(x - x')$.

4. The complex amplitude of a Gaussian beam is typically written as

$$\varphi_G(q,z) = A(z)e^{ik\frac{q^2}{2Q(z)}}e^{-i\zeta(z)},$$

where $\zeta(z)$ denotes the Gouy phase and $Q(z)$ is the complex beam parameter, depending on the wave front radius curvature and the beam width; k is the light wavenumber in the medium. Verify that on propagating through an $ABCD$ system the Gaussian beam remains exactly Gaussian, and that specifically the complex parameter changes to

$$Q(z_o) = \frac{AQ(z_i)+B}{CQ(z_i)+D},$$

provided that the same medium is in the object and image spaces.

References

[1] Some basic titles concerned with the formalization of paraxial wave optics by a quantization-like process of paraxial ray optics are listed in [9] of ch. 4. As an alternative to the quantization procedure, the direct group theoretical approach to the parabolic wave equation is presented in

[1.1] D. Stoler, "Operator method in physical optics", J. Opt. Soc. Am. **71**, 334-339 (1981).

[1.2] H. Bacry and M. Cadilhac, "Metaplectic group and Fourier optics", Phys. Rev. A **23**, 2533-2536 (1981).

[1.3] H. Bacry, "Group theory and paraxial optics", in Proc. XIIIth Intern. Coll. on Group Theor. Meth. in Phys., W.W. Zachary (ed.) (World Sc. Publ., Singapore, 1984), pp. 215-224.

[1.4] J. Ojeda-Castañeda and A. Noyola-Isgleas, "Differential operator for scalar wave equation", J. Opt. Soc. Am. A **5**, 1605-1609 (1988).

[1.5] G. Dattoli, S. Solimeno and A. Torre, "Algebraic view of the optical propagation in a nonhomogeneous medium", Phys. Rev. A **35**, 1668-1672 (1987).

[1.6] G. Dattoli, J. C. Gallardo and A. Torre, "An algebraic view to the operatorial ordering and its applications to optics", Riv. Nuovo Cim. **11**, 11, 1-79 (1988).

[1.7] G. Dattoli and A. Torre, "A general view to Lie algebraic methods in applied mathematics, optics and transport systems for charged beam accelerators", in *Dynamical Symmetries and Chaotic Behaviour in Physical Systems*, G. Maino, L. Fronzoni and M. Pettini (eds.) (World Sc. Publ., Singapore, 1991), pp. 21-89.

[2] For a basic treatment of the symplectic group in the general context of the group theory see [9] of ch. 1; specific investigations are in [11] of ch. 2. Studies on the metaplectic representation, firstly introduced by Weil [2.1], and on its relation to the symplectic geometry are presented in the titles below. We signalize the enlightening presentations in [2.6], and, within the optical context, in [2.4] and [2.8], also evidencing the wave-optics/quantum mechanics correspondence.

[2.1] A. Weil, "Sur certains groupes d'opérateurs unitaires", Acta Math. **111**, 143-211 (1964); also in *Collected Papers*, Vol. 3 (Springer-Verlag, Heidelberg, 1980), pp. 1-69.

[2.2] S. Sternberg, "Some recent results on the metaplectic representation", in *Group Theoretical Methods in Physics*, P. Kramer and A. Rieckers (eds.) (Springer-Verlag, Berlin, 1978), pp. 117-143.

[2.3] V. Guillemin and S. Sternberg, "The metaplectic representation, Weyl operators, and spectral theory", J. Functional Anal. **42**, 129-225 (1981).

[2.4] V. Guillemin and S. Sternberg, *Symplectic Techniques in Physics* (Cambridge Un. Press, New York, 1984), ch. 1, pp. 1-104.

[2.5] M. García-Bullé, W. Lassner and K.B. Wolf, "The metaplectic group within the Heisenberg-Weyl ring", J. Math. Phys. **27**, 29-36 (1986).

[2.6] R. G. Littlejohn, "The semiclassical evolution of wave packets", Phys. Rep. **138**, 198-291 (1986).

[2.7] R. Simon and N. Mukunda, "The two-dimensional symplectic and metaplectic groups and their universal cover", in *Symmetries in Science VI*, B. Gruber (ed.) (Plenum Press, New York, 1993), pp. 659-689.

[2.8] R. Simon and N. Mukunda, "Optical phase space, Wigner representation, and invariant quality parameters", J. Opt. Soc. Am. A **17**, 2440-2463 (2000).

[2.9] H. ter Morsche and P.J. Oonincx, "On the integral representations for metaplectic operators", J. Fourier Anal. & Appl. **8**, 245-257 (2002).

[3] For the diffraction theory the reader is directed to [3] of ch. 4, and to [6] below.

[4] The link between the ray-matrix and wave-operator formulations of first-order optics through the Huygens-Fresnel integral, firstly recognized by Collins [4.1], is clearly elucidated in the textbook by Siegman [4.2].

[4.1] S.A. Collins, "Diffraction-integral written in terms of matrix-optics", J. Opt. Soc. Am. **60**, 1168-1177 (1970).

[4.2] A.E. Siegman, *Lasers* (Univ. Science Book, Mill Valley, 1986), ch. 20, pp. 777-782.

[5] For a discussion of the concept of duality in relation with optical systems, see

[5.1] A. Papoulis, "Dual optical systems", J. Opt. Soc. Am. **58**, 653-654 (1968).

[5.2] H.J. Butterweck, "General theory of linear, coherent, optical data-processing systems", J. Opt. Soc. Am. **67**, 60-70 (1977).

[5.3] H.J. Butterweck, "Principles of optical data-processing',' in *Progress in Optics*, Vol. XIX, E. Wolf (ed.) (North-Holland, Amsterdam, 1981), ch. 4, pp. 211-280.

[5.4] A. Lohmann, "Ein neues Dualitätsprinzip in der Optik',' Optik **11**, 478-488 (1954). For the English translation, see: A. Lohmann, "Duality in optics',' Optik **89**, 93-97 (1992).

[6] For a mathematics- and/or physics-oriented treatment of the Fourier transform the reader may consult the textbooks in [15] and [16] of ch. 4. In particular, for a discussion of the Fresnel and Fraunhofer diffraction regimes see:

[6.1] J.W. Goodman, *Introduction to Fourier Optics* (McGraw-Hill, New York, 1968), ch. 4, pp. 57-76.

[6.2] A. Papoulis, *Systems and Transforms with Applications in Optics* (McGraw-Hill, New York, 1968), ch. 9, pp. 315-345.

[6.3] J.D. Gaskill, *Linear Systems, Fourier Transforms and Optics* (Wiley & Sons, New York, 1978), pp. 361-385.

[7] An essential bibliography about the fractional Fourier transform is in [8] of ch. 3. We quote below the paper by Namias, in which the fractional Fourier transform is introduced through an eigenvalue equation like (5.4.42), and those by Agarwal and Simon, and Dattoli, Torre and Mazzacurati, the former dealing with an interpretation of the fractional Fourier transform in terms of the time evolution operator of the quantum oscillator, and the latter with an extension of such an interpretation to an in general quadratic Hamiltonian. Papers concerned with fractional Fourier transforms of complex order and generalized forms of the fractional Fourier transform are also suggested. For a discussion of the fractional Fourier transform within the general context of real and complex linear canonical transforms, see [10].

[7.1] S.V. Namias, "The fractional order Fourier transform and its application to quantum mechanics", J. Inst. Maths. Applics. **25**, 241-265 (1980).

[7.2] G.S. Agarwal and R. Simon, "A simple realization of fractional Fourier transformation and relation to harmonic oscillator Green's function',' Opt. Comm. **110**, 23-26 (1994).

[7.3] G. Dattoli, A. Torre and G. Mazzacurati, "An alternative point of view to the theory of fractional Fourier transform", IMA J. Appl. Math. **60**, 215-224 (1998).

[7.4] C.-C. Shih, "Optical interpretation of a complex-order Fourier transform", Opt. Lett. **20**, 1178-1180 (1995).

[7.5] L.M. Bernardo and O.O.D. Soares, "Optical fractional order Fourier transforms with complex orders", Appl. Opt. **35**, 3163-3166 (1996).

[7.6] J. Hua, L. Liu and G. Li, "Extended fractional Fourier transform", J. Opt. Soc. Am. A **14**, 3316-3322 (1997).

[8] A Gaussian beam related bibliography can be found in [13] of ch. 2.

[9] For a discussion of the Ehrenfest theorem within the quantum mechanical context see the textbooks in [10] of ch. 4. The presentation in § 5.5 is inspired by that in

[9.1] D. Marcuse, *Light Transmission Optics* (Van Nostrand-Reinhold, Princeton, 1972), ch. 3, pp. 107-109.

[10] Following the suggestion in [10.1], the presentation in § 5.6 rephrases within the optical context the general considerations concerning the integral transform aspects of the construction of unitary representations of linear canonical transformations of the Heisenberg-Weyl algebra operators $\widehat{\mathsf{q}}$, $\widehat{\mathsf{p}}$ on the Hilbert space $\mathcal{L}^2(\mathbb{R})$ of square-integrable functions. In this connection, we suggest the titles below. See also [2.4]. An enlightening explanation of the π-phase shift suffered by a light wave passing through a conjugate point is given in [10.10].

[10.1] M. Nazarathy and J. Shamir, "First-order optics - a canonical operator representation: lossless systems", J. Opt. Soc. Am. **72**, 356-364 (1982).

[10.2] L. Infeld and J. Plebanski, "On a certain class of unitary transforamtions",Acta Phys. Polon. **14**, 41-75 (1954).

[10.3] V. Bargmann, "On a Hilbert space of analytic functions and an associated integral transform. I", Comm. Pure & Appl. Math. **14**, 187-214 (1961); II. A family of related function spaces application to distribution theory, ibid., **20**, 1-101 (1967).

[10.4] M. Moshinsky and C. Quesne, "Linear canonical transformations and their unitary representations", J. Math. Phys. **12**, 1772-1780 (1971); C. Quesne and M. Moshinsky, "Canonical transformations and matrix elements", *ibid.*, 1780-1783.

[10.5] M. Moshinsky, "Canonical transformations and quantum mechanics", SIAM J. Appl. Math. **25**, 1772-1780 (1973).

[10.6] K.B. Wolf, "Canonical transforms. I. Complex linear transforms", J. Math. Phys. **15**, 1295-1301 (1974); II. Complex radial transforms", *ibid.*, 2102-2111.

[10.7] P. Kramer, M. Moshinsky and T.H. Seligman, "Complex extensions of canonical transformations and quantum mechanics", in *Group Theory and Its Applications*, Vol. 3, E.M. Loebl (ed.) (Academic Press, New York, 1975), pp. 249-332.

[10.8] K.B. Wolf, *Integral Transforms in Science and Engineering* (Plenum Press, New York, 1979), ch. 9, pp. 381-416.

[10.9] K.B. Wolf, *Geometric Optics On Phase Space* (Springer, Berlin, 2004), pp. 277-282.

[10.10] V. Guillemin and S. Sternberg, *Geometric Asymptotics* (Am. Math. Soc., Providence, 1977), ch. 1, pp. 1-19.

6

The Wigner Distribution Function: Analytical Evaluation

6.1 Introduction

The Wigner distribution function was originally introduced by Wigner [1] as the *simplest* quantum analogue of the classical phase space distribution function permitting to find probabilities and expectation values of quantum operators using as much classical language and methods as allowed. The Wigner method establishes a rule to associate a c-number function in phase-space with every operator being a function of position and momentum operators. Thus, the expectation values of quantum mechanical observables can be calculated in the same mathematical form as the averages of the classical statistical mechanics rather than through the operator formalism of quantum mechanics.

As an illustrative example, we consider a particle moving along a line.

In classical mechanics the dynamical state of the particle at any instant of time is fully described by the position $q(t)$ and the momentum $p(t)$ of the particle [2]. The motion of the particle is generated by the classical Hamiltonian

$$H(q,p) = \frac{p^2}{2m} + V(q), \tag{6.1.1}$$

according to which the classical equations of motion obtain as

$$\frac{dq}{dt} = \frac{p}{m}, \quad \frac{dp}{dt} = -V'(q) \tag{6.1.2}$$

m denoting the mass of the particle and $V(q)$ the potential it experiences ($V'(q) \equiv dV/dq$). If the initial conditions $q_0 = q(0)$ and $p_0 = p(0)$ are known, the evolution of the state of the particle is uniquely determined by the solution $(q(t),\, p(t))$ of (6.1.2), which individualizes as well a definite trajectory in the particle phase-space.

It may happen however that the initial state of the particle be not specified through the definite values q_0 and p_0 of the position and momentum of the particle at $t = 0$, but rather through the joint probability distribution of q and p at $t = 0$. In that case, the dynamics of the particle will not be determined by Hamilton's equations (6.1.2) for $q(t)$ and $p(t)$, but by Liouville's equation for the phase-space distribution function $\rho_{\mathrm{C}}(q, p; t)$,

$$\frac{\partial}{\partial t}\rho_{\mathrm{C}}(q, p; t) = \{\rho_{\mathrm{C}}, H\} = -\frac{p}{m}\frac{\partial}{\partial q}\rho_{\mathrm{C}} + V'(q)\frac{\partial}{\partial p}\rho_{\mathrm{C}}. \tag{6.1.3}$$

Clearly, $\rho_{\mathrm{C}}(q, p; t)$ means the probability density for the particle to be at the phase-space point (q, p) at time t. Hence, $\rho_{\mathrm{C}}(q, p; t) \geq 0$ and $\iint \rho_{\mathrm{C}}(q, p; t)dqdp = 1$ at any t, the integration being over the inherent phase space of the particle. If the probability distribution of the dynamical variables q and p at the initial time is known: $\rho_{\mathrm{C}}(q, p; 0) = \rho_0(q, p)$, the probability distribution $\rho_{\mathrm{C}}(q, p; t)$ at any subsequent time is uniquely determined by the solution of Eq. (6.1.3).

In other words, in classical mechanics the information contained into the canonically conjugate variables q and p, representing the particle state as a point in phase-space, can be transferred into the distribution function $\rho_{\mathrm{C}}(q, p; t)$, representing the probability for every point in phase space to be a possible state for the particle. Accordingly, the evolution of the particle state can uniquely be accounted for by the motion of the phase plane point $(q(t), p(t))$, or by the evolution of the distribution function $\rho_{\mathrm{C}}(q, p; t)$.

We note that Liouville's equation has been derived in § 2.2.1 within the context of the statistical ensemble description of the propagation of a bundle of light rays, the relevant dynamical function $\rho(q, p; z)$ being interpreted as the phase-space ray density at every reference z plane. In effect, $\rho(q, p; z)dqdp$ yields the number of rays at z in the elemental volume $dqdp$ around the phase-space point (q, p); hence at each z plane $\iint_{\mathcal{V}} \rho(q, p; z)dqdp = \mathcal{N}$ (the total number of rays in the beam), the integration involving the invariant phase space representative volume $\mathcal{V}$ of the bunch of rays. It is evident that, within the mechanical-optical basic correspondence of the temporal coordinate t to the axial coordinate z, the distribution functions $\rho_{\mathrm{C}}(q, p; t)$ and $\rho(q, p; z)$ are equivalent, apart from the relative normalization settings. The probabilistic picture inherent in $\rho_{\mathrm{C}}(q, p; t)$ implicitly involves, in fact, the ensemble picture of a huge number of identical systems, i.e. particles with mass m, distributed at each time t over all the possible dynamical states (q, p) according to the relevant weighting function $\rho_{\mathrm{C}}(q, p; t)$.

Equation (6.1.3) corresponds to (6.1.2) when a statistical ensemble picture of the system dynamics is concerned. Accordingly, the average value of any dynamical function $f(q, p)$ is calculated by the integral

$$\langle f \rangle_{\mathrm{C}}(t) = \iint \rho_{\mathrm{C}}(q, p; t) f(q, p)dqdp, \tag{6.1.4}$$

the integration involving the entire phase space domain pertaining to the system [2].

Integral (6.1.4) formally expresses the nature of the statistical description of classical mechanics, where the mean values of dynamical variables are obtained by averaging over uniquely determined processes. This clearly mirrors the fact that in classical mechanics the position and momentum of the particle as well as any function of q and p, and hence the probability distribution $\rho_C(q, p; t)$, can exactly be known at any time, the relevant time evolution being uniquely determined by deterministic laws of evolution; the whole uncertainty in classical mechanics is indeed contained in the form of the initial probability distribution of the dynamical variables specifying the given system.

In contrast, in quantum mechanics the simultaneous specification of the position and momentum of the particle with arbitrary precision is impossible. The dynamical state of the particle is described by specifying at each time t the wave function $\psi(q, t)$, which determines the probability density of the position coordinate according to [3]

$$\mathfrak{Q}_Q(q, t) = |\,\psi(q, t)\,|^2, \tag{6.1.5}$$

and separately the probability density of the momentum coordinate through the Fourier transform $\widetilde{\psi}(p, t)$ according to

$$\mathfrak{P}_Q(p, t) = |\,\widetilde{\psi}(p, t)\,|^2. \tag{6.1.6}$$

In addition, the knowledge of $\psi(q, t)$ enables us to find the expectation value of any observable $\widehat{f}(\widehat{\mathsf{q}}, \widehat{\mathsf{p}})$ in the state ψ by the scalar product

$$\langle\,\widehat{f}\,\rangle_Q(t) \equiv \langle\,\psi(q, t)|\,\widehat{f}(\widehat{\mathsf{q}}, \widehat{\mathsf{p}})\,|\,\psi(q, t)\,\rangle = \int \psi^*(q, t)\widehat{f}(\widehat{\mathsf{q}}, \widehat{\mathsf{p}})\psi(q, t)dq. \tag{6.1.7}$$

In particular, if $\widehat{f}$ depends only on the coordinate operator $\widehat{\mathsf{q}}$, i.e., $\widehat{f}(\widehat{\mathsf{q}}, \widehat{\mathsf{p}}) = \widehat{h}(\widehat{\mathsf{q}})$, we have

$$\langle\,\widehat{f}\,\rangle_Q(t) = \int \mathfrak{Q}_Q(q, t)h(q)dq, \tag{6.1.8}$$

whereas, if $\widehat{f}$ depends only on the momentum operator $\widehat{\mathsf{p}}$, namely: $\widehat{f}(\widehat{\mathsf{q}}, \widehat{\mathsf{p}}) = \widehat{g}(\widehat{\mathsf{p}})$, one obtains

$$\langle\,\widehat{f}\,\rangle_Q(t) = \int \mathfrak{P}_Q(p, t)g(p)dp, \tag{6.1.9}$$

where $h(q)$ and $g(p)$ are the classical functions corresponding to the quantum operators $\widehat{h}(\widehat{\mathsf{q}})$ and $\widehat{g}(\widehat{\mathsf{p}})$.

The evolution of the dynamical state of the particle is determined by the Schrödinger equation for the wave function $\psi(q, t)$,

$$i\hbar\frac{\partial}{\partial t}\psi(q, t) = \widehat{\mathsf{H}}(\widehat{\mathsf{q}}, \widehat{\mathsf{p}})\psi(q, t), \tag{6.1.10}$$

with $\widehat{\mathsf{H}}(\widehat{\mathsf{q}},\widehat{\mathsf{p}}) = \widehat{\mathsf{p}}^2/2m + \widehat{V}(\widehat{\mathsf{q}})$ being the quantum Hamiltonian operator corresponding to the classical function (6.1.1) [3].

As noted above, a classical joint probability distribution $\rho_{\mathrm{C}}(q,p;t)$ can be devised whose transformation with time is ruled by Liouville's equation (6.1.3), thus allowing for a fully probabilistic-like approach to classical dynamical problems without explicitly resorting to the description in terms of the position and momentum coordinates, and hence to Hamilton's equations.

It is natural to investigate the possibility of elaborating a classical-like description of the dynamical state of the quantum system (the point particle in the present discussion) and its evolution without explicitly resorting to the wave function, and hence to the Schrödinger equation. In fact, despite the impossibility of measuring position and momentum simultaneously and precisely, one can legitimately think of the simultaneous and precise values of position and momentum as existing entities, and accordingly conceive a quantum phase space, where a "probability" distribution for any pair of q and p values could be defined and used in a classical-like fashion to characterize and predict the quantum state and its evolution (Fig. 6.1). In other words one can legitimately question about the possibility of defining a joint probability distribution $\rho_{\mathrm{Q}}(q,p;t)$, which gives the probability that the coordinate and momentum of the particle have the values q and p at time t. If such a distribution function could be defined to correspond to the state $\psi(q,t)$, then the expectation value of any operator function of $\widehat{\mathsf{q}}$ and $\widehat{\mathsf{p}}$ in that state, typically calculated by (6.1.7), could as well be obtained by a (q,p) integration similar to the classical expression (6.1.4); namely,

$$\langle \widehat{f} \rangle_{\mathrm{Q}}(t) = \iint \rho_{\mathrm{Q}}(q,p;t) f(q,p)\, dq\, dp, \tag{6.1.11}$$

where $f(q,p)$ should represent the classical function corresponding to the quantum observable $\widehat{f}(\widehat{\mathsf{q}},\widehat{\mathsf{p}})$, just as $h(q)$ and $g(p)$ in (6.1.8) and (6.1.9) in relation with $\widehat{h}(\widehat{\mathsf{q}})$ and $\widehat{g}(\widehat{\mathsf{p}})$. Remarkably, there is no ambiguity in the correspondence between the quantum operators $\widehat{h}(\widehat{\mathsf{q}})$ and $\widehat{g}(\widehat{\mathsf{p}})$ and the classical functions $h(q)$ and $g(p)$, as they respectively depend only on $\widehat{\mathsf{q}}$ and $\widehat{\mathsf{p}}$. In contrast, the correspondence between the quantum function $\widehat{f}(\widehat{\mathsf{q}},\widehat{\mathsf{p}})$ of the non-commuting operators $\widehat{\mathsf{q}}$ and $\widehat{\mathsf{p}}$ and the classical function $f(q,p)$ of the commuting variables q and p is not unique. Classical mechanics operates with algebraic functions, whereas quantum mechanics works with operators, and there is no unique rule to translate from one mathematical context to the other. If the operators $\widehat{\mathsf{q}}$ and $\widehat{\mathsf{p}}$ are indiscriminately replaced by the scalars q and p, different operator functions $\widehat{f}(\widehat{\mathsf{q}},\widehat{\mathsf{p}})$ come to correspond to the same c-function $f(q,p)$. The ambiguity is of the order of $\hbar$ (see Problem 1 and § 8.2.3). We may therefore guess that phase space distributions in quantum mechanics are not unique and that their definition is closely connected with the correspondence rule set up

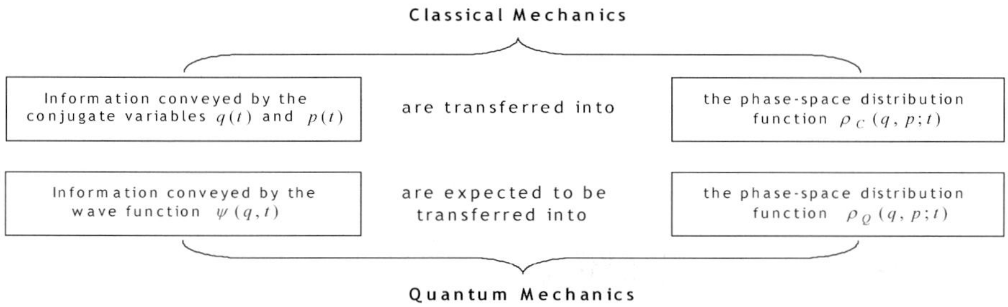

FIGURE 6.1. Paralleling the mapping of the Hamiltonian view into the Liouvillian view of classical mechanics, which transfers the information about the system state from the canonically conjugate variables $q(t)$, $p(t)$ into the phase-space distribution function $\rho_{\mathrm{C}}(q,p;t)$, the phase-space formulation of quantum mechanics is expected to map the Schrödinger view into the Wigner view, transferring the information about the system state conveyed by the wave function $\psi(q,t)$ into a suitably defined phase-space distribution function $\rho_{\mathrm{Q}}(q,p;t)$.

between quantum observables and classical functions.

It is of course desirable that $\rho_{\mathrm{Q}}(q,p;t)$ be in the classical limit the phase-space distribution function $\rho_{\mathrm{C}}(q,p;t)$. Also, in view of its role as a joint probability density, it should be real and nonnegative for all values of q and p:

$$\rho_{\mathrm{Q}}(q,p;t) \geq 0, \tag{6.1.12}$$

and give the proper probabilities for the coordinate q and momentum p, if integrated respectively over p and q:

$$\int \rho_{\mathrm{Q}}(q,p;t)dp = |\,\psi(q,t)\,|^2, \qquad \int \rho_{\mathrm{Q}}(q,p;t)dq = |\,\widetilde{\psi}(p,t)\,|^2. \tag{6.1.13}$$

The integrals above are known in the probability theory as the *marginal* distributions of the given (q,p) distribution $\rho_{\mathrm{Q}}(q,p;t)$.

Further conditions are usually required to be satisfied by $\rho_{\mathrm{Q}}(q,p;t)$, which however do not lead to a unique determination of ρ_{Q}. The Wigner distribution function is one of the possible distributions $\rho_{\mathrm{Q}}(q,p;t)$, that can be elaborated to satisfy the required conditions at a large extent. It benefits from the greatest mathematical simplicity and elegance. It is typically used as a joint probability density, although a strict interpretation in this sense is not possible as it may in general take on negative values in some regions of the quantum phase space, thus violating the usual axioms of probability theory[(a)] .

[a] The reader interested in a discussion of phase-space distribution functions in the light of the instance for *extended probabilities*, which are not necessarily positive, may consult the stimulating review: W. Muckenheim, "A review of extended probabilities", Phys. Rep. **133**, 337-401 (1986).

As already noted, the Wigner distribution function was first introduced by Wigner for a system of many particles in order to study the quantum corrections to thermodynamics functions and then applied advantageously to a large variety of quantum mechanical problems [4]. It was reintroduced by Ville in the signal analysis [5] and more recently by Bastiaans [6] in the optical context where it found wide applications within both the space-frequency and time-frequency domains. Naturally the space-frequency (as well time-frequency) Wigner distribution function in optics (as well in general in signal theory) is mathematically identical to the quantum Wigner distribution function of position and momentum. The quantum mechanical results have suitably been rephrased within the optical context, with the signal amplitude and relevant frequency spectrum replacing the position and momentum quantum wave functions. The interpretation of the optical Wigner distribution function in terms of the signal energy distribution has also been attempted paralleling that of the position-momentum distribution of the quantum theory in terms of a probability distribution. As in the case of the quantum theory, however, the strict interpretation of the Wigner function as an energy distribution is not possible as it in general takes on negative values in the phase space.

The Wigner distribution function represents a particularly useful tool in quantum optics as well. The Wigner-space picture yields an intuitive visualization of the dynamics of coherent and squeezed states, which can be generated from the vacuum state by phase-space rotation, translation and squeeze operators, all having their geometrical ray-optics analogs [7].

It has been already emphasized the analogy between light and magnetic optics, which provides the foundations for sharing both concepts and methods. Accordingly the application of the Wigner picture to magnetic optics has also been suggested, as, for instance, in relation to the description and analysis of the optical characteristics of synchrotron radiation sources [8].

Section 6.2 introduces the formal definition of the *optical* Wigner distribution function, whose properties are described in detail in Sect. 6.3. Several examples are illustrated in Sect. 6.4 in order to create the feeling for the hybrid nature of the Wigner function, which inherently incorporates, and hence can display, the basic features of the ray-optics description of light signals within and through the seemingly irreconcilable context and methods of wave optics.

The above discussion has had a pure illustrative purpose. We considered in fact the simple case of a system with one degree of freedom in a pure state. More in general a system with many degrees of freedom and in a mixed state may be considered, thus demanding for a certain number of pairs of canonically conjugated variables as well as for the density matrix formalism, for which the interested reader is addressed to [1.1, 4, 7]. Finally, we note that, although our discussion has concerned with the position and momentum operators, the Wigner representation can as well be elaborated for any pair of complete sets of mutually incompatible Hermitian operators [9].

6.2 The optical Wigner distribution function: basic concepts

We consider a time-harmonic $1D$ optical signal, represented in the space and spatial frequency domains, respectively, by the complex amplitude $\varphi(q,z)$ and the respective Fourier transform $\widetilde{\varphi}(\kappa,z)$. Linked by the reciprocal Fourier relations (§ 4.3.1.2),

$$\begin{aligned}\varphi(q,z) &= \tfrac{1}{\sqrt{2\pi}}\int \widetilde{\varphi}(\kappa,z)e^{i\kappa q}d\kappa,\\ \widetilde{\varphi}(\kappa,z) &= \tfrac{1}{\sqrt{2\pi}}\int \varphi(q,z)e^{-i\kappa q}dq,\end{aligned} \tag{6.2.1}$$

$\varphi(q,z)$ and $\widetilde{\varphi}(\kappa,z)$ are understood as square-integrable complex-valued functions of the Fourier conjugate variables q and κ for any value of the real parameter z. We will also consider signals, which are not normalizable in the common sense of the energy-like integral, i.e., by the requirement that $\int|\varphi(q,z)|^2\,dq < \infty$. As noted in § 4.3.2, such signals may be handled as limiting processes of normalizable signals or by resorting to the generalized function method. The impulse and the pure harmonic signal, respectively represented by the delta and the complex exponential, are prime examples of such idealized signals.

The Wigner distribution function $\mathcal{W}(q,\kappa)$ of the $1D$ optical signal $\varphi(q)$ is defined as a function of the Fourier conjugate variables q and κ by the integral in the space domain [6]

$$\mathcal{W}(q,\kappa) = \tfrac{1}{2\pi}\int \varphi^*(q-\tfrac{q'}{2})\varphi(q+\tfrac{q'}{2})e^{-i\kappa q'}dq'. \tag{6.2.2}$$

A basically identical integral in the frequency domain, which then involves the signal spectrum $\widetilde{\varphi}(\kappa)$, is also given as an alternative definition, i.e.

$$\mathcal{W}(q,\kappa) = \tfrac{1}{2\pi}\int \widetilde{\varphi}^*(\kappa-\tfrac{\kappa'}{2})\widetilde{\varphi}(\kappa+\tfrac{\kappa'}{2})e^{i\kappa' q}d\kappa'. \tag{6.2.3}$$

As usual all the integral in the chapter without limits of integration are intended from $-\infty$ to $+\infty$.

The equivalence of the integrals (6.2.2) and (6.2.3) can easily be proved. In fact, expressing the amplitudes $\varphi^*(q-\frac{q'}{2})$ and $\varphi(q+\frac{q'}{2})$ in (6.2.2) in terms of the signal spectrum according to (6.2.1), we obtain

$$\mathcal{W}(q,\kappa) = \tfrac{1}{2\pi}\iint \widetilde{\varphi}^*(\kappa')\widetilde{\varphi}(\kappa'')e^{i(\kappa''-\kappa')q}[\tfrac{1}{2\pi}\int e^{iq'(\frac{\kappa'+\kappa''}{2}-\kappa)}dq']d\kappa' d\kappa''. \tag{6.2.4}$$

As the bracketed expression above turns to be $\delta[\frac{\kappa'+\kappa''}{2}-\kappa]$, the double integral over frequency simplifies in the form (6.2.3). Inversely, using the Fourier integral for $\widetilde{\varphi}$ in (6.2.3), we regain the q-domain definition (6.2.2).

The dependence of the signal amplitude φ and the Fourier transform $\tilde{\varphi}$ on the axial coordinate z is implied in the integrals (6.2.2) and (6.2.3), which are understood as performed at $z = const.$ planes. Hence the Wigner distribution function depends on z as well, but this dependence will in general not be displayed. Also, a suitable subscript, referring to the specific signal at hand, will be attached to the basic symbol $\mathcal{W}(q, \kappa)$, when needed for clarity's sake.

There is an evident formal similarity between the space and frequency domain definitions (6.2.2) and (6.2.3), so that one can be determined from the other by interchanging the roles of q and κ. This feature is in conformity with the desired hybrid nature of the Wigner distribution function, which is expected to describe the signal in space and frequency simultaneously.

It may be instructive to give a closer look at the symmetric product $\varphi^*(q - \frac{q'}{2})\varphi(q + \frac{q'}{2})$, which is central to the evaluation of the Wigner integral. To help visualization we consider a real and nonnegative signal. It is then apparent that the Wigner distribution function at point q is obtained by adding up terms consisting of the product of the signal amplitudes at points symmetrically disposed on the left and on the right of q. With this view in mind, after the manner of [10.2, 5.5, 5.7], we can gain a visual account of some general properties of the Wigner distribution function inspecting the simple waveforms, depicted in Fig. 6.2. Both signals extend over the finite interval (Q_a, Q_b), one signal showing a gap as the other a sinusoidal perturbance for a short interval over an otherwise uniform amplitude. Thinking in optical terms, the waveforms depicted may be thought of as obtained by a uniform illumination of two constant amplitude transmittance slits, spatially separated by an opaque section in Fig. 6.2.a) and by a sinusoidal amplitude transmittance grating in Fig. 6.2.b). Evidently, the Wigner distribution functions of both signals vanish outside the interval (Q_a, Q_b). In fact, at any point $q \leq Q_a$ the signal on the right overlaps over the zero signal on the left, and similarly at any point $q \geq Q_b$ the signal on the left overlaps over the zero signal on the right, thus yielding in both cases no contribution to the integral (6.2.2). On the other hand, in regard to the waveform of Fig. 6.2.a), we can see that the Wigner distribution function is not zero through the whole gap, i.e., for the interval (Q_c, Q_d). At point q_1, for instance, the amplitudes outside the interval (Q_c, Q_d) contribute a non zero overlap to the Wigner integral (6.2.2). In fact, as a general property, the Wigner distribution function is limited to the same interval of the signal, but is not necessarily zero at points where the signal vanishes. Similarly, the harmonic perturbance appearing in the waveform of Fig. 6.2.b) is seen to yield no contribution to the Wigner integral at the sampled points q_1 and q_3, where the sinusoid overlaps over the vanishing, respectively, left and right part of the signal. In contrast, it contributes to the Wigner integral at point q_2, having here a nonzero overlap with the signal amplitude at the left.

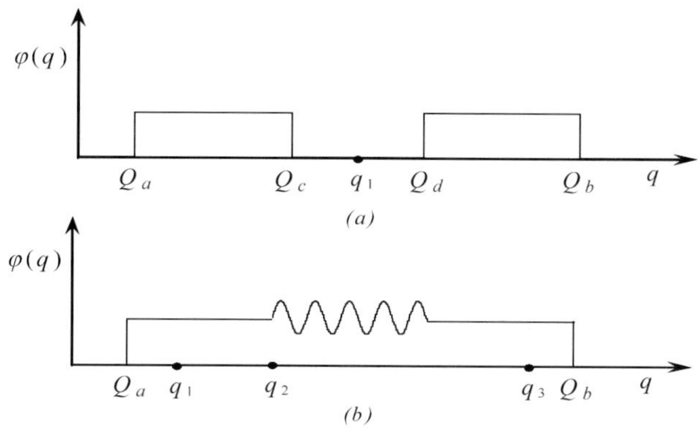

FIGURE 6.2. *(a)* The Wigner distribution function of the depicted signal does not vanish throughout the gap extending from Q_c to Q_d. The left and right part of the signal about the point q_1, for instance, have a non zero overlap, thus yielding a non zero value of $\mathcal{W}$. *(b)* The sinusoidal perturbance extending through the short interval (Q_c, Q_d) does not contribute to the Wigner function at the points q_1 and q_3, whilst it does at the point q_2. (Inspired by [5.7])

Of course, the same considerations hold for the frequency domain, due to the symmetry between the space and frequency domain definitions.

Interestingly, by an easy manipulation, expressions (6.2.2) and (6.2.3) are recast in the equivalent forms

$$\begin{aligned}\mathcal{W}(q,\kappa) &= \tfrac{1}{\pi}\int \varphi^*(2q-q')\varphi(q')e^{-2i\kappa(q'-q)}dq',\\ \mathcal{W}(q,\kappa) &= \tfrac{1}{\pi}\int \widetilde{\varphi}^*(2\kappa-\kappa')\varphi(\kappa')e^{2iq(\kappa'-\kappa)}d\kappa',\end{aligned} \tag{6.2.5}$$

clearly evidencing the convolution nature of the Wigner integrals. In particular,

$$\begin{aligned}\mathcal{W}(q,0) &= \tfrac{1}{\pi}\int \varphi^*(2q-q')\varphi(q')dq',\\ \mathcal{W}(0,\kappa) &= \tfrac{1}{\pi}\int \widetilde{\varphi}^*(2\kappa-\kappa')\varphi(\kappa')d\kappa',\end{aligned} \tag{6.2.6}$$

saying that the profiles of the Wigner distribution function on the q and κ axes equal (apart from $\frac{1}{\pi}$) the autoconvolutions of the signal and its spectrum.

The Wigner distribution function can equally be defined as a function of the canonically conjugate variables q and p. In fact, being $\kappa = k_0 p$ (as usual, k_0 denotes the vacuum wave number), the integrals in (6.2.2) and (6.2.3) can be recast in terms of p and $\widetilde{\varphi}(p)$ according to (see also Eq. (4.6.3))

$$\begin{aligned}\mathcal{W}(q,p) &= \tfrac{1}{2\pi}\int \varphi^*(q-\tfrac{q'}{2})\varphi(q+\tfrac{q'}{2})e^{-ik_0pq'}dq'\\ &= \tfrac{k_0}{2\pi}\int \widetilde{\varphi}^*(p-\tfrac{p'}{2})\widetilde{\varphi}(p+\tfrac{p'}{2})e^{ik_0p'q}dp',\end{aligned} \tag{6.2.7}$$

where $\varphi(q)$ and $\widetilde{\varphi}(p)$ are linked by the Fourier integrals (4.6.3).

We will mainly use definitions (6.2.2) and (6.2.3) in terms of the pair (q, κ).

6.2.1 *The Wigner distribution of light signals: simple examples.*

We give a primary illustration of the valence of the Wigner distribution function to convey an intermediate signal description between the pure space and pure spatial-frequency descriptions. To this end, we consider some simple examples; additional examples are discussed in § 6.4.

Firstly we derive the Wigner distribution functions of the prime Fourier pair, i.e. the point-like and the pure harmonic signals. The q-representation of a point source located at the point q_0 is provided by the δ-function,

$$\varphi_{q_0}(q) = \delta(q - q_0), \tag{6.2.8}$$

whose Wigner distribution function $\mathcal{W}_{q_0}(q, \kappa)$ comes to be a δ-function as well:

$$\mathcal{W}_{q_0}(q, \kappa) = \tfrac{1}{2\pi}\delta(q - q_0), \tag{6.2.9}$$

independent of κ and fully concentrated along the line parallel to the κ-axis at $q = q_0$; all the frequencies are present only at this point, with no contributions from other points. This exactly corresponds to the ray-optics picture of a point source, the light rays emanating from the position q_0 in all directions. The relevant frequency spectrum $\widetilde{\varphi}(\kappa) \propto e^{-i\kappa q_0}$ evidences that all the frequencies are present in the signal, whereas the distribution $\mathcal{W}_{q_0}(q, \kappa)$ adds the precise information about where the frequencies are present, i.e. at $q = q_0$.

Correspondingly, the spatial harmonic

$$\varphi_{\kappa_0}(q) = \tfrac{1}{\sqrt{2\pi}} e^{i\kappa_0 q}, \tag{6.2.10}$$

is described in ray optics as a collection of parallel rays all propagating in the direction $p_0 = \kappa_0/k_0$, whichever be the spatial coordinate. The Wigner distribution function $\mathcal{W}_{\kappa_0}(q, \kappa)$ is easily obtained as

$$\mathcal{W}_{\kappa_0}(q, \kappa) = \tfrac{1}{2\pi}\delta(\kappa - \kappa_0), \tag{6.2.11}$$

which reproduces the ray picture of the optical disturbance (6.2.10), and accordingly individualizes on the (q, κ) plane a straight line parallel to the q-axis through the point $\kappa = \kappa_0$. The frequency spectrum $\widetilde{\varphi}_{\kappa_0}(\kappa) = \delta(\kappa - \kappa_0)$ reveals that only the frequency κ_0 is present in the signal, whilst the function $\mathcal{W}_{\kappa_0}(q, \kappa)$ provides the further information that κ_0 is present at all points.

It is also interesting to consider a spherical wave, propagating in a uniform medium. As noted in § 5.2.1, a spherical wave is represented in the linear

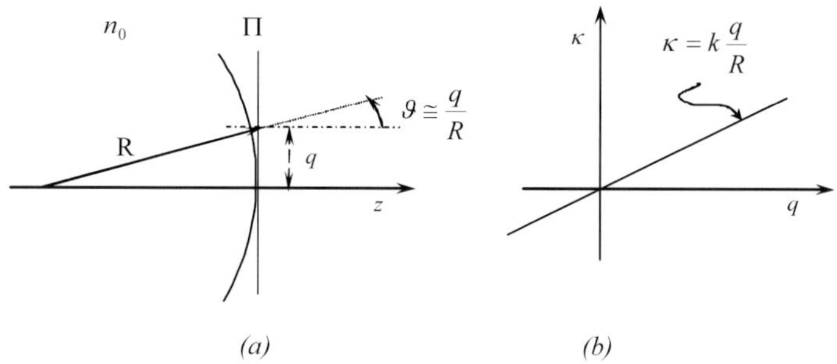

FIGURE 6.3. The Wigner distribution function of *(a)* a spherical wave paraxially crossing the reference plane Π at q into the direction $\vartheta \simeq q/R$ is *(b)* a line delta, whose trace in the Wigner plane (q, κ) is the straight line $\kappa = kq/R$.

planar model by the quadratic-phase waveform

$$\varphi_s(q) = e^{i\frac{k}{2R}q^2}, \tag{6.2.12}$$

where R denotes the curvature radius of the parabolic wavefront at the z-plane, which $\varphi_s(q)$ is referred to, and $k = n_0 k_0$ is the wave number in the medium. Carrying on the calculations prescribed by (6.2.2), the Wigner distribution function of the signal (6.2.12) takes the form

$$\mathcal{W}_s(q, \kappa) = \delta(\kappa - k\tfrac{q}{R}), \tag{6.2.13}$$

yielding a definite relation of the space coordinate q to the frequency κ (i.e. the direction of propagation) of the wave. At the point q in the reference plane one finds the only frequency $\kappa = \frac{kq}{R}$, in accord with the ray optics representation of the spherical wave, as the ray crossing the reference plane at q propagates in the paraxial direction $\vartheta \simeq \frac{q}{R}$. We recall that in the wave-optical picture the rays are understood as the normals to the wavefronts. The support[b] of $\mathcal{W}_s$ in the Wigner plane (q, κ) is evidently the straight line $\kappa = \frac{kq}{R}$ (Fig. 6.3).

As a conclusive example, due to its relevance as both a mathematical model and an analysing and synthesizing tool for optical as well as not strictly optical signals, we consider the Gabor function

$$\varphi_G(q) = (\tfrac{1}{\pi a^2})^{1/4} e^{-\frac{1}{2a^2}(q-q_0)^2 + i\kappa_0(q-\frac{q_0}{2})}, \tag{6.2.14}$$

where a is a positive quantity. As emphasized in § 4.6.4, the above function describes an optical signal with an intensity concentrated around the point q_0

[b] We recall that the *support* of a function is defined as the region where the function takes on non negligible values.

through a Gaussian profile, whose width is measured by a. The frequency spectrum $\widetilde{\varphi}_{\mathsf{G}}$ has the same analytical form as φ_{G}, the Fourier conjugate variables q and κ being interchanged. In fact, we obtain the Gaussian amplitude

$$\widetilde{\varphi}_{\mathsf{G}}(\kappa) = (\tfrac{a^2}{\pi})^{1/4} e^{-\frac{a^2}{2}(\kappa-\kappa_0)^2 - iq_0(\kappa-\frac{\kappa_0}{2})}, \tag{6.2.15}$$

centered around the frequency κ_0, whose width is measured by $1/a$.

The Wigner distribution function $\mathcal{W}_{\mathsf{G}}(q,\kappa)$ reads as

$$\mathcal{W}_{\mathsf{G}}(q,\kappa) = \tfrac{1}{\pi} e^{-\frac{1}{a^2}(q-q_0)^2 - a^2(\kappa-\kappa_0)^2}. \tag{6.2.16}$$

Interestingly it combines the features of φ_{G} and $\widetilde{\varphi}_{\mathsf{G}}$; in fact, we see that

$$\mathcal{W}_{\mathsf{G}}(q,\kappa) = |\varphi_{\mathsf{G}}(q)|^2 |\widetilde{\varphi}_{\mathsf{G}}(\kappa)|^2. \tag{6.2.17}$$

Notably, it is positive everywhere in the (q,κ) plane. Specifically, it is Gaussian in both q and κ around the space-frequency point (q_0,κ_0) through widths approximately of a and $1/a$, thus conveying the view of the Gabor signal as a kind of a "wave package" located at position q_0 and spatial frequency κ_0.

6.3 The Wigner distribution function: basic properties

In this section we describe in some detail a number of properties of the Wigner distribution function [10]. We will see that $\mathcal{W}(q,\kappa)$ is a real-valued function, invariant to translations and reversible. The signal can be reconstructed from it up to a constant phase factor. The Wigner distribution function yields the correct marginal distributions for any arbitrary signal and has finite support property. However, in general it does not behave as a proper energy density function, because it is not everywhere positive through the space-frequency domain. Also, being a function of two variables, derived from a function of one variable, i.e. the signal amplitude or the corresponding frequency spectrum, it must satisfy certain conditions, and thereby not every phase space function configures as the Wigner distribution function of some signal.

As remarked in the Introduction, the formalism related to the Wigner distribution function is widely used within the quantum mechanical context as well as within the general signal theory. Therefore the characterizations, we report below to identify the properties of the Wigner distribution, resort to the language of both probabilistic and signal theory.

Firstly, we note that by application of Schwarz's inequality to the Wigner integrals (6.2.2) and (6.2.3), an estimation of the upper limit for the absolute value of the Wigner distribution function obtains as

$$|\mathcal{W}(q,\kappa)| \leq \tfrac{\|\varphi\|^2}{\pi}. \tag{6.3.1}$$

6.3.1 *Reality and symmetry property*

The Wigner distribution function is always real irrespective of that the function φ is real or complex:

$$\mathcal{W}_{\varphi}(q,\kappa) = \mathcal{W}^{*}_{\varphi}(q,\kappa) \qquad \forall\, \varphi. \tag{6.3.2}$$

It is also invariant with respect to space and frequency reflection of φ, i.e.

$$\begin{array}{llllll} \text{if} & \varphi(q) & \rightarrow & \varphi(-q) & \text{then} & \mathcal{W}_{\varphi}(q,\kappa) \rightarrow \mathcal{W}_{\varphi}(q,-\kappa), \\ \text{if} & \widetilde{\varphi}(\kappa) & \rightarrow & \widetilde{\varphi}(-\kappa) & \text{then} & \mathcal{W}_{\varphi}(q,\kappa) \rightarrow \mathcal{W}_{\varphi}(-q,\kappa). \end{array} \tag{6.3.3}$$

Besides, the Wigner distribution of a real function, $\varphi(q) = \varphi^{*}(q)$, is an even function of κ, as the Wigner distribution of a function having a real frequency spectrum, $\widetilde{\varphi}(\kappa) = \widetilde{\varphi}^{*}(\kappa)$, is an even function of q. Since a real function possesses a spectrum, which turns into the complex conjugate under frequency reversal: $\widetilde{\varphi}(-\kappa) = \widetilde{\varphi}^{*}(\kappa)$, and similarly a real spectrum corresponds to a function with the reflection property: $\varphi(-q) = \varphi^{*}(q)$[c], we can synthesize the symmetry properties of the Wigner distribution function in relation to the symmetry properties of the signal and its frequency spectrum as

$$\begin{array}{lll} \text{if } \varphi \text{ real: } \widetilde{\varphi}(-\kappa) = \widetilde{\varphi}^{*}(\kappa) & \text{then} & \mathcal{W}_{\varphi}(q,-\kappa) = \mathcal{W}_{\varphi}(q,\kappa), \\ \text{if } \widetilde{\varphi} \text{ real: } \varphi(-q) = \varphi^{*}(q) & \text{then} & \mathcal{W}_{\varphi}(-q,\kappa) = \mathcal{W}_{\varphi}(q,\kappa). \end{array} \tag{6.3.4}$$

6.3.2 *Shift and modulation property*

The Wigner distribution function is shifted in space and frequency just as are shifted the signal and its frequency spectrum. The reader can verify that

$$\begin{array}{llllll} \text{if} & \varphi(q) & \rightarrow & \varphi(q-q_0) & \text{then} & \mathcal{W}_{\varphi}(q,\kappa) \rightarrow \mathcal{W}_{\varphi}(q-q_0,\kappa); \\ \text{if} & \widetilde{\varphi}(\kappa) & \rightarrow & \widetilde{\varphi}(\kappa-\kappa_0) & \text{then} & \mathcal{W}_{\varphi}(q,\kappa) \rightarrow \mathcal{W}_{\varphi}(q,\kappa-\kappa_0). \end{array} \tag{6.3.5}$$

As illustrated in § 4.6.1, a space shift of the signal corresponds to a linear modulation of the spectrum, and conversely a frequency shift of the spectrum corresponds to a linear modulation of the signal; namely

$$\begin{array}{llllll} \text{if} & \varphi(q) & \rightarrow & \varphi(q-q_0) & \text{then} & \widetilde{\varphi}(\kappa) \rightarrow \widetilde{\varphi}(\kappa)e^{i\kappa q_0}; \\ \text{if} & \widetilde{\varphi}(\kappa) & \rightarrow & \widetilde{\varphi}(\kappa-\kappa_0) & \text{then} & \varphi(q) \rightarrow \varphi(q)e^{i\kappa_0 q}. \end{array} \tag{6.3.6}$$

Combining space and frequency shifts, we may see that the Wigner distribution of the space and frequency shifted function

$$\psi(q) = \varphi(q-q_0)e^{i\kappa_0 q} \tag{6.3.7}$$

[c] Functions with the reflection property: $f(-x) = f^{*}(x)$ are referred to as *Hermitian*, or also *symmetric*, on account of the consequent symmetry of the modulus: $|f(-x)| = |f(x)|$.

is just the Wigner distribution of the original function $\varphi(q)$ equally translated through the (q, κ) plane to the point (q_0, κ_0) as

$$\mathcal{W}_\psi(q, \kappa) = \mathcal{W}_\varphi(q - q_0, \kappa - \kappa_0). \tag{6.3.8}$$

The behaviour of the Wigner distribution function under the space and frequency shifts of the signal, exemplified by (6.3.8), is reported as the *space-* and *frequency-shift invariance.*

6.3.3 *Finite support property*

The Wigner distribution function reflects the space and frequency range limitations of the signal.

In fact, if the function is limited to a certain space interval and vanishes outside, it follows from (6.2.2) that the Wigner distribution function is restricted to the same interval. Likewise, if the frequency spectrum is nonzero only over a finite frequency interval, the Wigner integral (6.2.3) is different from zero only within the same frequency interval. In symbols, we can write

$$\begin{array}{llll} \text{if} & \varphi(q) = 0, \ q \notin (q_a, q_b) & \text{then} & \mathcal{W}_\varphi(q, \kappa) = 0, \ q \notin (q_a, q_b); \\ \text{if} & \widetilde{\varphi}(\kappa) = 0, \ \kappa \notin (\kappa_a, \kappa_b) & \text{then} & \mathcal{W}_\varphi(q, \kappa) = 0, \ \kappa \notin (\kappa_a, \kappa_b). \end{array} \tag{6.3.9}$$

The above express a distinctive property of the Wigner distribution, which is reported as the *finite* (*space* and *spectral*) *support property* [5.7]. It is not a property common to every space-frequency distribution; the ambiguity function, for instance, is an example of a signal representation that extends through space and frequency intervals larger than those pertinent to the signal (§ 7.4.4).

The finite support property of the Wigner distribution has been already evidenced in Sect. 6.2, when illustrating, in relation with Fig. 6.2, the visual interpretation of the Wigner integral (6.2.2) at one point in terms of the overlap of the portion of the signal to the left over the portion to the right.

6.3.4 *Non linearity*

The Wigner distribution function is defined as a bilinear functional of the signal or its frequency spectrum, $\varphi(q)$ and $\widetilde{\varphi}(\kappa)$ entering indeed quadratically in the relevant integrals (6.2.2) and (6.2.3). The bilinearity of the Wigner distribution function in relation with the signal parallels the inherent bilinearity of the signal energy, and so it supports the desirable role of $\mathcal{W}(q, \kappa)$ as energy distribution. Thus, the Wigner distribution function of a linear superposition of a certain number of signals is not simply the linear superposition of the Wigner distribution functions of the individual signal components.

Suppose, for instance, to sum the functions $\varphi_1(q)$ and $\varphi_2(q)$ into

$$\varphi(q) = \varphi_1(q) + \varphi_2(q). \tag{6.3.10}$$

From definition (6.2.2) it follows that the Wigner distribution of φ reads as

$$\begin{aligned}\mathcal{W}_{\varphi}(q,\kappa) &= \mathcal{W}_{\varphi_1}(q,\kappa) + \mathcal{W}_{\varphi_2}(q,\kappa) \\ &\quad +2\Re\{\tfrac{1}{2\pi}\int \varphi_1^*(q-\tfrac{q'}{2})\varphi_2(q+\tfrac{q'}{2})e^{-i\kappa q'}dq'\},\end{aligned} \tag{6.3.11}$$

whose non linearity in relation to superposition (6.3.10) is clearly manifested by the term enclosed between brackets (the symbol $\Re$ signifying the real part of the argument), which adds to the two signal-components arising terms $\mathcal{W}_{\varphi_1}$ and $\mathcal{W}_{\varphi_2}$. This term is referred to as the *cross*-(or, *mixed*-)Wigner distribution function of the functions φ_1 and φ_2. Denoted by $\mathcal{W}_{\varphi_1,\varphi_2}(q,\kappa)$, it generalizes integral (6.2.2) as displayed in (6.3.11), i.e by the expression [4.4, ?]

$$\mathcal{W}_{\varphi_1,\varphi_2}(q,\kappa) = \tfrac{1}{2\pi}\int \varphi_1^*(q-\tfrac{q'}{2})\varphi_2(q+\tfrac{q'}{2})e^{-i\kappa q'}dq', \tag{6.3.12}$$

which, of course, with $\varphi_1 = \varphi_2 = \varphi$ turns into the expression (6.2.2), often referred to as the *auto*-(or, *self*-)Wigner distribution function of φ (see § 9.3.2).

Needless to say, $\mathcal{W}_{\varphi_1,\varphi_2}(q,\kappa)$ writes in terms of the pertinent frequency spectra $\tilde{\varphi}_1(\kappa)$ and $\tilde{\varphi}_2(\kappa)$ in the plain form

$$\mathcal{W}_{\varphi_1,\varphi_2}(q,\kappa) = \tfrac{1}{2\pi}\int \tilde{\varphi}_1^*(\kappa-\tfrac{\kappa'}{2})\tilde{\varphi}_2(\kappa+\tfrac{\kappa'}{2})e^{i\kappa' q}d\kappa'. \tag{6.3.13}$$

The integrals in (6.3.12) and (6.3.13) can be given the same visual interpretation as the auto-Wigner integrals (6.2.2) and (6.2.3). Clearly, the *folding* of the signal at position q (or frequency κ) involves one signal (or frequency spectrum) component on the left and the other component on the right.

Thus, we can express the Wigner distribution function of the composed signal (6.3.10) in terms of the *auto* and *cross*-Wigner distribution functions of the two components φ_1 and φ_2 according to

$$\mathcal{W}_{\varphi}(q,\kappa) = \mathcal{W}_{\varphi_1}(q,\kappa) + \mathcal{W}_{\varphi_2}(q,\kappa) + \mathcal{W}_{\varphi_1,\varphi_2}(q,\kappa) + \mathcal{W}_{\varphi_2,\varphi_1}(q,\kappa), \tag{6.3.14}$$

which turns into expression (6.3.11) on account of the easily provable property of the cross Wigner distribution function that $\mathcal{W}_{\varphi_1,\varphi_2} = \mathcal{W}^*_{\varphi_2,\varphi_1}$.

As we will see in relation with Moyal's formula, the cross-Wigner distribution function is used within the quantum mechanical context as a tool to investigate the interference between quantum mechanical wave functions, and accordingly the occurrence of transitions between different quantum states. The quantum issue is in a sense paralleled by the wave-optical issue of determining the efficiency of light launching into waveguides, for the purpose of optimizing the transfer of energy from the incident beam into a definite mode of the guiding structure.

A brief comment on the cross-term behaviour

Here we detail the interplay between the self and cross-terms in relation to the overall distribution, after the manner of [10.3, 10.4].

Relation (6.3.14) can easily be generalized to the case of a signal composed by a certain number, say N, of components located at different regions in the space-frequency plane. In fact, if

$$\varphi(q) = \sum_{j=1}^{N} a_j \varphi_j(q), \tag{6.3.15}$$

then

$$\mathcal{W}_\varphi(q,\kappa) = \sum_{j=1}^{N} |a_j|^2 \mathcal{W}_{\varphi_j}(q,\kappa) + 2 \sum_{\substack{j=1 \\ j>l}}^{N} \sum_{l=1}^{N} \Re\{a_j^* a_l \mathcal{W}_{\varphi_j,\varphi_l}(q,\kappa)\}. \tag{6.3.16}$$

Each signal component φ_j contributes to $\mathcal{W}_\varphi$ with its own signal term $\mathcal{W}_{\varphi_j}$, and each pair of signal components (φ_j, φ_l) produces a corresponding cross term, given by the real part of the cross Wigner distribution function $\mathcal{W}_{\varphi_j,\varphi_l}$ of the signal pair. Thus, the Wigner distribution function of the multicomponent signal (6.3.15) will display N signal terms and $\binom{N}{2}$ cross terms (customarily termed in the signal theory context as *interference* terms).

Is there some relation between the cross and self terms, which might enable us to foresee the behaviour of the former from that of the latter? Notably, the squared modulus of the cross Wigner distribution function $\mathcal{W}_{\varphi_j,\varphi_l}(q,\kappa)$ is expressible in terms of the auto Wigner distribution functions of the involved signals, i.e., $\mathcal{W}_{\varphi_j}$ and $\mathcal{W}_{\varphi_l}$. Explicitly, after some algebra one finds

$$|\mathcal{W}_{\varphi_j,\varphi_l}(q,\kappa)|^2 = \tfrac{1}{2\pi} \iint \mathcal{W}_{\varphi_j}(q-\tfrac{q'}{2}, \kappa-\tfrac{\kappa'}{2}) \mathcal{W}_{\varphi_l}(q+\tfrac{q'}{2}, \kappa+\tfrac{\kappa'}{2}) dq' d\kappa', \tag{6.3.17}$$

giving $\mathcal{W}_{\varphi_j,\varphi_l}$ as the space and frequency convolution of the Wigner distributions of the individual signals.

The $2D$ integral on the right reproduces with respect to the space and frequency variables the structure of the cross-Wigner integral (6.3.12) at zero frequency. Hence, resorting to the hinge-like picture of the self-Wigner integral (6.2.2) elaborated in § 6.2, it can be visualized as arising at any point (q,κ) in the Wigner plane by summing up the contributions of $\mathcal{W}_{\varphi_j}$ and $\mathcal{W}_{\varphi_l}$ respectively at points lying symmetrically to the chosen point (q,κ). Thus, if $\mathcal{W}_{\varphi_j}$ and $\mathcal{W}_{\varphi_l}$ sharp around the points (q_1,κ_1) and (q_2,κ_2) in the (q,κ) plane, the cross contribution $\mathcal{W}_{\varphi_j,\varphi_l}(q,\kappa)$ develops mainly in the surroundings of the midway point $(q_{\mathrm{m}},\kappa_{\mathrm{m}})$, with $q_{\mathrm{m}} = \frac{q_1+q_2}{2}$ and $\kappa_{\mathrm{m}} = \frac{\kappa_1+\kappa_2}{2}$.

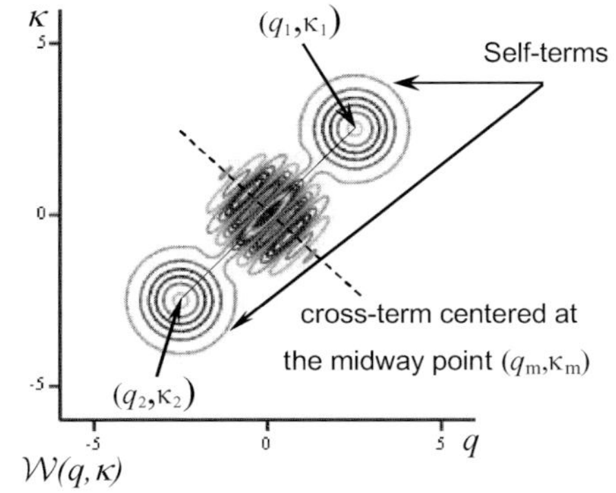

FIGURE 6.4. Geometry in the Wigner plane of the self and cross terms in the Wigner distribution function of the sum of two identical signals centered on (q_1, κ_1) and (q_2, κ_2).

Specifically, let us consider a signal $\psi(q)$ composed by two identical copies of a model signal $\varphi(q)$. We suppose that $\varphi(q)$ is mainly concentrated around the origin of the space-frequency plane, and that the two signal replicas are shifted in space and frequency to the points (q_1, κ_1) and (q_2, κ_2). So we write

$$\psi(q) = \varphi(q - q_1)e^{i\kappa_1 q} + \varphi(q - q_2)e^{i\kappa_2 q}, \tag{6.3.18}$$

and accordingly,

$$\begin{aligned} \mathcal{W}_\psi(q, \kappa) &= \mathcal{W}_\varphi(q - q_1, \kappa - \kappa_1) + \mathcal{W}_\varphi(q - q_2, \kappa - \kappa_2) \\ &+2\cos[\kappa_\mathrm{d}(q - q_\mathrm{m}) - q_\mathrm{d}(\kappa - \kappa_\mathrm{m}) + \kappa_\mathrm{d} q_\mathrm{m}]\mathcal{W}_\varphi(q - q_\mathrm{m}, \kappa - \kappa_\mathrm{m}). \end{aligned} \tag{6.3.19}$$

Here, $\mathcal{W}_\varphi$ is the Wigner distribution function of the model signal φ; q_m and κ_m are as defined before, whilst q_d, κ_d are the difference coordinates: $q_\mathrm{d} = q_1 - q_2$ and $\kappa_\mathrm{d} = \kappa_1 - \kappa_2$. The cross-term above has been obtained from an appropriate manipulation of the cross-Wigner integral (6.3.12) explicitly involving the two spatially and spatial frequency shifted versions of the model signal.

The geometry of the terms entering (6.3.19) is illustrated in Fig. 6.4. As expected, the two signal contributions in $\mathcal{W}_\psi$ are located at the regions in the space-frequency plane where the relative signals are located, whilst the cross contribution lies midway between the signal terms. Moreover, the envelope of the cross term shapes just as the Wigner distribution $\mathcal{W}_\varphi$ of the model signal with a double amplitude, but is modulated by an oscillation occurring transversely to the line through the centres of the self terms, with a frequency governed by the separation between those centres. Although established in the specific case of a signal like (6.3.18), i.e., composed by two identical waveforms, the behaviour exemplified in Fig. 6.4 is quite generally displayed by the self-

and cross-terms of the Wigner distribution function of a composed signal, even when the signal terms have not the same form. Specifically, the graph in Fig. 6.4 refers to a Gaussian-shaped model signal (see § 6.4.5).

It is evident from (6.3.19) that even though the signal terms are positive everywhere through the space-frequency plane, as it occurs in the only case of a Gaussian-shaped signal, the oscillating behaviour of the cross term unavoidably makes the Wigner distribution function of the composed signal to be negative over a certain region of the (q, κ) plane.

Accordingly, the occurrence of the cross terms, that may point to energy in space-frequency regions where the signal does not exist, and also may be a factor of generating negative values in the Wigner function, makes rather difficult the interpretation of the Wigner distribution function as an energy distribution over space and frequency. Hence an appropriate smoothing in both the space and frequency directions is often carried out to reduce the magnitude of the cross-terms contribution. Formally, this is obtained by convolving the Wigner distribution function with a *smoothing kernel* $\mathcal{L}(q, \kappa)$, i.e.,

$$\mathcal{W}(q, \kappa) \longrightarrow \mathcal{W}_{\text{smoothed}}(q, \kappa) = \iint \mathcal{W}(q', \kappa')\mathcal{L}(q - q', \kappa - \kappa')dq'd\kappa', \quad (6.3.20)$$

which, of course, must optimally be designed according to the requirements of the problem of interest in order to reduce the disturbing cross terms without affecting too much the signal terms, and so in general without altering too much the information content carried by the original Wigner distribution.

Indeed, although mostly often undesired, the cross terms carry however significative information regarding the relations between the signal components, as shown by (6.3.17), which traces the modulus of the cross term back to the corresponding signal terms. For instance, they allow one to distinguish between a meaningful and a fictitious decomposition of a signal into components. In fact, if one expresses a signal as the sum of two parts as

$$\psi(q) = a\varphi(q) + (1 - a)\varphi(q),$$

the relevant Wigner distribution function turns to be

$$\mathcal{W}_{\psi}(q, \kappa) = |a|^2\mathcal{W}_{\varphi}(q, \kappa) + |1-a|^2\mathcal{W}_{\varphi}(q, \kappa) + 2\Re[a^*(1-a)]\mathcal{W}_{\varphi}(q, \kappa) = \mathcal{W}_{\varphi}(q, \kappa), \quad (6.3.21)$$

and no oscillations arise in the (fictitious) cross distribution as a modulation of the signal generated envelope $\mathcal{W}_{\varphi}$. Thus, inspection of the structure of the Wigner distribution function can be an effective tool to individualize multicomponent signals, i.e., signals which in the space-frequency planes are well delineated into definite (and possibly intersecting) regions [10.5, 5.5, 5.7]. Indeed, reliable suppression or reduction of the cross-terms must wisely be designed in order to preserve the information about the various independent space-frequency "entities", which may compose the observed signal.

6.3.5 *Marginal distributions and inversion formulas*

The Wigner integrals (6.2.2) and (6.2.3) manifestly display a Fourier transform structure. Expression (6.2.2) in fact conveys the Wigner distribution function as the Fourier transform of the product $\varphi(q+\frac{q'}{2})\varphi^*(q-\frac{q'}{2})$, regarded as a function of q' with q taken as a parameter. So, inverse transforming $\mathcal{W}(q,\kappa)$ to the q'-domain, we obtain

$$\varphi(q+\tfrac{q'}{2})\varphi^*(q-\tfrac{q'}{2}) = \int \mathcal{W}(q,\kappa)e^{i\kappa q'}d\kappa, \tag{6.3.22}$$

which, redefining the arguments of φ and φ^*, is recast in the form

$$\varphi(q)\varphi^*(q') = \int \mathcal{W}(\tfrac{q+q'}{2},\kappa)\, e^{i\kappa(q-q')}d\kappa. \tag{6.3.23}$$

In a like manner, by (6.2.3) the Wigner distribution function relates also to the inverse Fourier transform of the product $\widetilde{\varphi}(\kappa+\frac{\kappa'}{2})\widetilde{\varphi}^*(\kappa-\frac{\kappa'}{2})$ with respect to κ', thus enabling the Fourier integral of $\mathcal{W}(q,\kappa)$ to yield

$$\widetilde{\varphi}(\kappa+\tfrac{\kappa'}{2})\widetilde{\varphi}^*(\kappa-\tfrac{\kappa'}{2}) = \int \mathcal{W}(q,\kappa)e^{-i\kappa' q}dq, \tag{6.3.24}$$

which by changing the variables writes in a form similar to (6.3.23), i.e.,

$$\widetilde{\varphi}(\kappa)\widetilde{\varphi}^*(\kappa') = \int \mathcal{W}(q,\tfrac{\kappa+\kappa'}{2})e^{-iq(\kappa-\kappa')}dq. \tag{6.3.25}$$

Evidently, setting $q'=q$ and $\kappa'=\kappa$ respectively in (6.3.23) and (6.3.25), we obtain on one side the square moduli of $\varphi(q)$ and $\widetilde{\varphi}(\kappa)$, and on the other side the integrals of $\mathcal{W}(q,\kappa)$ over the κ- and q-variable, namely

$$|\varphi(q)|^2 = \int \mathcal{W}(q,\kappa)\,d\kappa, \quad |\widetilde{\varphi}(\kappa)|^2 = \int \mathcal{W}(q,\kappa)\,dq. \tag{6.3.26}$$

The integral of $\mathcal{W}$ over the frequency variable at position q yields the energy density of the signal at q. Correspondingly, integrating $\mathcal{W}$ over the space variable at the spatial frequency κ gives the energy density spectrum of the signal at κ. The projections of the Wigner distribution onto the q- and κ-axis, mathematically represented by the integrals in (6.3.26), are respectively named the *space* and *frequency marginal distributions* (or simply, *marginals*) of $\mathcal{W}(q,\kappa)$. Indeed, they yield the desired quantities in conformity with the basic properties (6.1.13), one should expect in general for a phase-space distribution having the definite role of an energy density distribution.

In accord with the marginal relations, the signal energy contained within a certain space interval $[q_i,q_o]$ is given by the integral of $\mathcal{W}(q,\kappa)$ over the vertical strip $-\infty<\kappa<+\infty$, $q_i\le q\le q_o$ in the q-κ plane, as correspondingly the content of energy within the frequency interval $[\kappa_i, \kappa_o]$ is obtained by integrating the Wigner distribution function over the horizontal strip $-\infty < q < +\infty$, $\kappa_i \leq \kappa \leq \kappa_o$. Evidently the integration of $\mathcal{W}(q, \kappa)$ over the whole q-κ plane provides the total energy according to

$$\iint \mathcal{W}(q,\kappa)\, d\kappa dq = \int |\varphi(q)|^2 dq = \int |\widetilde{\varphi}(\kappa)|^2 d\kappa \qquad (6.3.27)$$

Equations (6.3.26) and (6.3.27) establish a link of the rather abstract Wigner function with the measurable intensity and power spectrum of the signal, and so with the total energy carried by the optical perturbance. As we will see in Chapter 9, they are crucial to establish that the moments of the Wigner distribution function are the moments of the signal and its frequency spectrum; moments offer an alternative and effective tool to convey as well as to extract information about the signal, as for instance in pattern recognition processes.

Notably it is possible to reconstruct the space and frequency representations $\varphi(q)$ and $\widetilde{\varphi}(\kappa)$ of the signal from the Wigner function. In fact, with assigned values of $q' = q_0$ and $\kappa' = \kappa_0$, (6.3.23) and (6.3.25) give the inversion formulas

$$\begin{aligned} \varphi(q) &= \frac{1}{\varphi^*(q_0)} \int \mathcal{W}(\tfrac{q+q_0}{2}, \kappa) e^{i\kappa(q-q_0)}\, d\kappa, \\ \widetilde{\varphi}(\kappa) &= \frac{1}{\widetilde{\varphi}^*(\kappa_0)} \int \mathcal{W}(q, \tfrac{\kappa+\kappa_0}{2})\, e^{-iq(\kappa-\kappa_0)} dq, \end{aligned} \qquad (6.3.28)$$

from which $\varphi(q)$ and $\widetilde{\varphi}(\kappa)$ may be recovered up to a constant phase factor, since only the magnitudes of $\varphi^*(q_0)$ and $\widetilde{\varphi}^*(\kappa_0)$ can be known from the squared moduli in (6.3.26). One may conveniently choose $q_0 = 0$ and $\kappa_0 = 0$, if of course $\varphi^*(0)$ and $\widetilde{\varphi}^*(0)$ are not zero. Accordingly, the signal amplitude $\varphi(q)$ is retrieved from the inverse Fourier transform with respect to κ of the Wigner distribution at position $\frac{q}{2}$. Likewise, the frequency spectrum $\widetilde{\varphi}(\kappa)$ results from the Fourier transform of $\mathcal{W}(q, \frac{\kappa}{2})$ with respect to q.

The inversion formulae (6.3.28) offer evidence that the Wigner distribution contains as much information as the signal itself, being the signal uniquely recoverable from it, up to a constant phase. In other terms, we may say that the expectations pictorially expressed in Fig. 6.1 are satisfactorily fulfilled by the Wigner distribution function. The information about the system state are equivalently displayed by the wave function and the Wigner distribution, on due account of course of the intrinsic limitations of the Wigner view, inherent in the inhibited interpretation of $\mathcal{W}(q, \kappa)$ as a true energy density distribution.

From the applicative viewpoint, relations (6.3.28) are the mathematical expressions for the inverse problems of signal reconstruction and synthesis through the Wigner distribution function, and possibly suitable synthesis functions. Signal reconstruction and synthesis are basically concerned with an estimate of the signal waveform $\varphi(q)$ from the knowledge of some other signal representation related to φ, with the possible aid of auxiliary tools. In fact, the reader may easily prove that synthesis of a signal is also possible by use of a "key" function $\phi(q)$ (as it is commonly said in the context of radar/sonar applications), in accord with the relation

$$\varphi(q) = \iint \phi(2q' - q)\mathcal{W}_{\phi,\varphi}(q', \kappa')e^{i2\kappa'(q-q')}dq'd\kappa', \tag{6.3.29}$$

holding if the key function is normalized to unity. $\mathcal{W}_{\phi,\varphi}$ is the cross-Wigner distribution function defined by (6.3.12). Thus, according to Eqs. (6.3.28) and (6.3.29), it is possible to synthesize a signal having the desired space-frequency characteristics. Constructing a Wigner distribution function, or a cross-Wigner distribution function in conjunction with a suitable key function, with the desired attributes, one can obtain the signal which produces that auto-, or cross-Wigner distribution, and hence has the desired space-frequency properties.

The inference of the system wave function from the Wigner distribution function is of a great relevance in quantum mechanics/optics. From $\mathcal{W}_\varphi(q, \kappa)$, which is typically reconstructed by quantum tomography (see § 8.4.3), we can numerically construct not only the squared modulus of the wave function, but the very complex wave function of the system under investigation (perhaps a particle). The phase indeterminacy of the inferred wave function is, of course, irrelevant, since, as we know, any wave function can be multiplied by an arbitrary phase factor without altering the physical state.

A brief comment on partially coherent signals

As remarked in § 6.3.4 when illustrating the effects of the cross-terms, the bilinear dependence of $\mathcal{W}(q, \kappa)$ on the signal amplitude may be a disadvantage in some applications as, for instance, in pattern recognition, since the nonlinearity generated terms may reduce the readability of the signal pattern. In contrast, it can reveal advantageous in other applications where the field enters the theory quadratically, as in the theory of partial coherence. In fact, the concept of Wigner distribution function with the relevant formalism has been extended from deterministic to stochastic signals, without increasing the dimensionality of the formulae. Roughly speaking, it seems natural, retracing the lines of the definition of the Wigner distribution in quantum mechanics, to replace in the defining integral (6.2.2) the product $\varphi^*(q_1)\varphi(q_2)$, appropriate to completely coherent signals, comparable to the pure states of quantum statistics, by the ensemble average $\langle\varphi^*(q_1)\varphi(q_2)\rangle$, suitable to describe partially coherent signals, comparable to the quantum statistical mixtures. In effect, as the complex amplitude $\varphi(q)$ is the proper descriptor of coherent signals, the appropriate descriptors for randomly fluctuating perturbances are known, from the optical coherence theory, to be the *coherence function* $\Gamma(q_1, q_2, \tau)$ and the *cross-spectral density* $\widetilde{\Gamma}(q_1, q_2, \omega)$ [11]. The former describes the spatial correlation of the temporally stationary stochastic signal $\psi(q,t)$ at pairs of positions q_1 and q_2 as a function of time τ through the ensemble average[(d)]

$$\Gamma(q_1,q_2,\tau) = \langle\psi^*(q_1,t)\psi(q_2,t+\tau)\rangle, \tag{6.3.30}$$

whilst the latter describes the spatial correlation of $\psi(q,t)$ as a function of frequency ω, arising from the temporal Fourier transform of $\Gamma(q_1,q_2,\tau)$, i.e.

$$\widetilde{\Gamma}(q_1,q_2,\omega) = \int \Gamma(q_1,q_2,\tau)e^{-i\omega\tau}d\tau. \tag{6.3.31}$$

Definition (6.2.2) contains by itself the product of the signal waveforms in a form resembling that in (6.3.30). Hence, within the spirit of (6.2.2), it is natural to define the Wigner distribution function for partially coherent light by the Fourier-like integral of the cross spectral density as [12]

$$\mathcal{W}(q,\kappa) = \tfrac{1}{2\pi}\int \widetilde{\Gamma}(q-\tfrac{q'}{2},q+\tfrac{q'}{2},\omega)e^{-i\kappa q'}dq'. \tag{6.3.32}$$

The Wigner optics of partially coherent light is beyond the purposes of the book; the interested reader is directed to [12]. Further discussion is in § 7.5.

Definition (6.3.32) is similar to that suggested by Walther as *generalized radiance* to link up radiometry with the theory of partial coherence [13.1], and later investigated and refined by Wolf [13.2].

6.3.6 *Nonpositivity*

It has been emphasized in the introductory notes that in order for a space-frequency distribution function to be properly interpreted as the signal energy density in space and frequency, it must be positive defined over all the space-frequency domain and, once integrated over one of the variables, it must yield the energy density with respect to the other variable.

According to (6.3.26), the Wigner distribution function satisfies the marginal requirements as its space and frequency marginal distributions just equal the space and frequency marginal distributions, $|\varphi(q)|^2$ and $|\widetilde{\varphi}(\kappa)|^2$, of the signal.

Unfortunately, the Wigner distribution function is not necessarily nonnegative through the whole (q,κ) plane; in general it equally assumes positive and negative numerical values. Therefore, even though relations (6.3.26) might suggest the interpretation of the Wigner distribution function of a signal φ as the

---

[d] The ensemble average, signified by the sharp brackets, is understood over an ensemble of different realizations of the optical disturbance. Equivalently, under the customary hypothesis of ergodicity of the light field, it can as well be interpreted as the time average

$$\langle f(t,..)\rangle \equiv \lim_{T\to\infty}\tfrac{1}{2T}\int_{-T}^{T} f(t,..)dt.$$

energy distribution of φ in space and frequency, actually a strict interpretation of $\mathcal{W}(q,\kappa)$ as an energy density distribution is not possible. Thus information about how the energy of the signal is distributed in q and κ can be gained from the marginal distributions of $\mathcal{W}(q,\kappa)$, but this does not enable the Wigner distribution to be truly interpreted as an energy density distribution, although it is commonly treated and manipulated as an energy density distribution.

As Wigner showed, bilinear distributions, which have the correct marginal distributions and are at the same time nonnegative defined for arbitrary state functions, do not exist [1.2]. The only waveforms for which the Wigner distribution function is everywhere nonnegative are represented by exponentials of quadratic polynomials, i.e., Gaussian functions [10.6]. Expression (6.2.16), we obtained for the Wigner distribution function $\mathcal{W}_{\mathsf{G}}(q,\kappa)$ of the Gabor amplitude (6.2.14), is in agreement with this assertion (see also § 6.4.5). The existence of space-frequency distributions that satisfy the marginal conditions and are as well positive for arbitrary signals has been demonstrated. They are not bilinear in the signal amplitude or frequency spectrum; also, a method to generate them within the general class of mixed space-frequency distributions has been elaborated [10.7]. Notably, among all the bilinear distributions satisfying the marginal properties, the Wigner distribution function is the closest to being positive, in the sense that it requires the least amount of Gaussian smoothing to become nonnegative everywhere in the space-frequency plane [10.12].

Smoothing as a tool to obtain a positive phase-space distribution function

We will not enter the details of the positivity question, addressing the reader to § 8.2.2 for a more accurate account. However, we can gain an immediate feeling for the nonpositivity of the Wigner distribution function, examining its value at the origin of the (q,κ) plane, which explicitly writes as

$$\mathcal{W}(0,0) = \tfrac{1}{\pi}\int_{-\infty}^{\infty} \varphi^*(-q)\varphi(q)dq = \tfrac{2}{\pi}\Re[\int_{0}^{\infty} \varphi^*(-q)\varphi(q)dq]. \qquad (6.3.33)$$

For illustrative purposes, we consider the real function $\varphi(q) = q$. By (6.3.33), the corresponding Wigner distribution attains at $(0,0)$ a negative value, being

$$\mathcal{W}_{\varphi}(0,0) = -\tfrac{2}{\pi}\int_{0}^{\infty} q^2 dq < 0. \qquad (6.3.34)$$

In general, for any signal such that $q\varphi(q) > 0$ for all q or $q\varphi(q) < 0$ for all q it happens that $\mathcal{W}_{\varphi}(0,0) < 0$ in both cases (Problem 3).

In § 6.4 we will calculate the Wigner distribution function of a certain number of optical waveforms. As we will see, for all of them, except that for the Gaussian waveform, the Wigner distribution function takes on negative values over some regions of the space-frequency plane.

The smoothing of $\mathcal{W}$ by an appropriate kernel

$$\mathcal{W}(q,\kappa) \longrightarrow \mathcal{W}_{\text{smoothed}}(q,\kappa) = \iint \mathcal{W}(q',\kappa')\mathcal{G}(q-q',\kappa-\kappa')dq'd\kappa', \quad (6.3.35)$$

has thoroughly been investigated as a possible tool to yield a nonnegative distribution function. The smoothing procedure (6.3.35) parallels, and actually can complement, (6.3.20), entailed to reduce the cross-terms contribution.

The Gaussian shape for the smoothing kernel $\mathcal{G}(q,\kappa)$ is a natural choice:

$$\mathcal{G}(q,\kappa) = \frac{\sqrt{\beta\gamma}}{2\pi}e^{-\frac{1}{2}(\gamma q^2+\beta\kappa^2)}, \quad (6.3.36)$$

where, of course, γ and β are supposed to be positive. For convenience's sake, the multiplying factor above is set to have $\iint \mathcal{G}(q,\kappa)dqd\kappa = 1$.

It has been proved that if the Gaussian kernel (6.3.36) is as wide as or wider than the minimum uncertainty wave packet, i.e., if

$$\beta\gamma \leq 4, \quad (6.3.37)$$

the resulting distribution (6.3.35) is nonnegative everywhere in the phase-plane [10.14]. As we will mention in § 6.3.10, a $2D$ space-frequency Gaussian distribution like (6.3.36), satisfying the equality in (6.3.37), turns to be itself a Wigner distribution function; precisely the Wigner distribution function of a real Gaussian signal, whose spatial and spatial frequency extensions are faithfully reproduced into the respective extensions of the Wigner distribution (see also §§ 6.2.1-6.4.5). Thus, the smoothing by a minimum uncertainty $2D$ Gaussian kernel acquires the meaning of the convolution with another Wigner distribution function, i.e., that of the Gaussian amplitude

$$\varphi_{\mathcal{G}}(q) = (\tfrac{2}{\pi\beta})^{1/4}e^{-\frac{q^2}{\beta}}. \quad (6.3.38)$$

Then, by (6.3.36) the double convolution integral (6.3.35) explicitly writes

$$\begin{aligned}\mathcal{W}_{\text{smoothed}}(q,\kappa) &= \frac{\sqrt{\beta\gamma}}{2\pi}\iint \mathcal{W}(q',\kappa')e^{-\frac{\gamma}{2}(q-q')^2-\frac{\beta}{2}(\kappa-\kappa')^2}dq'd\kappa' \\ &= \frac{\sqrt{\beta\gamma}}{2\pi^2}\iiint \varphi^*(q'-q'')\varphi(q'+q'')e^{-\frac{\gamma}{2}(q-q')^2-\frac{\beta}{2}(\kappa-\kappa')^2}e^{-2i\kappa' q''}dq''dq'd\kappa'.\end{aligned} \quad (6.3.39)$$

Firstly we integrate by κ' as

$$\sqrt{\beta}\int e^{-\frac{\beta}{2}(\kappa-\kappa')^2}e^{-2i\kappa' q''}d\kappa' = \sqrt{2\pi}e^{-\frac{2}{\beta}q''^2-2i\kappa q''}, \quad (6.3.40)$$

and then redefine the integration variable to finally obtain

$$\mathcal{W}_{\text{smoothed}}(q,\kappa) = \frac{\sqrt{2\pi\gamma}}{4\pi^2}e^{-\frac{\gamma}{2}q^2}\iint f(x)f^*(y)e^{\xi xy}dxdy, \quad (6.3.41)$$

where we have set

$$f(x) \equiv \varphi(x)e^{-(\beta\gamma+4)\frac{x^2}{8\beta}+(\gamma q-2i\kappa)\frac{x}{2}}, \quad \xi \equiv \tfrac{1}{4\beta}(4-\beta\gamma). \tag{6.3.42}$$

Expanding the cross exponential in (6.3.41) as

$$e^{\xi xy} = \sum_{n=0}^{\infty} \tfrac{1}{n!}\xi^n x^n y^n, \tag{6.3.43}$$

$\mathcal{W}_{\text{smoothed}}(q,\kappa)$ can be recast into

$$\mathcal{W}_{\text{smoothed}}(q,\kappa) = \tfrac{\sqrt{2\pi\gamma}}{4\pi^2}e^{-\frac{\gamma}{2}q^2}\sum_{n=0}^{\infty}\tfrac{1}{n!}\xi^n\left|\int x^n f(x)dx\right|^2. \tag{6.3.44}$$

Evidently, the sign of $\mathcal{W}_{\text{smoothed}}$ depends on that of ξ; indeed, if ξ is non-negative the smoothed Wigner distribution turns to be positive. It is evident then positive ξ's amount to relation (6.3.37).

Generalization of the above result to obliquely oriented Gaussians, formally described by

$$\mathcal{G}(q,\kappa) = \tfrac{(\beta\gamma-4\alpha^2)^{1/2}}{2\pi}e^{-\frac{1}{2}(\gamma q^2-2\alpha q\kappa+\beta\kappa^2)}, \tag{6.3.45}$$

the degree of obliqueness of the Gaussian "bell" with respect to the phase-plane axes being measured by α, has been proved under the condition that

$$\beta\gamma \leq \alpha^2 + 4. \tag{6.3.46}$$

Needless to say, this inequality corresponds to (6.3.37) for vanishing values of the obliquity parameter α. Of course, in order that (6.3.45) be a $2D$ Gaussian, $\beta,\gamma > 0$ and $\beta\gamma - 4\alpha^2 > 0$. Also, note that again $\iint \mathcal{G}(q,\kappa)dqd\kappa = 1$.

The reader may verify the above assertion retracing the steps of the proof previously elaborated in regard to the kernel (6.3.36). In particular, if the parameters in (6.3.45) are chosen to satisfy the equality in (6.3.46), the resulting $2D$ Gaussian function is interpretable as a Wigner distribution function; specifically, the Wigner distribution function of the complex Gaussian amplitude

$$\varphi_{\mathcal{G}}(q) = (\tfrac{2}{\pi\beta})^{1/4}e^{-\frac{q^2}{\beta}(1-i\frac{\alpha}{2})}. \tag{6.3.47}$$

Evidently, one may regard (6.3.38) as the amplitude profile of a Gaussian beam at the waist with beam radius equal to $\sqrt{\beta}$, and accordingly (6.3.47) as the amplitude profile of a Gaussian beam at a plane different from the waist, where specifically the wave front curvature is $\frac{\alpha}{k\beta}$ and the beam radius is $\sqrt{\beta}$.

The smoothing by a minimum uncertainty Gaussian has been originally introduced by Husimi in his extensive investigation of the formal properties of the density matrix [10.13]. Accordingly, the phase-space distribution resulting from the convolution process (6.3.35) by a minimum uncertainty Gaussian is frequently reported as the *Husimi distribution function*. It is one of the basic phase-space distributions of quantum optics, corresponding to the normal ordering of the boson creation and annihilation operators.

The smoothing procedure implied by (6.3.35) translates in an elegant and adaptable form the naïve idea of coarsely replacing the value of the Wigner distribution function at each point in the phase plane with the value of the integral of $\mathcal{W}$ over a properly chosen region around that point. In a sense this is inspired by the uncertainty relation of Fourier analysis, which is commonly intended to make the "point" in phase plane a physically meaningless entity. The single points are not practically accessible, the minimum size of the phase-plane region that can be resolved being defined by the uncertainty product. So, it makes sense to replace the meaningless values of the Wigner function at points in phase plane with the relevant physically meaningful averages taken over accessible regions, possibly of minimum allowed area (see also § 9.4.4).

Of course, a judicious choice of the smoothing distribution, and so of the size of the integration region, must be performed in relation with the problem at hand. In fact, a very narrow windowing distribution in space would allow one to closely inspect the Wigner distribution function behaviour in space, but in turn would reduce the accuracy of the frequency inspection. Likewise, a very narrow distribution in frequency would allow one to inspect closely the Wigner distribution function behaviour in frequency, reducing in turn the resolvibility in space. The minimum uncertainty Gaussian distribution represents the optimum compromise to conciliate the accuracy requirements with the limitations imposed by the uncertainty relation. We will further face with the complementarity of the resolution in q and κ in relation with the windowing-like procedure in § 7.2, when describing the spectrogram, which grounds on a windowing procedure, and so on a double convolution like (6.3.35).

Finally, it should be mentioned that an even more general result has been established in regard to the convolution with the Wigner distribution function of any normalized signal [10.15]. This result straightforwardly follows from the Moyal relation we discuss next. In fact, on account of the Moyal relation and the shift property of the Wigner distribution function in response to the space and frequency shifts of the signal, we can relate the smoothing of the Wigner distribution function of the signal of interest with the Wigner distribution function of any other signal to the squared modulus of the overlap integral of the two signals (more precisely, one of the two signals is spatially and spatial frequency shifted). It is worth noting that the overlap integral of two waveforms, say φ and ψ, can be interpreted as the result of a measure operation on the "state" φ through a measurement apparatus described by ψ. In the light of this interpretation, the smoothing of the Wigner distribution function mathematically reflects the operative coarse-graining, by which any measure performed on the signal is invariably affected by the measuring apparatus.

6.3.7 *Moyal formula*

The Moyal formula relates the overlap of two functions to the overlap of the corresponding Wigner distribution functions. We consider two functions φ_1 and φ_2 [4.2, 12.1, 4.4, 5.3]. The inner product of their respective Wigner distribution functions $\mathcal{W}_{\varphi_1}$ and $\mathcal{W}_{\varphi_2}$ writes as

$$(\mathcal{W}_{\varphi_1}, \mathcal{W}_{\varphi_2}) = \iint \mathcal{W}^*_{\varphi_1}(q,\kappa)\mathcal{W}_{\varphi_2}(q,\kappa)dqd\kappa, \tag{6.3.48}$$

the asterisk being actually uninfluential because of the realness of $\mathcal{W}_{\varphi_1}$. Inserting the explicit forms of $\mathcal{W}_{\varphi_1}$ and $\mathcal{W}_{\varphi_2}$ according to (6.2.2), after an easy algebra we obtain the expression

$$(\mathcal{W}_{\varphi_1}, \mathcal{W}_{\varphi_2}) = \tfrac{1}{2\pi} \iint \varphi_1(q+\tfrac{q'}{2})\varphi_1^*(q-\tfrac{q'}{2})\varphi_2(q-\tfrac{q'}{2})\varphi_2^*(q+\tfrac{q'}{2})dqdq', \tag{6.3.49}$$

which, by a suitable change of variables, turns into the form as the square modulus of the inner product of φ_1 and φ_2. Explicitly, we have

$$(\mathcal{W}_{\varphi_1}, \mathcal{W}_{\varphi_2}) = \tfrac{1}{2\pi} \left| \int dq\varphi_1^*(q)\varphi_2(q) \right|^2 = \tfrac{1}{2\pi} \left|(\varphi_1, \varphi_2)\right|^2, \tag{6.3.50}$$

thus linking the overlap of the functions φ_1 and φ_2 to the overlap of the respective Wigner distribution functions.

Result (6.3.50) says that the average of one Wigner distribution function with another Wigner distribution function is always nonnegative. In particular, if $(\varphi_1, \varphi_2) = 0$, the overlap integral of $\mathcal{W}_{\varphi_1}$ and $\mathcal{W}_{\varphi_2}$ vanishes:

$$\iint \mathcal{W}^*_{\varphi_1}(q,\kappa)\mathcal{W}_{\varphi_2}(q,\kappa)dqd\kappa = 0 \quad \text{if} \quad (\varphi_1, \varphi_2) = 0. \tag{6.3.51}$$

As the equality above cannot be satisfied if both $\mathcal{W}_{\varphi_1}$ and $\mathcal{W}_{\varphi_2}$ are positive-valued everywhere in the Wigner plane (q,κ), it represents a further proof of the fact that the Wigner distribution function can attain negative values.

By use of Schwarz's inequality: $|(\varphi_1, \varphi_2)| \leq \|\varphi_1\| \cdot \|\varphi_2\|$, and relation (6.3.27), Eq. (6.3.50) gives

$$\iint \mathcal{W}_{\varphi_1}(q,\kappa)\mathcal{W}_{\varphi_2}(q,\kappa)dqd\kappa \leq \tfrac{1}{2\pi} \iint \mathcal{W}_{\varphi_1}(q,\kappa)dqd\kappa \iint \mathcal{W}_{\varphi_2}(q,\kappa)dqd\kappa, \tag{6.3.52}$$

which can be regarded as a kind of Schwarz's inequality for the Wigner distribution. The equality holds if and only if $\varphi_1 = \varphi_2 = \varphi$, or more in general if the two functions differ by a complex constant; in that case, we obtain

$$\|\mathcal{W}_\varphi\|^2 = \iint \mathcal{W}^2_\varphi(q,\kappa)dqd\kappa = \tfrac{1}{2\pi}\|\varphi\|^4 = \tfrac{1}{2\pi}\left(\iint \mathcal{W}_\varphi(q,\kappa)dqd\kappa\right)^2, \tag{6.3.53}$$

which relates the norm of the function to that of the corresponding Wigner distribution function. In a sense, (6.3.53) may be thought of as a kind of Parseval's relation for the Wigner transform, in the light of its interpretation as the Fourier transform of the signal intensity or the power spectrum.

We emphasize that Moyal's relation, in the form reported in (6.3.50), is a particular case of the more general formula

$$(\mathcal{W}_{\varphi_1,\phi_1},\mathcal{W}_{\varphi_2,\phi_2}) = \frac{1}{2\pi}\int \varphi_1^*(q)\varphi_2(q)dq \int \phi_1(q)\phi_2^*(q)dq = \frac{1}{2\pi}(\varphi_1,\varphi_2)(\phi_1,\phi_2)^*, \tag{6.3.54}$$

which involves four signals and hence two cross-Wigner distribution functions. The reader is invited to prove the above formula using definition (6.3.12) for the cross-Wigner functions $\mathcal{W}_{\varphi_1,\phi_1}$ and $\mathcal{W}_{\varphi_2,\phi_2}$.

Moyal's relation directly introduces us into a new quantum mechanical interpretation of the scalar product of two wave functions in Hilbert space, as proposed in [10.22, 10.23]. For illustrative purposes, we may consider two quantum particles, whose state being respectively described by the wave functions, φ_1 and φ_2. The scalar product of the two wave functions gives the projection of one state on the other. Moyal's formula substitutes the scalar product of φ_1 and φ_2, i.e., the overlap in the configuration space, by the overlap of the corresponding Wigner distribution functions in phase space. So, on one side it confirms the equivalence of the quantum mechanics formulation in terms of Schrödinger wave function and Wigner distribution function. On the other side, since in quantum mechanics the Wigner distribution function $\mathcal{W}_\varphi$ can be interpreted as the probability that the particle in state φ has a specific coordinate and a specific momentum, the product $\mathcal{W}_{\varphi_1}\mathcal{W}_{\varphi_2}$, entering (6.3.50), is the probability that the two particles, in the considered states, have the same specific position and the same specific momentum. Therefore, the integral of that product over the whole phase space is a measure of the proximity of the two states, φ_1 and φ_2, in phase space, and so of the "propensity" of the two particles "to have the same positions and the same momenta" [10.22]. Also, shifting one of the states, say φ_2 in both position and momentum by specific amounts, the relevant Wigner distribution function is equally shifted in position and momentum, and hence the scalar product in phase space is a measure of the "propensity" of the two particle, in the considered states, "to have positions and momenta differing" just by those specific amounts [10.23]. This interpretation in terms of "proximity of states in phase plane" gives further insight into the measurement process, since one of the particles, prepared in the state φ_1, can be understood as a part of the apparatus for measuring certain properties of the other particle in the state φ_2.

Finally, as the reader can verify, the inner product of two functions equals the phase space integral of the cross Wigner distribution function of the two functions (see § 6.3.4 and § 9.3.2):

$$\int \varphi_1^*(q)\varphi_2(q)dq = \iint \mathcal{W}_{\varphi_1,\varphi_2}(q,\kappa)dqd\kappa. \qquad (6.3.55)$$

Therefore, Moyal's formula can be formulated in the alternative form

$$\tfrac{1}{2\pi}\left|\int \varphi_1^*(q)\varphi_2(q)dq\right|^2 = (\mathcal{W}_{\varphi_1},\mathcal{W}_{\varphi_2}) = \tfrac{1}{2\pi}\left|\iint \mathcal{W}_{\varphi_1,\varphi_2}(q,\kappa)dqd\kappa\right|^2. \qquad (6.3.56)$$

In the light of the discussion above, (6.3.56) relates the "proximity" of the two states φ_1, φ_2 to the phase space integral of the cross-Wigner distribution function of the two states. We have learned in § 6.3.4 that the cross Wigner distribution function expresses the cross-term contribution to the Wigner distribution function of the (possible) superposition of the two states; in a sense it is a measure of the "amount" of interference [10.24, 7].

The analogue in optics is found in the launching of an optical signal into a mode-supporting structure like a waveguide [10.25]-[10.28]. Evidently, the scalar product $\int \varphi_1^*(q)\varphi_2(q)dq$ relates to the coupling coefficient between the incoming signal, say φ_1, and a given mode, φ_2, of the waveguide. By Moyal's formula, such a coupling coefficient comes to be interpretable in the light of the overlap of the Wigner distribution functions of the signal and the waveguide mode, and so in the light of the "proximity" of the signal and the mode to occupy the same region in the space-angle plane. In a sense, it is a measure of the signal-mode interfering area in phase plane. This view has as well practical advantages since the phase plane representation may straightforwardly incorporates the effects of the finite angular aperture of the receiving structure. Also, it may be easier to obtain information about the Wigner distribution function rather than about the spatial distribution of the signal at the plane of interest. Besides, the Wigner distribution function, unlike the signal amplitude, benefits from very simple transfer laws (see § 8.3). Hence, it can be measured at any plane and then propagated up to the desired plane. Finally, as far as paraxial propagation is concerned, phase plane ellipses, pertaining to the signal and the waveguide modes, can be individualized in terms of the second-order moments of the respective Wigner distribution functions. Accordingly, the optimization of the coupling efficiency of the light launching into the waveguide acquires the intuitive visualization in terms of the matching of the phase-plane emittance of the incident signal to the phase-plane acceptance of the receiver's mode.

6.3.8 *Scaling property*

Under linear scaling the Wigner distribution function reproduces in its space and frequency dependence the scaling laws for the signal and its spectrum.

Suppose in fact to scale the signal $\varphi(q)$ to

$$\phi(q) = \sqrt{\mathfrak{a}}\,\varphi(\mathfrak{a}q), \tag{6.3.57}$$

with $\mathfrak{a}$ positive real parameter. The square root factor is required to keep the normalization of the scaled signal $\phi(q)$ the same as for the original signal $\varphi(q)$.

The Fourier spectrum scales inversely as the signal according to (§ 4.6.1)

$$\widetilde{\phi}(\kappa) = \tfrac{1}{\sqrt{\mathfrak{a}}}\,\widetilde{\varphi}(\tfrac{\kappa}{\mathfrak{a}}). \tag{6.3.58}$$

Likewise, by (6.2.2) we obtain for $\mathcal{W}_{\phi}(q, \kappa)$ the scaling relation

$$\mathcal{W}_{\phi}(q, \kappa) = \mathcal{W}_{\varphi}(\mathfrak{a}q, \tfrac{\kappa}{\mathfrak{a}}), \tag{6.3.59}$$

which expresses the *scale invariance* of the Wigner distribution function scaling in space and frequency the same way as the signal and its spectrum.

6.3.9 *Convolution and multiplication property*

The Wigner distribution function of the convolution or product of two functions is given respectively by the space domain or frequency domain convolution of the individual Wigner distributions of the two functions.

Given the function $\varphi(q)$ as the convolution integral

$$\varphi(q) = \int \varphi_1(q - q')\varphi_2(q')\, dq' \tag{6.3.60}$$

of the functions $\varphi_1(q)$ and $\varphi_2(q)$, the relevant Wigner distribution function can be easily evaluated making use of the κ-domain definition (6.2.3). Recalling from § 4.6.1 that the frequency spectrum $\widetilde{\varphi}(\kappa)$ relates to the spectra $\widetilde{\varphi}_1$ and $\widetilde{\varphi}_2$ of the convolved functions φ_1 and φ_2 by the product

$$\widetilde{\varphi}(\kappa) = \sqrt{2\pi}\widetilde{\varphi}_1(\kappa)\,\widetilde{\varphi}_2(\kappa), \tag{6.3.61}$$

we obtain

$$\mathcal{W}_{\varphi}(q, \kappa) = 2\int \widetilde{\varphi}_1^*(\kappa - \kappa')\widetilde{\varphi}_1(\kappa + \kappa')\widetilde{\varphi}_2^*(\kappa - \kappa')\widetilde{\varphi}_2(\kappa + \kappa')e^{2i\kappa' q}\, d\kappa'. \tag{6.3.62}$$

Then, applying relation (6.3.25) to the products $\widetilde{\varphi}_1^*(\kappa - \kappa')\widetilde{\varphi}_1(\kappa + \kappa')$ and $\widetilde{\varphi}_2^*(\kappa-\kappa')\widetilde{\varphi}_2(\kappa+\kappa')$, we end up with the convolution of the Wigner distributions $\mathcal{W}_{\varphi_1}$ and $\mathcal{W}_{\varphi_2}$ in the space variable; namely

$$\mathcal{W}_{\varphi}(q, \kappa) = 2\pi \int \mathcal{W}_{\varphi_1}(q - q', \kappa)\mathcal{W}_{\varphi_2}(q', \kappa)\, dq'. \tag{6.3.63}$$

Similarly, given $\phi(q)$ as the product of the functions $\phi_1(q)$ and $\phi_2(q)$, i.e.

$$\phi(q) = \phi_1(q)\,\phi_2(q), \tag{6.3.64}$$

it is convenient to use definition (6.2.2) and relation (6.3.23) to determine the relevant Wigner distribution function. The resulting expression,

$$\mathcal{W}_{\phi}(q,\kappa) = \int \mathcal{W}_{\phi_1}(q,\kappa-\kappa')\mathcal{W}_{\phi_2}(q,\kappa')\, d\kappa', \qquad (6.3.65)$$

is just the convolution of the Wigner distributions $\mathcal{W}_{\phi_1}$ and $\mathcal{W}_{\phi_2}$ along the frequency dimension. We note that the Fourier spectrum of (6.3.64) is the convolution of the spectra $\widetilde{\phi}_1$ and $\widetilde{\phi}_2$ of the functions ϕ_1 and ϕ_2 (§ 4.6.1)

$$\widetilde{\phi}(\kappa) = \tfrac{1}{\sqrt{2\pi}} \int \widetilde{\phi}_1(\kappa-\kappa')\widetilde{\phi}_2(\kappa')\, d\kappa'. \qquad (6.3.66)$$

A slightly more general result accounts for a possible scaling factor in (6.3.64). If the function $\phi(q)$ is obtained as

$$\phi(q) = \phi_1(q)\, \phi_2(\mathfrak{a}q), \qquad (6.3.67)$$

with $\mathfrak{a} > 0$, the Wigner distribution $\mathcal{W}_{\phi}(q,\kappa)$ still comes to be the frequency domain convolution of $\mathcal{W}_{\phi_1}$ and $\mathcal{W}_{\phi_2}$, the arguments of the latter being appropriately scaled according to (6.3.59), thus giving

$$\mathcal{W}_{\phi}(q,\kappa) = \tfrac{1}{\mathfrak{a}} \int \mathcal{W}_{\phi_1}(q,\kappa-\kappa')\mathcal{W}_{\phi_2}(\mathfrak{a}q, \tfrac{\kappa'}{\mathfrak{a}})\, d\kappa'. \qquad (6.3.68)$$

Note that the Fourier spectrum of (6.3.67) reads

$$\widetilde{\phi}(\kappa) = \tfrac{1}{\sqrt{2\pi}|\mathfrak{a}|} \int \widetilde{\phi}_1(\kappa-\kappa')\widetilde{\phi}_2(\tfrac{\kappa'}{\mathfrak{a}})\, d\kappa'. \qquad (6.3.69)$$

Expressions (6.3.63) and (6.3.65) of the Wigner distributions $\mathcal{W}_{\varphi}(q,\kappa)$ and $\mathcal{W}_{\phi}(q,\kappa)$ relative to the convolutions (6.3.60) and (6.3.66) make evident that the convolution is transmitted from the functions to the corresponding Wigner distributions, and involves the space or frequency dimension according to whether the functions or the respective Fourier transforms are convolved.

Interestingly, the convolutions (6.3.63) and (6.3.65) may be interpreted as the space-frequency representations of the transformations described by products (6.3.61) and (6.3.64). Thinking in physical terms, the latters may describe filtering operations respectively in the frequency and space domains, with one function representing the signal and the other the filter function. Thus, both the frequency and spatial filtering manifest in the q-κ plane through the convolution of the Wigner distributions of the signal and the filter function along the space and frequency dimension, respectively. Within the optical context, relations (6.3.60) and (6.3.67) may be seen as describing the transfer properties of space-invariant and diffraction-free optical systems, which, as noted in § 4.6.3, act as frequency and spatial filters for optical signals. Convolution (6.3.60) and product (6.3.64) express therefore the transformation laws for optical signals under propagation through a homogeneous medium and a thin lens, one of the functions representing the input waveform, the other one the line spread function of the medium or correspondingly the phase modulation function of the lens. Accordingly, convolutions (6.3.63) and (6.3.65) may be regarded as the transfer laws of the Wigner distribution function for free-propagation and lens transfer, as they relate the Wigner distribution function of the input and the output signals. We will recover them in § 8.3.1 when investigating the transfer laws for the Wigner distribution function under propagation through a free-medium section and a thin lens.

6.3.10 *Representability*

As a natural conclusion to this section, we comment on the *representability* of the Wigner distribution function.

Evidently, not every function of space and frequency can be understood as a Wigner distribution function, because there may not exist a signal, from which it might be generated. Thus, a $2D$ function of space and frequency is said to be Wigner *representable* or *realizable*, if it is generated by some signal.

An explicit condition can be established as a mean to ascertain whether a space-frequency function is representable as a Wigner distribution function or not [4.13, 6.7, 10.21]. Basically, it grounds on the compatibility of relations (6.3.23), (6.3.25) and (6.3.28); indeed, looking at (6.3.23) and (6.3.25), we see that a function of q and κ is a Wigner distribution if and only if the integrals on the right-hand sides result into the separable forms appearing in the left-hand sides. Specifically, we can phrase the necessary and sufficient condition for a function of the Fourier-conjugate variables $\mathcal{W}(q,\kappa)$ to be a representable Wigner distribution in terms of the function itself through the relation [4.13, 6.7]

$$\begin{aligned} &\mathcal{W}(q_0+\tfrac{q}{2},\kappa_0+\tfrac{\kappa}{2})\mathcal{W}(q_0-\tfrac{q}{2},\kappa_0-\tfrac{\kappa}{2}) \\ &=\tfrac{1}{2\pi}\iint \mathcal{W}(q_0+\tfrac{q'}{2},\kappa_0+\tfrac{\kappa'}{2})\mathcal{W}(q_0-\tfrac{q'}{2},\kappa_0-\tfrac{\kappa'}{2})e^{-i(\kappa q'-q\kappa')}dq'd\kappa'. \end{aligned} \tag{6.3.70}$$

which must hold for any q_0 and κ_0 (see Problem 5).

In practical terms, in order to establish whether a function is representable or not, we assume the function, we are dealing with, be a Wigner distribution function of some signal, which can be found from the inverse formula (6.3.28). Then, we calculate the Wigner distribution of the derived signal and compare it with the original function. Evidently, if we recover the same function, the distribution at hand is a representable Wigner distribution function. In § 6.4.6 we will illustrate an example of a function of q and κ, which is not a representable Wigner distribution function. As we will see, the features of the possible generating signal, derived from the inverse formula (6.3.28), are in contrast with relations (6.3.23) and (6.3.26).

Also, we know that the Wigner distribution function attains always negative values somewhere in the q-κ plane, except when it is generated by a Gaussian signal. Thus, any nonnegative $2D$ function can immediately be recognized as a non representable Wigner distribution function, unless it is in the Gaussian form (6.3.45) with the relevant parameters satisfying the equality in (6.3.46).

6.4 The Wigner distribution function of light signals: further examples

We will further illustrate the potential of the Wigner distribution function to act as a *local* frequency spectrum, simultaneously displaying the space and frequency features of the signal. We will consider typical waveforms[e] and derive the analytical expression of the corresponding Wigner distribution functions and/or show on graphs the $3D$ images of it [5, 10].

We will start by considering signals limited within a finite-length interval in space or frequency. The mathematical representation for such waveforms is the rectangular function, whose dual representation is the sinc-function.

Then, resorting to the prime Fourier-transform pair, i.e., the monofrequency and δ-like signals, discussed in § 6.2.1, we will firstly consider a cosine signal, arising as the sum of two spatial harmonics of equal and opposite frequencies. The obvious generalization is evidently the infinite (numerable) sum of pure spatial harmonics, which for suitable coefficients can be regarded as the Fourier series of spatially periodic wavefields. As a basic example of periodic signal, we will consider the comb function. It is of great relevance in connection with the analysis of optical effects, like, for instance, the Talbot and Lau effects, produced by proper illumination of suitably arranged periodic structures.

Frequency-modulated and amplitude-modulated signals will then be considered. In the first case, we will start describing the pure quadratic phase signals, which are the basic waveforms of first-order wave optics, naturally arising in fact in the Fresnel approximation as model function for the response function of optical systems as well as for the phase front of the incoming disturbance. As a natural continuation, the pure polynomial-phase signals are briefly commented on, being seen as a tool for investigations of the optical propagation properties beyond the Fresnel approximation.

As a primary example of amplitude-modulated signal, we will consider the real Gaussian amplitude which provides, for instance, the model function for the transverse profile of the fundamental mode of the field oscillations in a stable resonator laser as well as of the quantum mechanical harmonic oscillator.

---

[e] A detailed description of many typical waveforms of relevance in optics is [14].

The Hermite-Gaussian beams will also be discussed as well as the more recently introduced Hermite-cosh-Gaussian modes. We will give a numerical account of the superGaussian envelope waveforms, which may model the lateral profile of beams from high-gain unstable resonator lasers suitably equipped with superGaussian transmittance or reflectivity mirrors.

It is worth noting that, since the Wigner distribution function is defined by an infinite integral of a supposedly continuous function, its exact evaluation is in general impossible. Indeed, for almost all of the waveforms, we will consider in what follows, an exact expression of the respective Wigner distribution function can be obtained. This is so, except in the case of the quartic and quintic polynomial phase signals, and in the case of the superGaussian amplitude, for which a numerical account is given. Of course, this implies a discretization of the signal and so a discretization of the resulting distribution. We shall not dwell on the questions related to the definition of a discrete version of the Wigner distribution function and on the consequence of this discretization, for which the reader is directed to [5.3, 5.4]. In fact, it is our purpose on showing the $3D$ graphs and the relevant phase-plane contour plots of the aforementioned signals, only to give a feeling for their behaviour; no exactitude is needed. In this connection, we emphasize that the optical display of the Wigner distribution function provides an alternative effective tool for its evaluation; this will be described in the next chapter (§§ 7.6-7-9).

6.4.1 *The Wigner phase-plane*

It has been remarked in § 4.6.5 that as the Hamiltonian conjugate variables q, p span the geometrical optical phase space, where the light-ray dynamics can be pictorially displayed and effectively investigated, likewise the Fourier conjugate variables q, κ span the wave-optical phase space. It has also been noted that the matrix relation connecting the ray variables (q, p) at different planes along the propagation axis, can equally be extended to the wave-optical pair (q, κ), of course, in a purely formal fashion.

The Wigner distribution function composes both the spatial and frequency features of the signal into a single representation. So it is tailored to be that "entity" in terms of which we can afford a phase-plane visualization of the signal propagation properties, as we will do indeed in the examples below. However, such a picture in the phase plane (that we will also refer to as the *Wigner phase-plane*) suffers from an inherent limitation arising from the uncertainty relation of Fourier analysis, which, as it is commonly interpreted, makes "points" in the phase plane inaccessible and hence values of the joint representation at points in phase plane not measurable, and so meaningless. A critical comment on this commonly held notion will be addressed in § 9.4.4.

We know that the geometrical optical phase-space has a symplectic structure. The Wigner phase-space has its own structure as well, that an accurate investigation has revealed to be conform to an affine metric [4.6, 4.10], being invariant under unimodular linear transformations (a subgroup, indeed, of symplectic transformations, under which the geometrical-optical phase space is invariant). Unimodular linear transformations preserve straightness, mapping straight lines into straight lines (see § 6.4.4).

Within the paraxial approximation, amounting to symplectic transformations of the ray variables which actually are linear (represented,as they are, by the ray matrices), the difference between the geometrical-optical phase-space and the Wigner phase-space becomes imperceptible (apart from, of course, that arising from the indeterminacy relation). Also, it is beyond the purposes of our studies here and accordingly we shall not dwell on it any further.

Let us finally remark that, as also pointed out when dealing with geometrical-optical phase-plane, in order that the worked relations be meaningful we must think of the Wigner plane axes q and κ as suitably scaled (possibly by some characteristic length of the problem at hand), so that they have the same dimensionality. It is in the light of this implicit normalization that the mathematical relations, that will subsequently be derived, must be understood. Accordingly, no mention will be given to the units in relation to the values assigned to the parameters in the graphs (which will appear just as numbers), the specific units being, of course, determined in accord with the chosen normalization settings.

6.4.2 *Finite-extent waveforms in space or frequency*

Rectangular signal

A unit-magnitude uniform signal, limited to the space interval of length $2Q$, may be described as

$$\varphi_{\mathrm{R}}(q) = \mathrm{rect}(\tfrac{q}{2Q}) \equiv \begin{cases} 1 & |q| \leq Q, \\ 0 & \text{otherwise}, \end{cases} \tag{6.4.1}$$

where Q is a positive number. The frequency spectrum $\widetilde{\varphi}(\kappa)$ behaves as a sinc-function, being

$$\widetilde{\varphi}_{\mathrm{R}}(\kappa) = \sqrt{\tfrac{2Q^2}{\pi}}\,\mathrm{sinc}(\kappa Q). \tag{6.4.2}$$

Using definition (6.2.2), it is easy to work out for the Wigner distribution function $\mathcal{W}_{\mathrm{R}}(q,\kappa)$ of φ_{R} the explicit expression

$$\mathcal{W}_{\mathrm{R}}(q,\kappa) = \begin{cases} \frac{2Q}{\pi}(1-\frac{|q|}{Q})\mathrm{sinc}[2Q\kappa(1-\frac{|q|}{Q})] & |q| \leq Q, \\ 0 & |q| > Q. \end{cases} \tag{6.4.3}$$

As expected by virtue of the finite support property (§ 6.3.3), being the function (6.4.1) concentrated within the space interval $-Q \leq q \leq Q$, the

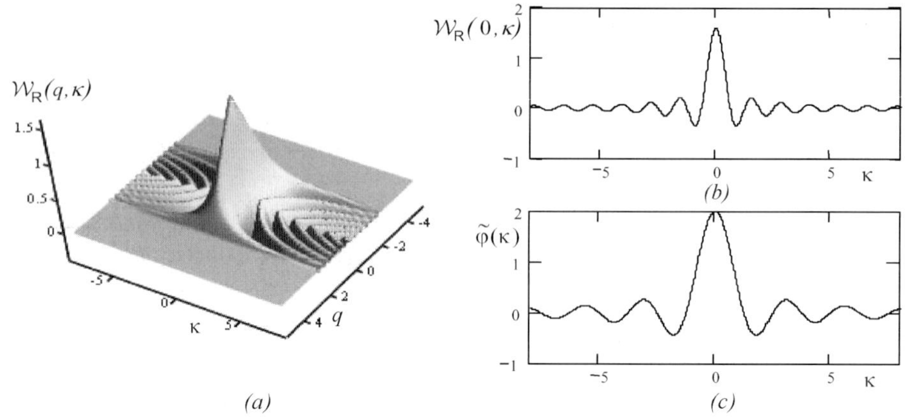

FIGURE 6.5. *(a)* 3*D* plot of the Wigner distribution function of a rectangle-signal of width $2Q = 5$, *(b)* relevant profile vs. κ at $q = 0$ and *(c)* Fourier transform of the rectangle-signal.

relevant Wigner distribution function comes to be concentrated in the phase plane into the vertical strip through the same space interval $-Q \leq q \leq Q$. Here, $\mathcal{W}_R(q, \kappa)$ displays a sinc-like shape along the frequency dimension. For each $q \in (-Q, Q)$, the main lobe is centered at $\kappa = 0$, and its width increases with increasing values of $|q|$, whilst the height decreases to zero. In particular, $\mathcal{W}_R(q, \kappa)$ attains its maximum at the space-frequency point $(0, 0)$ with the value: $\mathcal{W}_R(0, 0) = 2Q/\pi$, in accord with the limit expressed in (6.3.1). Notably, $\mathcal{W}_R(q, \kappa)$ is negative in certain regions of the q-κ plane. Finally, the zero value loci of the distribution (6.4.3) are the symmetrical hyperbolas

$$\kappa(1 - \tfrac{|q|}{Q}) = \tfrac{n\pi}{2Q}, \tag{6.4.4}$$

corresponding to integer values of $n = \pm 1, \pm 2, ...$

Figure 6.5.a) shows the 3*D* plot of $\mathcal{W}_R(q, \kappa)$ corresponding to a given value of Q. The sinc-like behaviour versus κ is evidenced in Fig. 6.5.b), showing the profile of the distribution versus κ at $q = 0$, which can be compared with the frequency spectrum $\widetilde{\varphi}(\kappa)$ traced in Fig. 6.5.c).

Sinc signal

The natural counterpart of the rectangle signal is sinc-amplitude signal,

$$\varphi_S(q) = \sqrt{\tfrac{2}{\pi Q^2}}\,\mathrm{sinc}(\tfrac{q}{Q}), \tag{6.4.5}$$

which in turn has the finite-extent frequency spectrum,

$$\widetilde{\varphi}_S(\kappa) = \mathrm{rect}(\tfrac{\kappa}{2K}), \qquad K \equiv \tfrac{1}{Q}. \tag{6.4.6}$$

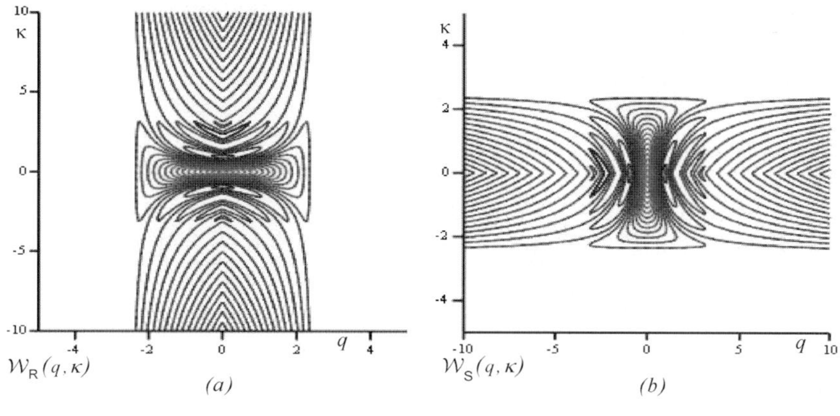

FIGURE 6.6. Contour plots of the Wigner distribution functions of *(a)* the rectagle-waveform (6.4.1) ($2Q = 5$) and *(b)* the sinc-waveform (6.4.5) ($2K = 5$).

Using (6.2.3) the relevant Wigner distribution function $\mathcal{W}_{\rm S}(q,\kappa)$ obtains as

$$\mathcal{W}_{\rm S}(q,\kappa) = \begin{cases} \frac{2K}{\pi}(1-\frac{|\kappa|}{K})\text{sinc}[2Kq(1-\frac{|\kappa|}{K})] & |\kappa| \leq K, \\ 0 & |\kappa| > K, \end{cases} \tag{6.4.7}$$

which evidently has the same behaviour with respect to the space variable as (6.4.3) with respect to the frequency variable. Figure 6.6 shows the (q,κ)-contour images of the distributions (6.4.3) and (6.4.7) for the same numerical value of Q and K; we may see the characteristic structure of nested symmetric hyperbolas, which are rotated by $\frac{\pi}{2}$ one to the other, being confined respectively within a vertical and a horizontal strip. Rectangle and sinc signals relate to each other by the Fourier integral, and accordingly the respective Wigner charts in the q-κ plane relate to each other by a$\frac{\pi}{2}$-rotation. Resorting to the results of § 6.2.1, we note that the same $\frac{\pi}{2}$-rotation in the q-κ plane relates the Wigner distribution functions of the impulse and the spatial harmonic, which are a Fourier-transform pair as well.

Modulated and shifted rectangular signal

As noted in § 6.3.2, a modulation of the rectangular signal (6.4.1) by the harmonic factor $e^{i\kappa_0 q}$ according to

$$\varphi(q) = \text{rect}(\tfrac{q}{2Q})e^{i\kappa_0 q} \tag{6.4.8}$$

produces a frequency shift of the corresponding Wigner distribution (6.4.3) as

$$\mathcal{W}_\varphi(q,\kappa) = \mathcal{W}_{\rm R}(q,\kappa-\kappa_0). \tag{6.4.9}$$

The above behaves as (6.4.3), apart from a shift of the center of gravity from the phase-plane point $(0, 0)$ to the point $(0, \kappa_0)$.

Translating signal (6.4.8) along the q line by q_0 as

$$\varphi(q) = \mathrm{rect}(\tfrac{q-q_0}{2Q})e^{i\kappa_0 q}, \tag{6.4.10}$$

leads to

$$\mathcal{W}_\varphi(q, \kappa) = \mathcal{W}_\mathrm{R}(q - q_0, \kappa - \kappa_0), \tag{6.4.11}$$

whose behaviour again replicates that of distribution (6.4.3) with the centroid being shifted to the point (q_0, κ_0) in the space-frequency plane.

Signals (6.4.8) and (6.4.10) have the same sinc^2-like power spectrum. In fact, their frequency spectra, respectively given by

$$\widetilde{\varphi}(\kappa) = \sqrt{\tfrac{2Q^2}{\pi}}\,\mathrm{sinc}[Q(\kappa - \kappa_0)], \quad \widetilde{\varphi}(\kappa) = \sqrt{\tfrac{2Q^2}{\pi}}\,\mathrm{sinc}[Q(\kappa - \kappa_0)]e^{-i(\kappa-\kappa_0)q_0}, \tag{6.4.12}$$

simply differ for the phase function $e^{-i(\kappa-\kappa_0)q_0}$. The corresponding square magnitudes shape therefore for both signals as the sinc^2-function

$$|\widetilde{\varphi}(\kappa)|^2 = \tfrac{2Q^2}{\pi}\,\mathrm{sinc}^2[Q(\kappa - \kappa_0)], \tag{6.4.13}$$

which evidently contains no information about the location in space of the energy density. In contrast, the vertical strips pertinent to the nonzero values of the Wigner distribution functions (6.4.9) and (6.4.11) have different locations in the (q, κ)-plane, as they retain memory, unlike the power spectra, of the space shift of (6.4.10) with respect to (6.4.8).

Thinking in optical terms, we note that the rectangle-shaped function (6.4.1) as well as the spatially shifted version $\mathrm{rect}((q - q_0)/2Q)$ in (6.4.10) may describe the transmittance of a slit of height $2Q$, located in an opaque screen with the center on the z axis or transversely shifted to it by q_0. The finite-length complex-sinusoids (6.4.8) and (6.4.10) may accordingly describe the light distribution immediately behind the screen, when it is illuminated on the other side by the plane wave $e^{i\kappa_0 q}$, and similarly, (6.4.1) may be seen as arising from a normally incident illuminating wave. In this connection, we recall that the far-field distribution of a plane wave issuing from a rectangular aperture like (6.4.1) is well known to have a sinc-like profile.

6.4.3 *Monofrequency and impulse related waveforms*

Cosine signal

The Wigner distribution of a cosine signal, described by the amplitude

$$\varphi(q) = \varphi_0 \cos(\kappa_0 q) = \tfrac{\varphi_0}{2}e^{i\kappa_0 q} + \tfrac{\varphi_0}{2}e^{-i\kappa_0 q}, \tag{6.4.14}$$

φ_0 being in general a complex constant, is easily obtained making use of property (6.3.11) and result (6.2.11). In fact, as the cosine wave can be represented as the sum of two plane waves of frequencies κ_0 and $-\kappa_0$, the Wigner distribution function obtains as

$$\mathcal{W}_\varphi(q,\kappa) = \frac{|\varphi_0|^2}{4}\{\delta(\kappa-\kappa_0)+\delta(\kappa+\kappa_0)+2\delta(\kappa)\cos(2\kappa_0 q)\}. \tag{6.4.15}$$

It displays two q-independent self-terms concentrated along the lines $\kappa = \kappa_0$ and $\kappa = -\kappa_0$ in the q-κ plane, and a q-dependent cross-term concentrated on the line $\kappa = 0$. In accord with the general considerations of § 6.3.4, the cross term exhibits an oscillatory behaviour at the frequency $2\kappa_0$; it accounts in fact for the sinusoidal oscillations of the signal power with q occurring at the frequency $2\kappa_0$ around the stable level $\frac{|\varphi_0|^2}{2}$. Also, it is sharp at $\kappa = 0$; this is the only value of frequency, where the spectrum $\widetilde{\varphi}(\kappa) \propto \delta(\kappa-\kappa_0)+\delta(\kappa+\kappa_0)$, which sharpens at $\kappa = \kappa_0$ and $\kappa = -\kappa_0$, can be folded to yield a nonzero overlap of the left over the right part, and so a nonzero cross-term.

Sum of two spatial harmonics. Fourier series

As a more general case, we may consider the sum of two spatial harmonics of frequencies κ_1 and κ_2, weighted by the complex coefficients φ_{0_1} and φ_{0_2}:

$$\varphi(q) = \varphi_{0_1} e^{i\kappa_1 q} + \varphi_{0_2} e^{i\kappa_2 q}. \tag{6.4.16}$$

In that case, the Wigner distribution function comes to be

$$\begin{aligned}\mathcal{W}_\varphi(q,\kappa) &= |\varphi_{0_1}|^2\delta(\kappa-\kappa_1) + |\varphi_{0_2}|^2\delta(\kappa-\kappa_2) + \\ &+ 2|\varphi_{0_1}\varphi_{0_2}|\cos[(\kappa_1-\kappa_2)q+\tau]\delta(\kappa - \tfrac{\kappa_1+-\kappa_2}{2})\end{aligned} \tag{6.4.17}$$

where τ denotes the phase difference between the two waves at $q = 0$, i.e.

$$e^{i\tau} = \frac{\varphi_{0_1}\varphi_{0_2}^*}{|\varphi_{0_1}\varphi_{0_2}|}. \tag{6.4.18}$$

Distribution (6.4.17) is concentrated along the auto-terms lines at the two component frequencies κ_1 and κ_2, and along the cross-term line at the mean frequency $\frac{\kappa_1+\kappa_2}{2}$. Like the signal power, the cross term manifests an oscillating behaviour at the difference frequency $\kappa_1 - \kappa_2$ with a constant phase fixed by the initial phase difference τ between the two component waves.

The graphs in Fig. 6.7 show the Wigner distribution function of a cosine amplitude and the relevant space-frequency contour image. Specifically, Fig. 6.7.a) pertains to an unlimited cosine signal. The contributions from the self- and the cross-terms, respectively located at the component frequencies and at the mean frequency, are evident. In turn, Fig. 6.7.b) refers to a finite-length cosine envelope. The delta spikes of Fig. 6.7.a), typical of unlimited sinusoids, widen into sinc-like profiles, proper of finite extent sinusoids.

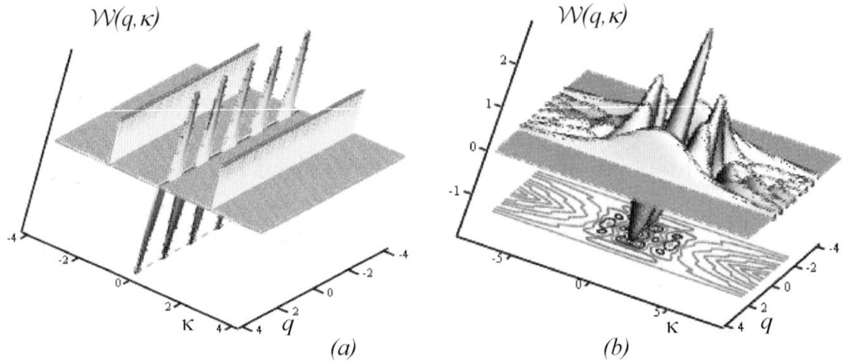

FIGURE 6.7. $3D$ view with the relevant (q,κ)-plane contour plots of the Wigner distribution functions of *(a)* the cosine signal $\cos(2q)$, and of *(b)* the same cosine signal limited to the finite-length interval $-2 \leq q \leq 2$.

We invite the reader to generalize expression (6.4.17) relevant to a signal composed only by two harmonics to the infinite series of harmonics

$$\varphi(q) = \sum_{n=-\infty}^{\infty} \varphi_n e^{i\kappa_n q}. \tag{6.4.19}$$

where φ_n's and κ_n's are assigned constants [10.29].

Notably, waveform (6.4.19) can be interpreted as a Fourier series describing for instance a periodic signal: $\varphi(q) = \varphi(q+d)$ (Problem 8). As is well known, in that case, the frequencies κ_n of the component harmonics in the series can be expressed as integer multiples of the *fundamental* frequency κ_0: $\kappa_n = n\kappa_0$, κ_0 being determined by the spatial period as $\kappa_0 = \frac{2\pi}{d}$. Besides, the weights φ_n of the various harmonics take the meaning as the Fourier coefficients, linked to the signal waveform $\varphi(q)$ by the integrals

$$\varphi_n = \tfrac{1}{d}\int_{-d/2}^{d/2} \varphi(q) e^{-in\kappa_0 q} dq. \tag{6.4.20}$$

Sum of two impulses

The Wigner distribution function of a signal, that arises as the superposition of two impulses, located at $q = q_1$ and $q = q_2$,

$$\varphi(q) = \delta(q - q_1) + \delta(q - q_2), \tag{6.4.21}$$

is formally similar to the Wigner distribution (6.4.17) of the sum of two spatial harmonics, with the variables q and κ being exchanged. Explicitly we find

$$\mathcal{W}_\varphi(q,\kappa) = \tfrac{1}{2\pi}\{\delta(q-q_1) + \delta(q-q_2) + 2\delta(q - \tfrac{q_1+q_2}{2})\cos[(q_1-q_2)\kappa]\}, \tag{6.4.22}$$

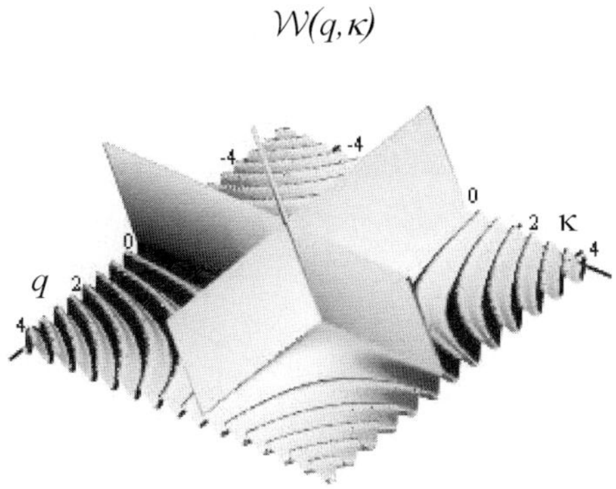

FIGURE 6.8. $3D$ view of the Wigner distribution function of the sum of an impulse and a spatial harmonic, having $\varphi_1 = \varphi_2 = 1$, $q_1 = 0.5$ and $\kappa_2 = 1$ (see Problem 9).

which shows two contributions at the lines $q = q_1$ and $q = q_2$, arising from the self-terms, and the cross-terms contribution, placing values along the intermediate lines $q = \frac{q_1+q_2}{2}$ in the space-frequency plane.

The reader may deduce the expression of the Wigner distribution function of the sum of a δ-function and a spatial harmonic, whose $3D$ plot is shown in Fig. 6.8 for given values of the relevant parameters (see Problem 9).

Train of δ-like waveforms

The sum of two δ-like waveforms in (6.4.21) can be generalized to comprise the infinite sequence of impulses

$$\varphi(q) = \sum_{m=-\infty}^{\infty} \delta(q - md), \tag{6.4.23}$$

which is one of the various representation reported in the literature for the *comb* function [14]; d denotes the spacing between any two consecutive δ functions. Graphically the series (6.4.23) is usually depicted as an infinite array of unit-height spikes located at positions $q = md$, $m = 0, \pm 1, \pm 2, ...$ (Fig. 6.9.a)).

Clearly, $\varphi(q)$ is periodic and hence admits the Fourier series representation

$$\varphi(q) = \frac{1}{d} \sum_{n=-\infty}^{\infty} e^{in\kappa_0 q}, \qquad \kappa_0 = \frac{2\pi}{d}. \tag{6.4.24}$$

The Wigner distribution writes therefore as

$$\mathcal{W}_{\text{comb}}(q, \kappa) = \frac{1}{d^2} \sum_{n=-\infty}^{\infty} \sum_{n'=-\infty}^{\infty} \delta(\kappa - \tfrac{n+n'}{2}\kappa_0) e^{-i(n-n')\kappa_0 q}, \tag{6.4.25}$$

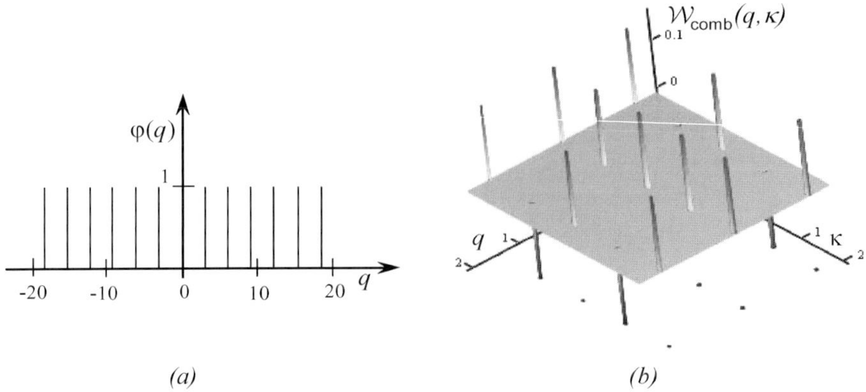

FIGURE 6.9. *(a)* Schematic of the comb function with $d = \pi$, and *(b)* 3D view of the relative Wigner distribution function with the corresponding lattice of points on the (q, κ)-plane.

which with a bit of manipulation can be recast into

$$\mathcal{W}_{\text{comb}}(q,\kappa) = \frac{1}{2d} \sum_{n'=-\infty}^{\infty} \frac{2}{d} e^{i2n'\kappa_0 q} \sum_{m=-\infty}^{\infty} \delta(\kappa - m\tfrac{\kappa_0}{2}) e^{-im\kappa_0 q}. \tag{6.4.26}$$

The first series above is just the Fourier series of the comb function with spatial period $\frac{d}{2}$: $\sum_{n'=-\infty}^{\infty} \frac{2}{d} e^{i2n'\kappa_0 q} = \sum_{n=-\infty}^{\infty} \delta(q - n\frac{d}{2})$. Then, exploiting the well known property of the comb function to extract the values of a function at positions regularly spaced by the period as

$$e^{-im\kappa_0 q} \cdot \sum_{n=-\infty}^{\infty} \delta(q - n\tfrac{d}{2}) = \sum_{n=-\infty}^{\infty} e^{-imn\kappa_0 \frac{d}{2}} \delta(q - n\tfrac{d}{2}), \tag{6.4.27}$$

we end up with the expression

$$\mathcal{W}_{\text{comb}}(q,\kappa) = \frac{1}{2d} \sum_{n=-\infty}^{\infty} \sum_{m=-\infty}^{\infty} (-)^{nm} \delta(q - n\tfrac{d}{2}) \delta(\kappa - m\tfrac{\kappa_0}{2}). \tag{6.4.28}$$

We note that the product of the two delta functions can also be embedded into the two-dimensional δ-function [14]

$$\delta(q - n\tfrac{d}{2}) \delta(\kappa - m\tfrac{\kappa_0}{2}) = \delta(q - n\tfrac{d}{2}, \kappa - m\tfrac{\kappa_0}{2}). \tag{6.4.29}$$

Distribution (6.4.28) identifies a lattice of delta functions with q- and κ-spacings respectively of $\frac{d}{2}$ and $\frac{\kappa_0}{2}$. On graphs the δ functions are visualized as spikes of height $\frac{1}{2d}$ located at the points $(n\frac{d}{2}, m\frac{\kappa_0}{2})$; so, each of them has the volume $\frac{1}{2d}$. However, $\mathcal{W}_{\text{comb}}(q,\kappa)$ cannot be interpreted as a $2D$ comb function, as the sign of the various δ functions depends on the lattice points and hence on the integers n and m. The δ functions located at points identified by an even n or m are positive, whilst those placed at points identified by an odd n have alternate signs with m and vice versa (Fig. 6.9.b)).

The comb function reveals a useful mathematical tool when dealing, for instance, with the sampling of a signal at regular intervals. As an example, we may consider an opaque screen containing a certain number of very narrow identical slits, regularly spaced by d. The screen is uniformly illuminated on one side by a collimated light beam, whose optical axis, for formal convenience, may be supposed to pass through some one of the slits. Then, as the amplitudes contributed by the individual slits are of the same magnitudes, the signal amplitude $\psi(q)$ on the other side of the screen can be represented by

$$\psi(q) = \mathcal{A} \sum_{m=-N}^{N} \delta(q - md), \tag{6.4.30}$$

where the constant $\mathcal{A}$ relates to the illuminating amplitude through the transmittance of the slits and the sum extends over all the $2N+1$ slits comprised within the beam width $2w_0$, i.e., $Nd \leq w_0$.

Due to the sampling property of the comb function, we can also write

$$\psi(q) = \mathcal{A}\,\text{rect}(\tfrac{q}{2w_0}) \sum_{m=-\infty}^{\infty} \delta(q - md), \tag{6.4.31}$$

thus conveniently expressing the transmitted amplitude as the product of the comb function and the rectangle function. Exploiting the windowing property (6.3.65) and the proper expressions (6.4.3) and (6.4.28) for $\mathcal{W}_{\text{R}}$ and $\mathcal{W}_{\text{comb}}$, the reader may deduce the explicit expression of the Wigner distribution function of the finite δ-train (6.4.31). As earlier mentioned, periodic structure are of relevance in connection with investigations of self-imaging phenomena, in which the lateral periodicity of the wave field is mirrored into a longitudinal periodicity as well. The Wigner distribution function-based description yields a clear, almost visualizable, account of the properties of the phenomenon [10.30, 10.29]. As we will see in § 8.3.4, the correspondence of the lateral to the longitudinal periodicity finds a plain interpretation in the light of the propagation properties of the Wigner distribution function.

6.4.4 *Frequency-modulated waveforms*

Quadratic phase signal

The quadratic-phase signal is described by the purely imaginary Gaussian

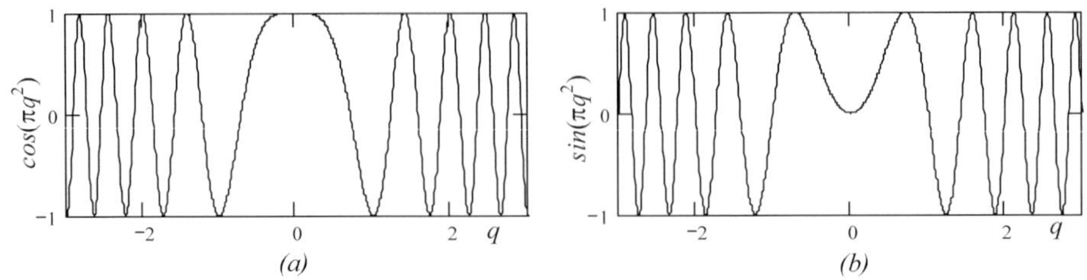

FIGURE 6.10. *(a)* Real part and *(b)* imaginary part of the chirp $e^{i\pi q^2}$.

$$\varphi(q) = e^{i\beta\frac{q^2}{2}}, \tag{6.4.32}$$

with β being a real number.

It should be mentioned, perhaps, that, in the language of signal theory, waveform (6.4.32) is referred to as a *chirp signal* (or chirp, for short). Roughly speaking, chirps are in general intended as waveforms of the type

$$\varphi(q) = a(q)e^{i\phi(q)}, \tag{6.4.33}$$

where the amplitude function $a(q)$ is supposed to produce a smooth modulation of the rapid oscillations produced by the phase function $\phi(q)$. The quadratic-phase signal (6.4.32) is a particular case of the *linear chirp*, which having constant amplitude and quadratic polynomial phase designates in general a linearly frequency modulated signal. Needless to say, waveform (6.4.32) is of great relevance in optics. In fact, we have found in Sect. 5.2 that it is relates to the line spread of the first-order optical models of both filtering and modulating systems, like free-medium sections and thin lenses. So, we are exclusively concerned with linear chirps.

Waveform (6.4.32) exhibits an oscillatory behaviour at the local frequency $\mathfrak{K}(q)$, which, given by the derivative of the phase: $\mathfrak{K}(q) = \frac{d}{dq}(\beta\frac{q^2}{2}) = \beta q$, changes linearly with q. As an example, the real and imaginary parts of the chirp $e^{i\pi q^2}$, i.e. $\cos\pi q^2$ and $\sin\pi q^2$, are plotted in Fig. 6.10. Both $\cos\pi q^2$ and $\sin\pi q^2$ oscillate at an increasing frequency and their first lobes lie within the intervals $|q| < \frac{1}{\sqrt{2}}$ and $|q| < 1$, respectively. Signal (6.4.32) is represented in the spatial frequency domain by a quadratic-phase signal as well, the Fourier transform being, in fact, $\widetilde{\varphi}(\kappa) = \sqrt{i^{\text{sgn}(\beta)}/|\beta|}e^{-i\kappa^2/2\beta}$, with $\text{sgn}(\beta) \equiv \beta/|\beta|$.

An easy calculation leads to the following expression

$$\mathcal{W}_{\text{chirp}}(q,\kappa) = \delta(\kappa - \beta q) = \delta(\kappa - \mathfrak{K}(q)), \tag{6.4.34}$$

for the Wigner distribution function of (6.4.32). It is totally concentrated along the local frequency curve and accordingly it identifies on the q-κ plane a straight line through the origin making the angle $\theta = \tan^{-1}\beta$ with the q-axis. In § 6.2.1, the Wigner distribution function of the spherical wave (6.2.12) has been found to be confined along a δ-line, making the angle $\theta = \tan^{-1}(k_0/R)$ (R is now intended to be scaled by the refractive index medium n_0) with the q-axis in the space-frequency plane. Indeed, pertinently with the linear approximation (i.e. small spatial and angular extensions across the reference plane), the amplitude (6.2.12), representative of the spherical wave with curvature radius R at the observation point q, can be assimilated to a chirp signal with coefficient $\beta = k_0/R$. In the same approximation, a thin lens is known to produce a modulation of the signal amplitude by the chirp $e^{-ik_0 q^2/2f}$.

More in general, a linear chirp accounts for both space and frequency shifts with respect to the envelope (6.4.32) according to

$$\varphi_{\text{chirp}}(q) = e^{i\frac{\beta}{2}(q-q_0)^2 + i\kappa_0 q}. \tag{6.4.35}$$

the local frequency $\mathfrak{K}(q) = \kappa_0 + \beta(q - q_0)$ being still linear in q. The representative line of the distribution (6.4.34) in the phase plane is then translated to pass through the point (q_0, κ_0). We obtain in fact

$$\mathcal{W}_{\text{chirp}}(q, \kappa) = \delta(\kappa - \kappa_0 - \beta(q - q_0)) = \delta(\kappa - \mathfrak{K}(q)). \tag{6.4.36}$$

As noted in § 6.2.1, the Wigner distribution of both an impulse and a monofrequency signal configures as a δ-line. We have found here that a δ-line form as well characterizes the Wigner distribution function of a chirp. Due to the relevance that these signals have in optics, we summarize the results concerning the respective Wigner distributions into the table below with the visual aid of Fig. 6.11, showing the pertinent supports in the q-κ plane.

$\varphi(q)$	$\mathcal{W}_\varphi(q, \kappa)$	$\widetilde{\varphi}(\kappa)$
$\delta(q - q_0)$	$\frac{1}{2\pi}\delta(q - q_0)$	$\frac{e^{-iq_0\kappa}}{\sqrt{2\pi}}$
$\frac{e^{i\kappa_0 q}}{\sqrt{2\pi}}$	$\frac{1}{2\pi}\delta(\kappa - \kappa_0)$	$\delta(\kappa - \kappa_0)$
$e^{i\frac{\beta}{2}(q-q_0)^2 + i\kappa_0 q}$	$\delta(\kappa - \kappa_0 - \beta(q - q_0))$	$\sqrt{\frac{i\beta/\|\beta\|}{\|\beta\|}}\, e^{-\frac{i}{2\beta}(\kappa-\kappa_0)^2 - iq_0(\kappa-\kappa_0)}$

(6.4.37)

In all cases the support of the Wigner distribution in the q-κ plane is a straight line, respectively disposed vertical through the point $q = q_0$, horizontal through the point $\kappa = \kappa_0$, and inclinated by the angle $\theta = \tan^{-1}\alpha$ to the q-axis through the point (q_0, κ_0). As already remarked, the impulse and the harmonic wave are Fourier conjugate to each other, and the supports of the respective Wigner distribution functions relate to each other by a $\pi/2$-rotation in the q-κ plane. It is worth noting that from chirp (6.4.35) one can recover the

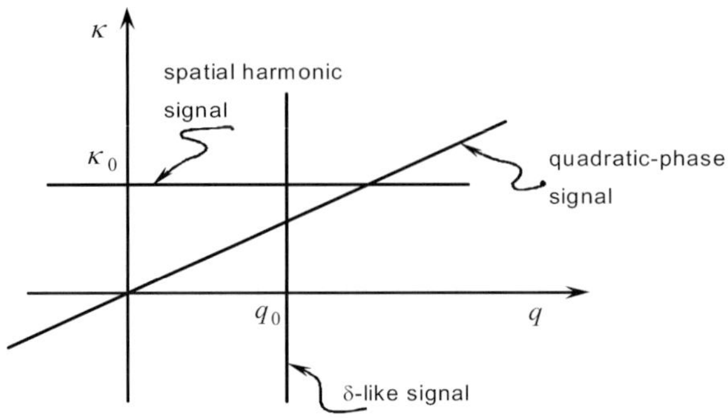

FIGURE 6.11. Supports of the Wigner distribution functions of a δ-like, a spatial harmonic and a quadratic-phase signal.

impulse and the spatial harmonic, sending respectively $\alpha \to \infty$ and $\alpha \to 0$; in the phase plane picture of Fig. 6.11, in fact, the representative line for chirp (6.4.35) rotates to the vertical line $q = q_0$ with α approaching $+\infty$ and to the horizontal line $\kappa = \kappa_0$ with $\alpha \to 0$.

Resorting to the operator formalism of Chapter 4, we recall that the δ function and the complex exponential are the eigenfunctions of the position and momentum (and so, spatial frequency) operators, $\widehat{\mathsf{q}}$ and $\widehat{\mathsf{p}}$ (and so, $\widehat{\kappa}$). Accordingly, the quadratic-phase signal (6.4.35) can be understood as the eigenfunction of the operator $a\widehat{\mathsf{q}} + b\widehat{\mathsf{p}}$, obtained by a linear transformation of $\widehat{\mathsf{q}}$ and $\widehat{\mathsf{p}}$ for real values of the coefficients a and b. In all cases the Wigner distribution function is a δ-function, whose argument identifies a line in the Wigner plane. Since linear transformations of $\widehat{\mathsf{q}}$ and $\widehat{\mathsf{p}}$ leave linear functions linear, lines remain lines under linear transformations of the Wigner plane. As we know, linear transformations are basic to first-order optics. This means that within the framework of first-order optics the impulse, the spatial harmonic and the chirp are the only "states", whose Wigner distribution function is a δ-line, which in addition remains a δ-line under any first-order optical transformation. As earlier noted, this property identifies the affine geometry of the Wigner phase-plane.

We invite the reader to calculate the Wigner distribution function of the sum of two chirps and of a chirp limited to a finite extent interval (see Fig. 6.12, and Problem 9). It is interesting to compare the Wigner distribution function obtained for the finite-extent chirp (Fig. 6.12.b)), which may optically represent a spherical wave issuing from a rectangular aperture, with the Wigner distribution function of (6.4.1), which as noted may optically represent a unit-amplitude plane wave issuing from a rectangular aperture.

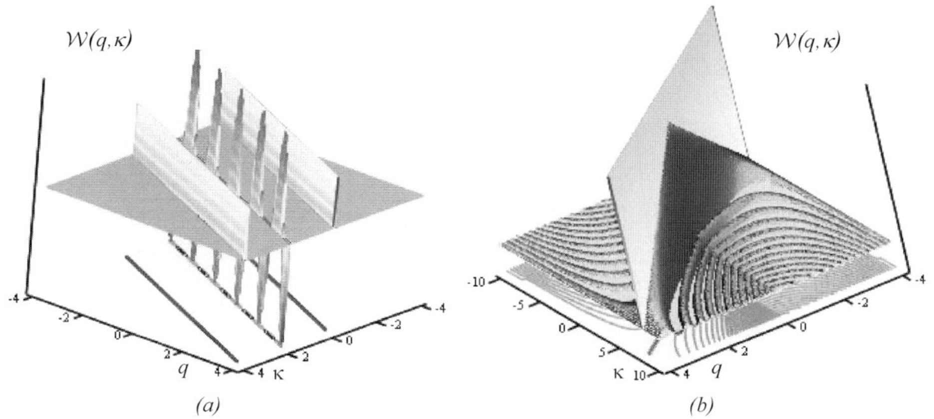

FIGURE 6.12. $3D$ image and relevant (q, κ)-contour plots of the Wigner distribution functions of *(a)* the sum of the chirps $e^{i(q^2+4q)}$ and $e^{i(q^2-4q)}$, and of *(b)* the chirp e^{iq^2} limited to the finite extent interval $-4 \leq q \leq 4$. Here, it is also shown the δ-function wall, with the relevant straight trace on the (q, κ)-plane, representative of the Wigner distribution function of the unlimited chirp e^{iq^2}.

Cubic phase signal

The cubic phase signal is represented by the complex amplitude

$$\varphi_{\text{cubic}}(q) = e^{i\beta\frac{q^3}{3}}, \tag{6.4.38}$$

which displays a local frequency quadratically varying with q as $\mathfrak{K}(q) = \beta q^2$. Inserting waveform (6.4.38) into the Wigner integral (6.2.2), we obtain

$$\mathcal{W}_{\text{cubic}}(q, \kappa) = \frac{1}{\pi} \int e^{i[\frac{2}{3}\beta q'^3 + 2(\mathfrak{K}(q)-\kappa)]} dq', \tag{6.4.39}$$

which, on due account of the integral representation of the Airy function [18]

$$Ai(\frac{x}{(3a)^{1/3}}) = \frac{(3a)^{1/3}}{2\pi} \int e^{i(ax'^3+xx')} dx', \tag{6.4.40}$$

is recast in the closed form

$$\mathcal{W}_{\text{cubic}}(q, \kappa) = (\tfrac{4}{\beta})^{1/3} Ai((\tfrac{4}{\beta})^{1/3}(\mathfrak{K}(q) - \kappa)). \tag{6.4.41}$$

As shown in Fig. 6.13.a), the Airy function oscillates for negative arguments whilst shows a fast decay for positive arguments. Therefore, distribution (6.4.41) is expected to have a fast decay at the convex side of the local frequency curve $(q, \mathfrak{K}(q))$ in the q-κ plane and in contrast an oscillatory trend at

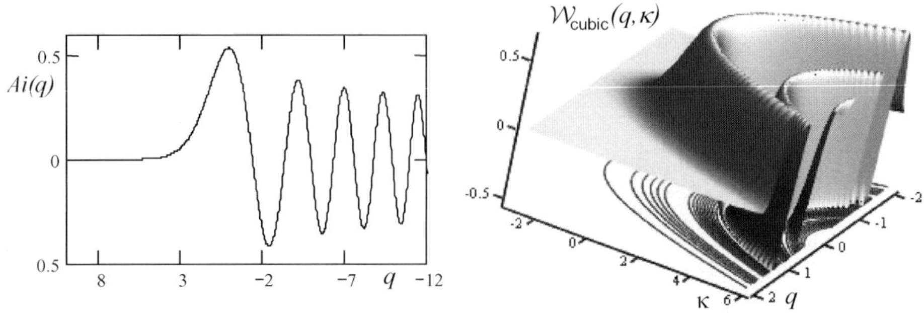

(a) (b)

FIGURE 6.13. *(a)* Airy function $Ai(x)$, and *(b)* $3D$ image of the Wigner distribution function $\mathcal{W}_{\text{cubic}}(q,\kappa)$ of the cubic phase signal $e^{i\frac{2q^3}{3}}$ with the relevant q-κ contour image.

the concave side of the curve. This is the behaviour we may observe inspecting the surface graph and the corresponding contour plots of $\mathcal{W}_{\text{cubic}}$, shown in Fig. 6.13.b) for a given value of β. Evidently, unlike $\mathcal{W}_{\text{chirp}}(q,\kappa)$, the Wigner distribution function of a cubic-phase signal does not develop exclusively around the local frequency line $\kappa = \mathfrak{K}(q)$, although it remains mainly concentrated about that line. In Chapter 9 we will see that the local frequency (i.e. the derivative of the signal phase) is the first-order local moment of the Wigner distribution function in the frequency variable, and hence in a sense it represents the local center of gravity of the distribution with respect to the frequency dimension. It is then highly desirable that the Wigner function as a sort of space-frequency energy density distribution be mainly concentrated near the local frequency line in the space-frequency plane, around which the signal energy is expected to be located. This is indeed the typical behaviour of the Wigner distribution function as we will see in relation to other signals as well [10.4, 5.7].

Polynomial phase signals

Figure 6.14 shows the trend of the Wigner distribution functions of the pure phase signal

$$\varphi_n(q) = e^{i\beta\frac{q^n}{n}}, \qquad n \geq 2 \tag{6.4.42}$$

with in particular $n = 4$ and $n = 5$, whose local frequency lines are respectively given by the cubic curve $\mathfrak{K}(q) = \beta q^3$ and the quartic curve $\mathfrak{K}(q) = \beta q^4$. Note that an exact expression of the Wigner distribution function for the polynomial phase signal (6.4.42) is not obtainable.

We see that, like $\mathcal{W}_{\text{cubic}}$, $\mathcal{W}_{\text{quartic}}(q,\kappa)$ and $\mathcal{W}_{\text{quintic}}(q,\kappa)$ model essentially in conformity to the local frequency line in the q-κ plane, displaying a rapid

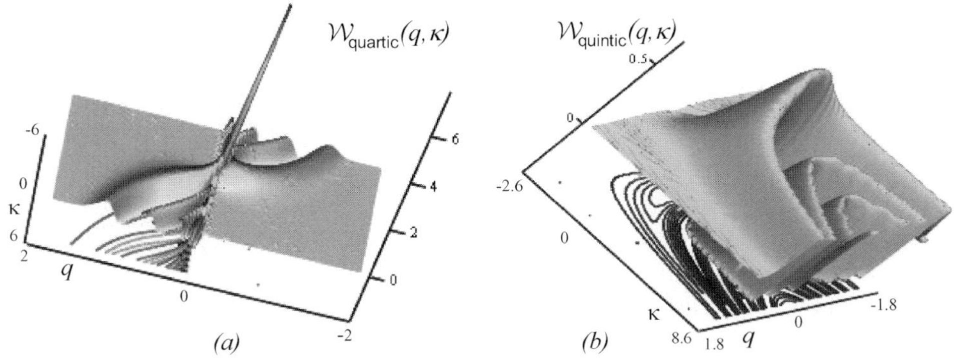

FIGURE 6.14. 3*D* image with the relevant q-κ contour image of the Wigner distribution functions of *(a)* the quartic phase signal $e^{i\frac{q^4}{2}}$ and *(b)* the quintic phase signal $e^{i\frac{2q^5}{5}}$.

decay at the convex side of that curve and an oscillatory trend at the concave side. Note also that for even powers the Wigner distribution function at the origin of the q-κ plane comes to be a δ-point, being $\mathcal{W}_{2n}(0,\kappa) = \delta(\kappa)$.

The n-th power phase signals of the type (6.4.42) are a particular case of the more general polynomial phase signals represented by

$$\varphi_{\text{pol}}(q) = e^{i\sum_{j=1}^{N}\beta_j q^j}, \tag{6.4.43}$$

Due to the Weierstrass theorem, allowing for a uniform polynomial approximation, with properly chosen coefficients, of any continuous function over a closed interval, waveforms of the type (6.4.43) may model the response function of different kinds of systems. From the optical viewpoint, higher-order powers than the second in the phase function in (6.4.43) may signify the corrections to the parabolic approximation of the distance of the image to the object, or also to the exact phase of the response function of a pure phase-modulation system. The Wigner distribution function may accordingly offer an effective tool for testing the departure of the response function of a thin lens from ideality, and thereby for studying the relative wave aberrations. In fact, an ideal lens should affect the signal by a pure quadratic-phase modulation, imaging for instance a unit-amplitude normally incident plane wave into the spherical wave front signal $e^{-ik_0q^2/2f}$ (apart from a constant phase factor accounting for the real thickness of the lens). If we could produce and display the Wigner distribution function of the transmitted wave, we might check the presence of aberrations produced by the lens, inspecting the deviations of the Wigner chart from the rectilinear pattern, expected for a pure quadratic-phase modulation (see § 7.8)[12.3, 10.32].

Sinusoidal phase signal

As a final example of frequency modulated signals, we consider the sinusoidally frequency modulated envelope

$$\varphi(q) = e^{i[\kappa_0 q + \frac{\gamma}{2}\sin(\kappa_f q)]}. \tag{6.4.44}$$

The local frequency of the signal has an oscillatory variation with q about the carrier frequency as $\mathfrak{K}(q) = \kappa_0 + \kappa_f \gamma \cos(\kappa_f q)/2$. Hence κ_f is the modulation frequency and γ specifies the maximum excursion of the phase variation.

By (6.2.2), we obtain for the Wigner distribution function the integral

$$\mathcal{W}_\varphi(q,\kappa) = \frac{1}{\pi}\int e^{-2i(\kappa-\kappa_0)q'} e^{i\gamma\cos(\kappa_f q)\sin(\kappa_f q')} dq', \tag{6.4.45}$$

which can be further handled by resorting to the well known identity

$$e^{ix\sin\phi} = \sum_{n=-\infty}^{\infty} e^{in\phi} J_n(x), \tag{6.4.46}$$

where J_n is the Bessel function of first kind of order n [18]. Thus, we can write

$$\begin{aligned} \mathcal{W}_\varphi(q,\kappa) &= \frac{1}{\pi}\sum_{n=-\infty}^{\infty} J_n(\gamma\cos(\kappa_f q)) \int e^{-2i(\kappa-\kappa_0-\frac{n\kappa_f}{2})q'} dq' \\ &= \sum_{n=-\infty}^{\infty} J_n(\gamma\cos(\kappa_f q))\delta(\kappa-\kappa_0-\tfrac{n\kappa_f}{2}), \end{aligned} \tag{6.4.47}$$

which shows a line structure with horizontal lines through the points $\kappa = \kappa_0 + n\kappa_f/2$, equally spaced by half of the modulation frequency. The contribution on each of these lines varies periodically with q through magnitudes governed by the Bessel functions.

Sinusoidal phase signals like (6.4.44) are appropriate to describe the light distribution transmitted by a sinusoidal phase grating, whose transmittance function $\mathsf{t}_{\rm ph}(q)$ may indeed be written as

$$\mathsf{t}_{\rm ph}(q) = e^{i\frac{\gamma}{2}\sin(\kappa_f q)}. \tag{6.4.48}$$

If the grating is illuminated on one side by a unit-amplitude, normally incident plane wave, the light distribution on the other side is given by (6.4.48). We also mention, as counterpart of (6.4.48), the sinusoidal amplitude grating, whose transmittance $\mathsf{t}_{\rm am}(q)$ is mathematically represented by the function

$$\mathsf{t}_{\rm am}(q) = 1 + \cos(\kappa_f q). \tag{6.4.49}$$

Interestingly, whereas the transmittance function (6.4.48) transforms an incident plane wave into the infinite superposition of diffracted waves $J_n(\frac{\gamma}{2})e^{in\kappa_f q}$, the transmittance (6.4.49) splits the incident wave into three waves: the undiffracted wave and the $\pm$1st order diffracted waves. The reader may calculate the Wigner distribution function of the transmittance (6.4.49) (see also § 8.3.4)

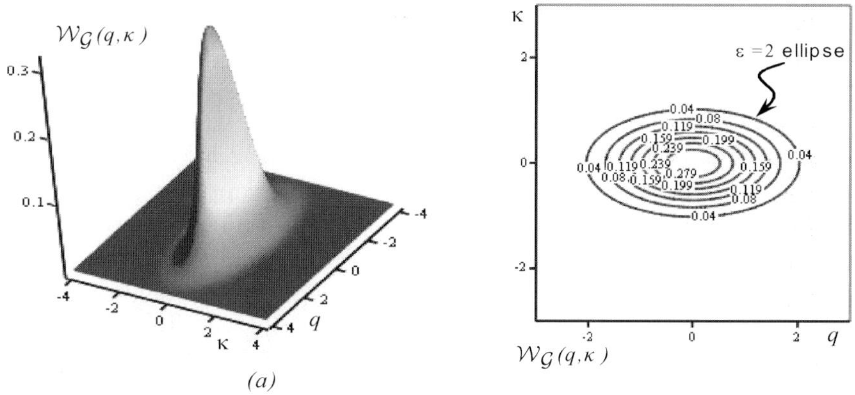

FIGURE 6.15. *(a)* 3D plot and *(b)* relevant q-κ elliptical contours of the Wigner distribution function $\mathcal{W}_{\mathcal{G}}(q,\kappa)$ of the Gaussian waveform $(\frac{1}{2\pi})^{1/4}e^{-q^2/4}$.

6.4.5 *Amplitude-modulated waveforms*

Gaussian signal

As a primary example of amplitude modulated waveforms, we consider the real Gaussian signal

$$\varphi_{\mathcal{G}}(q) = (\tfrac{2}{\pi w^2})^{1/4} e^{-\frac{1}{w^2}q^2}, \tag{6.4.50}$$

where w is a positive quantity, representing the e^{-2} intensity extension of the signal. The corresponding Wigner distribution function $\mathcal{W}_{\mathcal{G}}(q,\kappa)$ is easily obtained in the form

$$\mathcal{W}_{\mathcal{G}}(q,\kappa) = \tfrac{1}{\pi} e^{-\frac{2}{w^2}q^2 - \frac{w^2}{2}\kappa^2}, \tag{6.4.51}$$

which exhibits a Gaussian profile in both space and frequency dimensions in the (q,κ) plane. It is seen that $\mathcal{W}_{\mathcal{G}}$ is the same as the Wigner distribution function $\mathcal{W}_{\mathrm{G}}$ of the Gabor signal (6.2.14), on due account of the response of the Wigner distribution function to space and frequency shifts of the signal. The Gabor amplitude (6.2.14) results in fact from the plain Gaussian amplitude (6.4.50) through a suitable sequence of space and frequency shifts.

Figure 6.15.a) shows the familiar "bell" shape of $\mathcal{W}_{\mathcal{G}}(q,\kappa)$, with the relevant space-frequency contours being pictured in 6.15.b). Described by the equation

$$\tfrac{2}{w^2}q^2 + \tfrac{w^2}{2}\kappa^2 = \varepsilon, \tag{6.4.52}$$

for arbitrary positive numbers ε, the level curves are identified as ellipses, centered on the phase plane origin with axes determined by the signal parameter w and the assigned number ε. The ellipse corresponding to the value $\varepsilon = 2$, and hence to the value $\frac{1}{\pi}e^{-2} \sim 0.04$ of $\mathcal{W}_{\mathcal{G}}$, is marked in the figure. It relates to the concepts of phase space *emittance* and *acceptance*. We have earlier touched on these concepts in § 2.5, when illustrating the geometric-optics phase-plane view of the evolution of the Gaussian ray-density (2.2.83), whose space-frequency contour image consists as well of nested ellipses (Fig 2.13). It has been mentioned there that the concepts of emittance and acceptance provide a valid tool in regard, for instance, to the design of optical communication channels, being instrumental to determine the efficiency of the coupling of the light from the source to the receiver or more generally from one component to another in the channel [10.25]-[10.28]. We also emphasized that the concepts of phase-space *emittance* and *acceptance* are familiar as well to the optics of charged particle beams, being accordingly used in the design of transport channels for particle accelerators. In this connection, we note that the phase plane of magnetic optics is spanned by the conjugate variables (x, x'), representing respectively the small transverse deviation and the relevant slope of the "real" path of the charged particle with respect to the *design trajectory* of the transport channel. Evidently, (x, x') parallel the pair (q, p) of geometrical optics.

Complex Gaussian amplitude

The more general complex Gaussian amplitude function

$$\varphi_{\mathcal{G}}(q) = (\tfrac{2}{\pi w^2})^{1/4} e^{-\frac{1}{w^2}(q-q_0)^2 + i\frac{\beta}{2}(q-q_0)^2 + i\kappa_0 q}, \tag{6.4.53}$$

is obtained from a quadratic phase modulation of the real Gaussian amplitude (6.4.50), further shifted along the space dimension by q_0. In optical terms, waveform (6.4.50) may accounts for the q-profile of the amplitude distribution of a Gaussian beam at its waist; we may think, for instance, of the transverse envelope of a TEM_{00} mode from a stable unconfined large Fresnel-number optical resonator at the position of its minimum lateral extension [15]. Correspondingly, (6.4.53) may describe the q-profile of the amplitude distribution of a Gaussian beam at the position along its own optical axis where it exhibits w as beam radius and β as wavefront curvature, scaled by the light wavenumber; possibly, the optical axis of the beam may not be aligned with the chosen z axis thus accounting for the space and frequency shifts displayed in (6.4.53).

Convolving in the frequency variable the Wigner distribution functions (6.4.51) and (6.4.34) of the real Gaussian amplitude and the chirp, we obtain for the Wigner distribution function of the envelope (6.4.53) the expression

$$\mathcal{W}_{\mathcal{G}}(q, \kappa) = \tfrac{1}{\pi} e^{-\frac{2}{w^2}(q-q_0)^2 - \frac{w^2}{2}[\kappa-\kappa_0-\beta(q-q_0)]^2} = |\varphi_{\mathcal{G}}(q)|^2 |\widetilde{\varphi}_{\mathcal{G}}(\kappa)|^2. \tag{6.4.54}$$

The elliptical contours of $\mathcal{W}_{\mathcal{G}}(q, \kappa)$ are traced in Fig. **??**, the relevant equation being

$$\tfrac{2}{w^2}(q - q_0)^2 + \tfrac{w^2}{2}[\kappa - \kappa_0 - \beta(q - q_0)]^2 = \varepsilon. \tag{6.4.55}$$

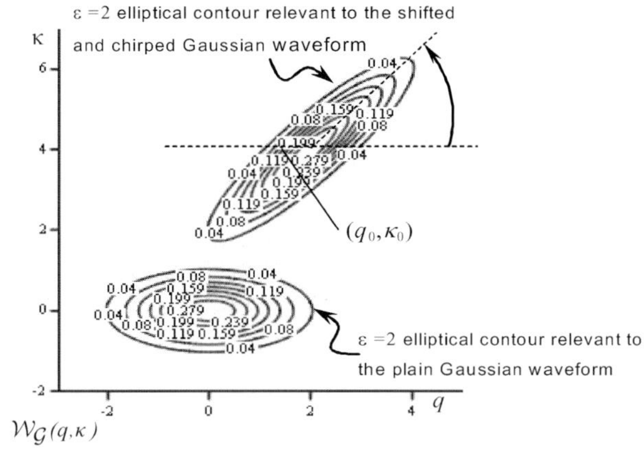

FIGURE 6.16. The elliptical contours of the Wigner distribution function of the chirped and shifted Gaussian signal (6.4.53) may be seen as resulting from a translation and a κ-shear of the elliptical contours of the Wigner distribution function of the basic Gaussian amplitude (6.4.50) (see Fig. 6.15). In the graph, $w = 2$, $q_0 = 2$, $\kappa_0 = 4$.

A comparison with the ellipses relevant to the Wigner distribution function (6.4.51), also traced in the figure, allows one to recognize a translation of the ellipse center from the origin to the phase-plane point (q_0, κ_0), and a κ-shear, determined by the curvature parameter β. With increasing the value of w the chirp dominates over the Gaussian amplitude, which widens almost uniformly over the q-axis. On graphs, with $w \to +\infty$ the phase-plane ellipse collapses into the representative line of the chirp signal, i.e. the line delta $\delta(\kappa - \kappa_0 - \beta(q - q_0))$. Likewise, for small values of w the Gaussian is almost like a δ-function, and so the representative phase-plane ellipse collapses into the vertical line $\delta(q - q_0)$ proper of a point source.

Notably, distributions (6.4.51) and (6.4.54) are positive through the whole phase plane, in accord with the Hudson theorem [10.6], which, as mentioned in § 6.3.6, identifies the Gaussian signals as the only waveforms whose Wigner distribution function is positive everywhere in the space-frequency plane.

Sum of two Gaussian signals

As a further example of the non linearity of the Wigner distribution function and accordingly of the behaviour of the cross-terms entering the expression of $\mathcal{W}(q, \kappa)$ when the superposition of signals is involved, we consider a signal composed by two identical Gaussian signals, which are however shifted one to the other in both space and frequency. So, we consider the amplitude

$$\varphi(q) = \left(\tfrac{2}{\pi w^2}\right)^{1/4} e^{-\frac{1}{w^2}(q-q_1)^2 + i\kappa_1 q} + \left(\tfrac{2}{\pi w^2}\right)^{1/4} e^{-\frac{1}{w^2}(q-q_2)^2 + i\kappa_2 q}. \qquad (6.4.56)$$

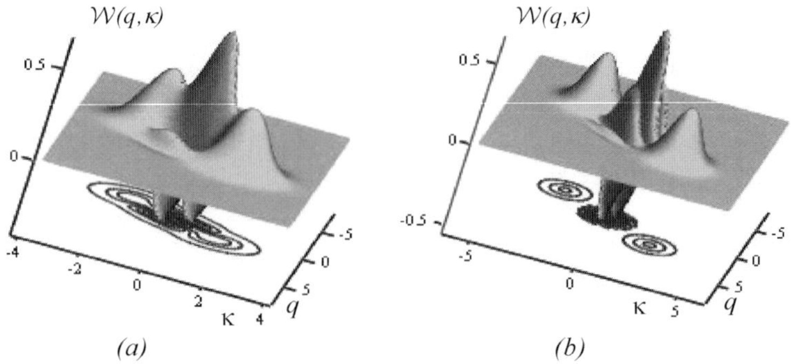

FIGURE 6.17. 3*D* image and relevant space-frequency contour plots of the Wigner distribution function of the signal (6.4.56), composed of two Gaussian amplitudes with *(a)* $w = 1.5$, $q_1 = -q_2 = 1.5$, $\kappa_1 = -\kappa_2 = 1.5$, and *(b)* $q_1 = -q_2 = 2.5$, $\kappa_1 = -\kappa_2 = 2.5$.

After some lengthy algebra we end up with the explicit expression of the Wigner distribution function which writes

$$\begin{array}{l} \mathcal{W}_\varphi(q,\kappa) = \mathcal{W}_\mathcal{G}(q-q_1,\kappa-\kappa_1) + \mathcal{W}_\mathcal{G}(q-q_2,\kappa-\kappa_2) \\ +2\cos[(q-q_\mathrm{m})\kappa_\mathrm{d} - (\kappa-\kappa_\mathrm{m})q_\mathrm{d} - \delta]\mathcal{W}_\mathcal{G}(q-q_\mathrm{m},\kappa-\kappa_\mathrm{m}). \end{array} \quad (6.4.57)$$

Here, $\mathcal{W}_\mathcal{G}$ is the Wigner distribution function (6.4.51) of the real Gaussian amplitude (6.4.50); q_m, κ_m and q_d, κ_d denote the mean and difference space and frequency coordinates, whilst δ is a constant phase as

$$q_\mathrm{m} = \tfrac{q_1+q_2}{2},\ \kappa_\mathrm{m} = \tfrac{\kappa_1+\kappa_2}{2};\quad q_\mathrm{d} = q_1 - q_2,\ \kappa_\mathrm{d} = \kappa_1 - \kappa_2;\ \delta = q_\mathrm{m}\kappa_\mathrm{d}. \quad (6.4.58)$$

In accord with the considerations of § 6.3.4, distribution (6.4.57) displays two self-terms, mainly located around the space-frequency centroids of the two component signals, i.e. (q_1, κ_1) and (q_2, κ_2). The cross term mainly develops around the central point $(q_\mathrm{m}, \kappa_\mathrm{m})$ with the Gaussian profile being sinusoidally modulated; the oscillations occur orthogonally to the line connecting the two signal centers (q_1, κ_1) and (q_2, κ_2). As it can be seen comparing the surface plots of Figs. 6.17.a) and 6.17.b), corresponding to the same value of w but different values of the space and frequency shifts, the visibility of the cross-term oscillations through the overlap of the self-terms is governed by the relative value of the phase-plane distance between the two self-terms centers to the respective space and frequency extents, fixed by the signal parameter w.

Interestingly, graphs like those in Fig. 6.17 are encountered in quantum optics when dealing with the superposition of states. We may mention, for instance, the quantum state obtained superposing two coherent states of the same quantum number and a given phase difference (the Schrödinger cat state). Of course, the occurrence of the cross term in the Wigner distribution function of such a state is a consequence of the superposition and the bilinearity of the Wigner integral. Accordingly, the cross-term shape is a manifestation of the separation of the two states in the pertinent phase-plane (i.e., the q-p oscillator phase-plane), ultimately determined by the relative value of the phase difference to the quantum number. Thus, as noted in § 6.3.7, the cross-term behaviour in the Wigner surface, yields a key-tool to individualize the occurrence of overlap, and hence of interference and transitions, between the individual states in the superposition state. Also, it is closely associated with the features of the relevant photon statistics [10.24].

Hermite-Gaussian beams

As is well known, the Gaussian signal (6.4.53) plays a significant role in optics as both an effective signal and a convenient model signal. Gaussian beams are introduced in optics in several different ways; for instance, as solutions to the paraxial Helmholtz equation in uniform and quadratic index media or as fundamental modes of optical stable resonators. The basic features of signal (6.4.53) are the paraboloidal profile of the wavefronts and the Gaussian shape of the intensity distribution; the former is accounted for by the complex quadratic exponential, the parameter β identifying the curvature of the wavefront, and the latter is accounted for by the real quadratic exponential, the parameter w fixing the beam radius.

The Gaussian envelope is not the only solution to the aforementioned optical problems. The other solutions, known as Hermite-Gaussian beams (or modes), have the paraboloidal wavefronts as (6.4.53), but do not exhibit a plain Gaussian distribution of the intensity across the axis as the Gaussian envelope is modulated by Hermite polynomials [15]. We do not dwell on the mathematical and physical properties of the Hermite-Gaussian modes, but simply consider the basic real Hermite-Gaussian amplitude,

$$\varphi_{\mathcal{HG}_n}(q) = \left(\frac{\sqrt{2}}{2^n n! \sqrt{\pi} w}\right)^{1/2} H_n\left(\frac{\sqrt{2}q}{w}\right) e^{-\frac{1}{w^2}q^2}, \tag{6.4.59}$$

and calculate the relevant Wigner distribution function. Here $H_n(\cdot)$ denotes the Hermite polynomial of degree n and the constant factor is fixed to satisfy the condition: $||\varphi_{\mathcal{HG}}||^2 = 1$ [18]. With $n = 0$ we recover the *fundamental* Gaussian waveform (6.4.50), as $H_0(x) = 1$. In Fig. 6.18.a) the profiles (6.4.59) are graphed vs. q/w for some values of n; note that the Hermite polynomials are even or odd functions according to whether n is even or odd.

It should be mentioned, perhaps, that (6.4.59) may be interpreted as the position representation of the energy eigenstates of the quantum mechanical harmonic oscillator, and so of the Fock states of quantum optics (see also § 5.4.2). Indeed, within the quantum mechanical picture, the argument $\frac{q^2}{w^2/2}$ +

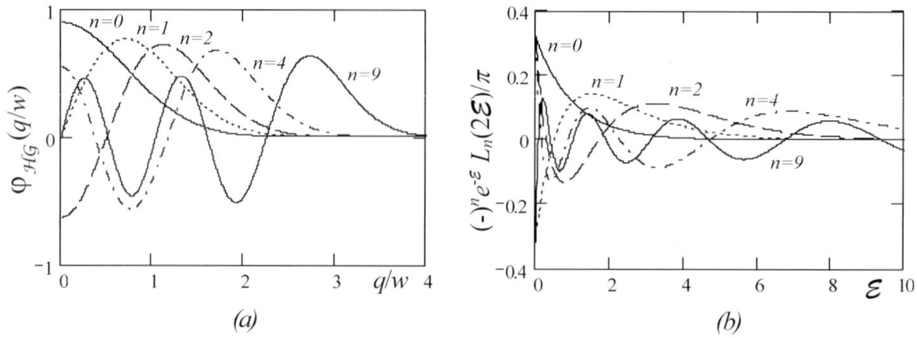

FIGURE 6.18. *(a)* Hermite-Gaussian beam profiles vs. q/w for some beam orders n, and *(b)* corresponding $\mathcal{W}_{\mathcal{HG}_n}$-profiles vs. $\mathcal{E} = \frac{2q^2}{w^2} + \frac{w^2\kappa^2}{2}$.

$\frac{w^2k_0^2p^2}{2}$ of the exponential factor, entering the Wigner distribution function $\mathcal{W}_{\mathcal{G}}$ as well as $\mathcal{W}_{\mathcal{HG}_n}$ below, can be understood as the energy function of the harmonic oscillator, which for given input data individualize in the q-p oscillator phase-plane the familiar elliptical contours like those in Fig. 6.15.b).

The reader is invited to compute the frequency spectrum $\widetilde{\varphi}_{\mathcal{HG}_n}(\kappa)$. In this connection, we note that the Hermite-Gaussian functions (6.4.59) are eigenfunctions of the Fourier transform operator $\widehat{\mathcal{F}}$, forming indeed an orthonormal and complete set. This can readily be seen using dimensionless variables in both the space and frequency domains, i.e., setting

$$\frac{\sqrt{2}q}{w} \leftarrow x \rightarrow \frac{w\,\kappa}{\sqrt{2}}$$

in the two dual domains. Then, in the x coordinate one finds

$$\widehat{\mathcal{F}}\varphi_{\mathcal{HG}_n}(x) = (-i)^n\varphi_{\mathcal{HG}_n}(x), \qquad (6.4.60)$$

since in the dimensionless variable $x = \frac{\kappa w}{\sqrt{2}}$ the Fourier transform $\widetilde{\varphi}_{\mathcal{HG}_n}(x)$ takes the same form as the wave function $\varphi_{\mathcal{HG}_n}(x)$ in the corresponding variable $x = \sqrt{2}\frac{q}{w}$. As mentioned in §§ 5.4.1 and 5.4.5, the eigenvalue equation (6.4.60) has been the key-point for introducing the fractional Fourier transform operator $\widehat{\mathcal{F}}^{\alpha}$, defined in fact through a generalization of (6.4.60) to eigenvalues evenly separated by a *fraction* of the imaginary unit; namely,

$$\widehat{\mathcal{F}}^{\alpha}\varphi_{\mathcal{HG}_n}(x) = (-i)^{n\alpha}\varphi_{\mathcal{HG}_n}(x). \qquad (6.4.61)$$

We use (6.2.2), so that after rearranging the variables we obtain

$$\mathcal{W}_{\mathcal{HG}_n}(q,\kappa) = \frac{1}{2^n n!\sqrt{\pi}}e^{-\frac{2q^2}{w^2}-\frac{w^2}{2}\kappa^2}\int H_n\left(\frac{\sqrt{2}q}{w}+i\kappa w-y\right)H_n\left(\frac{\sqrt{2}q}{w}-i\kappa w+y\right)e^{-y^2}dy. \qquad (6.4.62)$$

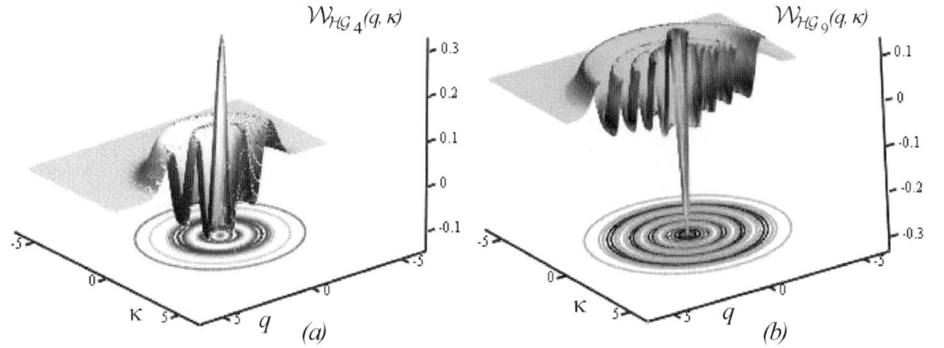

FIGURE 6.19. Surface plots and q-κ contour plots of the Wigner distribution function of the Hermite-Gaussian signal (6.4.59) with $w = 1.5$ and order *(a)* $n = 4$ and *(b)* $n = 9$.

Taking advantage of the parity property of the Hermite polynomials: $H_n(-x) = (-)^n H_n(x)$ and the integral [19]

$$\int H_n(y+x)H_n(y+z)e^{-y^2}dy = 2^n n!\sqrt{\pi}L_n(-2xz), \tag{6.4.63}$$

where $L_n(\cdot)$ denotes the Laguerre polynomial of degree n [18], we end up with the nice expression

$$\mathcal{W}_{\mathcal{HG}_n}(q,\kappa) = \frac{(-)^n}{\pi}e^{-\frac{2q^2}{w^2}-\frac{w^2}{2}\kappa^2}L_n[2(\tfrac{2q^2}{w^2}+\tfrac{w^2\kappa^2}{2})]. \tag{6.4.64}$$

Expression (6.4.51) for the Wigner distribution function of the fundamental Gaussian mode (6.4.50) follows from (6.4.64) with $n = 0$, being $L_0(x) = 1$. In Fig. 6.18.a) the amplitudes (6.4.59) are graphed vs. q/w for some values of n.

Notably, on graphs the contour plots of $\mathcal{W}_{\mathcal{HG}_n}$ reproduce the elliptical structures (6.4.52) pertinent to $\mathcal{W}_{\mathcal{G}}$, with the basic difference being that, unlike $\mathcal{W}_{\mathcal{G}}$, the $\mathcal{W}_{\mathcal{HG}_n}$'s take on negative values in some regions of the space-frequency plane in accord with the order n and the value of the argument of the Laguerre polynomial. In Fig. 6.18.b) we have plotted $\mathcal{W}_{\mathcal{HG}_n}$ as a function of the quantity $\mathcal{E} = \frac{2q^2}{w^2} + \frac{w^2\kappa^2}{2}$. We see that at the phase-plane origin

$$\mathcal{W}_{\mathcal{HG}_n}(0,0) = \frac{(-)^n}{\pi}, \tag{6.4.65}$$

turning to be positive or negative according to whether the mode order is even or odd. Also, we see that $\mathcal{W}_{\mathcal{HG}_n}$ oscillates versus $\mathcal{E}$ roughly up to $\mathcal{E} \sim 2n+1$, and then for greater values decays rather quickly. Accordingly, as it can be seen

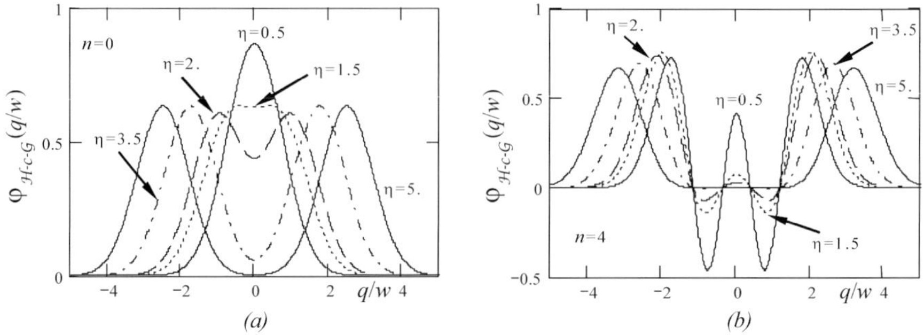

FIGURE 6.20. Hermite-cosh-Gaussian beam profiles vs. q/w for specified values of the parameter η for the modes of order *(b)* $n = 0$ and *(b)* $n = 4$.

in the exemplary plots of Fig. 6.19, relevant to the orders $n = 4$ and $n = 9$, $\mathcal{W}_{\mathcal{HG}_n}(q,\kappa)$ is mainly concentrated over the region of the space-frequency plane where the signal itself is concentrated, i.e. roughly $\sqrt{2n+1}\frac{w}{2}$ and $\frac{\sqrt{2n+1}}{w}$ across the origin respectively along the q and κ axes. In addition, we may note that $\mathcal{W}_{\mathcal{HG}_n}(q,\kappa)$ oscillates inside the elliptical contour identified as well by Eq. (6.4.52) for the value of $\varepsilon = 2(2n+1)$, whilst decaying rather fast outside.

Hermite-cosh-Gaussian signals

A new set of beam solutions of the parabolic wave equation has been recently elaborated, generally reported as *sinusoidal Hermite-Gaussian* modes [16]. Such solutions are in the form of sinusoidal functions of a possibly complex argument times a conventional Hermite-Gaussian beam factor.

We consider here one of the simplest forms of these new solutions, namely, the Hermite-cosh-Gaussian beam, which is represented by the real amplitude

$$\varphi_{\mathcal{H}\cdot c\cdot\mathcal{G}_n}(q) = \left(\frac{2\sqrt{2}}{2^n n!\sqrt{\pi}w\,\xi_n(\Omega)}\right)^{1/2} H_n\left(\frac{\sqrt{2}q}{w}\right)e^{-\frac{1}{w^2}q^2}\cosh(\Omega q), \tag{6.4.66}$$

where Ω is a real parameter, and the factor $\xi_n\ (\Omega)$ conforms to the normation condition for unity, and hence

$$\xi_n(\Omega) = 1 + e^{\frac{\Omega^2 w^2}{2}} L_n(-\Omega^2 w^2). \tag{6.4.67}$$

The Hermite-cosh-Gaussian modes can optically be produced, for instance, by passing an Hermite-Gaussian mode from a stable resonator laser through an aperture having a cosh-amplitude transmission function.

Noting that $\cosh(\Omega q) \propto e^{\Omega q} + e^{-\Omega q}$, we may interpret the beams (6.4.66) as resulting from the superposition of two conventional Hermite-Gaussian modes

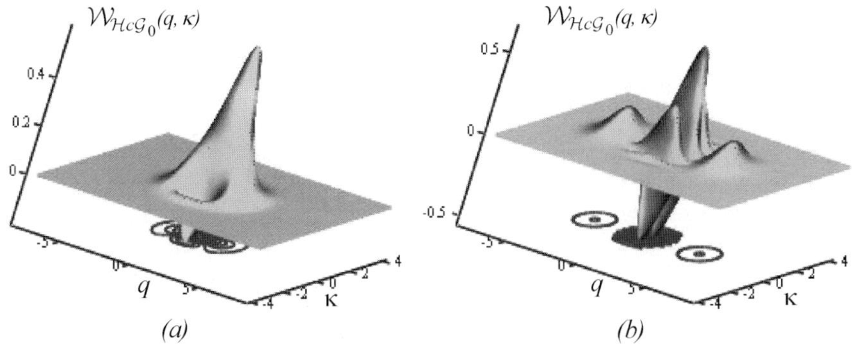

FIGURE 6.21. Surface plots and relevant q-κ contour plots of the Wigner distribution functions of the cosh-Gaussian signal (6.4.66) with $w = 1.5$ for *(a)* $\eta = 2$ and *(b)* $\eta = 5$.

weighted by the factors $e^{\Omega q}$ and $e^{-\Omega q}$. It becomes evident therefore that it is the value of Ω in relation to that of the Gaussian beam radius w to determine the overall shape of the resulting beam. In fact, in Fig. 6.20 we have plotted the amplitude profiles versus q/w of the fundamental mode ($n = 0$) and of an higher order mode ($n = 4$) for increasing values of the parameter $\eta = \Omega w$. Referring to the fundamental mode, we see that the profile firstly becomes flat, closely resembling the shape of super-Gaussian field distributions, that are greatly desired in high-power lasers due to their potential of efficiently extracting energy from the amplifying medium (see next paragraph). With increasing η, a center deep appears which becomes deeper and larger with increasing values of η, thus finally modelling the overall distribution into a double-lobe envelope. This is basically the effect acted by the cosh-factor in (6.4.66) for higher order modes; the near-axis oscillations of the conventional Hermite-Gaussian modes are more or less strongly reduced in favour of a magnification of the lateral lobes, in which then the beam energy comes to be concentrated. Hence we may guess that the Wigner distribution function will come to resemble that of the sum of two Gaussian-like signals.

It is an easy task to verify that (6.4.66) can be recast as

$$\varphi_{\mathcal{H}\cdot c\cdot\mathcal{G}_n}(q) = \left(\frac{\sqrt{2}e^{\eta^2/2}}{2^n n! \sqrt{\pi} w\, \xi_n(\Omega)}\right)^{1/2} H_n\left(\frac{\sqrt{2}q}{w}\right)\left\{e^{-\left(\frac{q}{w}-\frac{\eta}{2}\right)^2} + e^{\left(\frac{q}{w}-\frac{\eta}{2}\right)^2}\right\}, \qquad \eta = \Omega w. \tag{6.4.68}$$

Accordingly, carrying on the calculations, we explicitly find

$$\mathcal{W}_{\mathcal{H}\cdot c\cdot\mathcal{G}_n}(q,\kappa) = \frac{1}{\xi_n(\Omega)}\{\mathcal{W}_{\mathcal{H}\mathcal{G}_n}(q,\kappa)\cosh(2\Omega q) + \Re[\mathcal{W}_{\mathcal{H}\mathcal{G}_n}(q,\kappa - i\Omega)]\}. \tag{6.4.69}$$

Figure 6.21 shows the $3D$ image with the relevant contour plots in the q-κ

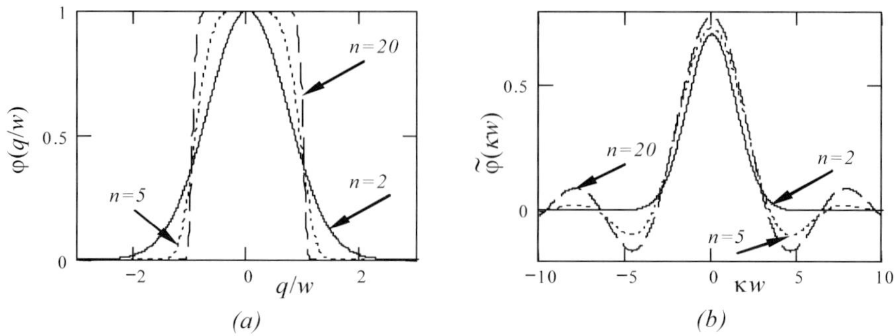

FIGURE 6.22. Profiles of *(a)* the superGaussian distributions (6.4.70) vs. q/w for $n = 2$, 5 and 20, and of *(b)* the corresponding frequency spectra vs. κw.

plane of the Wigner distribution function (6.4.69) for the fundamental mode in correspondence with two values of the parameter η.

SuperGaussian signal

We consider the superGaussian signal, which can plainly be represented by the amplitude function

$$\varphi(q) = \varphi_0 e^{-\left(\frac{|q|}{w}\right)^n}, \tag{6.4.70}$$

n being a positive integer; with $n = 2$ we regain the Gaussian signal (6.4.50).

To aide intuition, Fig. 6.22 shows the profiles of the signal (6.4.70) and the corresponding Fourier transform in the cases $n = 2$, 5, 20. The spatial profile is basically flat through a certain interval and smoothly goes to zero, with the profile width and the degree of smoothness being governed by the parameter w and the order n. Correspondingly, in the frequency domain the distribution displays a hybrid behaviour between a Gaussian and a sinc function.

SuperGaussian amplitude distributions like (6.4.70) are produced by a wide class of laser systems, as, for instance, lasers with high-gain, saturated active media as well as lasers equipped with variable reflectivity mirror cavities [17].

The Wigner distribution function of the signal (6.4.70) for $n > 2$ cannot be analytically calculated. Figure 6.23 shows the trend of the Wigner distribution function of the signal (6.4.70) for two values of n. As expected, the Wigner distribution function tends to be mainly concentrated within a region of the q-κ plane, whose extensions relate to the widths of the energy distribution of the signal in the q and κ domains, exhibiting oscillations inside and fast decay outside. For a given w with increasing the order n the spatial profile of the signal (6.4.70) approaches that of a rectangular signal and, accordingly, the Wigner distribution function comes to display a sinc-like behavior with respect to the frequency variable.

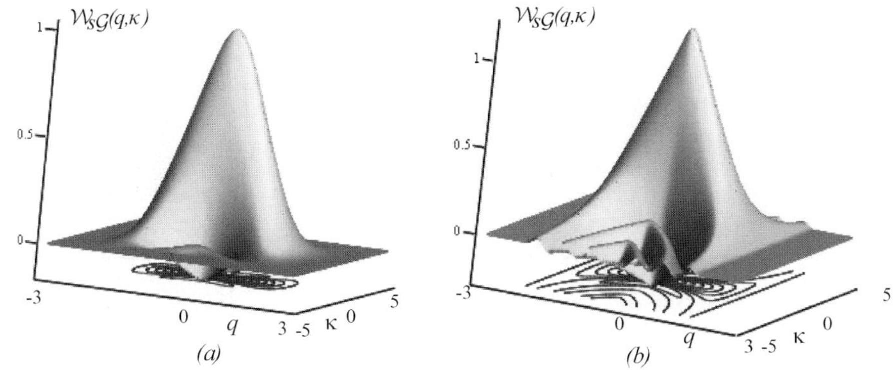

FIGURE 6.23. $3D$ images and q-κ contour plots of the Wigner distribution functions of the superGaussian amplitude (6.4.70) with $w = 2$, and *(a)* $n = 5$ and *(b)* $n = 13$.

6.4.6 *An example of Wigner nonrepresentability*

As an example of Wigner nonrepresentability, we consider a function $\mathsf{W}(q, \kappa)$ attaining a constant value over a rectangular region of the (q, κ) plane, i.e.

$$\mathsf{W}(q, \kappa) = \begin{cases} 1 & |q| \leq Q \; ; \; |\kappa| \leq K \\ 0 & \text{otherwise.} \end{cases} \tag{6.4.71}$$

We assume $\mathsf{W}(q, \kappa)$ be a Wigner distribution function arising from a signal, whose amplitude, by the inversion formula (6.3.28), would obtain in the form

$$\varphi(q) = \begin{cases} \frac{2K}{\varphi^*(0)} \operatorname{sinc}(Kq) & |q| \leq 2Q \\ 0 & |q| > 2Q. \end{cases} \tag{6.4.72}$$

The absolute value of $\varphi^*(0)$ could be inferred from (6.3.26), which, indeed, on account of (6.4.71), gives

$$|\varphi(q)|^2 = \begin{cases} 2K & |q| \leq Q \\ 0 & |q| > Q, \end{cases} \tag{6.4.73}$$

and hence $|\varphi^*(0)|^2 = 2K$.

We see that results (6.4.72) and (6.4.73) are in contrast to one another. In fact, (6.4.72) yields for the signal magnitude $|\varphi(q)|$ a q-dependent expression vanishing outside the interval $-2Q \leq q \leq 2Q$. In contrast, (6.4.73) gives a constant value for $|\varphi(q)|$, which also would differ from zero within the smaller interval $-Q \leq q \leq Q$. Thus, we may argue that the $2D$ function (6.4.71) is

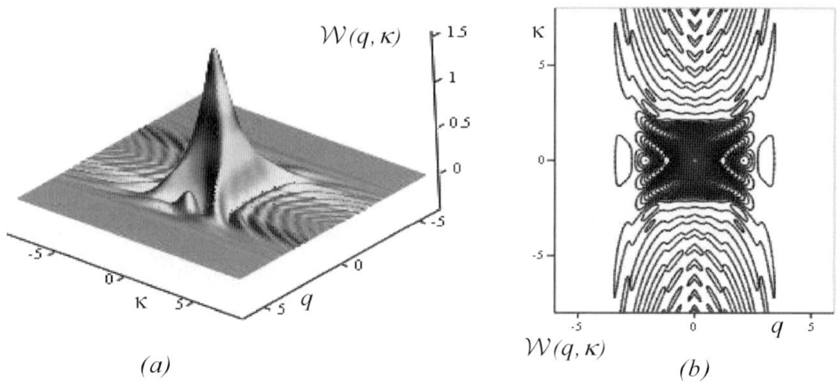

FIGURE 6.24. *(a)* $3D$ view and *(b)* relevant contour plots of the Wigner distribution function of the signal (6.4.74) with $Q = 1$ and $K = 2$.

not a representable Wigner function. In fact, the reader may easily verify that the Wigner distribution function of the signal

$$\varphi(q) = \sqrt{2K}\text{sinc}(Kq)\text{rect}\,(\frac{q}{4Q}) \tag{6.4.74}$$

does not reproduce the original function (6.4.71). It can be expressed as the convolution over κ of two sinc functions, precisely a sinc function like (6.4.3) corresponding to the rect-signal and a sinc function like (6.4.7) corresponding to the sinc-signal. Indeed, the reader may easily work out the analytical expression of $\mathcal{W}_\varphi(q,\kappa)$ involving a definite integral over the frequency variable; also, it turns out to be confined within the vertical strip $|q| \leq 2Q$. As an example, Fig. 6.24.a) shows the $3D$ image of the Wigner distribution function of the waveform (6.4.74) with $Q = 1$ and $K = 2$. Even though it tends to be mainly concentrated within a rectangular region in the (q,κ) plane, as clearly evidenced by the corresponding contour plots of Fig. 6.24.b), it extends indefinitely through the κ-direction; also, it does not attain a constant value over the region of maximum localization.

We will seen in Chapter 9 that the space and spatial extensions of the signal reflect into the corresponding extensions of the Wigner distribution function. So, (6.4.71) would correspond to a signal that should be limited to a finite interval in both the space and frequency domain in contrast to the width principle of Fourier analysis. This would immediately suggest the nonrepresentability of $2D$ function like (6.4.71) as Wigner distribution functions.

6.5 Summary

After introducing the formal definition of the optical Wigner distribution function, which indeed draws its root in the issue of expressing quantum mechanics in a form as similar as possible to classical mechanics, we have evaluated and displayed on graphs the Wigner distribution functions, and the relevant Wigner charts in the (q, κ) plane, of a large variety of signals, which are of interest in optics as well as in quantum and signal theory.

We have gained a feeling for the potential of the Wigner distribution function to convey information about both the space and frequency features of the signal in accord as well with the inherent limitations, dictated by the uncertainty relation, concerning the entire space and the entire frequency extents of the Wigner distribution function. Remarkably, it has been apparent that in general the Wigner distribution function takes on negative values over some regions in the phase plane. This feature prevents us from interpreting the Wigner distribution function as a true space-frequency energy distribution even though its marginals correctly convey the energy distributions of the signal separately in space and in frequency.

In the next chapter we will reconsider the Wigner representation, and the related issue of composing into a single "entity" the descriptions separately conveyed by the signal amplitude and the Fourier spectrum, within the context of signal theory, thus paralleling the quantum-mechanics oriented presentation given here.

Problems

1. *(a)* Prove that the quantum observables $\widehat{f}_1(\widehat{\mathsf{q}}, \widehat{\mathsf{p}}) = (\widehat{\mathsf{q}}\widehat{\mathsf{p}} + \widehat{\mathsf{p}}\widehat{\mathsf{q}})/2$ and $\widehat{f}_2(\widehat{\mathsf{q}}, \widehat{\mathsf{p}}) = \widehat{\mathsf{q}}\widehat{\mathsf{p}}$ differ by $-i\hbar/2$. However, they would turn into the same classical function $f = qp$ if we trivially replaced the operators $\widehat{\mathsf{q}}$ and $\widehat{\mathsf{p}}$ with the classical variable q and p. *(b)* Verify that the same occurs with the observables $\widehat{f}_1(\widehat{\mathsf{q}}, \widehat{\mathsf{p}}) = \widehat{\mathsf{p}}\widehat{\mathsf{q}}^2\widehat{\mathsf{p}}$ and $\widehat{f}_2(\widehat{\mathsf{q}}, \widehat{\mathsf{p}}) = (\widehat{\mathsf{q}}^2\widehat{\mathsf{p}}^2 + \widehat{\mathsf{p}}^2\widehat{\mathsf{q}}^2)/2$, which differ by $\hbar^2$. Similarly both would turn into the same classical function if one simply replaced $\widehat{\mathsf{q}} \rightarrow q$ and $\widehat{\mathsf{p}} \rightarrow p$.

2. Show that *(a)* if two functions $\varphi_1(q)$ and $\varphi_2(q)$ vanish outside the space intervals (Q_a, Q_b) and (Q_c, Q_d), respectively, then the corresponding cross-Wigner distribution function turns to be zero outside the vertical strip $\frac{Q_a+Q_c}{2} \leq q \leq \frac{Q_b+Q_d}{2}$ in the phase-plane, and likewise *(b)* if the frequency spectra of two functions vanish outside the frequency intervals (κ_a, κ_b) and (κ_c, κ_d), respectively, the relevant cross-Wigner distribution function is nonzero only within the horizontal strip $\frac{\kappa_a+\kappa_c}{2} \leq \kappa \leq \frac{\kappa_b+\kappa_d}{2}$ in the phase-plane.

3. Verify that the $\mathcal{W}_\varphi(0, 0) < 0$ for functions $\varphi(q)$, such that $q\varphi(q) > 0$ or $q\varphi(q) < 0$ for all q over the real line. Check this property for the power functions: $\varphi(q) = q^n$, distinguishing between odd and even exponents.

4. Prove that smoothing the Wigner distribution function of a given signal $\varphi(q)$ by the Wigner distribution function of any normalized signal, say $\psi(q)$, the resulting distribution $\mathcal{W}_{\text{smoothed}}(q, \kappa)$ still satisfies the energy relation, i.e,

$$\iint \mathcal{W}_{\text{smoothed}}(q, \kappa) dq d\kappa = \iint \mathcal{W}(q, \kappa) dq d\kappa,$$

whilst the marginals are lost, since projecting $\mathcal{W}_{\text{smoothed}}(q,\kappa)$ on the q and κ axis one obtains the average of the power spectrum and the energy density of φ through the power spectrum and the energy density of ψ, according to

$$\int \mathcal{W}_{\text{smoothed}}(q,\kappa)dq = \int |\widetilde{\varphi}(\kappa')|^2|\widetilde{\psi}(\kappa-\kappa')|^2 d\kappa',$$
$$\int \mathcal{W}_{\text{smoothed}}(q,\kappa)d\kappa = \int |\varphi(q')|^2|\psi(q-q')|^2 dq'.$$

5. Prove relation (6.3.70) as the necessary and sufficient condition for a given function $\mathcal{W}(q,\kappa)$ to be a representable Wigner distribution function. (Hint: relation (6.3.70) is straightforwardly obtained applying definition (6.2.2) for $\mathcal{W}(q,\kappa)$; correspondingly, if $\mathcal{W}(q,\kappa)$ behaves according to (6.3.70), the signal $\varphi(q)$, derived through (6.3.28), also satisfies relation (6.3.23), thus proving the representability of the given $\mathcal{W}(q,\kappa)$.)

6. Verify that the $2D$ Gaussian function

$$\mathcal{W}(q,\kappa) = e^{-\frac{1}{2}(\gamma q^2 - 2\alpha q\kappa + \beta\kappa^2)},$$

for which $\beta, \gamma > 0$ and $\beta\gamma - 4\alpha^2 > 0$, is Wigner representable only if the parameters α, β and γ satisfy the equality

$$\beta\gamma = \alpha^2 + 4.$$

7. Write down the expression of the Wigner distribution function of the double rectangle function

$$\varphi(q) = \tfrac{1}{2}\{\mathrm{rect}(\tfrac{q-q_1}{2Q_1})e^{i\kappa_1 q} + \mathrm{rect}(\tfrac{q-q_2}{2Q_2})e^{i\kappa_2 q}\}.$$

Show that it is zero for $q \le \frac{q_1-Q_1+q_2-Q_2}{2}$ and $q \ge \frac{q_1+Q_1+q_2+Q_2}{2}$, and that in the limit $Q_1 = Q_2 \to \infty$ turns into the expression (6.4.17) relevant to the superposition of two harmonics with in particular $\varphi_{0_1} = \varphi_{0_2} = \frac{1}{2}$.

8. Calculate the Wigner distribution function of the rectangle-wave function

$$\varphi(q) = \sum_{n=-\infty}^{\infty} \tfrac{1}{2Q}\mathrm{rect}(\tfrac{q-nd}{2Q}),$$

formed by an infinite sequence of rectangle waveform, $2Q$ wide and spaced by d. (Hint: express $\varphi(q)$ as a Fourier series on account of its periodicity by d.) Verify that in the limit $Q \to 0$, where the rectangle–wave function turns into the comb-function, the obtained expression for $\mathcal{W}_\varphi(q,\kappa)$ turns into (6.4.25).

9. Calculate the Wigner distribution function for the composed signals

$$\begin{aligned}
\varphi(q) &= \varphi_1\delta(q-q_1) + \tfrac{\varphi_2}{\sqrt{2\pi}}e^{i\kappa_2 q},\\
\varphi(q) &= \tfrac{\varphi_0}{2}e^{i\alpha_1\frac{q^2}{2}+i\kappa_{01}q} + \tfrac{\varphi_0}{2}e^{i\alpha_2\frac{q^2}{2}+i\kappa_{02}q},\\
\varphi(q) &= \mathrm{rect}(\tfrac{q}{2Q})e^{i\alpha\frac{q^2}{2}+i\kappa_0 q}.
\end{aligned}$$

In particular, show that with α_1 approaching α_2: $\alpha_1 \to \alpha_2$, we regain the expression (6.4.34) for a single chirp.

References

[1] We quote below the fundamental papers by Wigner concerned with the Wigner method. In [1.1] the Wigner distribution function is introduced in an almost heuristic manner in connection with the problem of determining the quantum correction to thermodynamic functions. In the later paper [1.2] the formal procedure to elaborate an in general quantum phase-space distribution on account of the specific requirements and inherent limitations is described.

[1.1] E.P. Wigner, "On the quantum correction for thermodynamic equilibrium", Phys. Rev. **40**, 749-759 (1932).

[1.2] E.P. Wigner, "Quantum-mechanical distribution function revisited", in *Perspectives in Quantum Theory*, W. Yourgrau and A. van der Merwe (eds.) (MIT Press, Cambridge, 1971), pp. 25-36.

[2] A few titles on classical statistical mechanics are given in [4] of ch. 2. We signalize the textbook by Balescu, wherein the Wigner method is described and used.

[2.1] R. Balescu, *Equilibrium and Nonequilibrium Statistical Mechanics* (John Wiley & Sons, New York, 1975), ch. 3, pp. 89-104.

[3] Titles on quantum mechanics are listed in [10] of ch. 4. The textbook by Blum is specifically concerned with the density matrix formalism.

[3.1] K. Blum, *Density Matrix Theory and Applications* (Plenum Press, New York, 1981).

[4] The literature concerned with the Wigner representation in the quantum mechanical framework is considerably rich. Only a restricted list of exemplary papers and reviews is given here. A comprehensive analysis of the Wigner distribution function in quantum mechanics was firstly presented in the book by Kim and Noz [4.12]. See also the review paper by Dragoman [4.11], offering an overall investigation of the correspondence between the phase-space description of classical optics and quantum mechanics, and the fascinating presentations in [2.1], and, within the context of quantum electrodynamics, in [4.13]. Additional references are suggested in [1] of ch. 8. For investigations about the structure of the (quantum) Wigner phase space we address the reader to the enlightening analysis in [4.6].

[4.1] H.J. Groenewold, "On the principles of elementary quantum mechanics", Physica **12**, 405-460 (1946).

[4.2] J.E. Moyal, "Quantum mechanics as a statistical theory", Proc. Cambridge Phil. Soc. **45**, 99-124 (1949).

[4.3] H. Mori, I. Oppenheim and J. Ross, "Some topics in quantum statistics. The Wigner function and transport theory", in *Studies in Statistical Mechanics*, J. de Boer and G.E. Uhlenbeck (eds.) (North-Holland, Amsterdam, 1962), Vol. 1, pp. 212-298.

[4.4] N.G. de Bruijn, "A theory of generalized functions, with applications to Wigner distribution and Weyl correspondence", Nieuw Archief voor Wiskunde (3), Vol XXI, 205-280 (1973).

[4.5] M.V. Berry, "Semi-classical mechanics in phase-space: a study of Wigner's function", Phil. Trans. Roy. Soc. **287**, 237-271 (1977).

[4.6] N.L. Balazs, "Weyl's association, Wigner's function and affine geometry", Physica **102A**, 236-254 (1980).

[4.7] J.P. Dahl, "The Wigner function", Physica **114A**, 439-444 (1982).

[4.8] R.F. O'Connell, "The Wigner distribution function - 50th birthday", Found. Phys. **13**, 83-92 (1983).

[4.9] V.I. Tatarskií, "The Wigner representation of quantum mechanics", Sov. Phys. Usp. **26**, 311-327 (1983).

[4.10] N.L. Balazs and B.K. Jennings, "Wigner's function and other distribution functions in the Mock phase spaces", Phys. Rep. **104**, 347-391 (1984).

[4.10] M. Hillery, R.F. O'Connell, M.O. Scully and E.P. Wigner, "Distribution functions in physics: fundamentals", Phys. Rep. **106**, 121-167 (1984).

[4.11] D. Dragoman, "Phase space correspondence between classical optics and quantum mechanics", in *Progress in Optics*, Vol. 43, E. Wolf (ed.) (Elsevier, Amsterdam, 2002), ch. 5, pp. 433-496.

[4.12] Y.S. Kim and M.E. Noz, *Phase Space Picture of Quantum Mechanics* (World Scientific, Singapore, 1991).

[4.13] S. R. De Groot and L.G. Suttorp, *Foundations of Electrodynamics* (North-Holland, Amsterdam, 1972), ch. VI-VII, pp. 317-406. The properties of the Weyl transform and the Wigner distribution function are specifically described in he appendix of ch. VI, pp. 341-364.

[5] A few titles concerned with the Wigner representation in relation to the signal theory are listed here. See also [4]-[6] of ch. 7 in relation to the various aspects of the joint signal representation theory.

[5.1] J. Ville, "Théorie et applications de la notion de signal analitique", Cables et Transmissions **2A**, 61-74 (1948).

[5.2] A.W. Rihaczek, "Signal energy distribution in time and frequency", IEEE Trans. Inf. Theory, **IT-14**, 369-374 (1968).

[5.3] T.C.A.M. Claasen and W.F.G. Mecklenbräuker, "The Wigner distribution - a tool for time-frequency signal analysis. I: continuous-time signals", Philips J. Res. **35**, 217-250 (1980); II: discrete-time signals", *ibid.*, 276-300; III: relations with other time-frequency signal transformations", *ibid.*, 372-389.

[5.4] F. Peyrin and R. Prost, "A unified definition for the discrete-time, discrete-frequency, and discrete-time/frequency Wigner distributions", IEEE Trans. on Acoust., Speech and Signal Process. **34**, 858-867 (1986).

[5.5] L. Cohen, "Time-frequency distributions-A Review", Proc. IEEE **77**, 941-981 (1989).

[5.6] F. Hlawatsch and G.F. Boudreaux-Bartels, "Linear and quadratic time-frequency signal representations," IEEE Signal Process. Mag., 21-27, April 1992.

[5.7] L. Cohen, *Time-Frequency Analysis* (Prentice-Hall, Englewood Cliffs, 1995).

[5.8] *The Wigner Distribution: Theory and Applications in Signal Processing*, W. Mecklenbräuker and F. Hlawatsch (eds.) (Elsevier, Amsterdam, 1997).

[6] The pioneering papers by Bastiaans introducing the Wigner representation in optics and investigating many related theoretical and applicative issues are quoted below. We also address the reader to the review papers [6.8] and [6.9], and to the JOSA issue [6.10].

[6.1] M.J. Bastiaans, "Wigner distribution function applied to optical signals and systems", Opt. Comm. **25**, 26–30 (1978).

[6.2] M.J. Bastiaans, "The Wigner distribution function and Hamilton's characteristics of a geometrical-optical system", Opt. Comm. **30**, 321-326 (1979).

[6.3] M.J. Bastiaans, "Wigner distribution function and its application to first-order optics", J. Opt. Soc. Am. **69**, 1710-1716 (1979).

[6.4] M.J. Bastiaans, "The Wigner distribution function and its applications to optics", in *Optics in Four Dimensions-1980,* AIP Conf. Proc. ***65***, M.A. Machado and L.M. Narducci (eds.) (AIP, New York, 1981), ch. 9, pp. 292-312.

[6.5] M.J. Bastiaans, "Signal description by means of a local frequency spectrum", in *Transformations in Optical Signal Processing,* W.T. Rhodes, J.R. Fienup and B.E.A. Saleh (eds.) SPIE **373**, 49-62 (1981).

[6.6] M.J. Bastiaans, "Use of the Wigner distribution function in optical problems", in *1984 European Conference in Optics, Optical Systems and Applications,* SPIE **492**, 251-262 (1984).

[6.7] M.J. Bastiaans, "Application of the Wigner distribution function in optics", in *The Wigner Distribution: Theory and Applications in Signal Processing,* W. Mecklenbräuker and F. Hlawatsch (eds.) (Elsevier, Amsterdam, 1997), pp. 375-426.

[6.8] G. Cristóbal, C. Gonzalo and J. Bescós, "Image filtering and analysis through the Wigner distribution", in *Advances in Electronics and Electron Physics,* Vol. 80, P.W. Hawkes (ed.) (Academic Press, San Diego, 1991), pp. 309-397.

[6.9] D. Dragoman, "The Wigner distribution function in Optics and Optoelectronics", in *Progress in Optics,* Vol. XXXVII, E. Wolf (ed.) (Elsevier, Amsterdam, 1997), ch. 1, pp. 1-56.

[6.10] *Wigner Distribution and Phase Space in Optics,* G.W. Forbes, V.I. Man'ko, H.M. Ozaktas, R. Simon and K.B. Wolf (eds.), J. Opt. Soc. Am. A **17** (12), 2273-2542 (2000).

[7] The Wigner distribution function is one of the basic tool of quantum optics, from which the whole class of s-parameterized distribution functions can be obtained. We signalize the seminal papers by Cahill and Glauber [7.1] and some fundamental textbooks on quantum optics, dealing as well with the Wigner representation, specifically addressed in the quoted pages.

[7.1] K.E. Cahill and R.J. Glauber, "Ordered expansions in boson amplitude operators", Phys. Rev. **177**, 1857-1881 (1969); "Density operators and quasiprobability distributions", *ibid.*, 1882-1902.

[7.2] W.H. Louisell, *Quantum Statistical Properties of Radiation* (John Wiley & Sons, New York, 1973), ch. 3, pp. 168-176.

[7.3] L. Mandel and E. Wolf, *Optical Coherence and Quantum Optics* (Cambridge U. Press, New York, 1995), pp. 541-542.

[7.4] U. Leonardt, *Measuring the Quantum State of Light* (Cambridge U. Press, New York, 1997), ch. 3, pp. 37-54.

[7.5] W.P. Schleich, *Quantum Optics in Phase Space* (Wiley-VCH, Berlin, 2001), ch. 3, pp. 67-98.

[8] Presentations concerned with the application of the Wigner representation to electron optics in both synchrotron radiation and microscope transfer function related issues are in

[8.1] K.-J. Kim, "Brightness, coherence and propagation characteristics of synchrotron radiation", Nucl. Instr. & Meth. in Phys. Res. A **246**, 71-76 (1986).

[8.2] V.M. Castaño, "A generalized display of the contrast transfer function in transmission electron microscopy", Optik **81**, 35-37 (1988).

[8.3] G. Dattoli, C. Mari and A. Torre, "A note on Wigner-phase space and electron beam transport", ENEA RT/INN/91/30.

[8.4] V.M. Castaño, P. Santiago, A. Gómez, "Image displacements in transmission electron microscopy: a phase space approach", Optik **95**, 31-33 (1993).

[8.5] F. Ciocci, G. Dattoli, A. Torre and A. Renieri, Insertion Devices for Synchrotron Radiation and Free Electron Laser (World Sc., Singapore, 2000), ch. 5, pp. 167-178.

[9] Generalization of the phase-space formulation of quantum mechanics to any set of non commuting conjugate operators are elaborated in the papers below. Some titles concerned with similar elaborations within the context of signal theory can be found in [5] of ch.7.

[9.1] M.O. Scully and L. Cohen, "Quasi-probability distributions for arbitrary operators", in *The Physics of Phase Space*, Y.S. Kim and W.W. Zachary (eds.) (Springer-Verlag, New York, 1987), pp. 253-260.

[9.2] A. Pimpale and M. Razavy, "Quantum-mechanical phase space: a generalization of Wigner phase-space formulation to arbitrary coordinate systems", Phys. Rev. **38**, 6046-6054 (1988).

[9.3] J.A. Vaccaro and D.T. Pegg, "Wigner function for number and phase", Phys. Rev. A **41**, 5156-6054 (1990).

[10] Description at diverse depths, and within different contexts, of the general properties of the Wigner distribution function, from both theoretical and applicative viewpoints, can be found in [1], [4], [8] and in the papers below. Additional titles can be found in the forthcoming chapters.

[10.1] R.F. O'Connell and E. P. Wigner, "Quantum mechanical distribution functions: conditions for uniqueness", Phys. Lett. A **83**, 145-148 (1981).

[10.2] L. Cohen, "On a fundamental property of the Wigner distribution", IEEE Trans. on Acoust., Speech, and Signal Processing, ASSP-**35**, 559-561 (1987).

[10.3] F. Hlawatsch, "Interference terms in the Wigner distribution", in *Digital Signal Processing-84*, V. Cappellini and A.G. Constantinides (eds.) (Elsevier, Amsterdam, 1984), pp. 363-367.

[10.4] A.J.E.M. Janssen, "On the locus and spread of pseudo-density functions in the time-frequency plane", Philips J. Res. **37**, 79-110 (1982).

[10.5] L. Cohen, "What is a multicomponent signal?", Proc. IEEE , ICASSP–92 **5**, 113-116 (1992).

[10.6] Hudson, R.L., "When is the Wigner quasi-probability density non-negative?", Rep. Math. Phys. **6**, 249-252 (1974).

[10.7] L. Cohen and Y.I. Zaparovanny, "Positive quantum joint distributions", J. Math. Phys. **21**, 794-796 (1980).

[10.8] R.F. O'Connell and E. P. Wigner, "Some properties of non-negative quantum-mechanical distribution function ", Phys. Lett. A **85**, 121-126 (1981).

[10.9] F. Soto and P. Claverie, "When is the Wigner function of multi-dimensional systems non-negative?", J. Math. Phys **24**, 97-100 (1983).

[10.10] A.J.E.M. Janssen, "A note on Hudson's theorem about functions with non-negative Wigner distributions", SIAM J. Math. Anal. **15**, 170-176 (1984).

[10.11] A.J.E.M. Janssen, "Positivity properties of phase-plane distribution functions", J. Math. Phys. **25**, 2240-2252 (1984).

[10.12] A.J.E.M. Janssen and T.A.C.M. Claasen, "On positivity of time-frequency distributions", IEEE Trans. Acoust., Speech, Signal Processing, ASSP-**32**, 81-94 (1985).

[10.13] K. Husimi, "Some formal properties of the density matrix", Proc. Phys. Math. Soc. Japan **22**, 264-314 (1940).

[10.14] N.D. Cartwright, "A non-negative Wigner-type distribution", Physica **83A**, 210-212 (1976).

[10.15] P. Bertrand, J.P. Doremus, B. Izrar, V.T. Nguyen and M.R. Feix, "Obtaining non-negative quantum mechanical distribution function", Phys. Lett. A **94**, 415-417 (1983).

[10.16] T.B. Smith, "A comment on the smoothed Wigner function", Phys. Lett. A **127**, 75-78 (1988).

[10.17] D. Dragoman, "The origin of negative values of the Wigner distribution function", Optik **111**, 179-183 (2000).

[10.18] R. Jagannathan, R. Simon, E.C.G. Sudarshan and R. Vasudevan, "Dynamical maps and nonnegative phase-space distribution functions in quantum mechanics", Phys. Lett. A **120**, 161-164 (1987).

[10.19] F.J. Narcowich, "Conditions for the convolution of two Wigner distributions to be itself a Wigner distribution", J. Math. Phys. **29**, 2036-2041 (1988).

[10.20] H.H. Szu and J.A. Blodgett, "Wigner distribution and ambiguity function", in *Optics in Four Dimensions-1980,* AIP Conf. Proc. ***65***, M.A. Machado and L.M. Narducci (eds.) (AIP, New York, 1981), pp. 355-381.

[10.21] F.J. Narcowich and R.F. O'Connell, "Necessary and sufficient conditions for a phase-space function to be a Wigner distribution", Phys. Rev. A **34**, 1-6 (1986).

[10.22] Y. Aharonov, D.Z. Albert and C.K. Au, "New interpretation of the scalar product in Hilbert space", Phys. Rev. Lett. **47**, 1029-1031 (1981).

[10.23] R. F. O'Connell and A.K. Rajagopal, "New interpretation of the scalar product in Hilbert space", Phys. Rev. Lett. **48**, 525-526 (1981).

[10.24] V. Buzek and P.L. Knight, "Quantum interference, superposition states of light, and nonclassical effects", in *Progress in Optics*, Vol. XXXIV, E. Wolf (ed.) (Elsevier, Amsterdam, 1995), ch. 1, pp. 1-158.

[10.25] D. Onciul, "Efficiency of light launching into waveguides", Optik **96**, 20-24 (1994).

[10.26] D. Onciul, "Waveguide launching efficiency for multimode and partially coherent light sources", Optik **97**, 75-77 (1994).

[10.27] D. Dragoman, "Phase space representation of modes in optical waveguides", J. Mod. Opt. **42**, 1815-1823 (1995).

[10.28] D. Dragoman, "Wigner distribution function representation of the coupling coefficient", Appl. Opt. **35**, 6758-6763 (1995).

[10.29] M. Testorf and J. Ojeda-Castañeda, "Fractional Talbot effect: analysis in phase space", J. Opt. Soc. Am. A **13**, 119-125 (1996).

[10.30] J. Ojeda-Castañeda and E.E. Sicre, "Quasi ray-optical approach to longitudinal periodicities of free and bounded wavefields", Opt. Acta **32**, 17-26 (1985).

[10.31] K.-H. Brenner and A.W. Lohmann, "Wigner distribution function display of complex 1D signals", Opt. Comm. **42**, 310-314 (1982).

[10.32] A.W. Lohmann, J. Ojeda-Castañeda and N. Streibl, "The influence of wave aberrations on the Wigner distribution", Opt. Appl. **13**, 465-471 (1983).

[11] The concept of the mutual coherence function was firstly introduced in the seminal paper by Wolf [11.1]. Some representative textbooks are listed here. See also [7.2] and [7.3].

[11.1] E. Wolf, "A macroscopic theory of interference and diffraction of light from finite sources. II. Fields with a spectral range of arbitrary width", Proc. Roy. Soc. (London) **230**, 246-265 (1955).

[11.2] M. Born and E. Wolf, *Principles of Optics* (Pergamon, Oxford, 6th,1980), ch. 10, pp. 491-555.

[11.3] M.J. Beran and G.B. Parrent, Jr., *Theory of Partial Coherence* (Prentice-Hall, Englewwod Cliffs, 1964).

[11.4] J.W. Goodman, *Statistical Optics* (John Wiley & Sons, New York, 1985).

[11.5] B.E.A. Saleh and M.C. Teich, *Fundamentals of Photonics* (John Wiley & Sons, New York, 1991), ch. 10, pp. 342-383.

[12] For the Wigner-representation of partially coherent optical fields the reader may consult [6.4], [6.6] and

[12.1] M.J. Bastiaans, "The Wigner distribution function of partially coherent light", Optica Acta **28**, 1215-1224 (1981).

[12.2] M.J. Bastiaans, "Applications of the Wigner distribution function to partially coherent light", J. Opt. Soc. Am. A **3**, 1227-1238 (1986).

[12.3] Brenner, K.-H. and J. Ojeda-Castañeda, "Ambiguity function and Wigner distribution applied to partially coherent imagery", Optica Acta **31**, 213-223 (1984).

[13] A list of papers investigating the connection between the phase-space distribution functions and the theory of partial coherence is given in [9] of ch. 2.

[13.1] E. Walther, "Radiometry and coherence", J. Opt. Soc. Am. **58**, 1256-1259 (1968); *ibid.*, J. Opt. Soc. Am. **63**, 1219-1226 (1973).

[13.2] E. Wolf, "Coherence and radiometry", J. Opt. Soc. Am. **68**, 6-14 (1978).

[14] J.D. Gaskill, *Linear Systems, Fourier Transforms and Optics* (Wiley & Sons, New York, 1978), ch. 3, pp. 40-98.

[15] As to Gaussian and Hermite-Gaussian beams see [13] of ch. 2.

[16] For the basic treatment of Hermite-cosh-Gaussian beams we suggest

[16.1] L.W. Casperson, D.G. Hall and A.A. Tovar, "Sinusoidal-Gaussian beams in complex optical systems", J. Opt. Soc. Am. A **14**, 3341-3348 (1997).

[16.2] L.W. Casperson and A.A. Tovar, "Hermite-sinusoidal-Gaussian beams in complex optical systems", J. Opt. Soc. Am. A **15**, 954-961 (1998).

[16.3] A.A. Tovar and L.W. Casperson, "Production and propagation of Hermite-sinusoidal-Gaussian laser beams", J. Opt. Soc. Am. A **15**, 2425-2432 (1998).

[16.4] A. Belafhal and M. Ibnchaikh, "Propagation properties of Hermite-cosh-Gaussian laser beams", Opt. Comm. **186**, 269-276 (2000).

[17] For the basic treatment of superGaussian beams we suggest

[17.1] S. De Silvestri, P. Laporta, V. Magni and O. Svelto, "Solid-state laser unstable resonators with tapered reflectivity mirrors: the supergaussian approach", IEEE J. Quant. Electron. **24**, 1172-1177 (1988).

[17.2] C. Palma and V. Bagini, "Propagation of super-Gaussian beams", Opt. Comm. **111**, 6-10 (1994).

[18] W. Magnus and F. Oberhettinger, *Formulas and Theorems for the Special Functions of Mathematical Physics* (Chelsea, New York, 1949).

[19] I.S. Gradshteyn and I.M. Ryzhik, *Table of Integrals, Series and Products* (Academic Press, New York, 1965), p. 838, # 7.377.

7

The Wigner Distribution Function: Optical Production

7.1 Introduction

We have learned in the previous chapters that an optical signal can be fully and globally represented by the complex amplitude $\varphi(q)$ as well as by the relevant Fourier transform $\widetilde{\varphi}(\kappa)$. The former conveys information about the spatial distribution of the signal and the latter about the frequency content.

However, on performing the Fourier analysis we loose all information about the space localization of a given frequency component. in contrast, we may wish to know the very position at which a given frequency component occurs. In addition, Fourier analysis is rather "uneconomical" and highly unstable with respect to perturbation. In fact, one needs to add an infinite series of alternating contributions even to reproduce a uniform signal over a finite space interval. Also, if one adds a small-amplitude term to a linear superposition of sinusoidal waveforms, the spatial distribution of the overall signal changes very little, whilst the frequency spectrum is strongly modified. This is due to the global character of the Fourier transform.

What we wish indeed is to construct a local frequency representation of a signal. The instance is then how to infer a local information from knowledge of φ and $\widetilde{\varphi}$, how to join the two parallel information conveyed by φ and $\widetilde{\varphi}$ in order to obtain an "intersection" of information, enabling us to know the position at which given frequencies are present.

This is the same question as was approached in the Introduction of Chapter 6 within the quantum-mechanical context, even though in a slightly different formulation. In that context, the Wigner distribution function emerged as one of the possible (and most convenient) solutions to the posed problem. Here the question is posed within the context (and accordingly with the inherent terminology) of signal theory, where indeed the Wigner distribution function was reintroduced (by Ville) some 15 years after its original introduction (by Wigner). The most popular example of a joint signal representation in the

FIGURE 7.1. A musical score is basically a joint representation of a signal (from Satie's Pièces Froides-Danses de Travers, no. 3).

inherent time-frequency domain is the musical score [1] (Fig. 7.1).

Then we may say that we are looking for a method to construct the "musical score" of an optical signal. As proved by the various examples illustrated in the previous chapter, the Wigner distribution function can be considered as a kind of such a desired musical score. We give here a rather naïve demonstration of this on the basis of operative considerations.

It is, in fact, clear that in order to individualize the local features of a signal one may inspect it piece by piece with an analyzing function. This leads to the formalization of the concepts of sliding-window Fourier transform and spectrogram, as described in Sect. 7.2. As an external probe unavoidably influences the result, one may prefer to use the signal itself or its mirrored version as analyzing function, thus respectively obtaining the ambiguity function and the Wigner distribution function. Both are then intrinsic transforms, actuating the former a correlative-type inspection of the signal as the latter a convolutive-type inspection. A brief account of the properties of the ambiguity function and its relationship to the Wigner distribution function is given in Sect. 7.4.

Spectrogram, Wigner distribution function and ambiguity function are all representatives of a comprehensive and wider class of joint (bilinear and not) signal representations, on which we comment rather briefly in Sect. 7.3.

The intrinsic non linearity of the Wigner distribution and ambiguity functions naturally frame them within the context of the partial coherence theory, as the product function $\varphi^*(q - \frac{q'}{2})\varphi(q + \frac{q'}{2})$, entering both the Wigner and ambiguity function integrals, is interpretable in the light of the second-order statistics of the wave field, as discussed in Sect. 7.5.

Finally, the optical production of the Wigner distribution function is discussed in a certain detail, firstly framing it within the fascinating issue of optical computing (Sect. 7.6) and then detailing the optical architectures that have been devised to display the Wigner integral for both real (Sect. 7.7) and complex (Sect. 7.8) $1D$ signals. Illustration of the optical implementation of the Gaussian smoothed Wigner distribution (Sect. 7.9) closes the chapter.

7.2 The sliding-window Fourier transform

The simplest way one can devise to answer the question "which frequencies pertain to which position" is that of sectioning the entire spatial distribution into a certain number of separate portions of some length and then Fourier analyse each selected portion to ascertain the frequencies present in that portion. The collection of all the obtained *local* spectra provides us with a kind of a joint space-frequency description of the signal. The reliability and the sharpness of the resulting joint representation is determined by the details of the sectioning procedure. However, they are inherently limited since the "identity" of the original signal is at a certain extent lost when sectioning is operated; also, the position and frequency resolutions of the obtained representation are intimately related to the extent of the various signal sections, that we have created, through the uncertainty relation of Fourier analysis.

More specifically, such an intermediate description can operatively be realized by windowing the signal $\varphi(q)$ by a suitable function $\mathsf{w}(q)$ and then Fourier transforming the windowed signal for running locations of $\mathsf{w}(q)$. Namely, for a given position q we multiply the signal amplitude $\varphi(q')$ by the conjugate and shifted version of the window function $\mathsf{w}^*(q'-q)$, and then take the Fourier transform of the product $\varphi(q')\mathsf{w}^*(q'-q)$. Shifting the window function $\mathsf{w}(q)$ continuously over the signal, we create a function of q and κ, which allows us to estimate the frequency content of the signal at different positions within the relevant space domain. The obtained *sliding-window Fourier transform* [2]

$$\mathfrak{F}_{\varphi}(q,\kappa;\mathsf{w}) = \frac{1}{\sqrt{2\pi}} \int \varphi(q')\mathsf{w}^*(q'-q)e^{-i\kappa q'}dq', \tag{7.2.1}$$

can in fact be seen as a local frequency spectrum of the signal (Fig. 7.2). The window function w may be chosen rather arbitrarily; however, being the procedure implied by (7.2.1) aimed at gaining information about the space localization of the frequency components of the signal, it is convenient that $\mathsf{w}(q)$

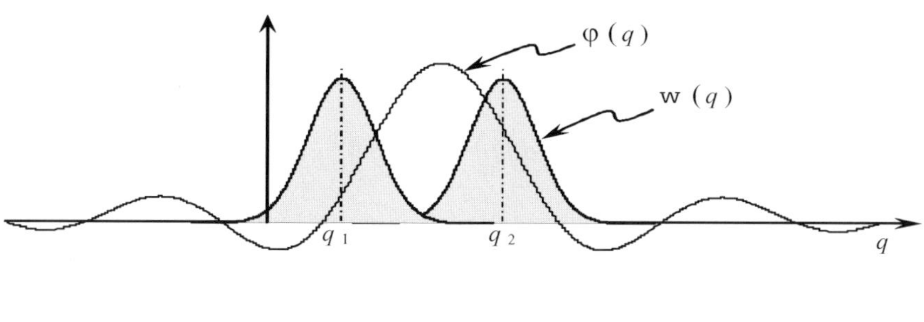

FIGURE 7.2. Sliding-window Fourier spectrum as a local frequency spectrum: moving the window function $\mathsf{w}(q)$ over the waveform $\varphi(q)$ to be analysed, definite sections of $\varphi(q)$ are selected whose frequency content can then be estimated upon Fourier transformation.

be relatively "narrow". Thus, the integration limits in (7.2.1) are in practice set according to the chosen position and the width of the window function.

Evidently, $\mathfrak{F}_\varphi(q,\kappa;\mathsf{w})$ is linear and in general complex valued. It preserves frequency shifts in the signal, and space shifts up to a modulation, thus allowing for a uniform analysis of the signal features over the whole (q,κ) plane (see Problem 7.1). In addition, the signal can uniquely be reconstructed from the relevant windowed spectrum (7.2.1). In fact, transforming back (7.2.1) to the space domain results into the obvious inversion formula:

$$\varphi(q')\mathsf{w}^*(q'-q) = \frac{1}{\sqrt{2\pi}\mathsf{w}^*(0)}\int \mathfrak{F}_\varphi(q,\kappa;\mathsf{w})e^{i\kappa q}d\kappa. \tag{7.2.2}$$

Alternatively, one may transform back (7.2.1) to the space domain, multiply by the window and then integrate over the space variable, thus finally obtaining

$$\varphi(q) = \frac{1}{\sqrt{2\pi}}\iint \mathfrak{F}_\varphi(q',\kappa';\mathsf{w})[\mathsf{w}(q-q')e^{i\kappa' q}]dq'd\kappa', \tag{7.2.3}$$

the window function being supposed to have a unit energy: $\int |\mathsf{w}|^2(q)dq = 1$. Interestingly, in (7.2.3) the signal at position q is expressed as a weighted superposition of what can be regarded as basis signals $s_{q',\kappa'}(q) = \mathsf{w}(q-q')e^{i\kappa' q}$, centered on the point (q',κ') in the phase plane, if as usual the window function is centered around the origin in both the space and frequency domain. A more general form of the *synthesis* formula (7.2.3) is described in Problem 7.1.

The magnitude squared of the sliding-window spectrum, namely

$$\mathcal{S}_\varphi(q,\kappa;\mathsf{w}) = |\mathfrak{F}_\varphi(q,\kappa;\mathsf{w})|^2 \tag{7.2.4}$$

yields the desired space-frequency representation of the signal conveying information about the energy density distribution in the spatial position and frequency. In the signal theory, $\mathcal{S}_\varphi(q,\kappa;\mathsf{w})$ is named the *spectrogram*. However, a certain variety of names exists. Thus, for instance, $\mathfrak{F}_\varphi$ is as well called the *complex spectrogram* and $\mathcal{S}_\varphi$ the *physical spectrum* of $\varphi(q)$.

It is worth observing that the sliding-window spectrum entangles the signal and the window, and hence unavoidably the properties of the signal are admixed with those of the window function. As an example, note that integrating the spectrogram (7.2.4) over the entire (q,κ) plane one obtains

$$\iint \mathcal{S}_\varphi(q,\kappa;\mathsf{w})dqd\kappa = ||\varphi(q)||^2||\mathsf{w}(q)||^2, \tag{7.2.5}$$

i.e., the product of the total energies of the signal and the window function. Analogously, as the reader may easily verify, the marginal distributions of the spectrogram do not relate to the "pure" energy densities of the signal, the energy density of the window being as well involved in the relevant expressions. Relation (7.2.5) makes evident the interchangeability of $\varphi(q)$ and $\mathsf{w}(q)$ as "signal" and "window function". It is in fact evident that the signal and the

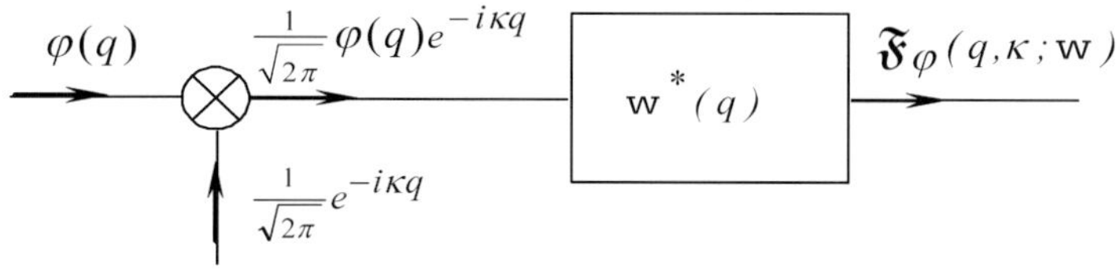

FIGURE 7.3. The windowed Fourier spectrum can be seen as a filtering operation of the linearly frequency modulated amplitude $\frac{1}{\sqrt{2\pi}}\varphi(q)e^{-i\kappa q}$ by the conjugate spectrum of the window function.

window play essentially symmetric roles; it is in a certain measure ambiguous whether a window $\mathsf{w}(q)$ is used to compute the spectrogram of a signal $\varphi(q)$ or inversely a window $\varphi(q)$ is used to compute the spectrogram of a signal $\mathsf{w}(q)$. Indeed, one has that $\mathfrak{F}^*_{\mathsf{w}}(q,\kappa;\varphi) = e^{i\kappa q}\mathfrak{F}_\varphi(-q,-\kappa;\mathsf{w})$. One may remedy the inherent ambiguity in the roles of the signal and the window, using the signal itself, or its mirrored version, as window. This leads to the ambiguity function, or the Wigner distribution function, as we will see in § 7.2.3.

Finally we note that the short-space Fourier transform $\mathfrak{F}_\varphi(q,\kappa;\mathsf{w})$ can well be interpreted as the "output" from a space-invariant system, whose spread function is basically $\mathsf{w}^*(q)$, with the input being the linearly frequency modulated amplitude $\frac{1}{\sqrt{2\pi}}\varphi(q)e^{-i\kappa q}$. The modulation of $\varphi(q)$ serves to shift the frequency spectrum of $\varphi(q)$ at frequency κ to 0 frequency. Accordingly, the windowed-Fourier transforming operation implied by $\mathfrak{F}_\varphi(q,\kappa;\mathsf{w})$ is understood as a filtering operation of the shifted spectrum of $\varphi(q)$ by the conjugate spectrum of the window function (Fig. 7.3).

7.2.1 *The frequency-domain view of* $\mathfrak{F}_\varphi(q,\kappa;\mathsf{w})$

The windowed Fourier transform (7.2.1) is amenable to a windowing-like interpretation in the frequency domain as well. In fact, being the Fourier transform of the product $\psi_q(q') = \varphi(q')\mathsf{w}^*(q'-q)$, regarded as a function of the running position q' for a fixed q, $\mathfrak{F}_\varphi(q,\kappa;\mathsf{w})$ comes to be the convolution of the spectra of φ and the shifted window. Hence, the alternative expression

$$\mathfrak{F}_\varphi(q,\kappa;\mathsf{w}) = \frac{1}{\sqrt{2\pi}}e^{-i\kappa q}\int \widetilde{\varphi}(\kappa')\widetilde{\mathsf{w}}^*(\kappa'-\kappa)e^{i\kappa' q}d\kappa' \tag{7.2.6}$$

immediately follows. The integral over κ' signifies evidently a windowing operation in the frequency domain with the window function being the very Fourier transform of $\mathsf{w}(q)$. Then it can be interpreted as a *sliding-window space transform*, and accordingly denoted as

$$\mathfrak{F}_{\widetilde{\varphi}}(q,\kappa;\widetilde{\mathsf{w}}) = \frac{1}{\sqrt{2\pi}}\int \widetilde{\varphi}(\kappa')\widetilde{\mathsf{w}}^*(\kappa'-\kappa)e^{i\kappa' q}d\kappa'. \tag{7.2.7}$$

Expression (7.2.6), we may suitably rewrite as

$$\mathfrak{F}_{\varphi}(q, \kappa; \mathsf{w}) = e^{-i\kappa q}\mathfrak{F}_{\widetilde{\varphi}}(q, \kappa; \widetilde{\mathsf{w}}), \tag{7.2.8}$$

shows that windowing the signal $\varphi(q)$ in space by a function $\mathsf{w}(q)$ automatically implies windowing the spectrum $\widetilde{\varphi}(\kappa)$ by the Fourier transform $\widetilde{\mathsf{w}}(\kappa)$. It signifies the equivalence of the windowing operation in space and frequency if, of course, the window functions in the two domains are just a Fourier transform pair. Either case, we obtain the same space-frequency distribution $\mathcal{S}_{\varphi}$, which accordingly can be used to inspect the frequency properties of the signal at given positions and inversely the space properties at given frequencies.

The observation above makes evident an intrinsic limit of the space-frequency representation based on a windowing-procedure. In order to achieve good space localization of the frequency properties a narrow analysis window $\mathsf{w}(q)$ in space is required. Likewise, for a good frequency localization of the space properties a narrow window $\widetilde{\mathsf{w}}(\kappa)$ in frequency would be needed. Of course, the requirements for a "narrow" window is rather imprecise, the precise definition depending on the applications. We know that the extensions of a function and its Fourier transform can not be arbitrarily chosen, being linked by the uncertainty relation. We can not have narrow windows in both the space and the frequency domains. The sliding-window Fourier transform is indeed affected by a basic resolution trade off as improving the space resolution by narrowing the window results in a loss of frequency resolution and vice versa. Criteria for fixing the optimal window function can be set up according to the problem under investigation.

7.2.2 *Some examples*

The spectrogram depends evidently on the choice of the window function. For illustration purposes we calculate the spectrograms of some elementary signals, using rectangular and/or Gaussian windows; accordingly, the various parameters in the graphs are specified simply by numbers, units being implicit. As noted in § 4.6.4, Gaussian functions have the notable property of self-reproducing under Fourier transform with the relevant space-bandwidth product taking the minimum value. As we will see in § 7.2.4, the spectrogram of a signal by a Gaussian window is also interpretable as a Gaussian-smoothed Wigner distribution function (see also § 6.3.6).

Firstly we compare the spectrograms of the quadratic phase signal

$$\varphi_{\text{chirp}}(q) = e^{i\frac{\beta}{2}q^2}, \tag{7.2.9}$$

obtained using a rectangular and a Gaussian window.

A unit-energy rectangular function is, of course, described by

$$\mathsf{w}_{\text{R}}(q) = \begin{cases} \frac{1}{\sqrt{2Q_{\mathsf{w}}}}, & |q| \leq Q_{\mathsf{w}}, \\ 0 & |q| > Q_{\mathsf{w}}. \end{cases} \tag{7.2.10}$$

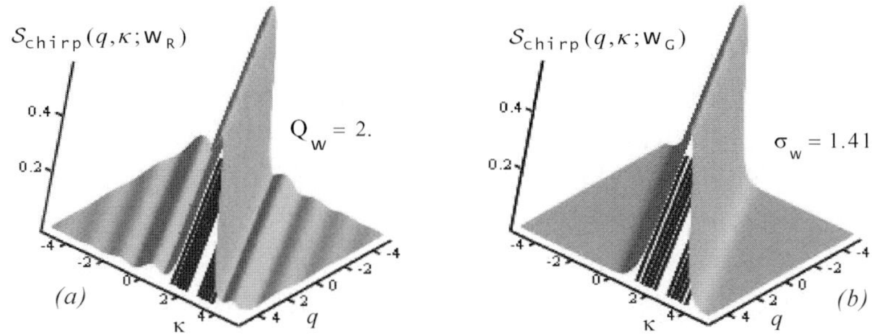

FIGURE 7.4. Spectrograms of the chirp $\varphi_{chirp}(q) = e^{i\frac{1}{4}q^2}$ using *(a)* the rectangular window (7.2.10) and *(b)* the Gaussian window (7.2.13) with the respective optimal window widths.

A direct calculation leads to the expression

$$\mathfrak{F}_{chirp}(q,\kappa;\mathsf{w}_R) = \sqrt{\frac{i}{8\beta Q_w}}\, e^{-i\frac{\kappa^2}{2\beta}}\{\mathrm{erf}(\tfrac{\kappa-\beta q}{\sqrt{2i\beta}} + \sqrt{\tfrac{\beta}{2i}}Q_w) - \mathrm{erf}(\tfrac{\kappa-\beta q}{\sqrt{2i\beta}} - \sqrt{\tfrac{\beta}{2i}}Q_w)\}, \tag{7.2.11}$$

for the windowed spectrum of (7.2.9) using (7.2.10). Here, $\mathrm{erf}(x)$ is the error function defined by [13]

$$\mathrm{erf}(x) = \frac{2}{\sqrt{\pi}}\int_0^x e^{-t^2}dt. \tag{7.2.12}$$

It can be guessed from (7.2.11) that the spectrogram $\mathcal{S}_{chirp}(q,\kappa;\mathsf{w}_R)$ of the chirp (7.2.9) develops around the local frequency line $\kappa = \beta q$ (recall the line-delta $\delta(\kappa-\beta q)$ obtained for the Wigner distribution function). The frequency resolution is optimal for a window width roughly estimated as $Q_w \sim \sqrt{2/\beta}$ (see Problem 7.2). In Fig. 7.4.a) the $3D$ graph of $\mathcal{S}_{chirp}(q,\kappa;\mathsf{w}_R)$ is shown in the case $\beta = 0.5$ for the corresponding optimal window width $Q_w = 2$.

Correspondingly, a unit-energy Gaussian window shapes as

$$\mathsf{w}_G(q) = (\frac{1}{\pi\sigma_W^2})^{\frac{1}{4}} e^{-\frac{q^2}{2\sigma_W^2}}. \tag{7.2.13}$$

With this analysis window the spectrogram of (7.2.9) explicitly writes as

$$\mathcal{S}_{chirp}(q,\kappa;\mathsf{w}_G) = \frac{1}{\sqrt{\pi}\sigma} e^{-\frac{(\kappa-\beta q)^2}{\sigma^2}}, \tag{7.2.14}$$

with σ being

$$\sigma^2 = \frac{1+\beta^2\sigma_W^4}{\sigma_W^2}. \tag{7.2.15}$$

It develops as well around the line $\kappa = \beta q$ in the (q,κ) plane with a frequency localization depending on β and σ_W. It is in fact Gaussian in the variable

$\kappa-\beta q$, and would turn into a line-delta $\delta(\kappa-\beta q)$ with $\sigma \longrightarrow 0$, a value however that is not reachable. The best we can do is to choose σ_{W} in order to minimize σ, which requirement is fulfilled with $\sigma_{\mathsf{W}} = \frac{1}{\sqrt{\beta}}$. A $3D$ view of $\mathcal{S}_{\text{chirp}}(q, \kappa; \mathsf{w}_{\text{G}})$ for the chirp $\beta = 0.5$ is given in Fig. 7.4.b) in correspondence with the optimal value $\sigma_{\mathsf{W}} = 1.41$. Evidently the Gaussian window accommodates for a better frequency localization with respect to the rectangle window.

An accurate account of several window functions typically used in spectrogram related computations, and in general in Fourier spectrum estimations, is given in [2.2], [2.3] and [6.3]. In particular, in [2.3] the relation between the spectrogram-based signal synthesis and the Gabor signal expansion is elucidated. The reader may guess the nature of such a relation from (7.2.3), which as noted expresses the signal as a weighted superposition of basis signals in the space/spatial frequency plane like $s_{q',\kappa'}(q) = \mathsf{w}(q-q')e^{i\kappa' q}$. With a Gaussian amplitude taken as a synthesis window in the basis signals $s_{q',\kappa'}(q)$, we basically obtain the Gabor elementary signals $\varphi_{\text{G}}(q)$, reported in § 4.6.4 in the form

$$\varphi_{\text{G}}(q) = \left(\frac{1}{\pi a^2}\right)^{1/4} e^{-\frac{1}{2a^2}(q-q_0)^2 + i\kappa_0(q-\frac{q_0}{2})}, \tag{7.2.16}$$

with a being the $\frac{1}{e}$-intensity width. As said in § 4.6.4, $\varphi_{\text{G}}(q)$ has the smallest space-bandwidth product permitted by the uncertainty relation of Fourier analysis. The signal expansion suggested in (7.2.3) is however in terms of basis signals that are continuously placed in the (q, κ) plane, whereas Gabor's signal expansion is in terms of basis signals that are discretely placed in the (q, κ) plane. Indeed, in practical problems the "continuum" of the space-frequency plane is replaced by a lattice of points, made by rectangular cells conveniently sized to have the minimum area in accord with the uncertainty relation.

The basis signals - or *logons* - was originally chosen by Gabor as a set of Gaussian-weighted complex exponentials like (7.2.16) [3]. Each logon is located in only one cell of the lattice and appropriately sized to be highly concentrated in its own cell, thus minimizing the overlap with the other logons, and accordingly yielding a "quantum of information" independent of those provided by the other basis signals. Thus the number of logons for an effective signal representation turns to be equal to the (finite) number of degrees of freedom of the signal. Interestingly, in the light of the ray-like picture of the Gaussian-weighted complex exponential signals, the relevant rms widths in space and frequency multiplying to the minimum allowed value, Gabor's signal expansion is amenable to the nice interpretation as representing the optical signal in the form of a superposition of a certain number of "rays" appearing in discrete positions in the space and with discrete directions, uniquely determined by the definite Gabor lattice in the pertinent space-frequency domain [3.3]. The accurate investigations of the Gabor signal expansion variedly developed in the literature have generalized the original Gabor's expansion to rather arbitrarily shaped elementary signals and to partially coherent light as well as to discrete-time signals and oversampling [3].

Relation (7.2.3) signifies the link of the windowed Fourier transform-based signal representation with the Gabor signal expansion, the coefficients of the latter being obtained as sampling values of the former at the definite space-frequency lattice points. Gabor's signal expansion represents indeed a primary example in a discrete view of a joint space-frequency signal representation.

As an example of the resolution trade-off affecting the spectrogram, it may be instructive to consider the sum of a monofrequency signal and an impulse [4.7]. We take as analysis window the unit-energy Gaussian (7.2.13). Firstly, we note that for the spatial harmonic $\varphi_{\kappa 0}(q) = \frac{1}{\sqrt{2\pi}} e^{i\kappa_0 q}$ one obtains

$$\mathcal{S}_{\kappa 0}(q, \kappa; \mathsf{w}_{\mathrm{G}}) = \frac{\sigma_{\mathsf{W}}}{2\pi\sqrt{\pi}} e^{-(\kappa-\kappa_0)^2\sigma_{\mathsf{W}}^2}, \tag{7.2.17}$$

and correspondingly for the point-like signal $\varphi_{q0}(q) = \delta(q - q_0)$, one has

$$\mathcal{S}_{q0}(q, \kappa; \mathsf{w}_{\mathrm{G}}) = \frac{1}{2\pi\sqrt{\pi}\sigma_{\mathsf{W}}} e^{-\frac{(q-q_0)^2}{\sigma_{\mathsf{W}}^2}}. \tag{7.2.18}$$

The reader may compare the above expressions with those for the corresponding Wigner distribution functions being two delta-lines at $\kappa = \kappa_0$ and $q = q_0$, respectively. The spectrogram (7.2.17) turns into a delta line: $\mathcal{S}_{\kappa 0}(q, \kappa; \mathsf{w}_{\mathrm{G}}) \to \frac{1}{2\pi}\delta(\kappa - \kappa_0)$, for $\sigma_{\mathsf{W}} \to \infty$, whilst in turn $\mathcal{S}_{q0}(q, \kappa; \mathsf{w}_{\mathrm{G}}) \to \frac{1}{2\pi}\delta(q - q_0)$ with $\sigma_{\mathsf{W}} \to 0$. Thus, the spectrogram $\mathcal{S}_{\mathrm{h}+\delta}(q, \kappa; \mathsf{w}_{\mathrm{G}})$ of the signal

$$\varphi(q) = \delta(q - q_0) + \frac{1}{\sqrt{2\pi}} e^{i\kappa_0 q}, \tag{7.2.19}$$

displays a good frequency resolution when the signal is windowed by a "wide" Gaussian at the expense of the space resolution (Fig. 7.5.a)). Conversely a good spatial resolution is attained by using a "narrow" Gaussian, thus reducing the frequency localization (Fig. 7.5.b)). An intermediate situation is reached by an intermediate value of σ_{W} (Fig. 7.5.c)). It is interesting to compare the distribution shown in Fig. 7.5.c) with that representing the Wigner distribution function as plotted in Fig. 6.8. Note that $\mathcal{S}_{\mathrm{h}+\delta}(q, \kappa; \mathsf{w}_{\mathrm{G}})$ presents as well a cross term, which is oscillatory in κ with a period governed by σ_{W} and q_0 and modulated by a double Gaussian. Hence it is significative mostly around the point (q_0, κ_0) in the phase plane inside an elliptical contour approximately of width $2\sqrt{2}/\sigma_{\mathsf{W}}$ along the κ direction and $2\sqrt{2}\sigma_{\mathsf{W}}$ along the q direction.

We close this section briefly commenting on the behaviour of the cross terms that unavoidably enter the spectrogram when a multicomponent signal is considered. In fact, although the short-space Fourier transform is a linear transform, the spectrogram is evidently a bilinear transform, and hence, like the Wigner distribution function, has cross-terms. In particular, for a two-component signal like

$$\varphi(q) = \varphi_1(q) + \varphi_2(q), \tag{7.2.20}$$

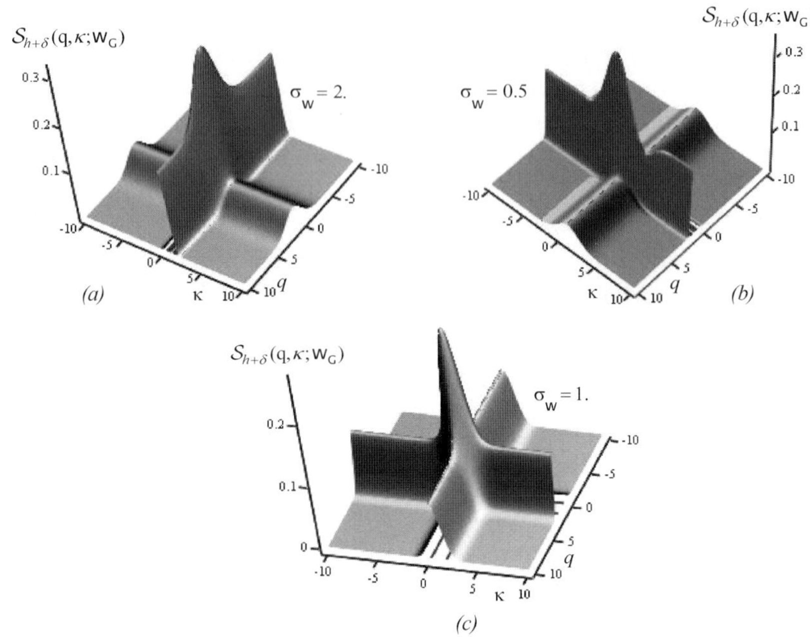

FIGURE 7.5. Spectrograms of the signal $\varphi(q) = \delta(q-1) + \frac{1}{\sqrt{2\pi}} e^{iq}$ using the Gaussian window (7.2.13) with different widths σ_{w}'s.

we have

$$\mathcal{S}_\varphi(q,\kappa;\mathsf{w}) = |\mathfrak{F}_{\varphi 1}(q,\kappa;\mathsf{w})|^2 + |\mathfrak{F}_{\varphi 2}(q,\kappa;\mathsf{w})|^2 + 2\Re\{\mathfrak{F}_{\varphi 1}(q,\kappa;\mathsf{w})\mathfrak{F}^*_{\varphi 2}(q,\kappa;\mathsf{w})\}, \tag{7.2.21}$$

which clearly shows that the cross-term contribution adds to the auto terms $|\mathfrak{F}_{\varphi 1}(q,\kappa;\mathsf{w})|^2$ and $|\mathfrak{F}_{\varphi 2}(q,\kappa;\mathsf{w})|^2$ through an interference-like mechanism.

Then, let us consider a spatial harmonic starting from $q = 0$; namely,

$$\varphi_+(q) = \begin{cases} e^{i\kappa_0 q} & q \geq 0, \\ 0 & \text{elsewhere.} \end{cases} \tag{7.2.22}$$

Taking as a window function the unit-energy rectangle waveform (7.2.10), a straightforward calculation leads to

$$\mathcal{S}_{\varphi_+}(q,\kappa;\mathsf{w}_{\mathrm{R}}) = \begin{cases} \frac{\sin^2[(\kappa-\kappa_0)(q+Q_{\mathsf{w}})/2]}{\pi Q_{\mathsf{w}}(\kappa-\kappa_0)^2}, & -Q_{\mathsf{w}} < q \leq Q_{\mathsf{w}}, \\ \frac{\sin^2[(\kappa-\kappa_0)Q_{\mathsf{w}}]}{\pi Q_{\mathsf{w}}(\kappa-\kappa_0)^2}, & q > Q_{\mathsf{w}}. \end{cases} \tag{7.2.23}$$

It is evident that $\mathcal{S}_{\varphi_+}$ is non zero even for values $q < 0$, specifically over the interval $-Q_{\mathsf{w}} < q \leq 0$, where the signal does not exist. As expected, the frequency and space localization depends on Q_{w} through an inverse relation.

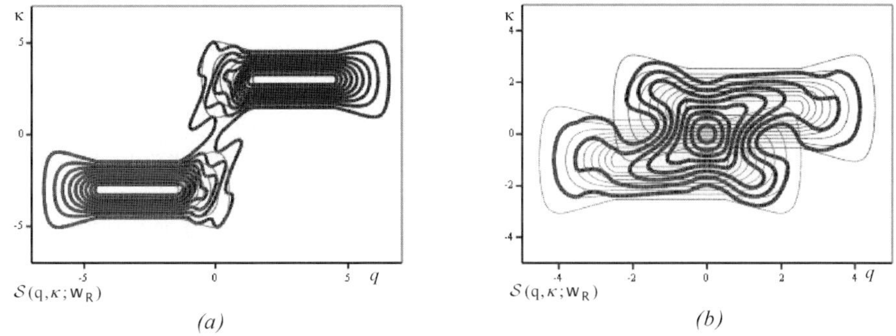

FIGURE 7.6. Spectrograms of the signal (7.2.25) with $Q_1 = Q_2 = 3$. *(a)* $\kappa_{0_1} = -\kappa_{0_2} = 3$; $q_{0_1} = -q_{0_2} = 3$. *(b)* $\kappa_{0_1} = -\kappa_{0_2} = 1$; $q_{0_1} = -q_{0_2} = 1$. The subtle lines represent the spectrograms of the individual signal components.

The reader is suggested to calculate the spectrogram of a complex sinusoidal signal limited to a segment of width $2Q$, i.e.,

$$\varphi_Q(q) = \begin{cases} e^{i\kappa_0 q} & |q| \leq Q, \\ 0 & \text{elsewhere}, \end{cases} \tag{7.2.24}$$

using again the rectangular window (7.2.10). It turns out to vanish outside the vertical strip $-Q_w - Q < q \leq Q_w + Q$, and be mostly concentrated within the frequency interval centered on κ_0 of width $\frac{2\pi}{Q_w}$.

The graphs shown in Fig. 7.6 are relevant to the spectrogram of the two-component signal

$$\varphi(q) = \varphi_{Q_1}(q) + \varphi_{Q_2}(q), \tag{7.2.25}$$

each component being a shifted version of the finite-length harmonic waveform (7.2.24), namely

$$\varphi_{Q_j}(q) = \begin{cases} e^{i\kappa_{0_j}(q-q_{0_j})} & |q - q_{0_j}| \leq Q_j, \\ 0 & \text{elsewhere}. \end{cases} \qquad j = 1, 2. \tag{7.2.26}$$

Two sets of values of the involved parameters have been considered; again, a rectangular window has been used ($Q_w = 1.5$).

The displayed graphs are aimed at providing a certain feeling for the mechanism of the cross-terms in the spectrogram, also in comparison with that of the Wigner distribution function (see § 6.3.4). In fact, by inspecting the various contour plots in connection with the corresponding set of parameter values, we may infer that in general the cross-terms between the signal components have a significant effect on the overall shape of the spectrogram of a multi-component signal only if there is an effective overlap of the self-terms, which ultimately depends on the signal structure and the window used [4.6, 2.5]. Then, if the auto-terms are sufficiently far apart in the space-frequency plane, the cross-terms are significantly attenuated and masked by the self-terms. In contrast, as noted in § 6.3.4, the cross-terms of the Wigner distribution function always occur unattenuated midway between each pair of signal components whether the corresponding auto-terms overlap or not.

A great variety of further examples of spectrograms can be found in [4.7] (ch. 7, pp. 93-112), by which this subsection is partly inspired.

7.2.3 *From the short-space spectrum to the Wigner distribution and ambiguity function*

A closer look at the expression (7.2.1) of $\mathfrak{F}_\varphi(q, \kappa; \mathsf{w})$ reveals a certain similarity with the cross-Wigner distribution function (6.3.12) and better yet with the cross-ambiguity function (7.4.10) [12.5, 4.5]. The similarity becomes more evident if we rewrite the expression (7.2.1) in a symmetric form by the simple change of variable $q' \to q''$, being $q' = q'' + q/2$. The integral in (7.2.1) turns into

$$\mathfrak{F}_\varphi(q, \kappa; \mathsf{w}) = \tfrac{1}{\sqrt{2\pi}} e^{-i\kappa q/2} \int \varphi(q' + \tfrac{q}{2}) \mathsf{w}^*(q' - \tfrac{q}{2}) e^{-i\kappa q'} dq', \tag{7.2.27}$$

clearly resembling the expression (7.4.10) of the cross-ambiguity function of the signal and the window. In particular, choosing as a window function the signal itself, i.e., $\mathsf{w}(q) = \varphi(q)$, the integral above just reproduces that defining the auto-ambiguity function (7.4.1). The self-windowed spectrum becomes indeed

$$\mathfrak{F}_\varphi(q, \kappa; \varphi(q)) = \sqrt{2\pi} e^{-i\kappa q/2} \mathcal{A}_\varphi(q, \kappa). \tag{7.2.28}$$

On the other hand, with the choice $\mathsf{w}(q) = \varphi(-q)$, we obtain

$$\mathfrak{F}_\varphi(q, \kappa; \varphi(-q)) = \tfrac{1}{\sqrt{2\pi}} e^{-i\kappa q/2} \int \varphi(\tfrac{q}{2} + q') \varphi^*(\tfrac{q}{2} - q') e^{-i\kappa q'} dq', \tag{7.2.29}$$

the integral being immediately recognized as $\pi \mathcal{W}_\varphi(q/2, \kappa/2)$. Hence,

$$\mathfrak{F}_\varphi(q, \kappa; \varphi(-q)) = \sqrt{\tfrac{\pi}{2}} e^{-i\kappa q/2} \mathcal{W}_\varphi(\tfrac{q}{2}, \tfrac{\kappa}{2}). \tag{7.2.30}$$

Thus, we can synthesize the windowing-operation picture of the Wigner distribution and ambiguity function in relation with the short-space Fourier transform through the diagram

$$\boxed{\begin{array}{ccc} & \mathfrak{F}_\varphi(q, \kappa; \mathsf{w}) & \\ \mathsf{w}(q) = \varphi(q) \;\swarrow & & \searrow\; \mathsf{w}(q) = \varphi(-q) \\ \mathcal{A}_\varphi(q, \kappa) & & \mathcal{W}_\varphi(\tfrac{q}{2}, \tfrac{\kappa}{2}) \end{array}} \tag{7.2.31}$$

which outlines the basic correspondences, disregarding the precise mathematical relations demanding for the cross quadratic phase-term $e^{-i\kappa q/2}$ and some constant factors.

It appears evident that a direct correspondence holds between the space-frequency variables of $\mathfrak{F}_{\varphi}$ and $\mathcal{A}_{\varphi}$, representing in either case the position and frequency difference-coordinates of the signal and its spectrum to the corresponding sliding replicas. In contrast, the space-frequency variables of the Wigner distribution function represent the position and frequency centre-coordinates of the signal and its spectrum to their mirrored sliding replicas.

The considerations above support the interpretation of the Wigner distribution function suggested in § 6.2, and anticipate a possible interpretation of the ambiguity function, on which we will comment in more detail in Sect. 7.4.

7.2.4 *The spectrogram as a smoothed Wigner distribution function*

As previously noted, the sliding-window Fourier transform can also be related to the smoothed Wigner distribution function when a gaussian-shaped window is involved in. To see this, we resort to the Moyal formula (6.3.50) which expresses a definite link between the overlap integral of two signals and that of the respective Wigner distribution functions, i.e.,

$$\frac{1}{2\pi}\left|\int \varphi_1^*(q)\varphi_2(q)dq\right|^2 = \iint \mathcal{W}_{\varphi_1}(q,\kappa)\mathcal{W}_{\varphi_2}(q,\kappa)dqd\kappa. \tag{7.2.32}$$

The defining integral (7.2.1) of the short-space spectrum is interpretable as the inner product of the signal $\varphi(q')$, whose Wigner distribution function is $\mathcal{W}_{\varphi}(q',\kappa')$, and the shifted and frequency modulated window function $\mathsf{w}(q'-q)e^{i\kappa q'}$, whose Wigner distribution function is then equally shifted in space and frequency as $\mathcal{W}_{\mathsf{w}}(q'-q,\kappa'-\kappa)$. Accordingly, interpreting the integral on the left of (7.2.32) as a sliding-window Fourier transform with φ_2 being the signal φ and φ_1 the window-related waveform $\mathsf{w}(q'-q)e^{i\kappa q'}$, Moyal's relation can be written as

$$\mathcal{S}_{\varphi}(q,\kappa;\mathsf{w}) = \iint \mathcal{W}_{\mathsf{w}}(q'-q,\kappa'-\kappa)\mathcal{W}_{\varphi}(q',\kappa')dq'd\kappa'. \tag{7.2.33}$$

The double integral above is evidently the convolution of the Wigner distribution function of the signal with that of the window. Therefore, due to the arbitrariness of the window function, we may conclude that the convolution of any two Wigner distribution functions can be interpreted as a spectrogram. The reader may deduce (7.2.30) from (7.2.33) with $\mathsf{w}(q) = \varphi(-q)$.

Interestingly, the convolution in (7.2.33) can also be viewed as resulting from a windowing operation in the phase plane of the Wigner distribution function of the signal, whose value at the phase-plane point (q',κ') could be picked up by a rather narrow distribution $\mathcal{W}_{\mathsf{w}}$. Thus, the windowing operation in the space domain of a given signal, or in the frequency domain of the relevant Fourier spectrum, corresponds to a windowing operation in the space-frequency plane of the Wigner distribution function of the signal. In the light of this view, the resolution trade-off of the spectrogram is clearly evidenced, as we know that the support of the Wigner distribution function exactly reproduces that of the signal in space and of the relevant spectrum in frequency. Also, we see from (7.2.33) that the windowing operation in the Wigner plane by any Wigner distribution function produces a space-frequency distribution which is everywhere positive through the whole (q, κ) plane.

In particular, taking a Gaussian window, whose Wigner distribution function is as well Gaussian in both space and frequency, one obtains on the right side of (7.2.33) the *Gaussian-smoothed* Wigner distribution function of $\varphi(q)$, which, as described in § 6.3.6, is obtained by the convolution of the Wigner distribution function of the signal with a space-frequency Gaussian distribution, having appropriately related space and frequency rms widths. Evidently, relation (7.2.33) establishes that the Gaussian-smoothed Wigner distribution function is everywhere positive through the whole (q, κ) plane, in accord with the discussion of § 6.3.6. In § 7.9 we will describe the optical implementation of the Gaussian-smoothed Wigner distribution.

According to (7.2.33) the Gaussian-smoothed Wigner distribution function can be interpreted as the spectrogram of the signal by a Gaussian window, and then be related to the projection of the signal upon a Gabor-like waveform, i.e., a Gaussian-weighted complex exponential.

7.3 The Wigner distribution function and the general class of space-frequency signal representations

The spectrogram, the Wigner distribution function and the ambiguity function are representatives of a wide class of bilinear space-frequency distribution functions, which are related to each other through linear transformations. More in general, they are cast into a common analytical framework, including as well nonbilinear signal representations, which has been devised by Cohen [4.1] within the context of quantum mechanics and then applied to time-frequency as well as space-frequency signal analysis [4].

It is worth noting that presently the "strategies" underlying the joint space-frequency (as well time-frequency) signal analysis develop through two channels. On one hand, the linear approach is inspired by the Fourier-like philosophy, and so it attempts to directly decompose the signal into components, which are simultaneously localized in space and frequency. Formally it can be regarded as a linear mapping between the set of the $1D$ complex functions of spatial position or frequency, $\varphi(q)$ or $\widetilde{\varphi}(\kappa)$, to the set of $2D$ in general complex-valued functions of position and frequency:

$$\widehat{\mathfrak{L}} : \varphi(q) \in \mathcal{L}^2(\mathbb{R}) \longrightarrow \mathfrak{L}_\varphi(q, \kappa; \mathfrak{X}). \tag{7.3.1}$$

As earlier said, Gabor's signal expansion is the primary example of such a linear approach, the analysing components being the highly space- and frequency-localized Gaussian-weighted complex sinusoids. The windowed short-space Fourier transform is conceptually based on the same decomposition strategy, performing the Fourier analysis of the signal as it appears when seen through an ensemble of identical appropriately chosen windows shifted in position with respect to each other. The symbol $\mathfrak{X}$ generally signifies the auxiliary tools, as, for instance, the analysing or the synthesizing window function, which pertain to the specific transform and/or the signal under examination.

On the other hand, the bilinear approach attempts to combine the instantaneous power $|\varphi(q)|^2$ and the power spectral density $|\widetilde{\varphi}(\kappa)|^2$ of the signal in a single $2D$ distribution such that it could indicate the energy content of the signal simultaneously in position and frequency. Formally, it realizes a mapping between the pairs of $1D$ complex functions of spatial position or frequency to the set of $2D$ in general complex-valued functions of position and frequency:

$$\widehat{\mathfrak{B}} : (\varphi(q), \varphi(q)) \in \mathcal{L}^2(\mathbb{R}) \times \mathcal{L}^2(\mathbb{R}) \longrightarrow \mathfrak{B}_\varphi(q, \kappa; \mathfrak{X}). \tag{7.3.2}$$

The Wigner distribution function is the primary example of such a quadratic approach, and as noted in § 7.2.3 it can be seen as arising from a windowing-like procedure with the ensemble of analysing windows being provided by replicas of the signal itself, mirrored and shifted in position. Indeed, the scheme (7.2.31) exemplifies the general relation holding between the linear $\mathfrak{L}_\varphi$ and the bilinear $\mathfrak{B}_\varphi$ joint representations, with the former turning into the latter when the waveform to be analysed is itself used as the analysing tool. It is shown that any bilinear signal representation is obtainable as a linear transformation, and also under more specific conditions as a linearly filtered version, of the Wigner distribution function of the signal to be analysed (Problem 6).
Again, $\mathfrak{X}$ quite generally denotes any auxiliary tool that can be used to make the mapping (7.3.2) adjustable and selectable in accord with the specific requirements inherent in the problem under examination. More in general, $\mathfrak{X}$ may depend on the signal itself, thus allowing for higher-order representations. Evidently, this provides a further degree of freedom in the light of the optimal "design" of the mapping (7.3.2) in relation with the signal and the specific processing/analysis requirements.

In order to gain a feeling for the "position" of the Wigner distribution within the general class of the joint signal representations, we report below the unifying formal expression from which all the space-frequency distributions can be generated. Only a few comments are given, addressing the reader to [4] and [5] for a detailed treatment of the subject and the several related topics.

7.3.1 The general class of space-frequency signal representations

All the space-frequency representations of the $1D$ signal $\varphi(q)$ can be obtained from the triple integral [4]

$$\mathcal{C}_\varphi(q,\kappa;\Psi) = \frac{1}{4\pi^2}\iiint \varphi^*(q'-\tfrac{\eta}{2})\varphi(q'+\tfrac{\eta}{2})\Psi(\nu,\eta)e^{-i(\nu q'+\kappa\eta-\nu q)}dq'd\eta d\nu, \tag{7.3.3}$$

where the auxiliary function $\Psi(\nu,\eta)$, referred to as the *kernel* [4.3], is basic to individualize the different representations associable with the signal $\varphi(q)$.

The Wigner integral corresponds to the identity kernel

$$\Psi_{\mathcal{W}}(\nu,\eta) = 1. \tag{7.3.4}$$

The spectrogram also belongs to the generalized class of signal representations (7.3.3). It is individualized by the kernel

$$\Psi_{\mathcal{S}}(\nu,\eta) = \int \mathsf{w}(q-\tfrac{\eta}{2})\mathsf{w}^*(q+\tfrac{\eta}{2})e^{i\nu q}dq, \tag{7.3.5}$$

defined by the window function $\mathsf{w}(q)$ through an ambiguity-like integral form.

The alternative expression for $\mathcal{C}_\varphi(q,\kappa;\Psi)$ in terms of the signal spectrum $\widetilde{\varphi}(\kappa)$ holds as well:

$$\mathcal{C}_\varphi(q,\kappa;\Psi) = \frac{1}{4\pi^2}\iiint \widetilde{\varphi}^*(\kappa'-\tfrac{\nu}{2})\widetilde{\varphi}(\kappa'+\tfrac{\nu}{2})\Psi(\nu,\eta)e^{i(\eta\kappa'+\nu q-\eta\kappa)}d\kappa'd\nu d\eta, \tag{7.3.6}$$

as it may easily be verified by expressing the signal amplitude through the inverse Fourier integral: $\varphi(q) = \frac{1}{\sqrt{2\pi}}\int\widetilde{\varphi}(\kappa)e^{i\kappa q}d\kappa$.

The properties of a distribution $\mathcal{C}_\varphi$ are reflected into specific constraints on the corresponding kernel Ψ. Hence, it is possible to analyse the properties of a particular representation in a systematic way once the kernel is given; also, one can choose those kernels that produce space-frequency distributions satisfying the desired properties for the specific problem at hand. For example, the distributions satisfying the marginal requirements are generated by kernels having the simple property: $\Psi(\nu,0) = \Psi(0,\eta) = 1$ (Problem 3.a). In that case, the one-dimensional (i.e., in space and frequency) energy densities of the signal directly relate to the frequency and space integrals of $\mathcal{C}_\varphi(q,\kappa,\Psi)$ according to

$$\begin{aligned}\int \mathcal{C}_\varphi(q,\kappa,\Psi)d\kappa &= |\varphi(q)|^2,\\ \int \mathcal{C}_\varphi(q,\kappa,\Psi)dq &= |\widetilde{\varphi}(\kappa)|^2.\end{aligned} \tag{7.3.7}$$

As a consequence, the signal energy can be derived by integrating the joint distribution over the pertinent space-frequency domain. However, as also noted in § 6.3.5 when commenting on the marginal properties of the Wigner distribution function, properties (7.3.7) do not warrant the interpretation of the relative distribution as a space-frequency energy density at each point in the space-frequency plane, being also such a pointwise interpretation commonly intended as precluded by the fundamental resolution limitation imposed by the uncertainty product.

Reality is as well another desirable property. Real distributions arise from kernels satisfying the relation: $\Psi(\nu,\eta) = \Psi^*(-\nu,-\eta)$. Evidently, both $\Psi_{\mathcal{W}}$ and $\Psi_{\mathcal{S}}$ generate real distributions, but whilst $\Psi_{\mathcal{W}}$ generates a distribution with the proper marginal properties, as confirmed by Eqs. (6.3.26), $\Psi_{\mathcal{S}}$ does not. The spectrogram is an example of a signal representation that, as opposed to $\mathcal{W}$, is positive but does not yield the correct marginal distributions, its marginals being not exactly equal to the energy densities of the signal.

Although not shown in the notation, the kernel $\Psi(\nu,\eta)$ may depend explicitly on the space and frequency variables; it may also be a functional of the signal itself. Independence of the kernel of space and frequency assures the invariance of the corresponding distribution under shifts of the signal in space and frequency (Problem 3). Hence, the analysis characteristics of the signal representation do not change with position and frequency, the signal being then uniformly analyzed with the same space resolution and the same frequency resolution at various space-frequency points through the pertinent domain. Also, we have already noted that the space and frequency-shift invariance is essential to guarantee the correspondence of the space and frequency variables of the distribution to those of the signal and its spectrum respectively. Independence of the kernel of the signal gives the associated distribution a quadratic dependence on both the signal and the spectrum; in that case, the distribution is said to be bilinear. As seen in § 6.4, the Wigner distribution function is both bilinear and space and frequency-shift invariant, in accord with the mathematical form (7.3.4) of the relative kernel $\Psi_{\mathcal{W}}$. It has also been noted in § 6.3.6 in relation with the positiveness of the Wigner distribution function that, as a general property, bilinear joint distributions cannot be positive and simultaneously satisfy the marginal requirements (see also § 8.2.2).

Invertibility is of course a further important issue to be assessed. For signal-independent kernels the following inversion formula is readily deduced:

$$\varphi(q) = \frac{1}{2\pi\varphi^*(0)} \iiint \frac{\mathcal{C}_\varphi(q',\kappa';\Psi)}{\Psi(\nu',q)} e^{i\kappa' q - i(q'-\frac{q}{2})\nu'}\, dq'\, d\kappa'\, d\nu'. \tag{7.3.8}$$

It is therefore possible to recover the signal from the relevant distribution $\mathcal{C}_\varphi$ (up to a constant phase-factor) only if the division by the kernel is allowed for all values of ν', q. The reader may easily regain the inversion formulas for the spectrogram and the Wigner distribution function from this general relation.

A strictly related topic is concerned with the representability of a $2D$ function, i.e. with the individualization of the condition under which a $2D$ function could be considered as the space-frequency representation of a physical signal. This is the mathematical formulation of the practical problem of signal synthesis from a space-frequency distribution, i.e., a surface over the space-frequency plane. In a typical space-frequency representation-based scheme for signal synthesis, the space-frequency representation is modified (e.g., by masking or smoothing) in order to fit some desired positional and spectral properties. This may result into a "nonvalid" space-frequency distribution which then is no longer the space-frequency representation of any signal. We have addressed such a problem specifically in relation with the Wigner distribution function, for which a definite condition for representability can analytically be formulated (see § 6.3.10; for the spectrogram, see Problem 1.c). Indeed, as far as a bilinear mapping over space-frequency domains is concerned, a general theory and solution for signal synthesis problem can be elaborated. In more general situations ad hoc procedures must be devised. The interested reader may consult the related literature [5].

7.4 The ambiguity function

The *ambiguity function* $\mathcal{A}(\widetilde{q},\widetilde{\kappa})$ is defined much the same way as the Wigner distribution function. It mirrors the Wigner distribution function in the windowed Fourier transform arising scheme (7.2.31). It has found wide applications in radar and sonar as useful measure of simultaneous range and velocity target information [6]. Its application in the optical context has firstly been discussed by Papoulis [7.1], and later investigated by many authors [7]. The properties of the ambiguity function are very well understood; also, several optical setups have been implemented to work as optical processors for displaying the ambiguity function for both real and complex signals [8]. Due to the formal similarity between $\mathcal{W}$ and $\mathcal{A}$, optical architectures have been devised to operate as processors for both the Wigner distribution and ambiguity functions with minor or no changes in the hardware. In the paragraphs below we will evidence the analogies and differences between the Wigner and ambiguity integrals, emphasizing their interpretation as Fourier transforms of the symmetric products $\varphi^*(q-\frac{q'}{2})\varphi(q+\frac{q'}{2})$ and $\widetilde{\varphi}^*(\kappa-\frac{\kappa'}{2})\widetilde{\varphi}(\kappa+\frac{\kappa'}{2})$. We will see that the direction of the transform, q/q' or κ/κ', is central to determine the meaning, the interpretation and the use of $\mathcal{W}$ and $\mathcal{A}$ as signal descriptors over the space-frequency plane.

7.4.1 *Basic definitions*

The ambiguity function $\mathcal{A}(\widetilde{q},\widetilde{\kappa})$ is defined as a complex cross-correlation of the signal with a version of itself which is both spatially and spatial-frequency shifted[(a)]. It arises therefore as a function of the Fourier conjugate variables

---

[a] The (*complex*) *cross-correlation* function $\gamma_{f,g}(x)$ of two deterministic signals represented by the functions $f(x)$ and $g(x)$ is defined by the integral [10.15, 10.16]

$\widetilde{q}$ and $\widetilde{\kappa}$ through the space-domain integral

$$\mathcal{A}(\widetilde{q},\widetilde{\kappa}) = \tfrac{1}{2\pi}\int \varphi^*(q' - \tfrac{\widetilde{q}}{2})\varphi(q' + \tfrac{\widetilde{q}}{2})e^{-i\widetilde{\kappa}q'}dq', \qquad (7.4.1)$$

as well as through the frequency-domain integral

$$\mathcal{A}(\widetilde{q},\widetilde{\kappa}) = \tfrac{1}{2\pi}\int \widetilde{\varphi}^*(\kappa' - \tfrac{\widetilde{\kappa}}{2})\widetilde{\varphi}(\kappa' + \tfrac{\widetilde{\kappa}}{2})e^{i\kappa'\widetilde{q}}d\kappa'. \qquad (7.4.2)$$

Unlike the Wigner distribution function, which is always real, the ambiguity function is in general complex.

It is worth mentioning some differences in terminology concerning the ambiguity function. We may report, for instance, the alternative definition

$$\chi(\widetilde{q},\widetilde{\kappa}) = \tfrac{1}{2\pi}\int \varphi^*(q' - \widetilde{q})\varphi(q')e^{-i\widetilde{\kappa}q'}dq', \qquad (7.4.3)$$

reproducing the integral pertinent to the windowed Fourier transform (7.2.1), and hence being equivalent to (7.4.1) within a linear phase term. Also, the complex conjugates of (7.4.1) or (7.4.3) as well as the respective absolute values, or squared absolute values, are often referred to as the ambiguity function.

As said earlier, the ambiguity function has primarily been used in radar applications as a tool for predicting the ability of a radar signal to measure the range and the velocity of a target via Doppler shift [6]. In this context, it is properly defined as a function of the dual variables (τ,ν) (the "time" equivalents of the "space" pair $(\widetilde{q},\widetilde{\kappa})$, pertaining to the optical context); τ represents the time-delay of an echo produced by the target as ν the Doppler frequency shift due to the target motion.

7.4.2 *Relating the Wigner distribution and ambiguity functions*

Notably the Wigner distribution and the ambiguity function are related by a double Fourier-like transform relation, being

$$\begin{aligned}\mathcal{W}(q,\kappa) &= \tfrac{1}{2\pi}\iint \mathcal{A}(\widetilde{q},\widetilde{\kappa})e^{-i(\kappa\widetilde{q}-q\widetilde{\kappa})}d\widetilde{q}d\widetilde{\kappa},\\ \mathcal{A}(\widetilde{q},\widetilde{\kappa}) &= \tfrac{1}{2\pi}\iint \mathcal{W}(q,\kappa)e^{i(\kappa\widetilde{q}-q\widetilde{\kappa})}dqd\kappa.\end{aligned} \qquad (7.4.4)$$

$$\gamma_f(x) = \int f(x')g^*(x'-x)dx' = \int f(x'+x)g^*(x')dx'.$$

In general, $f(x)$ and $g(x)$ may signify both the signal waveforms and their spectra; correspondingly, $\gamma_{f,g}(x)$ may represent the crosscorrelation of the signals or their spectra.

In particular, if $g(x) = f(x)$ one obtains the (*complex*) *autocorrelation* function γ_f of f. At the origin the modulus $|\gamma_f(x)|$ attains its maximum value, which is just the norm of f:

$$|\gamma_f(x)| \le \gamma_f(0) = \|f\|^2\,.$$

Resorting to the notation of § 4.6.1 for the Fourier transform, we can write (7.4.4) in the elegant and enlightening form

$$\begin{aligned} \mathcal{W}(q,\kappa) &= \widehat{\mathcal{F}}_{(\widetilde{q},\kappa)}\widehat{\mathcal{F}}^{-1}_{(q,\widetilde{\kappa})}\mathcal{A}(\widetilde{q},\widetilde{\kappa}), \\ \mathcal{A}(\widetilde{q},\widetilde{\kappa}) &= \widehat{\mathcal{F}}^{-1}_{(\widetilde{q},\kappa)}\widehat{\mathcal{F}}_{(q,\widetilde{\kappa})}\mathcal{W}(q,\kappa). \end{aligned} \tag{7.4.5}$$

The subscripts on the Fourier operator symbols signify the pair of Fourier conjugate variables involved into the respective Fourier transformation. Indeed, considering a $2D$ function $h(x,\eta)$ of a pair of Fourier conjugate variables (x,η), the operator $\widehat{\mathcal{F}}_{(x,\widetilde{\eta})}$ acts on $h(x,\eta)$ as a Fourier transformation with respect to x, thus turning $h(x,\eta)$ into the function $g(\widetilde{\eta},\eta)$ according to

$$\widehat{\mathcal{F}}_{(x,\widetilde{\eta})}\,h(x,\eta) = \tfrac{1}{\sqrt{2\pi}}\int dx e^{-ix\widetilde{\eta}}h(x,\eta) = g(\widetilde{\eta},\eta). \tag{7.4.6}$$

Evidently, by the inverse operation

$$\widehat{\mathcal{F}}^{-1}_{(x,\widetilde{\eta})}\,g(\widetilde{\eta},\eta) = \tfrac{1}{\sqrt{2\pi}}\int d\widetilde{\eta} e^{ix\widetilde{\eta}}g(\widetilde{\eta},\eta\xi) = h(x,\eta), \tag{7.4.7}$$

one can regain $h(x,\xi)$. In a like manner, the pair of operators $\widehat{\mathcal{F}}_{(\widetilde{x},\eta)}$ and $\widehat{\mathcal{F}}^{-1}_{(\widetilde{x},\eta)}$ establishes a Fourier-like relation between the functions $h(x,\eta)$ and $f(x,\widetilde{x})$, the variables involved in the Fourier transformation being, of course, $\widetilde{x}$ and η.

Accordingly, the Wigner distribution function may be thought as resulting from the ambiguity function by an inverse Fourier transform with respect to the frequency variable $\widetilde{\kappa}$, being turned into q, and then by a Fourier transform with respect to the space variable $\widetilde{q}$, being turned into κ. The relation for $\mathcal{A}(\widetilde{q},\widetilde{\kappa})$ has a similar interpretation, being it obtained from $\mathcal{W}(q,\kappa)$ by a Fourier transform with respect to q and an inverse Fourier transform with respect to κ. Of course, the order of the two transformations in Eqs. (7.4.5) is uninfluential.

The Fourier integrals (7.4.4) evidence the duality relation holding between $\mathcal{W}$ and $\mathcal{A}$, as exemplified by the variable-to-variable correspondences:

$\mathcal{W}$-space variable q	$\longleftrightarrow$	$\mathcal{A}$-frequency variable $\widetilde{\kappa}$
$\mathcal{A}$-space variable $\widetilde{q}$	$\longleftrightarrow$	$\mathcal{W}$-frequency variable κ

(7.4.8)

showing that in the ambiguity function space-frequency plane the "horizontal" axis corresponds to the frequency variable $\widetilde{\kappa}$, whilst the "vertical" axis corresponds to the space variable $\widetilde{q}$. Therefore, it would be more pertinent to write $\mathcal{A}(\widetilde{\kappa},\widetilde{q})$ instead of $\mathcal{A}(\widetilde{q},\widetilde{\kappa})$. We prefer postponing the frequency variable to the space variable, thus writing $\mathcal{A}(\widetilde{q},\widetilde{\kappa})$.

From (7.4.5), taking account of Parseval's and Moyal's relations, it follows that (see also Problem 4.a)

$$\iint |\mathcal{A}(\widetilde{q},\widetilde{\kappa})|^2\, d\widetilde{q}d\widetilde{\kappa} = \iint \mathcal{W}^2(q,\kappa)dqd\kappa = \tfrac{\|\varphi\|^4}{2\pi}. \tag{7.4.9}$$

For future use, we write below the expression of the cross-ambiguity function $\mathcal{A}_{\varphi_1,\varphi_2}(\widetilde{q},\widetilde{\kappa})$ relevant to the pair of functions $\varphi_1(q)$ and $\varphi_2(q)$:

$$\mathcal{A}_{\varphi_1,\varphi_2}(\widetilde{q},\widetilde{\kappa}) = \frac{1}{2\pi}\int \varphi_1^*(q' - \tfrac{\widetilde{q}}{2})\varphi_2(q' + \tfrac{\widetilde{q}}{2})e^{-i\widetilde{\kappa}q'}dq', \tag{7.4.10}$$

which evidently parallels the cross-Wigner integral (6.3.12). Interestingly, whereas $\mathcal{W}_{\varphi_1,\varphi_2}(q,\kappa)$ is commutative in the signals, being $\mathcal{W}_{\varphi_2,\varphi_1}(q,\kappa) = \mathcal{W}^*_{\varphi_1,\varphi_2}(q,\kappa)$, the cross-ambiguity function is not, since

$$\mathcal{A}_{\varphi_2,\varphi_1}(\widetilde{q},\widetilde{\kappa}) = \mathcal{A}^*_{\varphi_1,\varphi_2}(-\widetilde{q},-\widetilde{\kappa}) \tag{7.4.11}$$

Equation (7.4.10) supports the consideration in § 7.2.3 concerning the relation between the windowed Fourier transform and the cross-ambiguity function.

7.4.3 *Energetic and correlative classes of the bilinear space-frequency signal representations*

Within the general class of space-frequency distributions represented by (7.3.3) with signal-independent kernels, the ambiguity integrals (7.4.1) and (7.4.2) are obtained in correspondence with the kernel

$$\Psi_{\mathcal{A}}(\nu,\eta) = 2\pi\delta(\eta - \widetilde{q})\delta(\nu - \widetilde{\kappa}). \tag{7.4.12}$$

Since $\Psi_{\mathcal{A}}$ depends on both the space and frequency variables, the ambiguity function is not invariant under shifts of the signal in space and frequency. In fact, translating the signal or the spectrum affects the function $\mathcal{A}(\widetilde{q},\widetilde{\kappa})$ by a linear phase factor (Problem 4.b). In addition, unlike the Wigner distribution, the ambiguity function does not satisfy the marginal properties. It satisfies the so called *correlative* marginal properties expressed by the relations

$$\begin{aligned}\int \mathcal{A}(\widetilde{q},\widetilde{\kappa})d\widetilde{\kappa} &= \varphi(\tfrac{\widetilde{q}}{2})\varphi^*(-\tfrac{\widetilde{q}}{2}),\\ \int \mathcal{A}(\widetilde{q},\widetilde{\kappa})d\widetilde{q} &= \widetilde{\varphi}(\tfrac{\widetilde{\kappa}}{2})\widetilde{\varphi}^*(-\tfrac{\widetilde{\kappa}}{2}).\end{aligned} \tag{7.4.13}$$

We also see that

$$\begin{aligned}\mathcal{A}(\widetilde{q},0) &= \frac{1}{2\pi}\int \varphi(q')\varphi^*(q' - \widetilde{q})dq',\\ \mathcal{A}(0,\widetilde{\kappa}) &= \frac{1}{2\pi}\int \widetilde{\varphi}(\kappa')\widetilde{\varphi}^*(\kappa' - \widetilde{\kappa})d\kappa',\end{aligned} \tag{7.4.14}$$

showing that the profiles of $\mathcal{A}$ along the $\widetilde{q}$ and $\widetilde{\kappa}$ axes just equal (apart from the factor $\frac{1}{2\pi}$) the autocorrelation functions of the signal and its spectrum (see footnote on p. 358). Then, the energy densities of the signal obtain by Fourier transforming the zero-space and zero-frequency profiles of $\mathcal{A}$; namely,

$$\begin{aligned}|\varphi(q)|^2 &= \int \mathcal{A}(0,\widetilde{\kappa})e^{i\widetilde{\kappa}q}d\widetilde{\kappa}\ ,\\ |\widetilde{\varphi}(\kappa)|^2 &= \int \mathcal{A}(\widetilde{q},0)e^{-i\kappa\widetilde{q}}d\widetilde{q}.\end{aligned} \tag{7.4.15}$$

Moreover, it follows from (7.4.14) that

$$\mathcal{A}(0,0) = \frac{\|\varphi\|^2}{2\pi}. \tag{7.4.16}$$

The ambiguity function at the origin is proportional to the total energy of the signal, and hence $\mathcal{A}(0,0) > 0$. In contrast, the Wigner distribution function at the origin may attain a negative value, as evidenced in § 6.3.6. More strictly, by Schwarz's inequality we find

$$|\,\mathcal{A}(\widetilde{q},\widetilde{\kappa})\,| \leq \mathcal{A}(0,0), \tag{7.4.17}$$

and hence the modulus of $\mathcal{A}$ is always greatest at the origin of the $(\widetilde{q},\widetilde{\kappa})$ plane.

Since one usually considers unit-energy signals, relations (7.4.9) and (7.4.16) take the φ-independent forms

$$\iint |\,\mathcal{A}(\widetilde{q},\widetilde{\kappa})\,|^2 d\widetilde{q}d\widetilde{\kappa} = \frac{1}{2\pi}, \tag{7.4.18}$$

and

$$|\,\mathcal{A}(\widetilde{q},\widetilde{\kappa})\,| \leq \mathcal{A}(0,0) = \frac{1}{2\pi}. \tag{7.4.19}$$

Relation (7.4.18) states that the volume under the ambiguity surface is independent of the signal waveform. Also, it shows that the ambiguity function must distribute throughout the $(\widetilde{q},\widetilde{\kappa})$ plane; it cannot be concentrated arbitrarily close to the origin. According to (7.4.19) a definite "portion" of $\mathcal{A}(\widetilde{q},\widetilde{\kappa})$ must remain in the proximity of the origin. Therefore we may argue that if the region over which the ambiguity function is significantly different from zero enlarges, then $\mathcal{A}(\widetilde{q},\widetilde{\kappa})$ becomes more nearly concentrated near the origin, the "remaining" portion of $\mathcal{A}$ having to distribute over an enlarged region. Thus, the ambiguity function becomes more sharply peaked as it is spread out over a relatively larger domain in the $(\widetilde{q},\widetilde{\kappa})$ plane. We will illustrate this behaviour in the next paragraph in connection with the Hermite-Gaussian beams, the differences with the Wigner distribution function becoming then evident.

Relations (7.4.15) parallel the marginal relations holding for the Wigner distribution function. They are central to mark the differences between $\mathcal{W}$ and $\mathcal{A}$. In fact, relations (6.3.26) express the *energetic* nature of the Wigner distribution function, which seeks to combine the spatial and spectral energy densities of the signal, we recover through the relevant marginal distributions. Correspondingly, relations (7.4.13)-(7.4.14) express the *correlative* nature of the ambiguity function, which seeks to combine the spatial and spectral autocorrelations of the signal, we recover from the ambiguity profiles along the phase-plane axes. As noted in [4.6], $\mathcal{W}$ and $\mathcal{A}$ may be regarded within the general class of the bilinear space-frequency representations (7.3.3) as the "founders" of two dual subclasses, whose members share their respective energetic and correlative character (see Problem 6). In addition, it is evident that we are not allowed to interpret the space and frequency variables of the ambiguity function $\widetilde{q}$, $\widetilde{\kappa}$ as corresponding to those of the signal, as we loosely do with the space and frequency variables of the Wigner distribution. As already emphasized in § 7.2.3 and here evidenced by (7.4.14), the ambiguity function variables $(\widetilde{q}, \widetilde{\kappa})$ are properly interpreted as the *space lag* and the *frequency lag*; we will further discuss this point § 7.5.1. Therefore, unlike the Wigner plane, the $(\widetilde{q}, \widetilde{\kappa})$ plane can not be identified with the signal phase plane. The space and frequency variables proper to $\mathcal{W}$ and $\mathcal{A}$ are linked by the duality relation implied by the Fourier integrals (7.4.4), and displayed in (7.4.8).

As a conclusion, a few words on the invertibility of the mapping (7.4.1), and the representability of a space-frequency distribution as the ambiguity function of some signal. The reader may easily verify that the signal amplitude can be recovered from the ambiguity function up to a constant phase factor as

$$\varphi(q) = \frac{1}{\varphi^*(0)} \int \mathcal{A}(\widetilde{q}, \widetilde{\kappa}) e^{\frac{i}{2} q \widetilde{\kappa}} \, d\widetilde{\kappa}, \tag{7.4.20}$$

a dual formula holding for the signal spectrum.

Finally, carrying on the various steps suggested in § 6.3.10 in connection with the representability condition (6.3.70) for $\mathcal{W}$, it can be shown that the necessary and sufficient condition for a function of the Fourier-conjugate variables $(\widetilde{q}, \widetilde{\kappa})$ to be interpretable as the ambiguity function of some signal is expressible through the relation

$$\begin{aligned} &\mathcal{A}(\widetilde{q} + \tfrac{\widetilde{q}_0}{2}, \widetilde{\kappa} + \tfrac{\widetilde{\kappa}_0}{2}) \mathcal{A}^*(\widetilde{q} - \tfrac{\widetilde{q}_0}{2}, \widetilde{\kappa} - \tfrac{\widetilde{\kappa}_0}{2}) \\ &= \frac{1}{2\pi} \iint \mathcal{A}(\widetilde{q}' + \tfrac{\widetilde{q}_0}{2}, \widetilde{\kappa}' + \tfrac{\widetilde{\kappa}_0}{2}) \mathcal{A}^*(\widetilde{q}' - \tfrac{\widetilde{q}_0}{2}, \widetilde{\kappa}' - \tfrac{\widetilde{\kappa}_0}{2}) e^{-i(\widetilde{\kappa}' \widetilde{q} - \widetilde{q}' \widetilde{\kappa})} d\widetilde{q}' d\widetilde{\kappa}'. \end{aligned} \tag{7.4.21}$$

which must hold for any $\widetilde{q}_0$ and $\widetilde{\kappa}_0$.

7.4.4 *Comparing the ambiguity function and the Wigner distribution function through some examples*

In order to gain an intuitive insight into the differences between the Wigner distribution and the ambiguity function, we may calculate $\mathcal{A}(\widetilde{q}, \widetilde{\kappa})$ for the simple signals in the table (6.4.39), i.e., the δ-like, monofrequency and chirp-like signals. We easily obtain

$$\begin{array}{lcll} \delta(q - q_0) & \longrightarrow & \mathcal{A}_{q_0}(\widetilde{q}, \widetilde{\kappa}) & = \frac{1}{2\pi} \delta(\widetilde{q}) e^{-iq_0 \widetilde{\kappa}}, \\ \frac{1}{\sqrt{2\pi}} e^{i\kappa_0 q} & \longrightarrow & \mathcal{A}_{\kappa_0}(\widetilde{q}, \widetilde{\kappa}) & = \frac{1}{2\pi} \delta(\widetilde{\kappa}) e^{i\kappa_0 \widetilde{q}}, \\ e^{i\frac{\beta}{2}(q-q_0)^2 + i\kappa_0 q} & \longrightarrow & \mathcal{A}_{\mathsf{chirp}}(\widetilde{q}, \widetilde{\kappa}) & = e^{i\widetilde{\kappa}(\kappa_0 - \beta q_0)} \delta(\widetilde{\kappa} - \beta \widetilde{q}). \end{array} \tag{7.4.22}$$

In all cases the ambiguity function is confined along a straight line in the $(\widetilde{q}, \widetilde{\kappa})$ plane, respectively the $\widetilde{\kappa}$-axis, the $\widetilde{q}$-axis and the line $\widetilde{\kappa} = \alpha \widetilde{q}$. The absolute

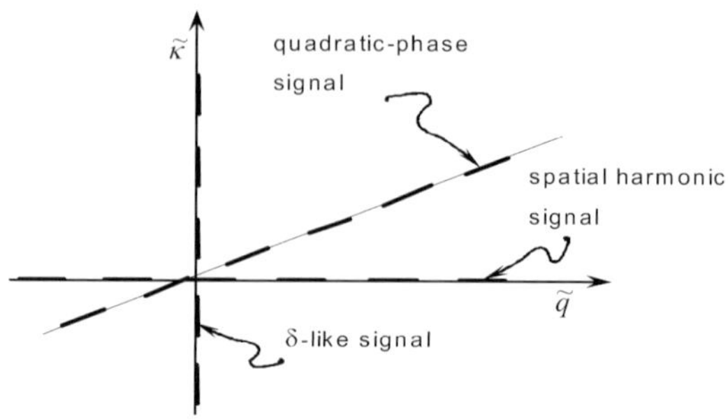

FIGURE 7.7. Supports of the ambiguity function of the δ-like, monofrequency and quadratic phase signals. The dashed line appearance is aimed at evoking the sinusoidal behaviour of the ambiguity function over the inherent supports.

values of $\mathcal{A}_{q_0}$, $\mathcal{A}_{\kappa_0}$ and $\mathcal{A}_{\text{chirp}}$ are insensible to frequency modulations and space shifts of the signal, which simply manifest through linear phase terms. In Fig. 7.7 the supports of $\mathcal{A}_{q_0}$, $\mathcal{A}_{\kappa_0}$ and $\mathcal{A}_{\text{chirp}}$ in the proper $(\widetilde{q}, \widetilde{\kappa})$ plane are depicted, paralleling Fig. 6.11 that illustrates the supports of the corresponding Wigner distributions $\mathcal{W}_{q_0}$, $\mathcal{W}_{\kappa_0}$ and $\mathcal{W}_{\text{chirp}}$ in the respective (q, κ) plane.

For future reference, we give the expression of the ambiguity function $\mathcal{A}_{\text{R}}(\widetilde{q}, \widetilde{\kappa})$ of a rectangle-signal of width $2Q$:

$$\mathcal{A}_{\text{R}}(\widetilde{q}, \widetilde{\kappa}) = \begin{cases} \frac{Q}{\pi}(1 - \frac{|\widetilde{q}|}{2Q})\text{sinc}[Q\widetilde{\kappa}(1 - \frac{|\widetilde{q}|}{2Q})] & |\widetilde{q}| \leq 2Q, \\ 0 & |\widetilde{q}| > 2Q. \end{cases} \tag{7.4.23}$$

which the reader may interestingly compare with the corresponding expression (6.4.3) of the Wigner distribution function $\mathcal{W}_{\text{R}}(q, \kappa)$. We see that $\mathcal{A}_{\text{R}}(\widetilde{q}, \widetilde{\kappa})$ is confined within the vertical strip $-2Q < \widetilde{q} \leq 2Q$, with the zero value loci being the symmetric hyperbolas (Fig. 7.8.a))

$$\widetilde{\kappa}(1 - \tfrac{|\widetilde{q}|}{2Q}) = \frac{n\pi}{Q}, \qquad |\widetilde{q}| \leq 2Q, \quad n = 1, 2, .. \tag{7.4.24}$$

It is also instructive to calculate the ambiguity function of the real Gaussian amplitude

$$\varphi_{\mathcal{G}}(q) = (\tfrac{2}{\pi w^2})^{1/4} e^{-\frac{1}{w^2} q^2}, \tag{7.4.25}$$

for which we obtain

$$\mathcal{A}_{\mathcal{G}}(\widetilde{q}, \widetilde{\kappa}) = \tfrac{1}{2\pi} e^{-\frac{1}{4}(\frac{2\widetilde{q}^2}{w^2} + \frac{w^2}{2}\widetilde{\kappa}^2)}. \tag{7.4.26}$$

Like $\mathcal{W}_{\mathcal{G}}(q, \kappa)$ (see Eq. (6.4.51)), it displays a Gaussian profile along both the $\widetilde{q}$ and $\widetilde{\kappa}$ axis, but with variances $\sigma^2_{\widetilde{q}} = w^2$ and $\sigma^2_{\widetilde{\kappa}} = \frac{4}{w^2}$ [b]. Roughly speak-

[b] The variances for a $2D$ function are defined by double integrals arising as an obvious generalization of (4.6.35) appropriate to a $1D$ function (see § 9.2.2).

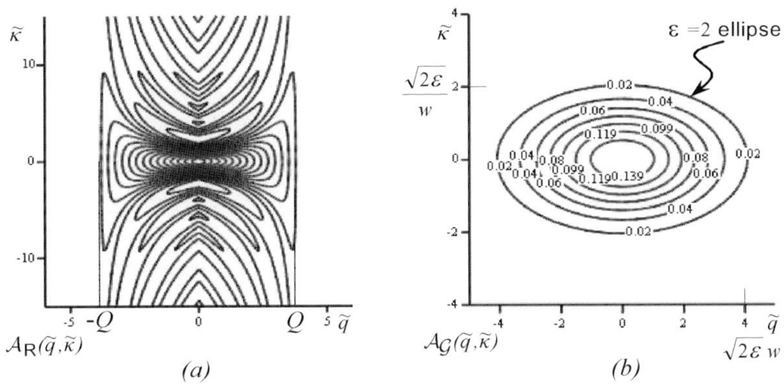

FIGURE 7.8. Contour plots of the ambiguity function of *(a)* the rectangle signal $\varphi_{\rm R}(q) = {\rm rect}(\frac{q}{4})$ and *(b)* the Gaussian waveform $\varphi_{\mathcal{G}}(q) = (\frac{1}{2\pi})^{1/4} e^{-\frac{q^2}{4}}$.

ing, we may say that $\mathcal{W}_{\mathcal{G}}$ extends over the phase plane significantly within the rectangle $[-\frac{w}{2}, \frac{w}{2}] \times [-\frac{1}{w}, \frac{1}{w}]$, thus reproducing the (approximate) extensions of $\varphi_{\mathcal{G}}(q)$ in the space and frequency domains. In contrast, $\mathcal{A}_{\mathcal{G}}$ extends in the pertinent $(\widetilde{q}, \widetilde{\kappa})$ plane significantly over the larger rectangle $[-w, w] \times [-\frac{2}{w}, \frac{2}{w}]$, the relevant level curves being in fact the ellipses $\frac{\widetilde{q}^2}{w^2} + \frac{w^2 \widetilde{\kappa}^2}{4} = 2\varepsilon$, which evidently for any given value of ε have double axes with respect to the corresponding elliptical contours of $\mathcal{W}_{\mathcal{G}}(q, \kappa)$ (Eq. (6.4.52), Fig. 7.8.b)).

To help intuition, in Fig. 7.8 we show the contour plots of $\mathcal{A}_{\rm R}(\widetilde{q}, \widetilde{\kappa})$. and $\mathcal{A}_{\mathcal{G}}(\widetilde{q}, \widetilde{\kappa})$. In this connection, note that the ambiguity function is in general complex; therefore the modulus $|\mathcal{A}|$ or the squared modulus $|\mathcal{A}|^2$ is ordinarily plotted. In particular, the ambiguity function of the rectangle and Gaussian signals as well as of all the waveforms we consider in this paragraph are real. Hence they can directly be shown in a $3D$ plot or in a phase-plane plot.

Examples (7.4.23) and (7.4.26) illustrate the general property of the ambiguity function, whose support in the pertinent space-frequency plane has $\widetilde{q}$ and $\widetilde{\kappa}$ extensions double than those of the signal and its spectrum, respectively (Problem 6). We may gain an intuitive justification for this property, by resorting to the naive picture, elaborated in § 6.2.1 in relation with the Wigner distribution function, and further supported by the considerations developed in § 7.2.3. As in § 6.2.1, we consider a real and nonnegative signal. Thus, on computing the Wigner integral at a point we may think of folding the part of the signal to the left over to the right, the resulting overlap contributing to the Wigner integration. Correspondingly, on computing the ambiguity integral at a point we may think of shifting the signal to the chosen point, the overlap with the amoved replica of the signal contributing to the ambiguity function

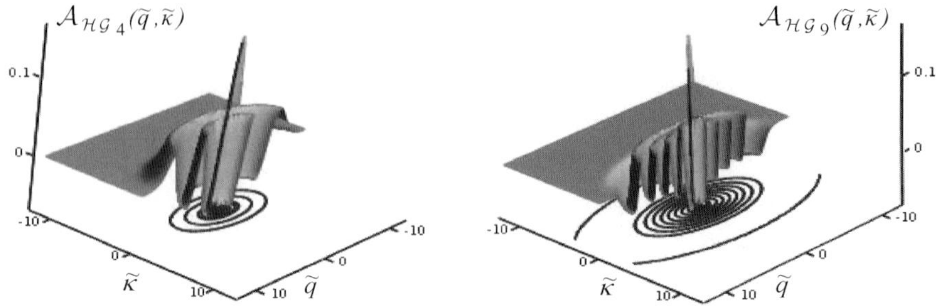

FIGURE 7.9. $3D$ graphs and relevant phase-plane plots of the ambiguity function of the Hermite-Gaussian waveform (7.4.27) of order $n = 4$ and $n = 9$.

integration. So, whilst the folding gives zero just at the extremes of the signal interval, the overlap vanishes at double of the extremes. The ambiguity function is in fact a measure of the degree of similarity between the signal and its replica being shifted in space and frequency one to the other.

We compute now the ambiguity function of the Hermite-Gaussian beam amplitudes, which can be given in the form (6.4.59), i.e.,

$$\varphi_{_{\mathcal{HG}_n}}(q) = (\tfrac{\sqrt{2}}{2^n n! \sqrt{\pi} w})^{1/2} H_n(\tfrac{\sqrt{2}q}{w}) e^{-\frac{1}{w^2}q^2}, \tag{7.4.27}$$

Following the lines of the calculation leading to (6.4.66), we obtain

$$\mathcal{A}_{\mathcal{HG}_n}(\widetilde{q}, \widetilde{\kappa}) = \tfrac{1}{2\pi} e^{-\frac{1}{2}(\frac{\widetilde{q}^2}{w^2} + \frac{w^2}{4}\widetilde{\kappa}^2)} L_n[\tfrac{\widetilde{q}^2}{w^2} + \tfrac{w^2 \widetilde{\kappa}^2}{4}], \tag{7.4.28}$$

where $L_n(\cdot)$ denotes the Laguerre polynomial of degree n [13].
The rms widths of $\varphi_{_{\mathcal{HG}_n}}$ and its Fourier transform $\widetilde{\varphi}_{_{\mathcal{HG}_n}}$ are respectively measured by the variances of $|\varphi_{_{\mathcal{HG}_n}}|^2$ and $|\widetilde{\varphi}_{_{\mathcal{HG}_n}}|^2$, being $\sigma^2_{_{q,n}} = \frac{w^2}{4}(2n+1)$ and $\sigma^2_{_{\kappa,n}} = \frac{1}{w^2}(2n+1)$. Hence, for increasing values of the order n the widths of the space and frequency domains of $\varphi_{_{\mathcal{HG}_n}}$ increase, and so do as well the supports of $\mathcal{W}_{\mathcal{HG}_n}$ and $\mathcal{A}_{\mathcal{HG}_n}$. The reader may guess that the support of $\mathcal{A}_{\mathcal{HG}_n}$ in the relevant $(\widetilde{q}, \widetilde{\kappa})$ plane involves a region with rms $\widetilde{q}$ and $\widetilde{\kappa}$ extensions just given by $4\sigma_{_{q,n}}$ and $4\sigma_{_{\kappa,n}}$. Therefore in accord with the considerations suggested by (7.4.18) and (7.4.19), with increasing n the ambiguity function becomes more sharply peaked about the origin in the $(\widetilde{q}, \widetilde{\kappa})$ plane whilst accommodating its invariant volume over a larger region. Figure 7.9 shows the $3D$ graphs with the phase-plane plots of the ambiguity function of the Hermite Gaussian amplitudes (7.4.27) of order $n = 4$ and $n = 9$, that can be compared with the corresponding plots of the Wigner distribution function in Fig. 6.19.

Examples (7.4.23), (7.4.26) and (7.4.28) are in accord with the relation:

$$\mathcal{W}(q, \kappa) = 2\mathcal{A}(2q, 2\kappa), \tag{7.4.29}$$

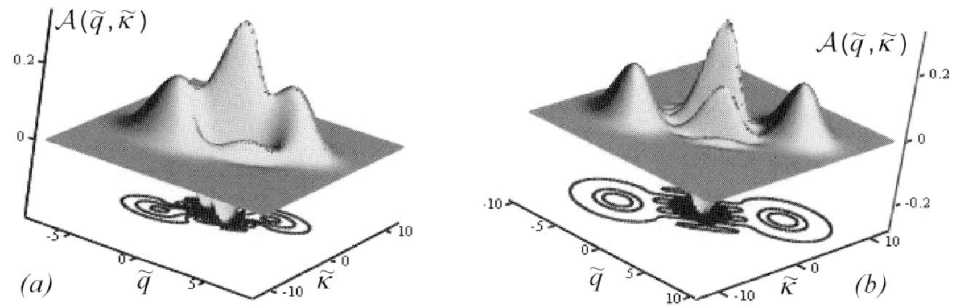

FIGURE 7.10. $3D$ graphs and relevant contour plots of the ambiguity function of the two component signal (7.4.30) with $w = 1.5$ and *(a)* $q_1 = -q_2 = 1.5$, $\kappa_1 = -\kappa_2 = 1.5$; *(b)* $q_1 = -q_2 = 2.5$, $\kappa_1 = -\kappa_2 = 2.5$.

holding between the Wigner distribution and ambiguity functions in the case of an even signal, $\varphi(-q) = \varphi(q)$, in which case in fact the ambiguity function turns to be real.

Finally, to close this tracking shot on the characteristics of the ambiguity function we illustrate the behaviour of the cross-terms in the ambiguity function, considering a two component signal. Such a behaviour is usefully compared with that of the cross-terms in the Wigner distribution function. Due to its analytical computability, we consider the sum of two identical Gaussian amplitudes being both spatially and spatial-frequency shifted. Namely,

$$\varphi(q) = \varphi_1(q) + \varphi_2(q) \tag{7.4.30}$$

where specifically

$$\varphi_j(q) = (\tfrac{2}{\pi w^2})^{1/4} e^{-\frac{1}{w^2}(q-q_j)^2 + i\kappa_j q}, \quad j = 1, 2 \tag{7.4.31}$$

In general the ambiguity function of a two component waveform is composed by four terms, i.e.,

$$\mathcal{A}_\varphi(\widetilde{q},\widetilde{\kappa}) = \mathcal{A}_{\varphi 1}(\widetilde{q},\widetilde{\kappa}) + \mathcal{A}_{\varphi 2}(\widetilde{q},\widetilde{\kappa}) + \mathcal{A}_{\varphi_1,\varphi_2}(\widetilde{q},\widetilde{\kappa}) + \mathcal{A}_{\varphi_2,\varphi_1}(\widetilde{q},\widetilde{\kappa}), \tag{7.4.32}$$

where according to (7.4.11) $\mathcal{A}_{\varphi_2,\varphi_1}(\widetilde{q},\widetilde{\kappa}) = \mathcal{A}^*_{\varphi_1,\varphi_2}(-\widetilde{q},-\widetilde{\kappa})$. The reader is invited to work out the explicit expression of $\mathcal{A}_\varphi$ in the general case of arbitrary spatial and frequency shifts: q_1, q_2, κ_1 and κ_2.

Here we report, and comment on, the expression of $\mathcal{A}_\varphi$ relevant to the case of identical and opposite shifts, i.e., $q_1 = -q_2 = q_0$ and $\kappa_1 = -\kappa_2 = \kappa_0$. The pictorial view of $\mathcal{A}_\varphi$ in that case is offered by Fig. 7.10 for specific values of q_0, κ_0 and w. The self-terms $\mathcal{A}_{\varphi 1}$ and $\mathcal{A}_{\varphi 2}$ shape both as $\mathcal{A}_g$, apart from a linear phase factor accounting for the space shift and the frequency modulation of

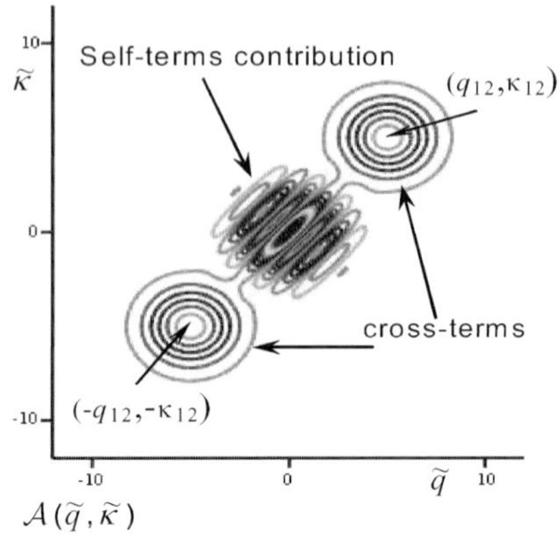

FIGURE 7.11. Cross- and self-terms geometry of the ambiguity function.

the respective signals, and so are both concentrated around the origin in the $(\widetilde{q},\widetilde{\kappa})$ plane. Accordingly, they sum to

$$[\mathcal{A}_{\varphi 1}+\mathcal{A}_{\varphi 2}](\widetilde{q},\widetilde{\kappa}) = \frac{1}{\pi}e^{-\frac{1}{2}(\frac{\widetilde{q}^2}{w^2}+\frac{w^2}{4}\widetilde{\kappa}^2)}\cos(\kappa_0\widetilde{q}-q_0\widetilde{\kappa}) = 2\mathcal{A}_{\mathcal{G}}(\widetilde{q},\widetilde{\kappa})\cos(\kappa_0\widetilde{q}-q_0\widetilde{\kappa}), \tag{7.4.33}$$

which evidently displays an oscillatory behaviour, governed by q_0 and κ_0, and modelled by the Gaussian envelope as (7.4.26).

In contrast, the cross-terms $\mathcal{A}_{\varphi_1,\varphi_2}$ and $\mathcal{A}_{\varphi_2,\varphi_1}$ yield in general two separate contributions, concentrated respectively on the points $(q_2-q_1,\kappa_2-\kappa_1)$ and $(q_1-q_2,\kappa_1-\kappa_2)$ in the $(\widetilde{q},\widetilde{\kappa})$ plane. In the particular case of equal and opposite shifts we are dealing with, one obtains, in fact,

$$\begin{aligned}\mathcal{A}_{\varphi_1,\varphi_2}(\widetilde{q},\widetilde{\kappa}) &= \frac{1}{2\pi}e^{-\frac{1}{2}[\frac{1}{w^2}(\widetilde{q}+2q_0)^2+\frac{w^2}{4}(\widetilde{\kappa}+2\kappa_0)^2]} = \mathcal{A}_{\mathcal{G}}(\widetilde{q}+2q_0,\widetilde{\kappa}+2\kappa_0),\\ \mathcal{A}_{\varphi_2,\varphi_1}(\widetilde{q},\widetilde{\kappa}) &= \frac{1}{2\pi}e^{-\frac{1}{2}[\frac{1}{w^2}(\widetilde{q}-2q_0)^2+\frac{w^2}{4}(\widetilde{\kappa}-2\kappa_0)^2]} = \mathcal{A}_{\mathcal{G}}(\widetilde{q}-2q_0,\widetilde{\kappa}-2\kappa_0).\end{aligned} \tag{7.4.34}$$

Both terms have the same Gaussian envelope as $\mathcal{A}_{\mathcal{G}}(\widetilde{q},\widetilde{\kappa})$, and are symmetrically localized in the space-frequency plane around the points $(-2q_0,-2\kappa_0)$ and $(2q_0,2\kappa_0)$. The contribution from each cross-term is half that of the self-terms arising contribution (7.4.33).

Figure 7.11 schematically synthesizes the above described behaviour of the ambiguity function, which is as noted rather general. Indeed, Figs. 7.10 and 7.11 parallel Figs. 6.17 and 6.4, which are relevant to the Wigner distribution function. Specifically, Figs. 7.11 and 6.4 evidence the complementary geometry of the ambiguity and Wigner distribution function cross-terms. In fact, in the case of the ambiguity function the signal terms are located around the origin in the $(\widetilde{q},\widetilde{\kappa})$ plane, whilst the cross terms are separately centered at

$(-q_{12}, -\kappa_{12})$ and (q_{12}, κ_{12}), q_{12} and κ_{12} being the space and frequency difference coordinates between the central coordinates of the component signal. On the contrary, as illustrated in § 6.3.4, the self-terms of the Wigner distribution function are concentrated at the same space-frequency points around which the signal components occur, whereas the cross-terms contribution is oscillatory around a point located midway between the self-terms centres. Notably such a complementarity may effectively been used to filter out the undesired cross-terms contributions in the Wigner distribution function, since from the viewpoint of the ambiguity function an appropriate weighting function sharply concentrated around the origin of the space-frequency plane may allow the signal terms to pass whilst reducing or suppressing the cross terms contributions.

7.5 Understanding the Wigner and ambiguity functions from the viewpoint of the mutual intensity function

Here we further emphasize the mathematical interpretation of the Wigner distribution and ambiguity function as $1D$ Fourier transforms of $2D$ functions, formed by the product of shifted versions of the signal or its spectrum, which ultimately can be interpreted as the mutual intensity and spectrum. Both $\mathcal{W}$ and $\mathcal{A}$ are seen to arise from a proper sequence of a multiplication and a Fourier transform process, thus suggesting a possible optical implementation for the Wigner distribution and ambiguity functions display. Also, the forthcoming discussion conveys the significance of the Wigner distribution and ambiguity functions within the partial coherence theory (see also § 6.3.5).

7.5.1 *Signal and spectrum generated functions $\pi_s(q,\widetilde{q})$ and $\pi_f(\kappa,\widetilde{\kappa})$*

Consider two signals $\varphi_1(q)$ and $\varphi_2(q)$, with the respective Fourier spectra $\widetilde{\varphi}_1(\kappa)$ and $\widetilde{\varphi}_2(\kappa)$. We regard $\varphi_1(q)$ and $\varphi_2(q)$ as two independent, entities and accordingly associate to each its own space and frequency coordinates. Thus, we write $\varphi_1(q_1)$ and $\widetilde{\varphi}_1(\kappa_1)$, and correspondingly $\varphi_2(q_2)$ and $\widetilde{\varphi}_2(\kappa_2)$.
On describing the relations between the two signals in both the space and frequency domains, it is rather natural to introduce the centre and difference coordinates:

$$\begin{aligned} q &= \frac{q_1+q_2}{2}, \quad \kappa = \frac{\kappa_1+\kappa_2}{2}; \\ \widetilde{q} &= \widetilde{q}_2 - \widetilde{q}_1, \quad \widetilde{\kappa} = \widetilde{\kappa}_2 - \widetilde{\kappa}_1. \end{aligned} \tag{7.5.1}$$

Evidently, the original coordinates are recovered by inverting (7.5.1) as

$$\begin{aligned} q_1 &= q - \frac{\widetilde{q}}{2}, \quad \kappa_1 = \kappa - \frac{\widetilde{\kappa}}{2}; \\ q_2 &= q + \frac{\widetilde{q}}{2}, \quad \kappa_2 = \kappa + \frac{\widetilde{\kappa}}{2}. \end{aligned} \tag{7.5.2}$$

Also, in order to individualize the interrelations between the two signals, it is natural to consider the space-domain product $\varphi_1(q_1)\varphi_2(q_2)$ as well as the frequency-domain product $\widetilde{\varphi}_1(\kappa_1)\,\widetilde{\varphi}_2(\kappa_2)$. On account of (7.5.2), such space- and frequency-domain products can equivalently be interpreted as functions of the centre and difference variable pairs $(q,\widetilde{q})$ and $(\kappa,\widetilde{\kappa})$, respectively.

Of course, we may take as independent signals any waveform $\varphi(q)$ and its complex conjugate $\varphi^*(q)$. Hence, the aforementioned space-domain product turns to be [4.3, 12.5, 7.4, 4.8]

$$\pi_s(q,\widetilde{q}) = \varphi^*(q-\tfrac{\widetilde{q}}{2})\varphi(q+\tfrac{\widetilde{q}}{2}), \tag{7.5.3}$$

manifestly expressed in terms of the centre and difference spatial coordinates q and $\widetilde{q}$.

We can then rewrite the space-domain integral defining the Wigner distribution in the form

$$\mathcal{W}(q,\kappa) = \tfrac{1}{2\pi}\int \pi_s(q,\widetilde{q})e^{-i\kappa\widetilde{q}}d\widetilde{q} = \tfrac{1}{\sqrt{2\pi}}\widehat{\mathcal{F}}_{(\widetilde{q},\kappa)}[\pi_s(q,\widetilde{q})], \tag{7.5.4}$$

relating the Wigner distribution function to the Fourier transform of the product function $\pi_s(q,\widetilde{q})$ with respect to the difference space-coordinate $\widetilde{q}$.

Replicating the same structure as (7.5.3) in terms of the frequency representation $\widetilde{\varphi}(\kappa)$, we introduce the $2D$ product [4.3, 12.5, 7.4, 4.8]

$$\pi_f(\kappa,\widetilde{\kappa}) = \widetilde{\varphi}^*(\kappa-\tfrac{\widetilde{\kappa}}{2})\varphi(\kappa+\tfrac{\widetilde{\kappa}}{2}), \tag{7.5.5}$$

being a function of the centre and difference (spatial) frequency variables κ and $\widetilde{\kappa}$. In terms of (7.5.5) the frequency-domain integral concerning the Wigner distribution function writes as

$$\mathcal{W}(q,\kappa) = \tfrac{1}{2\pi}\int \pi_f(\kappa,\widetilde{\kappa})e^{i\widetilde{\kappa}q}d\widetilde{\kappa} = \tfrac{1}{\sqrt{2\pi}}\widehat{\mathcal{F}}^{-1}_{(q,\widetilde{\kappa})}\,\pi_f(\kappa,\widetilde{\kappa}), \tag{7.5.6}$$

i.e., apart from the uninfluential factor $\frac{1}{\sqrt{2\pi}}$, as the inverse Fourier transform of $\pi_f(\kappa,\widetilde{\kappa})$ with respect to the difference frequency variable $\widetilde{\kappa}$.

Similar relations can be established between the ambiguity function $\mathcal{A}(\widetilde{q},\widetilde{\kappa})$ and the $2D$ functions π_s and π_f. In fact, rewriting the ambiguity integrals (7.4.1) and (7.4.2) in terms of π_s and π_f, i.e.,

$$\begin{aligned}\mathcal{A}(\widetilde{q},\widetilde{\kappa}) &= \tfrac{1}{2\pi}\int \pi_s(q,\widetilde{q})e^{-i\widetilde{\kappa}q}dq = \tfrac{1}{\sqrt{2\pi}}\widehat{\mathcal{F}}_{(q,\widetilde{\kappa})}\,\pi_s(q,\widetilde{q}),\\ \mathcal{A}(\widetilde{q},\widetilde{\kappa}) &= \tfrac{1}{2\pi}\int \pi_f(\kappa,\widetilde{\kappa})e^{i\kappa\widetilde{q}}d\kappa = \tfrac{1}{\sqrt{2\pi}}\widehat{\mathcal{F}}^{-1}_{(\widetilde{q},\kappa)}\,\pi_f(\kappa,\widetilde{\kappa}).\end{aligned} \tag{7.5.7}$$

we see that, the ambiguity function can be interpreted (apart from the factor $\frac{1}{\sqrt{2\pi}}$) as the Fourier transform of the space-domain product $\pi_s(q,\widetilde{q})$ with respect to the centre space-coordinate q and likewise as the inverse Fourier transform of the frequency-domain product $\pi_f(\kappa,\widetilde{\kappa})$ with respect to the centre

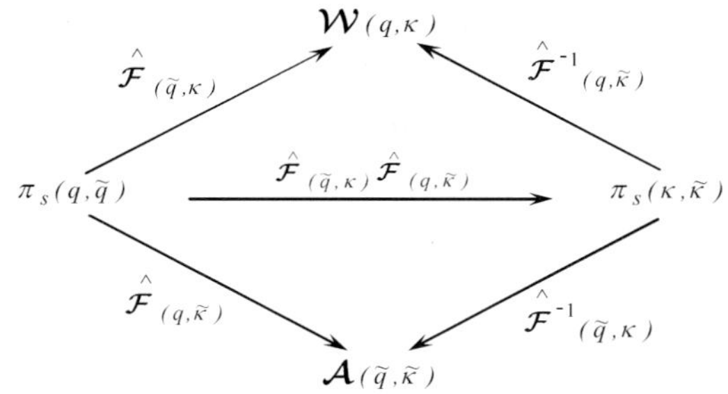

FIGURE 7.12. Schematic of the Fourier transform relation between the product functions $\pi_s(q,\widetilde{q})$, $\pi_f(\kappa,\widetilde{\kappa})$ and the Wigner distribution and ambiguity functions (inspired by [12.5]). The various space, frequency and hybrid $2D$ domains are then recognized.

frequency-coordinate κ. The diagram in Fig. 7.12 synthesizes the Fourier-like relations between π_s, π_f, and $\mathcal{W}$ and $\mathcal{A}$ [12.5, 7.4, 4.8].

Relations (7.5.4)-(7.5.7) are in evident accord with the established correspondences between the two sets of variables (q,κ) and $(\widetilde{q},\widetilde{\kappa})$ pertaining to the Wigner distribution function and the ambiguity function, sketched in (7.4.8).

With these correspondences in mind, we can write down the formal expression of the $2D$ Fourier transform of $\pi_s(q,\widetilde{q})$, namely

$$\widetilde{\pi}_s(\widetilde{\kappa},\kappa) = \frac{1}{2\pi}\iint \pi_s(q,\widetilde{q})e^{-i(\widetilde{\kappa}q+\kappa\widetilde{q})}dqd\widetilde{q}, \tag{7.5.8}$$

which is easily proved to equal the frequency-domain product $\pi_f(\kappa,\widetilde{\kappa})$:

$$\widetilde{\pi}_s(\widetilde{\kappa},\kappa) = \pi_f(\kappa,\widetilde{\kappa}). \tag{7.5.9}$$

The Fourier transform relation between the signal and spectrum generated functions π_s and π_f is also evidenced in the diagram of Fig. 7.12.

As a conclusion, we synthesize the relations between the space and frequency variables $(q,\widetilde{q},\kappa,\widetilde{\kappa})$ involved in the Wigner and ambiguity function representations of the signal. The space variables $(q,\widetilde{q})$ span the $2D$ domain of the signal-generated function $\pi_s(q,\widetilde{q})$ as the frequency variables $(\kappa,\widetilde{\kappa})$ span the $2D$ domain of the spectrum-generated function $\pi_f(\kappa,\widetilde{\kappa})$. The two subspaces are duals in accord with the Fourier transform relation (7.5.8). On the other hand, the space-frequency pairs (q,κ) and $(\widetilde{q},\widetilde{\kappa})$ span the space-frequency domains of the Wigner distribution and the ambiguity function, respectively. These hybrid domains as well are duals according to (7.4.4). Although both (q,κ) and $(\widetilde{q},\widetilde{\kappa})$ are pairs of dual variables through the Fourier transform relation holding between the signal and its spectrum, when framed within the Wigner and ambiguity function representation they acquire different roles, (q,κ) corresponding directly to the space and frequency variables of the signal, whilst $(\widetilde{q},\widetilde{\kappa})$ being associable with the space and frequency lags.

7.5.2 *Relating the Wigner distribution and ambiguity functions to the mutual intensity function*

The product structure of the function $\pi_s(q, \widetilde{q})$ suggests to relate the Wigner distribution and ambiguity functions to the mutual intensity function. We recall from the theory of partial coherence [9] that the mutual intensity function describes the spatial correlations of the space- and time-varying wave function $\psi(\mathbf{r}, t)$, representing the scalar optical disturbance at position $\mathbf{r}$ and time t, through the ensemble average (or equivalently, the infinite-time average)

$$\mathsf{J}(\mathbf{r}_1, \mathbf{r}_2) = \langle\, \psi^*(\mathbf{r}_1, t)\psi(\mathbf{r}_2, t)\,\rangle \tag{7.5.10}$$

for any pair of positions $\mathbf{r}_1 = (q_{x1}, q_{y1}, z_1)$ and $\mathbf{r}_2 = (q_{x2}, q_{y2}, z_2)$ across the region of interest. The sharp brackets above signify both the ensemble and the time-average, which are indeed interchangeable due to the hypothesized ergodicity of the optical wave field; and hence

$$\langle\, \psi^*(\mathbf{r}_1, t)\psi(\mathbf{r}_2, t)\,\rangle = \lim_{T\to\infty} \frac{1}{2T} \int_{-T}^{T} \psi^*(\mathbf{r}_1, t)\psi(\mathbf{r}_2, t)dt. \tag{7.5.11}$$

Evidently, $\mathsf{J}(\mathbf{r}_1, \mathbf{r}_2)$ yields the mutual coherence function $\Gamma(\mathbf{r}_1, \mathbf{r}_2; \tau)$ at null delay τ, and the field intensity for $\mathbf{r}_1 = \mathbf{r}_2$:

$$I(\mathbf{r}) = \langle |\, \psi(\mathbf{r},t)|^2\,\rangle. \tag{7.5.12}$$

Conforming with the formalism of the book, we consider distributions of the optical field across profile planes at fixed longitudinal z coordinates, so that, with the further simplification introduced by the single transverse-coordinate account, the general definition (7.5.10) turns into the simpler

$$\mathsf{J}(q_1, q_2) = \langle\, \psi^*(q_1, t)\psi(q_2, t)\,\rangle, \tag{7.5.13}$$

the dependence on z as a parameter being implied.

The mutual intensity function yields a complete description of the spatial correlation of the optical disturbance for quasi-monochromatic fields, whose temporal frequency spread $\Delta\omega$ is indeed small compared to the corresponding mean frequency $\overline{\omega}$: $\Delta\omega/\overline{\omega} \ll 1$. In that case, under the additional restriction that all the optical path differences $\Delta\mathcal{L}$ inherent in the optical system are much shorter than the temporal coherence length of the wavefield: $\Delta\mathcal{L} \ll 2\pi c/\Delta\omega$, the light may be considered to effectively posses complete temporal coherence. Hence the mutual coherence function $\Gamma(q_1, q_2; \tau)$ takes the form of a simple modulation of $\mathsf{J}(q_1, q_2)$ at the mean temporal frequency $\overline{\omega}$:

$$\Gamma(q_1, q_2; \tau) \simeq \mathsf{J}(q_1, q_2)e^{i\overline{\omega}\tau}. \tag{7.5.14}$$

Besides, under the above specified restrictions, the time-varying wave function $\psi(q, t)$, representing a nonmonochromatic field is shown to obey the same laws of propagation as the time-independent amplitude $\varphi(q)$ representing a monochromatic disturbance at temporal frequency $\overline{\omega}$. Evidently, this is crucial in order to establish the transformation relationship for the mutual intensity function under propagation through optical systems.

In particular, for quasi-monochromatic spatially coherent fields, the infinite-time average in (7.5.10) yields the simple product of the time-independent amplitude $\varphi(q)$ measured at the points of concern, q_1 and q_2; namely,

$$\mathsf{J}(q_1,q_2) = \varphi^*(q_1)\varphi(q_2). \tag{7.5.15}$$

In contrast, in the opposite (and idealized) limit of spatially incoherent fields, in which case light from each point across the plane of interest is assumed to be statistically independent of light from every other point, $\mathsf{J}(q_1,q_2)$ writes as

$$\mathsf{J}(q_1,q_2) = \langle|\,\psi(q_1,t)|^2\,\rangle\delta(q_1 - q_2) = I(q_1)\delta(q_1 - q_2), \tag{7.5.16}$$

where the intensity $I(q) = \langle|\,\psi(q,t)|^2\,\rangle$ turns into the square modulus of the time-independent amplitude for quasi-monochromatic fields. As noted in § 4.3.2, due to the restriction to $1D$ description, the physical quantities involved into our analysis may come not to have the right dimensions. Again, we invite the reader to disregard this accident, focussing attention rather on the mathematical relations between the involved quantities, the right dimensions being recoverable in a correct $2D$ analysis.

Relations (7.5.15) and (7.5.16), relevant to the opposite limits of spatially coherent and incoherent wavefields, are basic to demonstrate that optical processing under coherent illumination can be described as a linear mapping of complex amplitudes, whilst optical processing under incoherent illumination amounts to a linear mapping of real intensities. In the intermediate case of partially coherent light, it is necessary to establish the transformation relationships for the mutual intensity function under propagation from the object to the receiving plane of the optical set-up under examination. Indeed, it can easily be argued that the mutual intensity is in general expressible in the form

$$\mathsf{J}(q_1,q_2;z_o) = \iint dq_1'\,dq_2'\,\mathsf{J}(q_1',q_2';z_i)\mathsf{g}^*(q_1,q_1')\mathsf{g}(q_2,q_2'), \tag{7.5.17}$$

where $\mathsf{g}(q,q')$ denotes the line-spread function of the optical system, and z_i and z_o individualize as usual the location of the input (i.e., illuminating plane and the output (i.e., observing) plane.

In this connection, we mention that in [7.2] it has been shown that the paraxial propagation into the Fraunhofer region and the thin-lens transfer to the back focal plane affect the mutual intensity function by a double Fourier-like transform. In fact, recalling that the far-field distribution of a freely propagated object amplitude looks likes the imperfect Fourier transform of the object itself (see § 5.4.3), one finds

$$\varphi_o(q, z_i + \ell) \simeq \sqrt{\tfrac{k}{i\ell}}\,e^{i\frac{k}{2\ell}q^2}\,\widetilde{\varphi}_i(\tfrac{kq}{\ell}, z_i) \tag{7.5.18}$$

where, as required by the Fraunhofer condition, $\ell \gg \frac{kw^2}{2}$, $2w$ being the maximum width of the object distribution; $k = n_0 k_0$ is the light wavenumber in the medium. Thus for a quasi-monochromatic wave field $\psi(q,t)$ one can write

$$\psi_o(q, z_i + \ell; t) \simeq \sqrt{\tfrac{k}{i\ell}}\, e^{i\frac{k}{2\ell}q^2}\, \widetilde{\psi}_i(\tfrac{kq}{\ell}, z_i; t - \tfrac{\ell}{c}). \tag{7.5.19}$$

Inserting the above expression into the infinite-time average in (7.5.10) with ψ_o, and accordingly $\widetilde{\psi}_i$, being taken at two different points q_1 and q_2, we obtain

$$\begin{aligned} \mathsf{J}(q_1,q_2; z_i + \ell) &= \tfrac{k}{\ell} e^{i\frac{k}{2\ell}(q_2^2 - q_1^2)} \langle\, \widetilde{\psi}_i^*(\tfrac{kq_1}{\ell}, z_i; t - \ell/c)\widetilde{\psi}_i(\tfrac{kq_2}{\ell}, z_i; t - \tfrac{\ell}{c})\,\rangle \\ &= \tfrac{k}{2\pi\ell} e^{i\frac{k}{2\ell}(q_2^2 - q_1^2)} \iint dq' dq'' \mathsf{J}(q',q''; z_i) e^{-i\frac{k}{\ell}(q_2 q'' - q_1 q')}, \end{aligned} \tag{7.5.20}$$

where the second step has been obtained by expressing the product between brackets back to the spatial domain through the Fourier relation. Then, the mutual intensity functions at planes z_i and $z_i + \ell$ (maybe, the object and image planes) are related by a double Fourier transform relation. A similar relation can be drawn in the case as well of the propagation to the back focal plane of a thin lens, with ℓ taking then the meaning of the lens focal length.

After this preliminary and qualitative digression on the mutual intensity function and its relevance in the optical system theory under partial coherent illumination, we turn back to the question of the relation between the Wigner distribution and ambiguity functions, and the mutual intensity function.

A closer look at the product-like form of $\mathsf{J}(q_1,q_2)$ in the case of spatially coherent light immediately suggests a link with the $2D$ product function $\pi_s(q,\widetilde{q})$, which is basic to the definition of $\mathcal{W}$ and $\mathcal{A}$. It is in fact evident that we can define the mutual intensity function in terms of the center and difference coordinates q and $\widetilde{q}$. Thus, using the same symbol J for the mutual intensity we can rewrite (7.5.10) in terms of q and $\widetilde{q}$ as

$$\mathsf{J}(q,\widetilde{q}) = \langle\, \psi^*(q - \tfrac{\widetilde{q}}{2}, t)\psi(q + \tfrac{\widetilde{q}}{2}, t)\,\rangle, \tag{7.5.21}$$

thus yielding just the signal generated function $\pi_s(q,\widetilde{q})$ in the case of strictly coherent wave fields:

$$\mathsf{J}(q,\widetilde{q}) = \varphi^*(q - \tfrac{\widetilde{q}}{2})\varphi(q + \tfrac{\widetilde{q}}{2}) = \pi_s(q,\widetilde{q}) \qquad \text{for coherent signals.}$$

The Wigner distribution and ambiguity functions are accordingly interpretable as the $1D$ Fourier transforms of the mutual intensity function $\mathsf{J}(q,\widetilde{q})$ with respect to the difference and centre spatial coordinates $\widetilde{q}$ and q:

$$\boxed{\begin{array}{ccc} & \mathsf{J}(q,\widetilde{q}) & \\ \widehat{\mathcal{F}}_{(\widetilde{q},\kappa)} \swarrow & & \searrow \widehat{\mathcal{F}}_{(q,\widetilde{\kappa})} \\ \mathcal{W}(q,\kappa) & & \mathcal{A}(\widetilde{q},\widetilde{\kappa}) \end{array}} \tag{7.5.22}$$

This naturally introduces the Wigner distribution and ambiguity functions within the context of the partial coherence theory.

Formally, one has

$$\mathcal{W}(q,\kappa) = \frac{1}{2\pi}\int \mathsf{J}(q,\widetilde{q})e^{-i\kappa\widetilde{q}}d\widetilde{q} = \frac{1}{2\pi}\int \langle\psi^*(q-\tfrac{\widetilde{q}}{2},t)\psi(q+\tfrac{\widetilde{q}}{2},t)\rangle e^{-i\kappa\widetilde{q}}d\widetilde{q}, \quad (7.5.23)$$

and correspondingly

$$\mathcal{A}(\widetilde{q},\widetilde{\kappa}) = \frac{1}{2\pi}\int \mathsf{J}(q,\widetilde{q})e^{-i\widetilde{\kappa}q}dq = \frac{1}{2\pi}\int \langle\psi^*(q-\tfrac{\widetilde{q}}{2},t)\psi(q+\tfrac{\widetilde{q}}{2},t)\rangle e^{-i\widetilde{\kappa}q}dq. \quad (7.5.24)$$

We note that, referring to the pertinent discussion in § 6.3.5, one can recover expression (7.5.23) for $\mathcal{W}$ in terms of the mutual intensity $\mathsf{J}(q,\widetilde{q})$ from expression (6.3.33) giving $\mathcal{W}$ in terms of the cross spectral density $\widetilde{\Gamma}(q_1,q_2;\omega)$, since under the restrictions inherent in (7.5.14) $\widetilde{\Gamma}(q_1,q_2;\omega)$ turns into

$$\widetilde{\Gamma}(q_1,q_2;\omega) = \mathsf{J}(q_1,q_2)\delta(\omega-\overline{\omega}). \quad (7.5.25)$$

Remarkably, relating the Wigner distribution and ambiguity function to the mutual intensity function has the further valence of offering a tool to deduce the transformation laws for J under paraxial propagation through optical systems from those for the Wigner distribution or the ambiguity function. In turn, from the transfer relationships for J under paraxial propagation into the Fraunhofer region or thin-lens transfer to the back focal plane (for strictly coherent light), it is possible to deduce the corresponding transfer laws for the Wigner distribution and ambiguity functions.

As a conclusion, we rewrite (7.5.20) in terms of the centre and difference coordinates q and $\widetilde{q}$, thus obtaining

$$\mathsf{J}(q,\widetilde{q};z_i+\ell) = \frac{k}{2\pi\ell}e^{i\frac{k}{\ell}q\widetilde{q}}\iint dq'd\widetilde{q}'\,\mathsf{J}(q',\widetilde{q}';z_i)e^{-i\frac{k}{\ell}(\widetilde{q}q'+q\widetilde{q}')}, \quad (7.5.26)$$

which shows that the transformation law of $\mathsf{J}(q,\widetilde{q})$ for the Fraunhofer diffraction is a pure $2D$ Fourier transform with respect to q and $\widetilde{q}$.

Also, the integral above can directly be expressed in terms of the Wigner distribution and ambiguity function as

$$\begin{aligned}\mathsf{J}(q,\widetilde{q};z_i+\ell) &= \tfrac{k}{\ell}e^{i\frac{k}{\ell}q\widetilde{q}}\textstyle\int dq'\mathcal{W}(q',\frac{kq}{\ell};z_i)e^{-i\frac{k}{\ell}\widetilde{q}q'}\\ &= \tfrac{k}{\ell}e^{i\frac{k}{\ell}q\widetilde{q}}\textstyle\int d\widetilde{q}'\mathcal{A}(\widetilde{q}',\frac{k\widetilde{q}}{\ell};z_i)e^{-i\frac{k}{\ell}q\widetilde{q}'},\end{aligned} \quad (7.5.27)$$

thus showing that $\mathcal{W}$ and $\mathcal{A}$ may naturally enter optical propagation problems under both coherent and partially coherent illumination, with $\mathcal{W}$ or $\mathcal{A}$ coming into play according to whether it is the centre or the difference coordinate to have relevance for the specific problem under examination.

In Sect. 8.3 when illustrating the paraxial propagation properties of the Wigner distribution function, we will see that $\mathcal{W}(q,\kappa)$, defined in terms of wave optics, is closely related to the ray-optical description of light propagation, being in fact interpretable as the "amplitude" of geometrical rays seen from the viewpoint of wave optics. Hence, relations like (7.5.27) signify a valuable link between the wave- and ray-optics tools and representations.

7.5.3 *The ambiguity function and the optical transfer function*

The analysis presented in the preceding paragraph has evidenced the significance of the Wigner distribution and ambiguity function within the context of coherent [7.1] as well as partially coherent [7.2]-[7.4] image formation, as conveyed by the basic transfer law (7.5.26), which may involve $\mathcal{W}$ and $\mathcal{A}$ according to the features of the optical setup at hand.

Here, following [7.5] we will illustrate the link between the ambiguity function and the optical transfer function, which is basic to the space-invariant image formation and evaluation theory under strictly incoherent illumination, representing the spatial frequency response of the optical system to an incoherent object ([10.13], ch. 6, pp. 113–125; [10.15], ch. 11, pp. 483-504). Figure 7.13 shows the typical optical geometry for image formation, with the input or object plane and the output or image plane being related by the lens law:

$$\frac{1}{d_{\text{O}}} + \frac{1}{d_{\text{I}}} - \frac{1}{f} = 0. \tag{7.5.28}$$

Without entering the details of the theory of the optical transfer function, we will simply report the relevant definition as the normalized auto-correlation of the generalized pupil function $\mathsf{P}(\xi)$. Explicitly, one has

$$\mathrm{H}(u) = \frac{\int \mathsf{P}^*(\xi - \frac{u}{2})\mathsf{P}(\xi + \frac{u}{2})dx}{\int |\mathsf{P}(\xi)|^2 dx}, \tag{7.5.29}$$

where ξ and u denote normalized spatial frequency variables to the coherent cut-off frequency $\kappa_{\text{cutoff}} = \frac{ka}{d_{\text{I}}}$, $2a$ being the lens aperture diameter and $k = n_0 k_0$ the light wavenumber in the medium.

The generalized pupil function $\mathsf{P}(\xi)$ is defined as

$$\mathsf{P}(\xi) = P(\xi)e^{-ik[W(\xi)+\varepsilon\xi^2]} \equiv \mathsf{P}_0(\xi)e^{-ik\varepsilon\xi^2} \tag{7.5.30}$$

The complex pupil function $\mathsf{P}_0(\xi) = P(\xi)\exp[-ikW(\xi)]$ accounts for the effective extent of the lens aperture and the aberrations. In fact, $W(\xi)$ is the aberration function, whilst the simple pupil function $P(\xi)$ is defined to be unity on the opening and zero otherwise; in terms of the normalized frequency variable ξ one has

$$P(\xi) = \begin{cases} 1 & |\xi| \leq 1, \\ 0 & \text{otherwise.} \end{cases} \tag{7.5.31}$$

The quadratic phase term, which is of main concern to us, originates from an out-of-focus of the image plane location with respect to the focus condition

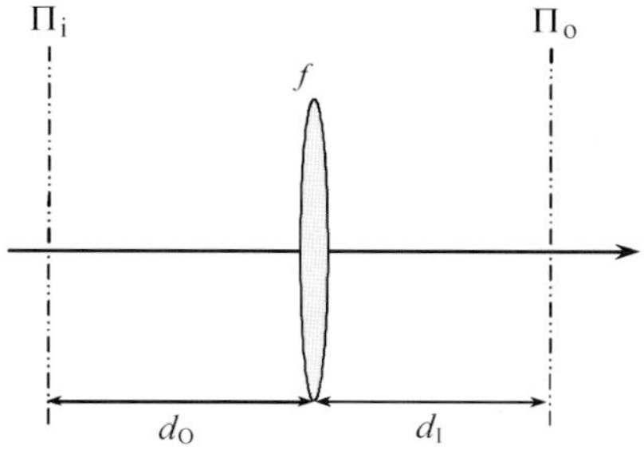

FIGURE 7.13. Typical geometry for image formation by a thin lens.

expressed by (7.5.28). Indeed, the parameter ε yields the maximum path-length error due to a departure of the relative positions of the object and image planes from the imaging condition (7.5.28), being

$$\varepsilon = \frac{a^2}{2}(\frac{1}{f} - \frac{1}{d_I} - \frac{1}{d_O}). \tag{7.5.32}$$

Evidently, since $|\mathsf{P}(\xi)|^2 = P^2(\xi)$, it follows that

$$\int |\mathsf{P}(\xi)|^2 d\xi = \int_{-1}^{1} P^2(\xi) d\xi = 2. \tag{7.5.33}$$

Then, inserting (7.5.30) and (7.5.33) into (7.5.29) and evidencing the dependence of the optical transfer function on the defocusing error ε, we can write

$$\mathrm{H}(u;\delta) = \tfrac{1}{2}\int \mathsf{P}_0^*(\xi - \tfrac{u}{2})\mathsf{P}_0(\xi + \tfrac{u}{2})e^{-i2\delta u\xi}d\xi, \qquad \delta = k\varepsilon. \tag{7.5.34}$$

The integral above has the very form of the ambiguity integral for the "signal" $\mathsf{P}_0(\xi)$, with u and $2\delta u$ being placed as the spatial and frequency variables of the ambiguity function. Accordingly, we can relate the monochromatic optical transfer function to the ambiguity function of the complex pupil P_0 as

$$\mathrm{H}(u;\delta) = \pi\mathcal{A}_{\mathsf{P}_0}(u, 2\delta u). \tag{7.5.35}$$

Resorting then to the expression (7.4.23) for the ambiguity function of a slit of width $2Q$, the ambiguity function $\mathcal{A}_{\mathsf{P}_0}(u,v)$ of the complex pupil $\mathsf{P}_0(\xi)$, in the absence of aberrations (i.e., $\mathsf{P}_0(\xi) = P(\xi)$) and in the normalized set of variables we are working with, readily obtains as

$$\mathcal{A}_{\mathsf{P}_0}(u,v) = \begin{cases} \frac{1}{\pi}(1 - \frac{|u|}{2})\mathrm{sinc}[v(1 - \frac{|u|}{2})] & |u| \le 2, \\ 0 & |u| > 2. \end{cases} \tag{7.5.36}$$

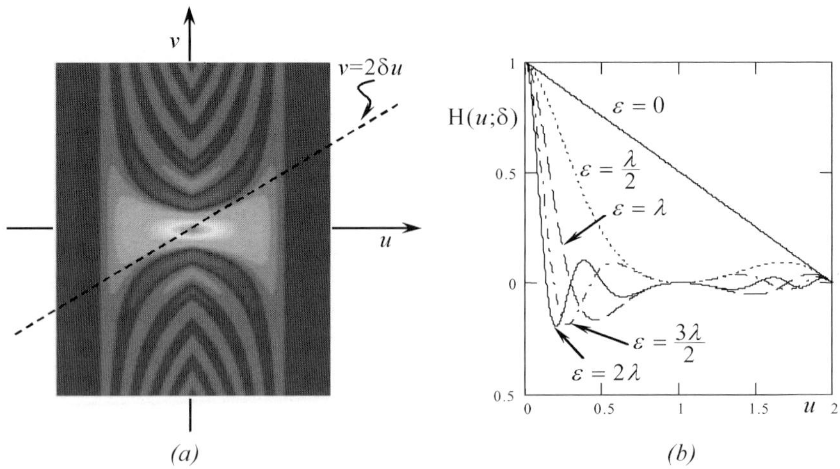

FIGURE 7.14. *(a)* Planar plot of the function $\pi\mathcal{A}_{\mathsf{P}_0}(u,v)$ representing the optical transfer function of the system with varying defocus parameter δ. *(b)* Profiles of $\pi\mathcal{A}_{\mathsf{P}_0}(u,v)$ along the straight line $v=2\delta u$ for definite values of $\delta=k\varepsilon$. (Inspired by [7.5].)

Clearly, on relating $\mathcal{A}_{\mathsf{P}_0}$ to H the reading of $\mathcal{A}_{\mathsf{P}_0}$ in the pertinent (u,v) plane must be performed along the straight line $v=2\delta u$. The value of $\mathcal{A}_{\mathsf{P}_0}$ at the point $(u,2\delta u)$ yields the value of the out-of-focus optical transfer function corresponding to the normalized frequency u and to the normalized defocus $\delta=k\varepsilon$, also obtained graphically as the tangent of the polar angle β of the line $v=2\delta u$ to the u axis: $2\delta=\tan\beta$. In particular in the condition of focus (7.5.28), amounting to $\delta=0$, the values of $\mathcal{A}_{\mathsf{P}_0}$ along the u axis yield the values of the in-focus optical transfer function [7.3]. Also, the intersections of the straight line $v=2\delta u$ with the symmetric hyperbolas

$$v=\frac{2n\pi}{2-|u|},\quad |u|\leq 2,\ n=1,2,.. \tag{7.5.37}$$

which as noted in § 7.4.4 are the zero value loci of function (7.5.36), determine the points at which the optical transfer function $\mathrm{H}(u;\delta)$ vanishes.

In Fig. 7.14.a) the planar plot of $\pi\mathcal{A}_{\mathsf{P}_0}(u,v)$ is reported, with a straight line $v=2\delta u$, for a chosen value of δ, being shown along which the reading of the optical transfer function for that value of the defocus must be performed. Finally, the profiles of $\pi\mathcal{A}_{\mathsf{P}_0}(u,v)$ along the lines $v=2\delta u$ for different values of the defocusing parameter δ, i.e., ε, are shown in Fig. 7.14.b).

Interestingly, if $\mathsf{P}_0(\xi)$ is modelled by a symmetrical function, an equivalent description in terms of the Wigner distribution function can be elaborated in virtue of relation (7.4.29).

Finally, it is worth mentioning that the ambiguity function-based account of the optical transfer function can effectively be used to obtain the polychromatic optical transfer function. In fact, since longitudinal chromatic aberrations, which affect imaging with polychromatic light, amount to a wavelength-dependent defocus parameter $\varepsilon(\lambda)$, planar plots like that in Fig. 7.14.a) can be interpreted as polar displays of all the monochromatic optical transfer functions of the system, and hence used to calculated the relevant polychromatic optical transfer function.

7.6 Optical production of the Wigner distribution function: general considerations

The optical production of the Wigner distribution function is a particular case of the more general issue concerning the realization of coherent optical computing systems. Noncoherent optical techniques for signal processing have been developed as well. As noted before, coherent illuminated optical processors operate on typically complex amplitudes, whereas incoherent illuminated optical processors operate on intensities, and so on inherently real signals. Conforming with the limited purpose of the present analysis, we consider here coherently illuminated optical processors.

Optical processors may be thought of as analog computers, which by use of optical architectures are capable of performing a certain class of operations [10]. Optical systems can inherently handle $2D$ data, displayed on the surface of the object or image plane. They can therefore operate simultaneously and independently on two separate channels as opposed to serially as in digital processing, which employs electronic devices having only time as an independent variable. The Fresnel and Fourier transforms are prime examples of mathematical operations that can be implemented by very simple optical systems. As noted in § 5.2, the propagation of coherent light between two planes in a homogeneous and isotropic medium provides a means to perform the $2D$ Fresnel transform. Similarly, the propagation between the focal planes (front to back) of a positive spherical lens provides a means to perform the $2D$ Fourier transform. In both cases, obviously, the respective optical setups must satisfy the Fresnel approximation, demanding for small extensions of the light distribution in both the object and observation regions with respect to the propagation distance.

The Fresnel and Fourier transformations are specific realizations of a general integral transform-like relation, expressing through the mathematical form of a superposition integral the relation between the input and output signals for linear systems. We recall, in fact, from the introductory considerations of § 4.3.2 that for a $2D$ linear system the input $\Phi_i(q_1, q_2)$ and the output $\Phi_o(q_1, q_2)$ are linked through the $2D$ superposition integral

$$\Phi_o(q_1, q_2) = \iint \mathfrak{G}(q_1, q_2; q_1', q_2')\Phi_i(q_1', q_2')dq_1' dq_2', \tag{7.6.1}$$

the point-spread function $\mathfrak{G}(q_1, q_2; q_1', q_2')$ being the mathematical model for the physical transmission properties of the system. As in § 4.3.2, q_1 and q_2 denote Cartesian coordinates in the reference planes, transverse to the optical z axis, the same symbols being used for both the input and output coordinates. The point spread function $\mathfrak{G}(q_1, q_2; q_1', q_2')$ signifies the response of the system at position (q_1, q_2) in the output plane to a delta signal $\delta(q_1'' - q_1')\delta(q_2'' - q_2')$, applied at position (q_1', q_2') in the input plane.

A special case is when the response function $\mathfrak{G}(q_1, q_2; q_1', q_2')$ factors as

$$\mathfrak{G}(q_1, q_2; q_1', q_2') = \mathsf{g}_1(q_1, q_1')\mathsf{g}_2(q_2, q_2'), \tag{7.6.2}$$

displaying independent responses of the system in orthogonal directions. This has been the central hypothesis of the considerations developed in the previous chapters, having enabled us to turn from a $2D$ to a $1D$ description (i.e., in terms of a single profile coordinate) of the light propagation. As noted a number of times, factorizations like (7.6.2) apply to stigmatic and simple astigmatic systems. Clearly, the arrangements for the optical implementation of the Fresnel transform, namely, a straight section of a homogeneous medium, as well as of the Fourier transform, i.e., a spherical lens or a set of cylindrical lenses, allow for the decomposition of the point spread function $\mathfrak{G}(q_1, q_2; q_1', q_2')$ into two line spread functions $\mathsf{g}(q, q')$ as in (7.6.2). Accordingly, for each of the two Cartesian directions we may write, in the Fresnel approximation,

$$\mathsf{g}_{\textsf{Fresnel}}(q, q') = \sqrt{\frac{k}{2\pi i d}}\, e^{i\frac{k}{2d}(q-q')^2}, \qquad \mathsf{g}_{\textsf{Fourier}}(q, q') = \sqrt{\frac{k}{2\pi i f}}\, e^{-i\frac{k}{f}qq'}. \tag{7.6.3}$$

Here, k denotes the wavenumber in the medium, as d and f specify respectively the section length and the lens focal length[c].

Also interesting, and pertinent to the forthcoming discussion, is the imaging transformation, which is described by a separable impulse response in the form as (7.6.2), each direction being acted according to

$$\mathsf{g}_{\textsf{imaging}}(q, q') = \sqrt{\tfrac{1}{\mathcal{M}}}\, \delta(q' - \mathcal{M}q), \tag{7.6.4}$$

where $\mathcal{M}$ denotes the magnification factor. The optical implementation of an imaging transform may consist of a simple two-lens system; a sequence of two single-lens Fourier transformers (a pure magnifier), for instance, performs an imaging transformation, with magnification proportional to the ratio of the lens focal lengths: $\mathcal{M} = -f_2/f_1$ (see § 3.4.3 and Figs. 3.9-10).

The spread functions $\mathsf{g}_{\textsf{Fourier}}(q, q')$ and $\mathsf{g}_{\textsf{imaging}}(q, q')$ above pertain to optical setups that realize an exact Fourier transforming or imaging relation between

---

[c] We evidence that the expressions of the line spread functions for the Fresnel and Fourier transformations given in §§ 5.2.1-2 contain the wavenumber k_0 in the vacuum space, and accordingly the reduced section and focal lengths. Of course, they are the same as those in (7.6.3), where the refractive index is absorbed into the wavenumber k in the medium, and hence d and f signify the physical section and focal lengths.

the input and output planes. In general a nonexact Fourier transforming or imaging relation between the input and output planes gives rise to a spatially varying quadratic phase factor in the output signal that is not present in the input signal. If, for instance, the signal is out of the front focal plane, the lens produces the Fourier transform of the input with a quadratic phase factor depending on the signal mislocation. Similarly, if in the double-lens arrangement of an imaging system the back focal plane of the first lens is not perfectly superimposed on the front focal plane of the second, the imaged distribution presents a quadratic phase factor depending on the mislocation of the two planes relatively one to the other.

7.6.1 *The Wigner transform as an optical 2D linear transformation*

It has been already emphasized that the Wigner integrals represent a bilinear transformation of $1D$ signals. The general expression for a *bilinear* [11.1] transformation of the signal $\varphi(q)$ to $\psi(q)$ has the form of the double integral

$$\psi(q) = \iint \varphi^*(q_1')\varphi(q_2')\mathfrak{Q}(q; q_1', q_2')dq_1' dq_2', \tag{7.6.5}$$

where the kernel $\mathfrak{Q}(q; q_1', q_2')$ may be interpreted as the response of the system to two impulses applied at positions q_1' and q_2'. Although not explicitly evidenced in the notation, the kernel $\mathfrak{Q}(q; q_1', q_2')$ may depend on some other parameters and hence also the function $\psi(q)$ will. Evidently the Wigner distribution and ambiguity function are in conformity with the scheme (7.6.5), both converting a $1D$ function of space or frequency into a $2D$ function of space and frequency together. It is also evident that self-correlation and self-convolution are further examples of bilinear transformations. Moreover, we mention that a bilinear transformation like (7.6.5) relates the optical intensity $\psi(q)$ in the image plane to the optical transmittance $\varphi(q)$ of a transparency illuminated by partially coherent light, the kernel accounting then for both the response of the system and the coherence degree of the illuminating light. Relation (7.5.20) with $q_1 = q_2$ illustrates the possible bilinearity of the response of an optical system to partially coherent illumination. Indeed, it has been evidenced in §§ 6.3.5 and 7.5.2 that the Wigner distribution naturally finds an eminent place within the partial coherence theory, also due to its intrinsic bilinearity. As observed in [11.1], Eq. (7.6.5) describes more optical systems and processes than the linear transformation equation (7.6.1), as a consequence of the quadratic relation between the optical intensity and the optical field amplitude.

The reader may readily verify that the Wigner transform is recovered from the general expression (7.6.5) with

$$\mathfrak{Q}_{\mathcal{W}}(q; q_1', q_2') = \tfrac{1}{\pi}\delta(q_1 + q_2 - 2q)e^{i\kappa(q_1 - q_2)}, \tag{7.6.6}$$

whereas the ambiguity function is obtained setting

$$\mathfrak{Q}_{\mathcal{A}}(\widetilde{q}; q_1', q_2') = \frac{1}{2\pi}\delta(q_1 - q_2 + \widetilde{q})e^{-\frac{i}{2}\widetilde{\kappa}(q_1+q_2)}. \tag{7.6.7}$$

The spatial frequencies κ and $\widetilde{\kappa}$ enter the above kernels as parameters on which the Wigner distribution and the ambiguity function come then to depend. The functional forms of $\mathfrak{Q}_{\mathcal{W}}$ and $\mathfrak{Q}_{\mathcal{A}}$ manifestly show the relation of the space variables q and $\widetilde{q}$, of $\mathcal{W}$ and $\mathcal{A}$, to the center coordinate: $q = \frac{q_1+q_2}{2}$, and to the difference coordinate: $\widetilde{q} = q_2 - q_1$, respectively. We invite the reader to deduce the alternative expression of the transformation (7.6.5) in terms of the signal spectrum $\widetilde{\varphi}(\kappa)$, and accordingly the frequency-domain expressions of the kernels (7.6.6) and (7.6.7).

It is easy to recognize the possibility of relating the bilinear transformation (7.6.5) to a $2D$ linear operation, whose mathematical form is given by (7.6.1). In fact, we may arrange a $2D$ system whose spread function $\mathfrak{G}(q_1, q_2; q_1', q_2')$ be

$$\mathfrak{G}(q, q; q_1', q_2') = \mathfrak{Q}(q; q_1', q_2'). \tag{7.6.8}$$

Then, letting the $2D$ object signal $\Phi_i(q_1, q_2)$ be generated by the $1D$ waveform $\varphi(q)$ through the product

$$\Phi_i(q_1, q_2) = \varphi^*(q_1)\varphi(q_2), \tag{7.6.9}$$

we may observe the function $\psi(q)$ along the line $q_1 = q_2$ in the output plane of the system, namely

$$\psi(q) = \Phi_o(q, q). \tag{7.6.10}$$

Evidently, other choices for the system response and hence for the geometry of the output detection are allowed. Thus, for instance, with

$$\mathfrak{G}(q, 0; q_1', q_2') = \mathfrak{Q}(q; q_1', q_2'), \tag{7.6.11}$$

we may observe $\psi(q)$ along the line $q_2 = 0$ in the output plane, being

$$\psi(q) = \Phi_o(q, 0). \tag{7.6.12}$$

In § 7.5.1 we proved that the Wigner distribution and the ambiguity function can both be interpreted as $1D$ Fourier transforms of the $2D$ product function

$$\pi_s(q, \widetilde{q}) = \frac{1}{\sqrt{2\pi}}\varphi^*(q - \tfrac{\widetilde{q}}{2})\varphi(q + \tfrac{\widetilde{q}}{2}), \tag{7.6.13}$$

the former as Fourier transform with respect to the difference space coordinate $\widetilde{q}$ and the latter as Fourier transform with respect to the mean space coordinate q (see Fig. 7.12). Here we conveniently rewrite Eqs. (7.5.4) and (7.5.7) in the slightly different but equivalent forms

$$\begin{aligned} \mathcal{W}(q, \kappa) &= \frac{1}{\sqrt{2\pi}}\iint \pi_s(q', \widetilde{q}')e^{-i\kappa\widetilde{q}'}\delta(q' - q)dq'd\widetilde{q}', \\ \mathcal{A}(\widetilde{q}, \widetilde{\kappa}) &= \frac{1}{\sqrt{2\pi}}\iint \pi_s(q', \widetilde{q}')e^{-i\widetilde{\kappa}q'}\delta(\widetilde{q}' - \widetilde{q})d\widetilde{q}'dq'. \end{aligned} \tag{7.6.14}$$

Accordingly, by comparison with (7.6.1), it becomes evident that $\mathcal{W}(q,\kappa)$ or $\mathcal{A}(\widetilde{q},\widetilde{\kappa})$ can be computed optically by a $2D$ linear transformation where the product function $\pi_s(q,\widetilde{q})$ be assigned as the $2D$ object signal and the optical processor be designed to have the response function

$$\mathfrak{G}_{\mathcal{W}}(q,\widetilde{q};q',\widetilde{q}') = \frac{1}{\sqrt{2\pi}} e^{-i\kappa\widetilde{q}'}\delta(q'-q), \qquad (7.6.15)$$

to compute the Wigner distribution, or

$$\mathfrak{G}_{\mathcal{A}}(q,\widetilde{q};q',\widetilde{q}') = \frac{1}{\sqrt{2\pi}} e^{-i\widetilde{\kappa}q'}\delta(\widetilde{q}'-\widetilde{q}), \qquad (7.6.16)$$

to compute the ambiguity function. Clearly, the space variables q and $\widetilde{q}$ come to acquire the role as the orthogonal coordinates of the Cartesian frames in the input and output planes. It is evident that $\mathfrak{G}_{\mathcal{W}}$ describes an optical system capable to perform the Fourier transform in the q direction and the imaging transformation in the $\widetilde{q}$ direction, and similarly for $\mathfrak{G}_{\mathcal{A}}$ with the role of the two directions interchanged.

The optical implementation of the Wigner and/or ambiguity integrals requires therefore the following two processes:

i) a multiplication process to form the $2D$ product function $\pi_s(q,\widetilde{q})$ from the $1D$ signal $\varphi(q)$, and then

ii) a parallel imaging/Fourier transforming process involving the two orthogonal directions of the object $\pi_s(q,\widetilde{q})$; specifically, the q-axis direction is imaged whilst $\widetilde{q}$ is Fourier transformed in order to display $\mathcal{W}$ and vice versa, q is Fourier transformed whilst $\widetilde{q}$ is imaged to display $\mathcal{A}$.

It is worth emphasizing that since the ambiguity function is in general complex whilst the Wigner distribution function is always real, when the detection scheme of the display setup makes use of devices responding only to light intensity (like, for instance, a photographic film), information regarding the phase of the ambiguity function and correspondingly the negative regions of the Wigner distribution function are got lost.

The optical realization of the processes i) and ii) is illustrated in the following section. Within this respect, we mention that, due to its relevance to radar applications, the generation and display of the ambiguity function have received a great attention [8]. Since the Wigner and ambiguity integrals simply differ in the direction along which the signal-generated $2D$ product function $\pi_s(q,\widetilde{q})$ is Fourier transformed, the optical implementation of the Wigner transform has greatly benefited from the optical arrangements for the ambiguity function display [12.1]; also, optical setups operating as both Wigner distribution and ambiguity function processors have been arranged. In the considerations to follow, we shall refer specifically to the Wigner transform.

7.7 Wigner processor for $1D$ real signals: basic configurations

Since our main concern is to illustrate how the "mathematics" of the Wigner integral could be implemented by optical means, we describe the basic schemes of the display setups with emphasis on the concepts rather than on the technical details. The interested reader is directed to the literature for a more complete and update account of the subject, concerned also with the digital and hybrid implementations of the Wigner distribution as well as with the optical visualizations that have been proposed to display the $4D$ Wigner distribution function of $2D$ signals [12].

We begin considering nonnegative real signals, so that the multiplicative process to form the $2D$ product function $\pi_s(q,\widetilde{q})$ involves the symmetrically shifted and real versions, $\varphi(q-\frac{\widetilde{q}}{2})$ and $\varphi(q+\frac{\widetilde{q}}{2})$, of the signal. Complex signals are briefly considered in § 7.8.

7.7.1 π_s-*product function modelling setup for* $1D$ *real signals: two object transparencies*

Optically one handles two dimensions, the $2D$ object distribution on the input plane Π_i being transformed into $2D$ image distribution on the output plane Π_o. Mathematically, the signal in (7.6.5) is a function of q only and hence is one-dimensional in nature. Therefore, in order that the signal be optically processed it must properly be represented on the $2D$ plane of a proper transparency. The term *transparency* is generally applied to a variety of elements designed to record or materialize a light distribution. We shall not enter the relevant technical details; to fix mind, we may consider here as a recording-display device the density-modulated transparency, whose amplitude-transmittance function represents the means by which the informations contained in the real signal are recorded and then displayed under coherent light illumination.

There are two simple ways to realize a $2D$ representation of $1D$ signals. With in mind that q and $\widetilde{q}$ signify the two Cartesian directions in the $2D$ transparency plane, the signal waveform $\varphi(q)$ may be extended into the $\widetilde{q}$ direction using the unit function $\mathrm{I}(\widetilde{q})=1$, i.e.,

$$\varphi(q) \to \Phi(q,\widetilde{q}) = \varphi(q)\mathrm{I}(\widetilde{q}), \tag{7.7.1}$$

which is practically realized by a planar transparency that varies in the q-direction according to the signal amplitude and is constant in the $\widetilde{q}$-direction. As an example, in the simple case of a rectangle waveform, illustrated in Fig. 7.15.b), the transparency may be designed to have a non zero uniform transmittance through a vertical strip with the same width as the signal.

Alternatively, the distribution in the transparency plane may be concentrated along a single line, which is mathematically represented by the δ-function as

$$\varphi(q) \to \Psi(q,\widetilde{q}) = \varphi(q)\delta(\widetilde{q}), \tag{7.7.2}$$

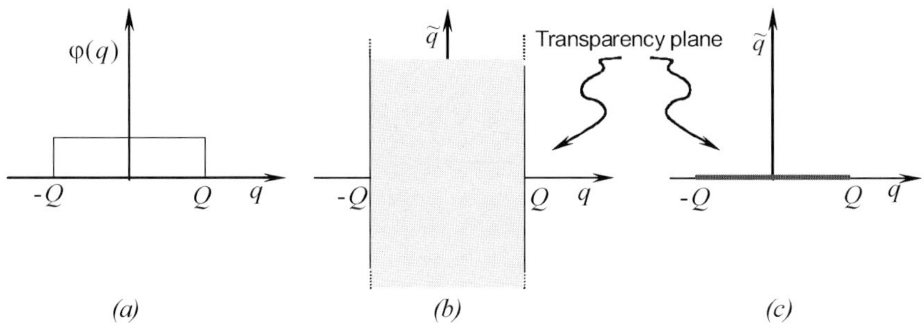

FIGURE 7.15. $2D$ representations of *(a)* a rectangle signal, obtained *(b)* by a nonzero uniform amplitude transmittance through the vertical strip $|q| \leq Q$ in the transparency plane, or *(c)* by use of a narrow slit of aperture $2Q$, selecting a single horizontal q-line in the strip in *(b)*.

and practically realized, for instance, by using a narrow slit to select the single horizontal q-line on the $2D$ representation (7.7.1) (Fig. 7.15.c)).

Once produced the $2D$ signal mask $\Phi(q, \widetilde{q})$, we must form the object signal $\pi_s(q, \widetilde{q})$ in the input plane Π_i of the optical system. Mathematically, π_s is given by the product of the symmetrically shifted versions of the signal $\varphi(q - \frac{\widetilde{q}}{2})$ and $\varphi(q + \frac{\widetilde{q}}{2})$. Optically π_s can be produced by multiplying two planar transmittances, one representing $\varphi(q - \frac{\widetilde{q}}{2})$ as the other $\varphi(q + \frac{\widetilde{q}}{2})$.

Typically the multiplication of two real signals can be accomplished optically by placing the two transparencies, representing the signals, adjacent to each other thus yielding as an equivalent transmittance the product of the two signals. Alternatively one of the signal, more precisely its $2D$ representation, can be transmitted on the transparency representing the other, thus again effecting the desired multiplicative operation of the two signals.

How can one form the planar transmittances of the shifted signals $\varphi(q - \frac{\widetilde{q}}{2})$ and $\varphi(q + \frac{\widetilde{q}}{2})$? Note that, if we have a transmittance $\Phi(q, \widetilde{q})$ of the signal $\varphi(q)$, the signal $\varphi(\alpha q + \beta \widetilde{q})$, where α and β are arbitrary real numbers, can be obtained by a coordinate transformation of the transmittance plane with respect to the fixed $(q, \widetilde{q})$ coordinate frame of the working plane, that is the input plane for the optical processor we intend to build. Rotating, for instance, the transparency $\Phi(q, \widetilde{q})$ counterclockwise by the angle θ about the origin, we produce in the reference plane the transmittance

$$\Phi_\theta(q, \widetilde{q}) = \varphi(q \cos\theta + \widetilde{q} \sin\theta). \tag{7.7.3}$$

In particular, a rotation by $\theta = 45°$ generates the scaled and shifted signal

$$\Phi_{45°}(q, \widetilde{q}) = \varphi(\tfrac{q + \widetilde{q}}{\sqrt{2}}), \tag{7.7.4}$$

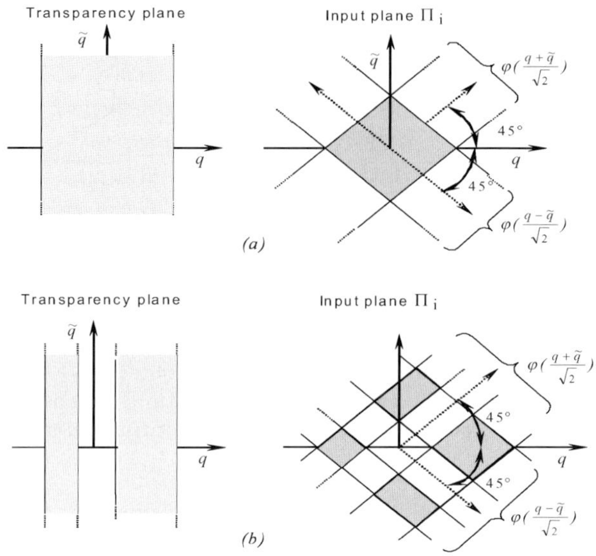

FIGURE 7.16. Resultant transmittance of the multiplication of two identical transparencies rotated by 90° relatively to each other in the case of *(a)* a single-rectangle signal and *(b)* a double-rectangle signal (inspired by [8.2]).

whereas a rotation by $\theta = -45°$ results into

$$\Phi_{-45°}(q, \widetilde{q}) = \varphi(\tfrac{q-\widetilde{q}}{\sqrt{2}}). \tag{7.7.5}$$

Evidently, apart from a scale factor, transmittances (7.7.4) and (7.7.5) display the same amount of shift and in the opposite directions, as required to form the product function $\pi_s(q, \widetilde{q})$. Then, we may place in the plane Π_i two identical transparencies, on which the $2D$ representation $\Phi(q, \widetilde{q})$ of the signal is recorded, one rotated by 45° and the other by −45°. Under uniform coherent illumination, the transmitted signal reproduces the desired product function $\pi_s(q, \widetilde{q})$ (or, some scaled version of it), from which the Wigner distribution function can be generated. Figure 7.16 illustrates the resultant transmittance of the multiplication of the two representations $\varphi(\frac{q-\widetilde{q}}{\sqrt{2}})$ and $\varphi(\frac{q+\widetilde{q}}{\sqrt{2}})$ for both a single and a double rectangle waveform.

Likewise, supposing to rotate by θ and dilate by m the axes, we are able to generate the transmittance

$$\Phi_{m,\theta}(q, \widetilde{q}) = \varphi(mq\cos\theta + m\widetilde{q}\sin\theta), \tag{7.7.6}$$

which can be made to fit the required shifts $q \pm \frac{\widetilde{q}}{2}$ by appropriately arranging m and θ. Indeed, requiring that $m\cos\theta = 1$ and $m\sin\theta = \pm\frac{1}{2}$, one obtains $m = \sqrt{\frac{5}{4}} \simeq 1$ and $\theta = \arccos(\frac{1}{m}) \simeq 26.6°$.

7.7.2 Optical processing of the product function $\pi_s(q, \widetilde{q})$ to display the Wigner distribution function

Once the $2D$ product function $\pi_s(q, \widetilde{q})$, or some scaled version of it, is available, it remains to arrange an optical system capable to perform the $1D$ Fourier transform of π_s with respect to the $\widetilde{q}$ direction to yield the Wigner distribution function. As we know, an optical Fourier transformer produces an output signal proportional to the Fourier transform of the input signal by a constant factor or a quadratic-phase term in the space coordinate. Here, the input signal is the field transmitted through the two transparencies arrangement under plane-wave illumination and hence is proportional to the product $\varphi(\frac{q-\widetilde{q}}{\sqrt{2}})\varphi(\frac{q+\widetilde{q}}{\sqrt{2}}) = \pi_s(\frac{q}{\sqrt{2}}, \sqrt{2}\widetilde{q})$. Then supposing that the $1D$ Fourier transform of the input signal with respect to $\widetilde{q}$ has been performed by an optical setup with effective focal length f, the resultant signal $\Phi_o(q, \widetilde{q})$ on the output plane should be proportional to the Fourier integral

$$\Phi_o(q, \widetilde{q}) \propto \int \varphi(\tfrac{q-\widetilde{q}'}{\sqrt{2}})\varphi(\tfrac{q+\widetilde{q}'}{\sqrt{2}})e^{-i\frac{k}{f}\widetilde{q}\widetilde{q}'}\,d\widetilde{q}'. \tag{7.7.7}$$

As a general property of optical Fourier transformers, the frequency variable κ of the transformation is related to the output coordinate $\widetilde{q}$ as (see § 5.4))

$$\kappa = \frac{k}{f}\widetilde{q}. \tag{7.7.8}$$

By changing the variable to $y = \sqrt{2}\widetilde{q}'$, the integral above turns into

$$\Phi_o(q, \widetilde{q}) \propto \int \varphi(\tfrac{q}{\sqrt{2}} - \tfrac{y}{2})\varphi(\tfrac{q}{\sqrt{2}} + \tfrac{y}{2})e^{-i\frac{\kappa}{\sqrt{2}}y}dy, \tag{7.7.9}$$

thus clearly revealing the relation of the output signal to the Wigner distribution of $\varphi(q)$, being in fact

$$\Phi_o(q, \widetilde{q}) \propto \mathcal{W}(\tfrac{q}{\sqrt{2}}, \tfrac{\kappa}{\sqrt{2}}). \tag{7.7.10}$$

Thus we may generate and display the scaled version of the Wigner distribution function of the given signal $\varphi(q)$ in the output plane of a $1D$ Fourier transforming optical processor. Notably, since the frequency variable of the displayed Wigner distribution comes to correspond directly to the output coordinate $\widetilde{q}$, the entire output plane of the optical processor is utilized, corresponding directly to the Wigner plane (q, κ).

Parallel imaging/Fourier transforming operations of the $2D$ real signal π_s

Central to this is the possibility of performing the Fourier transform in only one direction. As said earlier, the required functional form (7.6.15) of the impulse response reveals that the optical processor should act as a Fourier transformer in one direction and as a unity magnification imaging system in the orthogonal direction. Therefore, we must arrange an astigmatic optical system, capable of performing parallel operations in orthogonal directions; precisely, imaging in one dimension and Fourier transform in the other.

We describe the optical architectures that have been realized to perform such imaging/Fourier transforming parallel operations. Evidently cylindrical lenses are the natural candidates as components of an optical setup where the signal is affected differently in the two orthogonal directions. Various configurations, which employ cylindrical lenses in combination as well with spherical lenses, have been suggested [10.6].

Two-lens configurations. We begin our discussion considering the configuration depicted in Fig. 7.17.a) [8.2, 8.3, 8.7, 10.6], where we have assigned the variables q and $\widetilde{q}$ respectively to the horizontal and vertical axis in the reference planes. A cylindrical and a spherical lens, both having focal length f, are placed at distance f respectively from the input and the output plane; the two lenses are separated by $2f$ to each other, and the cylindrical lens is oriented with its cylindrical axis parallel to the $\widetilde{q}$-axis.

We may analyse the transformations acted on the input signal separately in the two orthogonal directions. Evidently, in the horizontal direction the signal "sees" a perfect imaging system, formed by two cascaded Fourier transformers, which then produces a negative unity magnification. In contrast, in the $\widetilde{q}$-direction only the spherical lens is active, being at distances $3f$ and f from the input and the output plane, respectively. The double-lens setup is therefore equivalent to an imperfect Fourier transformer, whose ray matrix has $D = -2$.

The impulse response of the system is exemplified by the function

$$\mathfrak{G}(q,\widetilde{q};q',\widetilde{q}') = \sqrt{\tfrac{ik}{2\pi f}}\, e^{-i\frac{k}{f}\widetilde{q}^2} e^{-i\frac{k}{f}\widetilde{q}\widetilde{q}'}\,\delta(q'+q), \tag{7.7.11}$$

and accordingly the input $\Phi_i(q,\widetilde{q}) = \varphi(\frac{q-\widetilde{q}'}{\sqrt{2}})\varphi(\frac{q+\widetilde{q}'}{\sqrt{2}})$ is transformed to

$$\Phi_o(q,\widetilde{q}) = \sqrt{\tfrac{i\pi k}{f}}\, e^{-i\frac{k}{f}\widetilde{q}^2}\, \mathcal{W}(-\tfrac{q}{\sqrt{2}}, \tfrac{\kappa}{\sqrt{2}}), \tag{7.7.12}$$

where, as already noted, the spatial frequency κ is readable along the output axis $\widetilde{q}$ through the scaling factor $\frac{k}{f}$: $\kappa = \frac{k\widetilde{q}}{f}$.
The system is seen to perform the desired parallel imaging and Fourier transforming operations, thus generating in the plane Π_o the scaled and q-inverted version of the Wigner distribution function of the assigned $1D$ signal $\varphi(q)$ with a quadratic phase factor in the vertical direction. This factor can be removed by placing a negative cylindrical lens immediately in front of the output plane.

Alternatively, in the astigmatic system of Fig. 7.17.b) the spherical and cylindrical lenses are in contact [10.3, 12.3]. The lenses have the same focal length f and are at distance f from both the input and output planes; as before, the cylindrical lens has its axis parallel to the $\widetilde{q}$ axis. Since both lenses

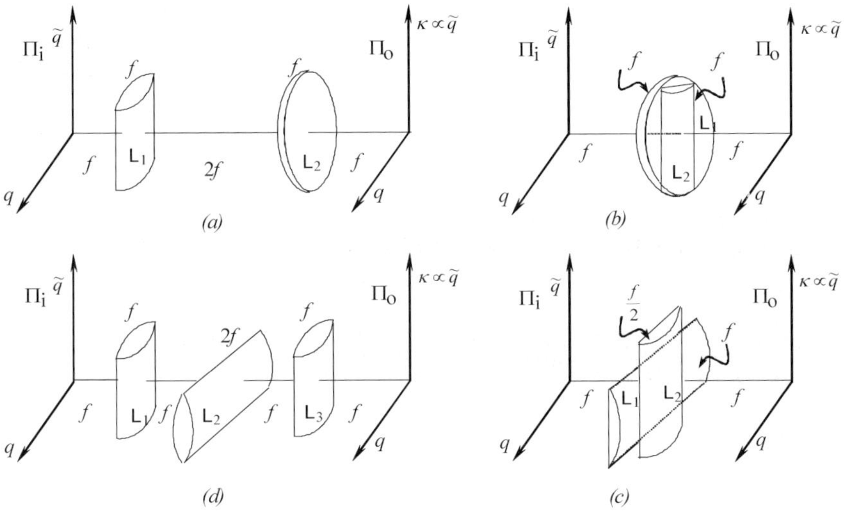

FIGURE 7.17. Schematic of astigmatic optical systems performing *(a)* parallel perfect imaging/imperfect Fourier transforming operations; *(b), (c)* parallel imperfect imaging/perfect Fourier transforming operations, and *(d)* parallel perfect imaging/perfect Fourier transforming operations.

are active in the horizontal direction, their combination is equivalent to a single lens with focal length f_{T} given by

$$\frac{1}{f_{\mathrm{T}}} = \frac{1}{f} + \frac{1}{f} = \frac{2}{f}. \tag{7.7.13}$$

Therefore the input signal $\Phi_i(q, \widetilde{q})$ experiences an imperfect imaging operation with a negative unity magnification and a curvature factor $e^{ikq^2/f}$. In contrast, in the vertical direction we have a perfect Fourier transforming arrangement.

The impulse response of the system is seen to be

$$\mathfrak{G}(q, \widetilde{q}; q', \widetilde{q}') = \sqrt{\tfrac{ik}{2\pi f}}\, e^{i\frac{k}{f}q^2}\, \delta(q' + q)\, e^{-i\frac{k}{f}\widetilde{q}\widetilde{q}'}. \tag{7.7.14}$$

It produces on the output plane a scaled and q-inverted version of the Wigner distribution function according to

$$\Phi_o(q, \widetilde{q}) = \sqrt{\tfrac{i\pi k}{f}}\, e^{i\frac{k}{f}q^2}\, \mathcal{W}\left(-\tfrac{q}{\sqrt{2}}, \tfrac{\kappa}{\sqrt{2}}\right), \tag{7.7.15}$$

with a quadratic phase factor in the horizontal direction, which can be removed by placing a positive cylindrical lens immediately in front of the plane Π_o.

Finally, a totally equivalent configuration can be realized by placing two cylindrical lenses, L_1 and L_2, of focal lengths f and $\frac{f}{2}$ back to back as sketched

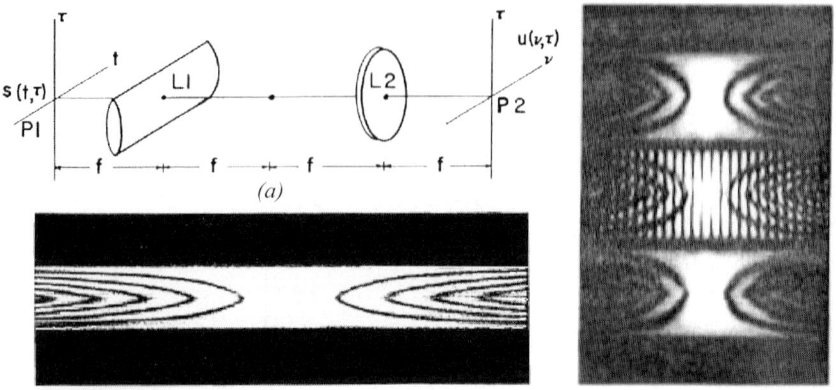

(b) (c)

FIGURE 7.18. Ambiguity function (modulus squared) display of *(b)* a single square pulse and *(c)* a double square pulse, obtained by *(a)* a sphero-cylindric doublet setup. (Reproduced with permission from [8.2].)

in Fig. 7.17.c). The lenses are at distance f from both the input and output plane and their axes are reciprocally perpendicular, with that of L_1 being parallel to the q axis.

Three-lens configuration. In order to avoid the presence of a spatially varying quadratic phase in the output, as occurs in the previously described processors using a spherocylindric doublet, we may use three cylindrical lenses arranged as illustrated in Fig. 7.17.d), thus creating exact imaging and Fourier transforming relations between the input and output planes [8.1, 10.5, 8.2, 10.6, 8.5]. Lenses L_1 and L_3 of focal length f are vertically oriented, separated by $2f$ and distanced by f respectively from plane Π_i on the right and plane Π_o on the left. Lens L_2 has a focal length double of that of L_1 and L_3; it is horizontally oriented and placed midway between L_1 and L_3.

Evidently, lens L_2 generates an exact Fourier transform relation for the $\widetilde{q}$ transformation, whilst the two-lens geometry in the horizontal direction produces a perfect imaging operation with negative unit magnification. The system response writes as

$$\mathfrak{G}(q, \widetilde{q}; q', \widetilde{q}') = \sqrt{\frac{ik}{4\pi f}} \delta(q' + q) e^{-i\frac{k}{2f}\widetilde{q}\widetilde{q}'}, \tag{7.7.16}$$

which then yields the output distribution

$$\Phi_o(q, \widetilde{q}) = \sqrt{\frac{i\pi k}{2f}} \mathcal{W}(-\tfrac{q}{\sqrt{2}}, \tfrac{\kappa}{\sqrt{2}}), \tag{7.7.17}$$

the frequency variable κ being related to the output coordinate $\widetilde{q}$ as $\kappa = \frac{k\widetilde{q}}{2f}$.

Figures 7.18.b)-c) show the ambiguity function display of a single- and a double-rectangle signal, as obtained by the two-lens astigmatic processor, sketched in part (a) of the figure [8.2]. The optical setup is practically the same as that pictured in Fig. 7.17.a), but rotated by 90° around the optical axis in order to perform the Fourier transform in the horizontal q-direction thus displaying the ambiguity function instead of the Wigner distribution function. The notation in Fig. 7.18 is in conformity with the time-frequency domain pertaining to the quoted paper. Thus, the pair (t, τ) corresponds to the pair $(q, \widetilde{q})$ relevant to the space-frequency domain of our analysis. Also, the function $s(t, \tau)$ correspond to the product function $\pi_s(q, \widetilde{q})$ as ν corresponds to $\frac{\widetilde{\kappa}}{2\pi}$. Finally, the function $u(\nu, \tau)$, which is the outcome of the displayed processor, is related to the scaled version of the asymmetric ambiguity function defined through the integral (7.4.3) without the factor $\frac{1}{2\pi}$.

7.7.3 π_s-product function modelling setups for $1D$ real signals: a single object transparency

The technique, described in § 7.7.1, to generate the shifted product $\pi_s(q, \widetilde{q})$ requires two identical transparencies that are placed close together in the plane Π_i and suitably rotated relatively to each other.

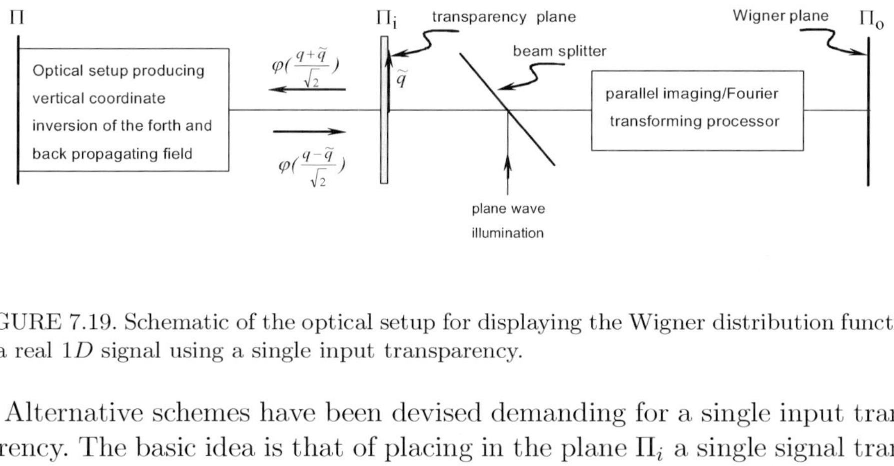

FIGURE 7.19. Schematic of the optical setup for displaying the Wigner distribution function of a real $1D$ signal using a single input transparency.

Alternative schemes have been devised demanding for a single input transparency. The basic idea is that of placing in the plane Π_i a single signal transparency properly rotated to produce any one of the shifted waveforms needed to produce the product $\pi_s(q, \widetilde{q})$, let us say $\varphi(\frac{q+\widetilde{q}}{\sqrt{2}})$. The transparency is then illuminated from the right by a plane wave reflected by a beam splitter (Fig. 7.19). The transmitted field is allowed to propagate forth and back through an optical setup suitably designed so that the $\widetilde{q}$-inverted field distribution is returned to Π_i, i.e., $\varphi(\frac{q-\widetilde{q}}{\sqrt{2}})$. Thus, after transmission through the $\varphi(\frac{q+\widetilde{q}}{\sqrt{2}})$-transparency, the desired product $\varphi(\frac{q-\widetilde{q}}{\sqrt{2}})\varphi(\frac{q+\widetilde{q}}{\sqrt{2}})$ is produced on the right of Π_i, which then can be input in any one of the previously described optical stages for the Wigner distribution function display. A three-lens setup, as that

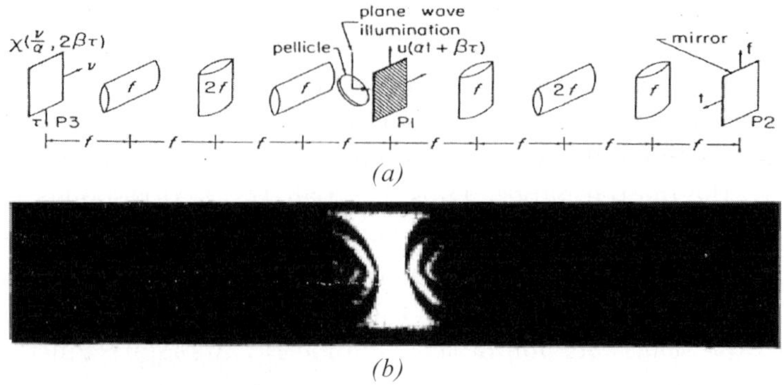

FIGURE 7.20. *(a)* A coherent processor for ambiguity function display using a single $1D$ input. *(b)* The ambiguity function (modulus squared) display of a single square pulse. (Reproduced with permission from [8.5].)

pictured in Fig. 7.16.d), and a totally reflecting mirror placed at Π yield, for instance, a suitable configuration to produce the vertical coordinate inversion of the object distribution from Π_i. In fact, it is evident that for the q direction the propagation from Π_i to the mirror and then back to Π_i is equivalent to the propagation through a cascade of four Fourier transforming systems, which produces a unit magnification imaging operation; hence, in the q direction the object distribution remains unaffected. In contrast, in the vertical direction lens L_2 generates a cascade of two Fourier transformers, which produces the desired coordinate inversion.

Figure 7.20 reproduces from [8.5] the single-input setup for ambiguity function display and the obtained squared modulus of the ambiguity function of a single square pulse. Here, $\chi(\nu, \tau)$ denotes the asymmetric ambiguity function defined by (7.4.3) with ν corresponding to $\frac{\widetilde{\kappa}}{2\pi}$, and the factor $\frac{1}{2\pi}$ being omitted. The quantities α and β specify the rotation angle of the transparency with respect to the reference plane: $\alpha = \cos\theta$ and $\beta = \sin\theta$. Evidently, the planes P_1, P_2 and P_3 correspond to the planes, Π_i, Π and Π_o of Fig. 7.19.

Another scheme is shown in Fig. 7.21, reproducing from [12.3] the single-input setup and the resulting Wigner distribution function display of a slit. The left arm of the processor comprises a $2D$ Fourier transformer and a roof top prism, whose reflection and one-coordinate inversion properties are sketched in the part (c) of the figure. As an overall effect of the propagation through the lens-prism arrangement, the rotated transparency $\varphi(\frac{q+\widetilde{q}}{\sqrt{2}})$ is imaged into the oppositely rotated transmittance $\varphi(\frac{q-\widetilde{q}}{\sqrt{2}})$. The pair (x, y) in [12.3] corresponds to the pair $(q, \widetilde{q})$ of our analysis as ν to $\frac{\widetilde{\kappa}}{2\pi}$; also, the Wigner distribution function $W(x, \nu)$ is defined by the expression (6.2.3) without the factor $\frac{1}{2\pi}$.

We observe that use of a single transparency representing the signal eliminates the problem of producing two identical object transparencies and of maintaining them aligned with respect to one another. However, part of the input power is wasted in passing through the beam splitter twice, and also the number of optical components in the processor is increased.

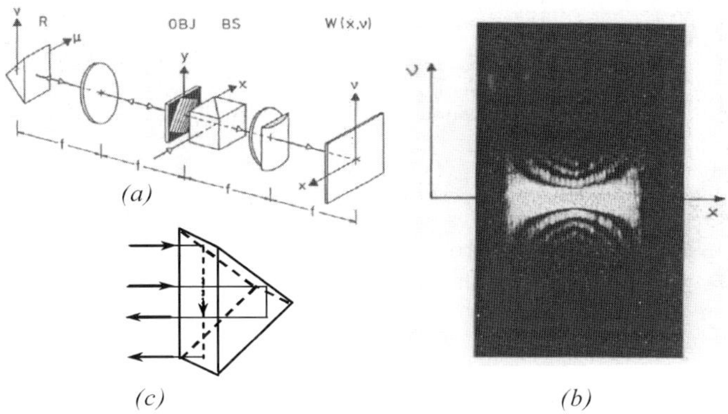

FIGURE 7.21. *(a)* Optical setup for displaying the Wigner distribution function of a real $1D$ signal using a single input transparency (BS: beam splitter, R: roof top prism), by which *(b)* the Wigner distribution function of a slit has been obtained. (Reprinted with permission from [12.3].) *(c)* Schematic of a roof top prism.

As suggested in [12.10], an appropriate polarization-sensitive setup can be devised to avoid the loss of power, normally consequent to the use of beam-splitting devices. The optical scheme, described in [12.10], makes use of a cubic polarizing beam splitter, which in combination with two retardation plates is able to convey the incident beam firstly into the π_s-product function modelling processor and then into the parallel imaging/Fourier transforming processor with an almost intact power. A polarizing beam-splitter cube normally separates the incident beam into two linearly polarized beams propagating at right angle to each other and with complementary states of polarization. The beam which straightly passes through the cube emerges as a p-polarized beam, the electric field oscillating in a plane parallel to the plane of incidence. The beam transmitted at the right angle to the incident beam emerges as an s-polarized beam with the plane of the electric vector being orthogonal to the incidence plane. The state of polarization of the incident light governs the distribution of the optical power among the two exiting beams.

In the experimental setup of [12.10], whose sensitive beam splitting section is reproduced in Fig. 7.22, the collimated linearly polarized coherent beam from a He-Ne laser is made to emerge from a properly oriented $\frac{\lambda}{2}$ plate as a pure p-polarized beam. Thus, it can be totally transmitted straightly by the polarizing cube into the π_s-function modelling section, which is designed in conformity

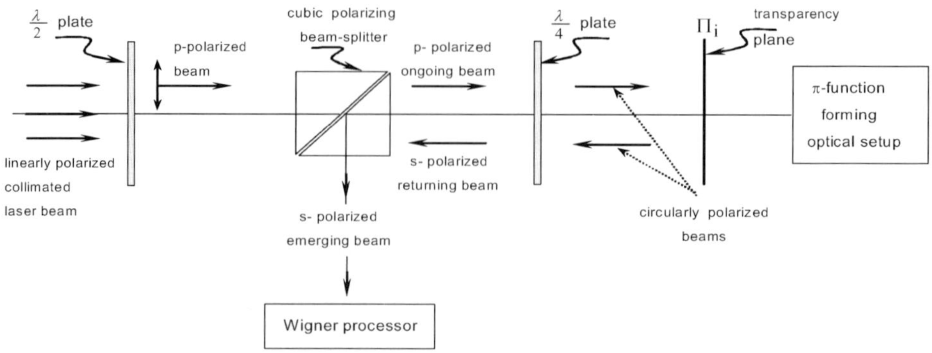

FIGURE 7.22. Optical scheme for the generation of the Wigner distribution function of a $1D$ or a $2D$ object using a single transparency. It exploits a suitable arrangement of polarization sensitive devices to convey almost the whole optical power of the incident beam into the Wigner processor (inspired by [12.10]).

with the above illustrated optical schemes. A $\frac{\lambda}{4}$ plate is inserted between the polarizing beam splitter cube and the object plane Π_i with its optic axis aligned at 45° to the vertical axis. Thus the initial incident p-polarized beam, passing twice through the $\frac{\lambda}{4}$ plate, returns to the polarizing beam-splitter cube as a pure s-polarized beam. Hence it emerges from the beam-splitter at 90° with respect to the returning beam and with almost the same input power as the original beam. Of course, the right angle section of the optical scheme is designed to perform the parallel imaging/Fourier transforming operation, which serves to generate the Wigner distribution function of the $1D$ object under examination as previously described.

Finally we note that the single-input setup is less flexible than the double-input arrangement since its use is limited to the display of the self-Wigner distribution function (and self-ambiguity function). Cross-Wigner distribution and cross-ambiguity function operations require, of course, two transparencies representing the two multiplying signals entering the pertinent integrals.

7.8 Wigner processor for $1D$ complex signals: basic configurations

In many practical situations $1D$ (or $2D$) complex signals are encountered. Evidently when dealing with complex signals one needs to produce the complex conjugate version of the signal in order to model the product function $\pi_s(q,\widetilde{q}) \propto \varphi^*(q - \frac{\widetilde{q}}{2})\varphi(q + \frac{\widetilde{q}}{2})$. We describe here the basics of some of the proposed methods to effect process i); of course, process ii) can be effected by the same optical arrangements, described in the foregoing paragraphs.

7.8.1 π_s-product function modelling setups for 1D complex signals

A certain number of π_s-function modelling arrangements for complex signals have been proposed and implemented [12].

A proposed method makes use of a (possibly computer generated) hologram of the signal, replacing the transparency at the object plane (OBJ) in the optical setup of Fig. 7.21.a) [12.3]. The hologram contains information about both the signal and its complex conjugate. Thus, by use of suitable masks, it is possible to select the information about φ or φ^*, as desired. We may suppose, for instance, that under coherent illumination in the first passage through the hologram (appropriately rotated according to the previous discussion) the reconstructed image of $\varphi(\frac{q-\widetilde{q}}{\sqrt{2}})$ is transmitted, which image is returned back as $\varphi(\frac{q+\widetilde{q}}{\sqrt{2}})$, whilst in the second passage the complex conjugate $\varphi^*(\frac{q-\widetilde{q}}{\sqrt{2}})$ is delivered. The obtained product function $\varphi^*(\frac{q-\widetilde{q}}{\sqrt{2}})\varphi(\frac{q+\widetilde{q}}{\sqrt{2}})$ is then input into the Wigner distribution function processor on the right of the object plane.

Alternatively, an interference based method can be used to effect the wanted phase conjugation of the complex signal. In fact, the interference pattern of the sum of two signals when yielded by an intensity recording material like, for instance, a holographic film, always contains, at its +/− first order, the product of the amplitude of one signal and the complex conjugate amplitude of the other and vice versa. Using therefore a non linear material, one can produce the required product function to be used for the Wigner distribution display of the original signal. Also, illuminating the complex object at a certain angle may improve the visibility of the final Wigner distribution function display.

Referring to Fig. 7.23.a), inspired by [12.3], we may suppose to illuminate the complex object representing the rotated signal $\varphi(\frac{q+\widetilde{q}}{\sqrt{2}})$ by a monochromatic plane wave at a certain angle θ_i with respect to the axis, thus producing the signal $\varphi(\frac{q+\widetilde{q}}{\sqrt{2}})e^{ik\widetilde{q}\sin\theta_i}$ behind the object plane. The signal is then separated in two parts at 90° apart by a beam splitter. One part is reflected back to the beam-splitter without change by a mirror. The other is reflected back as well, but inverted with respect to the vertical direction by a roof top prism; it is then formally represented as $\varphi(\frac{q-\widetilde{q}}{\sqrt{2}})e^{-ik\widetilde{q}\sin\theta_i}$. The two parts are again combined and added on some recording material yielding the modulus squared,

$$I(q,\widetilde{q}) \propto \left|\varphi(\tfrac{q+\widetilde{q}}{\sqrt{2}})e^{ik\widetilde{q}\sin\theta_i} + \varphi(\tfrac{q-\widetilde{q}}{\sqrt{2}})e^{-ik\widetilde{q}\sin\theta_i}\right|^2. \qquad (7.8.18)$$

Evidently it contains the self-terms $|\varphi(\frac{q+\widetilde{q}}{\sqrt{2}})|^2$ and $|\varphi(\frac{q-\widetilde{q}}{\sqrt{2}})|^2$, yielding two $\widetilde{q}$-mirrored contributions proportional to the Fourier transform of the instantaneous power $|\varphi(q)|^2$ at frequency $\kappa = \sqrt{2}k\frac{\widetilde{q}}{f}$, modulated by a q-depending

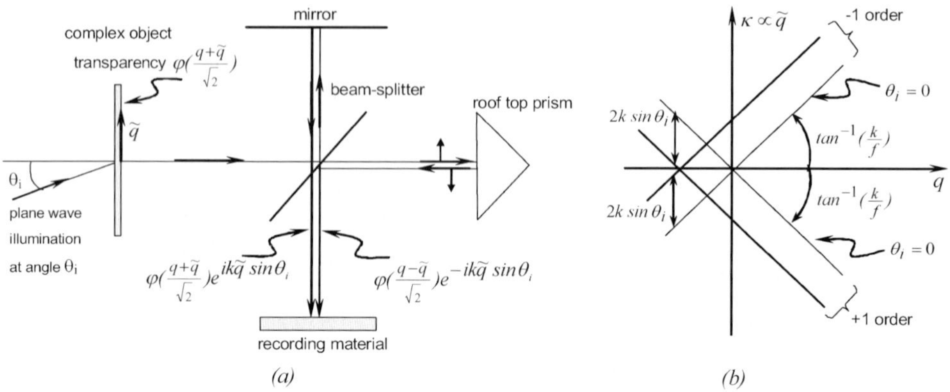

FIGURE 7.23. *(a)* Schematic for the Wigner display of a $1D$ complex signal. *(b)* The delta lines produced on the output plane by the $+/-$ order terms written on the recording film in the Wigner processor scheme of part *(a)* in the case of a lens-like quadratic phase waveform.

chirp. Besides, it contains the interference terms, each conveying just the desired complex product function multiplied by an exponential. Thus, each term produces a scaled version of the Wigner distribution function of the signal under examination. Precisely, the $+/-$ first order terms respectively yield

$$\varphi^*(\tfrac{q-\tilde{q}}{\sqrt{2}})\varphi(\tfrac{q+\tilde{q}}{\sqrt{2}})e^{2ik\tilde{q}\sin\theta_i} \longrightarrow \mathcal{W}_\varphi(-\tfrac{q}{\sqrt{2}}, \tfrac{\kappa-2k\sin\theta_i}{\sqrt{2}}), \tag{7.8.19}$$

and

$$\varphi(\tfrac{q-\tilde{q}}{\sqrt{2}})\varphi^*(\tfrac{q+\tilde{q}}{\sqrt{2}})e^{-2ik\tilde{q}\sin\theta_i} \longrightarrow \mathcal{W}_\varphi(-\tfrac{q}{\sqrt{2}}, -\tfrac{\kappa+2k\sin\theta_i}{\sqrt{2}}). \tag{7.8.20}$$

As noted, the spacing between the two distributions is controlled by the illumination angle θ_i. In Fig. 7.23.b) the expected view of the two interference-arising traces in the output plane is shown in the case of a $1D$ quadratic-phase waveform as that produced by a cylindrical aberration-free thin lens. It can be compared with the Wigner display of a cylindrical lens, shown in Fig. 7.24.b), which has been obtained using the setup sketched in Fig. 7.24.a), both reproduced from [12.3]; here, a holographic film has been employed as a recording device. In part (c) of the figure the Wigner distribution function of a third order phase object produced with the same setup is shown. The notation in the figure is as described in relation to Fig. 7.21.

Moreover, as described in [12.11], optical phase conjugation via a nonlinear four-wave mixing process can be used to model the product of the signal and its complex conjugate. To this end, a collimated monochromatic plane wave is separated into three beams, which are then suitably redirected to a third-order optical nonlinear material. As a result of the interaction of the three beams inside the material, a fourth beam is generated which carries the

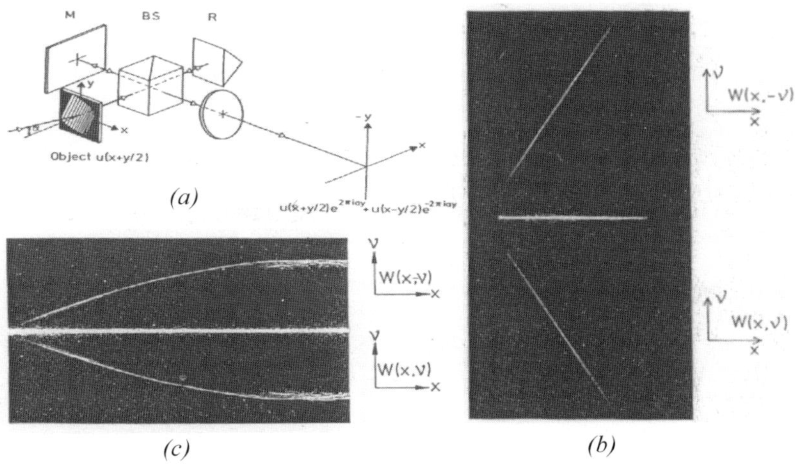

FIGURE 7.24. Wigner dispay for truly complex signals. M: mirror, BS: beam splitter, R: roof top prism. Wigner distribution function of *(b)* a cylindrical lens nad *(c)* a third order phase function, both produced with the setup in *(a)*. (Reprinted with permission from [12.3].)

product of the three ongoing beams, one of them entering the product through its complex conjugate version. Referring to the scheme of Fig. 7.25, adapted from [12.11] to the present discussion, we see that the complex object is inserted between the beam splitters B_1 and B_2. It may be supposed to be appropriately oriented in the input plane of the processor in order to produce the π_s-function factor $\varphi(\frac{q-\tilde{q}}{\sqrt{2}})$ upon illumination from the left by the undeflected portion of the ongoing monochromatic plane wave. The reflected portion of the incoming wave at B_1 is made to follow an appropriate path up to the nonlinear material. On the other hand, the signal $\varphi(\frac{q-\tilde{q}}{\sqrt{2}})$ is separated into two identical copies by the beam-splitter B_2. One copy is imaged without change to the nonlinear material along the path B_2-B_3-NLM. The other copy is conveyed through the path B2-M2-NLM, along which the required coordinate inversion is effected (by a Dove prism), thus reaching the non linear material in counterpropagating with respect to the unaltered illumination signal. The signal $s(q,\tilde{q})$ produced by the nonlinear material carries just the product

$$s(q,\tilde{q}) \propto \varphi(\tfrac{q+\tilde{q}}{\sqrt{2}})\varphi^*(\tfrac{q-\tilde{q}}{\sqrt{2}}),$$

travelling opposedly to the signal copy $\varphi(\frac{q-\tilde{q}}{\sqrt{2}})$. It can then be input into a strictly Wigner distribution function processor.

The details of the optical setup, described in [12.11], have been disregarded in the basic scheme of Fig. 7.25. The reader is directed to the quoted paper for a view of the complete scheme of the optical setup used to display the Wigner

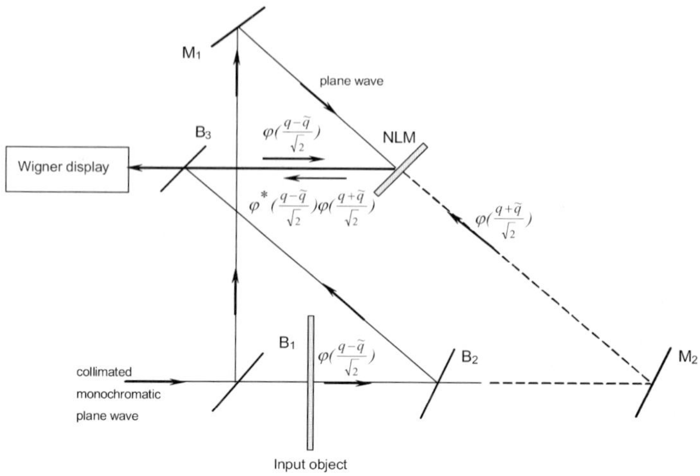

FIGURE 7.25. Schematic of the Wigner processor section (described in [12.11]) forming the π_s-product function for a complex signal.

distribution function and with slight modifications the ambiguity function.

7.9 The smoothed Wigner distribution function and the cross-ambiguity function: optical production

Following the suggestion of [12.2], we describe the optical implementation of the smoothed Wigner distribution function in particular by the Wigner distribution function of a Gaussian signal. The related discussion will further confirm the link, basically rooted in the Moyal relation, among all the concepts introduced and illustrated in the previous sections, i.e. the spectrogram, the (smoothed) Wigner distribution function and the (cross) ambiguity function.

We resort to the Gabor elementary signal (7.2.16), whose Wigner distribution function, as shown in § 6.2.1, shapes as a $2D$ real Gaussian centered at the point (q_0, κ_0) in the space-frequency plane with relative rms widths $\frac{a}{\sqrt{2}}$ and $\frac{1}{\sqrt{2}a}$ in the q and κ directions, respectively. Namely,

$$\mathcal{W}_{\mathsf{G}}(q,\kappa) = \frac{1}{\pi} e^{-\frac{(q-q_0)^2}{a^2} - a^2(\kappa-\kappa_0)^2}. \tag{7.9.1}$$

The overlap of any Wigner distribution function $\mathcal{W}_\varphi$ with $\mathcal{W}_{\mathsf{G}}$, i.e.,

$$\langle \mathcal{W}_\varphi \rangle_{\varphi_{\mathsf{G}}} \equiv \iint dq d\kappa \mathcal{W}_\varphi(q,\kappa)\mathcal{W}_{\mathsf{G}}(q,\kappa), \tag{7.9.2}$$

is evidently equivalent to the convolution of $\mathcal{W}_\varphi$ with the at the origin centered

$2D$ Gaussian $\frac{1}{\pi}e^{-\frac{q^2}{a^2}-a^2\kappa^2}$. In fact,

$$\langle \mathcal{W}_\varphi \rangle_{\varphi_{\mathsf{G}}}(q,\kappa;q_0,\kappa_0,a) = \frac{1}{\pi}\iint dq d\kappa \mathcal{W}_\varphi(q,\kappa) e^{-\frac{(q-q_0)^2}{a^2}-a^2(\kappa-\kappa_0)^2}, \tag{7.9.3}$$

where the dependence on the parameters of the Gabor signal q_0,κ_0 and a is explicitly displayed. Typically the convolution of $\mathcal{W}_\varphi$ with another appropriately chosen Wigner distribution function is reported as the smoothed Wigner distribution function. The smoothing has been suggested to overcome the problem of the in general non positiveness of the Wigner distribution function (see § 6.3.6) as well as that of the cross-terms contributions (see § 6.3.4).

In accord with the Moyal formula, the integral in Eq. (7.9.3) is related to the inner product of the signal $\varphi(q)$ with the Gabor function φ_{G}. Namely,

$$\langle \mathcal{W}_\varphi \rangle_{\varphi_{\mathsf{G}}}(q,\kappa;q_0,\kappa_0,a) = \frac{1}{2\pi}\left|\int \varphi(q)\varphi_{\mathsf{G}}(q)dq\right|^2. \tag{7.9.4}$$

Using the explicit expression of $\varphi_{\mathsf{G}}(q)$, a trivial manipulation of the integral above yields the enlightening relation,

$$\langle \mathcal{W}_\varphi \rangle_{\varphi_{\mathsf{G}}}(q,\kappa;q_0,\kappa_0,a) = 2\pi|\mathcal{A}_{\varphi_{\mathsf{G}},\varphi}(q_0,\kappa_0)|^2, \tag{7.9.5}$$

between the smoothed Wigner distribution function and the modulus squared of the cross-ambiguity function $\mathcal{A}_{\varphi_{\mathsf{G}},\varphi}$ of the Gabor signal and the signal under investigation, measured at the space and frequency mean values, q_0 and κ_0, of $|\varphi_{\mathsf{G}}(q)|^2$ and its spectrum $|\widetilde{\varphi}_{\mathsf{G}}(\kappa)|^2$. This result conforms with the analysis of § 7.2.3. Also, it is a particular case of the general result linking the convolution of two cross-Wigner distribution functions to the product of two cross-ambiguity functions, the involved signals in the former being crossly coupled in the latter (see [6.4] and Problem 10). As noted in § 7.2.3, the smoothed Wigner distribution function (7.9.2) can also be interpreted as a windowed Wigner distribution function. Therefore, the phase-space concentration of $\langle \mathcal{W}_\varphi \rangle_{\varphi_{\mathsf{G}}}$ is controlled by the windowing distribution $\mathcal{W}_{\mathsf{G}}$, and hence the position resolution is determined by the rms width of φ_{G} and similarly the frequency resolution is determined by the corresponding diffraction angle.

On account of relation (7.9.5), the implementation of the smoothed Wigner distribution function can be realized by an ambiguity function processor, and so by any one of the previously described optical setups with the roles of the q and $\widetilde{q}$ coordinates being interchanged. Also, two different signals are written on the two transparencies, which are rotated by 90° one to the other, as previously described. The signal under study is indeed written on one transparency, whilst for the Gaussian amplitude an appropriately rotated Gaussian transmission mask can be used or as well the Gaussian profile output from a laser.

Figure 7.26 reproduces from [12.2] the smoothed Wigner distribution function of a monofrequency signal of finite duration and a linearly increasing frequency signal. The symbols in these figures are as in Fig. 7.21.

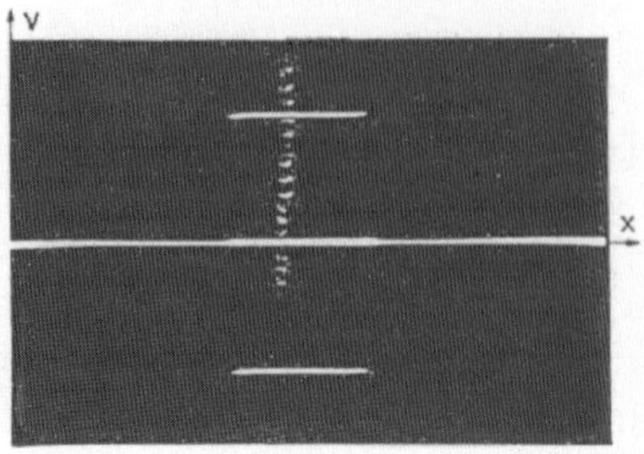

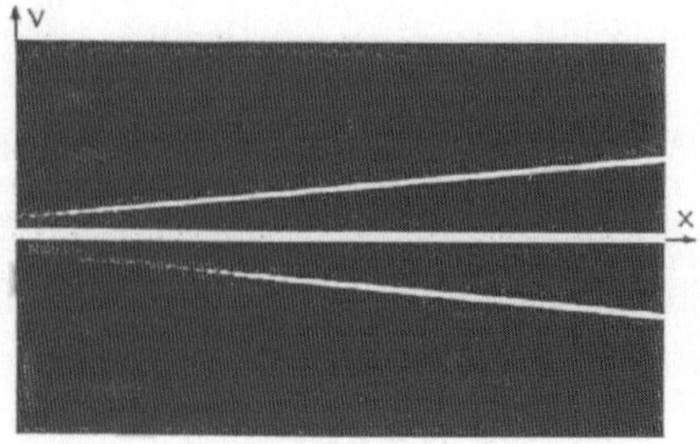

FIGURE 7.26. Smoothed Wigner distribution function of *(a)* a cosine signal of finite length and *(b)* a linearly incresing frequency signal. (Reprinted with permission from [12.2].)

7.10 Summary

The instance of the individualization of a method to construct the local frequency spectrum of a signal has been introduced. In this connection, the sliding-window Fourier transform and the spectrogram arise quite naturally as operating the piece-by-piece inspection of the signal with an external probe.

The ambiguity function and the Wigner distribution function come to be a kind of windowed Fourier transform, when the signal itself or its mirrored version are used as analyzing window. In spite of the similarity of the mathematical definitions, the information is encoded in the two signal representations very differently. As the Wigner distribution function is described as a kind of space-frequency convolution, the ambiguity function is in turn described as a kind of space-frequency correlation.

The Wigner distribution and ambiguity functions have been seen to be related by a Fourier transform to the mutual intensity function, thus introducing their conceptual and applicative potential into the partial coherence theory.

Finally, the optical architectures realized to implement the "mathematics" of the Wigner integral (and, by slight modifications, of the ambiguity function integral as well) have been described, restrictedly to $1D$ signals.

In the next chapter we will resort to the original quantum mechanical context, giving account of the formal concept of the Wigner distribution function on the basis of definite requirements formulated in the proper terminology of quantum mechanics. This completes the quantum mechanical view of the Wigner representation and evidences the peculiar role of (classical) optics comprising the instances of both quantum mechanics and signal theory.

Problems

1. *(a)* Verify that the sliding-window Fourier transform is invariant with respect to shifts of the signal in space (up to a modulation) and frequency; namely,

$$\begin{array}{llllll} \text{if} & \varphi(q) \longrightarrow \varphi(q-q_0), & \text{then} & \mathfrak{F}_\varphi(q,\kappa;\mathsf{w}) & \longrightarrow & \mathfrak{F}_\varphi(q-q_0,\kappa;\mathsf{w})\,e^{-i\kappa q_0}, \\ \text{if} & \varphi(q) \longrightarrow \varphi(q)e^{i\kappa_0 q}, & \text{then} & \mathfrak{F}_\varphi(q,\kappa;\mathsf{w}) & \longrightarrow & \mathfrak{F}_\varphi(q,\kappa-\kappa_0;\mathsf{w}). \end{array}$$

Compare with the corresponding relations for the Wigner distribution and the ambiguity function. (b) Show that the following representation of the signal,

$$\varphi(q) = \tfrac{1}{\sqrt{2\pi}} \iint \mathfrak{F}_\varphi(q',\kappa';\mathsf{w})[s(q-q')e^{i\kappa' q}]dq'd\kappa',$$

in terms of the relevant windowed spectrum is actuable for any function $s(q)$ such that $\int s(q)\mathsf{w}^*(q)dq = 1$. This representation generalizes (7.2.3), assigning complementary roles to the function $\mathsf{w}(q)$ as "analysis" window, and to $s(q)$ as "synthesis" window. As a natural choice, $s(q)$ is taken as $\mathsf{w}(q)$, thus leading to (7.2.3). *(c)* Prove that the condition for the representability of any function defined over the (q,κ) plane as a sliding-window spectrum of a signal $\varphi(q)$ writes as

$$\mathfrak{F}_\varphi(q,\kappa;\mathsf{w}) = \tfrac{1}{2\pi} \iint dq'd\kappa' \mathfrak{F}_\varphi(q',\kappa';\mathsf{w}) \int dq'' \mathsf{w}^*(q''-q)\mathsf{w}(q''-q')e^{-iq''(\kappa-\kappa')},$$

for any function $\mathsf{w}(q)$ such that $\int |\mathsf{w}|^2(q)dq = 1$.

2. Prove that the sliding-window Fourier transform $\mathfrak{F}_{\text{chirp}}(q,\kappa;\mathsf{w}_\text{R})$ of the chirp (7.2.9) with a unit-energy rectangular window can also be written as

$$\mathfrak{F}_{\text{chirp}}(q,\kappa;\mathsf{w}_\text{R}) = \sqrt{\tfrac{2i}{\pi\beta Q_\mathsf{w}}}\, e^{\frac{i}{2}(\beta q^2 - 2q\kappa)} \sum_{m=0}^{\infty} \tfrac{1}{(2m+1)!} \left(\sqrt{\tfrac{i\beta}{2}} Q_\mathsf{w}\right)^{2m+1} H_{2m}\left(\tfrac{\kappa-\beta q}{\sqrt{2i\beta}}\right),$$

where H_{2m} is the Hermite polynomial of degree $2m$ [13].

3. Prove that *(a)* if the kernel $\Psi(\eta,\nu)$ does not depend on the space and frequency variables, shifts of the signal in space or frequency result in corresponding shifts of the associated distribution, i.e.,

$$\varphi(q) \longrightarrow \varphi(q-q_0) \quad \text{yields} \quad C_\varphi(q,\kappa;\Psi) \longrightarrow C_\varphi(q-q_0,\kappa;\Psi), \quad \text{if } \Psi(\eta,\nu) \text{ is independent of } q,$$

$$\varphi(q) \longrightarrow \varphi(q)e^{i\kappa_0 q} \quad \text{yields} \quad C_\varphi(q,\kappa;\Psi) \longrightarrow C_\varphi(q,\kappa-\kappa_0;\Psi), \quad \text{if } \Psi(\eta,\nu) \text{ is independent of } \kappa,$$

and *(b)* if the kernel $\Psi(\eta,\nu)$ is signal-independent and depends on its arguments through the product: $\Psi(\eta,\nu) = \nu\eta$, the corresponding signal representation is scale invariant, and hence

$$\text{with } \varphi(q) \longrightarrow \sqrt{\mathfrak{a}}\varphi(\mathfrak{a}q),\ \mathfrak{a}>0 \quad \text{one has} \quad C_\varphi(q,\kappa;\Psi) \longrightarrow C_\varphi(\mathfrak{a}q,\tfrac{\kappa}{\mathfrak{a}};\Psi).$$

4. Verify that *(a)* the Moyal relation for the ambiguity function writes in the form

$$\iint d\widetilde{q}d\widetilde{\kappa}\, \mathcal{A}^*_{\varphi_1}(\widetilde{q},\widetilde{\kappa})\mathcal{A}_{\varphi_2}(\widetilde{q},\widetilde{\kappa}) = \tfrac{1}{2\pi}\left|\int \varphi_1^*(q)\varphi_2(q)dq\right|^2.$$

and *(b)* shifts of the signal in space and frequency result in linear phase factors in the ambiguity function, i.e.,

$$\begin{array}{llll} \text{if} & \psi(q) = \varphi(q-q_0), & \text{then} & \mathcal{A}_\psi(\widetilde{q},\widetilde{\kappa}) = e^{-i\widetilde{\kappa}q_o}\mathcal{A}_\varphi(\widetilde{q},\widetilde{\kappa}); \\ \text{if} & \widetilde{\phi}(\kappa) = \widetilde{\varphi}(\kappa-\kappa_0), & \text{then} & \mathcal{A}_\phi(\widetilde{q},\widetilde{\kappa}) = e^{i\widetilde{q}\kappa_o}\mathcal{A}_\varphi(\widetilde{q},\widetilde{\kappa}). \end{array}$$

5. Prove that the ambiguity function of *(a)* the sum of two rectangle-signals of the same width $2Q$ centered on the positions q_1 and q_2 writes as

$$\mathcal{A}(\widetilde{q},\widetilde{\kappa}) = e^{-i\kappa q_{\rm m}}\{2\mathcal{A}_{\rm R}(\widetilde{q},\widetilde{\kappa})\cos(\kappa\tfrac{q_{\rm d}}{2}) + \mathcal{A}_{\rm R}(\widetilde{q}+q_{\rm d},\widetilde{\kappa}) + \mathcal{A}_{\rm R}(\widetilde{q}-q_{\rm d},\widetilde{\kappa})\},$$

where $q_{\rm m} = (q_1 + q_2)/2$ and $q_{\rm d} = q_1 - q_2$; and *(b)* a diverging spherical wave issuing from a slit of width $2Q$, i.e., $\varphi(q) = {\rm rect}(\frac{q}{2Q})e^{i\frac{\beta}{2}q^2}$, is

$$\mathcal{A}(\widetilde{q},\widetilde{\kappa}) = \mathcal{A}_{\rm R}(\widetilde{q},\widetilde{\kappa}-\beta\widetilde{q}).$$

6. *(a)* Show that the representative $\mathcal{C}_\varphi(q,\kappa;\Psi)$ of the joint distributions (7.3.3) can be expressed in terms of the Wigner distribution function by a $2D$ convolution, like

$$\mathcal{C}_\varphi(q,\kappa;\Psi) = \tfrac{1}{2\pi}\iint dq'd\kappa'\Phi(q-q',\kappa-\kappa')\mathcal{W}_\varphi(q',\kappa'),$$

with

$$\Phi(q,\kappa) = \tfrac{1}{2\pi}\iint d\eta d\nu\Psi(\nu,\eta)e^{i(\nu q-\kappa\eta)}.$$

This means that any member of Cohen's class (7.3.3) can be obtained by a linear transformation of the Wigner distribution function of the signal at hand, with an "impulse response" $\Phi(q,\kappa)$ that is related to the kernel $\Psi(\nu,\eta)$ by a double Fourier transform. *(b)* Prove also the alternative expression

$$\mathcal{C}_\varphi(q,\kappa;\Psi) = \tfrac{1}{2\pi}\iint d\eta d\nu\Psi(\nu,\eta)e^{i(\nu q-\kappa\eta)}\mathcal{A}_\varphi(\eta,\nu),$$

linking the generic $\mathcal{C}_\varphi(q,\kappa;\Psi)$ to the ambiguity function $\mathcal{A}_\varphi$ through a product and a double Fourier-transform relation. *(c)* As an obvious consequence, verify that, defining the *dual* of $\mathcal{C}_\varphi(q,\kappa;\Psi)$ by the same double Fourier-like relation linking $\mathcal{W}$ and $\mathcal{A}$, i.e.,

$$\mathcal{D}_\varphi(\widetilde{q},\widetilde{\kappa};\Psi) = \tfrac{1}{2\pi}\iint \mathcal{C}_\varphi(q,\kappa;\Psi)e^{i(\kappa\widetilde{q}-q\widetilde{\kappa})}dqd\kappa,$$

if the kernel $\Psi(\nu,\eta)$ is position, frequency and signal independent, we obtain for $\mathcal{D}_\varphi$ the multiplicative form

$$\mathcal{D}_\varphi(\widetilde{q},\widetilde{\kappa};\Psi) = \Psi(\widetilde{\kappa},\widetilde{q})\mathcal{A}_\varphi(\widetilde{q},\widetilde{\kappa}).$$

This is a key relation for optimizing the smoothing of the Wigner distribution function aimed at suppressing the undesired cross-term contributions.

7. Prove that, if the signal is restricted to the space interval $(-Q,Q)$, the support of the ambiguity function extends through the $\widetilde{q}$ dimension from $-2Q$ to $2Q$, and so is double of that of the signal itself. A similar result holds in the frequency domain, the $\widetilde{\kappa}$-extension of $\mathcal{A}(\widetilde{q},\widetilde{\kappa})$ being double of that of the signal spectrum.

8. Deduce the transfer laws for the Wigner distribution and the ambiguity function under free-propagation into the Fraunhofer region exploiting relations (7.5.27), linking $\mathcal{W}$, $\mathcal{A}$ and J, and the transformation relation (7.5.26) of J for Fraunhofer diffraction.

9. As alternatives to the interpretations (7.5.4),(7.5.6) and (7.5.7)), verify by a simple manipulation, that the Wigner and ambiguity integrals can as well be considered as resulting from a process of multiplication, and a process of convolution in the case of the Wigner distribution and correlation in the case of the ambiguity function.

10. Prove that the cross-correlation of two cross-Wigner distribution functions equals the product of two cross-ambiguity functions, the involved signals in the former being correspondingly coupled in the latter as

$$\iint dq'd\kappa'\mathcal{W}^*_{\Phi_1,\Phi_2}(q'-q,\kappa'-\kappa)\mathcal{W}_{\varphi_1,\varphi_2}(q',\kappa') = 2\pi\mathcal{A}^*_{\Phi_1,\varphi_1}(q,\kappa)\mathcal{A}_{\Phi_2,\varphi_2}(q,\kappa).$$

References

[1] N.G. De Bruijn, "Uncertainty principles in Fourier analysis", in *Inequalities*, O.Shisha (ed.) (Academic Press, New York, 1967), pp. 57-71.

[2] Specific considerations on the windowed Fourier spectrum and the spectrogram within the general context of the joint (space-frequency or time-frequency) signal representations can be found in [4.3]-[4.8]. In particular, in [4.7], ch. 7, pp. 93-112, and in [4.8], ch. 3, pp. 63-65, the windowed Fourier transform is discussed from the viewpoint of the time-frequency and space-frequency signal representations, respectively. The space-frequency view of the sliding-window Fourier transform being interpreted as a local frequency spectrum is elaborated in detail in [2.3]. Several examples of windows extensively used in Fourier spectrum estimation related problems are illustrated in detail in [2.2], [2.3] and in [6.3], ch. 7, pp. 234-243. We also suggest reading [10.16], ch. 4, with a comparative description of four representative window functions being specifically addressed on pp. 160-163.

[2.1] J.B. Allen and L.R. Rabiner, "A unified approach to short-time Fourier analysis and synthesis", Proc. IEEE **65**, 1558-1564 (1977).

[2.2] F.J. Harris, "On the use of windows for harmonic analysis with the discrete Fourier transform", Proc. IEEE **66**, 51-83 (1978).

[2.3] M.J. Bastiaans, "Signal description by means of a local frequency spectrum", in *Transformations in Optical Signal Processing*, W.T. Rhodes, J.R. Fienup and B.E.A. Saleh (eds.) SPIE **373**, 49-62 (1981).

[2.4] S.H. Nawab and T.F. Quatieri, "Short-time Fourier transform", in *Advanced Topics in Signal Processing*, J.S. Lim and A.V. Oppenheim (eds.) (Prentice Hall, Englewood Cliffs, NJ, 1988), pp. 289-337.

[2.5] J. Jeong and W.J. Williams, "Mechanism of the cross-terms in spectrograms", IEEE Trans. Acoust., Speech, Signal Process. **40**, 2608-2613 (1992).

[3] The original papers by Gabor are quoted in [19] of ch. 4. For accurate investigations of Gabor's signal expansion in relation with the determination of the analysis function for a given synthesis function as well as the application of Gabor's concepts to discrete time signals and to the case of oversampling we suggest the papers below. Discussions about Gabor's signal expansion are also in [2.3], [4.5], [4.7], ch. 13, pp. 186-187, and in [4.8], ch. 3, pp. 65-67.

[3.1] M.J. Bastiaans, "Gabor's expansion of a signal into Gaussian elementary signals", Proc. IEEE **68**, 538-539 (1980).

[3.2] M.J. Bastiaans, "A sampling theorem for the complex spectrogram, and Gabor's expansion of a signal in Gaussian elementary signals", Opt. Eng. **20**, 594-598 (1981).

[3.3] M.J. Bastiaans, "Gabor's signal expansion and degrees of freedom of a signal", Optica Acta **29**, 1223-1229 (1982).

[3.4] M.J. Bastiaans, "Gabor's signal expansion applied to partially coherent light", Opt. Comm. **86**, 14-18 (1991).

[3.5] M.J. Bastiaans, "Gabor's signal expansion and Zak transform", Appl. Opt. **33**, 5241-5255 (1994).

[3.6] M.J. Bastiaans and M.C.W. Geilen, "On the discrete Gabor transform and the discrete Zak transform", Signal Process. **49**, 151-166 (1996).

[4] The general expression (7.3.3) for the joint signal representations was firsly given in [4.1] within a quantum mechanical context. The relations between some specific representatives of (7.3.3) with signal-independent kernels are illustrated at diverse depths in many papers and textbooks; a few of them are quoted below. In particular we signalize the comprehensive issue [4.9].

[4.1] L. Cohen, "Generalized phase space distribution functions", J. Math. Phys. **7**, 781-786 (1966).

[4.2] L. Cohen, "Quantization problem and variational principle in the phase-space formulation of quantum mechanics", J. Math. Phys. **17**, 1863-1866 (1976).

[4.3] T.C.A.M. Claasen and W.F.G. Mecklenbräuker, "The Wigner distribution - a tool for time-frequency signal analysis. III: relations with other time-frequency signal transformations", Philips J. Res. **35**, 372-389 (1980).

[4.4] L. Cohen, "Time-frequency distributions - A review", Proc. IEEE **77**, 941-981 (1989).

[4.5] L.D. Jacobson and H. Wechsler, "Joint spatial/spatial-frequency representation", Signal Process. **14**, 37-68 (1988).

[4.6] F. Hlawatsch and G.F. Boudreaux-Bartels, "Linear and quadratic time-frequency signal representations", IEEE Signal Process. Mag., 21-27, April 1992.

[4.7] L. Cohen, *Time-Frequency Analysis* (Prentice-Hall, Englewood Cliffs, 1995).

[4.8] H.M. Ozaktas, Z. Zalevsky and M. Alper Kutay, *The Fractional Fourier Transform with Applications in Optics and Signal Processing* (John Wiley & Sons Ltd., Chichester, 2001).

[4.9] *Time-Frequency Signal Analysis and Processing. A Comprehensive Reference.* B. Boashash (ed.) (Elsevier, Oxford, 2003).

[5] The literature on the joint signal representations is extremely rich, comprising as well specific investigations regarding, for instance, duality relation-based classification, optimization procedure for signal synthesis or kernel design, aimed at reducing the cross-terms contributions and at constructing positive distributions. Also, elaboration of new kernels to fulfil specific requirements, investigations of the relation between different signal representations, individualization of general operator procedure to generate joint signal representations of arbitrary variables, and so on. Here we give a short list of titles concerned with the above instances. A comprehensive account is offered by [4.9].

[5.1] H. Choi and W.J. Williams, "Improved time-frequency representation of multicomponent signal using exponential kernels", IEEE Trans. on Signal Process. **37**, 862-871 (1989).

[5.2] Y. Zhao, L.E. Atlas and R.J. Marks, II, "The use of cone-shape kernels for generalized time-frequency representation of nonstationary signals", IEEE Trans. on Signal Process. **38**, 1084-1091 (1990).

[5.3] F. Hlawatsch, "Duality and classification of bilinear time-frequency signal representations", IEEE Trans. on Signal Process. **39**, 1564-1574 (1991).

[5.4] F. Hlawatsch and W. Krattenthaler, "Bilinear signal synthesis", IEEE Trans. on Signal Process. **40**, 352-363 (1992).

[5.5] J. Jeong and W.J. Williams, "Kernel design for reduced interference distributions", IEEE Trans. on Signal Process. **40**, 401-412 (1992).

[5.6] A. Kumar, D.R. Fuhrmann, M. Frazier and B.D. Jawerth, "A new transform for time-frequency analysis", IEEE Trans. on Signal Process. **40**, 1697-1707 (1992).

[5.8] P.J. Loughlin, J. W. Pitton and L.E. Atlas, "Bilinear time-frequency representations: new insights and properties", IEEE Trans. on Signal Process. **41**, 750-766 (1993).

[5.9] R.G. Baraniuk and D.L. Jones, "A signal-dependent time-frequency representation: optimal kernel design", IEEE Trans. on Signal Process. **41**, 1589-1602 (1993).

[5.10] F. Hlawatsch and R.L. Urbanke, "Bilinear time-frequency representations of signals: the shift-scale invariant class", IEEE Trans. on Signal Process. **42**, 357-366 (1994).

[5.11] P.J. Loughlin, J. W. Pitton and L.E. Atlas, "Construction of positive time-frequency distributions", IEEE Trans. on Signal Process. **42**, 2697-2705 (1994).

[5.12] R.G. Baraniuk and L. Cohen, "On joint distribution of arbitrary variables", IEEE Signal Lett. **2**, 10-12 (1995).

[5.13] R.G. Baraniuk and D.L. Jones, "Unitary equivalence: a new twist on signal processing", IEEE Trans. Acoust., Speech, Signal Process. **43**, 2269-2282 (1995).

[5.15] F. Hlawatsch, T.G. Manickam, R.L. Urbanke and W. Jones, "Smoothed pseudo-Wigner distribution, Choi-Williams distribution and cone-kernel representation: ambiguity-domain analysis and experimental comparison", Signal Process. **43**, 149-168 (1995).

[5.16] R.G. Shenoy and T.W. Parks, "Wide-band ambiguity functions and affine Wigner distributions", Signal Process. **41**, 339-363 (1995).

[5.17] A.H. Costa and G.F. Boudreaux-Bartels, "Design of time-frequency representations using a multiform, tiltable exponential kernel", IEEE Trans. on Signal Process. **43**, 2283-2301 (1995).

[5.18] L. Cohen, "A general approach for obtaining joint representations in signal analysis-Part I: characteristic function operator method", IEEE Trans. on Signal Process. **44**, 1080-1090 (1996); Part II: general class, mean and local values, and bandwidth", *ibid.*, 1091-1098.

[5.20] R.G. Baraniuk, "Joint distributions of arbitrary variables made easy", Multdim. Sys. and Signal Process. **9**, 341-348 (1998).

[6] Many titles offer careful and thorough descriptions of the properties and the applicative potential of the ambiguity function within the context of the signal analysis. A few of them are quoted here. The reader is also addressed to [4.3]-[4.6], as well as to [4.7], where the ambiguity function is considered in relation with the characteristic function of the Wigner distribution function, and to [4.8], ch. 3, pp. 73-76. The Fourier transform duality of the Wigner distribution and ambiguity function is discussed in detail in [6.4]. Various aspects of the ambiguity function and its generalizations are as well investigated in some of refs. in [5].

[6.1] P. M. Woodward, *Probability and Information Theory with Applications to Radar* (Pergamon, New York, 1953).

[6.2] J.R. Klauder, "The design of radar signals having both high range resolution and high velocity resolution", Bell Syst. Tech. J. **39**, 809-820 (1960).

[6.3] A. Papoulis, *Signal Analysis* (McGraw-Hill, New York, 1977), ch. 8, pp. 284-295.

[6.4] H.H. Szu and J.A. Blodgett, "Wigner distribution and ambiguity function", in *Optics in Four Dimensions-1980,* AIP Conf. Proc. ***65***, M.A. Machado and L.M. Narducci (eds.) (AIP, New York, 1981), pp. 355-381.

[6.5] H.H. Szu and H.J. Caulfield, "The mutual time-frequency content of two signals", Proc. IEEE **72**, 902-908 (1984).

[7] Many applications of the ambiguity functions to optics have been devised starting from the seminal paper by Papoulis [7.1]. Below only the titles specifically concerned with the presentation in §§ 7.5.2-7.5.3 are reported.

[7.1] A. Papoulis, "Ambiguity function in Fourier optics", J. Opt. Soc. Am. **64**, 779-788 (1974).

[7.2] K. Dutta and J.W. Goodman, "Reconstructions of images of partially coherent objects from samples of mutual intensity", J. Opt. Soc. Am. **67**, 796-803 (1977).

[7.3] J.-P. Guigay, "The ambiguity function in diffraction and isoplanatic imaging by partially coherent beams", Opt. Comm. **26**, 136-138 (1978).

[7.4] K.-H. Brenner and J. Ojeda-Castañeda, "Ambiguity function and Wigner distribution applied to partially coherent imagery," Optica Acta **31**, 213-223 (1984).

[7.5] K.-H. Brenner, A.W. Lohmann and J. Ojeda-Castañeda, "The ambiguity function as a polar display of the OTF", Opt. Comm. **44**, 323-326 (1983).

[8] Among the wide literature concerned with the optical production of the ambiguity function we report a few representative papers, addressing the reader also to [6.4] and [12.4] for a comparative analysis of the ambiguity function implementation in relation respectively with the Wigner distribution function and the Cohen generalized space-frequency distribution functions.

[8.1] R.A.K. Said and D.C. Cooper, "Crosspath real-time optical correlator and ambiguity function processor", Proc. IEE **120**, 423-428 (1973).

[8.2] R.J. Marks II, J.F. Walkup and T.F. Krile, "Ambiguity function display: an improved coherent processor", Appl. Opt. **16**, 746-750 (1977). "Addendum", **16**, 1777 (1977).

[8.3] D. Casasent and B.V.K. Vijaya Kumar, "Optical image plane correlator for ambiguity surface computation", Appl. Opt. **18**, 1673-1678 (1979).

[8.4] D. Psaltis and D. Casasent, "Space-variant ambiguity function processor", Appl. Opt. **18**, 1869-1874 (1979).

[8.5] R. J. Marks II and M.W. Hall, "Ambiguity function display using a single $1D$ input", Appl. Opt. **18**, 2539-2540 (1979).

[8.6] T.C. Lee, J. Rebholz, P.N. Tamura and J. Lindquist, "Ambiguity processing by joint Fourier transform holography", Appl. Opt. **19**, 895-899 (1980).

[8.7] P.N. Tamura, J.J. Rebholz and T.C. Lee, "Ambiguity-function generation using passive optical τ-shift technique", Opt. Lett. **5**, 401-403 (1980).

[9] Basic titles on the partial coherence theory are suggested in [11] of ch. 6.

[10] The subject of optical processing comprises a plethora of techniques, differing from both the operating principles and the processing problems. Here a selection of titles is given in relation to both coherent [**?**]-[10.10] and incoherent [10.8]-[10.12] illumination, disregarding the specificity of the described applications. Some basic textbooks are also quoted; the application of the fractional Fourier transform-related methods to optical signal processing is accurately examinated in [4.8]. For an account of "digital optics" being the natural by-product of the theoretical and experimental work on optical computing see [10.19]-[10.22].

[10.1] K.G. Birch, "Spatial filtering in optical-data processing", Rep. Prog. Phys. **35**, 1265-1314 (1972).

[10.2] A. VanderLugt, "Coherent Optical processing", Proc. IEEE **62**, 1300-1319 (1974).

[10.3] W.T. Rhodes and J. M. Florence, "Frequency-variant optical signal analysis", Appl. Opt. **15**, 3073-3079 (1976).

[10.4] J.W.Goodman, "Operations achievable with coherent optical information processing systems", Proc. IEEE **65**, 29-38 (1977).

[10.5] R.J. Marks II, J.F. Walkup, M.O. Hagler and T.F. Krile, "Space-variant processing of $1D$ signals", Appl.Opt. **16**, 739-745 (1977).

[10.6] R.J. Marks II, and S.V. Bell, "Astigmatic coherent processor analysis", Opt. Eng. **17**, 167-169 (1978).

[10.7] J.F. Walkup, "Space-variant coherent optical processing", Opt. Eng. **19**, 339-346 (1980).

[10.8] W.T. Rhodes, "Incoherent spatial filtering", Opt. Eng. **19**, 323-330 (1980).

[10.9] W.T. Rhodes and A.A. Sawchuk, "Incoherent optical processing", in *Optical Information Processing Fundamentals*, S.H. Lee (ed.) (Springer, New York, 1981), pp. 69-110 .

[10.10] H. Bartelt and A.W. Lohmann, "Optical processing of one-dimensional signals", Opt. Comm. **42**, 87-91 (1982).

[10.11] T.E. Bell, "Optical computing: a field in flux", IEEE Spectrum, **23**, no. 8, 34-57 (1986).

[10.12] I. Glaser, "Information processing with spatially incoherent light", in *Progress in Optics*, Vol. XXV, E. Wolf (ed.) (North Holland, Amsterdam, 1987), ch. 5, pp. 389-509.

[10.13] J.W. Goodman, *Introduction to Fourier Optics* (McGraw-Hill, New York, 1968).

[10.14] G.L. Rogers, *Noncoherent Optical Processing* (Wiley, New York, 1977).

[10.15] J.D. Gaskill, *Linear Systems, Fourier Transforms and Optics* (Wiley & Sons, New York, 1978).

[10.16] A. VanderLugt, *Optical Signal Processing* (John Wiley & Sons, New York, 1992).

[10.17] J. Shamir, *Optical Systems and Processes* (SPIE, Bellingham, USA, 1999).

[10.18] *Digital Optical Computing*, special issue in Opt. Eng. **25**, 1-116 (1986).

[10.19] A.W. Lohmann, "What classical optics can do for the digital optical computer", Appl. Opt. **25**, 1543-1549 (1986).

[10.20] K.-H. Brenner, "Digital optical computing", Appl. Phys. B **46**, 111-120 (1988).

[10.21] N. Streibl, K.-H. Brenner, A. Huang, J. Jahns, J. Jewell, A.W. Lohmann, D.A.B. Miller, M. Murdocca, M.E. Prise and T. Sizer, "Digital optics", Proc. IEEE **77**, 1954-1968 (1989).

[10.22] H.M. Ozaktas and D.A.B. Miller, "Digital Fourier optics", Appl. Opt. **35**, 1212-1219 (1996).

[11] For an exhaustive analysis of the bilinear transformations in the light of their formal and applicative properties in the optical context as well as their relation to the Wigner distribution function and the ambiguity function, we suggest

[11.1] B.E.A. Saleh, "Bilinear processing of $1D$ signals by use of linear $2D$ coherent optical processors", Appl. Opt. **17**, 3408-3411 (1978).

[11.2] B.E.A. Saleh, "Optical bilinear transformations: general properties", Optica Acta **26**, 777-799 (1979).

[11.3] B.E.A. Saleh, "Linearity of quasi-homogeneous partially coherent imagery", J. Opt. Soc. Am. **72**, 1095-1096 (1982).

[11.4] B.E.A. Saleh, "Bilinear transformations in optical signal-processing", Proc. Soc. Photo-Opt. Inst. **373**, 125-134 (1984).

[11.5] J. Ojeda-Castañeda and E.E. Sicre, "Bilinear optical systems. The Wigner distribution function and ambiguity function representations", Optica Acta **31**, 255-260 (1984).

[12] The optical generation of the Wigner distribution function is documented in a very large variety of papers in relation with truly optical as well as not properly optical problems. Also quoted are some representative papers concerned with the Wigner display of $2D$ signals, which topic being not touched on in the text. See also the review papers [12.15], where a synthetic description of the setups for optical production of both the spatial and temporal Wigner distribution function is given, and [12.16], where an account of digital and hybrid implementations of the Wigner distribution function is presented.

[12.1] M.J. Bastiaans, "Wigner distribution function display: a supplement to ambiguity function display using a single $1D$ input", Appl. Opt. **19**, 192 (1980).

[12.2] H.O. Bartelt, K.-H. Brenner and A.W. Lohmann, "The Wigner distribution function and its optical production," Opt. Comm. **32**, 32-38 (1980).

[12.3] K.-H. Brenner and A.W. Lohmann, "Wigner distribution function display of complex 1D signals, " Opt. Comm. **42**, 310-314 (1982).

[12.4] G. Eichmann and B.Z. Dong, "Two-dimensional optical filtering of $1D$ signals," Appl. Opt. **21**, 3152-3156 (1982).

[12.5] R. Bamler and H. Glünder, "The Wigner distribution of two-dimensional signals. Coherent optical generation and display," Optica Acta **30**, 1789-1803 (1983).

[12.6] N. Subotic and B.E.A. Saleh, "Generation of the Wigner distribution function of two-dimensional signals by a parallel optical processor", Opt. Lett. **9**, 471-473 (1984).

[12.7] M. Conner and Y. Li, "Optical generation of the Wigner distribution of $2D$ real signals," Appl. Opt. **24**, 3825-3829 (1985).

[12.8] Ts. Mateeva and P. Sharlandjiev, "The generation of a Wigner distribution function of complex signals by spatial filtering", Opt. Comm. **57**, 153-155 (1986).

[12.9] T. Iwai, A.K. Gupta and T. Asakura, "Simultaneous optical production of the sectional Wigner distribution function for a two-dimensional object", Opt. Comm. **58**, 15-19 (1986).

[12.10] A.K. Gupta and T. Asakura, "New optical system for the efficient display of Wigner distribution function using a single object transparency", Opt. Comm. **60**, 265-268 (1986).

[12.11] Y. Li, G. Eichmann and M. Conner, "Optical Wigner distribution and ambiguity function for complex signals and images", Opt. Comm. **67**, 177-179 (1988).

[12.12] A.A. Rakes, C. Hester, R. Inguva and B.V.K. Kumar, "Wigner distribution function/ambiguity function processor", Proc. SPIE **936**, 260-269 (1988).

[12.13] G. Shabtay, D. Mendlovic and Z. Zalevsky, "Proposal for optical implementation of the Wigner distribution function", Appl. Opt. **37**, 2142-2144 (1998).

[12.14] K.F. Lee, F. Reil, S. Bali, A. Wax and J.E. Thomas, "Heterodyne measurement of Wigner distributions for classical optical fields", Opt. Lett. **24**, 1370-1372 (1999).

[12.15] D. Dragoman, "The Wigner distribution function in Optics and Optoelectronics," in *Progress in Optics*, Vol. XXXVII, E. Wolf, ed., (Elsevier, Amsterdam, 1997), ch. 1, pp.1-56.

[12.16] G. Cristóbal, C. Gonzalo and J. Bescós, "Image filtering and analysis through the Wigner distribution", in *Advances in Electronics and Electron Physics*, Vol 80, P.W. Hawkes (ed.) (Academic Press, San Diego 1991), pp. 309-397.

[13] W. Magnus and F. Oberhettinger, *Formulas and Theorems for the Special Functions of Mathematical Physics* (Chelsea, New York, 1949).

8

1D First-Order Optical Systems: Transfer Laws for the Wigner Distribution Function

8.1 Introduction

In Chapter 6 we have introduced the Wigner distribution function within a purely quantum mechanical context as an effective and appealing tool allowing for a formulation of quantum mechanics in terms of some kind of phase space formulation similar to that of classical mechanics. Using the Wigner method, it is possible to calculate a class of quantum mechanical averages in the same manner as the classical phase space distribution is used to calculate classical averages. As one merely multiplies the distribution function and the classical phase space function, whose average is desired, and then integrates over the entire phase space, likewise one merely multiplies the Wigner distribution function and the phase space function corresponding (by a precisely defined correspondence rule) to the quantum mechanical observable whose expectation value is desired, and then integrates over the entire phase space [1].

We have not dwelled in detail on the difference between the classical (symplectic) phase-space and the Wigner phase-space, which, as noted in § 6.4.1, has its own structure according to the inherent invariance laws. We have simply dealt with the Wigner phase-space as the mere Cartesian product of the eigenvalues q and p (or κ) of the coordinate and momentum (or spatial frequency) operators. In a sense, therefore, we have considered the Wigner representation like an extension of quantum mechanics into the "classical" phase space, focussing on the operative advantages of such a representation rather than on the structure of the manifold where it is defined (which is indeed object of interesting investigations). Naturally, whilst a point representation in phase space is allowed by the classical formalism, and so the motion can be viewed as a simple substitution of one point for another, the concept of moving point in phase space is replaced within the quantum mechanical formalism by a "quasi-probability" distribution which changes with time through non local transformations.

After importing the formal concept of the Wigner distribution function into optics [2], our attention has accordingly been focussed on the Wigner representation of optical signals, which yields a signal description displaying both the spatial and frequency features. Finally, in Chapter 7 we have framed the Wigner distribution function within the general context of signal theory in relation to the issue of individualizing the "musical score" of a signal.

In order to complete tracing the conceptual path to the definition of the Wigner integral, we return to the quantum mechanical context and show in Sect. 8.2 that the Wigner distribution function naturally reverts into the basic form when required to satisfy a certain set of well defined conditions suggested by the desired role it should play as a joint distribution of the conjugate variables q and p as well as by the desirable relation it might have to the classical phase-space distribution. The reader will easily recognize some of the properties described in § 6.3 for the optical Wigner distribution function in those considered in § 8.2 even though the motivations for, and the relevance of, the latter may sound slightly different due to the different context (quantum mechanical instead of wave optical) where the discussion of § 8.2 is placed.

Then, paralleling the "static" viewpoint of the description given in Chapter 6, where the Wigner distribution function has been calculated for a large variety of signals, in § 8.3 we approach the problem of the "dynamics" of the Wigner distribution function, and so we investigate how the Wigner distribution function of a travelling signal is modified by propagation. The ray-optics description of paraxial propagation in terms of symplectic matrices is recovered within the Wigner phase-space formalism, thus establishing an interesting similarity with the evolution of the phase-space ray-density distribution $\rho(q, p; z)$ (§ 2.2). The correspondence between the field distribution, or its frequency spectrum, and the Wigner distribution function is then reinforced. Not only we can substitute the parallel and interrelated space and frequency signal representations with the indirect but unique Wigner representation, but also we know how to "propagate" the Wigner distribution function through first-order optical systems, as it remarkably transforms by an affine transformation of its coordinates involving just the transfer matrices of ray-optics (Fig. 8.1).

As a completion of such a "dynamical" investigation, in § 8.5 we derive the transport equation for the Wigner distribution function, the wave-optical analog of the *quantum* Liouville equation. As expected, we find that for quadratic polynomial generated systems (and in particular for free-propagation systems), the quantum Liouville equation turns into the classical Liouville equation.

Moreover, some space (§ 8.4) is devoted to illustrate the relation between the Wigner distribution function and the fractional Fourier transform, which is of both conceptual and applicative interest, being basic to the phase-space tomographic investigation of complex wave fields, be they classical optical

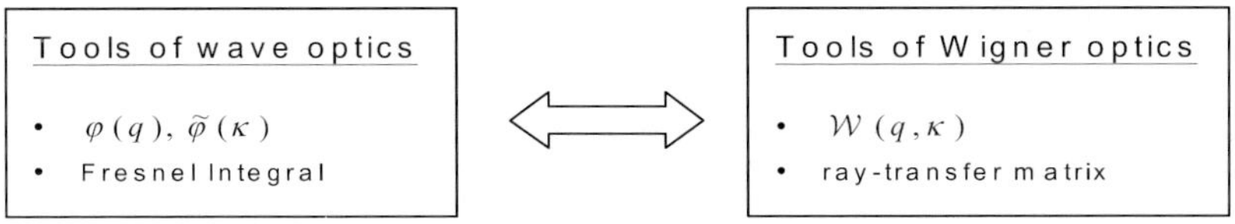

FIGURE 8.1. The identification of the propagation properties of the Wigner distribution function allows for a complete phase-space description of propagating signals.

disturbances or quantum matter-waves.

8.2 From the wave function to the phase space representation

We consider the one-dimensional motion of a single particle of mass m moving in a potential $V(q)$. Accordingly the relevant phase space is that of the particle position and momentum q and p, where both are treated as ordinary real variables instead of Hermitian operators. Our concern is to identify a mapping

$$\psi(q,t) \in \mathcal{L}^2(\mathbb{R}) \to \mathfrak{P}(q,p,t) \in \mathcal{L}^2(\mathbb{R}^2), \tag{8.2.1}$$

allowing for a unique correspondence of the particle wave function $\psi(q,t)$ to the function $\mathfrak{P}(q,p,t)$ of the positional and momentum coordinates q and p (and time t), we would like to interpret as giving the probability that the particle coordinate and momentum have the values q and p at time t.

8.2.1 *Identifying $\mathfrak{P}(q,p,t)$ as the Wigner distribution function*

The desired probabilistic interpretation of $\mathfrak{P}(q,p,t)$ naturally formalizes into a certain set of requirements. First of all, it is desirable that $\mathfrak{P}$ be real. This suggests relating $\mathfrak{P}(q,p,t)$ to the wave function $\psi(q,t)$, it represents, through the expectation value in the state ψ of some Hermitian operator depending on $\hat{\mathsf{q}}$ and $\hat{\mathsf{p}}$. It is therefore reasonable to write down the relation connecting the phase space distribution function $\mathfrak{P}(q,p,t)$ to the system wave function $\psi(q,t)$ in the form

$$\mathfrak{P}(q,p) = \iint K(q,p;x,x')\psi^*(x)\psi(x')dxdx', \tag{8.2.2}$$

with the kernel function $K(q,p;x,x')$ turning to have the basic property

$$K(q,p;x,x') = K^*(q,p;x',x), \tag{8.2.3}$$

in order that $\mathfrak{P}(q,p)$ be real. For formal convenience, the time t is not explicitly denoted in (8.2.2).

Further, it is natural to impose that the phase space representative $\mathfrak{P}(q,p)$ has the same invariances of the wave function, it corresponds to. Basically we require that $\mathfrak{P}$ be Galilei invariant, i.e. invariant for displacements in both the positional and momentum coordinates.

The invariance of $\mathfrak{P}$ for positional displacements of the wave function is expressed by the condition that

$$\text{if } \psi(q) \rightarrow \psi(q+Q), \quad \text{then } \mathfrak{P}(q,p) \rightarrow \mathfrak{P}(q+Q,p), \tag{8.2.4}$$

whatever Q be. In order to see how this reflects on the kernel of transformation (8.2.2), we note that $\psi(q+Q)$ has the distributional representation

$$\psi(q+Q) \rightleftharpoons \iint K(q,p;x,x')\psi^*(x+Q)\psi(x'+Q)dxdx'. \tag{8.2.5}$$

Equating the above integral to $\mathfrak{P}(q+Q,p)$, which writes

$$\mathfrak{P}(q+Q,p) = \iint K(q+Q,p;x,x')\psi^*(x)\psi(x')dxdx', \tag{8.2.6}$$

yields the following relation for the kernel

$$K(q,p;x,x') = K(q-Q,p;x-Q,x'-Q). \tag{8.2.7}$$

As Q denotes an arbitrary space displacement, which can as well be infinitesimal, relation (8.2.7) is seen to be equivalent to the differential equation

$$(\frac{\partial}{\partial q} + \frac{\partial}{\partial x} + \frac{\partial}{\partial x'})K(q,p;x,x') = 0. \tag{8.2.8}$$

This means that K, in addition to p, depends on the quantities q, x and x' only through the combinations $q - \frac{x+x'}{2}$ and $x - x'$, or equivalently through $q - x$ and $q - x'$.

Similarly, the invariance of $\mathfrak{P}$ with respect to a momentum displacement is expressed by the requirement that

$$\text{if } \widetilde{\psi}(p) \rightarrow \widetilde{\psi}(p+P), \quad \text{then } \mathfrak{P}(q,p) \rightarrow \mathfrak{P}(q,p+P), \tag{8.2.9}$$

whatever P be. Since a shift of the momentum representative $\widetilde{\psi}(p)$ is equivalent to a linear modulation of the wave function $\psi(q)$ according to

$$\widetilde{\psi}(p) \rightarrow \widetilde{\psi}(p+P) \rightleftharpoons \psi(q) \rightarrow \psi(q)e^{-iPq/\hbar}, \tag{8.2.10}$$

one easily obtains for $K(q,p;x,x')$ the relation

$$K(q,p;x,x') = K(q,p+P;x,x')e^{-iP(x-x')/\hbar}. \tag{8.2.11}$$

Since P is arbitrary, setting $P = -p$ we are able to further define the dependence of the kernel on the momentum variable in the form

$$K(q,p;x,x') = K(q,0;x,x')e^{ip(x-x')/\hbar}. \tag{8.2.12}$$

Also, according to the above considerations, the functional form of the kernel at $p = 0$ can be written as

$$K(q,0;x,x') = K_1(q - \tfrac{x+x'}{2}, x - x'), \tag{8.2.13}$$

or

$$K(q,0;x,x') = K_2(q - x, q - x'). \tag{8.2.14}$$

To fix mind, in the following we will use the first form so that

$$K(q,p;x,x') = K_1(q - \tfrac{x+x'}{2}, x - x')e^{ip(x-x')/\hbar}. \tag{8.2.15}$$

Evidently the Galilei invariance of $\mathfrak{P}(q,p)$, we have inquired above, corresponds to the space and frequency shift invariance of the optical Wigner distribution function described in § 6.3.2.

As emphasized in Chapter 6, it is desirable that $\mathfrak{P}(q,p)$, when integrated over q and p, yields the correct marginal probability distributions. Thus, integrating over p, we should obtain the proper probability distribution for the space coordinate, given by the square modulus of the wave function:

$$\int \mathfrak{P}(q,p)dp = |\psi(q)|^2. \tag{8.2.16}$$

Likewise, integrating over q we should obtain the probability distribution for the momentum variable, given by the square modulus of the Fourier transform of ψ:

$$\int \mathfrak{P}(q,p)dq = \left|\widetilde{\psi}(p)\right|^2 = \tfrac{1}{2\pi\hbar}\left|\int dq\psi(q)e^{-ipq/\hbar}\right|^2. \tag{8.2.17}$$

On account of relation (8.2.12), integrating (8.2.2) over p gives

$$\int \mathfrak{P}(q,p)dp = 2\pi\hbar \int dx K_1(q - x, 0)\,|\psi(x)|^2, \tag{8.2.18}$$

so that in order to meet condition (8.2.16) the kernel must satisfy the relation

$$K_1(\xi, 0) = \tfrac{1}{2\pi\hbar}\delta(\xi). \tag{8.2.19}$$

Similarly, requirement (8.2.17), which explicitly writes as

$$\begin{aligned}\iint dx dx' \int dq K_1(q - \tfrac{x+x'}{2}, x - x')e^{ip(x-x')/\hbar}\psi^*(x)\psi(x') \\ = \tfrac{1}{2\pi\hbar}\iint dx dx' e^{i\frac{p}{\hbar}(x-x')}\psi^*(x)\psi(x'),\end{aligned} \tag{8.2.20}$$

allows us to establish yet another relation for the kernel as

$$\int d\xi K_1(\xi, x - x') = \tfrac{1}{2\pi\hbar}. \tag{8.2.21}$$

However, $K(q,p;x,x')$ is still not uniquely defined.

As an additional simple request on the phase space distribution function $\mathfrak{P}(q,p)$, that might permit a unique correspondence of $\mathfrak{P}$ to ψ, we may demand that in the case of a free particle the equation of motion of $\mathfrak{P}$ be equal to the classical one for the phase space distribution $\rho(q,p;t)$, i.e. Liouville's equation for the classical free particle,

$$\frac{\partial}{\partial t}\rho(q,p;t) = -\frac{p}{m}\frac{\partial}{\partial q}\rho(q,p;t). \tag{8.2.22}$$

Let us therefore differentiate $\mathfrak{P}(q,p)$ with respect to t:

$$\frac{\partial}{\partial t}\mathfrak{P}(q,p) = \iint dx dx' K(q,p;x,x')\frac{\partial}{\partial t}[\psi^*(x)\psi(x')]. \tag{8.2.23}$$

Focusing on the derivative of the product, we can write

$$\frac{\partial}{\partial t}[\psi^*(x)\psi(x')] = \frac{i\hbar}{2m}[\psi^*(x)\frac{\partial^2}{\partial x'^2}\psi(x') - \psi(x')\frac{\partial^2}{\partial x^2}\psi^*(x)], \tag{8.2.24}$$

where use has been made of the Schrödinger equation for the free particle

$$\frac{\partial}{\partial t}\psi(q) = -\frac{\hbar^2}{2m}\frac{\partial^2}{\partial q^2}\psi(q). \tag{8.2.25}$$

Inserting (8.2.24) into (8.2.23), after integrating by parts we end up with

$$\frac{\partial}{\partial t}\mathfrak{P}(q,p) = \tfrac{i\hbar}{2m}\iint dx dx' \psi^*(x)\psi(x')[\frac{\partial^2}{\partial x'^2} - \frac{\partial^2}{\partial x^2}]K(q,p;x,x'). \tag{8.2.26}$$

Equating the t-derivative of $\mathfrak{P}$ to the q-derivative through the factor $-\frac{p}{m}$ we finally obtain

$$\frac{i\hbar}{2}[\frac{\partial^2}{\partial x'^2} - \frac{\partial^2}{\partial x^2}]K(q,p;x,x') = -p\frac{\partial}{\partial q}K(q,p;x,x'). \tag{8.2.27}$$

It is easy to express the derivative of K with respect to q, x and x' in terms of the derivatives of $K_1(\xi,\eta)$ with respect to the relative arguments $\xi = q - \frac{x+x'}{2}$ and $\eta = x - x'$. Thus, after a little calculation, Eq. (8.2.27) turns into

$$\frac{\partial^2}{\partial\xi\partial\eta}K_1(\xi,\eta) = 0, \tag{8.2.28}$$

which suggests $K_1(\xi,\eta)$ to depend separately on ξ and η; namely

$$K_1(\xi,\eta) = \alpha(\xi) + \beta(\eta). \tag{8.2.29}$$

Since according to Eq. (8.2.21) the integral of K_1 over ξ takes on a constant value, $\beta(\eta)$ is a constant function, and hence can be absorbed into $\alpha(\xi)$. Being therefore $K_1(\xi,\eta) = \alpha(\xi) = K_1(\xi,0)$, by (8.2.19) we immediately obtain that

$$\alpha(\xi) = \tfrac{1}{2\pi\hbar}\delta(\xi). \tag{8.2.30}$$

The kernel $K(q,p;x,x')$ comes to be unambiguously defined as

$$K_{\mathrm{W}}(q,p;x,x') = \tfrac{1}{2\pi\hbar}\delta(q - \tfrac{x+x'}{2})e^{ip(x-x')/\hbar}, \tag{8.2.31}$$

and hence the phase space distribution function takes the explicit expression

$$\mathrm{W}(q,p) = \tfrac{1}{2\pi\hbar}\int \psi^*(q - \tfrac{q'}{2})\psi(q + \tfrac{q'}{2})e^{-ipq'/\hbar}dq', \tag{8.2.32}$$

as firstly given by Wigner in his seminal paper (Eq. (5) in [3]).

Expression (8.2.32) establishes a unique correspondence of the state function $\psi(q)$ to the function $\mathrm{W}(q,p)$, defined in a classical-like manner in the c-number space (q,p), where q and p are real parameters turning into the classical coordinate and momentum in the limit $\hbar \to 0$. The link with the optical Wigner distribution function $\mathcal{W}(q,\kappa)$ defined through (6.2.2) is immediately recognized by noting that in the optical context the reduced wavelength $\frac{\lambda_0}{2\pi}$ corresponds to $\hbar$ and accordingly the spatial frequency κ plays the role as $\frac{p}{\hbar}$.

Naturally, in the presence of forces, $\mathrm{W}(q,p)$ does not satisfy the same equation as Liouville's equation for the classical phase space distribution $\rho(q,p;t)$,

$$\frac{\partial\rho}{\partial t} = -\frac{p}{m}\frac{\partial\rho}{\partial q} + V'(q)\frac{\partial\rho}{\partial p}, \tag{8.2.33}$$

except in the case when the potential $V(q)$ is linear or quadratic. We shall consider this aspect of the problem within the optical context. In fact, in § 8.5 we will show that in the Fresnel approximation (corresponding to the nonrelativistic regime of quantum mechanics, to which all the above considerations apply) the transport equation for the optical Wigner distribution function $\mathcal{W}(q,\kappa)$ shapes in the same form as the classical Liouville equation for $\rho(q,p;z)$, only if the refractive index function $n(q,z)$ depends quadratically on the off-axis distance. Within the mechanical-optical analogy the refractive index corresponds to the mechanical potential.

It is worth noting that the (q,p) representation of the quantum system as yielded by (8.2.32) is consistent with the assumption that the system state is completely described by the single wave function $\psi(q)$, i.e., that the system is in the *pure* state $\psi(q)$. The feasibility of such a univocal description by $\psi(q)$ is sometimes prevented by a lack of a precise knowledge of the system state. In many case, in fact, we may know only the probability that the system be in each of its own possible dynamical states. Accordingly the description of the system is suitably performed by means of the density matrix $\rho(q', q)$, which writes in the coordinate representation as[a]

$$\rho(q', q) = \sum_j w_j \psi_j^*(q')\psi_j(q), \tag{8.2.34}$$

the possible time dependence being again implied. The sum involves of course all the possible dynamical states of the system, precisely defined by the respective wave functions $\psi_j(q)$; the weight w_j specifies the probability that the system be in the state ψ_j. As an obvious consequence, the phase space distribution function comes to be expressed by the same integral form as in (8.2.32) with the single product $\psi^*(q - \frac{q'}{2})\psi(q + \frac{q'}{2})$ being replaced by the sum of weighted products $\psi_j^*(q - \frac{q'}{2})\psi_j(q + \frac{q'}{2})$, namely

$$\mathrm{W}(q, p) = \frac{1}{2\pi\hbar} \int \sum_j w_j \psi_j^*(q - \tfrac{q'}{2})\psi_j(q + \tfrac{q'}{2}) e^{-ipq'/\hbar} dq'. \tag{8.2.35}$$

In the optical context the density matrix related summation should be replaced by the coherence function, thus yielding the expression for the optical Wigner distribution function suitable to describe partially coherent light, as earlier noted in § 6.3.5. Notably in the literature the Wigner distribution function is mostly expressed in the form (8.2.35), involving the density matrix or operator, which is universally considered as the fundamental descriptor of a quantum system. Expression (8.2.32) is deduced from (8.2.35) in the case of a pure state.

The derivation of expression (8.2.32), presented here, is based on a definite set of requirements, that have allowed for a unique determination of the kernel of the transformation (8.2.2) relating the wave function to a distribution function in the (q, p) plane. In general many other reasonable requirements may be envisaged that can be restrictive for the kernel in (8.2.2); they can be variously combined to form several sets of conditions, minimal in number, each uniquely yielding the same Wigner kernel (8.2.31). The requirements we have examined here, i.e. Galilei invariance, free particle limit and correct marginals, are effective for providing a unique definition of the kernel of (8.2.2) in the form (8.2.31) only in the one dimensional case. The reader interested in the general instance of the necessary conditions for yielding a unique definition of the phase space distribution function may consult the refs. in [4], among which we signalize [4.3] offering an exhaustive investigation of the question.

---

[a] We refer to the density matrix by the customarily used symbol $\rho(q', q)$, hoping that this do not originate any confusion with the classical phase space distribution function, $\rho(q, p; t)$.

8.2.2 *The non positiveness of the Wigner distribution function*

The properties of the Wigner distribution function $\mathrm{W}(q,p)$, discussed above, are shared by any classical phase-space distribution. However, a classical phase space density is everywhere non negative, whilst the Wigner distribution function can in general attains negative values over some regions of the (q,p) space for some states. The non-positive property of the optical Wigner distribution function $\mathcal{W}(q,\kappa)$ has briefly been commented on in § 6.3.6, and verified in a certain number of examples in § 6.4. There we evidenced that only the quadratic phase signal (i.e., the chirp), which comprises as limiting case the impulse and the monofrequency signal, and the Gaussian signal have nonnegative Wigner distributions, respectively represented by a line delta and a Gaussian.

The non positiveness is not a peculiar property of the Wigner distribution function, but is a distinguishing feature of any bilinear real functional of the state vector $\psi(q)$ satisfying as well the marginal conditions. In fact, as proved in [4.1], there is no distribution function $\mathfrak{P}(q,p)$ of the type (8.2.2), i.e. bilinear in $\psi(q)$,which satisfies the marginal integrals,

$$\int \mathfrak{P}(q,p)dp = |\psi(q)|^2, \qquad \int \mathfrak{P}(q,p)dq = \left|\widetilde{\psi}(p)\right|^2, \tag{8.2.36}$$

for all wave functions $\psi(q)$ and is at the same time nonnegative

$$\mathfrak{P}(q,p) \geq 0, \tag{8.2.37}$$

for all ψ's and all values of q and p.

Following [4.1], we firstly consider a wave function $\psi(q)$, which is supposed to vanish at all points outside a limited interval $\mathbb{I}$ of the real line:

$$\psi(q) \neq 0 \qquad \text{only for } q \in \mathbb{I} \subset \mathbb{R}.$$

Let $\mathfrak{P}_\psi(q,p)$ be the distribution function corresponding to $\psi(q)$ through the bilinear functional (8.2.2). We suppose that $\mathfrak{P}_\psi$ is nonnegative definite. Also, we consider an observable $\widehat{g}(\widehat{\mathsf{q}})$, whose associated classical function is of course $g(q)$. We may suppose that $g(q)$ is zero inside the interval $\mathbb{I}$, and is nowhere negative in its own domain.

The expectation value of $\widehat{g}(\widehat{\mathsf{q}})$ in the state ψ is evidently zero, $g(q)$ and $|\psi(q)|^2$ yielding a null overlap integral; namely

$$\langle \widehat{g} \rangle = \int \psi^*(q)g(q)\psi(q)dq = 0. \tag{8.2.38}$$

On the other hand, in accord with the interpretation of $\mathfrak{P}_\psi(q,p)$, the expectation value $\widehat{g}$ can as well be calculated by the (q,p) integral

$$\langle \widehat{g} \rangle = \iint \mathfrak{P}_\psi(q,p)g(q)dqdp, \tag{8.2.39}$$

which then must vanish. The integrand function with respect to p above is evidently nonnegative, being the integral of a nonnegative function, i.e. the product of the nonnegative functions $\mathfrak{P}_\psi(q,p)$ and $g(q)$. Then it must be zero: $\int \mathfrak{P}_\psi(q,p)g(q)dq = 0$, and hence, $\mathfrak{P}_\psi(q,p)g(q) = 0$ as well. As a consequence, the distribution function $\mathfrak{P}_\psi(q,p)$ vanishes (except for a set of measure zero) for all values of q outside the domain $\mathbb{I}$ of ψ.

We consider now two wave functions, ψ_1 and ψ_2, which are supposed to vanish outside two disjoint bounded intervals, we denote respectively as $\mathbb{I}_1$ and $\mathbb{I}_2$. In symbols, we have

$$\psi_1(q) \neq 0 \text{ only for } q \in \mathbb{I}_1 \subset \mathbb{R} \quad \text{and} \quad \psi_2(q) \neq 0 \text{ only for } q \in \mathbb{I}_2 \subset \mathbb{R}.$$

Hence $\psi_1(q)\psi_2(q) = 0$.

Let $\mathfrak{P}_{\psi_1}(q,p)$ and $\mathfrak{P}_{\psi_2}(q,p)$ be the distribution functions associated to ψ_1 and ψ_2 by (8.2.2). We suppose that $\mathfrak{P}_{\psi_1}$ and $\mathfrak{P}_{\psi_2}$ are nonnegative for all values of q and p as well as that they give the correct marginal distributions when integrated with respect to q and p. Therefore, according to the previous discussion, $\mathfrak{P}_{\psi_1}$ and $\mathfrak{P}_{\psi_2}$ vanish outside the vertical strips in the (q,p) plane limited by the domains of the wave functions they represent, and hence $\mathfrak{P}_{\psi_1}$ vanishes outside the strip $\mathbb{I}_1 \times \mathbb{R}$ as $\mathfrak{P}_{\psi_2}$ vanishes outside the strip $\mathbb{I}_2 \times \mathbb{R}$.

We show that there exists an abundance of wave functions, we may construct as linear combinations of ψ_1 and ψ_2, whose phase-plane distribution functions fail to satisfy at the same time requirements (8.2.37) and (8.2.36). We may consider, for instance, the wave function $\psi(q)$ simply obtained as

$$\psi(q) = \psi_1(q) + \psi_2(q), \tag{8.2.40}$$

the relevant distribution function being denoted as $\mathfrak{P}_\psi(q,p)$. The assumption that $\mathfrak{P}_\psi(q,p)$ satisfies both requirements (8.2.37) and (8.2.36) leads to a contradiction. In fact, according to the general integral form definition (8.2.2), one finds that

$$\mathfrak{P}_\psi(q,p) = \mathfrak{P}_{\psi_1} + \mathfrak{P}_{\psi_2} + \mathfrak{P}_{\psi_1,\psi_2} + \mathfrak{P}_{\psi_2,\psi_1}, \tag{8.2.41}$$

which is the basic manifestation of the bilinearity of $\mathfrak{P}_\psi$ (see § 6.3.4); of course,

$$\mathfrak{P}_{\psi_1,\psi_2}(q,p) = \iint dx dx' K(q,p;x,x')\psi_1^*(x)\psi_2(x') = \mathfrak{P}^*_{\psi_2,\psi_1}(q,p). \tag{8.2.42}$$

Evidently, as $\mathfrak{P}_{\psi_1}$ and $\mathfrak{P}_{\psi_2}$ are nonzero only within their respective domains $\mathbb{I}_1 \times \mathbb{R}$ and $\mathbb{I}_2 \times \mathbb{R}$, in order that $\mathfrak{P}_\psi(q,p) \geq 0$ everywhere through the (q,p) plane the cross distribution $\mathfrak{P}_{\psi_1,\psi_2}$ must vanish, thus ensuring for $\mathfrak{P}_\psi$ the definitely nonnegative expression:

$$\mathfrak{P}_\psi(q,p) = \mathfrak{P}_{\psi_1} + \mathfrak{P}_{\psi_2}. \tag{8.2.43}$$

It is easy to verify that the above relation is in contrast with the assumption that $\mathfrak{P}_\psi$ yields the proper marginal distributions when either one of the variables is integrated over. In fact, on the ground of such an hypothesis, integrating $\mathfrak{P}_\psi(q,p)$ over q we should have

$$\int \mathfrak{P}_\psi(q,p)dq = \left|\widetilde{\psi}(p)\right|^2 = \left|\widetilde{\psi}_1(p)\right|^2 + \left|\widetilde{\psi}_2(p)\right|^2 + 2\Re[\widetilde{\psi}_1^*(p)\widetilde{\psi}_2(p)], \tag{8.2.44}$$

the last expression being in accord with (8.2.40) and the linearity of the Fourier transform operation, which implies that $\widetilde{\psi}(p) = \widetilde{\psi}_1(p) + \widetilde{\psi}_2(p)$. On the other hand, according to (8.2.43) the integration of $\mathfrak{P}_\psi(q,p)$ with respect to q gives

$$\int \mathfrak{P}_\psi(q,p)dq = \left|\widetilde{\psi}_1(p)\right|^2 + \left|\widetilde{\psi}_2(p)\right|^2, \tag{8.2.45}$$

since both $\mathfrak{P}_{\psi_1}$ and $\mathfrak{P}_{\psi_2}$ are supposed to yield the correct marginals. In order that (8.2.44) and (8.2.45) be equal, the product $\widetilde{\psi}_1^*(p)\widetilde{\psi}_2(p)$ should vanish. Such a condition could be satisfied only if the Fourier transforms $\widetilde{\psi}_1(p)$ and $\widetilde{\psi}_2(p)$ were nonzero over two nonoverlapping bounded intervals. But, being $\widetilde{\psi}_1$ and $\widetilde{\psi}_2$ the Fourier transforms of functions defined over bounded intervals, they can not be limited to finite intervals as well.

As a conclusion, we can state that phase space distribution functions involving the state bilinearly and having correct marginals must locally take on negative values for certain states. This raises the question about the interpretation of the $\mathfrak{P}_\psi$'s, which clearly cannot strictly represent the probability of finding the particle at q with momentum p.

The non positiveness of the (bilinear) quantum phase-space distribution functions is a plain manifestation of the departure of the joint position momentum description of quantum mechanics from the classical behaviour [5]. It is evident that not every phase space distribution function is realizable in the quantum mechanical case. As an example, we may consider a phase-plane point, described by the product of delta functions $\delta(q-q_0)\delta(p-p_0)$. Such a distribution is classically realizable, representing a single particle at q_0 with momentum p_0, which moves according to Hamilton's equations of motion, and thereby individualizes a definite trajectory in phase plane. Delta functions are evolved into delta functions by the classical Liouville equation (8.2.33). In contrast, the realizability of the point-like distribution function is inhibited within the quantum mechanical context by the uncertainty principle. Rephrasing, for example, properties (6.3.1) and (6.3.53) for the quantum distribution (8.2.32), we see that for any normalized pure state function ψ the corresponding Wigner distribution $\mathrm{W}(q,p)$ must satisfy the quantum constraints

$$\begin{aligned} |\mathrm{W}(q,p)| &\leq \tfrac{1}{\pi\hbar}, \\ \iint [\mathrm{W}(q,p)]^2 dq dp &= \tfrac{1}{2\pi\hbar}, \end{aligned} \tag{8.2.46}$$

which evidently prevent $\mathrm{W}(q,p)$ from being peaked on one single point of the (q,p) plane. Indeed, they yield a value of the order of $\hbar$ for the minimal area in the (q,p) plane over which the Wigner distribution function spreads out.

8.2.3 *Wigner distribution function and Weyl transform*

As already noted, the joint position-momentum description of a quantum system is dictated by the desire of treating the quantum theory within a framework similar as that of the classical statistical mechanics, thus making feasible, for instance, the calculation of the expectation values of quantum observables by a (q,p)-integration involving numerical-valued functions of q and p rather than by a q-integration involving operator-valued functions of $\widehat{\mathsf{q}}$ and $\widehat{\mathsf{p}}$.

The construction of the distribution function $\mathfrak{P}(q,p)$ pursues then the goal that the quantum mechanical expression $\left\langle \widehat{f}\right\rangle_{\mathrm{Q}}$ for the expectation value of the operator $\widehat{f}(\widehat{\mathsf{q}},\widehat{\mathsf{p}})$ in the state ψ,

$$\left\langle \widehat{f}\right\rangle_{\mathrm{Q}} = \int \psi^*(q)\widehat{f}(\widehat{q},\widehat{p})\psi(q)dq, \tag{8.2.47}$$

and the classical-like expression for the "average" of the scalar function $f(q,p)$ corresponding to $\widehat{f}(\widehat{\mathsf{q}},\widehat{\mathsf{p}})$,

$$\langle f\rangle = \iint \mathfrak{P}(q,p)f(q,p)dqdp, \tag{8.2.48}$$

$\mathfrak{P}(q,p)$ being treated as a classical probability density, yield the same value. We re-emphasize that integral (8.2.47) involves operators whereas (8.2.48) scalar functions; scalar functions may more easily be handled and investigated by both algebraic manipulations and intuitive perceptions. As in the previous paragraphs, we consider a system of one continuous degree of freedom, so that the pair of conjugate variables (q,p) suffices to span the inherent phase space.

The desired identity of (8.2.47) and (8.2.48), and the identification of the central quantity $\mathfrak{P}(q,p)$ is problematic due to the lack of a unique rule for associating a classical function of the conjugate variables q and p to a given quantum function of the conjugate operators $\widehat{\mathsf{q}}$ and $\widehat{\mathsf{p}}$, and vice versa. One cannot trivially replace the operators $\widehat{\mathsf{q}}$ and $\widehat{\mathsf{p}}$ in the expression of $\widehat{f}(\widehat{\mathsf{q}},\widehat{\mathsf{p}})$ by the scalars q and p, as so doing different operator functions would come to correspond to the same classical function. Therefore, the identity of the two expressions (8.2.47) and (8.2.48) would fail to occur, thus preventing the individualization of the probability distribution. A couple of examples illustrating this aspect of the problem has been given in Problem 6.1. As a further illustration, we may consider the exponential operator $e^{i\xi\widehat{\mathsf{q}}+i\eta\widehat{\mathsf{p}}}$. The simple substitution of $\widehat{\mathsf{q}}$ and $\widehat{\mathsf{p}}$ by q and p would yield the scalar exponential $e^{i\xi q+i\eta p}$, which according to the algebra of c-numbers can equally be written as

$$e^{i\xi q + i\eta p} = e^{i\xi q}e^{i\eta p} = e^{i\eta p}e^{i\xi q} = \tfrac{1}{2}(e^{i\xi q}e^{i\eta p} + e^{i\eta p}e^{i\xi q}).$$

In contrast, due to the non commutativity of $\widehat{\mathsf{q}}$ and $\widehat{\mathsf{p}}$: $[\widehat{\mathsf{q}}, \widehat{\mathsf{p}}] = i\hbar$, the operators $e^{i\xi\widehat{\mathsf{q}}+i\eta\widehat{\mathsf{p}}}$, $e^{i\xi\widehat{\mathsf{q}}}e^{i\eta\widehat{\mathsf{p}}}$, $e^{i\eta\widehat{\mathsf{p}}}e^{i\xi\widehat{\mathsf{q}}}$ and $\frac{1}{2}(e^{i\xi\widehat{\mathsf{q}}}e^{i\eta\widehat{\mathsf{p}}} + e^{i\eta\widehat{\mathsf{p}}}e^{i\xi\widehat{\mathsf{q}}})$ are clearly not equal. By the BCH formula (Eq. (2.2.60)) one has

$$e^{i\xi\widehat{\mathsf{q}}+i\eta\widehat{\mathsf{p}}} = e^{i\xi\widehat{\mathsf{q}}}e^{i\eta\widehat{\mathsf{p}}}e^{i\hbar\xi\eta/2} = e^{i\eta\widehat{\mathsf{p}}}e^{i\xi\widehat{\mathsf{q}}}e^{-i\hbar\xi\eta/2} = \tfrac{1}{2}(e^{i\xi\widehat{\mathsf{q}}+i\eta\widehat{\mathsf{p}}} + e^{i\eta\widehat{\mathsf{p}}}e^{i\xi\widehat{\mathsf{q}}})\sec(\tfrac{\hbar\xi\eta}{2}).$$

It is then apparent that the functional form of the phase space distribution function is closely related to the choice of the correspondence rule between operators and c-number functions. Conversely, each phase space distribution function, defined through suitable physical reasonings possibly suggested by the specific class of problems at hand, individualizes a well defined rule of association between classical and quantum observables.

Thus, the phase space representation of quantum mechanics, defining the correspondence (8.2.1) between quantum states and phase space distribution functions, establishes as well a one-to-one mapping between self-adjoint operators $\widehat{f}(\widehat{\mathsf{q}}, \widehat{\mathsf{p}})$ on the Hilbert space of the square-integrable complex-valued functions of the space coordinate and scalar functions $f(q,p)$ on the Cartesian product of the position and momentum coordinates:

$$f(q,p) \rightleftharpoons \widehat{f}(\widehat{\mathsf{q}}, \widehat{\mathsf{p}}). \tag{8.2.49}$$

As the joint q-p distribution $\mathfrak{P}(q,p)$ replaces the state function $\psi(q)$, similarly the c-number function $f(q,p)$ replaces the q-number function $\widehat{f}(\widehat{\mathsf{q}}, \widehat{\mathsf{p}})$. Note that in § 4.4.1 we have been concerned with the direction $f \to \widehat{f}$ in the association (8.2.49). Here we are concerned with the direction $\widehat{f} \to f$.

Considering in particular the Wigner distribution function $\mathrm{W}(q,p)$, let $f_{\mathrm{W}}(q,p)$ denote the Wigner phase plane function corresponding to the quantum operator $\widehat{f}(\widehat{\mathsf{q}}, \widehat{\mathsf{p}})$, which allows for the desired identity: $\left\langle \widehat{f} \right\rangle_{\mathrm{Q}} = \langle f \rangle$. Taking the explicit integral form (8.2.32) of $\mathrm{W}(q,p)$, the classical-like average (8.2.48) writes

$$\begin{aligned} \langle f \rangle_{(\mathrm{W})} &\equiv \iint \mathrm{W}(q,p) f_{\mathrm{W}}(q,p)\, dq dp \\ &= \tfrac{1}{2\pi} \iint dq dp \int d\tau \psi^*(q - \tfrac{\tau\hbar}{2})\psi(q + \tfrac{\tau\hbar}{2}) e^{-ip\tau} f_{\mathrm{W}}(q,p). \end{aligned} \tag{8.2.50}$$

Using the Fourier representation of $f_{\mathrm{W}}(q,p)$, namely

$$f_{\mathrm{W}}(q,p) = \tfrac{1}{2\pi} \iint d\xi d\eta \widetilde{f}_{\mathrm{W}}(\xi,\eta) e^{i\xi q + i\eta p}, \tag{8.2.51}$$

$\widetilde{f}_{\mathrm{W}}(\xi,\eta)$ being the Fourier transform of $f_{\mathrm{W}}(q,p)$, one can recast (8.2.50) into

$$\langle f \rangle_{(\mathrm{W})} = \tfrac{1}{2\pi} \iiint dq d\xi d\eta \psi^*(q - \tfrac{\eta\hbar}{2})\psi(q + \tfrac{\eta\hbar}{2}) \widetilde{f}_{\mathrm{W}}(\xi,\eta) e^{i\xi q}, \tag{8.2.52}$$

after noting that integration with respect to p gives $2\pi\delta(\tau-\eta)$.

It is convenient to appropriately redefine the involved variables in the integral on the right of (8.2.52) and then to rewrite the resulting integrand function in the operator form, thus obtaining

$$\begin{aligned}\langle f\rangle_{(\mathrm{W})} &= \iiint dq d\xi d\eta \psi^*(q)\psi(q+\eta\hbar)\widetilde{f}_{\mathrm{W}}(\xi,\eta)e^{i\xi q}e^{i\xi\eta\hbar/2} \\ &= \iiint dq d\xi d\eta \psi^*(q)\widetilde{f}_{\mathrm{W}}(\xi,\eta)e^{i\xi\eta\hbar/2}e^{i\xi\widehat{\mathsf{q}}}e^{i\eta\widehat{\mathsf{p}}}\psi(q).\end{aligned} \tag{8.2.53}$$

The last identity is an obvious consequence of the fact, that the shifted wave function $\psi(q+\eta\hbar)$ can be seen as resulting from the action of the displacement operator $e^{i\eta\widehat{\mathsf{p}}}$ according to

$$e^{i\eta\widehat{\mathsf{p}}}\psi(q) = e^{\eta\hbar\frac{d}{dq}}\psi(q) = \psi(q+\eta\hbar). \tag{8.2.54}$$

Finally, applying the BCH formula to recast the product of the two exponential operators above into a single exponential operator by the well know recipe:

$$e^{i\xi\widehat{\mathsf{q}}}e^{i\eta\widehat{\mathsf{p}}} = e^{-i\hbar\xi\eta/2}e^{i\xi\widehat{\mathsf{q}}+i\eta\widehat{\mathsf{p}}}, \tag{8.2.55}$$

we end up with

$$\langle f\rangle_{(\mathrm{W})} = \int dq\psi^*(q)\left[\frac{1}{2\pi}\iint d\xi d\eta \widetilde{f}_{\mathrm{W}}(\xi,\eta)e^{i\xi\widehat{\mathsf{q}}+i\eta\widehat{\mathsf{p}}}\right]\psi(q). \tag{8.2.56}$$

Equating the above expression with the quantum expectation value (8.2.47) allows us to define the relation between $\widehat{f}(\widehat{\mathsf{q}},\widehat{\mathsf{p}})$ and $f_{\mathrm{W}}(q,p)$ as

$$\widehat{f}(\widehat{\mathsf{q}},\widehat{\mathsf{p}}) = \frac{1}{2\pi}\iint d\xi d\eta \widetilde{f}_{\mathrm{W}}(\xi,\eta)e^{i\xi\widehat{\mathsf{q}}+i\eta\widehat{\mathsf{p}}}, \tag{8.2.57}$$

or equivalently, on account of (8.2.51),

$$\widehat{f}(\widehat{\mathsf{q}},\widehat{\mathsf{p}}) = \frac{1}{4\pi^2}\iiiint d\xi d\eta dq dp f_{\mathrm{W}}(q,p)e^{i\xi(\widehat{\mathsf{q}}-q)+i\eta(\widehat{\mathsf{p}}-p)}. \tag{8.2.58}$$

To invert (8.2.57) or (8.2.58) we evaluate the matrix element of the operator $\widehat{f}(\widehat{\mathsf{q}},\widehat{\mathsf{p}})$ between two eigenstates of the position operator corresponding, for instance, to the eigenvalues q' and q. Using the Dirac notation, we write

$$\left\langle q'\right|\widehat{f}(\widehat{\mathsf{q}},\widehat{\mathsf{p}})\left|q\right\rangle = \frac{1}{2\pi}\iint d\xi d\eta \widetilde{f}_{\mathrm{W}}(\xi,\eta)\left\langle q'\right|e^{i\xi\widehat{\mathsf{q}}+i\eta\widehat{\mathsf{p}}}\left|q\right\rangle, \tag{8.2.59}$$

where the inner product $\left\langle q'\right|e^{i\xi\widehat{\mathsf{q}}+i\eta\widehat{\mathsf{p}}}\left|q\right\rangle$ is evaluated as

$$\begin{aligned}&\left\langle q'\right|e^{i\xi\widehat{\mathsf{q}}+i\eta\widehat{\mathsf{p}}}\left|q\right\rangle \equiv \int dq''\delta(q''-q')\,e^{i\xi\widehat{\mathsf{q}}+i\eta\widehat{\mathsf{p}}}\delta(q''-q) \\ &= \int dq''\delta(q''-q')e^{i\hbar\xi\eta/2}e^{i\xi q''}\delta(q''+\hbar\eta-q) = e^{i\hbar\xi\eta/2}e^{i\xi q'}\delta(q'+\hbar\eta-q).\end{aligned} \tag{8.2.60}$$

Therefore, (8.2.59) turns into

$$\left\langle q'|\,\widehat{f}(\widehat{\mathsf{q}},\widehat{\mathsf{p}})|q\right\rangle = \frac{1}{2\pi}\int d\xi\,\widetilde{f}_{\mathrm{W}}(\xi,\tfrac{q-q'}{\hbar})\,e^{i\xi\frac{q+q'}{2}}, \tag{8.2.61}$$

more symmetrically written as

$$\left\langle q-\tfrac{\eta\hbar}{2}|\,\widehat{f}(\widehat{\mathsf{q}},\widehat{\mathsf{p}})|q+\tfrac{\eta\hbar}{2}\right\rangle = \frac{1}{2\pi}\int d\xi\,\widetilde{f}_{\mathrm{W}}(\xi,\eta)\,e^{i\xi q}. \tag{8.2.62}$$

It is immediate to obtain $\widetilde{f}_{\mathrm{W}}(\xi,\eta)$ through

$$\widetilde{f}_{\mathrm{W}}(\xi,\eta) = \int d\xi\,e^{i\xi' q}\left\langle q-\tfrac{\eta\hbar}{2}|\,\widehat{f}(\widehat{\mathsf{q}},\widehat{\mathsf{p}})|q+\tfrac{\eta\hbar}{2}\right\rangle, \tag{8.2.63}$$

and accordingly, by an inverse Fourier transformation, $f_{\mathrm{W}}(q,p)$ as

$$f_{\mathrm{W}}(q,p) = \int d\eta\,e^{i\eta p}\left\langle q-\tfrac{\eta\hbar}{2}|\,\widehat{f}(\widehat{\mathsf{q}},\widehat{\mathsf{p}})|q+\tfrac{\eta\hbar}{2}\right\rangle. \tag{8.2.64}$$

In the literature the function $f_{\mathrm{W}}(q,p)$ is reported as the *Weyl* (or *Wigner-Weyl*) *transform* of the quantum operator $\widehat{f}(\widehat{\mathsf{q}},\widehat{\mathsf{p}})$, $f_{\mathrm{W}}(q,p)$ and $\widehat{f}(\widehat{\mathsf{q}},\widehat{\mathsf{p}})$ being related through (8.2.57) and (8.2.64).

Remarkably, expression (8.2.57) mirrors the Weyl quantization procedure for associating a quantum operator with a classical function. The Weyl rule establishes in fact to decompose the classical function by harmonic analysis in exponential functions as in (8.2.51) and then to replace the exponential functions by exponential operators in which the quantum operators $\widehat{q}$ and $\widehat{p}$ take the place of the scalars q and p, thus leading to an expression of the form (8.2.57). In a way the "classical" orthogonal and complete set of harmonic c-number exponentials $e^{i\xi q+i\eta p}$ is replaced by the "quantum" orthogonal and complete set of harmonic exponential operators $e^{i\xi\widehat{\mathsf{q}}+i\eta\widehat{\mathsf{p}}}$ [6].

Therefore, Weyl transforms are related to quantum operators by a transformation which reproduces the Weyl rule. However, they do not obey the commutative algebra of classical functions. As an example, the Weyl transform $F_{\mathrm{W}}(q,p)$ of the quantum operator $\widehat{F}(\widehat{\mathsf{q}},\widehat{\mathsf{p}}) = \widehat{f}(\widehat{\mathsf{q}},\widehat{\mathsf{p}})\,\widehat{g}(\widehat{\mathsf{q}},\widehat{\mathsf{p}})$ is markedly different from the product of the Weyl transforms of $\widehat{f}(\widehat{\mathsf{q}},\widehat{\mathsf{p}})$ and $\widehat{g}(\widehat{\mathsf{q}},\widehat{\mathsf{p}})$; namely, $F_{\mathrm{W}}(q,p) \neq f_{\mathrm{W}}(q,p)g_{\mathrm{W}}(q,p)$. Moreover, $F_{\mathrm{W}}(q,p) \neq G_{\mathrm{W}}(q,p)$, with $G_{\mathrm{W}}(q,p)$ denoting the Weyl transform of the inverse order operator product $\widehat{G}(\widehat{\mathsf{q}},\widehat{\mathsf{p}}) = \widehat{g}(\widehat{\mathsf{q}},\widehat{\mathsf{p}})\,\widehat{f}(\widehat{\mathsf{q}},\widehat{\mathsf{p}})$ (Problem 1). Consequently, one cannot state in general that if

$$\widehat{f} \rightleftharpoons f_{\mathrm{W}}$$

the Weyl image of any function $H(\widehat{f})$ is then the function $H(f_{\mathrm{W}})$, i.e., that

$$H(\widehat{f}) \rightleftharpoons H(f_{\mathrm{W}}).$$

This feature of the Wigner phase space functions along with the non positiveness of the Wigner distribution unequivocally evidence the difference of the quantum phase space formulation to the classical stochastic theory [5.2].

8.3 First-order optical systems: propagation law for the Wigner distribution function

The Wigner distribution function is "propagated" through linear optical systems according to the ray matrix formalism of geometrical optics [2]. This is a plain consequence of the transfer relations, which for linear optical systems write in the form of superposition integrals whatever be the representation adopted to describe the input and output signals, i.e. the complex amplitude or the respective Fourier spectrum. As seen in Chapter 4, four different input-output relations can be written down for an optical system, all in the form of a superposition integral, the relevant kernels being completely determined by the system and the chosen signal representations. In particular, relations (4.3.13) and (4.6.19) characterize the transfer properties of optical systems in the space and the frequency domain, respectively. We will see that there is only one transfer relation for optical systems when the input and output signals are described in terms of the corresponding Wigner distribution functions.

8.3.1 *Transfer relations for the basic optical systems*

In order to gain an immediate insight into the propagation properties of the Wigner distribution function in relation to first-order optical systems, we deduce the input-output relations for the Wigner distribution function under propagation through the basic optical systems, we may directly associate with the second degree monomials in the ray-optical position and momentum variables or equivalently in the wave-optical position and momentum operators, as discussed in Chapters 2-5. The discussion is developed on the basis of the specific features of the optical system at hand. A more formal analysis is presented in the next subsection; accordingly the results of this paragraph are recovered within the context of the ray response of the system.

As earlier stated in relation to phase-plane graphs, the relevant axes are to be thought of as being suitably scaled in order to have the same dimensionality.

Section of homogeneous medium

The propagation through a homogeneous medium is described in the q-domain by the input-output relation (§ 5.2.1)

$$\varphi_o(q) = \sqrt{\frac{k_0}{2\pi i d}} \int e^{i\frac{k_0}{2d}(q-q')^2} \varphi_i(q')dq', \tag{8.3.1}$$

where d denotes the length of the section through which the input signal $\varphi_i(q)$ is freely propagated, factored by the reciprocal $\frac{1}{n_0}$ of the refractive index.

The transfer relation (8.3.1) involves the convolution of the input amplitude $\varphi_i(q)$ and the chirp $\mathfrak{g}(q) = \sqrt{\frac{k_0}{2\pi i d}} e^{ik_0 q^2/2d}$, which reproduces in proximity of the optical axis the circular profile of a spherical wave with curvature $\frac{1}{d}$. Hence, by result (6.3.62), the Wigner distribution function $\mathcal{W}_o(q,\kappa)$ of the output signal $\varphi_o(q)$ turns out to be the convolution in the space variable q of the Wigner distribution functions $\mathcal{W}_i(q,\kappa)$ and $\mathcal{W}_{\mathfrak{g}}(q,\kappa)$ of $\varphi_i(q)$ and $\mathfrak{g}(q)$; namely,

$$\mathcal{W}_o(q,\kappa) = 2\pi \int \mathcal{W}_{\mathfrak{g}}(q-q',\kappa)\mathcal{W}_i(q',\kappa)\,dq'. \tag{8.3.2}$$

Thus, on account of result (6.4.34), by which $\mathcal{W}_{\mathfrak{g}}(q,\kappa)$ takes the expression

$$\mathcal{W}_{\mathfrak{g}}(q,\kappa) = \tfrac{k_0}{2\pi d}\delta(\kappa - \tfrac{k_0}{d}q) = \tfrac{1}{2\pi}\delta(q - \tfrac{d}{k_0}\kappa), \tag{8.3.3}$$

we obtain the input-output relation for a free-medium section in the form

$$\mathcal{W}_o(q,\kappa) = \mathcal{W}_i(q - \tfrac{d}{k_0}\kappa, \kappa)\,. \tag{8.3.4}$$

Let us resort to the ray-optical picture of free propagation, according to which the ray-direction p remains unchanged whilst the ray-position changes in relation with the actual direction. The ray transfer matrix $\mathbf{T}(d)$ for free propagation along the reduced distance d reads, in fact, (Eq. (3.3.1))

$$\mathbf{T}(d) = \begin{pmatrix} 1 & d \\ 0 & 1 \end{pmatrix}. \tag{8.3.5}$$

As emphasized in § 4.6.5, the (q,κ)-representation of light rays allows for the ray-matrix picture as well, the off-diagonal elements of the ray-transfer matrix of the geometrical optical (q,p)-representation being suitably scaled by $1/k_0$ and k_0, respectively. It is therefore evident that the arguments of $\mathcal{W}_i$ in (8.3.4) relate to those of $\mathcal{W}_o$ through the inverse matrix of (8.3.5) in the (q,κ)-description, namely $\mathbf{T}^{-1}(d) = \left(\begin{smallmatrix} 1 & -d/k_0 \\ 0 & 1 \end{smallmatrix}\right)$. Gathering the Fourier conjugate variables (q,κ) into the column vector $\mathbf{u} = (q,\kappa)^{\top}$, we may recast the transfer relation (8.3.4) for a free-medium section into the compact form[b]

$$\mathcal{W}_o(\mathbf{u}) = \mathcal{W}_i(\mathbf{T}^{-1}\mathbf{u}), \tag{8.3.6}$$

the specification of the section length d being omitted for simplicity's sake.

---

[b] A remark is in order. When denoting the arguments of a function by a multicomponent column vector, the correct notation would require the use of the transpose vector, as for instance in Eq. (2.2.67). Thus, Eq. (8.3.6) should more precisely be written as

$$\mathcal{W}_o(\mathbf{u}^{\top}) = \mathcal{W}_i([\mathbf{T}^{-1}\mathbf{u}]^{\top}) = \mathcal{W}_i(\mathbf{u}^{\top}[\mathbf{T}^{-1}]^{\top}), \quad \mathbf{u}^{\top} = (q,\kappa).$$

For sake of formal convenience, we adopt in the following the shorthand notation of Eq. (8.3.6), thus implying transposition in the vector arguments on both sides.

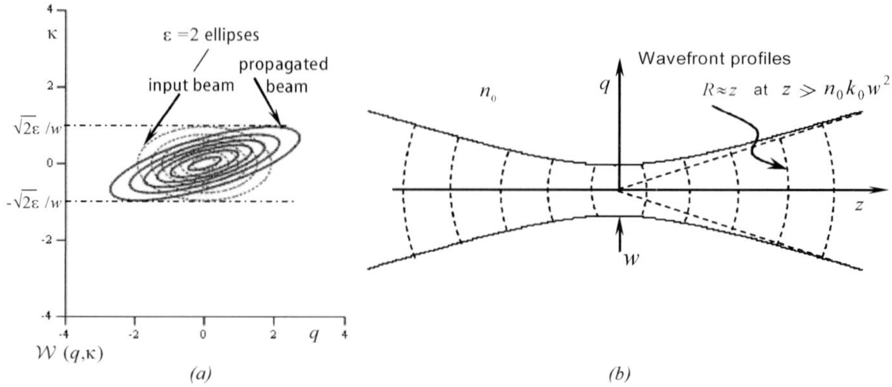

FIGURE 8.2. Phase-space and physical-space views of the transformation undergone by a Gaussian signal under free propagation, the latter being exemplified by the free propagation of a Gaussian beam from the waist. *(a)* q-shear of the Wigner chart of the signal $\varphi_{\mathcal{G}}(q) = (\frac{1}{2\pi})^{1/4} e^{-q^2/4}$ under free propagation. *(b)* At far distance from the waist, the wavefronts of a freely propagating Gaussian beam are like spherical wavefronts with curvature radius $R \sim z$.

In § 2.5 we noted that the free propagation has the graphical view in the geometrical-optics phase plane (q,p) as the horizontal shear of the ray distribution $\rho(q,p)$ (Fig. 2.13). Similarly the transfer relation (8.3.6) can be visualized in the Wigner plane (q,κ) as a horizontal shear of the Wigner chart, as shown by Fig. 8.2.a), where the Wigner charts of the Gaussian signal $\varphi_{\mathcal{G}}(q) = (\frac{1}{2\pi})^{1/4} e^{-q^2/4}$, before and after the propagation through the section, are depicted. The elliptical contours relevant to the Wigner distribution function $\mathcal{W}_i$ of the input signal are sheared along the q axis within the horizontal strip $-\frac{\sqrt{2\varepsilon}}{w} \leq \kappa \leq \frac{\sqrt{2\varepsilon}}{w}$. In fact, since the free propagation does not affect the ray direction, the κ (i.e. angular) extension of the signal is expected to remain unchanged. Also, the intersections with the q axis are unmoved, being representatives of rays (i.e. the normals to the beam wave fronts) that enter the section parallel to the optical axis and so propagate undeflected. Clearly, with increasing the section length $d \to \infty$, the ellipses tend to thin about the segment making the angle $\tan\beta = k_0/d$ with the q axis. As noted in § 6.4.5, real Gaussian waveforms may represent the q-profiles at the waist of the amplitude of Gaussian beams entering the section with waist radius w. At far distance d from the waist: $d \gg n_0 k_0 w^2$ and within the beam radius, the wavefronts of a Gaussian beam can be assimilated to portions of wave fronts of a spherical wave emanating from a point source at the waist, and so having a (reduced) curvature radius $R = d$ (Fig. 8.2.b)) [7]. Accordingly, the Wigner chart comes to be a straight line making the angle $\tan\beta = k_0/d$ with the q axis.

Thin lens

The propagation through a thin lens produces a quadratic phase modulation of the input amplitude $\varphi_i(q)$ according to (§ 5.2.2)

$$\varphi_o(q) = e^{-i\frac{k_0}{2f}q^2}\varphi_i(q), \tag{8.3.7}$$

with f being the lens focal distance. Having proved in § 6.3.9 that the Wigner distribution function of the product of two signals is the convolution in the frequency variable of the Wigner distributions of the two signals, we may write

$$\mathcal{W}_o(q,\kappa) = \int \mathcal{W}_{\mathfrak{f}}(q,\kappa-\kappa')\mathcal{W}_i(q,\kappa')\,d\kappa', \tag{8.3.8}$$

where $\mathcal{W}_{\mathfrak{f}}(q,\kappa)$ is the Wigner distribution function of the chirp $\mathfrak{f}(q) = e^{-ik_0q^2/2f}$, and so

$$\mathcal{W}_{\mathfrak{f}}(q,\kappa) = \delta(\kappa + \tfrac{k_0}{f}q). \tag{8.3.9}$$

It immediately follows that

$$\mathcal{W}_o(q,\kappa) = \mathcal{W}_i(q,\kappa + \tfrac{k_0}{f}q)\,. \tag{8.3.10}$$

Resorting again to the ray-optical picture, we recall that a thin lens is represented by the ray matrix (Eq. 3.3.2)

$$\mathbf{L}(f) = \begin{pmatrix} 1 & 0 \\ -1/f & 1 \end{pmatrix}, \tag{8.3.11}$$

with the off-diagonal entry being scaled into $-\frac{k_0}{f}$ when dealing with the spatial frequency κ instead of the momentum p. Thus, the transfer relation (8.3.10) can be recast in a form similar to (8.3.6), i.e.,

$$\mathcal{W}_o(\mathbf{u}) = \mathcal{W}_i(\mathbf{L}^{-1}\mathbf{u}), \tag{8.3.12}$$

involving the Wigner plane vector $\mathbf{u} = (q,\kappa)^\top$ and the inverse lens matrix in the (q,κ)-representation $\mathbf{L}^{-1}(f) = \begin{pmatrix} 1 & 0 \\ k_0/f & 1 \end{pmatrix}$. The specification of the lens focal distance f is omitted.

As seen in § 2.5, under the $\mathbf{L}$-transformation the phase-plane ray-distribution is sheared along the p axis. Likewise, the transformation (8.3.12) produces a vertical shear of the Wigner chart. As an example, Fig. 8.3.a) shows the κ shear induced by (8.3.12) on the Wigner chart of the Gaussian signal $\varphi_{\mathcal{G}}(q) = (\frac{1}{2\pi})^{1/4}e^{-\frac{q^2}{4}}$. The canonical ellipses of the initial distribution are sheared in the κ direction within the vertical strip $-w\sqrt{\frac{\varepsilon}{2}} \le q \le w\sqrt{\frac{\varepsilon}{2}}$. As the passage

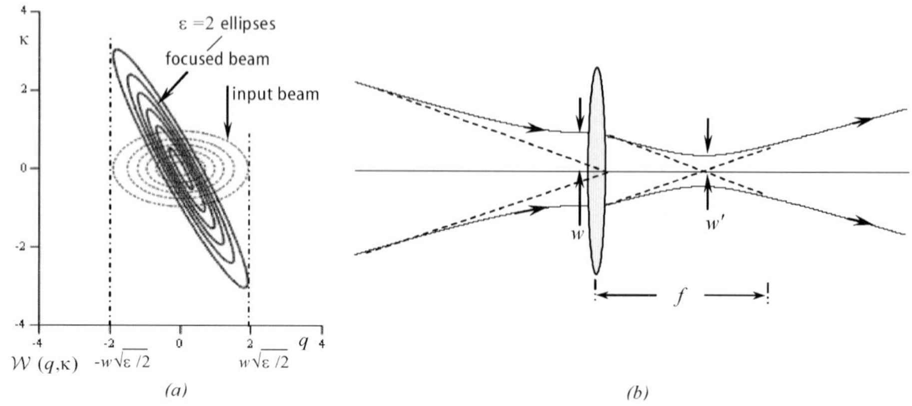

FIGURE 8.3. Phase-space and physical space views of the transformation undergone by a Gaussian signal under thin lens transfer, the latter being exemplified by the wavefront curvature transformation effected by a lens on a Gaussian beam. *(a)* κ-shear of the Wigner chart of the signal $\varphi_{\mathcal{G}}(q) = (\frac{1}{2\pi})^{1/4}e^{-q^2/4}$ under thin lens transfer. *(b)* Schematic of the focusing of a Gaussian beam by a positive thin lens.

through a thin lens has no effect on the ray position, the q extension of the signal does not change. Also, the intersections with the κ axis remain unchanged, being representatives of rays that passing through the center of the lens propagate undeflected. Increasing the value of the focusing power $\mathcal{P} = 1/f \to \infty$, the ellipses tend to thin about the segment making the angle $\tan\beta = -k_0/f$ with the κ axis. We recall that a lens placed at the waist of a Gaussian beam transforms the plane wavefront of the incoming beam into a spherical (more precisely, parabolic) one having the (reduced) curvature radius $R = -f$ (Fig. 8.3.b)) [7]. We know that the Wigner chart of the near-axis profile of a spherical wave is a straight line making the angle $\tan\beta = -k_0/R$ with the κ axis.

Note that relations (8.3.2) applies to (8.3.8) are representative respectively of shift-invariant and spreadless systems, as anticipated in § 6.3.9.

Ideal magnifier

The input-output relation for the ideal magnifier writes as (§ 5.3.3)

$$\varphi_o(q) = \frac{1}{\sqrt{m}}\varphi_i(\tfrac{q}{m}), \tag{8.3.13}$$

where m is the magnification factor.

Therefore, on account of result (6.3.59), we easily obtain the input-output relationship for the Wigner distribution function as

$$\mathcal{W}_o(q,\kappa) = \mathcal{W}_i(\tfrac{q}{m}, m\kappa)\,. \tag{8.3.14}$$

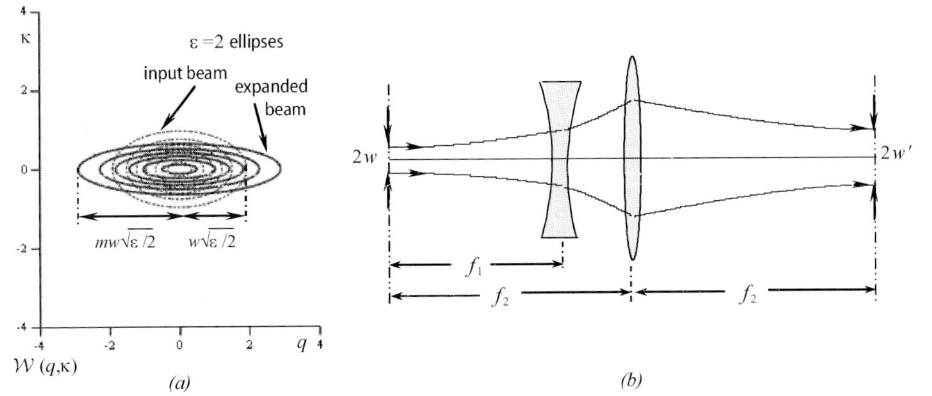

FIGURE 8.4. Phase-space and physical-space views of the transformation undergone by a Gaussian signal under scale transform, the latter being exemplified by the passage of a Gaussian beam through a telescopic system. *(a)* κ-shear of the Wigner chart of the signal $\varphi_{\mathcal{G}}(q) = (\frac{1}{2\pi})^{1/4}e^{-q^2/4}$ under propagation through a positive scaler with $m > 1$. *(a)* Schematic of the waist radius magnifying effect of a Galilean telescope on a Gaussian beam.

Evidently, being the system represented in ray optics by the transfer matrix (Eq. (3.4.26))

$$\mathbf{S}(m) = \begin{pmatrix} m & 0 \\ 0 & 1/m \end{pmatrix}, \tag{8.3.15}$$

for both the (q, p) and (q, κ) descriptions of the light ray, the above relation is equivalent to the following

$$\mathcal{W}_o(\mathbf{u}) = \mathcal{W}_i(\mathbf{S}^{-1}\mathbf{u})\,. \tag{8.3.16}$$

As shown in Fig. 8.4.a), the transformation (8.3.16) is depicted in the phase plane (q, κ) as a squeeze of the Wigner chart in the κ or q direction according to whether $m > 1$ or $m < 1$. In Fig. 8.4.b) the magnifying effect of a Galilean-like telescope on the beam radius of a Gaussian beam is sketched. Evidently, with the limiting values $m \to \infty$ and $m \to 0$, we recover the monofrequency and impulsive signals respectively, and accordingly the ellipses in Fig. 8.4.a) should crush over the horizontal and vertical axis, respectively. Phase-space and physical space views of the transformation undergone by a Gaussian signal under scale transform, the latter being exemplified by passage of a Gaussian beam through a telescopic system.

8.3.2 *The ray spread function*

We deduce now the transformation law for the Wigner distribution function under propagation through a linear optical system within a general context.

Let $\mathcal{W}(q,\kappa,z_o) \equiv \mathcal{W}_o(q,\kappa)$ and $\mathcal{W}(q,\kappa,z_i) \equiv \mathcal{W}_i(q,\kappa)$ denote the Wigner distribution functions of the signals $\varphi_i(q)$ and $\varphi_o(q)$ respectively entering and emerging from the optical system. As $\varphi_i(q)$ and $\varphi_o(q)$ are linked by the space domain transfer relation,

$$\varphi_o(q) = \int \mathsf{g}(q,q')\varphi_i(q')dq', \tag{8.3.17}$$

it is possible to express the Wigner distribution function $\mathcal{W}_o(q,\kappa)$ of the output signal in terms of the line-spread function $\mathsf{g}(q,q')$ of the system and the input signal amplitude $\varphi_i(q)$. In fact, according to the q-domain definition (6.2.2) of the Wigner distribution function, we write

$$\mathcal{W}_o(q,\kappa) = \tfrac{1}{2\pi}\iiint e^{-i\kappa q'}\mathsf{g}^*(q-\tfrac{q'}{2},q'')\mathsf{g}(q+\tfrac{q'}{2},q''')\varphi_i^*(q'')\varphi_i(q''')dq'dq''dq'''. \tag{8.3.18}$$

Using relation (6.3.23) to recover the Wigner distribution function of the input signal from the product $\varphi_i^*(q'')\varphi_i(q''')$, and suitably arranging the various integrals and variables entering the resulting expression, we end up with a definite relation of $\mathcal{W}_o(q,\kappa)$ to $\mathcal{W}_i(q,\kappa)$ in the form of the superposition integral

$$\mathcal{W}_o(q,\kappa) = \iint \mathsf{G}(q,\kappa,q',\kappa')\mathcal{W}_i(q',\kappa')dq'd\kappa'. \tag{8.3.19}$$

The kernel $\mathsf{G}(q,\kappa,q',\kappa')$ explicitly writes as

$$\mathsf{G}(q,\kappa,q',\kappa') = \tfrac{1}{2\pi}\iint \mathsf{g}^*(q-\tfrac{q''}{2},q'-\tfrac{q'''}{2})\mathsf{g}(q+\tfrac{q''}{2},q'+\tfrac{q'''}{2})e^{-i(\kappa q''-\kappa' q''')}dq''dq''', \tag{8.3.20}$$

and is seen to be the symplectic Fourier transform of the product $\mathsf{g}^*(q-\tfrac{q''}{2},q'-\tfrac{q'''}{2})\mathsf{g}(q+\tfrac{q''}{2},q'+\tfrac{q'''}{2})$ regarded as a function of q'' and q'''.

Naturally the transfer relation for the Wigner distribution function can also be worked out within the frequency domain representation of the system, using the pertinent input-output relation,

$$\widetilde{\varphi}_o(\kappa) = \int \mathsf{h}(\kappa,\kappa')\widetilde{\varphi}_i(\kappa')d\kappa'. \tag{8.3.21}$$

In fact, on account of definition (6.2.3) of $\mathcal{W}_o(q,\kappa)$ in terms of the Fourier spectrum $\widetilde{\varphi}_o(\kappa)$ and relation (6.3.25) for the product of two spectra at different frequencies, we obtain the superposition integral (8.3.19) with the kernel $\mathsf{G}(q,\kappa,q',\kappa')$ being expressed in terms of the wave-spread function $\mathsf{h}(\kappa,\kappa')$ in the same form as (8.3.20), namely

$$\mathsf{G}(q,\kappa,q',\kappa') = \tfrac{1}{2\pi}\iint \mathsf{h}^*(\kappa-\tfrac{\kappa''}{2},\kappa'-\tfrac{\kappa'''}{2})\mathsf{h}(\kappa+\tfrac{\kappa''}{2},\kappa'+\tfrac{\kappa'''}{2})e^{i(q\kappa''-q'\kappa''')}d\kappa''d\kappa'''. \tag{8.3.22}$$

The equality of the two expressions on the right of (8.3.20) and (8.3.22), implied by the use of the same symbol $\mathsf{G}(q, \kappa, q', \kappa')$, can rigorously be proved making use of the relations (4.6.20) of $\mathsf{g}(q, q')$ to $\mathsf{h}(\kappa, \kappa')$ or (4.6.21) of $\mathsf{h}(\kappa, \kappa')$ to $\mathsf{g}(q, q')$. The dual correspondence of the two expressions in (8.3.20) and (8.3.22) is evident. Accordingly, $\mathsf{G}(q, \kappa, q', \kappa')$ can also be interpreted as the inverse symplectic Fourier transform of the product $\mathsf{h}^*(\kappa - \frac{\kappa''}{2}, \kappa' - \frac{\kappa'''}{2})\mathsf{h}(\kappa + \frac{\kappa''}{2}, \kappa' + \frac{\kappa'''}{2})$ with respect to κ'' and κ'''.

The function $\mathsf{G}(q, \kappa, q', \kappa')$ is completely determined by the optical system through the point-spread function $\mathsf{g}(q, q')$ or the wave-spread function $\mathsf{h}(\kappa, \kappa')$; it provides a characterization of the system appropriate to the space-frequency representation of light signals conveyed by the Wigner representation. As the reader may verify, the hybrid input-output relations for the optical system, where the input and output signals are described in different domains, yield formal expressions for the kernel $\mathsf{G}(q, \kappa, q', \kappa')$, which involve the hybrid response functions in the same way as $\mathsf{g}(q, q')$ or $\mathsf{h}(\kappa, \kappa')$ in (8.3.20) and (8.3.22), the integrations being obviously performed over both the q and κ variables, and not exclusively in the space or frequency domain as in (8.3.20) and (8.3.22).

Notably, $\mathsf{G}(q, \kappa, q', \kappa')$ can be understood as the response of the system to a signal represented by the Wigner distribution function $\mathcal{W}_i(q, \kappa) = \delta(q - q_i)\delta(\kappa - \kappa_i)$, for which the transfer relation (8.3.19) would give $\mathcal{W}_o(q, \kappa) = \mathsf{G}(q, \kappa, q_i, \kappa_i)$. Of course, this interpretation has only a formal significance, since there does not exist any real optical signal, whose Wigner distribution function is a point delta in the (q, κ) plane, i.e., $\mathcal{W}_i(q, \kappa) = \delta(q - q_i)\delta(\kappa - \kappa_i)$. However, taking account of the geometrical optics representation of light rays, we may consider the nonrepresentable distribution $\mathcal{W}_i(q, \kappa) = \delta(q - q_i)\delta(\kappa - \kappa_i)$ as the mathematical representation of a single ray, entering the system at the position q_i with the optical momentum (i.e., direction) p_i, and hence with the frequency $\kappa_i = k_0 p_i$. The kernel $\mathsf{G}(q, \kappa, q', \kappa')$ is accordingly reported in the literature as the *ray-spread function* or the *ray response* of the system [2].

The ray spread function $\mathsf{G}(q, \kappa, q', \kappa')$ represents a sort of *trait d'union* between the ray and wave optical description of light propagation in the inherent paraxial and Fresnel approximations. In fact, on one side it marks the transformation law of the Wigner distribution function, which is defined in terms of the signal waveform $\varphi(q)$, and on the other side it reproduces the ray-optical picture of propagation, which is based on the transfer matrix formalism.

A little more about the ray-spread function

It is interesting to note that expressions (8.3.20) and (8.3.22) of $\mathsf{G}(q, \kappa, q', \kappa')$ have the structure of a "double" Wigner distribution function; hence, the function $\mathsf{G}(q, \kappa, q', \kappa')$ has all the properties of a Wigner function.

In fact, G is real. Also, it is easy to deduce the relations

$$\mathsf{g}(q,q')\mathsf{g}^*(q'',q''') = \frac{1}{2\pi}\iint \mathsf{G}(\tfrac{q+q''}{2},\kappa,\tfrac{q'+q'''}{2},\kappa')e^{i\kappa(q-q'')}e^{-i\kappa'(q'-q''')}d\kappa d\kappa',$$
$$\mathsf{h}(\kappa,\kappa')\mathsf{h}^*(\kappa'',\kappa''') = \frac{1}{2\pi}\iint \mathsf{G}(q,\tfrac{\kappa+\kappa''}{2},q',\tfrac{\kappa'+\kappa'''}{2})e^{-iq(\kappa-\kappa'')}e^{iq'(\kappa'-\kappa''')}dqdq', \tag{8.3.23}$$

which look very similar to (6.3.24) and (6.3.25). Indeed, we also obtain

$$|\mathsf{g}(q,q')|^2 = \frac{1}{2\pi}\iint \mathsf{G}(q,\kappa,q',\kappa')d\kappa d\kappa',\quad |\mathsf{h}(\kappa,\kappa')|^2 = \frac{1}{2\pi}\iint \mathsf{G}(q,\kappa,q',\kappa')dqdq', \tag{8.3.24}$$

and

$$\mathsf{g}(q,q') = \frac{1}{2\pi\mathsf{g}^*(0,0)}\iint \mathsf{G}(\tfrac{q}{2},\kappa,\tfrac{q'}{2},\kappa')e^{i(\kappa q-\kappa' q')}d\kappa d\kappa',$$
$$\mathsf{h}(\kappa,\kappa') = \frac{1}{2\pi\mathsf{h}^*(0,0)}\iint \mathsf{G}(q,\tfrac{\kappa}{2},q',\tfrac{\kappa'}{2})e^{-i(\kappa q-\kappa' q')}dqdq', \tag{8.3.25}$$

which formally resemble relations (6.3.26) and (6.3.28), respectively.

It is interesting to express the ray spread function of a composite system in terms of the ray spread functions of the various components. Referring to the sequence of N optical systems, considered in §§ 3.2 and 4.3.2 (Figs. 3.2 and 4.5) we may write for each pair of reference planes the input-output relation

$$\mathcal{W}_j(q,\kappa) = \iint \mathsf{G}_j(q,\kappa,q',\kappa')\mathcal{W}_{j-1}(q',\kappa')dq'd\kappa',\quad j=1,..,N \tag{8.3.26}$$

where $\mathcal{W}_{j-1}$ and $\mathcal{W}_j$ denote the Wigner distribution functions at the planes z_{j-1} and z_j, and accordingly $\mathsf{G}_j(q,\kappa,q',\kappa')$ is the ray response of the j-th element in the cascade.

Concatenating the N transfer relations (8.3.26) for $j=1$ to $j=N$, we can relate the Wigner distribution $\mathcal{W}_N(q,\kappa)$ on the output plane to the Wigner distribution $\mathcal{W}_0(q,\kappa)$ on the input plane in the form as (8.3.19), namely

$$\mathcal{W}_N(q,\kappa) = \iint \mathsf{G}(q,\kappa,q',\kappa')\mathcal{W}_0(q,\kappa)dq'd\kappa', \tag{8.3.27}$$

where $\mathsf{G}(q,\kappa,q',\kappa')$ has the meaning as the ray response of the overall system, and writes in terms of the ray responses of the system components as the $(N-1)$-dimensional integral

$$\mathsf{G}(q,\kappa,q',\kappa') = \int\cdots\int dq_{N-1}d\kappa_{N-1}dq_{N-2}d\kappa_{N-2}...dq_1d\kappa_1\mathsf{G}_N(q_N,\kappa_N,q_{N-1},\kappa_{N-1})$$
$$\cdot\mathsf{G}_{N-1}(q_{N-1},\kappa_{N-1},q_{N-2},\kappa_{N-2})...\mathsf{G}_1(q_1,\kappa_1,q',\kappa'). \tag{8.3.28}$$

The similarity with relation (4.3.18) for the impulse response of a compound system is apparent.

Finally, it is an easy task to verify that for a lossless optical system the ray spread function obeys the relation

$$\iint \mathsf{G}(q,\kappa,q',\kappa')dqd\kappa = 1. \tag{8.3.29}$$

8.3.3 *The ray response of an arbitrary ABCD system. Transformation law for the Wigner distribution function*

We consider now an arbitrary linear optical system, characterized by the ray-transfer matrix $\mathbf{M} = \begin{pmatrix} A & B \\ C & D \end{pmatrix}$ and accordingly by the point-spread function

$$\mathsf{g}(q, q') = \sqrt{\frac{k_0}{2\pi i B}} e^{i\frac{k_0}{2B}\left[Aq'^2 + Dq^2 - 2qq'\right]}. \tag{8.3.30}$$

Inserting the expression of $\mathsf{g}(q, q')$ into (8.3.20), after a simple algebra we obtain the following expression for the ray response

$$\mathsf{G}(q, \kappa, q', \kappa') = \frac{k_0}{4\pi^2 B} \iint dq'' dq''' e^{i\frac{k_0}{B}q''\left[Dq - q' - \frac{B}{k}\kappa\right]} e^{i\frac{k_0}{B}q'''\left[Aq' - q + \frac{B}{k}\kappa'\right]} \tag{8.3.31}$$

which turns into the product of delta functions according to

$$\mathsf{G}(q, \kappa, q', \kappa') = \delta(Aq' + \frac{B}{k_0}\kappa' - q)\delta(kCq' + D\kappa' - \kappa). \tag{8.3.32}$$

Resorting to the formalism of the Wigner plane vector $\mathbf{u} = (q, \kappa)^\top$, we can further recast the above expression into the vectorial form

$$\mathsf{G}(\mathbf{u}, \mathbf{u}') = \delta(\mathbf{M}\mathbf{u}' - \mathbf{u}). \tag{8.3.33}$$

Therefore, rewriting the relation (8.3.19) as

$$\mathcal{W}_o(\mathbf{u}) = \iint \mathsf{G}(\mathbf{u}, \mathbf{u}') \mathcal{W}_i(\mathbf{u}') d\mathbf{u}', \tag{8.3.34}$$

it is immediate to establish the transformation law of the Wigner distribution function under propagation through a first-order optical system in the form

$$\mathcal{W}_o(\mathbf{u}) = \mathcal{W}_i(\mathbf{M}^{-1}\mathbf{u}), \tag{8.3.35}$$

in accord with relations (8.3.6), (8.3.12) and (8.3.16). In formal terms the above relation states that the Wigner distribution function at a given phase plane point relevant to a transverse plane at the axial coordinate z is the same as the Wigner distribution function at the corresponding phase plane point relevant to another transverse plane, the correspondence being determined by the optical transformation, i.e. by the ray transfer matrix $\mathbf{M}$ connecting the chosen reference planes. Hence, the Wigner distribution function turns out to be invariant along the "paraxial geometrical ray-path". In particular, as the origin of the phase plane transforms into itself for any optical matrix, the Wigner distribution function at the origin of the phase plane is invariant:

$$\mathcal{W}_o(0, 0) = \mathcal{W}_i(0, 0). \tag{8.3.36}$$

It is worth emphasizing the formal similarity between the transfer relation (8.3.35) for the Wigner distribution function and the transfer relation (2.2.67) for the ray density distribution $\rho(q, p, z)$, appropriate to the geometrical optical description of ray propagation through first-order optical systems. Thus, though the Wigner distribution function cannot be interpreted as an energy density distribution, the resemblance of (8.3.35) to (2.2.67) confirms the role of the Wigner distribution function as a mathematical means for a joint space-spatial frequency description of the light perturbance evolution, which, yet conveniently, as far as first-order optics is concerned, can be dealt with by equations formally similar to those pertinent to the geometrical optics counterpart $\rho(q, p, z)$, legitimately interpretable as an energy density distribution. The Wigner distribution function appears therefore as an efficient and intuitively appealing tool to link the Fourier optics and the first-order ray optics, offering itself to be interpreted as the "amplitude" of a paraxial ray passing through the point (q, z) in the close vicinity of the optical axis with inclination directly deducible from the spatial frequency κ.

First-order geometrical optics is naturally associated with dynamical observables quadratically depending on the position coordinate q and its canonical conjugate, i.e., the momentum p. Correspondingly, in the wave optical context pertinently to the Fresnel approximation and the assumption of thin lenses and transmissive elements the observables quadratically depend on the position operator $\widehat{\mathsf{q}}$ and its canonical conjugate, i.e., the momentum operator $\widehat{\mathsf{p}}$ or the spatial frequency operator $\widehat{\kappa}$. Hence it is natural to presume that under these assumptions the Wigner distribution function formally behave as the ray density distribution. In fact, in § 8.5 we will see that in general the transport equation for the Wigner distribution function definitely differs from Liouville's equation (2.2.11), which is evidently the transport equation for the ray-density distribution $\rho(q, p, z)$. However, the former shapes in the same form as the latter in the case of quadratic Hamiltonian functions of (q, p) and correspondingly of $(\widehat{\mathsf{q}}, \widehat{\mathsf{p}})$, or, of course, in the geometrical optics limit $\lambda_0/2\pi \rightarrow 0$.

The transfer law (8.3.35) for the Wigner distribution function in a sense completes the issue approached in § 5.6, concerning the representation of linear canonical transformations (Fig. 8.5). We have seen in § 5.6 that linear canonical transformations, which map the conjugate operators $\widehat{\mathsf{q}}$, $\widehat{\mathsf{p}}$ into themselves through linear combinations preserving the Heisenberg commutation relations, find a possible representation in the Hilbert space $\mathcal{L}^2(\mathbb{R})$ of squared integrable wave functions in the form of integral transforms with quadratic phase kernels, depending on the coefficients of the transformation of $\widehat{\mathsf{q}}$ and $\widehat{\mathsf{p}}$. Correspondingly, in the wave-optical phase space, with the joint "state"-representation being conveyed by the Wigner distribution function, linear canonical transformations yield the affine transformation (8.3.35) of the coordinates of the Wigner distribution function through the same coefficients of the transforma-

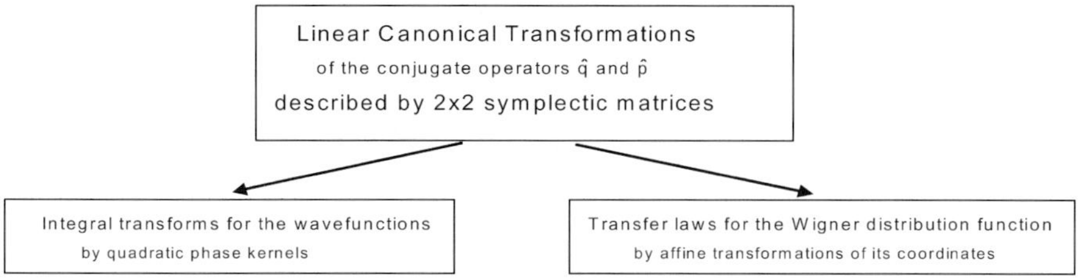

FIGURE 8.5. The transfer laws for the optical wavefunction and the Wigner distribution function are parallel representations of linear canonical transformations, in the Hilbert space $\mathcal{L}^2(\mathbb{R})$ and in the Wigner phase-plane (q, κ).

tion of $\widehat{\mathsf{q}}$ and $\widehat{\mathsf{p}}$. Thus, a definite link between the light signal representations as conveyed by ray optics and wave optics, with the respective mathematical machineries, is established in the Fresnel approximation. An alternative form of the (paraxial) transfer law for the Wigner distribution function in terms of the relevant second-order moments will be described in Chapter 9.

8.3.4 *Ray, wave and Wigner optics. I*

Relation (8.3.35) recovers to paraxial wave optics the practical and visually appealing potential of the matrix formalism of paraxial ray optics. It appears evident the utility of relation (8.3.35) for analyzing and synthesizing optical systems. In fact, whilst the propagation law for the wave field distribution or its spatial-frequency spectrum demands for the calculation of chirp-like integrals (that may reveal rather odd on a computational footing), the transfer law for the Wigner distribution function is based on the simple matrix-vector multiplication of ray-optics, thus offering a modular computational scheme, highly desirable for computer calculation. Hence, all the equivalence relations and decomposition theorems, described in Chapter 3, apply to *Wigner optics* as well. Also, the field distribution and its frequency spectrum can easily be obtained (by a $1D$ Fourier transform) from the Wigner distribution function.

Moreover, as noted in § 7.5.2, the Wigner representation is still valid when dealing with light of arbitrary coherence, the two-point field amplitude product in the Wigner integral being replaced by the mutual intensity, which is in fact the proper descriptor for partially coherent light (see also Problem 3).

As a final remark we emphasize that the matrices involved in Wigner optics calculation explicitly contains the wavelength of the light, by which, in fact, the off-diagonal entries of the ray-optics matrices are scaled. Accordingly, design or performance investigations of optical systems naturally display the wavelength dependence of the problem at hand [8]. This wavelength dependence can, of course, suitably be exploited when working, for instance, with polychromatic light to individualize the system parameters that can be adjusted in view of an achromatization of the system [8.3, 8.4].

"Ray"-optical description of longitudinal periodicities of optical wave fields

As a plain example of the practical manageability and conceptual transparency of the Wigner representation based approach to wave-optical propagation problems, we illustrate the application of the paraxial propagation rule for the Wigner distribution function to the investigation of longitudinal periodicities of optical wavefields, which manifest into the Talbot effect under coherent illumination, and into the Lau effect under partially coherent illumination [9]. Using the Wigner representation, it is possible to account for these inherently diffractive effects by a simple and visually appealing formalism, which indeed, due to the interpretation of the Wigner distribution function as the "amplitude" of geometrical rays, resorts to the ray-picture and formalism of paraxial geometrical optics. The presentation below follows that elaborated in [9.5], whose title is indeed rephrased in the section title above; the reader is directed to the quoted reference for a wider analysis.

We specifically focus on the transfer law for the Wigner distribution function, which, as demonstrated, writes in the form

$$\mathcal{W}(q,\kappa;z-z_i)=\mathcal{W}(q-\tfrac{(z-z_i)\kappa}{k_0},\kappa;z_i)\,, \tag{8.3.37}$$

for paraxial propagation from the plane at z_i to the plane at z. In the following, z_i is conveniently set to $z_i=0$.

Then, let us consider a coherently illuminated sinusoidal-amplitude grating in order to have on the grating plane (the $z=0$ plane) a wave amplitude like

$$\varphi(q)=1+\cos(\kappa_0 q), \tag{8.3.38}$$

which exhibits, of course, the spatial periodicity of the grating, $d=\frac{2\pi}{\kappa_0}$. The relevant Wigner distribution function is easily calculated to

$$\begin{aligned}\mathcal{W}(q,\kappa;0)=\delta(\kappa)[1+\tfrac{1}{2}\cos(2\kappa_0 q)]+\tfrac{1}{4}[\delta(\kappa-\kappa_0)+\delta(\kappa+\kappa_0)]\\ +[\delta(\kappa-\tfrac{\kappa_0}{2})+\delta(\kappa+\tfrac{\kappa_0}{2})]\cos(\kappa_0 q),\end{aligned} \tag{8.3.39}$$

whose support in the (q,κ) is pictorially sketched in Fig. 8.6.

At plane $z>0$, the Wigner distribution function changes to

$$\mathcal{W}(q,\kappa;z)=\mathcal{W}(q,\kappa;0)+\delta(\kappa-\tfrac{\kappa_0}{2})\cos[\kappa_0(q-\tfrac{\kappa_0 z}{2k_0})]+\delta(\kappa+\tfrac{\kappa_0}{2})\cos[\kappa_0(q+\tfrac{\kappa_0 z}{2k_0})] \tag{8.3.40}$$

which, as explicitly evidenced, differs from the distribution at the input plane $\mathcal{W}(q,\kappa;0)$ only along the two cross-terms lines $\kappa=\pm\kappa_0/2$, displaying there cos-like amplitudes with opposite z-depending initial phases.

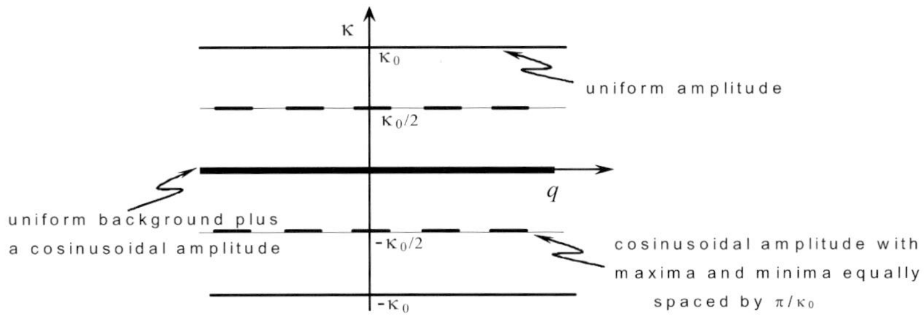

FIGURE 8.6. Phase-plane contours of the Wigner distribution function of the signal (8.3.38) showing the self- and cross-terms generated lines.

It is then evident that at planes z such that $\frac{\kappa_0^2 z}{2k_0} = 2m\pi$, which means

$$z = 2m\frac{d^2}{\lambda_0}, \tag{8.3.41}$$

where λ_0 is the vacuum wavelength of the incident light, the initial distribution is fully reproduced, and the field intensity as well. Recalling that according to our convention the propagation variable z is scaled by the refractive index n_0, it turns out that actually the physical length scales with the reciprocal of the wavelength in the medium, $\lambda = \lambda_0/n_0$. On the other hand, at planes

$$z = (2m+1)\frac{d^2}{\lambda_0}, \tag{8.3.42}$$

the amplitudes along the lines $\kappa = \pm\kappa_0/2$ have a contrast reversion. Finally, at planes

$$z = \tfrac{1}{2}(2m+1)\tfrac{d^2}{\lambda_0}, \tag{8.3.43}$$

the field intensity exhibits the lowest possible contrast. Notably the three formulae above just correspond to the locations of the positive-, negative- and zero-contrast Talbot planes; the axial distance $z_T = d^2/\lambda_0$ is in fact reported as the *Talbot distance.*

In contrast, if the grating (8.3.38) is incoherently illuminated, the description of the wave field must resort to the mutual intensity $\mathsf{J}(q,\widetilde{q};z)$, which as noted in § 7.5.2 is expressible in terms of the Wigner distribution function as

$$\mathcal{W}(q,\kappa;z) = \tfrac{1}{2\pi}\int \mathsf{J}(q,\widetilde{q};z)e^{-i\kappa\widetilde{q}}d\widetilde{q}, \tag{8.3.44}$$

where for the grating transmission function, we are considering, the mutual intensity turns out to specifically be

$$\mathsf{J}(q,\widetilde{q};0) = [1+\cos(\kappa_0 q)]^2\delta(\widetilde{q}). \tag{8.3.45}$$

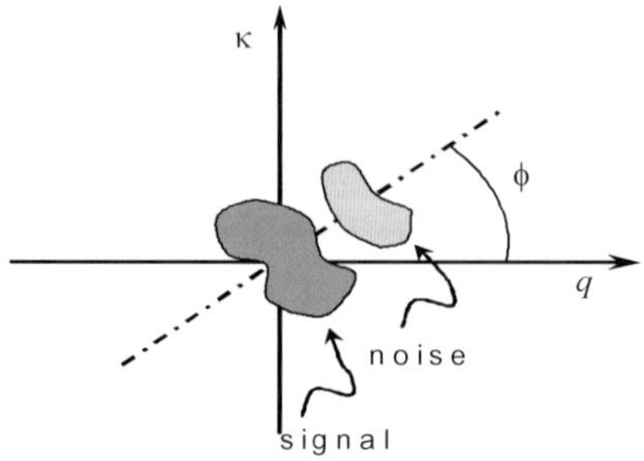

FIGURE 8.7. The overlap of the Wigner charts for the signal and the noise can be minimized if projected on axes at some angle to the Fourier axes.

The analysis of the longitudinal periodicities of the propagating wave field can clearly be performed in the same way as before, as suggested in Problem 2.

8.4 The Wigner distribution function and the optical Fourier transform: linking Fourier optics to Wigner optics

The fractional Fourier transform bridges the gap between Fourier and Wigner optics, that is, between optics described in terms of the complementary space and spatial frequency signal representations, the latter being conveyed by the Fourier transform, and optics described in terms of the joint space and frequency signal representation, as conveyed by the Wigner distribution function.

In fact, both the Wigner distribution function and the fractional Fourier transform yield hybrid signal representations, displaying space and spatial frequency features together. In accord with the analysis previously performed in connection with the fractional Fourier transform in both ray (§ 3.6.1) and wave (§ 5.4.1) optics, it is not surprising that, as we will see, performing the fractional Fourier transform of a signal manifests into a clockwise rotation of the corresponding Wigner chart in the Wigner- plane by an angle $\alpha\frac{\pi}{2}$. This correspondence between the Wigner distribution function and the fractional Fourier transform is formally expressed through the Radon transform.

The relation of the fractional Fourier transform to the Wigner distribution function was firstly asserted by Lohmann [10.2]. As a supposedly natural extension of the behaviour of the ordinary Fourier transform, the fractional Fourier transform of order α was defined as the transformation displayed by the function when the relevant Wigner distribution function is rotated by the angle $\alpha\frac{\pi}{2}$ in the inherent Wigner phase space. This relation has evidently applicative advantages, suggesting, for instance, optical designs for efficient signal multiplexing or for optimal filtering of signals from degradating noise terms [10.4, 10.7]. As sketched in Fig. 8.7, it may happen that the Wigner distribution functions of the signal and the noise overlap, when projected on both the space and spatial frequency axes, but have little or no overlap on some other direction, rotated by some angle with respect to the Fourier axes. Removal of the undesired signal components can then be actuated by using suitable amplitude masks after performing the appropriate fractional Fourier transform of the signal, and hence the appropriate rotation in the Wigner plane.

We describe in the forthcoming paragraphs this rotation based relation between the Wigner distribution function and the fractional Fourier transform. Firstly we examine the plain case of the ordinary Fourier transform.

8.4.1 *Fourier transform and Wigner distribution function: optical analog of the $\frac{\pi}{2}$-rotation in the Wigner phase plane*

In order to focus on the pure rotation part of the overall transformation induced by the optical Fourier transform in the Wigner phase plane, we explicitly scale the q and κ variables to the unitless $q\sqrt{\frac{k_0}{f}}$ and $\kappa\sqrt{\frac{f}{k_0}}$, although the familiar q and κ are still used to refer as well to the unitless coordinates. Note that the reciprocal scaling of q and κ by the same factor $\sqrt{\frac{k_0}{f}}$ does not affect the formal definition of the Wigner distribution function, being the wave function scaled as well according to the normation condition. Accordingly, the Fourier transform matrix $\mathbf{F}$ comes to correspond to the rotation matrix with angle $\frac{\pi}{2}$:

$$\mathbf{F} \xrightarrow[\text{unitless variables}]{} \mathbf{R}(\tfrac{\pi}{2}) = \begin{pmatrix} 0 & 1 \\ -1 & 0 \end{pmatrix}, \tag{8.4.1}$$

thus implying a direct relation between the optical Fourier transform $\widehat{\mathsf{F}}_f$ and the mathematical Fourier transform $\widehat{\mathcal{F}}$ (see Eq. (5.4.4)), apart from the factor $e^{-i\pi/4}$, which however has no effect on the Wigner distribution function. Hence, the transformation of the Wigner distribution function from $\mathcal{W}_i(q,\kappa)$ to

$$\mathcal{W}_o(q,\kappa) = \mathcal{W}_i(-\kappa, q) \tag{8.4.2}$$

manifests into a clockwise rotation by $\frac{\pi}{2}$ about the phase plane origin.

This $\frac{\pi}{2}$-rotation based relation in the Wigner plane between the Fourier transform and the Wigner distribution function has already been remarked when comparing the Wigner charts of some primary Fourier transform pairs, like the impulse and monofrequency signals (§ 6.4.4), or the rectangle and sinc-like signals (§ 6.4.2). In more detail, consider the point-like signal $\delta(q - q_0)$ with the relative Wigner distribution function

$$\varphi_{q_0}(q) = \delta(q - q_0) \quad \longrightarrow \quad \mathcal{W}_{q_0}(q,\kappa) = \tfrac{1}{2\pi}\delta(q - q_0), \tag{8.4.3}$$

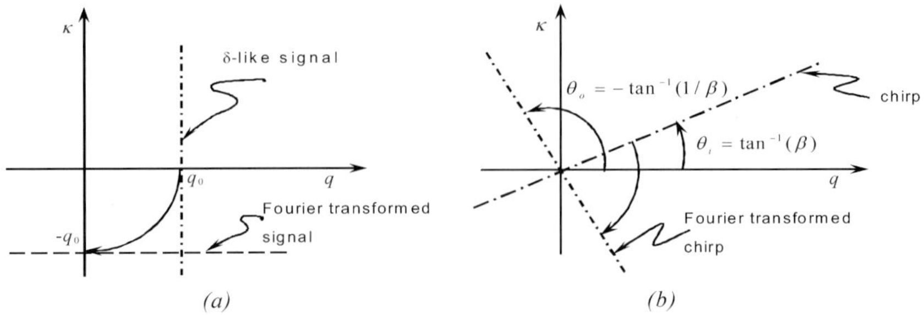

FIGURE 8.8. $\frac{\pi}{2}$-rotation of the Wigner charts of *(a)* a δ-like signal and *(b)* a chirp, under Fourier transformation.

whose support in the phase plane is the vertical line through the point $q = q_0$. The optical Fourier transform of $\varphi_{q_0}(q)$ is

$$[\widehat{\mathsf{F}}\varphi_{q_0}](q) \equiv \varphi_{\kappa_0}(q) = \frac{1}{\sqrt{2\pi i}} e^{iq_0 q}, \quad \kappa_0 = q_0 . \tag{8.4.4}$$

The Wigner distribution function is then

$$\mathcal{W}_{\kappa_0}(q,\kappa) = \frac{1}{2\pi}\delta(\kappa + q_0), \tag{8.4.5}$$

whose trace on the (q,κ) plane is a horizontal line through the point $\kappa = -q_0$, resulting evidently from a $\frac{\pi}{2}$ clockwise rotation of the $\mathcal{W}_{q_0}$-line (Fig. 8.8.a)).

Needless to say, the complementary view occurs in the case of the spatial harmonic, the horizontal line of the corresponding Wigner chart being rotated into a vertical line under Fourier transformation.

As a further example, we consider the pure chirp $e^{i\beta q^2/2}$ with the corresponding Wigner distribution function

$$\varphi_{\text{chirp}}(q) = e^{i\beta\frac{q^2}{2}} \quad \longrightarrow \quad \mathcal{W}_{\text{chirp}}(q,\kappa) = \delta(\kappa - \beta q), \tag{8.4.6}$$

the relevant support in the Wigner plane being the straight line $\kappa = \beta q$ (§ 6.4.4). The optical Fourier transform of $\varphi_{\text{chirp}}(q)$ is

$$[\widehat{\mathsf{F}}\varphi_{\text{chirp}}](q) = \frac{1}{\sqrt{\beta}} e^{-i\frac{q^2}{2\beta}}, \tag{8.4.7}$$

which, as expected, is still a chirp (see Eq. (5.2.2)). The Wigner distribution function is then

$$\mathcal{W}_{\widehat{\mathsf{F}}\varphi}(q,\kappa) = \frac{1}{\beta}\delta(\kappa + \frac{1}{\beta}q), \tag{8.4.8}$$

which clearly individualizes on the (q,κ) plane the straight line, $\kappa = -q/\beta$, orthogonal to the line $\kappa = \beta q$ (Fig. 8.8.b)).

Then, one can synthesize the (optical) Fourier transform/Wigner distribution function relation through the scheme

$$\boxed{\varphi(q) \xrightarrow[\text{unitless variables}]{\widehat{\mathsf{F}}} \tfrac{1}{\sqrt{i}}\,\widetilde{\varphi}(q) \iff \mathcal{W}_\varphi(q,\kappa) \xrightarrow[\text{unitless axes}]{\mathbf{R}(\frac{\pi}{2})} \mathcal{W}_\varphi[\left(\begin{smallmatrix}0 & 1\\ -1 & 0\end{smallmatrix}\right)\left(\begin{smallmatrix}q\\ \kappa\end{smallmatrix}\right)]}$$

8.4.2 *Fractional Fourier transform and Wigner distribution function: optical analog of the ϕ-rotation in the Wigner phase plane*

The fractional Fourier transforming operation produces a clockwise rotation by the angle ϕ of the Wigner distribution function in the (q, κ) plane. In fact, the optical fractional Fourier transform matrix $\mathbf{F}^\alpha$ corresponds to the rotation matrix by the angle $\phi = \alpha\pi/2$ in the (q, κ) plane with unitless axes:

$$\mathbf{F}^\alpha \xrightarrow[\text{unitless variables}]{} \mathbf{R}(\phi) = \begin{pmatrix} \cos\phi & \sin\phi \\ -\sin\phi & \cos\phi \end{pmatrix}. \tag{8.4.9}$$

Accordingly, the optical and mathematical fractional Fourier transforms $\widehat{\mathsf{F}}^\alpha_{f_S}$, $\widehat{\mathcal{F}}^{\,\alpha}$ come to differ simply by the phase factor $e^{-i\alpha\pi/4}$ (see Eq. (5.4.38)). So, as the signal $\varphi_i(q)$ is acted by the optical fractional Fourier transform operator:

$$\widehat{\mathsf{F}}^\alpha : \varphi_i(q) \to \varphi_o(q) = [\widehat{\mathsf{F}}^\alpha \psi_i](q),$$

the corresponding Wigner distribution function changes from $W_i(q, \kappa)$ to

$$\mathcal{W}_o^{(\alpha)}(q, \kappa) = \mathcal{W}_i([\mathbf{R}(\phi)]^{-1}\left(\begin{smallmatrix}q\\ \kappa\end{smallmatrix}\right)) = \mathcal{W}_i(q\cos\phi - \kappa\sin\phi,\ q\sin\phi + \kappa\cos\phi), \tag{8.4.10}$$

which is obtained rotating $\mathcal{W}_i(q, \kappa)$ clockwise by the angle $\phi = \alpha\frac{\pi}{2}$.

Evidently this mirrors the behaviour of the ray density distribution $\rho(q, p; z)$ (see Fig. 3.16). Also, it suggests an alternative and equivalent definition of the fractional Fourier transform as originally proposed in [10.2]; accordingly, the fractional Fourier transform of order α of the signal $\varphi_i(q)$ can implicitly be defined as the transformation experienced by $\varphi_i(q)$, when the associated Wigner distribution function is rotated in the (q, κ) plane by the α-th fraction of $\frac{\pi}{2}$. As learned in § 3.6.1, it turns out that such a Wigner rotation in phase plane can be accomplished by a simple bulk optics implementation consisting of free-medium lengths and thin lenses, which realize the proper three-stage transformation, i.e., the consecutive q, κ, and q shears or κ, q, and κ shears, yielding eventually an overall rotation of the Wigner distribution function. It was just by this reasoning that the fractional Fourier transform setups, Lohmann's type I and II, described in § 3.6.1, have been conceived.

Lohmann's definition parallels the definition given by Mendlovic and Ozaktas [10.1], modeled on the variation of the signal amplitude when propagating along a quadratic graded-index medium by a length proportional to the order

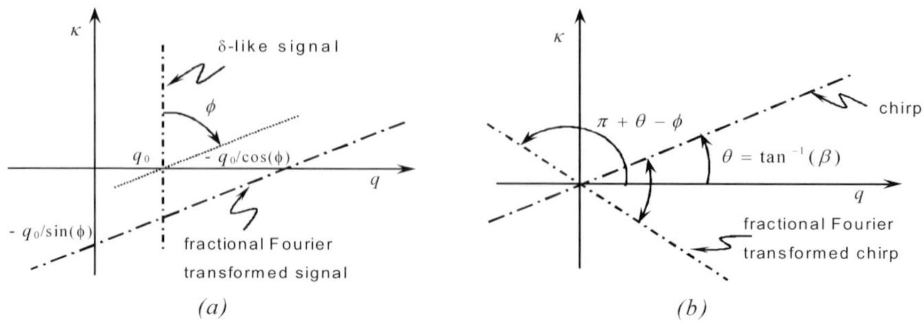

FIGURE 8.9. ϕ-rotation of the Wigner charts of *(a)* a δ-like signal and *(b)* a chirp, under fractional Fourier transformation.

of the transform, as discussed in § 3.5. The equivalence of the two definitions was proved in [10.3]. Basically it relies on the fact that, as remarked in § 8.3.3, linear canonical transformations, which are represented by symplectic matrices (in particular, 2×2 for one-dimensional systems), manifest in the signal-space representation through integral transforms having the mathematical form (5.3.4) and in the space-frequency representation through the transformation law (8.3.35) for the Wigner distribution function. The linear canonical transformation underlying the α-th order fractional Fourier transform is just the clockwise rotation with angle $\alpha\frac{\pi}{2}$ in the phase space of the appropriate conjugate variables. Yet again, first-order optics plainly provides a concrete realization of an abstract mathematical relation.

To help a comparison with the examples described in the previous section, we consider the δ-like signal $\delta(q - q_0)$ and the chirp $e^{i\beta q^2/2}$. Using the proper expression in Table (5.4.39) and the result of Problem 5.2, we find that the δ-like signal is fractionally Fourier transformed into the chirp

$$\delta(q - q_0) \xrightarrow[\text{unitless variables}]{\widehat{\mathsf{F}}^\alpha} \frac{1}{\sqrt{2\pi i \sin\phi}} e^{i\frac{q_0^2}{2}\sin\phi\cos\phi} e^{\frac{i}{2}(q\cos\phi - q_0)^2\cot\phi} e^{-iq_0 q\sin\phi}. \tag{8.4.11}$$

The Wigner distribution function is therefore a δ function, namely

$$\mathcal{W}_{\widehat{\mathsf{F}}^\alpha\delta}(q,\kappa) = \frac{1}{2\pi\sin\phi}\delta(\kappa - q\cot\phi + q_0\sin^{-1}(\phi)), \tag{8.4.12}$$

whose support results evidently from a rotation by ϕ of the input Wigner line $q = q_0$ (Fig. 8.9.a).

Finally, let us compare the Wigner distribution function $\delta(\kappa - \beta q)$ of the chirp $\varphi_{\text{chirp}}(q) = e^{i\beta q^2/2}$ with that of the fractionally Fourier transformed signal, which writes (see Table (5.4.39))

$$[\widehat{\mathsf{F}}^\alpha\varphi_{\text{chirp}}](q) = \frac{1}{\sqrt{\cos\phi + \beta\sin\phi}} e^{\frac{i}{2}q^2\frac{\beta - \tan\phi}{1 + \beta\tan\phi}}. \tag{8.4.13}$$

The Wigner distribution function is then

$$\mathcal{W}_{\widehat{\mathsf{F}}^{\alpha}\varphi_{\text{chirp}}}(q,\kappa) = \tfrac{1}{\cos\phi+\beta\sin\phi}\delta(\kappa - q\tfrac{\beta-\tan\phi}{1+\beta\tan\phi}), \tag{8.4.14}$$

which identifies a line through the phase plane origin making the angle $\theta_\alpha = \tan^{-1}[\frac{\beta-\tan\phi}{1+\beta\tan\phi}] = \pi+\theta-\phi$ with the q axis, thus confirming the expected clockwise ϕ-rotation of the input Wigner line (Fig. 8.9.b)); $\theta = \tan^{-1}(\beta)$.

Then, one can synthesize the relation between the (optical) fractional Fourier transform and the Wigner distribution function through the scheme

$$\boxed{\varphi(q) \xrightarrow[\text{unitless variables}]{\widehat{\mathsf{F}}^{\alpha}} \mathrm{e}^{-i\alpha\frac{\pi}{4}}\,\widetilde{\varphi}_\alpha(q) \Longleftrightarrow \mathcal{W}_\varphi(q,\kappa) \xrightarrow[\text{unitless axes}]{\mathbf{R}(\phi)} \mathcal{W}_\varphi[\begin{pmatrix}\cos\phi & -\sin\phi\\ \sin\phi & \cos\phi\end{pmatrix}\binom{q}{\kappa}]}$$

8.4.3 *Fractional Fourier transform and Radon-Wigner transform*

As a simple restatement of the Wigner rotation property, the connection between the Radon-Wigner transform and the fractional Fourier transform can be demonstrated [11.8], making it possible to transpose the signal processing capabilities of the Radon-Wigner transform into the fractional Fourier field, which benefits from easy digital computation and optical realization.

The $2D$ *Radon transform* is a set of $1D$ functions obtained by integrating a $2D$ function over all possible lines in the pertinent $2D$ domain. It is convenient to think of the Radon transform of a $2D$ function $v(x',y')$ as a collection of projections of $v(x',y')$ onto lines, each being identifiable as a coordinate axis into new frames (x,y)'s suitable rotated with respect to the original frame (x',y') (Fig. 8.10). The Radon transform consists then of two steps: rotation followed by projection, as expressed in the mathematical definition [11]

$$[\widehat{\mathcal{R}}v(x',y')](x,\phi) = \int_{\mathbb{R}} v([\mathbf{R}(\phi)]^{-1}\tbinom{x}{y})dy, \tag{8.4.15}$$

where $\mathbf{R}(\phi)$ denotes as usual the clockwise rotation matrix and (x,y) represent a coordinate system counterclockwise rotated by ϕ with respect to the original set (x',y'). For a fixed angle ϕ, the above yields integrals of $v(x',y')$ along lines parallel to the new y axis; it is evidently a $1D$ function, we denote as $[\widehat{\mathcal{R}}v]_\phi(x)$.

When applied to the Wigner distribution function the above definition straightforwardly leads to the squared modulus of the fractional Fourier transform. In fact, applying the definition (8.4.15) to the Wigner distribution function of a given signal $\varphi(q)$, i.e.,

$$[\widehat{\mathcal{R}}\mathcal{W}]_\phi(q) = \int_{\mathbb{R}} \mathcal{W}([\mathbf{R}(\phi)]^{-1}\tbinom{q}{\kappa})d\kappa, \tag{8.4.16}$$

we obtain the *Radon-Wigner transform* [11.6, 11.7]. Then, on account of the above established Wigner rotation property, the direct correspondence between the Radon-Wigner transform (8.4.16) and the squared modulus of the

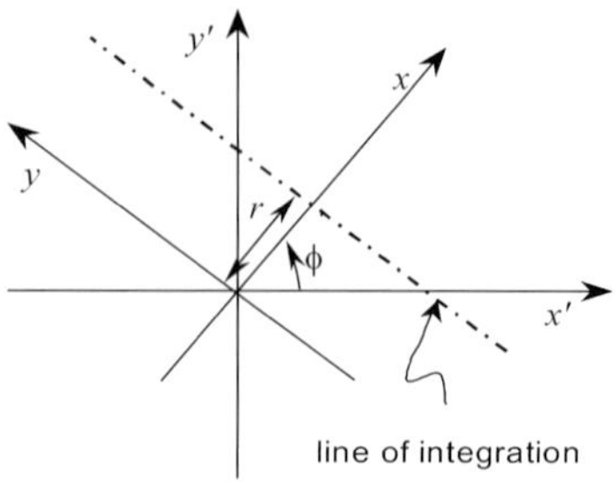

FIGURE 8.10. Geometry for the $2D$ Radon transform.

fractional Fourier transform of the signal $\varphi(q)$ immediately follows as [11.8, 11.7]

$$[\widehat{\mathcal{R}}_\phi \mathcal{W}](q) = \left|\widehat{\mathsf{F}}^\alpha \varphi\right|^2 (q) = |\widetilde{\varphi}_\alpha(q)|^2, \tag{8.4.17}$$

where obviously $\phi = \alpha\frac{\pi}{2}$ (and, we recall, we are still working with unitless axes in the Wigner plane).

Evidently the above relation has a considerable applicative relevance as it relates the fractional Fourier transform, which is implementable by simple optical means, to the Radon-Wigner transform, which is a basic tool for reconstruction of the Wigner distribution function and accordingly of the corresponding "signal" not only in optics but also, for instance, in signal processing and in measurements concerned with the quantum mechanical wave function. In other words, relation (8.4.17) suggests an easily practicable tactics for *tomographic* reconstruction of an "object".

In particular, if we take $\alpha = 0$ or $\alpha = 1$, we obtain on the left side of (8.4.17) the integral of the Wigner distribution function along a line in the $\frac{\pi}{2}$ and π-direction, respectively, and correspondingly on the right the amplitude squared of the signal and its Fourier spectrum. In other words, we recover respectively the space and the frequency marginals (Eqs. (6.3.26)), which indeed, as projections of the Wigner distribution function onto the q- and κ-axis, are particular realizations of the general relation (8.4.17) between the Radon-Wigner and fractional Fourier transforms. This also confirms the hybrid character of the fractional Fourier transform, which in a sense interpolates between the pure space and frequency signal representations; relation (8.4.17) conveys a geometrical view for such a hybrid character. In this connection, we mention that a space-order, i.e. (q, α), signal representation has been suggested as an alternative way to the space-frequency representation, with the intent of displaying signal features that may remain unrevealed by the space-frequency (as well as other) signal representation [10.5]. Also, an optical system to generate the (q, α) chart of a signal has been implemented, as described in [10.6].

Relation (8.4.17) can therefore be viewed as a "generalized marginal" involving projections of the Wigner distribution function onto arbitrary lines in the (q, κ) plane. Interestingly, as the space and frequency marginals (6.3.26) find their generalization in (8.4.17), likewise the inversion formulas (6.3.28) for the space and frequency signal representations find their generalization in [11.9]

$$\widetilde{\varphi}_\alpha(q)\widetilde{\varphi}^*_\alpha(0) = \int_{\mathbb{R}} \mathcal{W}(\tfrac{q}{2}\cos\phi - \kappa\sin\phi,\ \tfrac{q}{2}\sin\phi + \kappa\cos\phi)e^{i\kappa q}d\kappa, \tag{8.4.18}$$

that the reader may prove along the same line yielding (6.3.28) from (6.2.2).

Reconstruction of the Wigner distribution function from the Radon transform is made in principle possible by the *projection-slice* theorem [11.2]-[11.4]. This theorem states that the $1D$ Fourier transform of a projection of a $2D$ function yields one line through the $2D$ Fourier transform of the function itself, whose Fourier frequency space can therefore be sampled on a set of lines through the origin by transforming projections at different angles ϕ. To see this, we firstly note that a *slice* of a $2D$ function $f(x', y')$ is defined by

$$\text{slice}[f(x', y')](\xi) = f(\xi\cos\phi, \xi\sin\phi), \tag{8.4.19}$$

which evidently for fixed angle ϕ and variable $\xi \in \mathbb{R}$ selects a slice of the function along the line $(x' = \xi\cos\phi, y' = \xi\sin\phi)$ in the inherent plane (x', y'). Take now the $1D$ Fourier transform of the Radon transform (8.4.15),

$$\{\widehat{\mathcal{F}}[\widehat{\mathcal{R}}v]_\phi\}(\xi) = \frac{1}{\sqrt{2\pi}}\int dx e^{-i\xi x}\int_{\mathbb{R}} v([\mathbf{R}(\phi)]^{-1}\tbinom{x}{y})dy, \tag{8.4.20}$$

which, turning back to the old axes (x', y') in the double integral, yields

$$\begin{aligned}\{\widehat{\mathcal{F}}[\widehat{\mathcal{R}}v]_\phi\}(\xi) &= \tfrac{1}{\sqrt{2\pi}}\int dx' \int dy' v(x', y')e^{-i[x'\xi\cos\phi + y'\xi\sin\phi]} \\ &= \sqrt{2\pi}[\widehat{\mathcal{F}}v](\xi\cos\phi, \xi\sin\phi),\end{aligned} \tag{8.4.21}$$

i.e., the slice of the $2D$ Fourier transform of the original function $v(x', y')$ through the line $(\nu = \xi\cos\phi, \mu = \xi\sin\phi)$ in the Fourier plane (ν, μ).

Accordingly, reconstruction of the Wigner distribution function, and so of the corresponding signal, by tomographic measurements could in principle be achieved through the steps:

• take $1D$ projections of the $2D$ Wigner distribution function at successive angles ϕ covering at least $180°$, or equivalently take the fractional Fourier transform of the signal for orders corresponding to the sampled ϕ-range (this provides the so called (q, α) or Radon-Wigner display, of the signal);

• Fourier transform all projections, thus obtaining slices of the $2D$ Fourier transform of the Wigner distribution function at the same sampling angles ϕ in the relevant Fourier space;

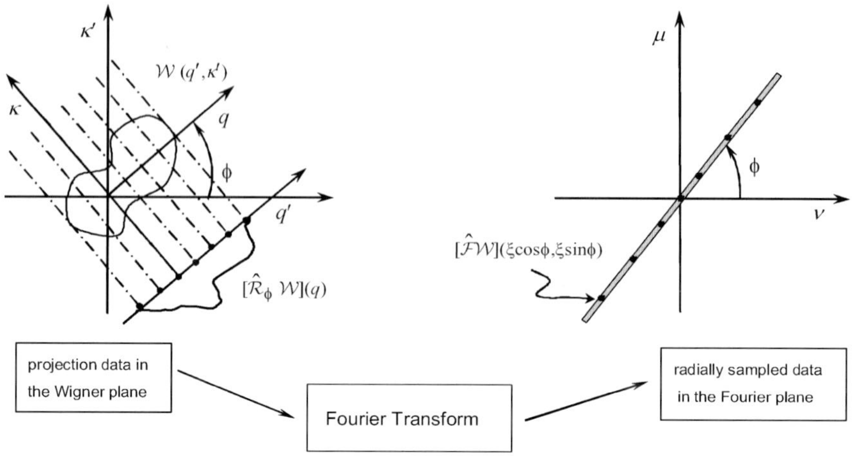

FIGURE 8.11. Schematic of the projection-slice theorem, according to which the collection of projection data in the (q', κ') plane generate by Fourier transform a collection of data in the Fourier plane along radial lines at the same angles as the projection axes in the (q', κ') domain (inspired by [11.5]).

• evaluate the $2D$ inverse Fourier transform of all obtained data to display the $2D$ Wigner distribution function, from which the signal amplitude can be reconstructed to within a phase constant.

Figure 8.11 illustrates the Fourier relation between the collection of projection data of $\mathcal{W}$, sampled at discrete intervals along the q-lines at the chosen angles ϕ, and the corresponding collection of Fourier resulting data, representing the $2D$ Fourier transform of $\mathcal{W}$ at discrete intervals along lines making in the relevant Fourier plane the same angles as the respective projection axes in the Wigner plane.

We mention that, in practice, the above sketched inversion scheme is rarely used because firstly all data must be obtained before reconstruction can begin and secondly $2D$ Fourier transform are quite expensive in computing time. An alternative widely applied method is the *filtered back projection*, by which the complete $2D$ reconstruction of the Wigner distribution function of the investigated "object" is obtained by adding modified versions of each of the projections taken over at least 180°. The modification amounts to convoluting the original projection with a spatial filter function, called the Radon or ramp filter, which is the same for all projections and compensate for the decreasing deal of information when moving radially far from the origin in the Fourier plane. Yet alternative methods have been devised for quantum-state tomography, since the filtering procedure used in the filtered back projection algorithm introduces a certain smearing of the data, which can hide some important features of the quantum states. The reader interested in this data-processing related issue may consult the relative literature [12].

Before closing this section, we wish to evidence a relation between the Fourier transform of the Radon-Wigner transform and the ambiguity function. On account of the Fourier-like relation of the Wigner distribution function $\mathcal{W}(q,\kappa)$ to the ambiguity function $\mathcal{A}(\widetilde{q},\widetilde{\kappa})$, it becomes evident that the slice of the $2D$ Fourier transform of the Wigner distribution function entering (8.4.21) is as well a slice of the ambiguity function $\mathcal{A}(-\widetilde{q},\widetilde{\kappa})$. Explicitly, we may write

$$\{\widehat{\mathcal{F}}[\widehat{\mathcal{R}}\mathcal{W}]_\phi\}(\xi) = \sqrt{2\pi}\mathcal{A}(-\xi\cos\phi, \xi\sin\phi). \tag{8.4.22}$$

Of course, the slice is oriented at the angle ϕ with respect to the horizontal axis, which in the ambiguity function plane turns out to be the $\widetilde{\kappa}$-axis. Finally, note that by (8.4.17) the above relation is equivalent to

$$[\widehat{\mathcal{F}}|\widetilde{\varphi}_\alpha(q)|^2](\xi) = \sqrt{2\pi}\mathcal{A}(-\xi\cos\phi, \xi\sin\phi), \tag{8.4.23}$$

which remarkably signifies a direct Fourier transform relation between measurable fractional Fourier transform intensities and ambiguity function slices in the relative $(\widetilde{q},\widetilde{\kappa})$ plane. Evidently, it suggests an alternative ambiguity-function based approach to phase-space tomographic reconstruction of "signals" to the Wigner distribution function based approach, represented by (8.4.17); in fact, we have learned in § 7. that the ambiguity function is linked through a $1D$ Fourier transform to the mutual intensity for both coherent and partially coherent optical wave fields (as well as to the density matrix in the quantum mechanical context).

A bit of mathematics about the Radon transform

Different and equivalent expressions can be given for the Radon transform of a $2D$ function. Let be $v(x',y')$ the function to be transformed. The Radon transform is the set of line integrals of $v(x',y')$ along all possible lines in the (x',y') plane:

$$\widehat{\mathcal{R}}\{v(x',y')\} = \int_\ell v(x',y')d\ell, \tag{8.4.24}$$

where $d\ell$ is the elemental Euclidean arc length along the line ℓ. Of course, the above is a $2D$ function, since a line in the (x',y') plane is specified by two parameters. Specifically, ℓ can be given in the slope intercept form or, equivalently, in the normal form; namely as

$$y' = \eta x' + \tau \quad \text{or} \quad x'\cos\phi + y'\sin\phi = r, \tag{8.4.25}$$

where ϕ is the polar angle and r the perpendicular distance from the origin to the line (see Fig. 8.10). Accordingly, the integral in (8.4.24) takes the form

$$[\widehat{\mathcal{R}}v](\eta,\tau) = \iint dx'dy'v(x',y')\delta(y' - (\eta x' + \tau)), \tag{8.4.26}$$

or,

$$[\widehat{\mathcal{R}}v](r,\phi) = \iint dx'dy'v(x',y')\delta(r - x'\cos\phi + y'\sin\phi). \tag{8.4.27}$$

Of course, if we define new axes (x, y) rotated by ϕ counterclockwise with respect to the original (x', y') so that $\binom{x'}{y'} = [\mathbf{R}(\phi)]^{-1}\binom{x}{y}$, we end up with the expression previously used to introduce the Radon-Wigner transform. Then, we may say that the Radon transform is a mapping from the (x', y') domain to the $2D$ domain individualized by the chosen line parameter sets. Needless to say, both (8.4.26) and (8.4.27) yield a $1D$ function if one of the line parameters are fixed, like, for instance, the slope η in (8.4.26) and the polar angle ϕ in (8.4.27). Also, if we define new axes (x, y) rotated by ϕ counterclockwise with respect to the old (x', y') so that $\binom{x'}{y'} = [\mathbf{R}(\phi)]^{-1}\binom{x}{y}$, we end up with the expression previously used to introduce the Radon-Wigner transform.

The Radon transform is linear; being

$$[\widehat{\mathcal{R}}(v_1 + v_2)](\eta,\tau) = [\widehat{\mathcal{R}}v_1](\eta,\tau) + [\widehat{\mathcal{R}}v_2](\eta,\tau), \tag{8.4.28}$$

and scales with the function according to

$$[\widehat{\mathcal{R}}v(\mathfrak{a}x', \mathfrak{b}y')](\eta,\tau) = \tfrac{1}{|\mathfrak{a}|}[\widehat{\mathcal{R}}v](\tfrac{\mathfrak{b}}{\mathfrak{a}}\eta, \mathfrak{b}\tau). \tag{8.4.29}$$

Interestingly, the Radon transform of the $1D$ convolution $h(x', y')$ of a $2D$ function $v(x', y')$ with a $1D$ function $t(y')$, i.e.,

$$h(x', y') = \int v(x', y' - y)t(y)dy, \tag{8.4.30}$$

turns out to be the $1D$ convolution of the Radon transform of $v(x', y')$ with $t(y')$ with respect to the variable τ; explicitly, we find

$$[\widehat{\mathcal{R}}h](\eta,\tau) = \int [\widehat{\mathcal{R}}v](\eta, \tau - \tau')t(\tau')d\tau'. \tag{8.4.31}$$

The reader may easily verify properties (8.4.28)-(8.4.31). Other properties can be found in the related literature in [11].

As an illustrative example, we calculate the Radon transform of the $2D$ symmetrical Gaussian

$$\mathcal{G}(x', y') = \tfrac{1}{2\pi\sigma^2} e^{-\frac{x'^2 + y'^2}{2\sigma^2}}. \tag{8.4.32}$$

Using definition (8.4.26), we obtain

$$\begin{aligned}[\widehat{\mathcal{R}}\mathcal{G}](\eta,\tau) &= \tfrac{1}{2\pi\sigma^2} \iint dx'dy' e^{-\frac{x'^2 + y'^2}{2\sigma^2}} \delta(y' - (\eta x' + \tau)) \\ &= \tfrac{1}{2\pi\sigma^2} e^{-\frac{\tau^2}{2\sigma^2}} \int dx' e^{-\frac{(1+\eta^2)x'^2}{2\sigma^2} x'^2 - \frac{\eta\tau}{\sigma^2} x'} = \tfrac{1}{\sqrt{2\pi}\Sigma(\sigma,\eta)} e^{-\frac{\tau^2}{2\Sigma^2(\sigma,\eta)}},\end{aligned} \tag{8.4.33}$$

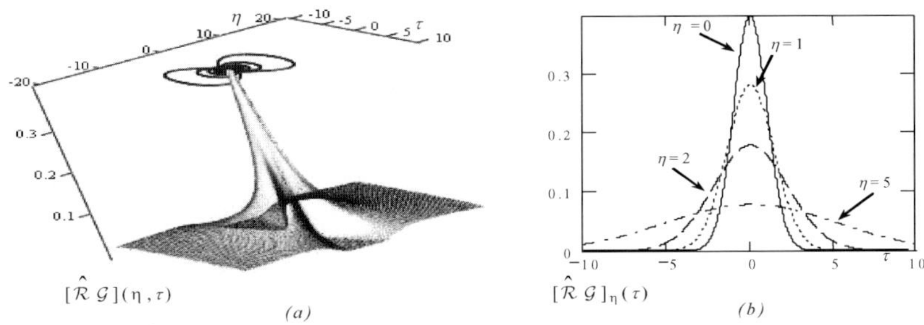

FIGURE 8.12. *(a)* $3D$ plot of the Radon transform of the Gaussian $\mathcal{G}(x', y') = \frac{1}{2\pi} e^{-(x'^2+y'^2)/2}$, and *(b)* corresponding projections at different line slopes.

whose $3D$ graph is shown in Fig. 8.12.a). For a fixed slope η, it turns into a $1D$ Gaussian in the intercept τ (Fig. 8.12.b)), whose width $\Sigma(\sigma, \eta)$ is determined by the slope η and the width σ of the original function according to

$$\Sigma(\sigma, \eta) = \sqrt{1 + \eta^2}\sigma. \tag{8.4.34}$$

Fractional Fourier transform based phase-space tomography

Relation (8.4.17) has a powerful applicative content. It tell us that it is possible to determine a complex wave field (i.e., its amplitude and phase as functions of position) by measurements of intensity only and, yet remarkably, using refractive optics only (so that less power field is lost). Also, although we have derived relation (8.4.17) within a purely (classical) optical context, it holds as well within a quantum mechanical context on due account, of course, of the frequently remarked analogy with the quantum-mechanical formulation of the phase-space representation. Thus, $\varphi(q)$ on one side may refer to a scalar coherent monochromatic classical electromagnetic wave or to a quantum-mechanical matter wave. Correspondingly, the line integral on the other side may signify the marginal distribution of the Wigner distribution function for the rotated phase-plane variable defined with respect to the rotation angle ϕ or the quasiprobability distribution for the rotated quadrature defined with respect the reference phase ϕ [12.1, 12.2]. Moreover, as briefly noted many times, also partial coherence as well as statistical mixture of quantum states can be accounted for by (8.4.17), with the Wigner distribution function being in that case expressed in terms of the mutual intensity or the density matrix.

As an example of the application potential of the fractional Fourier transform based approach to phase-space tomography, we synthesize below the detailed report of [12.4], where indeed that potential has been demonstrated (see also [12.8]). Figure 8.13.b) reproduces the scheme of the experimental setup de-

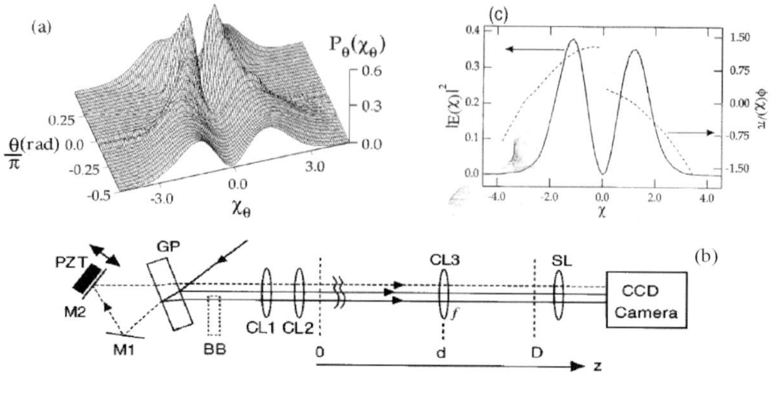

FIGURE 8.13. *(a)* Measured scaled intensity distribution for the fully coherent beam obtained by the experimental setup shown in *(b)*. *(c)* The intensity profile (solid curve) and the phase structure (dashed curves) of the reconstructed complex wave field obtained in the fully coherent case. When the intensity is near zero the phase is undefined and is not plotted (adapted with permission from [12.4]).

scribed in [12.4], implemented for reconstruction of a complex wave field $E(x, 0)$ located in the $z = 0$ plane (x is coordinate in the horizontal transverse direction). The output of a He-Ne lase, after being spatially filtered in a single-mode optical fiber, is transformed, by two surface reflections from a glass plate (GP) into two spatially separated, nearly parallel beams. "A constant relative phase is maintained between these beams so that together they represent a field with full spatial coherence." By passing through the two cylindrical lenses CL1 and CL2, the two beams are expanded in the vertical direction, thus yielding a field depending only on a single transverse coordinate.

The field then propagates through the cylindrical lens CL3, located at $z = d$ and oriented 90° with respect to both CL1 and CL2 so as to affect the field only along the horizontal direction. As noted in § 5.4.5 , an ABCD optical system can be regarded as performing a scaled fractional Fourier transform of scaled input with specific restrictions being imposed on the matrix elements if a real order transform is required as in the case we are dealing with. Therefore, varying d and D for fixed f amounts to varying the order of the fractional transform and so the rotation angle in the phase plane, which in [12.4] is denoted by θ. "For each combination of d and D, the intensity distribution $I(x, D, d, f)$ was digitized, normalized, and scaled to give the marginal distribution shown in" Fig. 8.13.a). The variable χ_θ denotes the rotated (and scaled to some characteristic length) phase-plane variable, and χ is obtained from χ_θ by a further conveniently θ-depending scaling.

A filtered back projection algorithm was used to reconstruct the Wigner distribution function, form which the complex wave field can be determined to within a constant phase (Fig. 8.13.c)).

For more details the reader is directly addressed to the quoted paper, where the reconstruction of the mutual intensity for a partially coherent wave field is also reported. In fact, the dotted line in Fig. 8.13.b) represents the path of a third beam, that was used to produce, by reflections by mirrors M1 and M2, phase randomization by a piezo-electric translator (PZ) and lower reflected beam block by a beam block (BB), "a field similar in intensity to the field reconstructed above, except that it is partially coherent."

Finally we mention that an alternative approach to the phase-space tomography based on the relation between the fractional Fourier transform and the ambiguity function (Eq. (8.4.23)) has been proposed in [12.5, 12.6].

8.5 Transport equation for the Wigner distribution function

Here we derive the transport equation for the Wigner distribution function, which in a sense parallels Liouville's equation for the ray-density distribution $\rho(q, p; z)$. It can be regarded as the continuous version of the input-to-output relations, discussed in § 8.3, which of course are recovered as solution to the transport equation for paraxial propagation.

Evidently, as the propagation of light rays involves a corresponding propagation of the associated ray density distribution, the propagation of light waves involves a corresponding propagation of the associated distribution function. In both cases, the equations for the phase space distributions are derived by the equations modelling respectively the ray propagation and the wave propagation, i.e. the Hamilton equations for the ray position and momentum variables, and the Helmholtz equation for the wave complex amplitude. In other words, we establish a relation between the evolution equation for the system and that for the associated phase space distribution function.

Firstly we devote a short paragraph to elaborate a bit of operator formalism, which will reveal its conceptual and practical convenience later when deriving the transport equation for the Wigner distribution function.

8.5.1 *"Wigner representation" of the position and momentum operators*

Let us start with the question concerned with the possibility of making the following identities

$$\begin{aligned} \mathcal{W}_{\varphi,q\psi}(q,\kappa) &= \widehat{\mathsf{Q}}\mathcal{W}_{\varphi,\psi}(q,\kappa), \\ \mathcal{W}_{\varphi,-i\frac{\partial}{\partial q}\psi}(q,\kappa) &= \widehat{\mathsf{K}}\mathcal{W}_{\varphi,\psi}(q,\kappa), \end{aligned} \tag{8.5.1}$$

to be true, individualizing definite expressions for the operators $\widehat{\mathsf{Q}}$ and $\widehat{\mathsf{K}}$, one may guess should depend on the position and frequency operators $\widehat{\mathsf{q}}$, $\widehat{\kappa}$.

Said in a different manner, within an operator picture one may consider the cross Wigner distribution function $\mathcal{W}_{\varphi,\psi}(q,\kappa)$ as the result of the operator $\widehat{\mathsf{W}}_\varphi$ acting on the Hilbert space of square-integrable functions for a chosen φ in the space, i.e.,

$$\widehat{\mathsf{W}}_\varphi : \psi \in \mathcal{L}^2(\mathbb{R}) \;\rightarrow\; \mathcal{W}_{\varphi,\psi}(q,\kappa) \;\in \mathcal{L}^2(\mathbb{R}^2). \tag{8.5.2}$$

Therefore, Eqs. (8.5.1) can be rephrased as

$$\begin{aligned} \widehat{\mathsf{W}}_\varphi[\widehat{\mathsf{q}}\psi](q,\kappa) &= \widehat{\mathsf{Q}}[\widehat{\mathsf{W}}_\varphi\psi](q,\kappa), \\ \widehat{\mathsf{W}}_\varphi[\widehat{\kappa}\psi](q,\kappa) &= \widehat{\mathsf{K}}[\widehat{\mathsf{W}}_\varphi\psi](q,\kappa). \end{aligned} \tag{8.5.3}$$

The above are reminiscent of relation (4.6.54), which is basic to the individualization of different representations of an operator. Indeed, the operators $\widehat{\mathsf{Q}}$ and $\widehat{\mathsf{K}}$ might be considered as the Wigner representation of the position and frequency operators $\widehat{\mathsf{q}}$ and $\widehat{\kappa}$ [14]. Apart from their interpretation, on which we shall not dwell, we shall regard the operators $\widehat{\mathsf{Q}}$ and $\widehat{\mathsf{K}}$ as inherently elegant and manageable calculational tools.

Taking the explicit expressions of the left hand sides of (8.5.3), one finds

$$\begin{aligned} &\widehat{\mathsf{W}}_\varphi[\widehat{\mathsf{q}}\psi](q,\kappa) = \tfrac{1}{2\pi}\int \varphi^*(q-\tfrac{q'}{2})[\widehat{\mathsf{q}}\psi](q+\tfrac{q'}{2})e^{-i\kappa q'}dq' \\ &= \tfrac{1}{2\pi}\int \varphi^*(q-\tfrac{q'}{2})(q+\tfrac{q'}{2})\psi(q+\tfrac{q'}{2})e^{-i\kappa q'}dq' \\ &= (q+\tfrac{i}{2}\tfrac{\partial}{\partial\kappa})\mathcal{W}_{\varphi,\psi}(q,\kappa) = \widehat{\mathsf{Q}}[\widehat{\mathsf{W}}_\varphi\psi](q,\kappa), \end{aligned} \tag{8.5.4}$$

and

$$\begin{aligned} &\widehat{\mathsf{W}}_\varphi[\widehat{\kappa}\psi](q,\kappa) = \tfrac{1}{2\pi}\int \varphi^*(q-\tfrac{q'}{2})[\widehat{\kappa}\psi](q+\tfrac{q'}{2})e^{-i\kappa q'}dq' \\ &= -\tfrac{i}{4\pi}\int \varphi^*(q-\tfrac{q'}{2})\{\tfrac{\partial}{\partial q}\psi(q+\tfrac{q'}{2}) + 2\tfrac{\partial}{\partial q'}\psi(q+\tfrac{q'}{2})\}e^{-i\kappa q'}dq' \\ &= (\kappa-\tfrac{i}{2}\tfrac{\partial}{\partial q})\mathcal{W}_{\varphi,\psi}(q,\kappa) = \widehat{\mathsf{K}}[\widehat{\mathsf{W}}_\varphi\psi](q,\kappa). \end{aligned} \tag{8.5.5}$$

Accordingly, the operators $\widehat{\mathsf{Q}}$ and $\widehat{\mathsf{K}}$ specialize as

$$\begin{aligned} \widehat{\mathsf{Q}} &= q + \tfrac{i}{2}\tfrac{\partial}{\partial\kappa}, \\ \widehat{\mathsf{K}} &= \kappa - \tfrac{i}{2}\tfrac{\partial}{\partial q}, \end{aligned} \tag{8.5.6}$$

which appear as a mixture of the q- and κ-representations respectively of the position and frequency operators $\widehat{\mathsf{q}}$, $\widehat{\kappa}$ (see § 4.6). Interestingly,

$$\left[\widehat{\mathsf{Q}},\widehat{\mathsf{K}}\right] = i. \tag{8.5.7}$$

The above operators have been originally introduced by Bopp, and recovered in more recent investigations relatively to the formulation of quantum mechanics in phase-space [14.3] or to the elaboration of phase-space operators aimed at simplifying the technical aspects of the Weyl transforms [14.2].

It is evident that for any power of $\widehat{\mathsf{q}}$ and $\widehat{\kappa}$ we find

$$\begin{aligned}\widehat{\mathsf{W}}_\varphi[\widehat{\mathsf{q}}^n\psi](q,\kappa) &= \widehat{\mathsf{Q}}^n[\widehat{\mathsf{W}}_\varphi\psi](q,\kappa) = \widehat{\mathsf{Q}}^n\mathcal{W}_{\varphi,\psi}(q,\kappa),\\ \widehat{\mathsf{W}}_\varphi[\widehat{\kappa}^n\psi](q,\kappa) &= \widehat{\mathsf{K}}^n[\widehat{\mathsf{W}}_\varphi\psi](q,\kappa) = \widehat{\mathsf{K}}^n\mathcal{W}_{\varphi,\psi}(q,\kappa).\end{aligned} \tag{8.5.8}$$

Hence, for any operator function $\widehat{\mathsf{O}}(\widehat{\mathsf{q}},\widehat{\kappa})$ of $\widehat{\mathsf{q}}$ and $\widehat{\kappa}$, which is supposed to be expressible as a series expansion of $\widehat{\mathsf{q}}$ and $\widehat{\kappa}$, the following relation is easily obtained

$$[\widehat{\mathsf{W}}_\varphi\widehat{\mathsf{O}}(\widehat{\mathsf{q}},\widehat{\kappa})\psi](q,\kappa) = \widehat{\mathsf{O}}(\widehat{\mathsf{Q}},\widehat{\mathsf{K}})\mathcal{W}_{\varphi,\psi}(q,\kappa). \tag{8.5.9}$$

Of course, the function $O(x,y)$ is the same of its arguments on both sides.

Notably, a pair of complex conjugate operators naturally correspond to the pair $(\widehat{\mathsf{Q}},\widehat{\mathsf{K}})$. In fact, the initial question concerning the feasibility of identities (8.5.1) can legitimately be concerned with the feasibility of the alternative identities:

$$\begin{aligned}\mathcal{W}_{q\psi,\varphi}(q,\kappa) &= \widehat{\mathsf{Q}}^*\mathcal{W}_{\psi,\varphi}(q,\kappa),\\ \mathcal{W}_{-i\frac{\partial}{\partial q}\psi,\varphi}(q,\kappa) &= \widehat{\mathsf{K}}^*\mathcal{W}_{\psi,\varphi}(q,\kappa),\end{aligned} \tag{8.5.10}$$

whose operator formulation is evidently

$$\begin{aligned}\widehat{\mathsf{W}}_{\cdot,\varphi}[\widehat{\mathsf{q}}\psi](q,\kappa) &= \widehat{\mathsf{Q}}^*[\widehat{\mathsf{W}}_{\cdot,\varphi}\psi](q,\kappa),\\ \widehat{\mathsf{W}}_{\cdot,\varphi}[\widehat{\kappa}\psi](q,\kappa) &= \widehat{\mathsf{K}}^*[\widehat{\mathsf{W}}_{\cdot,\varphi}\psi](q,\kappa).\end{aligned} \tag{8.5.11}$$

The operator $\widehat{\mathsf{W}}_{\cdot,\varphi}$ parallels $\widehat{\mathsf{W}}_\varphi$ (which should more properly written as $\widehat{\mathsf{W}}_{\varphi,\cdot}$), on account of the two possible ways by which a function can enter the Wigner integral. Hence by a direct calculation along the lines leading to (8.5.6), we obtain

$$\begin{aligned}\widehat{\mathsf{Q}}^* &= q - \tfrac{i}{2}\tfrac{\partial}{\partial\kappa},\\ \widehat{\mathsf{K}}^* &= \kappa + \tfrac{i}{2}\tfrac{\partial}{\partial q},\end{aligned} \tag{8.5.12}$$

which, as announced by the use of the asterisk, are the very complex conjugates of (8.5.6). Naturally relation (8.5.9) has its complex conjugate counterpart

$$[\widehat{\mathsf{W}}_{\cdot,\varphi}\widehat{\mathsf{O}}(\widehat{\mathsf{q}},\widehat{\kappa})\psi](q,\kappa) = \widehat{\mathsf{O}}(\widehat{\mathsf{Q}}^*,\widehat{\mathsf{K}}^*)\mathcal{W}_{\psi,\varphi}(q,\kappa). \tag{8.5.13}$$

8.5.2 *The transport equation for the Wigner distribution function*

We recall from § 4.2 the Helmholtz equation for the $1D$ picture of scalar wave propagation

$$(\frac{\partial^2}{\partial z^2} + \frac{\partial^2}{\partial q^2})\phi(q,z) + k^2(q,z)\phi(q,z) = 0, \tag{8.5.14}$$

where the possible weak inhomogeneity of the medium is accounted for by the position dependent wavenumber

$$k(q,z) = k_{\scriptscriptstyle 0} n(q,z). \tag{8.5.15}$$

Since only propagation in the positive z direction is considered, Eq. (8.5.14) can safely be "solved" for the z-derivative of the wave amplitude, thus yielding[c]

$$i\frac{\partial}{\partial z}\phi(q,z) = -\sqrt{k^2(\widehat{\mathsf{q}},z) - \widehat{\kappa}^2}\,\phi(q,z) \equiv \widehat{\mathcal{H}}(\widehat{\mathsf{q}},\widehat{\kappa})\phi(q,z). \tag{8.5.16}$$

The q-domain definition of the Wigner distribution function implies the integral

$$\mathcal{W}_\phi(q,\kappa;z) = \tfrac{1}{2\pi}\int \phi^*(q-\tfrac{q'}{2})\phi(q+\tfrac{q'}{2})e^{-i\kappa q'}dq', \tag{8.5.17}$$

the z-dependence of both ϕ and $\mathcal{W}_\phi$ being explicitly displayed. Differentiating $\mathcal{W}_\phi(q,\kappa;z)$ with respect to the propagation variable z, one obtains[d]

$$i\frac{\partial}{\partial z}\mathcal{W}_\phi(q,\kappa;z) = \mathcal{W}_{-i\frac{\partial}{\partial z}\phi,\phi}(q,\kappa;z) + \mathcal{W}_{\phi,i\frac{\partial}{\partial z}\phi}(q,\kappa;z), \tag{8.5.18}$$

which, on account of the wave equation (8.5.16) and the operator notation previously introduced, can be written as

$$i\frac{\partial}{\partial z}\mathcal{W}_\phi(q,\kappa;z) = [\widehat{\mathsf{W}}_\phi\,\widehat{\mathcal{H}}(\widehat{\mathsf{q}},\widehat{\kappa})\phi](q,\kappa;z) - [\widehat{\mathsf{W}}_{\cdot,\phi}\,\widehat{\mathcal{H}}(\widehat{\mathsf{q}},\widehat{\kappa})\phi](q,\kappa;z). \tag{8.5.19}$$

Therefore, by rules (8.5.9) and (8.5.13), we immediately obtain (dropping out the now unessential subscript)

$$i\frac{\partial}{\partial z}\mathcal{W}(q,\kappa;z) = [\widehat{\mathcal{H}}(\widehat{\mathsf{Q}},\widehat{\mathsf{K}}) - \widehat{\mathcal{H}}(\widehat{\mathsf{Q}}^*,\widehat{\mathsf{K}}^*)]\mathcal{W}(q,\kappa;z), \tag{8.5.20}$$

which represents the transport equation for the Wigner distribution function. It parallels the Helmholtz equation for the wave complex amplitude. It is reported as the *quantum Liouville equation.*

Explicitly, we write

$$\begin{aligned} i\tfrac{\partial}{\partial z}\mathcal{W}(q,\kappa;z) = \Big\{&\sqrt{k^2(q-\tfrac{i}{2}\tfrac{\partial}{\partial\kappa},z) - (\kappa+\tfrac{i}{2}\tfrac{\partial}{\partial q})^2} \\ &-\sqrt{k^2(q+\tfrac{i}{2}\tfrac{\partial}{\partial\kappa},z) - (\kappa-\tfrac{i}{2}\tfrac{\partial}{\partial q})^2}\Big\}\,\mathcal{W}(q,\kappa;z), \end{aligned} \tag{8.5.21}$$

which is useful to identify the behaviour of the Wigner distribution function under specific conditions. In particular, we are interested in determining the paraxial form of the transport equation for the Wigner distribution function.

---

[c] Note that this is nothing but Eq. (4.4.7) with the full expression of Hamiltonian operator (in terms of $\widehat{\kappa}$ instead of $\widehat{\mathsf{p}}$) being used.

[d] Naturally, it is assumed that the conditions under which the derivative of the integral can be turned into the integral of the derivative of the integrand function are satisfied [see, for instance, E.T. Whittaker and G.N. Watson, *A Course of Modern Analysis*, (Cambridge Univ. Press, 4th ed., 1965), ch. 4, p. 67].

8.5.3 Recovering the classical Liouville equation for quadratic refractive index

Let us expand the operators $(\kappa \pm \frac{i}{2}\frac{\partial}{\partial q})^2$ under the square roots in (8.5.21) above, thus obtaining

$$\sqrt{k^2(q \mp \tfrac{i}{2}\tfrac{\partial}{\partial \kappa}, z) - (\kappa \pm \tfrac{i}{2}\tfrac{\partial}{\partial q})^2} = \sqrt{k^2 - \kappa^2 \mp i\kappa\tfrac{\partial}{\partial q} + \tfrac{1}{4}\tfrac{\partial^2}{\partial q^2}}, \tag{8.5.22}$$

where, for formal convenience, the dependence of the wave number k on the operators $\widehat{\mathsf{Q}}$, $\widehat{\mathsf{Q}}^*$, and z is not explicitly shown on the right side. Then, according to the usual paraxial approximation expansion, we may write

$$\sqrt{k^2 - \kappa^2 \mp i\kappa\tfrac{\partial}{\partial q} + \tfrac{1}{4}\tfrac{\partial^2}{\partial q^2}} \simeq k_0 n(q \mp \tfrac{i}{2}\tfrac{\partial}{\partial \kappa}, z) + \tfrac{1}{2k_0 n_0(z)}[\tfrac{1}{4}\tfrac{\partial^2}{\partial q^2} \mp i\kappa\tfrac{\partial}{\partial q}], \tag{8.5.23}$$

which once inserted into (8.5.21) yields

$$\begin{aligned} \frac{\partial}{\partial z}\mathcal{W}(q,\kappa;z) &= -\tfrac{1}{k_0 n_0(z)}\kappa\frac{\partial}{\partial q}\mathcal{W}(q,\kappa;z) \\ &+ik_0[n(q + \tfrac{i}{2}\tfrac{\partial}{\partial \kappa}, z) - n(q - \tfrac{i}{2}\tfrac{\partial}{\partial \kappa}, z)]\mathcal{W}(q,\kappa;z), \end{aligned} \tag{8.5.24}$$

where evidently the first term accounts for the kinetic contribution whilst the second for the "potential" contribution. Also, expanding the refractive index function with respect to the transverse coordinate q and grouping the terms with equal powers, we end up with

$$\frac{\partial}{\partial z}\mathcal{W}(q,\kappa;z) = -\tfrac{1}{n_0(z)}p\frac{\partial}{\partial q}\mathcal{W}(q,\kappa;z) + \sum_{j=0}^{\infty}\tfrac{(-)^{j+1}}{(2j+1)!}\tfrac{1}{(2k_0)^j}n^{(2j+1)}\tfrac{\partial^{2j+1}}{\partial p^{2j+1}}\mathcal{W}(q,\kappa;z), \tag{8.5.25}$$

where $n^{(2j+1)} \equiv \frac{\partial^{2j+1}}{\partial q^{2j+1}}n(q,z)$. Note that we have turned from the Fourier conjugate pair (q,κ) to the Hamilton conjugate pair (q,p) to evidence the dependence on the wave number. In fact, the potential term turns into a series expansion in terms of the reduced wavelength $\frac{1}{k_0}$, the optical analog of $\hbar$.

Naturally in the geometrical optics limit: $\frac{1}{k_0} \to 0$, we recover the classical Liouville equation (as it should be appropriate to paraxial propagation, which amounts to the optical Hamiltonian $\frac{p^2}{2n_0(z)} - n(q,z)$), since only the $j = 0$ term survives in the series above. In general the refractive-index depending term is not expressible in the paraxial regime classical-like form, i.e, $\frac{\partial}{\partial q}n(q,z)$, except that in the case of free propagation: $n(q,z) = n_0(z)$, or quadratic refractive index: $n(q,z) = n_0(z) - \frac{1}{2}n_2(z)q^2$. In the latter case, in fact, the resulting equation can be written in the "classical" form

$$\frac{\partial}{\partial z}\mathcal{W}(q,\kappa;z) = -\{\mathcal{W}, H\}, \quad \mathcal{W}(q,\kappa;z_i) = \mathcal{W}_i(q,\kappa), \tag{8.5.26}$$

where H is the paraxial Hamiltonian of geometrical optics. It is therefore evident that the Wigner distribution comes to have a constant value along the geometrical optics ray-path:

$$\frac{d}{dz}\mathcal{W}(q,\kappa;z)=0, \qquad \text{along any "ray trajectory"} \tag{8.5.27}$$

and accordingly the transfer relations between two reference planes, we have obtained using the Fresnel integral, follow immediately, i.e.,

$$\mathcal{W}_o(\mathbf{u})=\mathcal{W}_i(\mathbf{M}^{-1}\mathbf{u}), \tag{8.5.28}$$

$\mathbf{M}$ being, of course, the optical ray-matrix with k_0-scaled off-diagonal entries.

Equation (8.5.28) means that, just as in the case of the ray-density distribution $\rho(q,p;z)$, we can run individual trajectories to determine the z-evolution of the Wigner distribution function and hence of the relevant signal. But this is true for the Wigner distribution function only for paraxial propagation under quadratic refractive index. In general, as the quantum Liouville operator involves derivatives of third and higher order, locality is lost. As far as the linear approximation is concerned, we may choose, possibly according to symmetry properties of the system, a phase-plane closed domain and follow the evolution of such domain as representative of the evolution of the Wigner distribution function (and so of the corresponding signal) under propagation through the system at hand. On account of the considerations of § 2.5, we may guess that phase-plane domains bounded by straight lines or by ellipses are very convenient to work with. Indeed, in the next chapter we will investigate the transformations effected on phase-plane domains, bounded by ellipses, by propagation through paraxial optical systems. We will recover the pictorial visualizations of the Wigner charts elucidated in § 8.3 in terms of the second-order moments of the Wigner distribution function.

8.6 Summary

We have investigated the paraxial propagation properties of the Wigner distribution function. Unlike the four possible input-output relations for optical systems when the signals are represented through the complex amplitude or the respective Fourier spectrum, there is only one transfer relation when the signals are represented through the Wigner distribution functions. It basically involves an affine transformation of the variables of the Wigner distribution function in the same form as that of the ray variables of geometrical optics. Thus, on one side the Wigner distribution function, being defined in terms of the optical wave function, captures all the wave aspects of light. On the other side, due to its transfer law, it retains in its dynamics the local features of ray propagation, local values of the output Wigner distribution function being determined by local values of the input Wigner distribution function.

Then, particular attention has been devoted to the relation between the Wigner distribution function and the fractional Fourier transform, which has been recognized to produce a rotation of the Wigner chart. Such a rotation based relation has been seen to be basic to the phase-space tomographic strategy for reconstruction of complex wave-fields, be they classical optical disturbances or quantum matter-waves.

Finally, the general form of the transport equation for the Wigner distribution function has been derived. It becomes the same as Liouville's equation for the classical phase space distribution function in the case of paraxial propagation through quadratic index profile media.

Problems

1. *(a)* Prove that the Weyl transform of the operator $\widehat{F}(\widehat{\mathsf{q}},\widehat{\mathsf{p}}) = \widehat{f}(\widehat{\mathsf{q}},\widehat{\mathsf{p}})\widehat{g}(\widehat{\mathsf{q}},\widehat{\mathsf{p}})$ is

$$F_{\mathrm{W}}(q,p) = \frac{1}{4\pi^2}\iiiint dq'dp'dq''dp'' f_{\mathrm{W}}(q',p')g_{\mathrm{W}}(q'',p'')e^{-i[(q'-q'')p-(p'-p'')q+p'q''-p''q']/\hbar},$$

where $f_{\mathrm{W}}(q,p)$ and $g_{\mathrm{W}}(q,p)$ are the Weyl transforms of $\widehat{f}(\widehat{\mathsf{q}},\widehat{\mathsf{p}})$ and $\widehat{g}(\widehat{\mathsf{q}},\widehat{\mathsf{p}})$ respectively. *(b)* Verify that $F_{\mathrm{W}}(q,p) \neq G_{\mathrm{W}}(q,p)$, where G_{W} is the Weyl transform of $\widehat{G}(\widehat{\mathsf{q}},\widehat{\mathsf{p}}) = \widehat{g}(\widehat{\mathsf{q}},\widehat{\mathsf{p}})\widehat{f}(\widehat{\mathsf{q}},\widehat{\mathsf{p}})$.

2. Prove that the the paraxial propagation law for the ambiguity function $\mathcal{A}(\widetilde{q},\widetilde{\kappa})$ writes as

$$\mathcal{A}_o(\widetilde{q},\widetilde{\kappa}) = \iint \mathsf{G}_{\mathcal{A}}(\widetilde{q},\widetilde{\kappa},\widetilde{q}',\widetilde{\kappa}')\mathcal{A}_i(\widetilde{q}',\widetilde{\kappa}')d\widetilde{q}'\,d\widetilde{\kappa}'.$$

with a definite relation between $\mathsf{G}_{\mathcal{A}}(\widetilde{q},\widetilde{\kappa},\widetilde{q}',\widetilde{\kappa}')$ and $\mathsf{G}(q,\kappa,q',\kappa')$, i.e. between the system responses to the incoming signal in the Wigner and ambiguity function representations; or equivalently as

$$\mathcal{A}_o(\widetilde{\mathbf{u}}) = \mathcal{A}_i(\mathbf{M}^{-1}\,\widetilde{\mathbf{u}}),$$

where $\mathbf{M}$ is the ray-matrix entering as well (8.3.35) and $\widetilde{\mathbf{u}} = (\widetilde{q},\widetilde{\kappa})^{\top}$ denotes vectors in the ambiguity function phase plane.

3. Using the propagation properties of the Wigner distribution function, prove that the propagation of an incoherent wave field, described by the mutual intensity function (see Eq. (7.5.16))

$$\mathsf{J}(q_1,q_2;z_i) = I(q_1;z_i)\delta(q_1-q_2),$$

through an *ABCD* optical system, is like that of a coherent wave field described by the complex amplitude

$$\varphi(q_1) = I(q_1;z_i)\mathsf{g}^*(0,q_1),$$

g^* being of course the spread function of the system. The above is known as the *Van Cittert-Zernike theorem.*

4. Find out the positive, negative and zero-contrast planes for a freely propagating wave field, described at the plane $z = 0$ by the mutual intensity (8.3.45), using the paraxial propagation properties of the Wigner distribution function. Take, in particular, $\kappa = \kappa_0$.

References

[1] A list of titles concerned with the Wigner distribution function and related issues within the quantum mechanical context, enriching [4] of ch.6, is below. For the quantum optics view of the Wigner representation see [12.9] and [12.10].

[1.1] T. Takabayasi, "The formulation of quantum mechanics in terms of ensemble in phase space", Progr. Theor. Phys. **11**, 341-373 (1954).

[1.2] G.A. Baker, "Formulation of quantum mechanics based on the quasi-probability distribution induced on phase space", Phys. Rev. **109**, 2198-2206 (1958).

[1.3] R. Kubo, "Wigner representation of quantum operators and its applications to electrons in a magnetic field", J. Phys. Soc. Japn 19, 2127-2139 (1964).

[1.4] K. Imre, E. Özizmir M. Rosenbaum and P.F. Zweifel, "Wigner method in quantum statistical mechanics", J. Math. Phys. **8**, 1097-1108 (1967).

[1.5] G.J. Ruggeri, "On phase-space description of quantum mechanics", Progr. Theor. Phys. **46**, 1703-1712 (1971).

[1.6] E.J. Heller, "Wigner phase space method. Analysis for semiclassical applications", J. Chem Phys. **65**, 1289-1298 (1976).

[1.7] D. Bohm and B.J. Hiley, "On a quantum algebraic approach to a generalized phase space", Found. Phys. **11**, 179-203 (1981).

[1.8] A. Royer, "Wigner function as the expectation value of a parity operator", Phys. Rev. A **15**, 449-450 (1977).

[1.9] M. Hillery, R.F. O'Connell, M. O. Scully and E.P. Wigner, "Distribution functions in physics: fundamentals", Phys. Rep. **106**, 121-167 (1984).

[1.10] R. Simon, E.C.G. Sudarshan and N. Mukunda, "Gaussian Wigner distributions: a complete characterization", Phys. Lett. **124**, 223-228 (1987).

[1.11] R. Simon, E.C.G. Sudarshan and N. Mukunda, "Gaussian-Wigner distributions in quantum mechanics and optics", Phys. Rev. Lett. A **36**, 3868-3880 (1987).

[1.12] A. Royer, "Wigner function in Liouville space: A canonical formalism", Phys. Rev. A **43**, 44-56 (1991).

[1.13] J.J. Wlodarz, "On quantum mechanical phase-space wave functions", J. Chem. Phys. **100**, 7478-7480 (1994).

[1.14] H.-W. Lee, "Theory and application of the quantum phase-space distribution functions", Phys. Rep. **259**, 147-211 (1995).

[2] As emphasized in ch. 6, the introduction of the Wigner distribution function in optics was suggested by Bastiaans, whose seminal papers and subsequent investigations, concerning as well the propagation properties of the Wigner distribution function in relation to first-order optical systems, are in [6] of ch. 6.

[3] E.P. Wigner, "On the quantum correction for thermodynamic equilibrium", Phys. Rev. **40**, 749-759 (1932).

[4] An accurate analysis of the conditions that uniquely lead to the formulation of the quantum phase space representative in the form of the Wigner distribution function can be found in

[4.1] E.P. Wigner, "Quantum-mechanical distribution functions-Revisited", in *Perspectives in Quantum Theory*, W. Yourgrau and A. van der Merwe (eds.) (MIT Press, Cambridge, 1971), pp. 25-36.

[4.2] R.F. O'Connell and E.P. Wigner, "Quantum-mechanical distribution functions: conditions for uniqueness", Phys. Lett. **83A**, 145-148 (1981).

[4.3] J.G. Krüger and A. Poffyn, "Quantum mechanics in phase space I. Unicity of the Wigner distribution function", Physica **85A**, 84-100 (1976).

[5] In relation with the negativeness of the Wigner distribution function see

[5.1] L. Cohen, "Generalized phase-space distribution functions", J. Math. Phys. **7**, 781-786 (1966).

[5.2] L. Cohen, "Can quantum mechanics be formulated as a classical probabilistic theory?", Philos. Sc. **33**, 317-322 (1966).

[5.3] M.D. Srinivas and E. Wolf, "Some nonclassical features of phase-space representations of quantum mechanics", Phys. Rev. D **11**, 1477-1485 (1975).

[5.4] R. Groblicki, "A generalization of Cohen's theorem", J. Math Phys. **24**, 841-843 (1983).

[5.5] D. Dragoman, "The origin of negative values of the Wigner distribution function", Optik **111**, 179-183 (2000).

[6] As to the Weyl formalism and its relation to the Wigner method see also [5.1, 5.3] and [1.4]. The paper by Shenoy and Parks [6.6] presents an elegant and interesting investigation of the relation between the Weyl correspondence (with the related Heisenberg group formalism) and the time-frequency analysis.

[6.1] C.L. Metha, "Phase-space formulation of the dynamics of canonical variables," J. Math. Phys. **5**, 677-686 (1964).

[6.2] B. Leaf, "Weyl transformation and the classical limit of quantum mechanics", J. Math. Phys. **9**, 65-72 (1968). "Weyl transform in nonrelativistic quantum mechanics", *ibid.*, 769-781.

[6.3] L. Cohen, "Quantization problem and variational principle in the phase-space formulation of quantum mechanics", J. Math. Phys. **17**, 1863-1866 (1976).

[6.4] M. Springborg, "Phase space functions and correspondence rules", J. Phys. A: Math. Gen. **16**, 535-542 (1983).

[6.5] A.M. Ozorio De Almeida, "The Weyl representation in classical and quantum mechanics", Phys. Rep. **295**, 265-342 (1998).

[6.6] R.G. Shenoy and T.W. Parks, "The Weyl correspondence and time-frequency analysis", IEEE Trans. Signal Process. **42**, 318-331 (1994).

[7] Basic references concerned with Gaussian beams are in [13] of ch. 2.

[8] As an illustration of the Wigner representation based approach to analysis and synthesis of optical systems on account as well of the wavelength dependence of the Wigner optics matrices, we suggest

[8.1] W.D. Furlan, G. Saavedra and J. Lancis, "Phase-space representations as a tool for the evaluation of the polychromatic OTF", Opt. Comm. **96**, 208-213 (1993).

[8.2] W.D. Furlan, G. Saavedra, E. Silvestre, P. Andrés and M. J. Yzuel, "Polychromatic axial behavior of aberrated optical systems: Wigner distribution function approach", Appl. Opt. **36**, 9146-9151 (1997).

[8.3] J. Lancis, E.E. Sicre, A. Pons and G. Saavedra, "Achromatic white self-imaging phenomenon: an appproach using the Wigner distribution function", J. Mod. Opt. **42**, 425-434 (1995).

[8.4] D. Wang, A. Pe'er, A.W. Lohmann and A.A. Friesem, "Wigner algebra as a tool for the design of achromatic optical processing systems", Opt. Eng. **39**, 3014-3024 (2000).

[9] An exhaustive bibliography on the Talbot-Lau effect, updated to 1997, is in [9.6]. Below are quoted the seminal papers by Talbot [9.1] and Lau [9.2], the basic review by Patorski [9.3] and the recent survey [9.7], where along with historical notes the potential of the Talbot effect for application to optical computing and optical measurement is emphasized. Papers [9.4, 9.5] are specifically related to the presentation in § 8.3.4.

[9.1] H. F. Talbot, "Facts relating to Optical Science", Phil. Mag. **9**, 401-407 (1836).

[9.2] E. Lau, "Beugungserscheinungen an Doppelrastern", Ann. Phys. **6**, 417-4237 (1948).

[9.3] K. Patorski, "The self imaging phenomenon and its applications", in *Progress in Optics*, Vol. XXVII, E. Wolf (ed.) (Elsevier, Amsterdam, 1989), ch. 1, pp. 1-108.

[9.4] A.W. Lohmann, J. Ojeda-Castañeda and N. Streibl, "Spatial periodicities in coherent and partially coherent fields", Opt. Acta **30**, 1259-1266 (1983).

[9.5] J. Ojeda-Castañeda and E.E. Sicre, "Quasi ray-optical approach to longitudinal periodicities of free and bounded wavefields", Opt. Acta **32**, 17-26 (1985).

[9.6] D. Ambrosini, "A bibliography about the Talbot-Lau effect" in http://dau.ing.univaq.it/omhat/Ref/talbib.htm.

[9.7] C. Zhou, W. Wang, E. Dai and L. Liu, "Simple principles of the Talbot effect", Optics & Phot. News **15** (11), 46-50 (2004), and references therein.

[10] A miscellanea of titles concerned with the applications of the fractional Fourier transform and its relation to the Wigner representation is given below. A basic reference is the comprehensive book [10.11]. Further refs. are in [5] of ch. 3.

[10.1] D. Mendlovic and H.M. Ozaktas, "Fractional Fourier transformations and their optical implementation: I", J. Opt. Soc. Am. A **10**, 1875-1880 (1993); II", *ibid.*, 2522-2531.

[10.2] A.W. Lohmann, "Image rotation, Wigner rotation and the fractional Fourier transform", J. Opt. Soc. Am. A **10**, 2181-2186 (1993).

[10.3] D. Mendlovic, H.M. Ozaktas and A.W. Lohmann, "Graded-index fibers, Wigner distirbution functions, and the fractional Fourier transforms", Appl. Opt. **33**, 6188-6193 (1994).

[10.4] H.M. Ozatkas, B. Barshan, D. Mendlovic and L. Onural, "Convolution, filtering, and multiplexing in fractional Fourier domain and their relation to chirp and wavelet transforms", J. Opt. Soc. Am. A **11**, 547-559 (1994).

[10.5] D. Mendlovic, Z. Zalevsky, R.G. Dorsch, Y. Bitran, A.W. Lohmann and H.M. Ozaktas, "New signal representation based on the fractional Fourier transform: definitions", J. Opt. Soc. Am. A**12**, 2424-2431 (1995).

[10.6] D. Mendlovic, R.G. Dorsch, A.W. Lohmann, Z. Zalevsky and C. Ferreira, "Optical illustration of a varied fractional-Fourier transform order and the Radon-Wigner display", Appl. Opt. **35**, 3925-3929 (1996).

[10.7] M.A. Kutay, H.M. Ozaktas, O. Arikan and L. Onural, "Optimal filtering in fractional Fourier domain", IEEE Trans. Signal Process. **45**, 1129-1143 (1997).

[10.8] A.W. Lohmann, D. Mendlovic and Z. Zalevsky, "Fractional transformations in optics," in *Progress in Optics*, Vol. XXXVIII, E. Wolf (ed.) (Elsevier, Amsterdam, 1997), ch. 4, pp. 263-342.

[10.9] A. Torre, "The fractional Fourier transform and some of its applications to optics," in *Progress in Optics*, Vol. 43, E. Wolf (ed.) (Elsevier, Amsterdam, 2002), ch. 7, pp. 521-586.

[10.10] T. Alieva and M.J. Bastiaans, "Wigner distribution and fractional Fourier transform" in *Time-Frequency Signal Analysis and Processing. A comprehensive Reference*, B. Boashash (ed.) (Elsevier, Oxford, 2003), sect. 4.8, pp.145-152.

[10.11] H.M. Ozaktas, Z. Zalevsky and M. A. Kutay, *The Fractional Fourier Transform with Applications in Optics and Signal Processing* (John Wiley & Sons Ltd, Chichester, 2001).

[11] Basic titles on the Radon transform, whose original application was mainly for medical computed tomography, are [11.1]-[11.5]. The classic paper by Radon, where he showed how a reconstruction from projections was possible, is reproduced in facsimile in [11.2], and in English translation in [11.3]. Formalization and applicative suggestions regarding the Radon-Wigner transform and its relation to the fractional Fourier transform are respectively in [11.6], [11.7] and [11.8]-[11.10].

[11.1] J. Radon, "Über die Bestimmung von Funktionen durch ihre Integralwerte längs gewisser Mannigfaltigkeiten", Ber. Saechs. Akad. Wiss. (Leipzig), Math-Phys. K **69**, 262-277 (1917).

[11.2] S. Helgason, *The Radon Transform*, Progress in Mathematics, Vol. 5 (Birkhaüser, Basel, 1980).

[11.3] S.R. Deans, *The Radon Transform and some of its Applications* (Wiley, New York, 1983).

[11.4] H. H. Barrett, "The Radon transform and its applications", in *Progress in Optics*, Vol. XXI, E. Wolf (ed.) (Elsevier, Amsterdam, 1984), ch. III, pp. 217-286.

[11.5] C.N. Guy, "The second revolution in medical imaging", Contemp. Phys. **37**, 15-45 (1996).

[11.6] J. C. Wood and D. T. Berry, "Radon transform of the Wigner spectrum", in *Advanced Signal Processing Algorithms, Architectures, and Implementations III*, F.T. Luk (ed.), Proc. SPIE 1770, 358-375 (1992).

[11.7] J. C. Wood and D. T. Berry, "Linear signal synthesis using the Radon-Wigner transform", IEEE Trans. Signal Process **42**, 2105-2111 (1994); "Radon transformation of time-frequency distributions for analysis of multicomponent signals", *ibid.*, 3166-3177.

[11.8] A.W. Lohmann and B.H. Soffer, "Relationships between the Radon-Wigner and fractional Fourier transforms", J. Opt. Soc. Am. A **11**, 1798-1801 (1994).

[11.9] D. Mustard, "The fractional Fourier transformation and the Wigner distribution", Austral. Math. Soc. Ser B **38**, 209-219 (1996).

[11.10] T. Alieva and M.J. Bastiaans, "Radon-Wigner transform for optical field analysis", Proc. Int. Conf. on Opt. and Optoelectron., Dec. 9-12, 1998, Derha, India, Vol. 1, pp. 132-135; http://www.sps.ele.tue.nl/members/M.J.Bastiaans/pdfs/icol98_5.pdf

[12] Since the pioneering paper by [12.8], where the tomographic strategy for the construction of the Wigner function was firstly suggested, the quantum tomography has found a great deal of applications, having also gained new insight from the

recognized Radon-Wigner/fractional Fourier transform relation. A few titles concerned with the phase-space tomographic reconstruction of complex wave-field are listed here. We direct the reader to [12.9] and [12.10] for an accurate account of quantum tomography. An enlightening survey of the status of art of the quantum states reconstruction based on the Radon-Wigner transform is in [12.8].

[12.1] K. Vogel and H. Risken, "Determination of quasiprobability distributions in terms of probability distributions for the rotated quadrature phase", Phys. Rev. A **40**, 2487-2489 (1989).

[12.2] D.T. Smithey, M. Beck, M.G. Raymer and A. Faridani, "Measurements of the Wigner distirbution and the density matrix of a light mode using homodyne tomography. Application to squeezed states and the vacuum", Phys. Rev. Lett. **70**, 1244-1247 (1993).

[12.3] M.G. Raymer, M. Beck and D.F. McAlister, "Complex wave-field reconstruction using phase-space tomography", Phys. Rev. Lett. **72**, 1137-1140 (1994).

[12.4] D.F. McAlister, M. Beck, L. Clarke, A. Mayer and M.G. Raymer, "Optical phase retrieval by phase space tomography and fractional-order Fourier transforms", Opt. Lett. **20**, 1181-1183 (1995).

[12.5] J. Tu and S. Tamura, "Wave field determination using tomography of the ambiguity function", Phys. Rev. E **55**, 1946-1949 (1997).

[12.6] J. Tu and S. Tamura, "Analytic relation for recovering the mutual intensity by means of intensity information", J. Opt. Soc. Am. A **15**, 202-206 (1998).

[12.7] D. Dragoman, M. Dragoman and K.-H. Brenner, "Tomographic amplitude and phase recovery of vertical-cavity surface-emitting laser by use of the ambiguity function", Opt. Lett. **27**, 1519-1521 (2002).

[12.8] J. Bertrand and P. Bertrand, "A tomographic approach to Wigner's function", Found. Phys. **17**, 397-405 (1987).

[12.8] M.G. Raymer, "Measuring the quantum mechanical wave function", Contemp. Phys. **38**, 343-355 (1997).

[12.9] U. Leonhardt, *Measuring the quantum state of light* (Cambridge Univ. Press, Cambridge, 1997), ch. 5, pp. 98-143.

[12.10] W.P. Schleich, *Quantum Optics in Phase Space* (Wiley-VCH, Berlin, 2001), pp. 143-146.

[13] For an accurate analysis of the transport equation for the Wigner distribution function we suggest the paper below. Readers interested in the quantum mechanical aspects of the evolution equation are directed to [1.9].

[13.1] H. Bremmer, "General remarks concerning theories dealing with scattering and diffraction in random media", Radio Sc. **8**, 511-534 (1973).

[13.2] M.J. Bastiaans, "Transport equations for the Wigner distribution function", Opt. Acta **26**, 1265-1272 (1979); "Transport equations for the Wigner distribution function in an inhomogeneous and dispersive medium", *ibid.*, 1333-1344.

[14] For the Wigner representation of number functions see [1.3], [1.9] and

[14.1] F. Bopp, "Statistische Mechanik bei Störung des Zustands eines physikalischen Systems durch die Beobachtung", in *Werner Heisenberg und die Physik unserer Zeit* (F. Vieweg, Braunschweig, 1961), 128-149.

[14.2] N.L. Balazs and H.C. Pauli, "Quantum-mechanical operators and phase-space operators", Z. Physik A **282**, 221-225 (1977).

[14.3] Go. Torres-Vega and J.H. Frederick,"Quantum mechanics in phase space: new approaches to the correspondence principle", J. Chem. Phys. 93, 8862-8874 (1990).

9

1D First-Order Optical Systems: Moments of the Wigner Distribution Function

9.1 Introduction

Although the Wigner distribution function of a signal φ cannot be interpreted as the energy distribution of φ in space and frequency, it provides an effective characterization of the signal. For a specific signal, indeed, we can evaluate (at least in principle) the Wigner distribution function and from the analysis of this function we can get an idea of how the energy of the signal is distributed in space and frequency. As an example, the energy of a Gaussian signal comes to spread out over all the (q, κ) plane, though remaining mainly concentrated within a region with definite rms extensions in both q and κ.

Alternatively, it is possible to characterize the signal without giving the complete Wigner representation, i.e. the values of the Wigner distribution function throughout the whole (q, κ) plane. Also, due to the uncertainty relation, such a point-wise description is commonly intended as meaningless. In effect, an equally effective characterization of the signal can be pursued by means of the moments of the Wigner distribution function, which, as we will see, are directly linked to the moments of the signal.

Moment methods for treating general problems in image manipulation and analysis have been pioneered by Hu [1.1], and later investigated and refined by many authors [1]. Object reconstruction is one of the basic instances addressed by the moment methods. The moments of a $2D$ function $f(x, y)$, as, for instance, the image irradiance distribution, are defined through the integrals

$$\mu_{rh} \propto \iint_{\mathfrak{R}} \nu_{rh}(x, y) f(x, y) dx dy. \tag{9.1.1}$$

The subindices r and h are positive integers. The sum $r + h$ defines the order of the moment. The functions $\{\nu_{rh}(x, y)\}_{r,h}$ are known as the basis set. The determination of the integration region $\mathfrak{R}$, the choice of the basis set $\{\nu_{rh}\}_{r,h}$

and accordingly the integration variables may be addressed in relation with the specific problem of concern. As demonstrated by Teague [1.3], for instance, $2D$ image moments with respect to Zernike polynomials have very convenient transformation properties in response to image transformations like scaling, translations or rotations. Also, an image has a unique approximate reconstruction in either real or Fourier space in terms of a finite set of such moments.

The simplest form of (9.1.1) takes as a basis set the monomials $\{x^r y^h\}_{r,h}$, thus yielding the *geometric moments* $\mathfrak{m}_{rh}$, that will be considered in what follows. Geometric moments have an intuitive, rather visible interpretation. In fact, the first-order spatial moments specify the location of the center of the field distribution, the second-order spatial moments characterize the size and the orientation of the signal, and higher-order moments determine further details of the field distribution, like the degree of symmetry possibly with respect to the centroid or the degree of peakedness.

Here, we shall be concerned with the geometric (from now on simply referred to as moments) of the Wigner distribution function. We may distinguish between *local* and *global* moments. The local moments are determined by treating the Wigner distribution as a function of the frequency κ for a given space variable q or as a function of the space variable q for a fixed frequency variable κ; therefore, they depend on q or κ, and give indication about the local behaviour of the Wigner distribution function. In contrast, the global moments are determined by considering the Wigner distribution as a function of both q and κ; accordingly, they are independent of both variables and provide global information about the Wigner distribution function.

In conformity with the linear approximation, which has underlied the presentations throughout the book, we limit ourselves to consider only first- and second-order moments. We will see that complex Gaussian signals are fully individualized by the second-order moments, the relevant Wigner distribution function being uniquely expressible in terms of them. When considering arbitrary non Gaussian signals, consideration of the second-order moments is enough to give a global characterization of the signal and of the transformations it experiences on passing through $ABCD$ optical systems. Also, they enable a pictorial visualization of the signal propagation in the Wigner plane, which propagation being mirrored into the phase-plane transformations of the variance-matrix ellipse.

Notably, the propagation laws for the first-order and the second-order moments directly involve the transfer matrix of linear ray-optics. Thus, we will see that, just rephrasing the results of § 5.5, the first-order moments, which represent the centroid of the Wigner distribution function, evolve in the same way as the ray-variables. Correspondingly, the second-order moments transform through a congruence transformation by the ray-transfer matrix, once properly embedded into the variance matrix. This enables a unique determi-

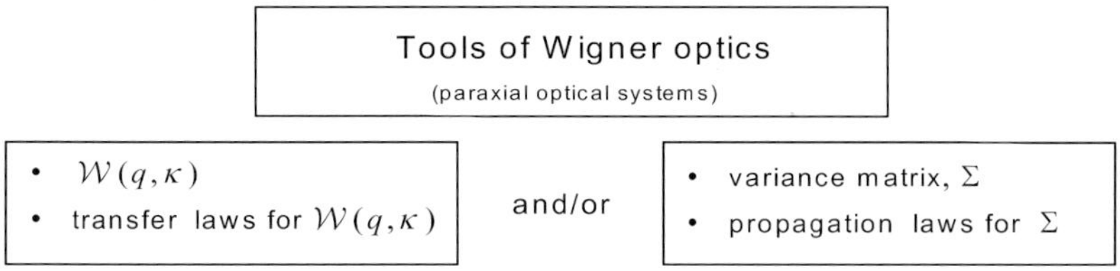

FIGURE 9.1. The second-order moments of the Wigner distribution function yield an effective description of the light propagation properties through paraxial optical systems.

nation of the second moments of the signal at the observation plane in terms of the second moments of the signal at the object plane.

Therefore, we may add to the "tools of Wigner optics" for the description of light propagation through paraxial optical systems, that comprise the Wigner distribution function and the relevant transfer laws (see Fig. 8.1), the second-order moments and the pertinent propagation laws (Fig. 9.1). This reinforces the link of Wigner optics to ray optics. It results from the presentation given in Chapter 8 that the Wigner distribution function retains in its kinematic structure the wave-optical description of signals. In contrast, in its dynamics it resorts to the transformation laws of ray-optics, its arguments being transformed as the geometrical ray variables. So, the results here will prove that such a ray-optical like characterization of the Wigner distribution function dynamics, in relation with paraxial optical systems, reflects as well in the propagation laws of its second-order moments through such systems.

Basic notions concerning the definition of the moments of a $1D$ function, not necessarily interpretable as a density function, and of the local and global moments of a $2D$ function, not necessarily interpretable as a density function, are given in Sect. 9.2. Section 9.3 develops some preliminary results, useful for the subsequent calculations. Section 9.4 addresses the specific calculation of the first-order and second-order, both local and global, moments of the Wigner distribution function. A variety of expressions are worked out, each having its counterpart in the signal analysis context. The results of Sect. 9.4 are then applied to Gaussian Wigner distribution functions (Sect. 9.5), which, as already remarked, are fully characterized by the relevant second-order moments. The variance matrix and the associated ellipse in the Wigner plane are introduced. The transfer laws for the variance matrix are finally deduced. Section 9.6 extends the results of Sect. 9.4 to arbitrary non Gaussian Wigner distribution functions. The propagation laws of the second-moments in arbitrary $ABCD$ systems are explicitly written down and visualized in terms of the transformations of the variance ellipse in the Wigner plane, specifically in relation to the propagation through a free-medium section, a thin lens, a beam expander and a Fourier transformer. Finally, a brief account of the higher-order moments and the inherent propagation laws is presented in Sect. 9.7.

9.2 Basic notions on moments

We give now the basic definitions of the moments of both one and two dimensional functions. Even though the defining relations formally reproduce those commonly introduced when dealing with density functions, the interpretation of the resulting quantities may come to be not the same, since we consider functions, that in general cannot be understood as density functions [2].

The reader will recognize that some of the definitions below have been used in previous analyses when dealing with the space and/or frequency extension of a signal or a phase-plane distribution, as, for instance, in § 4.6.4 in connection with the optical uncertainty relation, which indeed will be reformulated in terms of the second-order moments of the Wigner distribution function.

9.2.1 *Moments of a one-dimensional function*

Consider a real function $\mathcal{D}(q)$ defined over the real line: $q \in \mathbb{R}$, which is supposed to have a positive average, namely

$$\mathfrak{m}_0 = \int \mathcal{D}(q)dq > 0, \tag{9.2.2}$$

also called the *zero-order moment* of $\mathcal{D}$. As usual, unless otherwise specified, integrals are intended over the whole domain of the integrand function.

In general, $\mathcal{D}(q)$ is not assumed to be positive for all the values of q in its domain, and so it can not be interpreted as a density function. However, resorting to the strategy commonly applied when dealing with density functions, one may define the *first-order moment* of $\mathcal{D}$ by the usual integral:

$$\mathfrak{m}_1 = \frac{1}{\mathfrak{m}_0} \int q\mathcal{D}(q)dq. \tag{9.2.3}$$

Evidently, $\mathfrak{m}_1$ locates the *center of gravity* of $\mathcal{D}$, i.e., the value of q for which, thinking of $\mathcal{D}$ as a mass distribution along a straight line and allowing for negative masses where $\mathcal{D}(q) < 0$, the distribution is perfectly balanced.

Likewise the second-order *ordinary* and *central* moments, $\mathfrak{n}_2$ and $\mathfrak{m}_2$, write as

$$\mathfrak{n}_2 = \frac{1}{\mathfrak{m}_0} \int q^2\mathcal{D}(q)dq, \quad \mathfrak{m}_2 = \frac{1}{\mathfrak{m}_0} \int (q - \mathfrak{m}_1)^2\mathcal{D}(q)dq = \mathfrak{n}_2 - \mathfrak{m}_1^2\,. \tag{9.2.4}$$

If $\mathcal{D}(q)$ is nonnegative, the above can be considered as the variances of $\mathcal{D}$ respectively about the origin $q = 0$ and the centroid $\mathfrak{m}_1$. In fact, within the intuitive analogy of $\mathcal{D}$ with a mass distribution we can say that $\mathfrak{n}_2$ and $\mathfrak{m}_2$ give indication about the spread of the masses along the wire with respect to the origin and the center of gravity $\mathfrak{m}_1$, respectively. If $\mathcal{D}$ goes negative somewhere, this interpretation must be handled with care, since then, as we will see, both $\mathfrak{n}_2$ and $\mathfrak{m}_2$ may become negative. Notably, regarding $\mathfrak{m}_2$ as a function of $\mathfrak{m}_1$, the expression (9.2.3) just furnishes the value of $\mathfrak{m}_1$ for which $\mathfrak{m}_2$ is minimized.

The higher-order moments $\mathfrak{n}_r$ and $\mathfrak{m}_r$, $r > 2$, are then

$$\mathfrak{n}_r = \tfrac{1}{\mathfrak{m}_0} \int q^r \mathcal{D}(q) dq, \qquad \mathfrak{m}_r = \tfrac{1}{\mathfrak{m}_0} \int (q - \mathfrak{m}_1)^r \mathcal{D}(q) dq. \tag{9.2.5}$$

In particular, the central moment of order r relate to the ordinary moments of order zero up to r. In fact, using the binomial expansion $(q - \mathfrak{m}_1)^r = \sum_{h=0}^{r} \binom{r}{h} (-)^{r-h} q^h \mathfrak{m}_1^{r-h}$ in the expression for $\mathfrak{m}_r$ and integrating term by term, we obtain

$$\mathfrak{m}_r = \sum_{h=0}^{r} \binom{r}{h} (-)^{r-h} \mathfrak{n}_h \mathfrak{m}_1^{r-h}, \tag{9.2.6}$$

where specifically $\mathfrak{n}_0 = 1$ and $\mathfrak{n}_1 = \mathfrak{m}_1$. With $r = 2$ we regain the relation of $\mathfrak{m}_2$ to $\mathfrak{n}_2$ and $\mathfrak{m}_1$, as shown in (9.2.4).

Also, we note that by an appropriate translation of the coordinate origin to $\mathfrak{m}_1$ the central moments can be made to coincide with the ordinary moments.

The symbol q used here to denote the independent variable does not strictly relate to the space coordinate of optics, signifying in general a real quantity. Thus, the definitions above can be applied to the complex amplitude and the frequency spectrum of an optical signal as well. In that case, the pertinent $\mathcal{D}$-function identifies as the absolute square $|\varphi(q)|^2$ or $|\widetilde{\varphi}(\kappa)|^2$, which actually have both the conceptual valence and the mathematical properties to be considered as density functions. Besides, as we will see in the next section, the moments of $|\varphi(q)|^2$ and $|\widetilde{\varphi}(\kappa)|^2$ can be interpreted in terms of the expectation values of the position and frequency operators, and their powers, in the "state" φ.

Averages in the space and frequency domains as expectation values of the operators $\widehat{\mathsf{q}}$ and $\widehat{\kappa}$ and their powers

We show that the moments of $|\varphi(q)|^2$ and $|\widetilde{\varphi}(\kappa)|^2$ are interpretable as expectation values of the position and spatial-frequency operators $\widehat{\mathsf{q}}$ and $\widehat{\kappa}$, and their powers in the "state" φ. This conforms to the parallelism between the geometric/wave and classical/quantum correspondences. Also, the seemingly abstract operator picture, we have used in connection with the wave-optical description of the light propagation properties of paraxial optical systems, gains both calculational and physical merit revealing a profitable tool for the characterization of optical signals and systems.

According to the common definition, the expectation, or average, value of an operator $\widehat{\mathsf{O}}$ in a given "state" is interpreted in the sense of the inner product

$$\langle \widehat{\mathsf{O}} \rangle = (\cdot\,, \widehat{\mathsf{O}}\,\cdot). \tag{9.2.7}$$

The dots can interchangeably be replaced by $\varphi(q)$ or $\widetilde{\varphi}(\kappa)$ with the involved operator, i.e. $\widehat{\mathsf{q}}$ and $\widehat{\kappa}$, being accordingly given in the q or κ representation forms. The integration implied by (9.2.7) will then be over the q- or κ-domain.

It is evident that the zero order moments of $|\varphi(q)|^2$ and $|\widetilde{\varphi}(\kappa)|^2$ identify the norm $\|\varphi\|^2 = \|\widetilde{\varphi}\|^2$, which may conveniently be set equal to one.

Then let us compute the mean value $\overline{\kappa}$ of the spatial frequency. The defining relation (9.2.3), which specifically writes as

$$\overline{\kappa} = \int \kappa\, |\widetilde{\varphi}(\kappa)|^2\, d\kappa, \tag{9.2.8}$$

is expressible as

$$\overline{\kappa} = (\widetilde{\varphi}\,, \widehat{\kappa}\widetilde{\varphi}). \tag{9.2.9}$$

The calculation of (9.2.9) requires the knowledge of the Fourier spectrum $\widetilde{\varphi}(\kappa)$. However, it can also be expressed in terms of the signal amplitude $\varphi(q)$ in a very enlightening form. In fact, unfolding the square modulus $|\widetilde{\varphi}(\kappa)|^2 = \widetilde{\varphi}^*(\kappa)\widetilde{\varphi}(\kappa)$ and making use of the Fourier integral, we can rewrite (9.2.8) as

$$\overline{\kappa} = \int \widetilde{\varphi}^*(\kappa)\kappa\widetilde{\varphi}(\kappa)d\kappa = \tfrac{1}{2\pi} \iiint d\kappa dq dq'\, \varphi^*(q)\varphi(q')e^{i\kappa q}\kappa e^{-i\kappa q'}dq'. \tag{9.2.10}$$

Recognizing that $\kappa e^{-i\kappa q'} = i\frac{d}{dq'}e^{-i\kappa q'}$, we can integrate by part with respect to q', thus finally obtaining

$$\overline{\kappa} = \iint \varphi^*(q)[-i\tfrac{d}{dq'}\varphi(q')]\delta(q'-q)dqdq' = \int \varphi^*(q)[-i\tfrac{d}{dq}\varphi(q)]dq. \tag{9.2.11}$$

On account of the representation (4.6.57) of $\widehat{\kappa}$, the above integral identifies the inner product

$$\overline{\kappa} = (\varphi\,, \widehat{\kappa}\varphi). \tag{9.2.12}$$

The first order moment of $|\widetilde{\varphi}(\kappa)|^2$ is then interpretable as the expectation value of the spatial-frequency operator $\widehat{\kappa}$, i.e.,

$$\overline{\kappa} = \langle \widehat{\kappa} \rangle. \tag{9.2.13}$$

The explicit calculation involves the wave function $\varphi(q)$ or the frequency spectrum $\widetilde{\varphi}(\kappa)$ according to our convenience, and correspondingly the operator $\widehat{\kappa}$ takes the form $-i\frac{\partial}{\partial q}$ or κ.

Likewise the average value $\overline{\kappa^n}$ of the n-th power κ^n can be calculated as

$$\overline{\kappa^n} \equiv \int \kappa^n\, |\widetilde{\varphi}(\kappa)|^2\, d\kappa = (\widetilde{\varphi}\,, \widehat{\kappa}^n\widetilde{\varphi}), \tag{9.2.14}$$

and alternatively as

$$\overline{\kappa^n} = \int \varphi^*(q)(-i\tfrac{d}{dq})^n \varphi(q)dq = (\varphi\,, \widehat{\kappa}^n \varphi). \tag{9.2.15}$$

Therefore, the n-th order moment of $|\widetilde{\varphi}(\kappa)|^2$ can be brought into relation with the operator $\widehat{\kappa}$ through the average value of $\widehat{\kappa}^n$, namely

$$\overline{\kappa^n} = \langle \widehat{\kappa}^n \rangle, \tag{9.2.16}$$

the choice for the wave function and operator representation being a matter of convenience in accord with the problem at hand.

Finally, one can straightforwardly infer that the average $\overline{f(\kappa)}$ of a frequency function $f(\kappa)$ in the *state* $\varphi(q)$, defined by the integral

$$\overline{f(\kappa)} \equiv \int f(\kappa)\,|\widetilde{\varphi}(\kappa)|^2\,d\kappa = (\widetilde{\varphi}, f(\widehat{\kappa})\widetilde{\varphi}), \tag{9.2.17}$$

whose inherent inner-product interpretation has been displayed, is also obtainable as the integral

$$\overline{f(\kappa)} = \int \varphi^*(q) f(\widehat{\kappa})\varphi(q)dq = (\varphi, f(\widehat{\kappa})\varphi), \tag{9.2.18}$$

and hence interpretable as the expectation value of the operator function $f(\widehat{\kappa})$ in the *state* $\varphi(q)$:

$$\overline{f(\kappa)} = \langle f(\widehat{\kappa})\rangle. \tag{9.2.19}$$

It has implicitly been assumed that the function $f(\kappa)$ be expressible through the Taylor series

$$f(\kappa) = \sum_{n=0}^{\infty} f_n \kappa^n = f_0 + f_1\kappa + f_2\kappa^2 + ..., \tag{9.2.20}$$

with $f_n = f^n(0)/n!$. Then, applying result (9.2.16) to the series above, which amounts to substituting the operator $\widehat{\kappa}$ for κ in each occurrence, we end up with the operator function $f(\widehat{\kappa})$,

$$f(\widehat{\kappa}) = \sum_{n=0}^{\infty} f_n \widehat{\kappa}^n, \tag{9.2.21}$$

in both (9.2.17) and (9.2.18), the appropriate operator forms being κ and $-i\frac{\partial}{\partial q}$, respectively.

The computational relevance of relations (9.2.16) and (9.2.19) is apparent. The calculation of the average $\overline{\kappa^n}$, or, more in general, of the average $\overline{f(\kappa)}$ of a frequency function, does not necessarily require the knowledge of the spectrum $\widetilde{\varphi}(\kappa)$. It can alternatively be calculated using the signal amplitude $\varphi(q)$, which is indeed simply manipulated through appropriate differentiations and then inserted into the integrals implied by (9.2.16), or (9.2.19).

Clearly, similar results hold as well for the average of the spatial variable q and, in general, of any spatial function $g(q)$. In fact, starting from the definitions

$$\overline{q^n} \equiv \int q^n \left|\varphi(q)\right|^2 dq, \quad \overline{g(q)} \equiv \int g(q) \left|\varphi(q)\right|^2 dq, \tag{9.2.22}$$

the alternative expressions can be worked out

$$\begin{aligned} \overline{q^n} &= \int \widetilde{\varphi}^*(\kappa)(i\tfrac{d}{d\kappa})^n \widetilde{\varphi}(\kappa) d\kappa = (\widetilde{\varphi}, \widehat{\mathsf{q}}^n \widetilde{\varphi}), \\ \overline{g(q)} &= \int \widetilde{\varphi}^*(\kappa) g(\widehat{\mathsf{q}}) \widetilde{\varphi}(\kappa) d\kappa = (\widetilde{\varphi}, g(\widehat{\mathsf{q}}) \widetilde{\varphi}), \end{aligned} \tag{9.2.23}$$

according to the expression of $\widehat{\mathsf{q}}$ in the κ representation: $\widehat{\mathsf{q}} = i\frac{\partial}{\partial\kappa}$. Therefore relations (9.2.16) and (9.2.19) for the frequency averages are paralleled by

$$\overline{q^n} = \langle \widehat{\mathsf{q}}^n \rangle, \quad \overline{g(q)} = \langle g(\widehat{\mathsf{q}}) \rangle, \tag{9.2.24}$$

for the spatial averages. Thus, as regards the calculation of spatial averages the frequency spectrum $\widetilde{\varphi}(\kappa)$ plays the same role as the spatial amplitude $\varphi(q)$, when dealing with the calculation of frequency averages.

Finally, note that, if even powers are involved in (9.2.15), the final expression is further simplified. In fact, with $n = 2m$, (9.2.15) yields

$$\overline{\kappa^{2m}} = \int \varphi^*(q)(-i\tfrac{d}{dq})^{2m} \varphi(q) dq, \tag{9.2.25}$$

which by an m-fold integration by parts recasts into

$$\overline{\kappa^{2m}} = \int [\widehat{\kappa}^m \varphi(q)]^* [\widehat{\kappa}^m \varphi(q)] dq = \int \left|(-i\tfrac{d}{dq})^m \varphi(q)\right|^2 dq = \|\widehat{\kappa}^m \varphi\|^2. \tag{9.2.26}$$

In particular, the average square frequency can be calculated by the integral

$$\overline{\kappa^2} = \int \left|\tfrac{d}{dq}\varphi(q)\right|^2 dq, \tag{9.2.27}$$

simply involving the first-order derivative of the spatial distribution.

The corresponding expression for $\overline{q^{2m}}$ is, of course,

$$\overline{q^{2m}} = \int \left|(i\tfrac{d}{d\kappa})^m \widetilde{\varphi}(\kappa)\right|^2 d\kappa = \|\widehat{\mathsf{q}}^m \widetilde{\varphi}\|^2. \tag{9.2.28}$$

In § 4.6.4 we have discussed the well known uncertainty inequality

$$\sigma_q \sigma_\kappa \geq \frac{1}{2} \tag{9.2.29}$$

for the product of the space and frequency widths of the optical wave function σ_q, σ_κ, which, according to the accepted variance measures (4.6.35), are just the second-order moments of $|\varphi(q)|^2$ and $|\widetilde{\varphi}(\kappa)|^2$, and hence the average values $(\varphi, \widehat{\mathfrak{q}}^2\varphi)$ and $(\varphi, \widehat{\kappa}^2\varphi)$. In effect, (9.2.29) is a consequence of the relation holding between the operators $\widehat{\mathfrak{q}}$ and $\widehat{\kappa}$ as expressed by the Lie bracket: $[\widehat{\mathfrak{q}}, \widehat{\kappa}] = i$. It is a particular case of a more general theorem holding for any pair of variables, whose associated operators do not commute, as it is for $\widehat{\mathfrak{q}}$ and $\widehat{\kappa}$. The proof of the uncertainty relation on the ground of the operator representation is elegant and simple; also it evidences the generality of this principle which applies to any pair of non commuting observables (see, for instance, [3.10], pp. 216-218).

9.2.2 *Moments of a two-dimensional function*

Moment notions apply as well to a $2D$ real function $\mathcal{B}(q, \kappa)$; of course, q and κ denote now in general real independent quantities, without any close reference to the Fourier pair of optics. Evidently, the variety of possibilities is greatly enriched as both *local* and *global* moments can be defined. The formers depend on q or κ, as the latter, being constant, characterize the function $\mathcal{B}(q, \kappa)$ in a global sense. It is not required that $\mathcal{B}(q, \kappa)$ be nonnegative over the relevant domain, but only that the local and global averages be positive.

The global average is obtained by integrating $\mathcal{B}(q, \kappa)$ over the whole q-κ plane to the constant value

$$\mathfrak{m}_{00} = \iint \mathcal{B}(q, \kappa) dq d\kappa > 0. \tag{9.2.30}$$

In turn, the local averages are determined by partial integration of $\mathcal{B}(q, \kappa)$ along the κ or q line, thus yielding the q and κ depending quantities

$$\mathfrak{k}_0(q) = \int \mathcal{B}(q, \kappa) d\kappa > 0, \quad \mathfrak{q}_0(\kappa) = \int \mathcal{B}(q, \kappa) dq > 0. \tag{9.2.31}$$

It is apparent that

$$\mathfrak{m}_{00} = \int \mathfrak{k}_0(q) dq = \int \mathfrak{q}_0(\kappa) d\kappa. \tag{9.2.32}$$

Local moments

In the light of the averages (9.2.31), one defines the local moments of $\mathcal{B}(q, \kappa)$ with respect to κ and q simply reproducing the previously introduced formulae. Thus, the local first-order moments are given by the integrals

$$\mathfrak{k}_1(q) = \tfrac{1}{\mathfrak{k}_0(q)} \int \kappa \mathcal{B}(q, \kappa) d\kappa, \quad \mathfrak{q}_1(\kappa) = \tfrac{1}{\mathfrak{q}_0(\kappa)} \int q \mathcal{B}(q, \kappa) dq, \tag{9.2.33}$$

and respectively interpreted as the κ-average value of $\mathcal{B}$ at q and the q-average value of $\mathcal{B}$ at κ.

In general, the local higher-order moments with respect to κ and q write as

$$\mathfrak{k}_r(q) = \tfrac{1}{\mathfrak{k}_0(q)} \int [\kappa - \mathfrak{k}_1(q)]^r \mathcal{B}(q, \kappa) d\kappa, \quad \mathfrak{q}_r(\kappa) = \tfrac{1}{\mathfrak{q}_0(\kappa)} \int [q - \mathfrak{q}_1(\kappa)]^r \mathcal{B}(q, \kappa) dq. \tag{9.2.34}$$

Global moments

Similarly, following (9.2.30), the global first-order moments are obtained by integration over the whole q-κ plane as

$$\mathfrak{m}_{01} = \frac{1}{\mathfrak{m}_{00}} \iint \kappa \mathcal{B}(q,\kappa) dq d\kappa, \qquad \mathfrak{m}_{10} = \frac{1}{\mathfrak{m}_{00}} \iint q \mathcal{B}(q,\kappa) dq d\kappa. \tag{9.2.35}$$

Evidently, they specify the location of the centroid of $\mathcal{B}(q,\kappa)$, i.e. of the point $(\mathfrak{m}_{10}, \mathfrak{m}_{01})$ on the space-frequency plane (q,κ) at which $\mathcal{B}$, thought of as a mass distribution over the (q,κ) plane, is in balance.

Then, limiting ourselves to consider the central moments, it is obvious that the second-order moments write

$$\begin{aligned}
\mathfrak{m}_{02} &= \frac{1}{\mathfrak{m}_{00}} \iint (\kappa - \mathfrak{m}_{01})^2 \mathcal{B}(q,\kappa) dq d\kappa, \\
\mathfrak{m}_{11} &= \frac{1}{\mathfrak{m}_{00}} \iint (q - \mathfrak{m}_{10})(\kappa - \mathfrak{m}_{01}) \mathcal{B}(q,\kappa) dq d\kappa, \\
\mathfrak{m}_{20} &= \frac{1}{\mathfrak{m}_{00}} \iint (q - \mathfrak{m}_{10})^2 \mathcal{B}(q,\kappa) dq d\kappa,
\end{aligned} \tag{9.2.36}$$

as a natural extension of $\mathfrak{m}_2$ in (9.2.4), which indeed is folded into three quantities in order to characterize the extension of $\mathcal{B}$ with respect to q and κ as well as its orientation in the (q,κ) plane.

In general, the global moments are marked by two integer numbers, being

$$\mathfrak{m}_{rh} = \frac{1}{\mathfrak{m}_{00}} \iint (q - \mathfrak{m}_{10})^r (\kappa - \mathfrak{m}_{01})^h \mathcal{B}(q,\kappa) dq d\kappa. \tag{9.2.37}$$

As the coordinate origin can always be chosen or made, by appropriate translations, to coincide with the centroid of $\mathcal{B}$, the first-order moments $\mathfrak{m}_{10}$ and $\mathfrak{m}_{01}$ can be dropped out from (9.2.37), thus yielding the simplified expressions

$$\mathfrak{m}_{rh} = \frac{1}{\mathfrak{m}_{00}} \iint q^r \kappa^h \mathcal{B}(q,\kappa) dq d\kappa. \tag{9.2.38}$$

9.3 Preliminaries to the calculation of the moments of the Wigner distribution function

It is evident that the evaluation of the moments of the Wigner distribution function requires the integration over q or/and κ of products of $\mathcal{W}(q,\kappa)$ by monomials like q^r, κ^h and $q^n \kappa^m$. Thus, for purposes of future convenience, we firstly consider in some detail products of the type

$$q^r \mathcal{W}(q,\kappa), \qquad \kappa^h \mathcal{W}(q,\kappa), \qquad q^n \kappa^m \mathcal{W}(q,\kappa), \tag{9.3.1}$$

and express them in a form, which will be usefully exploited when integrating (9.3.1) with respect to q or/and κ.

It turns out that the integration of the products (9.3.1) comes to involve the cross-Wigner distribution of functions, which relate to the signal amplitude and frequency spectrum through the coordinate and frequency operators $\widehat{\mathsf{q}}$ and $\widehat{\kappa}$. Therefore, we devote a later section to a brief description of the properties of the cross-Wigner distribution function.

Although the presentation in the chapter is limited to the second-order moments of the Wigner distribution function, the discussion below offers a description of a useful operatorial machinery for possible advanced uses. The reader may also recognize the link with the formalism illustrated in § 8.5.1.

9.3.1 *About the products $q^r\mathcal{W}(q,\kappa)$, $\kappa^h\mathcal{W}(q,\kappa)$ and $q^n\kappa^m\mathcal{W}(q,\kappa)$*

Let us start with the product $q\mathcal{W}(q,\kappa)$.

Using for $\mathcal{W}(q,\kappa)$ the definition (6.2.3) involving the signal amplitude $\varphi(q)$, we conveniently arrange the factors under the integral as

$$
\begin{aligned}
q\mathcal{W}(q,\kappa) = \tfrac{1}{2}\Big\{\tfrac{1}{2\pi}\int (q-\tfrac{q'}{2})\varphi^*(q-\tfrac{q'}{2})\varphi(q+\tfrac{q'}{2})e^{-i\kappa q'}dq' \\
+\tfrac{1}{2\pi}\int (q+\tfrac{q'}{2})\varphi^*(q-\tfrac{q'}{2})\varphi(q+\tfrac{q'}{2})e^{-i\kappa q'}dq'\Big\}.
\end{aligned}
\tag{9.3.2}
$$

Since

$$
\begin{aligned}
(q-\tfrac{q'}{2})\varphi^*(q-\tfrac{q'}{2}) = (\widehat{\mathsf{q}}\varphi)^*(q-\tfrac{q'}{2}),\\
(q+\tfrac{q'}{2})\varphi(q+\tfrac{q'}{2}) = (\widehat{\mathsf{q}}\varphi)(q+\tfrac{q'}{2}),
\end{aligned}
\tag{9.3.3}
$$

(9.3.2) can compactly be expressed in terms of the mixed Wigner distribution function, as defined in (6.3.12), thus yielding

$$
q\mathcal{W}(q,\kappa) = \tfrac{1}{2}\left[\mathcal{W}_{\widehat{\mathsf{q}}\varphi,\varphi}(q,\kappa) + \mathcal{W}_{\varphi,\widehat{\mathsf{q}}\varphi}(q,\kappa)\right] = \Re[\mathcal{W}_{\varphi,\widehat{\mathsf{q}}\varphi}(q,\kappa)]. \tag{9.3.4}
$$

The final expression above is due to the property of the cross-Wigner distribution: $\mathcal{W}_{\varphi_1,\varphi_2}(q,\kappa) = \mathcal{W}^*_{\varphi_2,\varphi_1}(q,\kappa)$, already mentioned in § 6.3.4 (see also below). Notably, the expression on the right-hand side is reminiscent of the rule for the derivative of a product.

It is an easy matter to obtain for $q^r\mathcal{W}(q,\kappa)$ the expression

$$
q^r\mathcal{W}(q,\kappa) = (\tfrac{1}{2})^r\sum_{j=0}^{r}\binom{r}{j}\,\mathcal{W}_{\widehat{\mathsf{q}}^{r-j}\varphi,\widehat{\mathsf{q}}^{j}\varphi}(q,\kappa), \tag{9.3.5}
$$

where, of course, according to the coordinate representation of the operator $\widehat{\mathsf{q}}$, simply implying $\widehat{\mathsf{q}}^h\varphi = q^h\varphi$, the term $\mathcal{W}_{\widehat{\mathsf{q}}^{r-j}\varphi,\widehat{\mathsf{q}}^{j}\varphi}(q,\kappa)$ explicitly reads

$$
\mathcal{W}_{\widehat{\mathsf{q}}^{r-j}\varphi,\widehat{\mathsf{q}}^{j}\varphi}(q,\kappa) = \tfrac{1}{2\pi}\int (q-\tfrac{q'}{2})^{r-j}\varphi^*(q-\tfrac{q'}{2})(q+\tfrac{q'}{2})^j\varphi(q+\tfrac{q'}{2})e^{-i\kappa q'}dq'. \tag{9.3.6}
$$

Correspondingly, the product $\kappa\mathcal{W}(q,\kappa)$ can be handled as follows:

$$\begin{aligned}\kappa\mathcal{W}(q,\kappa) &= \tfrac{1}{2\pi}\int \kappa\varphi^*(q-\tfrac{q'}{2})\varphi(q+\tfrac{q'}{2})e^{-i\kappa q'}dq' \\ &= \tfrac{1}{2\pi}\int \varphi^*(q-\tfrac{q'}{2})\varphi(q+\tfrac{q'}{2})(i\tfrac{d}{dq'}e^{-i\kappa q'})dq',\end{aligned} \tag{9.3.7}$$

which, after integrating by parts, yields

$$\kappa\mathcal{W}(q,\kappa) = \tfrac{1}{2}\left[\mathcal{W}_{\widehat{\kappa}\varphi,\varphi}(q,\kappa)+\mathcal{W}_{\varphi,\widehat{\kappa}\varphi}(q,\kappa)\right] = \Re[\mathcal{W}_{\varphi,\widehat{\kappa}\varphi}(q,\kappa)]. \tag{9.3.8}$$

Again, use has been done of the operatorial notation on specifying the functions entering the cross-Wigner distribution function. Of course, here $\widehat{\kappa}\varphi = -i\frac{d}{dq}\varphi$ according to the coordinate representation (4.6.58) of the frequency operator $\widehat{\kappa}$.

The formal similarity of expressions (9.3.4) and (9.3.8) for $q\mathcal{W}(q,\kappa)$ and $\kappa\mathcal{W}(q,\kappa)$, with the operators $\widehat{\mathsf{q}}$ and $\widehat{\kappa}$ interchanged, suggests to write the product $\kappa^h\mathcal{W}(q,\kappa)$ in the form reminiscent of (9.3.5), i.e.

$$\kappa^h\mathcal{W}(q,\kappa) = (\tfrac{1}{2})^h\sum_{j=0}^{h}\binom{h}{j}\mathcal{W}_{\widehat{\kappa}^{h-j}\varphi,\widehat{\kappa}^j\varphi}(q,\kappa), \tag{9.3.9}$$

where

$$\widehat{\kappa}^m\varphi = (-i)^m\frac{d^m}{dq^m}\varphi, \tag{9.3.10}$$

and hence $\mathcal{W}_{\widehat{\kappa}^{h-j}\varphi,\widehat{\kappa}^j\varphi}(q,\kappa)$ is explicitly given by

$$\mathcal{W}_{\widehat{\kappa}^{h-j}\varphi,\widehat{\kappa}^j\varphi}(q,\kappa) = \frac{i^{h-2j}}{2\pi}\int[\frac{d^{h-j}}{dq^{h-j}}\varphi]^*(q-\tfrac{q'}{2})[\frac{d^j}{dq^j}\varphi](q+\tfrac{q'}{2})e^{-i\kappa q'}dq'. \tag{9.3.11}$$

The realness of the right-hand sides of both (9.3.5) and (9.3.9) follows again from the conjugation property of the cross-Wigner distribution function in response to the interchange of the functions.

The convenience of the operatorial notation adopted in the previous relations, involving the mixed Wigner distribution function, manifests, among other things, in the fact that the obtained expressions (9.3.5) and (9.3.9) for the products $q^r\mathcal{W}(q,\kappa)$ and $\kappa^h\mathcal{W}(q,\kappa)$ can be recast in terms of the signal frequency spectrum $\widetilde{\varphi}(\kappa)$ according to

$$\begin{aligned}q^r\mathcal{W}(q,\kappa) &= (\tfrac{1}{2})^r\sum_{j=0}^{r}\tbinom{r}{j}\mathcal{W}_{\widehat{\mathsf{q}}^{r-j}\widetilde{\varphi},\widehat{\mathsf{q}}^j\widetilde{\varphi}}(\kappa,q), \\ \kappa^h\mathcal{W}(q,\kappa) &= (\tfrac{1}{2})^h\sum_{j=0}^{h}\tbinom{h}{j}\mathcal{W}_{\widehat{\kappa}^{h-j}\widetilde{\varphi},\widehat{\kappa}^j\widetilde{\varphi}}(\kappa,q),\end{aligned} \tag{9.3.12}$$

where, taking into account expressions (4.6.60) and (4.6.59) for the operators $\widehat{\mathsf{q}}$ and $\widehat{\kappa}$ in the frequency representation, we have

$$\widehat{\mathsf{q}}^m\widetilde{\varphi} = i^m\frac{d^m}{d\kappa^m}\widetilde{\varphi}, \qquad \widehat{\kappa}^m\widetilde{\varphi} = \kappa^m\widetilde{\varphi}. \tag{9.3.13}$$

Also, the cross Wigner distribution functions appearing on the right hand sides are understood as

$$\begin{aligned}\mathcal{W}_{\widehat{\mathsf{q}}^{r-j}\widetilde{\varphi},\widehat{\mathsf{q}}^{j}\widetilde{\varphi}}(\kappa,q) &= \tfrac{(-i)^{r-2j}}{2\pi}\int[\tfrac{d^{r-j}}{d\kappa^{r-j}}\widetilde{\varphi}]^*(\kappa-\tfrac{\kappa'}{2})[\tfrac{d^j}{d\kappa^j}\widetilde{\varphi}](\kappa+\tfrac{\kappa'}{2})e^{iq\kappa'}d\kappa',\\ \mathcal{W}_{\widehat{\kappa}^{h-j}\widetilde{\varphi},\widehat{\kappa}^{j}\widetilde{\varphi}}(\kappa,q) &= \tfrac{1}{2\pi}\int(\kappa-\tfrac{\kappa'}{2})^{h-j}\widetilde{\varphi}^*(\kappa-\tfrac{\kappa'}{2})(\kappa+\tfrac{\kappa'}{2})^j\varphi(\kappa+\tfrac{\kappa'}{2})e^{iq\kappa'}d\kappa'.\end{aligned} \tag{9.3.14}$$

Accordingly, it easy to see that the product $q^n\kappa^m\mathcal{W}(q,\kappa)$ comes to be

$$\begin{aligned}q^n\kappa^m\mathcal{W}(q,\kappa) &= (\tfrac{1}{2})^m\sum_{j=0}^{m}\tbinom{m}{j}q^n\mathcal{W}_{\widehat{\kappa}^{m-j}\varphi,\widehat{\kappa}^{j}\varphi}(q,\kappa)\\ &= (\tfrac{1}{2})^n\sum_{j=0}^{n}\tbinom{n}{j}\kappa^m\mathcal{W}_{\widehat{\mathsf{q}}^{n-j}\varphi,\widehat{\mathsf{q}}^{j}\varphi}(q,\kappa),\end{aligned} \tag{9.3.15}$$

which can further be written as

$$q^n\kappa^m\mathcal{W}(q,\kappa) = (\tfrac{1}{2})^{n+m}\sum_{j=0}^{n}\sum_{l=0}^{m}\tbinom{n}{j}\tbinom{m}{l}\mathcal{W}_{\widehat{\mathsf{q}}^{n-j}\,\widehat{\kappa}^{m-l}\varphi,\widehat{\mathsf{q}}^{j}\,\widehat{\kappa}^{l}\varphi}(q,\kappa). \tag{9.3.16}$$

In conclusion, we see that the products of the Wigner distribution function with the monomials q^r, κ^h and $q^n\kappa^m$ can be expressed in the form of sums of cross-Wigner distribution functions, involving the signal amplitude or the frequency spectrum, properly acted by powers of the operators $\widehat{\mathsf{q}}$ and $\widehat{\kappa}$. The advantages of such a representation result from that, as we will see below, when integrated with respect to the variables q or κ, the cross-Wigner distribution function behaves in the similar way as the self-Wigner distribution function.

As a useful exercise, we invite the reader to deduce the basic relations of this section, summarized by (9.3.16) within the Bopp operators formalism, presented in § 8.5.1. It is evident, in fact, that

$$\begin{aligned}q &= \tfrac{1}{2}(\widehat{\mathsf{Q}}+\widehat{\mathsf{Q}}^*),\\ \kappa &= \tfrac{1}{2}(\widehat{\mathsf{K}}+\widehat{\mathsf{K}}^*).\end{aligned} \tag{9.3.17}$$

Since the operators $\widehat{\mathsf{Q}}$ and $\widehat{\mathsf{Q}}^*$ commute with each other as well as $\widehat{\mathsf{K}}$ and $\widehat{\mathsf{K}}^*$, powers of q or κ simply follow the binomial rule.

9.3.2 *Basics of the cross-Wigner distribution function*

The cross-Wigner distribution function $\mathcal{W}_{\varphi_1,\varphi_2}(q,\kappa)$ of the two functions $\varphi_1(q)$ and $\varphi_2(q)$ has been introduced in § 6.3.4 through the q-domain convolution

$$\mathcal{W}_{\varphi_1,\varphi_2}(q,\kappa) = \tfrac{1}{2\pi}\int\varphi_1^*(q-\tfrac{q'}{2})\varphi_2(q+\tfrac{q'}{2})e^{-i\kappa q'}dq', \tag{9.3.18}$$

and, correspondingly, through the κ-domain convolution

$$\mathcal{W}_{\widetilde{\varphi}_1,\widetilde{\varphi}_2}(\kappa,q) = \tfrac{1}{2\pi}\int\widetilde{\varphi}_1^*(\kappa-\tfrac{\kappa'}{2})\widetilde{\varphi}_2(\kappa+\tfrac{\kappa'}{2})e^{iq\kappa'}d\kappa' = \mathcal{W}_{\varphi_1,\varphi_2}(q,\kappa), \tag{9.3.19}$$

involving the Fourier transforms $\widetilde{\varphi}_1(\kappa)$ and $\widetilde{\varphi}_2(\kappa)$. Needless to say, both $\varphi_1(q)$ and $\varphi_2(q)$ are supposed to be Lebesque integrable.

Also, as already noted, it results that

$$\mathcal{W}_{\varphi_1,\varphi_2}(q,\kappa) = \mathcal{W}^*_{\varphi_2,\varphi_1}(q,\kappa), \tag{9.3.20}$$

implying the realness of the (auto-)Wigner distribution function ($\varphi_1 = \varphi_2$).

Basics to the forthcoming considerations, are the counterparts of relations (6.3.22-28). In fact, we find

$$\begin{aligned} \varphi_1^*(q - \tfrac{q'}{2})\varphi_2(q + \tfrac{q'}{2}) &= \int \mathcal{W}_{\varphi_1,\varphi_2}(q,\kappa)e^{i\kappa q'}d\kappa, \\ \widetilde{\varphi}_1^*(\kappa - \tfrac{\kappa'}{2})\widetilde{\varphi}_2(\kappa + \tfrac{\kappa'}{2}) &= \int \mathcal{W}_{\varphi_1,\varphi_2}(q,\kappa)e^{-i\kappa' q}dq, \end{aligned} \tag{9.3.21}$$

which, like (6.3.22) and (6.3.24), relate the products on the left hand sides, regarded respectively as functions of q' and κ', to the Fourier transforms of the cross-Wigner distribution function $\mathcal{W}_{\varphi_1,\varphi_2}(q,\kappa)$ with respect to κ and q.

Further relations can be inferred from (9.3.21). Setting for instance $q' = 0$ and $\kappa' = 0$, we may gain the product $\varphi_1^*(q)\varphi_2(q)$ by integrating $\mathcal{W}_{\varphi 1,\varphi 2}$ over the frequency variable, namely

$$\varphi_1^*(q)\varphi_2(q) = \int \mathcal{W}_{\varphi_1,\varphi_2}(q,\kappa)d\kappa, \tag{9.3.22}$$

and correspondingly the product of the spectra $\widetilde{\varphi}_1^*(\kappa)\widetilde{\varphi}_2(\kappa)$ from the integration of $\mathcal{W}_{\varphi 1,\varphi 2}$ over the coordinate:

$$\widetilde{\varphi}_1^*(\kappa)\widetilde{\varphi}_2(\kappa) = \int \mathcal{W}_{\varphi_1,\varphi_2}(q,\kappa)dq. \tag{9.3.23}$$

Consequently, the inner product of the two functions $\varphi_1(q)$ and $\varphi_2(q)$, as well as of the corresponding Fourier transforms $\widetilde{\varphi}_1(\kappa)$ and $\widetilde{\varphi}_2(\kappa)$, simply equals the integral of $\mathcal{W}_{\varphi 1,\varphi 2}(q,\kappa)$ over the whole plane (q,κ), namely

$$(\varphi_1,\varphi_2) = (\widetilde{\varphi}_1,\widetilde{\varphi}_2) = \iint \mathcal{W}_{\varphi_1,\varphi_2}(q,\kappa)dqd\kappa = \iint \mathcal{W}_{\widetilde{\varphi}_1,\widetilde{\varphi}_2}(q,\kappa)dqd\kappa, \tag{9.3.24}$$

which gives (6.3.26) with $\varphi_1 = \varphi_2$.

Moreover, the inversion formulas for both the functions and the Fourier transforms can be inferred from the relations (9.3.21) as well. In particular, for the amplitudes $\varphi_1(q)$ and $\varphi_2(q)$ we find

$$\begin{aligned} \varphi_1(q) &= \tfrac{1}{\varphi_2^*(q_0)} \int \mathcal{W}^*_{\varphi_1,\varphi_2}(\tfrac{q+q_0}{2},\kappa)e^{-i\kappa(q-q_0)}d\kappa, \\ \varphi_2(q) &= \tfrac{1}{\varphi_1^*(q_0)} \int \mathcal{W}_{\varphi_1,\varphi_2}(\tfrac{q+q_0}{2},\kappa)e^{i\kappa(q-q_0)}d\kappa, \end{aligned} \tag{9.3.25}$$

indicating that both $\varphi_1(q)$ and $\varphi_2(q)$ can be reconstructed from $\mathcal{W}_{\varphi_1,\varphi_2}$ up to a constant phase.

9.4 Wigner distribution function: local and global moments.

We apply the notion of moments, as described in § 9.2.2, to the Wigner distribution function. Specifically, we consider the local moments of $\mathcal{W}(q,\kappa)$ with respect to the spatial frequency κ and the coordinate q in § 9.4.1 and § 9.4.2. The global moments and related issues are considered in §§ 9.4.3-9.4.5.

9.4.1 *Local moments in the angular variable*

As earlier remarked, the moments with respect to one variable depend on the other and therefore yield informations about the local (i.e., on a chosen plane parallel to one or the other direction) properties of the Wigner distribution function, as, for instance, the relative center and extension.

On applying the scheme of § 9.2.2, we are led to consider the average $\mathfrak{k}_0(q)$ of $\mathcal{W}$ with respect to the frequency, i.e. the marginal distribution

$$\mathfrak{k}_0(q) = \int \mathcal{W}(q,\kappa)d\kappa = |\varphi(q)|^2 . \tag{9.4.1}$$

which is, of course, nonnegative. Accordingly, we can define the local higher-order moments of the Wigner distribution function with respect to κ only at positions for which $\varphi(q) \neq 0$.

The first-order moment: $\mathfrak{k}_1(q)$

The first-order moment $\mathfrak{k}_1(q)$ is then given as

$$\mathfrak{k}_1(q) = \tfrac{1}{\mathfrak{k}_0(q)} \int \kappa \mathcal{W}(q,\kappa)d\kappa, \tag{9.4.2}$$

and interpreted as the average frequency of $\mathcal{W}$ at position q. Taking into account relation (9.3.8), we can write

$$\mathfrak{k}_1(q) = \tfrac{1}{2\mathfrak{k}_0(q)} \int [\mathcal{W}_{\widehat{\kappa}\varphi,\varphi}(q,\kappa) + \mathcal{W}_{\varphi,\widehat{\kappa}\varphi}(q,\kappa)]d\kappa. \tag{9.4.3}$$

Then, exploiting the property (9.3.22) of the cross-Wigner distribution function, after some algebra we end up with

$$\mathfrak{k}_1(q) = -\frac{i}{2}\left(\frac{1}{\varphi}\frac{d}{dq}\varphi - \frac{1}{\varphi^*}\frac{d}{dq}\varphi^*\right) = \Im\left[\frac{\varphi'(q)}{\varphi(q)}\right], \tag{9.4.4}$$

where $\varphi'(q) \equiv d\varphi/dq$ and the symbol $\Im$ denotes the imaginary part.

In order to gain a deeper insight into the above expression, it is convenient to represent the signal in the form

$$\varphi(q) = f(q)e^{i\phi(q)}, \tag{9.4.5}$$

both the envelope and the phase, $f(q)$ and $\phi(q)$, being real-valued functions. The formal expression (9.4.4) yields then the phase derivative of the signal:

$$\mathfrak{k}_1(q) = \frac{d}{dq}\phi(q). \tag{9.4.6}$$

It suggests to interpret $\mathfrak{k}_1(q)$ as the local direction of propagation of the signal. This is in conformity with the wave optics view of light rays as the normals to the wavefront of the signal, which hence are mathematically represented by the gradient (and so by the derivative in a $1D$ model) of the phase. Such an interpretation is also suggested by the defining expression (9.4.2), which within the picture of the signal as a beam of rays, may be understood as the local mean value of the ray directions $\kappa = k\vartheta$ in the beam, being obtained by averaging over all the possible ray directions, each weighted by the relevant ray "amplitude" $\mathcal{W}(q,\kappa)$. Within the time-frequency domain of the signal theory the time derivative of the phase is reported as the *instantaneous frequency* [3].

As an example, we may consider the plane wave

$$\varphi(q) = e^{i\kappa_0 q}, \tag{9.4.7}$$

whose local frequency comes to be the q-independent value (see § 6.2.1)

$$\mathfrak{k}_1(q) = \kappa_0. \tag{9.4.8}$$

Correspondingly, in the case of the pure chirp

$$\varphi(q) = \varphi_0 e^{i\beta\frac{q^2}{2}}, \tag{9.4.9}$$

we find

$$\mathfrak{k}_1(q) = \beta q, \tag{9.4.10}$$

in accord with the considerations of § 6.4.4 regarding the Wigner distribution function of a pure chirp.

Likewise, for the sum of two chirps, written as

$$\varphi(q) = \frac{\varphi_0}{2} e^{i\beta_1\frac{q^2}{2}} + \frac{\varphi_0}{2} e^{i\beta_2\frac{q^2}{2}}, \tag{9.4.11}$$

we find

$$\mathfrak{k}_1(q) = \tfrac{1}{2}(\beta_1 q + \beta_2 q), \tag{9.4.12}$$

which gives indeed the angle within the cone between $\vartheta_1 = \beta_1 q$ and $\vartheta_2 = \beta_2 q$.

Interestingly, the average frequency $\mathfrak{k}_1(q)$ of the Wigner distribution function of a real signal, i.e., $\phi(q) = 0$, is zero:

$$\mathfrak{k}_1(q) = 0 \qquad \text{for a real signal,} \tag{9.4.13}$$

which, in effect, emerges also from (9.4.2) on account of that the Wigner function of real signals is an even function of κ.

In addition, if $\varphi(q)$ is the product of two signals

$$\varphi(q) = \varphi_1(q)\varphi_2(q), \tag{9.4.14}$$

the average frequency of the Wigner distribution function turns out to be

$$\mathfrak{k}_1(q) = \mathfrak{k}_{1,\varphi_1}(q) + \mathfrak{k}_{1,\varphi_2}(q), \tag{9.4.15}$$

i.e. the sum of the average frequencies of the Wigner distribution functions $\mathcal{W}_{\varphi_1}$ and $\mathcal{W}_{\varphi_2}$ of the two signals in the product.

Thus, for the frequency-modulated chirp

$$\varphi(q) = \varphi_0 e^{i\beta\frac{q^2}{2}} e^{i\kappa_0 q}, \tag{9.4.16}$$

we have

$$\mathfrak{k}_1(q) = \beta q + \kappa_0. \tag{9.4.17}$$

The second-order moment: $\mathfrak{k}_2(q)$

According to the general expressions in (9.2.34), the second-order moment of $\mathcal{W}$ with respect to κ is given by

$$\mathfrak{k}_2(q) = \frac{1}{\mathfrak{k}_0(q)} \int [\kappa - \mathfrak{k}_1(q)]^2 \mathcal{W}(q,\kappa) d\kappa = \frac{1}{|\varphi(q)|^2} \int \kappa^2 \mathcal{W}(q,\kappa) d\kappa - \mathfrak{k}_1^2(q). \tag{9.4.18}$$

Using relation (9.3.9) for the product $\kappa^2\mathcal{W}(q,\kappa)$, namely

$$\kappa^2\mathcal{W}(q,\kappa) = \tfrac{1}{4}\{\mathcal{W}_{\widehat{\kappa^2}\varphi,\varphi}(q,\kappa) + 2\mathcal{W}_{\widehat{\kappa}\varphi,\widehat{\kappa}\varphi}(q,\kappa) + \mathcal{W}_{\varphi,\widehat{\kappa^2}\varphi}(q,\kappa)\}, \tag{9.4.19}$$

and property (9.3.22) of the cross-Wigner distribution function, it is easy to calculate $\mathfrak{k}_2(q)$ in the form

$$\mathfrak{k}_2(q) = -\frac{1}{4}\Big[\frac{d}{dq}\frac{\varphi'}{\varphi} + \frac{d}{dq}\frac{\varphi'^*}{\varphi^*}\Big] = -\frac{1}{2}\Re\Big[\frac{d}{dq}\frac{\varphi'(q)}{\varphi(q)}\Big]. \tag{9.4.20}$$

The above is interestingly compared with the expression (9.4.4) for $\mathfrak{k}_1(q)$.

Notably, since $\mathcal{W}(q,\kappa)$ is not always positive, the local second-order moment $\mathfrak{k}_2(q)$ is not necessarily positive. Thereby, as already remarked, it cannot be interpreted as the variance of $\mathcal{W}$ in the κ direction at a given position q.

Using for $\varphi(q)$ the representation (9.4.5) in terms of the envelope and phase, we readily obtain

$$\mathfrak{k}_2(q) = -\frac{1}{2}\frac{d}{dq}\Big[\frac{f'(q)}{f(q)}\Big], \tag{9.4.21}$$

thus showing that the local second-order moment is determined by the signal envelope $f(q)$. As before, we have set $f'(q) \equiv df/dq$.

The second-order moment $\mathfrak{k}_2(q)$ for the chirp (9.4.9), for instance, is zero.

In general, $\mathfrak{k}_2(q) = 0$ whenever the signal envelope has the form $f(q) = f_0 e^{\beta q}$ with f_0 and β arbitrary constants.

The additivity property, obeyed by the average frequency $\mathfrak{k}_1(q)$ when a product of signals is concerned, holds for the second-order moment as well. In fact, for the signal (9.4.14) we have

$$\mathfrak{k}_2(q) = \mathfrak{k}_{2,\varphi_1}(q) + \mathfrak{k}_{2,\varphi_2}(q). \tag{9.4.22}$$

As an example, let us consider the frequency-modulated Gaussian signal

$$\varphi_G(q) = (\tfrac{2}{\pi w^2})^{\frac{1}{4}} e^{-\frac{1}{w^2}q^2} e^{i\kappa_0 q}, \tag{9.4.23}$$

for which we find

$$\mathfrak{k}_1(q) = \kappa_0, \quad \mathfrak{k}_2(q) = \frac{1}{w^2}, \tag{9.4.24}$$

in accord with the claimed role of the parameter w ruling the spatial and frequency extents of the relevant Wigner distribution function (see § 6.4.5).

9.4.2 *Local moments in the space variable*

The local moments in the angular variable are paralleled by the local moments in the spatial variable. The average of $\mathcal{W}(q,\kappa)$ with respect to q is given by the square amplitude of the frequency spectrum $\widetilde{\varphi}(\kappa)$ (Eq. 6.3.26)

$$\mathfrak{q}_0(\kappa) = \int \mathcal{W}(q,\kappa)dq = |\widetilde{\varphi}(\kappa)|^2 . \tag{9.4.25}$$

Then, we may define the local moments in the space variable only for frequencies κ at which $\widetilde{\varphi}(\kappa) \neq 0$.

The first-order moment: $\mathfrak{q}_1(\kappa)$

The local first-order moment $\mathfrak{q}_1(\kappa)$ is therefore

$$\mathfrak{q}_1(\kappa) = \tfrac{1}{\mathfrak{q}_0(\kappa)} \int q\mathcal{W}(q,\kappa)dq, \tag{9.4.26}$$

which, on account of the picture of the signal as a collection of rays and of the Wigner distribution function $\mathcal{W}(q,\kappa)$ as the "ray amplitude", may be interpreted as the average ray position at the frequency (i.e. direction) κ.

It is convenient to use the expression (9.3.12) for the product $q\mathcal{W}(q,\kappa)$ in terms of the frequency spectrum $\widetilde{\varphi}(\kappa)$, so that, due to the relation (9.3.23) obeyed by the cross-Wigner distribution function, we obtain

$$\mathfrak{q}_1(\kappa) = \frac{i}{2}(\frac{1}{\widetilde{\varphi}}\frac{d}{d\kappa}\widetilde{\varphi} - \frac{1}{\widetilde{\varphi}^*}\frac{d}{d\kappa}\widetilde{\varphi}^*) = -\Im[\frac{\widetilde{\varphi}'(\kappa)}{\widetilde{\varphi}(\kappa)}], \tag{9.4.27}$$

where, of course, $\widetilde{\varphi}'(\kappa) \equiv d\widetilde{\varphi}/d\kappa$.

Expression (9.4.27) can be regarded as the dual of (9.4.4) for $\mathfrak{k}_1(q)$. The correspondence between $\mathfrak{q}_1(\kappa)$ and $\mathfrak{k}_1(q)$ is better evidenced, writing the spectrum in a form similar to (9.4.5), i.e.

$$\widetilde{\varphi}(\kappa) = h(\kappa)e^{i\psi(\kappa)}, \tag{9.4.28}$$

where $h(\kappa)$ and $\psi(\kappa)$ are real functions. Thus, $\mathfrak{q}_1(\kappa)$ equals the derivative of the phase of the spectrum

$$\mathfrak{q}_1(\kappa) = -\frac{d}{d\kappa}\psi(\kappa). \tag{9.4.29}$$

As an example, we consider the chirp (9.4.9), whose Fourier transform is

$$\widetilde{\varphi}(\kappa) = \varphi_0\sqrt{\frac{i}{\beta}}e^{-i\frac{\kappa^2}{2\beta}}, \quad (\beta > 0) \tag{9.4.30}$$

and hence

$$\mathfrak{q}_1(\kappa) = \frac{\kappa}{\beta}, \tag{9.4.31}$$

with an evident correspondence with (9.4.10).

In order to clarify the meaning of $\mathfrak{q}_1(\kappa)$ we may consider a space-invariant system (see § 4.6.3). The relevant input-output relation in the κ domain writes in the simple multiplicative form

$$\widetilde{\varphi}_o(\kappa) = \mathfrak{h}(\kappa)\widetilde{\varphi}_i(\kappa), \tag{9.4.32}$$

involving the spectra of the input and output signals, $\widetilde{\varphi}_o(\kappa)$ and $\widetilde{\varphi}_i(\kappa)$, and the transfer function $\mathfrak{h}(\kappa)$ of the system. We write $\mathfrak{h}(\kappa)$ as

$$\mathfrak{h}(\kappa) = h(\kappa)e^{i\psi(\kappa)}, \tag{9.4.33}$$

and suppose that both h and ψ vary sufficiently slow about some frequency κ_o to make reasonable the approximations $h(\kappa) \simeq h(\kappa_o)$ and $\psi(\kappa) \simeq \psi(\kappa_o) + \psi'(\kappa_o)\kappa_o$. Accordingly, on reconstructing the amplitude $\varphi_o(q)$ of the output signal, we obtain

$$\varphi_o(q) = h(\kappa_o)e^{i[\psi(\kappa_o)+\psi'(\kappa_o)\kappa_o]}\varphi_i(q + \psi'(\kappa_o)). \tag{9.4.34}$$

Also, we assume the input signal to have a spatial envelope $f_i(q)$ linearly modulated at the frequency κ_o, namely

$$\varphi_i(q) = f_i(q)e^{i\kappa_0 q}. \tag{9.4.35}$$

Then, the output amplitude (9.4.34) becomes

$$\varphi_o(q) = h(\kappa_o)f_i(q - \mathfrak{q}_1(\kappa_o))e^{i\kappa_0[q-\mathfrak{q}_p(\kappa_o)]}, \tag{9.4.36}$$

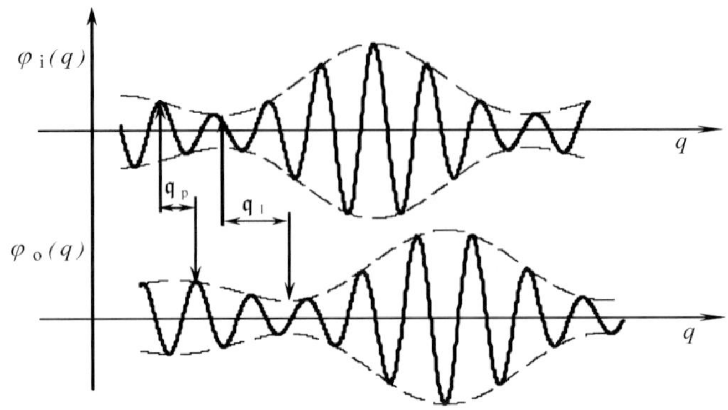

FIGURE 9.2. Typical diagram illustrating the meaning of $\mathfrak{q}_1$ and $\mathfrak{q}_p$ as envelope and phase shifts.

where we have exploited expression (9.4.29) for $\mathfrak{q}_1(\kappa)$ and denoted the ratio $\frac{\psi(\kappa_o)}{\kappa_0}$ as $\mathfrak{q}_p(\kappa_o) \equiv -\frac{\psi(\kappa_o)}{\kappa_0}$. Evidently, $\mathfrak{q}_1(\kappa_o)$ identifies the space-shift produced by the system on the envelope $f_i(q)$ of the input signal. On the other hand, the phase also is affected by a shift, specified by $\mathfrak{q}_p(\kappa_o)$. The familiar view illustrating the envelope and phase shifts, $\mathfrak{q}_1(\kappa_o)$ and $\mathfrak{q}_p(\kappa_o)$, is displayed in Fig. 9.2. These two shifts are equal only when the phase response $\psi(\kappa)$ of the system varies linearly with frequency. We recall that for time-dependent signals the quantities corresponding to $\mathfrak{q}_1$ and $\mathfrak{q}_p$, i.e. $-\frac{d}{dt}\psi(t)$ and $-\frac{\psi(t)}{t}$, are respectively called the *group delay* and the *phase delay* [3].

The second-order moment: $\mathfrak{q}_2(\kappa)$

The second-order moment $\mathfrak{q}_2(\kappa)$, defined by

$$\mathfrak{q}_2(\kappa) = \frac{1}{\mathfrak{q}_0(\kappa)} \int [q - \mathfrak{q}_1(\kappa)]^2 \mathcal{W}(q,\kappa)dq = \frac{1}{|\widetilde{\varphi}(\kappa)|^2} \int q^2 \mathcal{W}(q,\kappa)dq - \mathfrak{q}_1^2(\kappa), \tag{9.4.37}$$

takes the explicit expression

$$\mathfrak{q}_2(\kappa) = -\frac{1}{4}[\frac{d}{d\kappa}\frac{\widetilde{\varphi}'}{\widetilde{\varphi}} + \frac{d}{d\kappa}\frac{\widetilde{\varphi}'^*}{\widetilde{\varphi}^*}] = -\frac{1}{2}\Re[\frac{d}{d\kappa}\frac{\widetilde{\varphi}'(\kappa)}{\widetilde{\varphi}(\kappa)}], \tag{9.4.38}$$

after using relations (9.3.12) and (9.3.23). The correspondence with $\mathfrak{k}_2(q)$ is evident. Also, note that with the representation (9.4.28) of $\widetilde{\varphi}(\kappa)$ one finds

$$\mathfrak{q}_2(\kappa) = -\frac{1}{2}\frac{d}{d\kappa}[\frac{h'(\kappa)}{h(\kappa)}]. \tag{9.4.39}$$

Hence, the local second order moment $\mathfrak{q}_2(\kappa)$ depends only on the amplitude of the spectrum of the signal.

As an example, we consider the Gaussian signal (9.4.23), for which, being

$$\widetilde{\varphi}_{\mathcal{G}}(\kappa) = \left(\tfrac{w^2}{2\pi}\right)^{\frac{1}{4}} e^{-\frac{w^2}{4}(\kappa-\kappa_0)^2}, \tag{9.4.40}$$

Eqs. (9.4.29) and (9.4.39) yield

$$\mathfrak{q}_1(\kappa) = 0, \quad \mathfrak{q}_2(\kappa) = \tfrac{w^2}{4}. \tag{9.4.41}$$

9.4.3 *Global moments*

The global moments of the Wigner distribution function are obtained by integration over the whole space-frequency plane (q, κ) and hence they are constants, which characterize the Wigner distribution function in a global sense. We shall limit to consider the zero-, first- and second-order moments. A brief comment on the higher-order moments is given in § 9.7; a more accurate discussion is beyond the scope of the book. It is noteworthy the variety of expressions and relations, that, as we will see, can be deduced. This evidently conveys the effectiveness of the moments of the Wigner distribution function as inspection tools of the nature and the properties of a signal.

Thus, the average of $\mathcal{W}$ is given by

$$\mathfrak{m}_{00} = \iint \mathcal{W}(q, \kappa) dq d\kappa, \tag{9.4.42}$$

and, on account of (6.3.27), identified as the total power of the signal, being

$$\mathfrak{m}_{00} = \int |\varphi(q)|^2 \, dq = \int |\widetilde{\varphi}(\kappa)|^2 \, d\kappa. \tag{9.4.43}$$

Being the signal power always positive for all $\varphi \neq 0$, the global higher-order moments of the Wigner distribution function can safely be introduced.

The global first-order moments $\mathfrak{m}_{01}$ and $\mathfrak{m}_{10}$

The general definition (9.2.35) yields for the global average frequency $\mathfrak{m}_{01}$ the formal expression

$$\mathfrak{m}_{01} = \frac{1}{\mathfrak{m}_{00}} \iint \kappa \mathcal{W}(q, \kappa) dq d\kappa. \tag{9.4.44}$$

Integrating with respect to q, one obtains

$$\mathfrak{m}_{01} = \frac{1}{\|\widetilde{\varphi}\|^2} \int \kappa \, |\widetilde{\varphi}(\kappa)|^2 \, d\kappa = \overline{\kappa} = \langle \widehat{\kappa} \rangle, \tag{9.4.45}$$

which clearly shows that $\mathfrak{m}_{01}$ can be interpreted as the average frequency of the signal frequency spectrum (see § 9.2.1).

Alternatively, the integration with respect to κ, which by (9.4.2) gives $\int \kappa \mathcal{W}(q, \kappa) d\kappa = |\varphi(q)|^2 \, \mathfrak{k}_1(q)$, yields $\mathfrak{m}_{01}$ as the average of the local frequency

$\mathfrak{k}_1(q)$, with the local power $|\varphi(q)|^2$ acting as the pertinent density function. Explicitly, we have

$$\mathfrak{m}_{01} = \frac{1}{\|\varphi\|^2} \int \mathfrak{k}_1(q) \, |\varphi(q)|^2 \, dq = \overline{\mathfrak{k}_1(q)}. \tag{9.4.46}$$

Finally, according to the relation (9.4.6) appropriate to the representation (9.4.5) of the signal, the mean spatial frequency $\mathfrak{m}_{01}$ acquires the further meaning as the average of the phase derivative of the signal:

$$\mathfrak{m}_{01} = \frac{1}{\|f\|^2} \int \phi'(q) f^2(q) dq = \langle \phi'(q) \rangle . \tag{9.4.47}$$

Relations (9.4.45) and (9.4.46), implying the equality

$$\mathfrak{m}_{01} = \overline{\kappa} = \overline{\mathfrak{k}_1(q)}, \tag{9.4.48}$$

enable the equivalence of the weighted averages in the κ and q domains $\overline{\kappa}$ and $\overline{\mathfrak{k}_1(q)}$, according to which the carrier frequency of the signal equals the average of the local frequency of the signal. This link was firstly recognized by Franks within the context of the signal theory [3.8]. The discussion above evidences the formal and conceptual easiness through which that equivalence can be inferred within the formalism inherent in the Wigner representation.

Using expression (9.4.4) for $\mathfrak{k}_1(q)$, we can further recast (9.4.46) as

$$\mathfrak{m}_{01} = \frac{i}{2\|\varphi\|^2} \int (\varphi \tfrac{d}{dq} \varphi^* - \varphi^* \tfrac{d}{dq} \varphi) dq. \tag{9.4.49}$$

Similarly, the average space coordinate $\mathfrak{m}_{10}$, obtained through

$$\mathfrak{m}_{10} = \frac{1}{\mathfrak{m}_{00}} \iint q \mathcal{W}(q, \kappa) dq d\kappa, \tag{9.4.50}$$

turns out to be the centroid of the local power, since performing the integration with respect to κ we obtain

$$\mathfrak{m}_{10} = \frac{1}{\|\varphi\|^2} \int q \, |\varphi(q)|^2 \, dq = \overline{q} = \langle \hat{\mathfrak{q}} \rangle . \tag{9.4.51}$$

Also, as the integration with respect to q yields $\int q \mathcal{W}(q, \kappa) dq = |\widetilde{\varphi}(\kappa)|^2 \, \mathfrak{q}_1(\kappa)$, the alternative interpretation of $\mathfrak{m}_{10}$ as the mean value of the local first order moment $\mathfrak{q}_1(\kappa)$ is in accord with the resulting expression

$$\mathfrak{m}_{10} = \frac{1}{\|\widetilde{\varphi}\|^2} \int |\widetilde{\varphi}(\kappa)|^2 \, \mathfrak{q}_1(\kappa) d\kappa = \overline{\mathfrak{q}_1(\kappa)}. \tag{9.4.52}$$

Then, on account of (9.4.29), the first order moment $\mathfrak{m}_{10}$ is the weighted average of the phase derivative of the signal spectrum, i.e. $\mathfrak{m}_{10} = \langle -\psi'(\kappa) \rangle$.

Evidently, relations (9.4.51) and (9.4.52) imply the equality

$$\mathfrak{m}_{10} = \overline{q} = \overline{\mathfrak{q}_1(\kappa)}, \tag{9.4.53}$$

and so the equivalence of the two averages $\overline{q}$ and $\overline{\mathfrak{q}_1(\kappa)}$.

The global second-order moments $\mathfrak{m}_{02}$, $\mathfrak{m}_{20}$ and $\mathfrak{m}_{11}$

We start considering the global second-order moment in the frequency variable.

The global second-order moment $\mathfrak{m}_{02}$

As a direct transcription of the general definition in (9.2.36), the global second-order moment $\mathfrak{m}_{02}$ with respect to the frequency is formally given by

$$\mathfrak{m}_{02} = \frac{1}{\mathfrak{m}_{00}} \iint (\kappa - \mathfrak{m}_{01})^2 \mathcal{W}(q,\kappa) dq d\kappa. \tag{9.4.54}$$

Integrating with respect to q and using (9.4.45), we find

$$\mathfrak{m}_{02} = \frac{1}{\|\widetilde{\varphi}\|^2} \int (\kappa - \overline{\kappa})^2 \left|\widetilde{\varphi}(\kappa)\right|^2 d\kappa. \tag{9.4.55}$$

It is evident that $\mathfrak{m}_{02}$ is always non-negative and so can be interpreted as the spread of the Wigner distribution function in the κ-direction with respect to the centroid ($\mathfrak{m}_{10}$,$\mathfrak{m}_{01}$). It is also evident that $\mathfrak{m}_{02}$ is just the variance σ_κ^2 of the Fourier transformed signal $\widetilde{\varphi}(\kappa)$:

$$\mathfrak{m}_{02} = \sigma_\kappa^2. \tag{9.4.56}$$

Besides, in conformity with the formalism illustrated in § 9.2.1, we may relate $\mathfrak{m}_{02}$ to the expectation value of the squared operator $\widehat{\kappa}^2$:

$$\mathfrak{m}_{02} \rightleftharpoons \langle \widehat{\kappa}^2 \rangle, \tag{9.4.57}$$

and so to the matrix element in the state φ of the group generator $\widehat{\mathsf{K}}_-$ (§ 5.2).

Interestingly, the global moment $\mathfrak{m}_{02}$ relates to the local moments $\mathfrak{k}_1(q)$ and $\mathfrak{k}_2(q)$. In fact, on account of (9.4.2) and (9.4.18), it is easy to recast (9.4.54) as

$$\begin{aligned}\mathfrak{m}_{02} &= \frac{1}{\|\varphi\|^2} \int \mathfrak{k}_2(q) \left|\varphi(q)\right|^2 dq + \frac{1}{\|\varphi\|^2} \int [\mathfrak{k}_1(q) - \overline{\kappa}]^2 \left|\varphi(q)\right|^2 dq \\ &= \overline{\mathfrak{k}_2(q)} + \langle (\mathfrak{k}_1(q) - \overline{\kappa})^2 \rangle .\end{aligned} \tag{9.4.58}$$

We see that $\mathfrak{m}_{02}$ comprises two components: the first term represents the average $\overline{\mathfrak{k}_2(q)}$ of the local second-order moment $\mathfrak{k}_2(q)$, and the other is the variance of the local frequency $\mathfrak{k}_1(q)$. Within the representation (9.4.5) of the signal, it becomes evident that the magnitude and the phase of the signal yield separate contributions to $\mathfrak{m}_{02}$, on due account of expressions (9.4.21) and (9.4.6) relating $\mathfrak{k}_2(q)$ and $\mathfrak{k}_1(q)$ respectively to the envelope and the phase of the signal.

We have seen that the global moment $\mathfrak{m}_{01}$ is just equal to the average of the corresponding local moment $\mathfrak{k}_1(q)$. In contrast, $\mathfrak{m}_{02}$ can not be identified with the average of $\mathfrak{k}_2(q)$, unless the second average in (9.4.58) vanishes, as it happens, for instance, for signals displaying a phase linearly depending on κ.

Further expressions for the second-order moment $\mathfrak{m}_{02}$

The second-order moment $\mathfrak{m}_{02}$ is amenable to alternative expressions, which involve integrals over both the space and frequency domains. Besides being conceptually interesting and offering a challenge for their interpretation as well as a deeper insight into the interplay of the space and frequency variables, they provide useful tools for practical calculations.

Let us consider the frequency domain formula (9.4.55), yielding $\mathfrak{m}_{02}$ as the variance of $|\widetilde{\varphi}(\kappa)|^2$. By the Parseval theorem, it can be rewritten as

$$\mathfrak{m}_{02} = \frac{1}{\|\widetilde{\varphi}\|^2} \int |(\kappa - \overline{\kappa})\widetilde{\varphi}(\kappa)|^2 \, d\kappa = \frac{1}{\|\varphi\|^2} \int \left|\widehat{\mathcal{F}}^{-1}[(\kappa - \overline{\kappa})\widetilde{\varphi}](q)\right|^2 dq. \quad (9.4.59)$$

Here $\widehat{\mathcal{F}}^{-1}[(\kappa - \overline{\kappa})\widetilde{\varphi}](q)$ denotes the inverse Fourier transform of the function $(\kappa - \overline{\kappa})\widetilde{\varphi}(\kappa)$, for which an easy calculation gives

$$\widehat{\mathcal{F}}^{-1}[(\kappa - \overline{\kappa})\widetilde{\varphi}](q) = -ie^{i\overline{\kappa}q} \frac{d}{dq}[\varphi(q)e^{-i\overline{\kappa}q}]. \quad (9.4.60)$$

It is therefore immediate to write down the space domain formula for $\mathfrak{m}_{02}$ as

$$\mathfrak{m}_{02} = \frac{1}{\|\varphi\|^2} \int \left|\frac{d}{dq}[\varphi(q)e^{-i\mathfrak{m}_{01}q}]\right|^2 dq, \quad (9.4.61)$$

which relates $\mathfrak{m}_{02}$ to the norm of the derivative of the signal once filtered of the harmonic component $e^{i\mathfrak{m}_{01}q}$ at the carrier frequency $\mathfrak{m}_{01} = \overline{\kappa}$.

Moreover, expanding the derivative above we find

$$\mathfrak{m}_{02} = \frac{1}{\|\varphi\|^2} \int |\varphi'(q)|^2 \, dq - \mathfrak{m}_{01}^2 = \frac{\|\varphi'\|^2}{\|\varphi\|^2} - \mathfrak{m}_{01}^2. \quad (9.4.62)$$

Another interesting expression for $\mathfrak{m}_{02}$ is obtainable from (9.4.58) by use of expressions (9.4.6) and (9.4.21) for $\mathfrak{k}_1(q)$ and $\mathfrak{k}_2(q)$ in terms of the phase and the envelope of the signal. In fact, we end up with

$$\mathfrak{m}_{02} = \frac{1}{\|f\|^2} \int f'^{\,2}(q) dq + \frac{1}{\|f\|^2} \int [\phi'(q) - \overline{\kappa}]^2 f^{\,2}(q) dq. \quad (9.4.63)$$

It explicitly shows that, as noted before, the two averages in (9.4.58) depend one only on the envelope and the other only on the phase of the signal.

It is useful to comment on the two components of $\mathfrak{m}_{02}$ in (9.4.63). We consider a sample signal $\Phi(q)$, which is supposed to retain the envelope of $\varphi(q)$ while having zero-phase, namely $\Phi(q) = f(q)$. Being Φ real, the relevant spectrum $\widetilde{\Phi}(\kappa) = \widetilde{f}(\kappa)$ is symmetric, i.e., $\widetilde{\Phi}(-\kappa) = \widetilde{\Phi}^*(\kappa)$. Thus, the mean of $\left|\widetilde{\Phi}(\kappa)\right|^2$ is zero and hence the second term in (9.4.59) vanishes. Moreover, the corresponding variance turns out to be simply

$$\sigma^2(\widetilde{\Phi}) = \frac{1}{\|\widetilde{\Phi}\|^2} \int \kappa^2 \left|\widetilde{\Phi}(\kappa)\right|^2 d\kappa = \frac{1}{\|f\|^2} \int f'^2(q) dq, \quad (9.4.64)$$

the last equality being in accord with the Parseval theorem. Thus, the first term in (9.4.63) acquires the meaning as the variance $\sigma^2(\widetilde{\Phi})$ of the spectrum of an envelope-only equivalent signal of $\varphi(q)$. The other term is the variance of the phase derivative $\phi'(q)$, and so relates to a certain type of smoothness of $\phi(q)$. As said earlier, for a linearly varying phase, this term vanishes.

A further interesting relation is obtained by use of (9.4.6) and (9.4.21) in (9.4.58); namely,

$$\mathfrak{m}_{02} = \frac{1}{2\|\varphi\|^2} \int \{|\varphi'(q)|^2 - \tfrac{1}{2}[\varphi^* \tfrac{d^2}{dq^2}\varphi + \varphi \tfrac{d^2}{dq^2}\varphi^*]\} dq - \mathfrak{m}_{01}^2. \tag{9.4.65}$$

Interestingly, the integrand function can be seen as resulting from the action of the Kaiser-Teager energy operator $\widehat{\mathcal{E}}$ of signal analysis [4.6, 5], defined by

$$\widehat{\mathcal{E}}\varphi(q) = |\varphi'(q)|^2 - \tfrac{1}{2}[\varphi^* \frac{d^2}{dq^2}\varphi + \varphi \frac{d^2}{dq^2}\varphi^*]. \tag{9.4.66}$$

The global second-order moment $\mathfrak{m}_{20}$

The global second moment $\mathfrak{m}_{20}$ with respect to the space coordinate is

$$\mathfrak{m}_{20} = \frac{1}{\mathfrak{m}_{00}} \iint (q - \mathfrak{m}_{10})^2 \mathcal{W}(q,\kappa) dq d\kappa, \tag{9.4.67}$$

and can easily be recast into

$$\mathfrak{m}_{20} = \frac{1}{\|\varphi\|^2} \int (q - \overline{q})^2 |\varphi(q)|^2 dq, \tag{9.4.68}$$

showing that $\mathfrak{m}_{20}$ is a non-negative quantity, and accordingly interpretable as the spread of the Wigner distribution function along the q-direction with respect to the centroid $(\mathfrak{m}_{10}, \mathfrak{m}_{01})$. In addition, it is just the variance σ_q^2 of the local power $|\varphi(q)|^2$:

$$\mathfrak{m}_{20} = \sigma_q^2. \tag{9.4.69}$$

The relation of $\mathfrak{m}_{20}$ to the expectation value of the operator $\widehat{\mathsf{q}}^2$ is apparent:

$$\mathfrak{m}_{20} \rightleftharpoons \langle \widehat{\mathsf{q}}^2 \rangle, \tag{9.4.70}$$

which then, paralleling (9.4.57), relates $\mathfrak{m}_{20}$ to the matrix element in the state φ of the group generator $\widehat{\mathsf{K}}_+$ (see § 5.2). Again, the relation turns into an identity if one takes, as it is usually done, $\|\varphi\|^2 = 1$ and $\overline{q} = 0$.

Like $\mathfrak{m}_{02}$, also $\mathfrak{m}_{20}$ can be expressed in terms of the local moments $\mathfrak{q}_1(\kappa)$ and $\mathfrak{q}_2(\kappa)$ and thereby through integrals over the frequency domain. An easy manipulation of (9.4.67), on account of (9.4.26) and (9.4.37), leads in fact to

$$\begin{aligned} \mathfrak{m}_{20} &= \frac{1}{\|\widetilde{\varphi}\|^2} \int \mathfrak{q}_2(\kappa) |\widetilde{\varphi}(\kappa)|^2 d\kappa + \frac{1}{\|\widetilde{\varphi}\|^2} \int [\mathfrak{q}_1(\kappa) - \overline{q}]^2 |\widetilde{\varphi}(\kappa)|^2 d\kappa \\ &= \overline{\mathfrak{q}_2(\kappa)} + \langle (\mathfrak{q}_1(\kappa) - \overline{q})^2 \rangle, \end{aligned} \tag{9.4.71}$$

which parallels the corresponding expression (9.4.58) for $\mathfrak{m}_{02}$. Thus, $\mathfrak{m}_{20}$ can be separated into two components, one being the average $\overline{\mathfrak{q}_2(\kappa)}$ of the local second-order moment $\mathfrak{q}_2(\kappa)$ and the other the variance of the local centroid $\mathfrak{q}_1(\kappa)$. As we will see below, these two terms are related to the magnitude and phase components of the Fourier transformed signal $\widetilde{\varphi}(\kappa)$.

The second-order moment $\mathfrak{m}_{20}$ equals the average $\overline{\mathfrak{q}_2(\kappa)}$ only when the phase depending term in (9.4.71) vanishes. This happens, for instance, in the case where the phase of the spectrum $\widetilde{\varphi}(\kappa)$ is a linear function of κ.

Further expressions for the second-order moment $\mathfrak{m}_{20}$

The expressions for $\mathfrak{m}_{20}$ given below are basically the dual forms of those established for $\mathfrak{m}_{02}$. Then, we describe very little the details of the calculations, that as before ground on the Parseval theorem, connecting the space and frequency domains, and on the modulus and phase representation of the spectrum

We apply the Parseval theorem to the q-domain formula (9.4.68) for $\mathfrak{m}_{20}$. Thus, taking into account that the Fourier transform of $(q - \overline{q})\varphi(q)$ is

$$\widehat{\mathcal{F}}\left[(q-\overline{q})\varphi(q)\right](\kappa) = ie^{-i\overline{q}\kappa}\frac{d}{d\kappa}[\widetilde{\varphi}(\kappa)e^{i\overline{q}\kappa}], \tag{9.4.72}$$

we obtain

$$\mathfrak{m}_{20} = \frac{1}{\|\widetilde{\varphi}\|^2}\int \left|\frac{d}{d\kappa}[\widetilde{\varphi}(\kappa)e^{i\mathfrak{m}_{10}\kappa}]\right|^2 d\kappa, \tag{9.4.73}$$

which connects the second-order moment $\mathfrak{m}_{20}$ to the derivative of the spectrum, whose phase being normalized to the harmonic component $e^{i\mathfrak{m}_{10}\kappa}$.

In terms of the amplitude and phase of $\widetilde{\varphi}(\kappa)$ as given in (9.4.28), we find

$$\mathfrak{m}_{20} = \frac{1}{\|h\|^2}\int h'^2(\kappa)d\kappa + \frac{1}{\|h\|^2}\int [\psi'(\kappa) + \overline{q}]^2 h^2(\kappa)d\kappa, \tag{9.4.74}$$

which clearly separates the contributions from the magnitude $h(\kappa)$ and the phase $\psi'(\kappa)$ of the spectrum. The interpretation of these two contributions parallels that illustrated for the components of $\mathfrak{m}_{02}$ in (9.4.63). Consider a sample signal $\Psi(q)$, whose spectrum be $\widetilde{\Psi}(\kappa) = h(\kappa)$. Being $\widetilde{\Psi}(\kappa)$ real, $\Psi(q)$ is symmetric. Hence, the mean of $|\Psi(q)|^2$ is zero and the phase depending term in (9.4.74) vanishes. As to the first term, since $h'(\kappa) = \widetilde{\Psi}'(\kappa)$, we see that, applying again the Parseval formula, it can be understood as the variance of the local power $|\Psi(q)|^2$. Explicitly, we have

$$\sigma^2(\Psi) = \frac{1}{\|\Psi\|^2}\int q^2\left|\Psi(q)\right|^2 dq = \frac{1}{\|h\|^2}\int h'^2(\kappa)d\kappa. \tag{9.4.75}$$

Therefore, the second-order moment $\mathfrak{m}_{20}$ can be separated into the sum of two variances: the variance $\sigma^2(\Psi)$ of an equivalent signal of $\varphi(q)$, which retains the modulus of the spectrum, and the variance of the phase derivative $\psi'(\kappa)$ of the Fourier transformed signal $\widetilde{\varphi}(\kappa)$, which vanishes if $\psi(\kappa)$ is linear.

Finally, inserting the expressions (9.4.27) and (9.4.38) of $\mathfrak{q}_1(\kappa)$ and $\mathfrak{q}_2(\kappa)$ in (9.4.71), we obtain the dual form of (9.4.65), i.e.,

$$\mathfrak{m}_{02} = \frac{1}{2\|\widetilde{\varphi}\|^2} \int \{|\widetilde{\varphi}'(\kappa)|^2 - \frac{1}{2}[\widetilde{\varphi}^* \frac{d^2}{d\kappa^2}\widetilde{\varphi} + \widetilde{\varphi}\frac{d^2}{d\kappa^2}\widetilde{\varphi}^*]\} d\kappa - \mathfrak{m}_{10}^2, \qquad (9.4.76)$$

the function under the integral arising from the frequency spectrum $\widetilde{\varphi}(\kappa)$ when acted by the energy operator $\widehat{\mathcal{E}}$ according to the rule (9.4.66) [4.6, 5].

The global second-order moment $\mathfrak{m}_{11}$

The mixed second-order moment $\mathfrak{m}_{11}$ means the double integral

$$\mathfrak{m}_{11} = \frac{1}{\mathfrak{m}_{00}} \iint (q - \mathfrak{m}_{10})(\kappa - \mathfrak{m}_{01})\mathcal{W}(q,\kappa) dq d\kappa. \qquad (9.4.77)$$

Espanding the product $(q - \mathfrak{m}_{10})(\kappa - \mathfrak{m}_{01})$, and using (9.4.44) and (9.4.50), we end up with the equivalent expression

$$\mathfrak{m}_{11} = \frac{1}{\mathfrak{m}_{00}} \iint q\kappa\mathcal{W}(q,\kappa) dq d\kappa - \mathfrak{m}_{10}\mathfrak{m}_{01}, \qquad (9.4.78)$$

which can be further elaborated recasting the integrand in terms of the first order moments $\mathfrak{k}_1(q)$ or $\mathfrak{q}_1(\kappa)$. In fact, it is easy to obtain the q-domain formula

$$\mathfrak{m}_{11} = \frac{1}{\|\varphi\|^2} \int (q - \mathfrak{m}_{10})(\mathfrak{k}_1(q) - \mathfrak{m}_{01}) |\varphi(q)|^2 dq, \qquad (9.4.79)$$

and correspondingly the κ-domain formula

$$\mathfrak{m}_{11} = \frac{1}{\|\widetilde{\varphi}\|^2} \int (\mathfrak{q}_1(\kappa) - \mathfrak{m}_{10})(\kappa - \mathfrak{m}_{01}) |\widetilde{\varphi}(\kappa)|^2 d\kappa. \qquad (9.4.80)$$

Further expressions are obtained by explicitly calculating the integral over κ or q. In fact, noting that

$$\mathfrak{m}_{11} = \frac{1}{\mathfrak{m}_{00}} \int dq q [\int \kappa\mathcal{W}(q,\kappa) d\kappa] - \mathfrak{m}_{10}\mathfrak{m}_{01}, \qquad (9.4.81)$$

on account of (9.3.8) and the first of (9.3.21), we obtain

$$\mathfrak{m}_{11} = \frac{i}{2\mathfrak{m}_{00}} \int dq q [\varphi^{*\prime}(q)\varphi(q) - \varphi^*(q)\varphi'(q)] - \mathfrak{m}_{10}\mathfrak{m}_{01}. \qquad (9.4.82)$$

Similarly, since

$$\mathfrak{m}_{11} = \frac{1}{\mathfrak{m}_{00}} \int d\kappa\kappa [\int q\mathcal{W}(q,\kappa) dq] - \mathfrak{m}_{10}\mathfrak{m}_{01}, \qquad (9.4.83)$$

using (9.3.12) with $r = 1$ and the second of (9.3.21), we end up with

$$\mathfrak{m}_{11} = -\frac{i}{2\mathfrak{m}_{00}} \int d\kappa\kappa [\widetilde{\varphi}^{*\prime}(\kappa)\widetilde{\varphi}(\kappa) - \widetilde{\varphi}^*(\kappa)\widetilde{\varphi}'(\kappa)] - \mathfrak{m}_{10}\mathfrak{m}_{01}. \qquad (9.4.84)$$

The apex means, of course, derivation with respect to the specified argument.

Both (9.4.82) and (9.4.84) convey the relation of $\mathfrak{m}_{11}$ to the expectation value of the operator $\frac{1}{2}(\widehat{\mathsf{q}}\widehat{\kappa} + \widehat{\kappa}\widehat{\mathsf{q}})$:

$$\mathfrak{m}_{11} \rightleftharpoons \langle \tfrac{1}{2}(\widehat{\mathsf{q}}\widehat{\kappa} + \widehat{\kappa}\widehat{\mathsf{q}}) \rangle, \tag{9.4.85}$$

and so to the matrix element in the state φ of the group generator $\widehat{\mathsf{K}}_3$ (§ 5.2).

Conforming to the notation (9.4.69) and (9.4.56) used in relation with the space and frequency variances, the mixed moment is accordingly denoted as

$$\mathfrak{m}_{11} = \sigma_{q\kappa}. \tag{9.4.86}$$

9.4.4 *Local valence of the uncertainty relation*

An interesting set of inequalities can be deduced in relation with the second-order moments of the Wigner distribution function. The interested reader may consult the related literature [6.1, 6.2]. Here we comment on the global or local valence may be attributed to the basic inequality, involving the second-order moments $\mathfrak{m}_{20}$ and $\mathfrak{m}_{02}$, directly inferred from the width principle of Fourier analysis (§ 4.6.4) and the identification of $\mathfrak{m}_{20}$ and $\mathfrak{m}_{02}$ with the variances of the signal and its spectrum. Accordingly, using the common notation as in (9.4.69) and (9.4.56), we rewrite once again the standard form of the uncertainty relation, i.e.

$$\sigma_q \sigma_\kappa \geq \frac{1}{2}. \tag{9.4.87}$$

Of course, the quantities above refer to the whole spatial range and the whole frequency range of the signal. Notwithstanding. inequality (9.4.87) is commonly given a "local" valence, being intended as setting a limit to what can be done locally. It is therefore natural to inquire about the correctness and the general validity of interpreting (9.4.87) in terms of local properties of the signal. In other words, it is natural to inquire about the possibility of defining "local" quantities analogous to the global ones entering (9.4.87) and accordingly about the possibility of individualizing bounds on their product, that so might be interpreted as a local uncertainty product.

This issue is beautifully addressed in [6.3], where a *local uncertainty product* is suggested, which product is then shown to be upper bounded, under definite conditions, by the *global uncertatinty product* (9.4.87). Specific investigations are then presented within the time-frequency domain pertaining to the context of signal theory.

Following the suggestion of [6.3], we illustrate the question specifically in relation with the Wigner distribution function. Accordingly, let us go back to Eqs. (9.4.71) and (9.4.58) relating the global second-order moments to the local second-order moments. In view of such relations, one may ask whether inequality (9.4.87) may place a limit to some analogous product involving the local variances. Indeed, multiplying both sides of these equations we obtain

$$\sigma_q^2 \sigma_\kappa^2 = \overline{\mathfrak{q}_2(\kappa)\mathfrak{k}_2(q)} + \text{other terms}, \tag{9.4.88}$$

the "other terms" specifically contributing as

$$\begin{gathered}\text{other terms} = \overline{\mathfrak{q}_2(\kappa)} \left\langle (\mathfrak{k}_1(q) - \overline{\kappa})^2 \right\rangle \\ + \overline{\mathfrak{k}_2(q)} \left\langle (\mathfrak{q}_1(\kappa) - \overline{q})^2 \right\rangle + \left\langle (\mathfrak{q}_1(\kappa) - \overline{q})^2 \right\rangle \left\langle (\mathfrak{k}_1(q) - \overline{\kappa})^2 \right\rangle\end{gathered} \tag{9.4.89}$$

Evidently, the variances of the local centroids $\mathfrak{q}_1(\kappa)$ and $\mathfrak{k}_1(q)$ are positive quantities. As to the averages $\overline{\mathfrak{q}_2(\kappa)}$ and $\overline{\mathfrak{k}_2(q)}$ of the local variances, we may recognize inspecting the first terms in (9.4.74) and (9.4.63), that, although nothing can be said a priori on the sign of $\mathfrak{q}_2(\kappa)$ and $\mathfrak{k}_2(q)$, their averages turn out to be unequivocally positive quantities.

Therefore, equality (9.4.88) implies the inequality

$$\overline{\mathfrak{q}_2(\kappa)\mathfrak{k}_2(q)} \leq \sigma_q^2 \sigma_\kappa^2, \tag{9.4.90}$$

between the product of the averages of the local variances on the left and the product of the global variances on the right.

It seems natural to consider the former product as a local uncertainty product:

$$\overline{\mathfrak{q}_2(\kappa)\mathfrak{k}_2(q)} \quad \text{local uncertainty product}, \tag{9.4.91}$$

for which inequality (9.4.90) fixes an upper limit through (9.4.87).

Thus, we see that, although the global uncertainty product cannot be smaller than the lower limit set by (9.4.87), the local uncertainty product is allowed to be smaller than the global uncertainty product. Can it be arbitrarily small or does it have its own nonzero lower limit? Also, may inequality (9.4.90) stimulate a critical revision of the commonly held belief that the global uncertainty relation somehow limits the local quantities?

Leaving these fascinating questions open, as they are indeed, we address the reader to the related literature [6].

9.4.5 *Moments of the Wigner distribution function and ambiguity function*

It does exist an interesting relation between the moments of the Wigner distribution function and the derivatives of the ambiguity function, which is indeed implicit in the Fourier transform relation linking $\mathcal{W}$ to $\mathcal{A}$ [4.9].

In fact, just rewriting relation (7.4.16) we see that

$$\mathfrak{m}_{00} = 2\pi\mathcal{A}(0,0). \tag{9.4.92}$$

Also, taking account of the Fourier transform $\mathcal{W}$-$\mathcal{A}$ relation and of the obvious derivative relations

$$q^r e^{iq\widetilde{\kappa}} = (-i)^r \frac{\partial^r}{\partial \widetilde{\kappa}^r} e^{iq\widetilde{\kappa}}, \quad \kappa^h e^{-i\kappa\widetilde{q}} = i^h \frac{\partial^h}{\partial \widetilde{q}^h} e^{-i\kappa\widetilde{q}}, \tag{9.4.93}$$

it is easy to prove for the centroid of the Wigner distribution function the interesting relations

$$\mathfrak{m}_{10} = 2\pi i \left.\frac{\partial \mathcal{A}(\widetilde{q},\widetilde{\kappa})}{\partial \widetilde{\kappa}}\right|_{\widetilde{q},\widetilde{\kappa}=0}, \quad \mathfrak{m}_{01} = -2\pi i \left.\frac{\partial \mathcal{A}(\widetilde{q},\widetilde{\kappa})}{\partial \widetilde{q}}\right|_{\widetilde{q},\widetilde{\kappa}=0}. \tag{9.4.94}$$

Then, translating the origin of the ambiguity function phase plane $(\widetilde{q},\widetilde{\kappa})$ to the centroid $(\overline{q},\overline{\kappa})$ of the Wigner distribution function, we may express the second moments of $\mathcal{W}(q,\kappa)$ in terms of the second derivatives of $\mathcal{A}(\widetilde{q},\widetilde{\kappa})$ at the origin; namely,

$$\begin{aligned}
\mathfrak{m}_{20} &= -2\pi \left.\frac{\partial^2 \mathcal{A}(\widetilde{q},\widetilde{\kappa})}{\partial \widetilde{\kappa}^2}\right|_{\widetilde{q},\widetilde{\kappa}=0}, \quad \mathfrak{m}_{02} = -2\pi \left.\frac{\partial^2 \mathcal{A}(\widetilde{q},\widetilde{\kappa})}{\partial \widetilde{q}^2}\right|_{\widetilde{q},\widetilde{\kappa}=0} \\
\mathfrak{m}_{11} &= 2\pi \left.\frac{\partial^2 \mathcal{A}(\widetilde{q},\widetilde{\kappa})}{\partial \widetilde{q}\partial\widetilde{\kappa}}\right|_{\widetilde{q},\widetilde{\kappa}=0}
\end{aligned} \tag{9.4.95}$$

Evidently, the second-order moments of the Wigner distribution function around the origin are related to the second-order derivatives of the ambiguity function at the origin. The above displayed relations reproduce the general relations holding between the central moments of a distribution and the derivatives of the corresponding characteristic function at the origin, as established by a well known result of statistical theory [2]. Indeed, as earlier mentioned, the ambiguity function can be understood just as the characteristic function of the Wigner distribution function (see Problem 1).

9.5 Gaussian Wigner distribution functions: the variance matrix and its evolution

In order to gain a certain insight into the definitions given above, we consider Gaussian Wigner distributions. We have already considered Gaussian distributions in § 6.3.6 when illustrating the smoothing procedure acted on a Wigner distribution function in order to generate distributions being positive everywhere through the space-frequency plane. There, it was noted that a Gaussian distribution is Wigner representable under definite conditions for the relative parameters, the generating signal being in general complex Gaussian.

In § 6.4.5 we have calculated the Wigner distribution functions of both real and complex Gaussian amplitudes which have been noted to be interpretable as the amplitudes of a Gaussian beam respectively at the waist and at a plane far from the waist. We consider here the expression of a propagating Gaussian beam as it is usually represented, and focus our attention on the relevant Wigner distribution function. On calculating the moments, we will be able to put them in relation with the beam parameters and identify the relevant evolution law, which will be generalized to arbitrary non Gaussian Wigner distribution functions, and so to arbitrary non Gaussian (coherent) signals.

9.5.1 *Gaussian beams: review of the relevant parameters*

A Gaussian beam propagating in the z-direction, as obtained, for instance, from the lowest-order low-divergence solution of the parabolic wave equation in a free medium or in a quadratic index medium, is typically represented by the complex amplitude [7]

$$\varphi_{\mathcal{G}}(q,z) = \left(\frac{kQ_0}{i\pi Q^2(z)}\right)^{\frac{1}{4}} e^{i\left[\frac{k}{2Q(z)}q^2 - \zeta(z) + kz\right]}. \tag{9.5.1}$$

For completeness's sake, also the on-axis phase factor $e^{-i\zeta(z)+ikz}$ has been included in the expression above even though it is not functional to the calculations of the Wigner distribution function and the relevant moments, for which only the q-dependent part of the waveform has relevance.

Central to the representation (9.5.1) is the complex curvature $\frac{1}{Q(z)}$, which contains information about both the spatial width and the angular divergence of the beam, being

$$\frac{1}{Q(z)} = \frac{1}{R(z)} + i\frac{2}{kw^2(z)}. \tag{9.5.2}$$

As is well known, $R(z)$ is the curvature radius of the beam wavefront at z, which in accord with the paraxial approximation inherent in the waveform (9.5.1) has a parabolic profile; it, of course, determines the transverse deviation of the beam phase from the plane-wave phase. Correspondingly, $w(z)$ gives indication of the radial extent of the beam, being the off-axis distance where the beam intensity reduces to $e^{-2} \sim 0.135$ of its value on the axis.

As the beam propagates from z_i to z_o, the complex parameter Q transforms simply as

$$Q(z_o) = Q(z_i) + z_o - z_i. \tag{9.5.3}$$

It can therefore be understood as a kind of a complex curvature radius since it propagates just as the real curvature radius R_s of a spherical wave emanating from a point source at $z = 0$, for which, in fact, one has

$$R_s(z_o) = R_s(z_i) + z_o - z_i. \tag{9.5.4}$$

The analogy Q-R_s is better yet confirmed by the fact that in general, when propagating through an $ABCD$ system, the complex beam parameter changes according to the hyperbolic Möbius transformation (see also Problem 2.10)

$$Q(z_o) = n_{0_2} \frac{AQ(z_i)/n_{0_1} + B}{CQ(z_i)/n_{0_1} + D}, \tag{9.5.5}$$

which is the same suffered by the curvature radius R_s of a spherical wave propagating through the system from a point source at $z = 0$. As usual, n_{0_1} and n_{0_2} denote the refractive indices respectively in the object and the image space. Evidently relation (9.5.3) is a special case of the general law (9.5.5).

With the plane $z_i = 0$ placed at the beam waist, where by definition the beam wave front is plane: $R(0) = \infty$, and the beam radius is minimum: $w(0) = w_0$, the complex beam parameter Q at this plane is purely imaginary; namely,

$$Q(0) = -i\frac{kw_0^2}{2} = -iz_R. \tag{9.5.6}$$

Hence, in accord with (9.5.3), the beam parameters $R(z)$, $w(z)$ are seen to evolve along the propagation axis as

$$\begin{aligned} R(z) &= z(1 + \tfrac{z_R^2}{z^2}) \xrightarrow[z \gg z_R]{} z, \\ w^2(z) &= w_0^2(1 + \tfrac{z^2}{z_R^2}) \xrightarrow[z \gg z_R]{} \tfrac{4z^2}{k^2 w_0^2}. \end{aligned} \tag{9.5.7}$$

The *Rayleigh range* $z_R = \frac{kw_0^2}{2}$ (the waist area divided by λ), entering the above formulae, represents the axial distance from the waist at which the beam radius increases by a factor $\sqrt{2}$: $w(z_R) = \sqrt{2}w_0$, and so the beam area doubles: $\pi w^2(z_R) = \pi w_0^2$. The region from $-z_R$ to z_R can therefore be identified as the focal zone, or the region over which the beam is approximately collimated. The full distance between the two $w(z_R)$ beam radii is referred to as the *confocal parameter* of the beam. We note also that at $z = z_R$ the curvature radius takes the minimum value $R(z_R) = 2z_R$. Then, in the region $z \gg z_R$ the beam is similar to a spherical wave emating from a point source at $z = 0$, being indeed $\frac{1}{Q(z)} \xrightarrow{z \gg z_R} \frac{1}{R(z)} \sim \frac{1}{z}$. The two regions $z \ll z_R$ and $z \gg z_R$ are generally known as near-field region and far-field region respectively.

Another interesting parameter related to the shape and the propagation features of the beam (9.5.1) is the beam divergence angle $\theta(z)$ defined by

$$\theta(z) \equiv \frac{w(z)}{z} = \frac{w_0}{z}\sqrt{1 + \frac{z^2}{z_R^2}} \xrightarrow[z \gg z_R]{} \theta_\infty = \frac{\lambda}{\pi w_0}. \tag{9.5.8}$$

The far-field divergence angle θ_∞ represents the slope of the oblique asymptote of the beam radius dependence; it describes the spreading of the beam when propagating towards infinity. In accord with the width principle, θ_∞ and w_0 are reciprocal parameters, which multiply to the z-independent value λ/π.

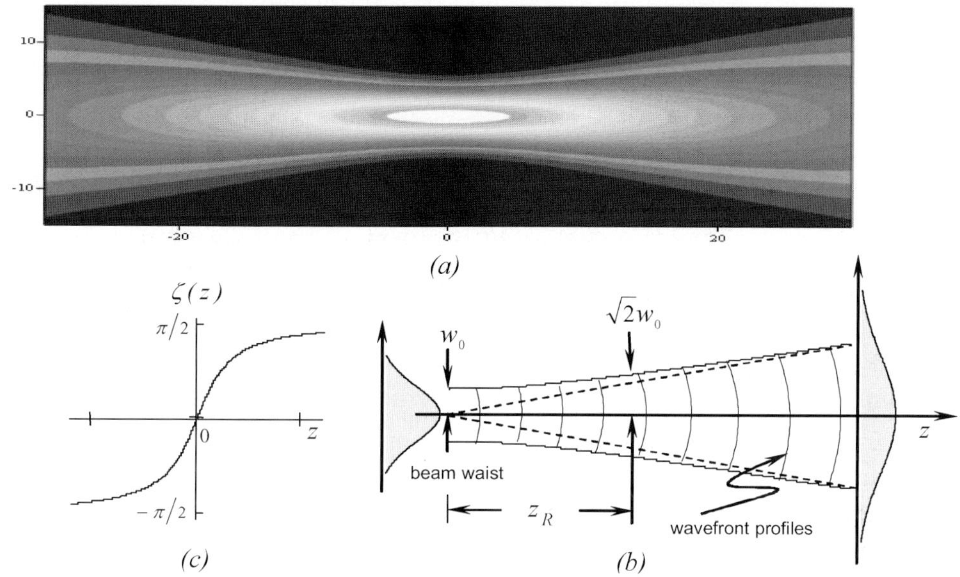

FIGURE 9.3. *(a)* Intensity of a Gaussian beam, *(b)* corresponding beam width $w(z)$ and wave front parabolic profiles, and *(c)* Guoy phase shift, as functions of the axial variable z.

Finally we recall that the Guoy phase

$$\zeta(z) = \tan^{-1}(\frac{z}{z_{\rm R}}), \tag{9.5.9}$$

which adds to the plane-wave phase kz to form the longitudinal phase of the beam, accounts for the excess delay of the beam wave front in comparison with a plane or a spherical wave. It ranges from $-\frac{\pi}{2}$ at $z = -\infty$ to $\frac{\pi}{2}$ at $z = \infty$, thus yielding a total excess delay of π as the beam travels from $-\infty$ to $+\infty$.

Figure 9.3 depicts the geometry of a propagating Gaussian beam.

9.5.2 *Gaussian beams: the matrix of second moments*

The Wigner distribution function of (9.5.1) is readily calculated to

$$\mathcal{W}_{\mathcal{G}}(q,\kappa;z) = \tfrac{1}{\pi} e^{-\frac{2q^2}{w^2(z)} - \frac{w^2(z)}{2}\left[\kappa - \frac{kq}{R(z)}\right]^2}. \tag{9.5.10}$$

The zero- and first-order global moments are of course

$$\begin{aligned} \mathfrak{m}_{00}(z) &= 1, \\ \mathfrak{m}_{10}(z) &= \mathfrak{m}_{01}(z) = 0, \end{aligned} \tag{9.5.11}$$

in accord with the normation condition and the on-axis propagation of (9.5.1).

Interestingly, for the global second-order moments we find

$$\begin{aligned}\mathfrak{m}_{20}(z) &= \frac{w^2(z)}{4},\\ \mathfrak{m}_{11}(z) &= \frac{kw^2(z)}{4R(z)} = \frac{z}{2z_{\mathrm{R}}},\\ \mathfrak{m}_{02}(z) &= \frac{1}{w^2(z)} + \frac{k^2w^2(z)}{4R^2(z)} = \frac{1}{w_0^2} = \frac{k^2\theta_\infty}{4}.\end{aligned} \tag{9.5.12}$$

As expected, the space second moment $\mathfrak{m}_{20}(z)$ relates to the local beam radius $w^2(z)$. Correspondingly, the angular second moment $\mathfrak{m}_{02}(z)$ is linked to the intrinsic beam property, i.e., the far field divergence θ_∞, through the light wave number, which is indeed the proper "constant" to turn from angles to spatial frequencies. Finally, we see that at the input plane $z = 0$, $\mathfrak{m}_{20}(0) = \frac{w_0^2}{4}$, which then multiplies with $\mathfrak{m}_{02}(0)$ to the minimum allowed by the optical uncertainty relation (9.2.29): $\mathfrak{m}_{20}(0)\mathfrak{m}_{02}(0) = \frac{1}{4}$.

It is convenient to arrange the above moments into a matrix scheme defining the *variance* (or, *second moment*) *matrix* $\mathbf{\Sigma}$ as

$$\mathbf{\Sigma}(z) \equiv \begin{pmatrix} \mathfrak{m}_{20}(z) & \mathfrak{m}_{11}(z) \\ \mathfrak{m}_{11}(z) & \mathfrak{m}_{02}(z) \end{pmatrix} = \begin{pmatrix} \frac{w^2(z)}{4} & \frac{kw^2(z)}{4R(z)} \\ \frac{kw^2(z)}{4R(z)} & \frac{1}{w^2(z)} + \frac{k^2w^2(z)}{4R^2(z)} \end{pmatrix}. \tag{9.5.13}$$

Accordingly the second moments will alternatively be denoted also as $\mathfrak{m}_{20} \equiv \Sigma_{qq}$, $\mathfrak{m}_{02} \equiv \Sigma_{\kappa\kappa}$ and $\mathfrak{m}_{11} \equiv \Sigma_{q\kappa}$.

Needless to say, $\mathbf{\Sigma}$ is defined as a real symmetric (and hence hermitian) matrix, whose off-diagonal elements are clearly unitless whilst the diagonal ones have dimensions respectively of length^2 and length^{-2}.

The determinant of $\mathbf{\Sigma}$, which is easily seen to be

$$\det \mathbf{\Sigma}(z) = \frac{1}{4}, \tag{9.5.14}$$

comes to be an invariant of the motion. The inverse of $\mathbf{\Sigma}$ is then

$$\mathbf{\Sigma}^{-1}(z) \equiv \frac{1}{\det \mathbf{\Sigma}} \begin{pmatrix} \mathfrak{m}_{02}(z) & -\mathfrak{m}_{11}(z) \\ -\mathfrak{m}_{11}(z) & \mathfrak{m}_{20}(z) \end{pmatrix}. \tag{9.5.15}$$

Also, it is worth mentioning that $\mathbf{\Sigma}$ is positive definite. Note that a Hermitian matrix $\mathbf{A}$ is said to be *positive definite* if for any vector $\mathbf{u}$ it results that [8]

$$(\mathbf{u}, \mathbf{A}\mathbf{u}) > 0. \tag{9.5.16}$$

It is a well established result of matrix theory that the necessary and sufficient condition for a Hermitian matrix $\mathbf{A}$ to be positive is that all leading principal minors of $\mathbf{A}$ are positive. Since $\mathfrak{m}_{02}(z) > 0$ (and as well $\mathfrak{m}_{20}(z) > 0$) and $\det \mathbf{\Sigma}(z) > 0$, the variance matrix $\mathbf{\Sigma}$ is evidently positive definite.

Notably, relation (9.5.14) implies the optical uncertainty relation (9.2.29). In fact, writing the explicit expression of the determinant of $\mathbf{\Sigma}$, i.e.,

$$\sigma_q^2 \sigma_\kappa^2 - \sigma_{q\kappa}^2 = \frac{1}{4}, \tag{9.5.17}$$

we obtain for the space-bandwidth product the relation

$$\sigma_q \sigma_\kappa = \frac{1}{2}\sqrt{1 + 4\sigma_{q\kappa}^2} \geq \frac{1}{2}, \tag{9.5.18}$$

which at a primary level confirms inequality (9.2.29). At a deeper level, it yields a more precise relation for the space and frequency signal widths product being linked to the beam-depending mixed moment $\sigma_{q\kappa}$ (see also Problem 4.8).

9.5.3 *Gaussian beams: propagation law of the variance matrix*

The utility of introducing the variance matrix is conveyed by the evident possibility of expressing the Wigner distribution function in a compact and elegant form in terms of $\mathbf{\Sigma}$. In fact, one may note that the exponent of (9.5.10) directly involves the entries of $\mathbf{\Sigma}$, i.e., the second moments, being

$$\tfrac{2q^2}{w^2(z)} + \tfrac{w^2(z)}{2}\left[\kappa - \tfrac{kq}{R(z)}\right]^2 = \tfrac{1}{2}\mathbf{u}^\top[\mathbf{\Sigma}(z)]^{-1}\mathbf{u}, \tag{9.5.19}$$

where as usual $\mathbf{u} = (q, \kappa)^\top$. Therefore, according to the normalization of Gaussian functions[(a)] we can finally write

$$\mathcal{W}_\mathcal{G}(q, \kappa; z) = \tfrac{1}{2\pi\sqrt{\det \mathbf{\Sigma}}}\, e^{-\frac{1}{2}\mathbf{u}^\top[\mathbf{\Sigma}(z)]^{-1}\mathbf{u}}, \tag{9.5.20}$$

thus relating the Wigner distribution function of the Gaussian beam (9.5.1) to the relevant variance matrix $\mathbf{\Sigma}$.

Remarkably, this allows us to obtain the propagation law for $\mathbf{\Sigma}$ and hence for the second moments. In fact, the paraxial transfer law for the Wigner distribution function signifies that

$$\mathcal{W}_\mathcal{G}(\mathbf{u}; z_o) = \mathcal{W}_\mathcal{G}(\mathbf{M}^{-1}\mathbf{u}; z_i), \tag{9.5.21}$$

where $\mathbf{M}$ is the ray-matrix connecting the planes at z_i and z_o between which the beam (9.5.1) is propagating. Recall that when dealing with the Wigner-plane pair (q, κ) instead of the geometrical-optics phase-plane pair (q, p), the off-diagonal entries of the optical matrix are scaled respectively by $\frac{1}{k_0}$ and k_0.

---

[a] Recall that for an n-dimensional Gaussian function $\mathcal{G}(\mathbf{u}) = e^{-\frac{1}{2}\mathbf{u}^\top\mathbf{A}\mathbf{u}}$, with $\mathbf{A}$ an $n \times n$ positive definite matrix, one has $\int \mathcal{G}(\mathbf{u})d\mathbf{u} = \frac{(2\pi)^{n/2}}{\sqrt{\det \mathbf{A}}}$.

In terms of the variance matrices $\mathbf{\Sigma}(z_i)$ and $\mathbf{\Sigma}(z_o)$ at the end-planes, relation (9.5.21) means

$$\frac{1}{\pi}e^{-\frac{1}{2}\mathbf{u}^\top[\mathbf{\Sigma}(z_o)]^{-1}\mathbf{u}} = \frac{1}{\pi}e^{-\frac{1}{2}[\mathbf{M}^{-1}\mathbf{u}]^\top[\mathbf{\Sigma}(z_o)]^{-1}[\mathbf{M}^{-1}\mathbf{u}]},$$

which implies of course the identity of the exponent. Hence, we end up with

$$\mathbf{\Sigma}(z_o) = \mathbf{M}\mathbf{\Sigma}(z_i)\mathbf{M}^\top, \tag{9.5.22}$$

which is the central relation of our analysis. Although it has been inferred in connection with the easily manageable Wigner distribution function of the signal (9.5.1), it holds true for any Wigner distribution function, and so optical disturbance (in the paraxial approximation). Once unfolded, it yields the explicit evolution laws for each of the second order moments, thus establishing the rules, by which the second-moment based phase-plane representative region associated with the beam changes during propagation, with the relevant area remaining constant, as stated by the invariance of $\det\mathbf{\Sigma}$.

We shall discuss relation (9.5.22) later in a more general context. Here we emphasize that it does yield the $ABCD$ law (9.5.5) for the complex beam parameter $Q(z)$. Note, in fact, that Q writes in terms of the $\mathbf{\Sigma}$-entries as

$$\frac{1}{Q} = \frac{\Sigma_{q\kappa} + i\sqrt{\det\mathbf{\Sigma}}}{k\Sigma_{qq}}, \tag{9.5.23}$$

or, equivalently

$$Q = \frac{k(\Sigma_{q\kappa} - i\sqrt{\det\mathbf{\Sigma}})}{\Sigma_{\kappa\kappa}}. \tag{9.5.24}$$

Of course, $\sqrt{\det\mathbf{\Sigma}} = \frac{1}{2}$ in the case of the fundamental Gaussian beam we are dealing with, however letting $\sqrt{\det\mathbf{\Sigma}}$ undefinite in the above expressions confer them a certain generality, that will reveal useful for future suggestions.

Inferring form (9.5.22) the relation of the matrix elements $\Sigma_{q\kappa}$ and $\Sigma_{\kappa\kappa}$ between the end-planes, i.e.,

$$\begin{aligned}\Sigma_{q\kappa}(z_o) &= ACk_0\Sigma_{qq}(z_i) + (AD+BC)\Sigma_{q\kappa}(z_i) + \tfrac{1}{k_0}BD\Sigma_{\kappa\kappa}(z_i),\\ \Sigma_{\kappa\kappa}(z_o) &= C^2k_0^2\Sigma_{qq}(z_i) + 2k_0CD\Sigma_{q\kappa}(z_i) + D^2\Sigma_{\kappa\kappa}(z_i),\end{aligned} \tag{9.5.25}$$

and noting from (9.5.24) that

$$\frac{2k\Sigma_{q\kappa}}{\Sigma_{\kappa\kappa}} = Q + Q^*, \quad \frac{-2ik\sqrt{\det\mathbf{\Sigma}}}{\Sigma_{\kappa\kappa}} = Q - Q^*, \quad \frac{k^2\Sigma_{qq}}{\Sigma_{\kappa\kappa}} = QQ^*, \tag{9.5.26}$$

after some algebra we recover the $ABCD$ transfer law (9.5.5) for the complex beam parameter under propagation from the plane at z_i to the plane at z_o being connected by the ray matrix $\mathbf{M} = \left(\begin{smallmatrix} A & B \\ C & D\end{smallmatrix}\right)$.

9.5.4 *Gaussian beams: the phase-plane ellipse of second moments*

Sections of the Wigner distribution function (9.5.20) at planes parallel to the (q, κ) plane individualize elliptical contours defined by the equation[(b)]

$$\tfrac{1}{2}\mathbf{u}^{\top}[\mathbf{\Sigma}(z)]^{-1}\mathbf{u} = \varepsilon. \tag{9.5.27}$$

In particular, the ellipse ($\varepsilon = 2$)

$$\tfrac{1}{4}\mathbf{u}^{\top}[\mathbf{\Sigma}(z)]^{-1}\mathbf{u} = 1, \tag{9.5.28}$$

occurs at an height which is 0.135 of the maximum height of the $3D$ plot of the Wigner distribution function.

Of course, any one of the ellipses (9.5.27) can be taken as a phase-plane representative region of the corresponding Wigner distribution function (and Gaussian beam) and hence used to inspect the modification suffered by the beam under propagation through the systen under investigation. Mantaining the correspondence with the analysis in previous chapters, we specifically consider the phase-plane ellipse identified by (9.5.28), which explicitly writes

$$\Sigma_{\kappa\kappa}(z)q^2 - 2\Sigma_{q\kappa}(z)q\kappa + \Sigma_{qq}(z)\kappa^2 = 1. \tag{9.5.29}$$

For a descriptiom of the geometrical properties of (9.5.29), the reader is directed to § 9.6.4, where the topic is illustrated in relation with the general case of an arbitrary non-Gaussian Wigner distribution function ($\sqrt{\det \mathbf{\Sigma}} \neq \frac{1}{2}$).

9.6 Propagation laws for the moments of the Wigner distribution function in first-order optical systems

Here we reformulate the propagation laws for the second moments of the Wigner distribution function in first-order optical systems just described in connection with Gaussian beams, within the general context pertaining to arbitrary non Gaussian optical signals [4.1, 4.2, 4.4].

Firstly we consider the first-order moments $\overline{q}$, $\overline{\kappa}$. In conformity to the vector notation $\mathbf{u} = (q,\kappa)^{\top}$, we may concisely write (9.4.44) and (9.4.50) into one form

$$\overline{\mathbf{u}}(z) = \iint \mathbf{u}\mathcal{W}(\mathbf{u}; z)d\mathbf{u}, \tag{9.6.1}$$

where, of course, $\iint ..d\mathbf{u} \equiv \iint ..dq d\kappa$.
As we know, the propagation of the signal through an $ABCD$ system implies a transformation of the first-moment vector (9.6.1), explicitly expressed by

$$\overline{\mathbf{u}}(z_o) = \iint \mathbf{u}\mathcal{W}(\mathbf{u}; z_o)d\mathbf{u} = \iint \mathbf{u}\mathcal{W}(\mathbf{M}^{-1}\mathbf{u}; z_i)d\mathbf{u}. \tag{9.6.2}$$

---

[b] Naturally the conic $\frac{1}{2}\mathbf{u}^{\top}[\mathbf{\Sigma}(z)]^{-1}\mathbf{u}$ is recognized as an ellipse since $\det \mathbf{\Sigma} > 0$.

Changing the variables from $\mathbf{u}$ to $\mathbf{v} = \mathbf{M}^{-1}\mathbf{u}$, for which $d\mathbf{u} = \det \mathbf{M} d\mathbf{v} = d\mathbf{v}$ due to the unimodularity of $\mathbf{M}$, we end up with

$$\overline{\mathbf{u}}(z_o) = \mathbf{M} \iint \mathbf{v}\mathcal{W}(\mathbf{v}; z_i) d\mathbf{v} = \mathbf{M}\overline{\mathbf{u}}(z_i). \tag{9.6.3}$$

As expected, the first-order moments of the Wigner distribution function evolve just as the ray-variables of paraxial ray-optics, in accord evidently with the results established in § 5.2 regarding the propagation law of the mean values, $\overline{q}$ and $\overline{\kappa}$, of the position and frequency operators.

In order to concisely write the second-order moments, we form the Kronecker product of the Wigner-plane vector $\mathbf{u}$ with the transpose vector $\mathbf{u}^\top$, thus obtaining the 2×2 symmetric matrix[(c)]

$$\mathbf{u} \otimes \mathbf{u}^\top = \begin{pmatrix} q \\ \kappa \end{pmatrix} (q \ \kappa) = \begin{pmatrix} q^2 & q\kappa \\ q\kappa & \kappa^2 \end{pmatrix}. \tag{9.6.4}$$

Accordingly, by (9.6.1) the product $\overline{\mathbf{u}} \otimes \overline{\mathbf{u}}^\top$ produces the matrix

$$\overline{\mathbf{u}} \otimes \overline{\mathbf{u}}^\top = \begin{pmatrix} \overline{q}^2 & \overline{q\kappa} \\ \overline{q\kappa} & \overline{\kappa}^2 \end{pmatrix}. \tag{9.6.5}$$

The dependence on the propagation variable z is implied in (9.6.4) and (9.6.5).

The variance matrix $\mathbf{\Sigma}$, introduced in (9.5.13) in connection with Gaussian beams, is therefore seen as a particular case of the general definition

$$\mathbf{\Sigma}(z) = \iint \mathbf{u} \otimes \mathbf{u}^\top \mathcal{W}(\mathbf{u}; z) d\mathbf{u} - \overline{\mathbf{u}} \otimes \overline{\mathbf{u}}^\top. \tag{9.6.6}$$

Typically it is assumed that $\overline{\mathbf{u}} = 0$, thus simplifying the working formulae.

In order to infer the transfer law for the variance matrix $\mathbf{\Sigma}(z)$ in first-order optical systems, we apply the propagation law for the Wigner distribution function, change the integration variables and then exploit the symplecticity of the ray-matrix $\mathbf{M}$ in the same way as on deriving (9.6.3). We finally find

$$\begin{aligned} \mathbf{\Sigma}(z_o) &= \iint \mathbf{u} \otimes \mathbf{u}^\top \mathcal{W}(\mathbf{u}; z_o) d\mathbf{u} = \iint \mathbf{u} \otimes \mathbf{u}^\top \mathcal{W}(\mathbf{M}^{-1}\mathbf{u}; z_i) d\mathbf{u} \\ &= \mathbf{M} \iint \mathbf{v} \otimes \mathbf{v}^\top \mathcal{W}(\mathbf{v}; z_i) d\mathbf{v} \mathbf{M}^\top, \end{aligned} \tag{9.6.7}$$

---

[c] The *Kronecker* (or *direct*) *product* of an $n \times m$ matrix $\mathbf{E}$ and an $n' \times m'$ matrix $\mathbf{F}$ produces the $nn' \times mm'$ matrix $\mathbf{E} \otimes \mathbf{F}$ by the rule [8]

$$\mathbf{E} \otimes \mathbf{F} \equiv \begin{pmatrix} e_{11}\mathbf{F} & \cdots & e_{1m}\mathbf{F} \\ \vdots & \vdots\vdots & \vdots \\ e_{n1}\mathbf{F} & \cdots & e_{nm}\mathbf{F} \end{pmatrix}.$$

Evidently, the Kronecker product is not commutative. It satisfies, among the others, the properties: *(a)* $\mathbf{E} \otimes \mathbf{F} \otimes \mathbf{G} = (\mathbf{E} \otimes \mathbf{F}) \otimes \mathbf{G} = \mathbf{E} \otimes (\mathbf{F} \otimes \mathbf{G})$; *(b)* $\mathbf{E} \otimes (\mathbf{F} + \mathbf{G}) = \mathbf{E} \otimes \mathbf{F} + \mathbf{E} \otimes \mathbf{G}$; *(c)* $(\mathbf{E} \otimes \mathbf{F})(\mathbf{H} \otimes \mathbf{G}) = \mathbf{EH} \otimes \mathbf{FG}$, provided that the ordinary products $\mathbf{EH}$ and $\mathbf{FG}$ may be formed; *(d)* $(\mathbf{E} \otimes \mathbf{F})^\top = \mathbf{E}^\top \otimes \mathbf{F}^\top$; *(f)* $(\mathbf{E} \otimes \mathbf{F})^{-1} = \mathbf{E}^{-1} \otimes \mathbf{F}^{-1}$, if $\mathbf{E}$ and $\mathbf{F}$ are non-singular.

which, after identifying the double integral as $\mathbf{\Sigma}(z_i)$, yields

$$\mathbf{\Sigma}(z_o) = \mathbf{M}\mathbf{\Sigma}(z_i)\mathbf{M}^{\top}, \tag{9.6.8}$$

reproducing, as expected, relation (9.5.22).

Relation (9.6.8) represents a congruence transformation[d]. It implies however a similarity transformation for the matrix $\mathbf{\Sigma}(z)\mathbf{J}$, for which, in fact, we firstly obtain

$$\mathbf{\Sigma}(z_o)\mathbf{J} = \mathbf{M}\mathbf{\Sigma}(z_i)\mathbf{M}^{\top}\mathbf{J}, \tag{9.6.9}$$

and then, being $\mathbf{M}^{\top}\mathbf{J} = \mathbf{J}\mathbf{M}^{-1}$,

$$\mathbf{\Sigma}(z_o)\mathbf{J} = \mathbf{M}\mathbf{\Sigma}(z_i)\mathbf{J}\mathbf{M}^{-1}. \tag{9.6.10}$$

Similarity transformations have the notable property of preserving eigenvalues and traces. Therefore, Eq. (9.6.10) may provide a useful supply of invariants. The interested reader may consult the related literature [4.2, 4.4, 4.7, 4.5]; here, we only detail the basic invariant yielded by the determinat of $\mathbf{\Sigma}$. It is, in fact, evident that, due to the unimodularity of the optical matrix $\mathbf{M}$,

$$\det \mathbf{\Sigma}(z_o) = \det \mathbf{\Sigma}(z_i), \tag{9.6.11}$$

as in the prevously described case of Gaussian Wigner distribution functions.

9.6.1 *The optical uncertainty relation. II*

The explicit form of (9.6.11) gives

$$\Sigma_{qq}\Sigma_{\kappa\kappa} - \Sigma^2_{q\kappa} = \text{const.} \tag{9.6.12}$$

In particular, choosing a transfer matrix $\mathbf{M}$ such that $\mathbf{\Sigma}$ comes to be diagonal: $\Sigma_{q\kappa} = 0$, we have

$$\det \mathbf{\Sigma} = \Sigma_{qq}\Sigma_{\kappa\kappa}, \tag{9.6.13}$$

which according to the width principle yields

$$\det \mathbf{\Sigma} \geq \tfrac{1}{4}. \tag{9.6.14}$$

Since, whichever be the optical matrix $\mathbf{M}$, the variance matrix resulting from (9.6.8) and obeyng (9.6.14) corresponds to a physical signal, i.e., the signal transformed by some other signal by propagation through the system represented by $\mathbf{M}$, relation (9.6.14) should hold for all variance matrices. This

---

[d] Given a matrix $\mathbf{A}$ and a matrix $\mathbf{B}$, the product $\mathbf{B}\mathbf{A}\mathbf{B}^{\top}$ realizes the *congruence transformation* of the matrix $\mathbf{A}$ by $\mathbf{B}$, whilst, if $\mathbf{B}$ is non-singular, the product $\mathbf{B}\mathbf{A}\mathbf{B}^{-1}$ realizes the *similarity transformation* of the matrix $\mathbf{A}$ by $\mathbf{B}$.

establishes a bound for the invariant value in (9.6.12), which then can be written as

$$\Sigma_{qq}\Sigma_{\kappa\kappa} \geq \frac{1}{4}(1 + 4\Sigma_{q\kappa}^2) \geq \frac{1}{4}. \qquad (9.6.15)$$

It represents the complete form of the optical uncertainty relation, involving a precise signal-dependent term (related to the mixed moment $\Sigma_{q\kappa}$) besides the signal-independent value $\frac{1}{4}$.

9.6.2 *Propagation laws for the second-order moments*

Unfolding the transfer law (9.6.8) for the variance matrix $\mathbf{\Sigma}(z)$, one gets the explicit input-output relations for the second-order moments, that can as well be arranged into a matrix scheme. In fact, it is straightforward to write

$$\begin{pmatrix} \Sigma_{qq} \\ \Sigma_{q\kappa} \\ \Sigma_{\kappa\kappa} \end{pmatrix}(z_o) = \begin{pmatrix} A^2 & \frac{2}{k_0}AB & \frac{1}{k_0^2}B^2 \\ k_0 AC & AD + BC & \frac{1}{k_0}BD \\ k_0^2 C^2 & 2k_0 CD & D^2 \end{pmatrix} \begin{pmatrix} \Sigma_{qq} \\ \Sigma_{q\kappa} \\ \Sigma_{\kappa\kappa} \end{pmatrix}(z_i). \qquad (9.6.16)$$

We will specifically examine the effect of some optical transformations on the moments later in relation with the phase-plane second-moment ellipse.

9.6.3 *Positive definiteness of* Σ

Firstly, let us note that relation (9.6.16) can be used to prove that the variance matrix $\mathbf{\Sigma}$ is positive definite. The first row, for instance, writes

$$A^2\Sigma_{qq}(z_i) + \frac{2}{k_0}AB\Sigma_{q\kappa}(z_i) + \frac{1}{k_0^2}B^2\Sigma_{\kappa\kappa}(z_i) = \Sigma_{qq}(z_o) > 0. \qquad (9.6.17)$$

Evidently the left-hand side above can be interpreted as the scalar product

$$\mathbf{u} \cdot \mathbf{\Sigma u}, \qquad (9.6.18)$$

with the vector $\mathbf{u}$ being given by the matrix entries $\mathbf{u} = (A, B)^\top$. Such a scalar product turns to be equal to a positive quantity, i.e., $\Sigma_{qq}(z_o)$. Then, being $\mathbf{u}$ arbitrary, one can state the positive definiteness of the moment matrix, according to the pertinent definition recalled in § 9.5.13 (Eq. (9.5.16)).

9.6.4 *The ellipse of second-order moments*

According to the results of § 8.5.3, as far as the linear approximation is concerned, the phase-plane dynamics of the Wigner chart mirrors that of the ray-density distribution in the geometrical optical phase-plane. This means that one may choose any closed domain in the Wigner plane to inspect the transformations experienced by the Wigner distribution function under propagation through first-order optical systems. As it can be inferred from the pictorial views presented in the preceding chapters (in full analogy with those pertaining to the "classical" ray-density distribution $\rho(q,p;z)$; see, for instance, § 2.5), it is convenient to work with regions delimited by straight lines or ellipses. In fact, δ-line distributions transform into δ-line distributions and Gaussian distributions transform into Gaussian distributions (§ 6.4.4-5), straight lines and ellipses respectively being the phase-plane equilevel contours pertaining to such distributions.

It is a common practice to adopt as "model" distribution the Gaussian distribution, and accordingly to focus on any one of the phase-plane elliptical regions individualized by the second-order moments according to (9.5.27), i.e.

$$\tfrac{1}{2}\mathbf{u}^{\top}[\mathbf{\Sigma}(z)]^{-1}\mathbf{u} = \varepsilon, \tag{9.6.19}$$

for real values of ε. In a sense we consider the Gaussian Wigner distribution function that best fits the Wigner distribution function under study, and hence the Gaussian signal that best fits the signal under examination.

To maintain the correspondence with the presentation in § 9.5.4, we will definitely examine the ellipse corresponding to the value $\varepsilon = 2$; namely,

$$\Sigma_{\kappa\kappa}(z)q^2 - 2\Sigma_{q\kappa}(z)q\kappa + \Sigma_{qq}(z)\kappa^2 = 4\epsilon^2, \tag{9.6.20}$$

where we have denoted by ϵ^2 the invariant value of the determinant of $\mathbf{\Sigma}$, i.e.

$$\det \mathbf{\Sigma} \equiv \epsilon^2. \tag{9.6.21}$$

The basic geometrical properties of (9.6.20) are enumerated below; in this respect, it is worth recalling once more that in order that the worked formulae be meaningful one must think of properly scaling the axes so that they have the same dimensionality. The reader can easily verify that (Fig. 9.4)

• (9.6.20) is a tilted ellipse centered on the origin of the phase plane.

• The points at which it intersects the q and κ axes are identified by the coordinates

$$q_{\text{int}} = \pm\frac{2\epsilon}{\sqrt{\Sigma_{\kappa\kappa}}}, \qquad \kappa_{\text{int}} = \frac{2\epsilon}{\sqrt{\Sigma_{qq}}}. \tag{9.6.22}$$

• The semiaxes $\mathfrak{a}_{\pm}$ of (9.6.20) are given by

$$\mathfrak{a}_{\pm} = \sqrt{\Sigma_{qq} + \Sigma_{\kappa\kappa} + 2\epsilon} \pm \sqrt{\Sigma_{qq} + \Sigma_{\kappa\kappa} - 2\epsilon}, \tag{9.6.23}$$

and disposed in conformity with the two solutions, spaced by $\frac{\pi}{2}$, of the equation

$$\tan(2\vartheta) = \frac{2\Sigma_{q\kappa}}{\Sigma_{qq}-\Sigma_{\kappa\kappa}}, \tag{9.6.24}$$

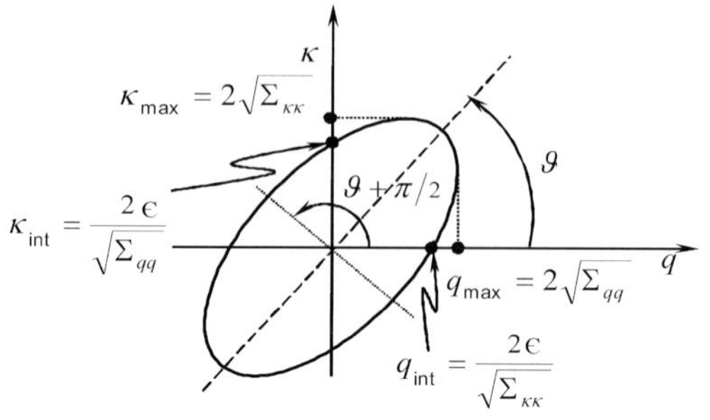

FIGURE 9.4. Schematic of the second-moment phase-plane ellipse for an arbitrary signal.

for the orientation of the ellipse axes.

• The maximum excursions along the coordinate axes are specified by the space and frequency variances according to

$$q_{\text{max}} = 2\sqrt{\Sigma_{qq}}, \qquad \kappa_{\text{max}} = 2\sqrt{\Sigma_{\kappa\kappa}}. \tag{9.6.25}$$

• The area of (9.6.20) can be brought into relation with det $\mathbf{\Sigma}$, being

$$\mathcal{A} = \pi \mathfrak{a}_+ \mathfrak{a}_- = 4\pi\epsilon. \tag{9.6.26}$$

The invariance of the determinant of $\mathbf{\Sigma}$ then relates to Liouville's invariance of phase-plane areas (see § 2.2.1 and § 8.5.3). The invariance of the area of the second-moment ellipse has practical consequences on the coupling efficiency when transferring optical power from one component to another in a given optical configuration. According to Liouville's theorem, there is no optical system that can reduce the area of the representative phase-plane domain of a given light beam, which, in fact, can be deformed but not transformed into a smaller area domain. Therefore, if the area of the representative ellipse of a given optical system (the *acceptance*, mentioned in § 2.5) is smaller than the ellipse area (the *emittance*, still mentioned in § 2.5) of the radiation that must be coupled to it, some of the optical power will be lost (Fig. 9.5). The design of optical devices, where light is coupled to some receiver, greatly benefits from the evaluation of the maximum coupling efficiency achievable. We see that basically it turns out to be a problem of matching of phase-plane areas.

As said earlier, the concepts of emittance and acceptance are familiar within the context of charged-particle optics. Indeed, the beam ellipse in Fig. 9.4 may equally represent the phase-plane ellipse of a charged beam travelling along a transport channel. In this respect, the beam ellipse is parameterized in terms of the second-order moments of the phase-plane beam distribution in full analogy with optical beams (typically this is done in beamlines), or in terms of

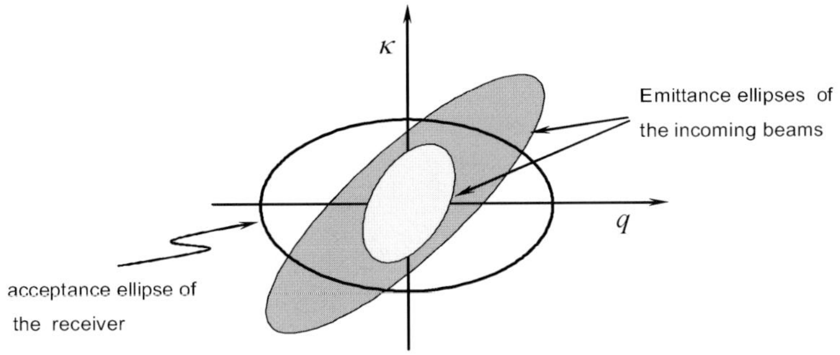

FIGURE 9.5. The optimization of the energy launching of a beam into a receiver requires the optimization of the matching of the phase-plane ellipse of the incoming beam to the phase-plane acceptance domain of the receiving structure (conveniently arranged through an elliptical shape as well). A beam ellipse both smaller and greater than the receiver ellipse produces uneconomical situations, implying in one case unneeded large sizes of the receiver and in the other case losses of the incident beam energy.

the Twiss parameters (typically in rings), linked to the second-order moments of the beam distribution by an emittance depending factor. Accordingly, the phase plane views of Fig. 9.6 may equally depict the transformations effected by definite magnetic components on the travelling particle beam in terms of the transformations of the corresponding phase-plane ellipse. The free-medium section corresponds in magnetic optics to a field-free section, and a thin lens to a thin quadrupole lens. Also, the Fourier transform-like matrix may accounts for the flight of the beam between the two focal planes of a magnetic quadrupole lens. As in light optics, the beam expander is a composite system.

9.6.5 *The second-moment ellipse and optical transformations*

We briefly investigate the effects of the elementary optical systems considered many times through the book on the second-order moments of the Wigner distribution function. The detailed investigation of such effects has been presented in Chapter 8. Here we shortly rephrase that analysis in terms of the second order moments, and so in terms of the transformations of the beam ellipse in the Wigner phase-plane (Fig. 9.6). We will see that each of the optical transformations that have been central to previous analyses, are characterized by the invariance or the skew invariance of one of the second order moments.

A diffraction region of length d leaves unchanged $\Sigma_{\kappa\kappa}$:

$$\Sigma_{\kappa\kappa}(z_o) = \Sigma_{\kappa\kappa}(z_i) \qquad \text{free-medium section,} \tag{9.6.27}$$

and hence leaves unchanged the maximum excursions $\kappa_{\max}$ along the κ-axis

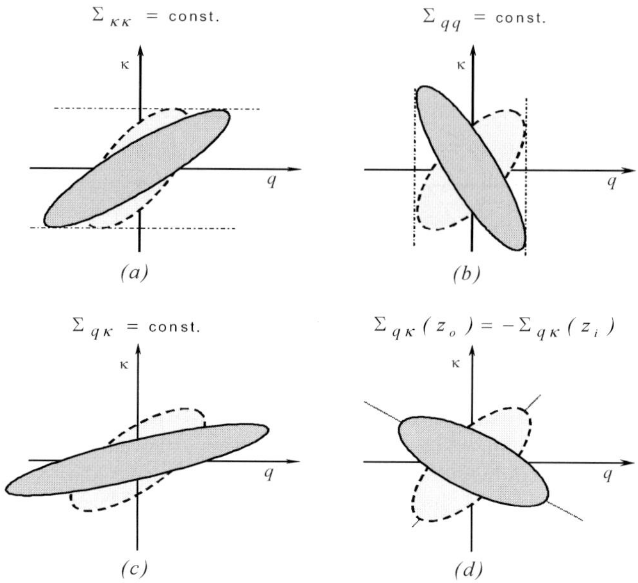

FIGURE 9.6. Phase-plane view of the transformations by which the second-moment ellipse is effected under propagation through *(a)* a free-medium section, *(b)* a thin lens, *(c)* a positive scaler, and *(d)* a Fourier transformer.

and the intersections q_{int} on the q-axis (Fig. 9.6.a)). Then, the $\Sigma_{\kappa\kappa}$-invariance becomes a distinctive characterization of the phase-plane area transformation acted by a free-propagation that has previously been identified as a q-shear.

Correspondingly, the propagation through a thin lens leaves unchanged the spatial variance of the signal, i.e.

$$\Sigma_{qq}(z_o) = \Sigma_{qq}(z_i) \qquad \text{thin lens.} \tag{9.6.28}$$

The maximum excursions $q_{\max}$ along the q-axis and the intersections κ_{int} with the κ-axis do not change, thus individualizing, of course, a p-shear (Fig. 9.6.b)).

As expected, a positive scaler is characterized by the invariance of the mixed moment

$$\Sigma_{q\kappa}(z_o) = \Sigma_{q\kappa}(z_i), \qquad \text{positive scaler} \tag{9.6.29}$$

whilst both $q_{\max}$ and q_{int} are scaled by the magnification factor m, and, reciprocally, $\kappa_{\max}$ and κ_{int} are scaled by the inverse $\frac{1}{m}$ (Fig. 9.6.c)).

Finally, a Fourier transformer basically carries the space and frequency moments one into the other, producing, as we know, a scaled $\frac{\pi}{2}$ rotation in the Wigner plane. Accordingly, the mixed moment is changed as (Fig. 9.6.d)).

$$\Sigma_{q\kappa}(z_o) = -\Sigma_{q\kappa}(z_i), \qquad \text{Fourier transformer.} \tag{9.6.30}$$

9.6.6 *Generalized curvature of a signal. Beam quality parameter*

Following the suggestion put forward in § 9.5.3 when dealing with Gaussian beams, whose parameters have been redefined in terms of the variance matrix entries, we introduce the generalized curvature parameter $\frac{1}{\mathsf{Q}}$ of an arbitrary non Gaussian beam through an obvious restatement of (9.5.23), i.e.

$$\frac{1}{\mathsf{Q}} = \frac{\Sigma_{q\kappa} + i\epsilon}{k\Sigma_{qq}} = \frac{\Sigma_{q\kappa}}{k\Sigma_{qq}} + i\frac{\epsilon}{k\Sigma_{qq}}. \tag{9.6.31}$$

Interpreting the real part of $\frac{1}{\mathsf{Q}}$ as the real curvature $\frac{1}{\mathsf{R}}$ of the beam wavefront, and relating the beam radius w to the second order moment Σ_{qq} in the same way as suggested by (9.5.12) for a Gaussian beam [7.4, 7.8], i.e.

$$\mathsf{w}^2 = 4\Sigma_{qq}, \tag{9.6.32}$$

we obtain for $\frac{1}{\mathsf{Q}}$ an expression formally similar to (9.5.2) relevant to a Gaussian beam:

$$\frac{1}{\mathsf{Q}} = \frac{1}{\mathsf{R}} + i\frac{4\epsilon}{k\mathsf{w}^2} = \frac{1}{\mathsf{R}} + i\frac{2M^2}{k\mathsf{w}^2}. \tag{9.6.33}$$

The factor M^2 introduced above (following a common practice) marks the difference with the beam radius w of a Gaussian beam, signifying, in fact, that

$$\mathsf{w}^2 = M^2 w^2. \tag{9.6.34}$$

In the literature related to laser beam characterizations, it is known as the M^2 *factor*. It is adopted as a measure of the (inverse) *beam quality,* with $M^2 = 1$ corresponding to a perfect Gaussian beam, and $M^2 > 1$ signifying a "poorer" quality [7.4, 7.6, 7.5]. Evidently, it is nothing but $(2\pi)^{-1}$ the area of the phase-plane beam ellipse.

Remarkably, the generalized beam curvature parameter $\frac{1}{\mathsf{Q}}$ obeys the same $ABCD$ law obeyed by the relevant parameter $\frac{1}{Q}$ of a Gaussian beam, as the reader can guess on account of the presentation given in § 9.5.3.

As an alternative to the M^2 factor, the beam quality is also defined as the *diameter-divergence product*, i.e. the product of the beam diameter at the focus with the far-field divergence angle (recall the space-bandwidth product mentioned in § 4.6.4, and the geometrical extent of a beam considered in § 2.2.4). It is evident that the concept of "beam quality" cannot be unique or universal, since the "quality" of a laser beam is evaluated in accord with the applications for which the laser is intended. Yet, the adopted definitions of beam quality yield useful criteria to characterize a laser beam, being basically a measure of how tightly it can be focused under certain conditions (as, for instance, with a limited beam divergence). A great variety of methods and instruments have been envisaged and implemented to yield a quantitative measure of the laser beam properties[7]. In the next section, we will describe a simple optical device, first introduced by Nemes [7.2] which is able to directly yield a measure of the beam diameter-divergence product.

9.6.7 *Phase-space beam analyzer*

Before describing the phase-space beam analyser [7.2, 7.9, 7.13], we give a few notes on the 4×4 ray-matrices of optical systems, that come into play when non separable optical components are present, as in the case below . We list here without proof the ray-transfer matrices for free propagation and thin lens operation (through different geometries) suitable to a $2D$ paraxial model of light propagation, in which indeed the ray is parameterized by the 4-component vector

$$\mathbf{u} = (q_x, q_y, p_x, p_y)^\top . \tag{9.6.35}$$

The reader may find an intuitive justification to the fact that, free propagation through a section of length d is accounted by the matrix

$$\mathbf{T}(d) = \begin{pmatrix} 1 & 0 & d & 0 \\ 0 & 1 & 0 & d \\ 0 & 0 & 1 & 0 \\ 0 & 0 & 0 & 1 \end{pmatrix}, \tag{9.6.36}$$

whilst cylindrical lenses with their axis respectively along the q_x and q_y direction (which then are active in the planes q_y-z and q_x-z) are described by[e]

$$\mathbf{L}(\mathcal{P}_y) = \begin{pmatrix} 1 & 0 & 0 & 0 \\ 0 & 1 & 0 & 0 \\ 0 & 0 & 1 & 0 \\ 0 & -\mathcal{P}_y & 0 & 1 \end{pmatrix}, \quad \mathbf{L}(\mathcal{P}_x) = \begin{pmatrix} 1 & 0 & 0 & 0 \\ 0 & 1 & 0 & 0 \\ -\mathcal{P}_x & 0 & 1 & 0 \\ 0 & 0 & 0 & 1 \end{pmatrix}. \tag{9.6.37}$$

Accordingly, a spherical lens amounts to the optical matrix

$$\mathbf{L}(\mathcal{P}) = \begin{pmatrix} 1 & 0 & 0 & 0 \\ 0 & 1 & 0 & 0 \\ -\mathcal{P} & 0 & 1 & 0 \\ 0 & -\mathcal{P} & 0 & 1 \end{pmatrix}. \tag{9.6.38}$$

Finally, a quadrupole lens, which consists of a negative and a positive cylinder lens, $-f$ and f, the respective axes being rotated one to the other by 90° and by 45° with respect to the q_x-axis [7.3], is described by the 4×4 matrix

$$\mathbf{Q}(\mathcal{P}) = \begin{pmatrix} 1 & 0 & 0 & 0 \\ 0 & 1 & 0 & 0 \\ 0 & -\mathcal{P} & 1 & 0 \\ -\mathcal{P} & 0 & 0 & 1 \end{pmatrix}, \quad \mathcal{P} = \frac{1}{f}. \tag{9.6.39}$$

We refer to Fig. 9.7 for the description of the phase-space beam analyzer, which as we will see is an effective device "to determine the one-dimensional

---

[e] According to our notation, we refer to lens matrices, in relation to both cylindrical and spherical lenses, in terms of the focal length f or the focusing power $\mathcal{P} = \frac{1}{f}$, and so by $\mathbf{L}(f)$ and $\mathbf{L}(\mathcal{P})$, the possible directional valence of both parameters being implied.

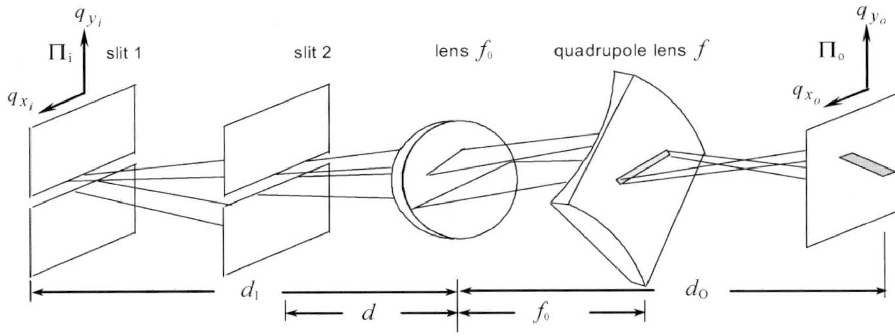

FIGURE 9.7. Schematic of the phase space beam analyzer. (Inspired by [7.9]).

beam parameter product (beam quality) by a single measurement." [7.9] A slit 1 is located at the input plane, having an aperture in the q_y direction sufficiently small to eliminate any wave field structure. It serves to reduce the incoming $2D$ optical wave to a line-like distribution, which then can be considered as a $1D$ input beam. A second slit acts as a collimator, selecting only rays with zero inclination in the q_y direction. In order that the subsequent spherical lens f_0 be reached by a collimated beam, the distance between the slit 2 and the spherical lens should be as small as possible (i.e., $d \sim 0$). Finally a quadrupole lens f is placed in the back focal plane of the spherical lens. The output plane is at distance d_O from the spherical lens, which in turn is at distance d_I from the input plane. Distances d_I and d_O fulfil the lens law for imaging:

$$\frac{1}{d_I} + \frac{1}{d_O} = \frac{1}{f_0}. \tag{9.6.40}$$

Therefore, in the absence of the quadrupole lens we would obtain the image of the slit on the observation plane.

We present here an account of the phase-space beam analyser by means of geometrical matrix optics, which is enough to gain a first undertanding. A deeper investigation on account as well of diffractive effects is given in [7.9].

The overall matrix of the system is, of course, obtained by multiplying the matrices of the various components, and so

$$\mathbf{M}_{\text{beam analyzer}} = \mathbf{T}(d_O - f_0)\mathbf{Q}(f)\mathbf{T}(f_0)\mathbf{L}(f_0)\mathbf{T}(d_I), \tag{9.6.41}$$

which explicitly reads

$$\mathbf{M}_{\text{beam analyzer}} = \begin{pmatrix} A & 0 & 0 & B \\ 0 & A & B & 0 \\ -1/f_0 & 0 & 1/A & -f_0/f \\ -1/f_0 & 0 & -f_0/f & 1/A \end{pmatrix}, \tag{9.6.42}$$

where

$$A = \frac{f_0 - d_O}{f_o}, \quad B = \frac{f_o(f_0 - d_O)}{f}. \tag{9.6.43}$$

We can now inspect the effect that the system in Fig. 9.7 produces on the rays from the input plane. By construction, the rays passing through the slit 1 are described by ray vectors like

$$\mathbf{u}_i = (q_{x_i}, 0, p_{x_i}, 0)^\top, \tag{9.6.44}$$

where q_{x_i} ranges over $\pm$ the radius of the incoming beam in the q_x direction and p_{x_i} over the $\pm$ the divergence angle in the q_x direction. The q_y and p_y components of the ray vector above have been set to zero, as a first approximation, since q_{y_i} is determined by the aperture, supposed small, of the slit 1, say a_1, i.e. $\Delta q_{y_i} \sim 2a_1$, and similarly p_{y_i} is determined by the aperture, still supposed small, of the slit 2, say a_2, and so $\Delta p_{y_i} \sim 2a_2/d_1$.

Limiting ourselves to consider only the spatial components of the ray-vectors representing the rays reaching the output plane, we see that

$$\begin{aligned} q_{x_o} &= A q_{x_i}, \\ q_{y_o} &= B p_{x_i}. \end{aligned} \tag{9.6.45}$$

Then the real appearance of the light distribution on the output screen is a direct (apart from the scaling factors A and B) phase-plane representation of the $1D$ distribution in the input plane. Evidently, one also obtains the beam diameter-divergence product.

9.6.8 *Ray, wave and Wigner optics. II*

The presentation given in Chapter 8 has revealed that the Wigner distribution function is wave-optical in its structure, resorting in its definition to the wave-optical descriptors of light wave fields. In contrast, the dynamics of the Wigner distribution function under propagation of the signal through first-order optical systems is determined by the transformation rules of the ray-variables in geometrical matrix optics. Thus, in full analogy with the classical ray-density, we can run the "individual phase-plane ray-trajectories" to follow the evolution of the Wigner distribution function under propagation through $ABCD$ optical systems. In fact, we have also seen in § 8.5.3 that, as far as the linear approxinmation (i.e. small inclination angles to and small distances off the axis) is concerned, the transport equation for the Wigner distribution function turns into the classical Liouville's equation for the ray-density distribution.

Here we have seen that the Wigner distribution function of a Gaussian beam is completely characterized by its second-order moments through a quadratic form. It identifies in the Wigner plane an elliptical shape, whose transformations mirror those of the signal under propagation through the system. Remarkably, the paraxial propagation laws for the second-order moments involve

FIGURE 9.8. The perfect beauty of Borromini's dome of Sant'Ivo alla Sapienza in Rome superbly conveys the aestethical and emotional valence of a spiral. (Picture taken by the author.)

the ray transfer matrix of geometrical optics in a form conforming to Liouville's invariance of phase-plane areas, whose formulation in terms of second-order moments yields indeed the optical uncertainty relation.

Indeed, this is not surprising since Gaussian beams, evolving from the minimum uncertainty amplitude, i.e. the real Gaussian signal, may be regarded as "ray-like beams". So it might be expected that the Wigner representation of a Gaussian beam operates a synthesis between ray and way optics, recovering the simple transformation rules and phase-plane depictions of the former but enriched by the intrinsic wave-optical nature of light distribution conveyed by the latter.

It is noticeable that the same holds true for arbitrary non Gaussian beams, by virtue of Liouville's theorem, holding true even for Wigner optics as far as the linear approximation is concerned (§ 8.5.3).

The Wigner representation amalgamates ray and wave optics, embedding the wave-optical features of light propagation into the ray-optics formalism.

Resorting to the beginning of our presentation started in Chapter 1 with the illustration of the geometrical matrix tools, we may say that, in a sense, we have moved along a spiral. So let us conclude our "path" with the view given in Fig. 9.8 of a charming architectural example of a spiral.

9.7 Higher-order moments of the Wigner distribution function

For completeness'sake, we trace the line of the investigation of higher-order moments of the Wigner distribution function and their propagation laws, following [4.5].

As suggested by the direct product scheme (9.6.4) yielding the second-moment matrix $\mathbf{\Sigma}$, we may embed the third-order moments into a matrix as well by further applying the Kronecker product operation to the ray vectors. Thus, we obtain the third-moment matrix $\mathbf{\Sigma}_3$ by the three-vector product

$$\mathbf{\Sigma}_3 = \langle \mathbf{u} \otimes \mathbf{u}^\top \otimes \mathbf{u} \rangle, \tag{9.7.1}$$

where the sharp brackets mean of course an average procedure over the phase plane through the Wigner distribution function as in (9.6.6), i.e.

$$\langle \cdots \rangle \equiv \iint \cdots \mathcal{W}(\mathbf{u}; z) d\mathbf{u}. \tag{9.7.2}$$

Explicitly, the third product above yields the 4×2 matrix

$$\mathbf{u} \otimes \mathbf{u}^\top \otimes \mathbf{u} = \begin{pmatrix} q^3 & q^2\kappa \\ q^2\kappa & q\kappa^2 \\ q^2\kappa & q\kappa^2 \\ q\kappa^2 & \kappa^3 \end{pmatrix}, \tag{9.7.3}$$

containing the powers of q and κ, q^3, $q^2\kappa$, $q\kappa^2$ and κ^3, pertaining to the formation of the desired third-order moments.

The propagation law for the matrix $\mathbf{\Sigma}_3$ can be obtained in the same way as before. In fact, changing the integration variable from $\mathbf{u}$ to $\mathbf{v} = \mathbf{M}^{-1}\mathbf{u}$ in the integral (9.7.2), implied by (9.7.1), we see that the three-vector product $\mathbf{u} \otimes \mathbf{u}^\top \otimes \mathbf{u}$ turns into

$$\mathbf{u} \otimes \mathbf{u}^\top \otimes \mathbf{u} \quad \xrightarrow{\mathbf{v}=\mathbf{M}^{-1}\mathbf{u}} \quad \mathbf{M}\mathbf{v} \otimes \mathbf{v}^\top\mathbf{M}^\top \otimes \mathbf{M}\mathbf{v}. \tag{9.7.4}$$

Repeated applications of the direct product property (see footnote on p. 500)

$$(\mathbf{E} \otimes \mathbf{F})(\mathbf{H} \otimes \mathbf{G}) = \mathbf{E}\mathbf{H} \otimes \mathbf{F}\mathbf{G}, \tag{9.7.5}$$

leads to the nice form

$$\mathbf{M}\mathbf{v} \otimes \mathbf{v}^\top\mathbf{M}^\top \otimes \mathbf{M}\mathbf{v} = (\mathbf{M} \otimes \mathbf{M})\mathbf{v} \otimes \mathbf{v}^\top \otimes \mathbf{v}\mathbf{M}^\top. \tag{9.7.6}$$

It is therefore evident that the propagation law for $\mathbf{\Sigma}_3$ writes as

$$\mathbf{\Sigma}_3(z_o) = (\mathbf{M} \otimes \mathbf{M})\mathbf{\Sigma}_3(z_i)\mathbf{M}^\top. \tag{9.7.7}$$

The procedure can be iterated, thus yielding for the n-th order moment matrix $\mathbf{\Sigma}_n$ the average of the n-vector direct product

$$\mathbf{\Sigma}_n = \langle \underbrace{\mathbf{u} \otimes \mathbf{u}^\top \otimes \mathbf{u} \otimes \mathbf{u}^\top \otimes \mathbf{u} \cdots}_{n\text{-times}} \rangle, \tag{9.7.8}$$

with $\mathbf{u}$ and $\mathbf{u}^\top$ being alternatingly involved into the sequence of products.

It is straightforwardly, although a little lengthy, using the properties of the Kronecker product, to verify that the matrix $\mathbf{\Sigma}_n$ propagates according to

$$\mathbf{\Sigma}_n(z_o) = (\mathbf{M}\otimes_{[\frac{n+1}{2}]})\mathbf{\Sigma}_n(z_i)(\mathbf{M}^\top\otimes_{[\frac{n}{2}]}). \tag{9.7.9}$$

Here, $[\frac{j}{2}]$ yields the closest integer to $\frac{j}{2}$, and

$$(\mathbf{M}\otimes_h) = \underbrace{\mathbf{M}\otimes\mathbf{M}\otimes\cdots\otimes\mathbf{M}}_{h\text{-times}} \tag{9.7.10}$$

denotes the h-th direct product of the ray matrix $\mathbf{M}$. Needless to say, with $n = 2$ we regain the propagation law (9.6.8) for the variance matrix $\mathbf{\Sigma}$, which indeed within the formalism of this section could be identified as $\mathbf{\Sigma} = \mathbf{\Sigma}_2$.

Let us separately consider the cases of n even and odd.

For n even: $n = 2j$, relation (9.7.9) yields the symmetric equation

$$\mathbf{\Sigma}_{2j}(z_o) = (\mathbf{M}\otimes_j)\mathbf{\Sigma}_{2j}(z_i)(\mathbf{M}^\top\otimes_j). \tag{9.7.11}$$

Note that $\mathbf{\Sigma}_{2j}$ is a $2^j \times 2^j$ matrix. It can also be seen as arising from

$$\mathbf{\Sigma}_{2j} = (\mathbf{\Sigma}_2\otimes_j). \tag{9.7.12}$$

Multiplying both sides by the $2^j \times 2^j$ matrix $(\mathbf{J}\otimes_j)$, using the relation $\mathbf{M}^\top = \mathbf{J}\mathbf{M}^{-1}\mathbf{J}^{-1}$ and repeatedly applying (9.7.5), we obtain on the right-hand side

$$\begin{aligned}[(\mathbf{M}\mathbf{\Sigma}_2(z_i)\mathbf{J}\mathbf{M}^{-1}\mathbf{J}^{-1})\otimes_j](\mathbf{J}\otimes_j) = [(\mathbf{M}\mathbf{\Sigma}_2(z_i)\mathbf{J}\mathbf{M}^{-1}\mathbf{J}^{-1}\mathbf{J})\otimes_j]\\ [(\mathbf{M}\mathbf{\Sigma}_2(z_i)\mathbf{J}\mathbf{M}^{-1})\otimes_j] = (\mathbf{M}\otimes_j)[\mathbf{\Sigma}_{2j}(z_i)(\mathbf{J}\otimes_j)](\mathbf{M}^{-1}\otimes_j).\end{aligned} \tag{9.7.13}$$

Finally, on due account of the further property of the direct product

$$(\mathbf{E}\otimes\mathbf{F})^{-1} = \mathbf{E}^{-1}\otimes\mathbf{F}^{-1}, \tag{9.7.14}$$

by which,

$$(\mathbf{M}^{-1}\otimes_j) = (\mathbf{M}\otimes_j)^{-1}, \tag{9.7.15}$$

the propagation law (9.7.11) is seen to be equivalent to the similarity transformation

$$\mathbf{\Sigma}_{2j}(z_o)(\mathbf{J}\otimes_j) = (\mathbf{M}\otimes_j)[\mathbf{\Sigma}_{2j}(z_i)(\mathbf{J}\otimes_j)](\mathbf{M}\otimes_j)^{-1}, \tag{9.7.16}$$

of the $2^j \times 2^j$ symmetric matrix $\mathbf{\Sigma}_{2j}(z_i)(\mathbf{J}\otimes_j)$ by the $2^j \times 2^j$ matrix $(\mathbf{M}\otimes_j)$.

A similar result, even though in a less direct form, can be worked out for the case of n odd. Indeed, with $n = 2j+1$ in (9.7.9) we obtain

$$\mathbf{\Sigma}_{2j+1}(z_o) = (\mathbf{M}\otimes_{j+1})\mathbf{\Sigma}_{2j+1}(z_i)(\mathbf{M}^\top\otimes_j), \tag{9.7.17}$$

for the $2^{j+1} \times 2^j$ matrix $\mathbf{\Sigma}_{2j+1}$.

Multiply both sides by $(\mathbf{J}\otimes_j)$, and repeatedly use (9.7.5), to finally write

$$\mathbf{\Sigma}_{2j+1}(z_o)(\mathbf{J}\otimes_j) = (\mathbf{M}\otimes_{j+1})\mathbf{\Sigma}_{2j+1}(z_i)(\mathbf{J}\otimes_j)(\mathbf{M}^{-1}\otimes_j), \tag{9.7.18}$$

both sides meaning a $2^{j+1} \times 2^j$ matrix. Correspondingly, taking the transpose of (9.7.17) and multiplying by $(\mathbf{J}\otimes_{j+1})$ we have

$$\mathbf{\Sigma}^\top_{2j+1}(z_o)(\mathbf{J}\otimes_{j+1}) = (\mathbf{M}\otimes_j)\mathbf{\Sigma}^\top_{2j+1}(z_i)(\mathbf{J}\otimes_{j+1})(\mathbf{M}^{-1}\otimes_{j+1}), \tag{9.7.19}$$

both sides implying of course a $2^j \times 2^{j+1}$matrix.
Finally, the product of (9.7.19) times (9.7.18) results into

$$\begin{aligned}&\mathbf{\Sigma}^\top_{2j+1}(z_o)(\mathbf{J}\otimes_{j+1})\mathbf{\Sigma}_{2j+1}(z_o)(\mathbf{J}\otimes_j)\\ &= (\mathbf{M}\otimes_j)[\mathbf{\Sigma}^\top_{2j+1}(z_i)(\mathbf{J}\otimes_{j+1})\mathbf{\Sigma}_{2j+1}(z_i)(\mathbf{J}\otimes_j)](\mathbf{M}\otimes_j)^{-1},\end{aligned} \tag{9.7.20}$$

which can be recognized as the similarity transformation of the $2^{j+1} \times 2^{j+1}$ matrix $\mathbf{\Sigma}^\top_{2j+1}(z_o)(\mathbf{J}\otimes_{j+1})\mathbf{\Sigma}_{2j+1}(z_o)(\mathbf{J}\otimes_j)$ by the direct product matrix $(\mathbf{M}\otimes_j)$.

As earlier noted, similarity transformations are very convenient to work with, since they preserve both eigenvalues and traces. Therefore, the identification of similarity transformations in relation with second-order as well as higher-order moments of light distributions provide an effective tool to individualize invariant quality parameters in term of which optical signals can suitably be characterized and accordingly optical systems effectively designed.

9.8 Summary

After introducing the basic notions of moments of $1D$ and $2D$ functions, we have focused our attention on the first- and second-order moments of the Wigner distribution function, for which a variety of significative relations have been established in general and then specialized to the case of the Wigner distribution function of a Gaussian signal, which is indeed $2D$ Gaussian. $2D$ Gaussian distribution function have relevance as well within the context of partial coherence theory.

In particular, a suitable matrix scheme has been envisaged enabling a compact and elegant formulation of the propagation laws for the second-order moments under propagation of the signal through $ABCD$ optical systems. A similarity transformation has been individualized, which involves the second-moment matrix and the ray transfer matrix of geometrical optics. From it a variety of interesting and useful relations can be deduced.

Of particular relevance is the invariance of the determinant of the variance matrix, interpretable as the area of the second-moment phase-plane ellipse. This has suggested the identification of suitable parameters for a meaningful definition (and operative procedure for the measure) of the "beam quality", as well as possible generalizations to partially coherent beams.

Problems

1. *(a)* Prove the general relation

$$\mathfrak{m}_{rs} = 2\pi i^{-r+s} \left. \frac{\partial^{r+s} \mathcal{A}(\widetilde{q}, \widetilde{\kappa})}{\partial \widetilde{q}^r \partial \widetilde{\kappa}^s} \right|_{\widetilde{q}, \widetilde{\kappa}=0},$$

holding between the central moments of the Wigner distribution function and the derivatives of the ambiguity function at the origin. It is implied above that the ambiguity plane origin has been translated at the mean values point of the Wigner distribution function. *(b)* Prove that the ambiguity function can be constructed by knowledge of the Wigner distribution function moments as

$$\mathcal{A}(\widetilde{q}, \widetilde{\kappa}) = \tfrac{1}{2\pi} \sum_{l,m=0}^{\infty} \frac{i^{l-m}}{l!m!} \mathfrak{m}_{ml} \widetilde{q}^l \widetilde{\kappa}^m,$$

thus also implying the reconstruction of the Wigner distribution function by knowledge (possibly gained by experimental observations) of its moments.

2. Verify that, the application of the energy operator $\widehat{\mathcal{E}}$ to a signal of the form (9.4.5), under the assumption that both the amplitude $f(q)$ and the local frequency $\mathfrak{k}_1(q) = \frac{d\phi}{dq}$ do not vary too rapidly, results into

$$\widehat{\mathcal{E}}\varphi(q) \simeq f^2 \mathfrak{k}_1^2,$$

i.e. the product of the squared amplitude and the squared (local) frequency of the signal.

3. Prove that for the Gaussian beam (9.5.1), the *local uncertainty product*, defined in (9.4.91), takes the explicit z-depending expression

$$\overline{\mathfrak{q}_2(\kappa)\mathfrak{k}_2(q)} = \frac{z_{\mathrm{R}}^2}{4(z^2 + z_{\mathrm{R}}^2)}.$$

Comment on it in the light of the considerations presented in § (9.4.4).

4. Prove that in order that, under propagation through an $ABCD$ system, the second-order moments of the light distribution do not change with respect to the respective input values, it must be

$$\Sigma_{qq}(z_i) = -\frac{B}{k_0} \frac{\epsilon}{\sin\beta}, \quad \Sigma_{\kappa\kappa}(z_i) = \frac{k_0 C\epsilon}{\sin\beta}, \quad \Sigma_{q\kappa}(z_i) = \frac{\epsilon}{\sin\beta} \frac{A-D}{2},$$

where $\cos\beta = \frac{A+D}{2}$. It is implied, of course, that $|A+D| < 2$ (*stability condition*).

5. Consider a light beam propagating through a quadratic index medium (see § 3.9). *(a)* Prove that the beam radius, defined by (9.6.32) evolves according to the Pinney equation

$$\frac{d^2\mathsf{w}}{dz^2} - \frac{16\epsilon^2}{n_0^2 k_0^2 \mathsf{w}^3} + \frac{n_2}{4n_0}\mathsf{w} = 0.$$

(Hint: use Eq. (9.6.8) for $\mathbf{\Sigma}$ and Eq. (1.7.18) for $\mathbf{M}$ to deduce the paraxial equations of motion for the second-order moments in a general form. Specialize the result to the case of propagation in a quadratic index medium.) *(b)* In particular, verify that with $n_2 = 0$ (free propagation) an evolution law for $\mathsf{w}(z)$ is obtainable in the similar form as for Gaussian beams, i.e.

$$\mathsf{w}^2(z) = \mathsf{w}_0^2(1 + \frac{z^2}{\mathsf{z}_{\mathrm{R}}^2}),$$

demanding for

$$\mathsf{z}_{\mathrm{R}} = \frac{z_{\mathrm{R}}}{M^2},$$

in accord with relation (9.6.34).

References

[1] Only a few titles on the moment methods in relation with image characterization, evaluation and manipulation, comprising the pioneering paper by Hu and the subsequent investigations by Teague, are listed here. The reader may find an exhaustive presentation in the textbook by Mukundan and Ramakrishnan.

[1.1] M.K. Hu, "Visual pattern recognition by moment invariants", IRE Trans. Info. Theory, IT-**8**, 179-187 (1962).

[1.2] M.R. Teague, "Optical calculation of irradiance moments", Appl. Opt. **19**, 1353-1356 (1980).

[1.3] M.R. Teague, "Image analysis via the general theory of moments", J. Opt. Soc. Am. **70**, 920-930 (1980).

[1.4] R. Mukundan and K.R. Ramakrishnan, *Moment Functions in Image Analysis: Theory and Applications* (World Sc., Singapore, 1998).

[2] The presentation in § 9.2 is beautifully covered in the quoted pages of [2.1]. Some representative titles concerned with the theory of general statistics are suggested.

[2.1] L. Mandel and E. Wolf, *Optical Coherence and Quantum Optics* (Cambridge Univ., New York, 1995), ch. 1, pp. 1-40.

[2.2] H. Cramér, *Mathematical Methods of Statistics* (Princeton Univ. Press, Princeton, 1946).

[2.3] M. Kendall and A. Stuart, *The Advanced Theory of Statistics* (Griffin, London, 1958).

[2.4] B.R. Martin, *Statistics for Physicists* (Academic Press, London, 1971).

[3] Concepts and definitions concerned with the moments (local and global) of a light distribution find their counterpart in the context of signal theory within the pertinent time-frequency domain. Some pertinent titles are suggested below. In particular we signalize the part I of the three-part paper by Claasen and Mecklenbräuker [3.1], where the moments of the Wigner distribution function and pertinent inequalities are considered. Well-known textbooks concerned with the signal theory and the relevant time-frequency representation issue are also listed.

[3.1] T.C.A.M. Claasen and W.F.G. Mecklenbräuker, "The Wigner distribution - a tool for time-frequency signal analysis. I: continuous-time signals", Philips J. Res. **35**, 217-250 (1980).

[3.2] S. C. Pohlig, "Signal duration and the Fourier transform", Proc. IEEE **68**, 629 (1980).

[3.3] L. Cohen, "Instantaneous mean quantities in time-frequency analysis", Proc. IEEE ICASSP-**88**, 2188-2191 (1988).

[3.4] L. Cohen and C. Lee, "Instantaneous frequency, its standard deviation and multicomponent signals", in *Advanced Algorithms and Architectures for Signal Processing. III* SPIE Vol. 975, 191-208 (1988).

[3.6] P.J. Loughlin and B. Tacer, "Instantaneous frequency and the conditional mean frequency of a signal", Signal Process. **60**, 153-162 (1997).

[3.7] B. Boashash, "Estimating and interpreting the instantaneous frequency of a signal. I: Fundamentals", Proc. IEEE **80**, 520-538 (1992); "II: Algorithms and applications", *ibid.* 540-568.

[3.8] L.E. Franks, *Signal Theory* (Prentice Hall, London, 1971).

[3.9] A. Papoulis, *Signal Analysis* (Mc-Graw Hill, New York, 1977).

[3.10] L. Cohen, *Time-Frequency Analysis* (Prentice-Hall, Englewood Cliffs, 1995).

[4] The papers by Bastiaans [4.1, 4.2], concerned with the definition of the second-order moments of the Wigner distribution function, relative propagation laws and related invariants, have been basic to subsequent investigations and generalizations [4.4, 4.5], also in relation with partially coherent beams [4.3]. The paper by Sheppard and Larkin [4.6] presents a variety of interesting relations involving the second-moments, some of which reproduced in § 9.4.3. The fascinating review by Simon and Mukunda [4.8] elucidates the relation between the ray, wave and Wigner optics formalisms, in relation as well to the partial coherence theory. The link of the second-moments of the Wigner distribution function to the second derivatives of the ambiguity function has been evidenced by Papoulis [4.9].

[4.1] M.J. Bastiaans, "Propagation laws for the second-order moments of the Wigner distribution function in first-order optical systems", Optik **82**, 173-181 (1989).

[4.2] M.J. Bastiaans, "Second-order moments of the Wigner distribution function in first-order optical systems", Optik **88**, 163-168 (1991).

[4.3] M.J. Bastiaans, "ABCD law for partially coherent Gaussian light, propagating through first-order optical systems", Opt. Quantum Electron. **24**, S1011-S1019 (1992).

[4.4] D. Onciul, "Invariance properties of general astigmatic beams through first-order optical systems", J. Opt. Soc. Am. A **10**, 295-299 (1993).

[4.5] D. Dragoman, "Higher-order moments of the Wigner distribution function in first-order optical systems", J. Opt. Soc. Am. A **11**, 2643-2646 (1994).

[4.6] C.J.R. Sheppard and K. G. Larkin, "Focal shift, optical transfer function, and phase space representation", J. Opt. Soc. Am. A **17**, 772-779 (2000).

[4.7] V.V. Dodonov and O.V. Man'ko, "Universal invariants of quantum-mechanical and optical systems", J. Opt. Soc. Am. A **17**, 2403-2410 (2000).

[4.8] R. Simon and N. Mukunda, "Optical phase space, Wigner representation, and invariant quality parameters", J. Opt. Soc. Am. A **17**, 2440-2463 (2000).

[4.9] A. Papoulis, "Ambiguity function in Fourier optics", J. Opt. Soc. Am. **64**, 779-788, (1974).

[5] The non linear energy operator $\widehat{\mathcal{E}}$ was originally developed by Teager and then systematically used by Kaiser. In the context of signal analysis, it has revealed an efficient tool for extracting amplitude and frequency modulation information in signals of the form $s(t) = a(t)\cos(\phi(t))$. As representative papers, we suggest

[5.1] H.M. Teager, "Some observations on oral air flow during phonation", IEEE Trans. Acoust., Speech, Signal Process., ASSP-**28**, 599-601 (1980).

[5.2] H.M. Teager and S.M. Teager, "Evidence for nonlinear sound production mechanisms in the vocal tract", in *Speech Production and Speech Modelling*, J. Hardcastle and A. Marchal (eds.) (Kluwer, Boston, 1990), pp. 241-261.

[5.3] J.F. Kaiser, "Some useful properties of Teager's energy operator", IEEE ICASSP-93, Vol. 3, 149-152 (1993).

[5.4] P. Maragos, J.F. Kaiser and T. F. Quatieri, "On amplitude and frequency demodulation using energy operators", IEEE Trans. Signal Process. **41**, 1532-1550 (1993).

[6] The richness of inequalities concerned with the moments of the Wigner distribution function is well conveyed by de Bruijn 's papers [6.1, 6.2], the second offering one of the first mathematical formalization of the Wigner representation. The presentation in § 9.4.4 is inspired by the paper by Loughlin and Cohen [6.3].

[6.1] N.G. de Bruijn, "Uncertainty principle in Fourier analysis", in *Inequalities*, O. Shisha (ed.) (Academic Press, New York, 1967), pp. 57-71.

[6.2] N.G. de Bruijn, "A theory of generalized functions, with applications to Wigner distribution and Weyl correspondence," Nieuw Archief voor Wiskunde **21**(3), 205-280 (1973).

[6.3] P.J. Loughlin and L. Cohen, "The uncertainty principle: global, local, or both?", IEEE Trans. Signal Process. **52**, 1218-1227 (2004).

[7] A reading list on Gaussian beams is in [13] of ch. 2. Here we list a few titles relevant to the presentation in §§ 9.6.6-9.6.7, and to definitions, applications and measurements of relevance to the individualization of suitable parameters for laser beams/devices characterization and design. See also the list of refs. in [7.1]

[7.1] http://www.stanford.edu/ siegman/beam_quality_ref_list.html.

[7.2] G. Nemes, I.E. Teodorescu and M. Nemes, "Phase space treatment of optical beams", Preprint LOP-43-1984, CIP, Bucharest(1984).

[7.3] M. Nemes and G. Nemes, "Optical Quadrupole", Rumanian Patent # 72789(1977).

[7.4] A. E. Siegman, "New developments in laser resonators", SPIE **1224**, 2-14 (1990).

[7.5] T.F. Johnston, Jr., "M^2 concept characterizes beam quality", Laser Focus, 173-183 (May 1990).

[7.6] M.W. Sasnett, "Propagation of multimode laser beams. The M^2 factor", in *The Physics and Technology of Laser Resonators*, D.R. Hall and P.E. Jackson (eds.) (Adam Hilger, Bristol, 1989).

[7.7] P.-A. Bélanger, "Beam propagation and the ABCD ray matrices", Opt. Lett. **16**, 196-198 (1991).

[7.8] A. E. Siegman, "Defining the effective radius of curvature for a nonideal optical beam", IEEE J. Quantum Electron. **QE-27**, 1146-1148 (1991).

[7.9] H. Weber, "Wave optical analysis of the phase space analyser", J. Mod. Opt. 39, 543-559 (1992).

[7.10] *Laser beam quality*, Special issue of Opt. Quantum Electron. **24**, H. Weber (ed.), S861-S1135 (1992).

[7.11] Y. Champagne, C. Baré and P.-A. Bélanger, "Method for direct measurement of the variance of laser beams", Opt. Lett. **19**, 505-507 (1994).

[7.12] Y. Champagne and P.-A. Bélanger, "Method for direct measurement of realistic second-moment propagation parameters for nonideal laser beams", Opt. Quantum Electron. **27**, 813-824 (1995).

[7.13] N. Hodgson and H. Weber, *Optical Resonators. Fundamentals, Advanced Concepts and Applications* (Springer, Berlin, 1997), § 2.6, pp. 92-104; ch. 23, pp. 605-617.

[7.14] J. Alda, "Laser and Gaussian beam propagation and transformation", in *Encyclopedia of Optical Engineering* (Dekker, New York, 2003).

[8] A. S. Deif, *Advanced Matrix Theory for Scientists and Engineers* (Abacus Press, Turnbridge Wells, 1982),as to positive definite matrices see ch. 4, pp. 103-106; as regards the Kronecker product see pp.184-185.

A

Lie Algebras and Lie Groups: Basic Notions

A *Lie algebra* $\mathcal{A}$ is a linear space over some field F, say, for instance, the field of real numbers $\mathbb{R}$ or of complex numbers $\mathbb{C}$, together with a binary operation $[\cdot,\cdot]$, which maps pair of elements in $\mathcal{A}$ into an element in $\mathcal{A}$:

$$[\cdot,\cdot]: \ (h,g) \in \mathcal{A} \times \mathcal{A} \longrightarrow [h,g] \in \mathcal{A}. \tag{A.1}$$

It is called the *Lie bracket* and satisfies the following properties:

- bilinearity:

$$[ah+bk,g] = a[h,g] + b[k,g], \quad [h,ag+bk] = a[h,g] + b[h,k], \tag{A.2}$$

for all $a,b \in F$ and $h,k,g \in \mathcal{A}$;

- anticommutativity:

$$[h,g] = -[g,h], \tag{A.3}$$

for all $h,g \in \mathcal{A}$;

- Jacobi identity:

$$[h,[k,g]] + [k,[g,h]] + [g,[h,k] = 0. \tag{A.4}$$

for all $h,k,g \in \mathcal{A}$.

It is evident that in general the Lie bracket is not associative, i.e.

$$[h,[k,g]] \neq [[h,k],g], \tag{A.5}$$

the associative law being replaced by the Jacobi identity.

The dimension of $\mathcal{A}$ is, of course, determined by the cardinality of the basis set $\{h_j\}_j$, which spans the underlying linear space.

As a concrete example of a Lie algebra, we may mention that the Euclidean space $\mathbb{R}^3$ becomes a three-dimensional Lie algebra with the Lie bracket taken as the cross product of vectors. Also, any associative algebra, with associative product fg (here simply denoted by justaposition), becomes a Lie algebra under the definition of the Lie bracket as the *commutator*, also called *Lie product*, of elements in the algebra:

$$[h,g] = hg - gh. \tag{A.6}$$

In a sense the commutator is a measure of how non-commutative the algebra is. The algebra of invertible real square matrices is made into a Lie algebra under the Lie product (A.6), involving the familiar row-by-column multiplication of matrices. Besides, in the text it is seen that the algebra of the $2D$ inhomogeneous linear polynomials (with respect to addition) acquires a Lie algebra structure under the Poisson bracket operation, as similarly does the algebra of $2D$ homogeneous quadratic polynomials.

A *homomorphism* σ between two Lie algebras $\mathcal{A}$ and $\mathcal{B}$ over the same field F:

$$\sigma : \mathcal{A} \longrightarrow \mathcal{B} \tag{A.7}$$

defines a linear mapping of $\mathcal{A}$ into $\mathcal{B}$, such that the Lie bracket in $\mathcal{B}$ of imaged elements of $\mathcal{A}$ is taken to the image of the Lie bracket in $\mathcal{A}$ of the original elements; namely:

$$[\sigma(h), \sigma(g)] = \sigma([h, g]) \quad \forall \ h, g \in \mathcal{A}. \tag{A.8}$$

If such a homomorphism is one-to-one (or, *faithful*), so that the inverse mapping exists, it is called an *isomorphism*. For all practical purposes, isomorphic Lie algebras are identical.

A *subalgebra* $\mathcal{S}$ in $\mathcal{A}$ is a linear subspace of $\mathcal{A}$, which is closed under Lie bracket of elements of $\mathcal{S}$ with elements of $\mathcal{S}$:

$$[\mathcal{S}, \mathcal{S}] \subset \mathcal{S}. \tag{A.9}$$

An *ideal* in $\mathcal{A}$ is a linear subspace $\mathcal{J}$ of $\mathcal{A}$, which is closed under Lie bracket of elements of $\mathcal{J}$ with elements of $\mathcal{A}$:

$$[\mathcal{J}, \mathcal{A}] \subset \mathcal{J}. \tag{A.10}$$

The *derived series* is represented by the sequence of subalgebras in $\mathcal{A}$ recursively obtained under Lie bracket at the same level of commutation:

$$\mathcal{A} \supset [\mathcal{A}, \mathcal{A}] \supset [[\mathcal{A}, \mathcal{A}], [\mathcal{A}, \mathcal{A}]] \supset [[[\mathcal{A}, \mathcal{A}], [\mathcal{A}, \mathcal{A}]], [[\mathcal{A}, \mathcal{A}], [\mathcal{A}, \mathcal{A}]]] \supset .... \tag{A.11}$$

Correspondingly, the *lower central series* is intended as the sequence of subalgebras in $\mathcal{A}$ obtained under repeated (right) Lie bracket with $\mathcal{A}$ itself:

$$\mathcal{A} \supset [\mathcal{A}, \mathcal{A}] \supset [\mathcal{A}, [\mathcal{A}, \mathcal{A}]] \supset [\mathcal{A}, [\mathcal{A}, , [\mathcal{A}, \mathcal{A}]]] \supset .... \tag{A.12}$$

A Lie algebra $\mathcal{A}$ is said *abelian* if $[\mathcal{A}, \mathcal{A}] = 0$; is said *solvable* if at a certain level of commutation the derived series terminates; and is said *nilpotent* if the lower central series becomes zero eventually. The Heisenberg-Weyl algebra $\{\widehat{\mathsf{q}}, \widehat{\mathsf{p}}, \widehat{\mathsf{l}}\}$ yields a significative example of a nilpotent algebra.

Also, a Lie algebra $\mathcal{A}$ is *simple* if it contains no nonzero ideals, whilst it is *semi-simple* if it contains no nonzero abelian ideals. A simple Lie algebra is semi-simple. It turns out that semi-simple Lie algebras are the direct sum of simple (non abelian) Lie algebras.

A Lie group $\mathcal{G}$ combines the continuous structure of a differentiable manifold with the basic properties of groups. A *group* is a set of elements in which it is defined a binary operation (the *group multiplication*), which maps any two elements in $\mathcal{G}$ into an element in $\mathcal{G}$ as well, and satisfies the properties:

- associativity:

$$H \cdot (G \cdot K) = (H \cdot G) \cdot K \tag{A.13}$$

for all $H, G, K \in \mathcal{G}$;

- existence of identity:

$$\exists\ I \in \mathcal{G} :\ \ I\ \cdot H = H \tag{A.14}$$

for all $H \in \mathcal{G}$;

- existence of unique inverse:

$$\exists\ H^{-1} \in \mathcal{G}\ :\ H^{-1} \cdot H = H \cdot H^{-1} = I. \tag{A.15}$$

for every $H \in \mathcal{G}$.

A familiar example is the group $\mathcal{P}_n$ of all permutations of n objects, which consists of $n!$ operations. Also, the collection of rotations of the circle in a plane through multiples of a given angle ϑ, say $\vartheta = \frac{2\pi}{n}$, with respect to a fixed point, constitutes a group made of n distinct operations.

In order that $\mathcal{G}$ be a Lie group, its group operation must be consistent with the continuous structure of the underlying manifold, and so it must differentiable as well. This naïvely means that if H and G in $\mathcal{G}$ multiply to $W \equiv H \cdot G$ in $\mathcal{G}$, then the group operation takes any elements J and K, respectively drawn from a neighborhood of H and a neighborhood of G, to the group element Y in a neighborhood of W.

As a simple example of Lie groups, we may mention that the set of real numbers, excluding 0, forms a Lie group with respect to the familiar multiplication of numbers. The set of translations of a straight line with respect to a reference point also constitutes a Lie group, each element of which being individualizable (i.e. parameterizable) by a real number (which specifies the shift from the identity). Also, the ensemble of rotations of the circle with respect to a fixed point by an angle ϑ, allowed to range continuously through the interval $0 \leq \vartheta \leq 2\pi$, generates a Lie group, whose elements can be parameterized just by the continuous angle ϑ. It is seen in the text that the group of the unimodular 2×2 real matrices (under the ordinary multiplication of matrices) forms a Lie group, and each matrix in it is parameterized by three real numbers, which can be differently interpreted according to the adopted decomposition. Another example is the group $GL(n)$ of $n \times n$ invertible matrices, or the relevant subgroup $O(n)$ of orthogonal matrices.

Roughly speaking, Lie algebras exponentiate to Lie groups. More precisely, the tangent space at the identity of a Lie group is always a Lie algebra. This Lie algebra determines the local structure of the Lie group through the exponential map. Considering, for illustrative purposes, a one-parameter Lie group $\mathcal{G}$, suppose that $G(t)$ represents a smooth curve in $\mathcal{G}$ passing through the group identity, e.g. $I = G(0)$. For values δt of the real parameter t infinitesimally close to 0, the corresponding element $G(\delta t)$ in $\mathcal{G}$ comes to be infinitesimally close to the identity $I = G(0)$, i.e.

$$G(\delta t) \approx I + \delta t G'(0), \tag{A.16}$$

thus establishing a mapping of elements $G'(0)$ in the Lie algebra to the infinitesimal generators $G(\delta t)$ of the Lie group. It is evident that the structure of the algebra is reflected locally into that of the group in a neighborhood of the identity.

Suppose now to take a finite number $t \in \mathbb{R}$. Of course, one can go infinitesimally close to 0, by properly fractioning t; set roughly, $\delta t = \frac{t}{m}$ and let the integer m go to infinity: $m \to \infty$. Since, by virtue of the closure property of the group, for any integer m, $[G(\delta t)]^m$ belongs to the group, we finally have

$$[G(\delta t)]^m \approx [I + \frac{t}{m} G'(0)]^m \approx e^{tG'(0)}, \tag{A.17}$$

thus establishing an exponential mapping through the real parameter t of elements $G'(0)$ in the Lie algebra to "finite" elements of the Lie group.

Historically, Lie algebras have emerged by considering elements in Lie groups infinitesimally close to the group identity.

Due to this intimate link between Lie algebras and Lie groups the characterization of Lie algebras mentioned above reflects in an analogous characterization (or, more precisely, classification) of Lie groups.

A representation of a Lie group $\mathcal{G}$ on a vector space V is a homomorphism of $\mathcal{G}$ into the group of automorphism of V (i.e. of transformation of V into itself). If a basis set is chosen for the vector space V, the representation can equally be intended as a homomorphism of $\mathcal{G}$ to the group $GL(n)$, and accordingly it provides a matrix representation of the group. In a sense, a group representation may yield a description of a group in terms of more "visible" objects.

Index